NEUTRINO '86

Proceedings of
The 12th INTERNATIONAL CONFERENCE ON NEUTRINO PHYSICS AND ASTROPHYSICS

SENDAI
JUNE 3–8, 1986

Editors

T. Kitagaki & H. Yuta
Tohoku University

Sponsored by
Tohoku University
National Laboratory for High Energy Physics
Research Institute for Fundamental Physics
Institute for Cosmic Ray Reseach

World Scientific

Published by

World Scientific Publishing Co Pte Ltd
P. O. Box 128, Farrer Road, Singapore 9128.

Library of Congress Cataloging-in-Publication data is available.

XIITH INTERNATIONAL CONFERENCE ON NEUTRINO PHYSICS AND ASTROPHYSICS

ISBN 9971-50-166-X

Printed in Singapore by Kim Hup Lee Printing Co. Pte. Ltd.

**The 12th International Conference on Neutrino Physics
and Astrophysics was partially supported by**

The Ministry of Education, Science
and Culture of Japan

The Japan World Exposition
Commemorative Fund

The Yamada Science Foundation

Foundation for Promotion of High
Energy Accelerator Science

Miyagi Prefectural Government

Sendai Municipal Government

CONFERENCE ORGANIZATION

International Advisory Committee

J. Bjorken	FNAL	L. Okun	Moscow
Kuang-Chao Chou	Beijing	D. Perkins	Oxford
H. Faissner	Aachen	V. Peterson	Hawaii
E. Fiorini	Milano	A. Pomansky	Moscow
C. Jarlskog	Stockholm	F. Reines	Irvine
D. Kiss	Budapest	F. Sciulli	Columbia
K. Kleinknecht	Mainz	G. Steigman	Newark
G. Marx	Budapest	G. Zatsepin	Moscow
M. Menon	Tata		

International Neutrino Conference Committee

H. Faissner	Aachen	J. Nilsson	Goteborg
E. Fiorini	Milano	V. Peterson	Hawaii
C. Jarlskog	Stockholm	A. Pomansky	Moscow
D. Kiss	Budapest	F. Reines	Irvine
T. Kitagaki	Tohoku	G. Zatsepin	Moscow
G. Marx	Budapest, Secretary		

Local Organizing Committee

T. Kitagaki	Tohoku, Chairman	K. Takahashi	KEK
M. Konuma	Keio	G. Takeda	Tohoku
T. Kotani	Osaka	A. Yamaguchi	Tohoku
S. Miyake	Tokyo	Y. Yamaguchi	INS
Y. Nagashima	Osaka	M. Yoshimura	KEK
S. Orito	Tokyo	H. Yuta	Tohoku
H. Sato	Kyoto		

FOREWORD

The twelfth International Conference on Neutrino Physics and Astrophysics was held at Sendai during the period of June 3 to June 8, 1986. The total number of participants was 230 — 160 from abroad and 70 from Japan — and 70 talks were presented.

The organizing committee chose the subjects of neutrino mass and mixing, cosmology and astrophysics, neutral current and charged current, weak bosons and standard theory, proton decay and lepton number non-conservation and future projects. Talks on these subjects were concentrated more on physics subjects rather than on each experiment or project. They put more weight on new data than on the preparation work and tried to cover the most recent exciting results till the last moment. The general structure of sessions was an inheritance from past neutrino conferences, especially the previous Dortmund Conference.

The conference place was the Sendai City Auditorium near the Nishi-Koen Park and Hirose-River. The environment, full of fresh green, was adequate for holding the International Neutrino Conference. We hope that all participants enjoyed the conference place, the Matsushima excursion, Sendai and Japan as well as the scientific program. The organizing committee thanks the International Advisory Committee for the allocation of quota and kind advice for organizing the program. The committee also thanks the International Neutrino Conference Committee for bringing the conference to Sendai, Japan and all members of Bubble Chamber Physics Laboratories, Tohoku University for their tireless efforts for running the conference smoothly.

T. Kitagaki
Chairman
Sendai, September 10, 1986

VI. FUTURE PROJECTS

Chairman: S. Miyake

Chairman: F. Reines

VII. LIST OF CONTRIBUTED PAPERS

VIII. LIST OF PARTICIPANTS

(*) no written version of talk contributed

THEORETICAL PERSPECTIVES ON NEUTRINO MASS

Lincoln Wolfenstein

Carnegie-Mellon University

Physics Department

Pittsburgh, PA 15213

There is today no confirmed experimental evidence or compelling theoretical argument for non-zero neutrino mass. In this talk I will assume neutrinos are massive and discuss a variety of theoretical ideas about the mass matrix (masses and mixings) for the three known types of neutrino.

I. Theory of Neutrino Mass

1. If neutrinos have a mass theorists would like them to be *Majorana particles*. Quarks and charged leptons are Dirac particles acquiring mass from a term that connects the left-handed fermion f_L to the right-handed f_R. On the other hand it is conceivable for neutrinos, because they have no charge, to acquire mass from a term (Majorana mass) that connects the left-handed neutrino ν_L to its anti-particle, which is right-handed, which I call ν_R^c. By assuming that neutrinos acquire their mass in this different way we hope to explain why their mass is very different in magnitude from quarks and charged leptons.

2. The Majorana mass term violates lepton number by two units. Thus in assuming Majorana neutrinos we assume that somewhere in the laws of physics lepton number is violated. It is interesting to assume that lepton number violation is associated with a *large mass scale M* and that the magnitude of the neutrino mass is inversely proportional to M. The Majorana mass term also violates weak isospin by one unit (ν_L *has* $I_3 = +\frac{1}{2}, \nu_R^c$ *has* $I_3 = -\frac{1}{2}$). If we assume that weak isospin is broken only by the vacuum expectation values v of Higgs doublets ϕ, then there must be an effective interaction of the form[1]

$$f M^{-1}(\bar{\nu}_L \nu_R^c \phi\phi)$$

where M^{-1} must be introduced for dimensional reasons. As a result

$$M_\nu = f v^2 / M \sim m_D^2 / M \tag{1.1}$$

where m_D is of the order of the Dirac mass of a normal fermion. (The last approximate equality in Eq. (1.1) corresponds to the fact that m_D is proportional to v.) Eq. (1.1) is the famous see-saw formula of Yanagida and Gell-Mann, Ramond, and Slansky.[2]

A more specific motivation for neutrino mass and for Eq. (1.1) comes from the grand-unified theory $SO(10)$. In this theory quarks and leptons are unified so that there exists a right-handed neutrino N_R in each generation. The neutrino necessarily obtains a Dirac mass m_D similar in magnitude to those of up quarks. The way out is to give N_R a large Majorana mass M so that the mass matrix takes the form

$$\begin{pmatrix} 0 & m_D \\ m_D & M \end{pmatrix} \tag{1.2}$$

where the zero corresponds to the Majorana mass of the left-handed neutrino. One then obtains, after diagonalizing, light left-handed neutrinos with a Majorana mass

$$m_\nu = m_D M^{-1} m_D \tag{1.3}$$

For more than one generation this is a matrix equation. It was in this context that the see-saw formula was first derived.

3. Within the context of $SO(10)$ it seems necessary for neutrinos to have a mass and to have a Majorana mass matrix of the form (1.3). However other solutions are possible if there exist *"extra" neutral leptons* beyond the two per generation in the minimal $SO(10)$. There are many possibilities for such extra particles. This is

most dramatically seen in supersymmetric E6, a group in which there is resurgent interest. The fermions in E6 are in the 27 representation ($27 = 16 + 5 + \bar{5} + 1$) giving three extra "neutrinos" per generation. Because of supersymmetry there are neutral gauginos, possibly as many as six, since E_6 contains six $U(1)$. (Among these are the "usual" gluino, zino, and photino.) Including three generations this would yield a 21 × 21 neutrino mass matrix. Even though many of the entries are likely to be zero this is still great fun.

Two sample examples illustrate the qualitatively new possibilities when there are extra neutrinos. (1) If there are three neutrinos per generation and there exists a conserved lepton number, the solution consists of a massless neutrino for each generation and a massive (possibly very massive) Dirac neutrino.[3] (2) If there are four neutrinos with a conservation law then one can get two pairs of Dirac neutrinos per generation, one being the light neutrino, which is now a Dirac particle.[4] There are many recent papers on neutrino mass matrices with extra neutrinos inspired by superstring theory. One example by Mohapatra and Valle[5] starts with a 4 × 4 matrix (considering only one generation) using 4 members of the 27 and contains an approximately conserved lepton number. Therefore, as in the previous case, there is a light Dirac neutrino and a heavy Dirac neutral in each generation. A small breaking of the lepton number converts the light neutrino to a pair of almost degenerate Majorana neutrinos with opposite CP eigenvalues. The effect is similar to that of a Dirac neutrino (highly suppressed neutrinoless double beta decay, for example) except for the existence of neutrino oscillations with a large oscillation length.

4. Returning to the standard scenario described by Eq. (1.3) we discuss prejudices concerning the *mass hierarchy* and *mixings* for three generations. The Majorana mass matrix in flavor space M_{ab} will be assumed to be real and symmetric, conserving CP. (Observations of CP violation in the neutrino mass matrix are extremely difficult although clearly of theoretical interest.[6]) The diagonalization is carried out using an orthogonal matrix U

$$(UMU^T)_{ab} = \delta_{ab} m_a \eta_a \tag{1.4}$$

The mass matrix is then described by 3 mixing angles, 3 masses, and two relative signs η_2/η_1 and η_3/η_1. The factor $\eta_a = \pm 1$ is the sign of the eigenvalue in Eq. (1.4) and the relative signs correspond to relative CP eigenvalues of the Majorana neutrinos.[7]

From GUTS we expect the neutrino Dirac mass m_D to scale like the generational mass μ. On the other hand there is no real guidance about the generation dependence in the heavy Majorana mass matrix M. If we assume M scales like $(\mu)^p$ then from Eq. (1.3), ignoring mixing, we expect for the light neutrino mass for generation k

$$m_k \approx \mu_k^{2-p}$$

If we choose μ_k as the charged lepton mass then

$$p = 0, \quad m(\nu_e) : m(\nu_\mu) : m(\nu_\tau) = m_e^2 : m_\mu^2 : m_\tau^2$$

$$p = 1, \quad m(\nu_e) : m(\nu_\mu) : m(\nu_\tau) = m_e : m_\mu : m_\tau \tag{1.5}$$

It is natural but not necessary in $SO(10)$ that mixings are small like quarks.[8] Thus the mixing angle between ν_μ and ν_e might be expected to be of the order of the Cabibbo angle and the mixing between ν_e and ν_τ to be especially small. Furthermore for the Fritzsch form of mass matrix one of the factors η_i/η_j is negative.[9] However there are special examples based on GUT ideas that do not satisfy these prejudices concerning masses and mixing angles.

5. An alternative approach for obtaining a Majorana neutrino mass strictly within $SU(2) \times U(1)$ without GUT is to introduce a *Higgs triplet* in addition to the doublet. There is no good motivation for doing this and in this case there is no guidance at all with respect to masses or mixings. The most interesting way to use the triplet is the Gelmini-Roncadelli model[10] in which the triplet VEV leads to the spontaneous violation of lepton number. There is then a massless goldstone boson called the majoron which has interesting consequences for cosmology.[11]

II. Cosmological Constraints

The big bang cosmology predicts that there should be a neutrino background similar to the microwave background. Assuming the neutrinos are not too heavy (less than a few Mev) there will be about 110 neutrinos/cc for each generation with a characteristic temperature of about 2^oK. If the neutrinos have a mass greater than 1ev they become the major component of the energy density of the universe. Given the Hubble constant one can then calculate the lifetime of the universe. Assuming conservatively that the universe is greater than 10^{10} years old one finds as discussed in Steigman's talk that

$$\Sigma m_\nu < 40ev \tag{2.1}$$

where this is the sum over all stable "light" neutrinos.

This limit on neutrino masses is much more severe than any empirical limits. If we accept the mass hierarchy of Eq. (1.5) this corresponds to a limit $m(\nu_\tau) < 40ev$ with $m(\nu_e)$ far too small to be relevant for beta-decay or double beta-decay. It is possible to evade the limit of Eq. (2.1) if the neutrinos with masses greater than 100ev are unstable. A great variety of laboratory experiments and cosmological analyses constrain this possibility. The results may be summarized[12,13] by saying that conventional decay mechanisms such as $\nu_2 \to \nu_1 + \gamma$ and $\nu_2 \to \nu_1 + e^+ + e^-$ are almost ruled out. Most interest has focused on exotic methods of getting rid of the heavy neutrinos. In the original Gelmini-Roncadelli model[10] the massive neutrinos annihilate into the massless majorons as soon as they become non-relativistic. As a result instead of relic massive neutrinos we have relic massless majorons. There has been some interest in using massive neutrinos for galaxy formation[14]; in this case it is necessary to get rid of the neutrinos after they become non-relativistic. This requires either (1) a new version of the majoron model in which neutrinos disappear by the decay $\nu_2 \rightarrow \nu_1 + majoron$ with the right lifetime[15] or (2) by giving the majoron a small mass to enhance the cross-section $\nu_2 + \nu_2 \rightarrow \nu_1 + \nu_1$.[16]

III The Solar Neutrino Problem

The classic experiment of Davis observes a neutrino flux significantly lower than calculations based on the standard solar model.[17] The ratio of the observed to calculated flux is about one-third, although given theoretical uncertainties the factor may be anywhere between 0.25 and 0.5. Furthermore it is possible that some of the observed flux may have other sources so that the factor may be even smaller. Here we shall assume that the factor is one-third and that the cause lies in some neutrino oscillation phenomenon rather than in a change in the solar model.

A simple possibility is that ν_e is an equal coherent mixture of three mass eigen-states. Very small values of $\triangle m^2 (\geq 10^{-10} ev^2)$ are sufficient to make the mixture incoherent when averaged over energy giving the reduction factor of one-third. However the large mixing angles, particularly between the first and third generation, are very unlikely from the theoretical arguments above. However, there exist theoretical models which give a result like this fairly naturally.[18]

Another possibility is to assume ν_e is an equal mixture of two mass eigenstates. This could give a reduction factor greater than $\frac{1}{2}$ if $\triangle m^2$ is so chosen that the earth-sun distance is half the oscillation length for neutrinos near the peak of the ^8B spectrum.[19] An interesting way to get such a mixture is to assume that ν_e is almost a Dirac particle.[5] This means that as a result of a small Majorana mass term the Dirac particle is split into two Majorana particles (with opposite CP eigenvalues) with a small mass difference. In this case the missing ν_e arrives as a sterile neutrino (anti-particle of ν_{eR}). In addition to the transformation to sterile neutrinos there may be some flavor oscillations causing further reduction in the ν_e flux.

Recently another possibility has been pointed out by Mikhaeyev and Smirnov [20] (MS). It is of particular interest in that it is consistent with our theoretical prejudices, particularly small mixing angles. In their model most of the ν_e are transformed into ν_μ or ν_τ during their passage through the sun. When neutrinos propagate through vacuum there is a phase change given by

$$e^{-i(M^2/2k)x}$$

where M is the mass and k the momentum. For two types of neutrinos there is a resulting oscillation with length

$$l_v = 4\pi k / \triangle M^2 = (2.5 E_\nu / \triangle M^2) \ meters \tag{3.1}$$

where in the last expression E_ν is in MeV and $\triangle M^2$ in $(ev)^2$. In matter [21] there is an additional phase change due to the index of refraction associated with forward scattering

$$e^{ik(n-1)x} - e^{-i(GN_e \sqrt{2})x}$$

where G is Fermi's constant, N_e is the electron number density, and we have considered only the *charged $-$ current* scattering of ν_e by electrons. (The neutral current interaction is the same for all generations and so gives only a common phase. The neutral current must be considered if we are interested in ν_e oscillating into sterile neutrinos; however (n-1) due to the neutral current vanishes for neutral matter composed only of electrons and protons so that there is only the small effect from the helium and heavier atoms in the sun.) A characteristic length can then be defined

$$l_o = \frac{2\pi}{\sqrt{2}GN_e} = 1.8 \cdot 10^7 m./\rho_e \tag{3.2}$$

where ρ_e is the electron density divided by Avagadro's number.

We now combine these two effects for the two generation case. Then we obtain for the x-dependent phase

$$e^{-i\Lambda x/2k}$$

$$\Lambda = \begin{pmatrix} M_{ee}^2 + 2\sqrt{2}GN_e k & M_{e\mu}^2 \\ M_{e\mu}^2 & M_{\mu\mu}^2 \end{pmatrix} \tag{3.3}$$

where the vacuum mixing angle θ is given by

$$\tan 2\theta = 2M_{e\mu}^2/(M_{ee}^2 - M_{\mu\mu}^2)$$

The MS idea is that at the surface of the sun ($N_e = 0$) with θ small the eigenstates are approximately ν_e and ν_μ with ν_μ the higher state assuming $M_{\mu\mu}^2 > M_{ee}^2$. In the core of the sun where the neutrinos are produced GN_e is the dominant term ($2\sqrt{2}kGN_e >> M_{\mu\mu}^2 - M_{ee}^2$) so that ν_e and ν_μ are approximate eigenstates with ν_e the higher state. Somewhere on the way out the diagonal elements become equal; that is, the levels appear to cross. This occurs for

$$l_v = l_o \cos 2\theta$$

$$E_\nu \rho_e = 7 \cdot 10^6 \triangle M^2 \cos 2\theta_v \tag{3.4}$$

Of course the $M_{e\mu}^2$ term keeps the levels from really crossing; if it is big enough (θ not too small) the adiabatic approximation applies and the emerging neutrinos remain in the upper eigenstate and thus leave the sun in almost a pure ν_μ state.

Detailed calculations[20,22] yield the result that a one-third reduction corresponds to values of $\triangle m^2$ and $sin^2 2\theta$ shown in Fig. 1 corresponding approximately to the equations

$$\triangle m^2 \simeq 10^{-4} (ev)^2 \ with \ sin^2 2\theta > 10^{-3} \tag{3.5}$$

$$sin^2(2\theta) \triangle m^2 \simeq 3 \cdot 10^{-8} (ev)^2 \tag{3.6}$$

We limit our discussion to values $sin\theta \leq 0.2$ so that vacuum oscillations between sun and earth are unimportant. For all the points lying between the two lines in Fig. 1 the reduction factor is greater than $\frac{1}{3}$. For points lying on the top curve (3.5) it is the higher energy ν_e that are eliminated so that a detector (such as ^{71}Ga) sensitive to $pp + Be^7$ neutrinos would detect the expected ν_e flux. On the other hand for points on the lower curve (3.6) with values of $\triangle m^2 \leq 2 \cdot 10^{-6}$ there is almost complete elimination of low energy ν_e so that the $pp + Be^7$ neutrinos would be more suppressed even than ^8B neutrinos.

IV Some Patterns of Masses and Mixings

We now look for patterns of neutrino masses and mixings that fit the Mikhaeyev-Smirnov scenario and the hierarchial mass relations of (1.5) with exponent p between 1 and 2. Since small masses are needed we will automatically fit the cosmological constraint of Eq. (2.1) and neutrinoless double beta-decay will be negligible. We assume that $\theta_{e\mu}$ is of order of the Cabibbo angle: $sin^2 2\theta_{e\mu} = 0.1 \ to \ 0.2$ and allow values of $sin^2 2\theta_{e\tau}$ between 10^{-2} and 10^{-3}. Three possible patterns are shown in Table 1. In each case $m(\nu_e) << m(\nu_\mu)$. The mass scale M corresponds to the see-saw formula of Eq. (1.1) for $m(\nu_\mu)$ with $m_D \approx 1 \ GeV$. The three cases have different experimental implications.

A. This corresponds to the lowest value of M and the largest neutrino masses. The high-energy solar ν_e are converted to ν_μ but the low-energy (Gallium detector)

ν_e arrive unsuppressed. The value of $(m^2(\nu_\tau) - m^2(\nu_\mu))$ is of the order 10^{-2} to $10(ev)^2$; for part of this range $\nu_\mu - \nu_\tau$ oscillations could be detected in accelerator experiments if the mixing angle is not too small. This is illustrated in Fig. 2 which shows that only a small portion of the allowed region is so far experimentally excluded. On the other hand $(m^2(\nu_\mu) - m^2(\nu_e)) = 10^{-4}(ev)^2$ which means that laboratory observations of $\nu_\mu - \nu_e$ oscillations are probably impossible.

B. This is the case in which ν_τ lies near the upper curve of **Fig. 1** and ν_μ lies near the lower curve. There are now three possibilities corresponding to relatively small shifts in the masses. For case B1 the major effect is ν_e being converted to ν_μ with almost complete suppression of the low energy ν_e. Case B3 is equivalent to case A for the solar neutrinos with ν_τ replacing ν_μ. Case B2 is interesting in that solar ν_e convert both to ν_τ and to ν_μ. While this case has not been worked out in detail it is clear that the high energy ν_e end up as ν_τ and the low energy as ν_μ. Thus the spectrum of the neutrinos detected by Davis is peaked at intermediate energies.

C. This corresponds to the largest value of M and the smallest neutrino masses. ν_τ lies on the lower curve in Fig. 1. For the larger values of $m(\nu_\tau)$ and the smaller values of θ the low energy (Ga detector) ν_e are unsuppressed but for the smallest values of $m(\nu_\tau)$ (larger θ) they are almost completely suppressed.

The attractiveness of the MS scenario is that starting with our theoretical prejudices we find that for M anywhere between 10^{11} and 10^{14} GeV it is quite natural to have a large suppression of solar neutrinos. In fact if masses and mixings calculated in various $SO(10)$ models[8] had been substituted in the equations of Ref. 21 and applied to the sun the result would have been a large suppression. Except for case A there is little hope for other experiments to find an effect of neutrino mass. On the other hand a variety of proposed solar neutrino experiments could help to clarify the situation.

As was pointed out earlier[21], experiments that observe neutrinos that have passed through the earth may need to take matter effects into account. For example, the parameter l_o of Eq. (3.2) is about 3×10^6 m for the core of the earth. Thus a search for neutrino oscillations with l_v comparable to the radius of the earth (as attempted by proton decay detectors) are sensitive to matter effects. If the adiabatic approximation were applicable to the earth there would, of course, be no matter effects since the entrance and exit conditions are the same. In fact this is not true since for the cases of interest the earth diameter and the matter oscillation length are comparable. If the parameters we have discussed are correct the only hope to see oscillations for neutrinos passing through the earth is if the mixing is enhanced by matter effects. For example for case A the mixing angle θ_m in the earth would be close to 45° for $E_\nu \sim 100$ to 300 MeV. This would enhance $\nu_e \rightarrow \nu_\mu$ but also $\nu_\mu \rightarrow \nu_e$; there would be no such effect for $\bar\nu_e$ or $\bar\nu_\mu$. Since the oscillation length in matter (for the case $\theta_m = 45^o$; i.e., $l_v = l_o$) is $l_v/sin\ 2\ \theta$ it is necessary that θ be not too small if significant oscillation is to take place.

Another interesting possibility is the effect of the earth on the solar neutrinos.[23] An example is case B1 with $\triangle m^2 \approx 3 \times 10^{-7}$ $(ev)^2$. In this case pp neutrinos are converted into ν_μ inside the sun but many of them are converted back to ν_e on passing through the earth. The mixing angle θ_m is close to 45° in the core for E_ν between 0.3 and 0.4 MeV. Thus the gallium detector might see the sun only at night!

V. ACKNOWLEDGEMENTS

This work was supported by the U.S. Department of Energy. The final version of this talk was written at the Aspen Center for Physics.

TABLE 1

Possible neutrino masses satisfying the MS scenario with $sin^2 2\theta_{e\mu} \sim 0.1$ and $sin^2 2\theta_{e\tau} \sim 10^{-2}$ *to* 10^{-3}. The last two columns indicate the transformation of high energy and low energy solar ν_e. OK indicates no suppression.

	$m(\nu_\tau)$	$m(\nu_\mu)$	M	High E ν_e	Low E ν_e
	(ev)	(ev)	(GeV)		
A	0.2 - 4	.01	10^{11}	$\to \nu_\mu$	OK
B1	0.02 - 0.2	5×10^{-4}		$\to \nu_\mu$	$\to \nu_\mu$
B2	.014	3×10^{-4}	$10^{12} - 10^{13}$	$\to \nu_\tau$	$\to \nu_\mu$
B3	.01	10^{-4}		$\to \nu_\tau$	OK
C	.002 - .008	$10^{-5} - 10^{-4}$	$10^{13} - 10^{14}$	$\to \nu_\tau$	$\to \nu_\tau^a$

[a]For the larger values of $m(\nu_\tau)$ the low energy ν_e are unsuppressed.

REFERENCES

1. S. Weinberg, Phys. Rev. Lett. 43, 1566 (1979); Phys. Rev. D22, 1694 (1980).

2. M. Gell-Mann, P. Ramond, and R. Slansky, private communication (1977) and in Supergravity, eds. P. van Nieuwehuizen and D. Freedman (North Holland, 1979). p. 317; T. Yanagida in Proc. Workshop on Unified Theory and Baryon Number in the Universe, eds. A. Sawada and A. Sugamoto, KEK, 1979.

3. D. Wyler and L. Wolfenstein, Nuc. Phys. B218, 205 (1983).

4. M. Roncadelli and D. Wyler, Phys. Lett. 133B, 325 (1983); P. Roy and O. Shanker, Phys. Rev. Lett. 52, 713 (1984).

5. R. N. Mohapatra and J. W. F. Valle, University of Maryland preprints 86-127, 86-164.

6. M. Doi et al, Phys. Lett 102B, 323 (1981); J. Schechter and J. W. F. Valle,

Phys. Rev. D22, 1227 (1980); S. M. Bilenky et al, Phys. Lett. 94B, 495 (1980).

7. L. Wolfenstein, Phys. Lett 107B, 77 (1983).

8. K. Kanaya, Prog. Theor. Phys. 64, 2278 (1980). R. Barbieri, D. V. Nanopoulos, G. Morchio, and F. Strocchi, Phys. Lett. 90B, 91 (1980); T. Yanagida and M. Yoshimura, ibid. 97B, 99 (1980); T. Goldman and G. J. Stephenson, Phys. Rev. D24, 236 (1981); K. Milton and K. Tanaka, Phys. Rev D23, 2087 (1981); S. Hama, K. Milton, S. Nandi, K. Tanaka, Phys. Lett. 97B, 221 (1980); K. Milton, S. Nandi, and K. Tanaka, Phys. Rev. D25, 800 (1982).

9. D. Chang and P. B. Pal, Phys. Rev. D26, 3113 (1982).

10. G. B. Gelmini and M. Roncadelli, Phys. Lett 99B, 411 (1981).

11. H. Georgi, S. L. Glashow, and S. Nussinov, Nucl. Phys. 193B, 297 (1981); G. B. Gelmini, S. Nussinov, and M. Roncadelli, Nucl. Phys. 209B, 157 (1982).

12. S. Sarkar and A. M. Cooper, Phys. Lett 148B, 347 (1984); D. Lindley, Ap. J. 294, 1 (1984).

13. For reviews see the talks of K. Sato at this conference and S. Sarkar at the Bari Europhysics Conference.

14. M. Turner, G. Steigman, and L. Krauss, Phys. Rev. Lett 52, 2090 (1984); G. Gelmini, D. Schramm, and J. F. Valle, Phys. Lett. 146, 311 (1984).

15. G. Gelmini and J. F. Valle, Phys. Lett. 142B, 181 (1984).

16. P. B. Pal, Phys. Rev. D 30, 2100 (1984).

17. For a review see Solar Neutrinos and Neutrino Astronomy (M. L. Cherry, W. A. Fowler, K. Lande, eds.), American Institute of Physics 1985.

18. L. Wolfenstein, Nuc. Phys. B 175, 93 (1980).

19. J. N. Bahcall and S. L. Frautschi, Phys. Lett. B29, 263 (1969).

20. S. P. Mikheyev and A. Smirnov, Nuov. Cim. 9C, 17 (1986).

21. L. Wolfenstein, Phys. Rev. D17, 2369 (1978).

22. S. P. Rosen and J. M. Gelb, Phys. Rev. D, to be published.

23. L. Wolfenstein, Paper submitted to XXIII International Conference on High Energy Physics, Berkeley, July 1986, Carnegie-Mellon Preprint CMU-HEP86-7.

FIGURE CAPTIONS

Fig. 1. Values of $(\Delta m^2, \sin^2 2\theta)$ that give a one-third reduction of the solar neutrino flux.[22] The top curves are for the ^{37}Cl detector and the bottom curves are for the ^{71}Ga detector. θ here represents mixing of ν_e with another neutrino.

Fig. 2. Present limits on $(\Delta m^2, \sin^2 2\theta)$ for $\nu_\mu - \nu_\tau$ mixing. The area to the left and below the curves is allowed. θ here represents the mixing of ν_μ and ν_τ. The dashed curve corresponds to $\sin^2 2\theta = 0.2$; we consider the region to the left of this curve of particular interest.

10⁻³
10⁻⁴
10⁻⁵
10⁻⁶
10⁻⁷
10⁻⁸
Δm²(eV)²
10⁻⁴
10⁻³
10⁻²
10⁻¹
10⁰
sin² 2θ
10⁻⁵
10⁻⁶
10⁻⁷
10⁻⁸
10⁻⁹
Δm²(eV)²
sin² 2θ

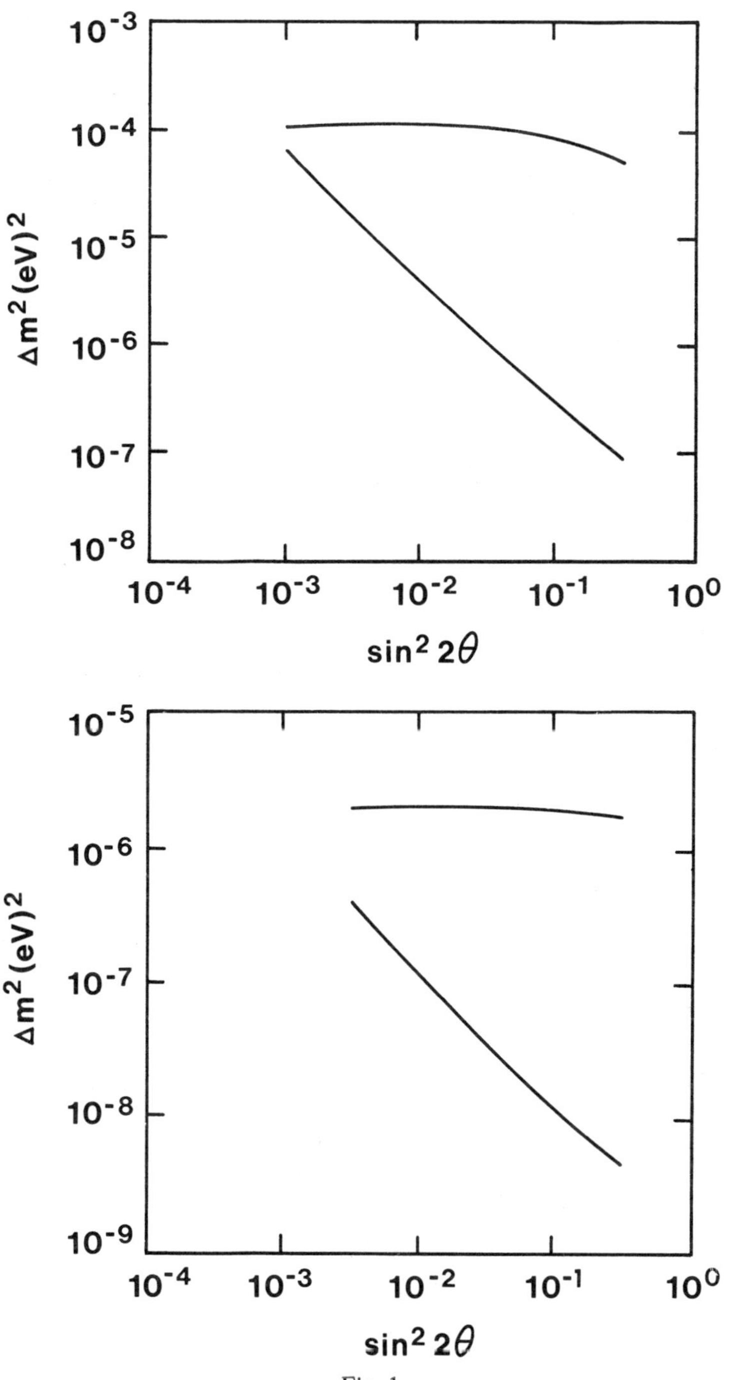

Fig. 1

$\nu_\mu \rightarrow \nu_x$
$\nu_\mu \rightarrow \nu_\tau$
$\Delta m^2 (eV^2)$
$\sin^2 2\theta$
10^3
10^2
10
1
10^{-1}
10^{-2}
0
0.5
1.0

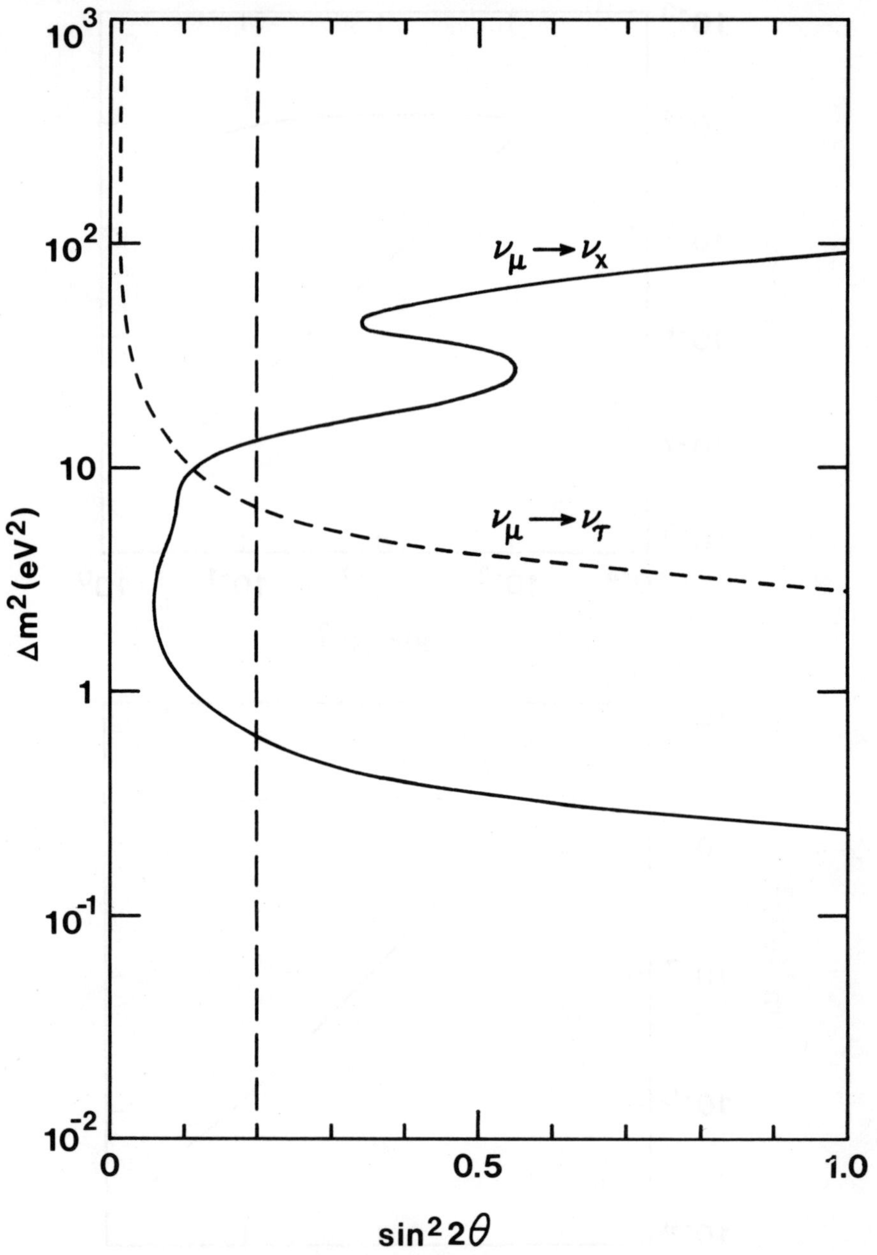

Fig. 2

THE NEUTRINO MASS FROM BETA SPECTRUM

Valentin Lubimov

Institute of Theoretical
and
Experimental Physics
Moscow,USSR

I. INTRODUCTION

This paper is the further development of the known investigation on the neutrino mass in beta-decay [I-3]. The results of a new series of measurements (ITEP-85) and the analysis of the beta-spectrum of tritium in a valine molecule are presented. The new cycle of measurements was carried out by the authors: S.Boris, A.Golutvin, L.Laptin, V. Lubimov, N.Myasoedov, V.Nagovizin, V.Nozik, E.Novikov, V.Soloshenko, I.Tihomirov, E.Tretjakov. The magneto-electrostatic spectrometer is described in [2]. We note that the beta-spectrum is scanned by electrostatic field variation while the focusing magnetic field was kept constant, i.e. the value of electron energy registered by the detector was fixed. This method of measurement has two considerable advantages:

I) The shape of the observed spectrum is virtually unaffected by the detector efficiency, which gives the possibility to widen the range of measurements. In the given cycle it amounts to 3.4 KeV.

2) There is a possibility to get rid of the main background component caused by the tritium contamination of the spectrometer parts. In the given set of measurements the electrons at the energy 2I.2 KeV were focused by the mag-

netic system, which exceeds the electron energy limit of
the tritium spectrum.

A proportional chamber with six channels is used as a
detector. The measurements are taken with a nonequipoten-
tional source, yielding a continuous focusing potential
secured by the backing made of weakly conducting glass. We
call such a source "pointlike", since the electrons from
the entire source surface are focused on a point of the
detector focal plane though they are actually emitted from
different geometrical points.

The employment of the pointlike source as well as the
multichannel detector makes it possible to use the whole
available focal plane of the spectrometer to the full. In
this way we reached the 20 eV resolution by narrowing the
entrance slit and regained intensity by increasing the
number of channels. With the improved resolution we can
explore the "mass-sensitive" 25 eV end portion of the
spectrum where the effect and background contributions are
nearly equal. The new spectrometer calibration was carried
out with more than 20 conversion lines of different gamma-
transitions of ^{169}Tm. The calibration coefficients are determined from the energy balance for a conversion electron,
the transition energies being known with high precision
(not employing the data on electron shell binding energy
for ^{169}Tm). From this equations we extracted the differences of binding energies for different shells, they are in
good agreement with [4], except for those involving the
shell M_3. We estimate the accuracy of the calibration
measurements as 3 eV. As before, for the evaluation of the
neutrino mass and the end-point energy we used the data
fitting procedure employing the model spectrum:

$$N^{th}(E) = AS(E); \qquad\qquad (I)$$

$$S(E) = \int dE' \, R(E,E') \left[I + \alpha_L (E' - E_o) \right] \left[I + \alpha_S (E' - E_o)^2 \right]$$

$$\sum_k W_k f_k (E', M_v, E_o);$$

$$f_k (E', M_v, E_o) = P E_t F(E') (E_o - E_k - E') \left[(E_o - E_k - E')^2 - M_v^2 \right]^{I/2};$$

(I)

The ITEP-85 investigation was aimed at thorough experimental analysis of the resolution function, involving its optical component and the component determined by the ionization losses in the source. The beta-spectrum measurements performed in a widened energy range are sensitive to the long tails of the function R, as well as to the possible heavy neutrino admixture in the mass below 3 KeV.

II. SPECTROMETER APPARATUS FUNCTIONS

The total response function of the spectrometer (TRF) is a sum of noninteracting electron spectrum, i.e. the optical resolution function (O.L.), and its convolution with the ionization loss spectrum (ILS) caused by the interaction of electrons with the source material (including the spectrum of backward scattering from the backing).

The O.L. measurement was carried out in the direct experiment free from any assumptions of the "shake off" behaviour. The shape of the L_I line (E_c - IO.6 KeV)(the intrinsic width L_I = 5.2 eV is minimal) was thoroughly measured in the accelerating field with the focusing energy E = 2I.2 KeV corresponding to the working cycle of measurements. The line obtained under these conditions is the fold of the sought optical line which corresponds to the working conditions: O.L.(2I.2) with ILS in Yb, with the "shake off" spectrum and a Lorentzian with Γ = 5.2 eV:

$$L_I(2I.2) = O.L.(2I.2) \otimes ILS \otimes "S-O" \otimes \Gamma$$

Then the same line was measured with "deceleration" at

E = 2 KeV:

$$L_I(2) = O.L.(2) \otimes ILS \otimes \text{"S-0"} \otimes \Gamma$$

When passing from "acceleration" to "deceleration" ILS, "shake off" and Γ remain unchangeable while O.L. getting narrower by an order of magnitude turning into the δ-function.

Thus, (see Fig.I) we get the sought optical resolution function involving the deconvolution procedure:

$$L_I(2I.2) \odot L_I(2) = O.L.(2I.2) \odot O.L.(2) \cong O.L.(2I.2)$$

(In reality this problem was solved without the mentioned above approximations made for simplicity). There were also other experimental ways of the O.L. determination [5]. In the subsequent analysis the spread resulted from different ways of the O.L. determination was taken into account in terms of systematic errors. Other contributions of TRF: ILS as well as the backward scattering spectrum has been studied experimentally [5]. Moreover the wide range fit interval 3.4 KeV has been treated as a sensitive test for the TRF shape adequacy.

The total resolution function (TRF) is shown in Fig.2.

III.MEASUREMENTS. RESULTS. DISCUSSIONS.

The scan of the experimental beta-spectrum was made in two runs. The first one was "standard": 4I points in 50 eV steps (A) and 6I points in 5 eV steps (B), exposition time being I00 sec and 300 sec. at each point correspondingly. On the whole the first run contained 34 series of double (A-B) measurements and covered the interval I.7 KeV of the beta-spectrum. The second run: 75px50eVx200sec contained I4 series and covered the interval 3.4 KeV. The total statistics is 55K/I00 eV at the end point region.

The shorter is the fitting interval from the end of the beta-spectrum the more insignificant is the influence of

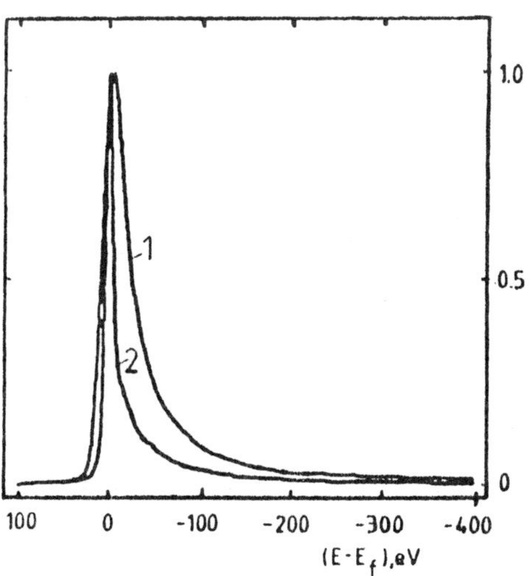

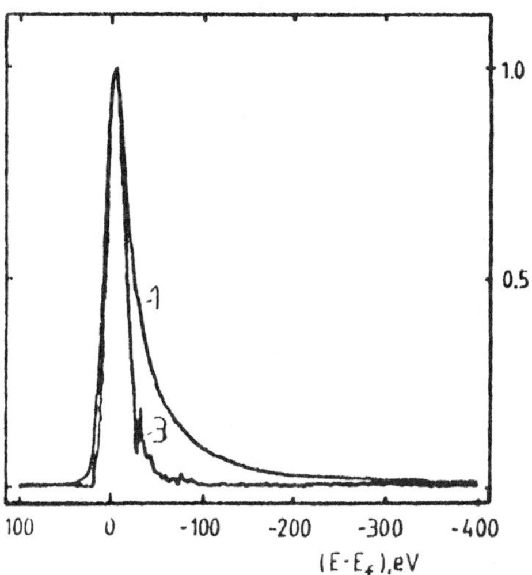

FIG. I. The O.L. determination in the direct experiment.

I – $L_I(21.2)$;   2 – $L_I(2)$;   3 – O.L.

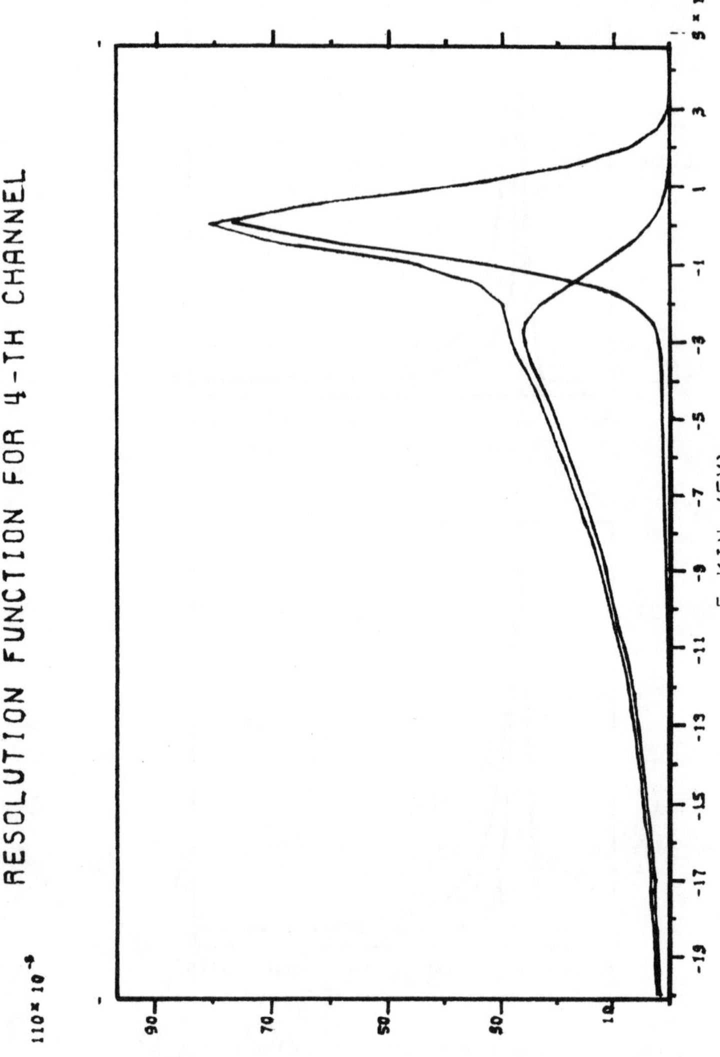

FIG.2. The total resolution function (TRF) for the working source. The contributions of the O.L. and IIS are shown.

the total resolution shape on the value of the obtained
parameters including those of physical significance such
as M_v and E_o. Really, the parameter values don't depend
at all on the TRF shape beyond the limits of the fitting
interval. On the contrary, the wider is the fitting interval the stronger is the dependence of the parameters on
the TRF form. Only when there is a complete adequacy of
TRF to the experimental spectrum the fitting parameters
have to be independent on the width of the fitting interval. The broadening of the beta-spectrum range of measurements up to 3.4 KeV made in the present work, appeared
to be extremely sensitive mode for choosing the right
shape of the TRF. Thus, choosing the shortest fitting interval (so far as statistics makes it possible) we get the
physical parameters (M_v and E_o) with the smallest systematic errors resulting from uncertainty in the **TRF** form.
At the same time the widest fitting interval serves to
test the adequacy of TRF.

Having at our disposal the checked form of O.L. and
ILS we recalculated the experimental spectra of the previous runs: ITEP-84 (Leipzig [3]). The results are given
in Table I.

The results of the ITEP-84 investigation are in a reasonable agreement with those given in the present work
(ITEP-85). The results of the joint data fit (ITEP-84 and
ITEP-85) are given in Table 2. The fit of data with the
neutrino mass set to zero is also presented in the table.
The shape of the curve at M=0 doesn't describe the experimental spectrum in any of the fit intervals, which can
be seen from the table. Shown in Fig.3 are the end point
region of the experimental and fitted spectra.

All the given results relate to the spectrum of final
states (SFS) for tritium in a valine molecule. This spectrum was calculated by Kaplan et al [6]. The mass diffe-

Table I

ΔE		1680		330	
				Standard O.L.; ILS-I; valine	
ITEP	T	M_ν	$E_O-18500$	M_ν	$E_O-18500$
85	1.98	$875\pm109(210)$	$80.9\pm0.6(2)$	$673\pm122(180)$	$79.3\pm0.8(2.5)$
84	2.10	$990\pm193(180)$	$81.5\pm1.1(2)$	$965+260(130)$	$81.4\pm1.4(2.5)$
	2.27	$1088\pm106(180)$	$81.8\pm0.6(2)$	$1094\pm152(150)$	$81.8\pm0.6(2.5)$
	2.82	$989\pm100(210)$	$80.9\pm0.6(2)$	$773\pm98(180)$	$79.1\pm0.8(2.5)$
average		$970\pm50(160)$	$81.2\pm0.3(1.7)$	$870\pm70(140)$	$80.3\pm0.4(2.3)$
$M_\nu = 30.3\pm(2)$; $\quad E_O = 18580.8\pm(4)^*$					

*) The absolute calibration error included. The sistematic errors are shown in braks. The total statistic: 130K/100eV.

Table 2

ΔE	3400	1680	330
M_ν	984 ± 50	966 ± 50	852 ± 66
E_O	81.0 ± 0.2	80.8 ± 0.2	80.0 ± 0.4
$\alpha_L \times 10^5$	-1.65 ± 0.12	-1.60 ± 0.12	-1.60 (Fix)
$\alpha_s \times 10^8$	0.13 ± 0.02	0.13 (Fix)	0.13 (Fix)
N	0.9896 ± 0.0016	0.9907 ± 0.0016	0.9991 ± 0.0039
x^2	$510/472$	$429/407$	$265/247$
$M_\nu = 0$			
E_O	76.4 ± 0.1	76.3 ± 0.1	74.0 ± 0.2
$\alpha_L \times 10^5$	-0.06 ± 0.10	-0.00 ± 0.08	0.00 (Fix)
$\alpha_s \times 10^8$	0.35 ± 0.02	0.35 (Fix)	0.35 (Fix)
N	1.0139 ± 0.0012	1.0144 ± 0.0012	1.0529 ± 0.0024
x^2	$1165/473$	$1032/408$	$475/248$

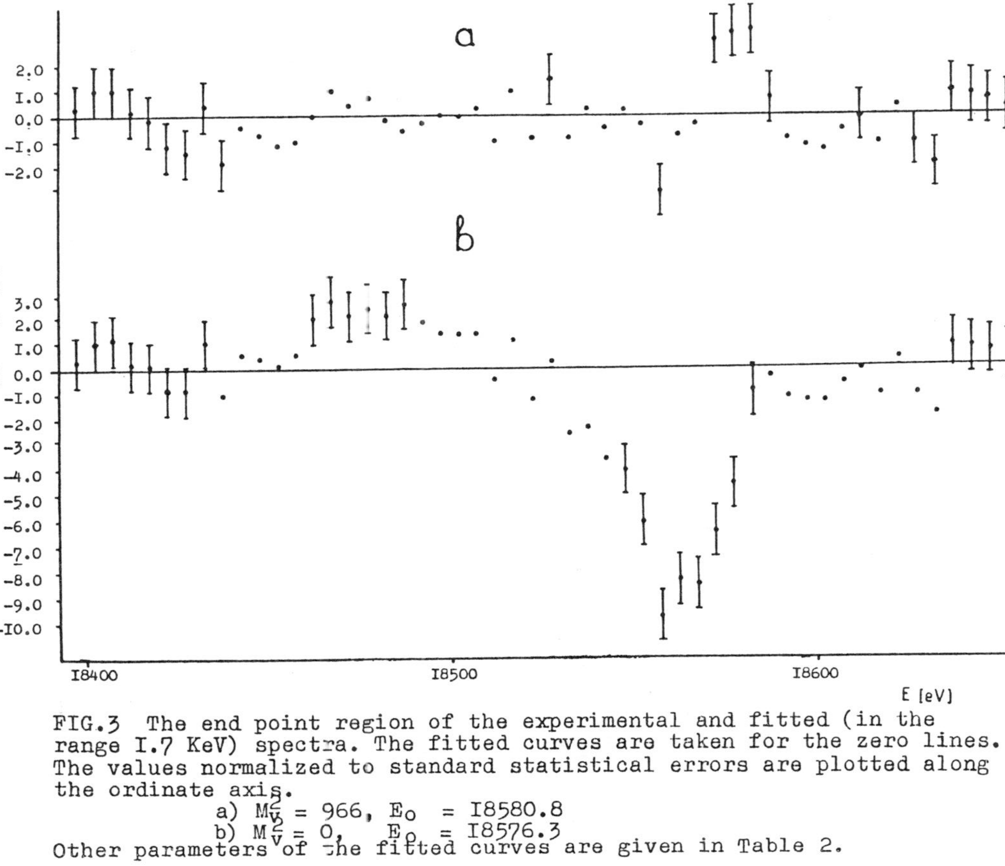

FIG.3 The end point region of the experimental and fitted (in the range I.7 KeV) spectra. The fitted curves are taken for the zero lines. The values normalized to standard statistical errors are plotted along the ordinate axis.

a) $M_\nu^2 = 966$, $E_0 = 18580.8$
b) $M_\nu^2 = 0$, $E_0 = 18576.3$

Other parameters of the fitted curves are given in Table 2.

rence of the neutral atoms of tritium and gelium can be
calculated from the fitted value of the parameter E_o,
employing the data [6]. The obtained from E_o value
$\Delta M_{T-He} = 18600.3 \pm 4$ is in an excellent agreement with
the mass measurements of the doublet T-He carried out by
Lipmaa et al [7] with the ICR-spectrometer:
$$\Delta M_{T-He} = 18599 \pm 2$$
It is common knowledge that the parameters M_v and
E_o depend on the adopted spectrum of final states (i.e.
model-dependent). Thus, for instance, for an extreme case
when the spectrum of final states is represented as
the δ-function (the bare nucleus decay), the fitted value
of E_o is reduced by 18 eV and becomes incompatible with
the Lipmaa data. The limits of the SFS variation agreeable
to the Lipmaa data (with the error taken into account in
our absolute calibration) lead to the interval:

$$17 < M_v < 40 \text{ eV}$$

REFERENCES

I. V.A. Lubimov et al., Phys.Lett. 1980, <u>v.94B</u>, p.266

2. S. Boris et al., HEP83, Intern. Europhysics Conf.
    on High energy physics (Brighton 1983), p.386

3. S. Boris et al., Proc. XXII Intern. Conf. on High
    energy physics (Leipzig 1984) <u>Vol.I</u>, p.259;
    Phys.Lett. 1985, <u>v.159B</u>, p.217

4. K. Sevier, ADNDT <u>v.24</u>, p.323, 1979.

5. V.Lubimov, A talk given at the XXI Moriond Conference.
    Tignes, January 1986.

6. I. Kaplan et al., DAN USSR 1984, <u>v.279</u>, p.IIIO

7. E. Lipmaa et al., Phys.Rev.Lett. 1985, <u>v.54</u>, p.285

AN UPPER LIMIT FOR THE ELECTRON ANTINEUTRINO MASS FROM THE ZURICH EXPERIMENT ON TRITIUM ß-DECAY

E. Holzschuh, M. Fritschi, W. Kündig, J.W. Petersen[a],
R.E. Pixley, and H. Stüssi

Physics Institute, University of Zurich
8001 Zurich, Switzerland

The endpoint region of the tritium ß-spectrum has been measured with 27 eV resolution, using a magnetic spectrometer. The tritium activity was implanted in to a thin layer of carbon. The neutrino mass determined is consistent with zero with an upper limit of 18 eV, which includes instrumental and statistical uncertainties as well as uncertainties due to energy loss in the source and the final electronic states.

1. Introduction

Measurement of the endpoint region in the tritium ß-spectrum still appears to be the most sensitive method to determine the rest mass m of the electron antineutrino. After the landmark paper of Bergkvist[1] in 1972 there was much interest in the subject only after a group at ITEP (Institute for Theoretical and Experimental Physics, Moscow) published the first of a series of papers[2] in 1980 claiming evidence for a nonzero value of m around 35 eV. This exciting result caused many groups to start new tritium experiments. Recently we have completed the analysis of our first measurements and have published the results.[3] So far we have no indication for a finite value of m larger than an upper limit of 18 eV. In this presentation we give a summary of the present status of our experiment.

The shape of the tritium ß-spectrum is sensitive to m only in the low intensity endpoint region which must be measured with high statistics and good resolution. There are three major sources for possible systematic errors common to most experiments of this type. These are shape and width of the spectrometer resolution function, energy loss when the ß-particles leave the source and the electronic final states. A fit to the measured ß-spectrum gives little indication for an inadequate representation of any of these distributions and independent knowledge is therefore of prime importance for a reliable determination of m.

2. Spectrometer

The instrument employed consisted of a toroidal field, magnetic spectrometer of the Tretyakov type[4] with 2662 mm average source detector distance, modified with a radial, electrostatic retarding field around the source. Figure 1 shows a cross section of the approximately cylindrically symmetric arrangement with some electron trajectories. The magnetic field is produced by 36 rectangular current loops (3,4). The electrons leaving the source (1) are decelerated by putting the source on positive high voltage with respect to a grid (2) near ground potential. The grid consists of 360 fine wires. The accepted solid angle and trajectories are limited by apertures (6,7).

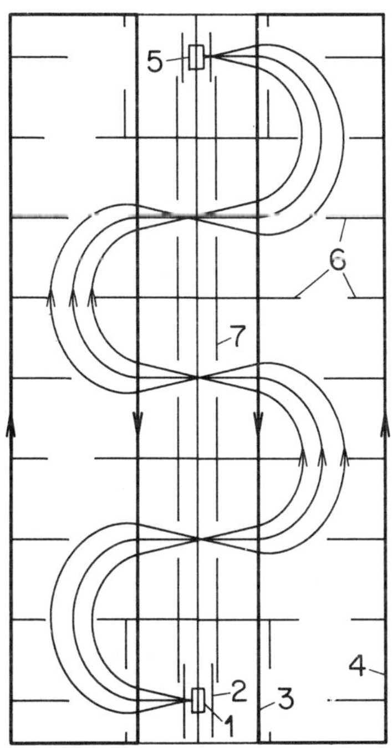

Fig.1.
Scale drawing of the spectrometer.

The detector (5) is a position sensitive proportional counter, also on high voltage (15 kV) with respect to a second grid of the same type as at the source. Position is determined by charge division, using a resistive anode wire. The diameter of the detector is 6 cm with 10 cm active length. The source assembly consists of 10 rings, each with 5 cm diameter and 1 cm width, i.e. the total active source surface is 157 cm². To compensate for the source length a gradient voltage, propotional to the spectrometer dispersion, is applied along the rings. Spectra are recorded by stepping the source high voltage while keeping the analysing energy, E_{mag}, of the magnetic spectrometer constant. For the present investigation we have E_{mag} = 2.2 keV, which gives 27 eV resolution (FWHM) and 1.1 eV/mm dispersion.

In summary the spectrometer has the following advantages: a) The modest relative resolution (1%) and the small dispersion in the magnetic part require a mechanical precision which can easily be achieved. Also the magnetic and electric fields are known analytically, allowing a reliable Monte-Carlo calculation of the resolution function. This is an important point since a resolution determination from measurements of conversion lines is complicated by shake-off effects.[5] b) The particles entering the detector have constant energy during a run, implying constant detection efficiency. c) A position sensitive detector has sufficient resolution, i.e. no energy defining slits are necessary and the luminosity is high. d) The resolution is simply determined by the magnet setting and can be chosen for a best compromise between high count rate and high resolution. The obvious disadvantage of the spectrometer is its sensitivity to tritium wall contamination, requiring sources of very high chemical stability. However, background from tritium on the walls is only a problem which reduces mass sensitivity and by itself does not cause systematic errors. In fact the background is a constant as a function of energy, since only the source potential is changed during a measurement which, of course, has no influence on the detector or on tritium on the walls.

3. Sources and Energy Loss Spectrum

The sources were prepared by implantation of T_2^+ ions into carbon, evaporated onto an Al backing. The thickness of the carbon layer was about 2000 Å. The implantation energies were in the range 185 to 400 eV per ion, giving typically an activity of 0.3 mCi/cm².

The dominant cause of energy loss of ß-particles traversing the carbon substrate is due to excitation of plasmons.[6] The probability for exactly n interactions is given by the Poission distribution

$$P_n(x) = 1/n! \ (x/\lambda)^n \ \exp(-x/\lambda), \tag{1}$$

where x is the layer thickness and λ denotes the mean free path. The deflection angle of an inelastically scattered electron is very small and completely negligible in our case. Assuming initially monoenergetic electrons, the energy distribution after the layer is given by

$$f(\Delta,x) = \sum_{n=o}^{\infty} P_n(x) \ \omega^{*n} , \tag{2}$$

where $\omega(\Delta)$ is the normalized energy loss spectrum for a single interaction with energy loss Δ and where

$$\omega^{*n} = \omega * \ldots * \omega \tag{3}$$

denotes the n-fold convolution of $\omega(\Delta)$ with itself. The no-loss term (n = 0) is defined by $\omega^{*0} = \delta(\Delta)$. For a distributed source Eq. 2 remains unchanged if P_n is replaced by the right side of Eq. 1 averaged over the source depth. This quantity was measured with 50 Å resolution (FWHM) using a nuclear recoil technique similar to that described in Ref. 7. Experimental values for λ and $\omega(\Delta)$ were taken from Ref. 6 and the energy loss spectra were determined using Eqs. 1 to 3.

4. Measurements and Data Analysis

The data so far available were obtained in four runs (R1 - R4) using three sources (T1 - T3). The measurements were performed with 27 eV resolution (FWHM) in an energy range of about 17.6 keV to 19.3 keV with 5 eV steps. To improve the mass sensitivity, this range was divided into several counting regions with more time spent near the endpoint. A single up-down pass required about one hour. The total measuring time was 27 days. Event data were recorded on magnetic tape and later binned into ß-spectra. Figure 2 shows the data from run R4 in the form of a Kurie plot.

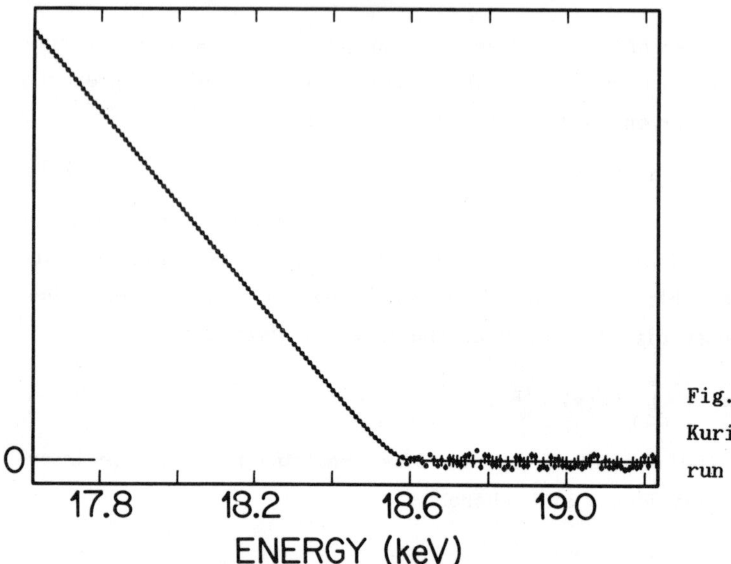

Fig.2. Kurie-plot from run R4.

The ß-spectrum was assumed to have the form

$$N(E) = A\ F(E)\ p\ E_t\ (1 + \alpha\varepsilon_o) \sum_i W_i\ \varepsilon_i^2 \sqrt{1 - m^2/\varepsilon_i^2}, \qquad (4)$$

where $\varepsilon_i = E_{oi} - E$. $F(E)$ is the Fermi function. E, E_t, and p are the electron kinetic energy, total energy and momentum, respectively. The sum runs over all electronic final states (FS) with branching fractions W_i and endpoints E_{oi}. Each term in the sum is set to zero for E above the point where the term first reaches zero. The E_{oi} are defined by $E_{oi} = E_{oo} - E_{ex,i}$, where E_{oo} corresponds to the electronic groundstate and is, for $m = 0$, the true endpoint of the spectrum. $E_{ex,i}$ are excitation energies measured from the groundstate. Backscattering in the source substrate is taken into account by the parameter α, which enters in this approximate form by a convolution of a constant backscattering distribution with a ß-spectrum proportional to ε_0^2. The fit function is obtained by convoluting Eq. (4) with the spectrometer resolution function and the energy loss spectrum and by adding a constant background term BG. A small correction factor ($<$ 7% variation) for the changing spectrometer acceptance as a function of retarding voltage has been included. The free parameters are α, E_{oo}, BG, an overall normalization A, and the neutrino mass squared m^2. The latter was allowed to take on nonphysical, negative values, in which case the square root in Eq. (4) was replaced by its first order expansion $|1 - m^2/(2\varepsilon_i|\varepsilon_i|)|$. This gave symmetric, parabola-like curves for χ^2 versus m^2. Our final result is essentially independent of this assumption.

Concerning the FS for our sources we argue as follows. The extremely small diffusion rate of hydrogen isotopes implanted in carbon[8] and the stability of our sources strongly suggest that C-T bonds are formed. Kaplan et al.[9] calculated the FS for a variety of molecules with C-T bonds and found very little variation in terms of the groundstate fraction W_0 and the mean excitation energy E_{ex}. By fitting various FS distributions we found that just these two quantities are important, whereas details of the FS influence the result very little. We thus take CH_3T, the simplest case, as an adequate representation for the FS of our sources. The corresponding fit results for the four spectra and the combined data set are listed in Tab. I. As can be seen, good fits were obtained and m^2 is compatible with zero within one sigma. The nonstatistical variation in the E_{oo} values is caused by uncertainties in the average source – detector distance mainly due to the use of different detectors. The m = 35 eV hypothesis was tested by fitting all data with fixed values m = 0 and m = 35 eV, giving χ^2 = 1370.2 and χ^2 = 1773.2, respectively, with 1370 degrees of freedom (DOF). As can be seen in Fig. 3, the latter case can be ruled out without question.

Possible systematic errors caused by uncertainties in spectrometer resolution, energy loss, and final states, were investigated by varying these distributions within estimated limits and fitting the data again.

It was found that the systematic error for m² should be smaller than 204 eV². For more details see Ref. 3. Assuming the best fitted value for m² would be exactly zero (and not −11 eV²) a statistical upper limit of 106 eV² for m² at the 95% confidence level was calculated. Adding the systematic and the statistical limit linearly we obtain our present result:

$$\mathbf{m^2 < 310 \ eV^2} \qquad \text{or} \qquad \mathbf{m < 18 \ eV.}$$

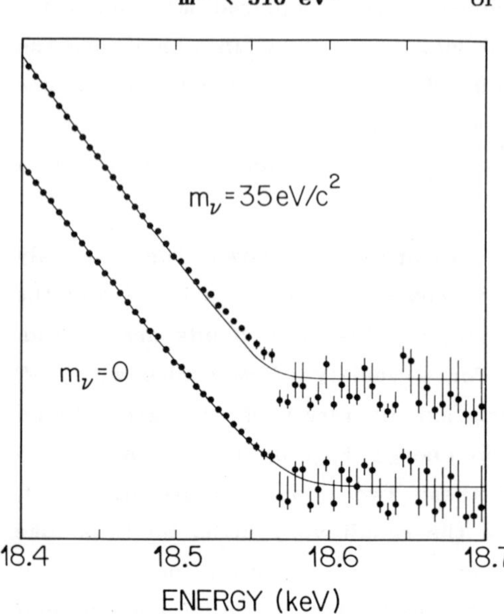

Fig. 3.
Kurie-plot of the data from run R4 fitted with m = 0 and m = 35 eV. All other parameter (E_{oo}, BG, A, α) are fitted. Notice the strong correlation of m² with the background in the fitted cuve.

Table I. Fit results for four runs with three different sources and the combined set of data using CH_3T final states. Errors indicated are one standard deviation. UL is a statistical upper limit at the 95% confidence level.

Run	Source	χ^2/DOF	$E_{oo}-18500$ (keV)	α (keV^{-1})	m² (eV²)	UL(95%) (eV²)
1	T1	328.8/337	82.6±0.3	0.018±0.004	+140±130	356
2	T2	345.6/356	84.5±0.2	0.020±0.002	−9±140	218
3	T2	349.9/356	84.4±0.2	0.022±0.002	−22±140	202
4	T3	343.7/317	77.6±0.2	0.021±0.002	−85± 93	70
all		1370.2/1369			−11± 63	95

5. Conclusion

No indication for a nonzero mass of the electron antineutrino was found, which is in strong contradiction to the result of Ref. 2. We see no possible source of error in our experiment large enough to account for this discrepancy. Presently we are preparing further measurements with new sources, in an attempt to reduce some of our systematic uncertainties.

References:

a)   Present address: CERN, 1211 Geneva, Switzerland.
1)   K.E. Bergkvist, Nucl. Phys. **39**, 317 (1972).
2)   V.A. Lubimov et al., Phys. Lett. B **94**, 266 (1980); Sov. Phys. JETP **54**, 616 (1981), (Zh. Eksp. Theor. Fiz. **81**, 1158 (1981)) and Phys. Lett. B **159**, 217 (1985).
3)   M. Fritschi, E. Holzschuh, W. Kündig, J.W. Petersen, R.E. Pixley, and H. Stüssi, Phys. Lett. B **173**, 485 (1986).
4)   E.F. Tretyakov, Izv. Akad. Nauk SSSR Ser. Fiz. **39**, 583 (1975).
5)   W. Kündig et al., Proc. of the Fourth Moriond Workshop, La Plagne, Savoie, France, 1984, ed. by J. Tran Thanh Van (Edition Frontières Paris, 1984), p. 261.
6)   R.E. Burge and D.L. Misell, Phil. Mag. **18**, 251, (1968).
7)   G.G. Ross et al., J. Nucl. Mat. **128**, 730 (1984).
8)   B.L. Doyle et al., J. Nucl. Mat. **103**, 513 (1981); V. Malka et al., Proc. Tritium Technology in Fission, Fusion and Isotopic Applications, American Nuclear Society National Topical Meeting, Dayton Ohio, 1980, p. 102.
9)   I.G. Kaplan, G.V. Smelov and V.N. Smutnyi, DAN USSR **279**, 1110 (1984); I.G. Kaplan, V.N. Smutnyi and Smelov, Phys. Lett. B **112**, 417 (1982); I.G. Kaplan, G.V. Smelov, and V.N. Smutnyi, Phys. Lett. B **161**, 389 (1985).

NEUTRINO MASS MEASUREMENT AT INS

H. Kawakami, K. Nisimura, T. Ohshima, S. Shibata,
Y. Shoji, I. Sugai, K. Ukai, T. Yasuda*
Institute for Nuclear Study, University of Tokyo

N. Morikawa, N. Nogawa
Isotope Center, University of Tokyo

T. Nagafuchi, F. Naito, T. Suzuki, H. Taketani
Tokyo Institute for Technology

M. Iwahashi
Kitasato University

K. Hisatake
Jissen Women's University

M. Fujioka
Cyclotron-Radioisotope Center, Tohoku University
and
Y. Fukushima, T. Matsuda, T. Taniguchi
National Laboratory for High Energy Physics, KEK

- Presented by T. Ohshima -

A first round of the neutrino mass measurement in ^{3}H β-decay at INS has been performed with the highest energy resolution of 10-16 eV in the world. An overall responce function of the spectrometer system was reliably obtained with a simple manner by using the same chemical structure for ^{3}H and reference ^{109}Cd sources. A preliminary result shows that the upper limit of the neutrino mass could be set as $m_\nu < 31$ eV with statistical and systematic errors.

* Present address : Northeastern University, Boston, MA 02115

The main apparatus of our experiment is an INS double focusing $\pi\sqrt{2}$ air-core spectrometer which has a magnetic field distribution proportinal to $1/\sqrt{r}$ and a central orbit radius of 75 cm. Its current stability is 2×10^{-5} and external magnetic field is cancelled down to 0.1 mGauss. Wide sources with a size of 25×100 mm^2 were used in combination with the spectrographic detection of β-ray, which greatly improved the luminosity. Focusing electric potential with a parabolic form was applied through the conductive source backing plate in order to compensate the optical aberration caused by the source expanse. A 40 cell proportional counter with slits of 0.5 mm wide was set on the focal plane of the spectrometer so that the energy range of $\sim$ 1600 eV around the ^{3}H end-point could be covered at one setting without changing the magnetic field strength. A momentum resolution $(\Delta p/p)$ of the system was examined by measuring several conversion and Auger lines of ^{169}Yb and ^{109}Cd. $\Delta p/p = 0.02\%$ was obtained for each individual line, which was equivalent to the energy resolution of

$$\Delta E = 8 \text{ eV}$$

for 20 keV β-rays. This is the highest energy resolution in the world so far achieved for the measurement of the ^{3}H β-spectrum.

We used tritium-labelled arachidic acid $(C_{20}H_{40}O_2)$, fatty acid, as a ^{3}H source material. A final state spectrum (FSS) in ^{3}H β-decay in this acid was calculated by Arafune et al. [1] with a self-consistent field molecular method. The reason to use this acid was as follows : Langmuir-Blodgett method [2] could be applied for preparation of the ^{3}H and reference ^{109}Cd sources. With this method, a very uniform source at the level of molecular size could be obtained and also the source thickness could be controlled in a unit of mono-molecular layer of $\sim$ 25 Å in thickness. Ag KLL Auger lines of ^{109}Cd, whose energies are around the ^{3}H end-point, were used as the reference β-rays to study the responce function, to calibrate absolute energy of ^{3}H β-spectrum and to monitor the stability of the system. In our case, both ^{3}H and ^{109}Cd sources are put in the same acid with the same chemical structure as illustrated in Fig.1.

The total responce function R_T in the measurement of the ^{3}H β-spectrum is decomposed into three components : the responce function given by the spectrometer optics (R_{opt}), the energy loss spectrum (EL) in the source substance and the backward scattering spectrum (BS) by the source backing plate. When the convolution procedure is expressed by ⊗ , the relation becomes

$$R_T = R_{opt} \otimes EL \otimes BS \tag{1}$$

The total responce function could be reliably obtained from the Auger spectra in the following way. The measured Auger spectrum includes the contribution from the natural line width (Γ) and the shake up /off effect (SUO), so that

$$\text{Measured Auger Spectrum} = R_{opt} \otimes EL \otimes BS \otimes \Gamma \otimes SUO ,$$
$$= R_T \otimes \Gamma \otimes SUO . \tag{2}$$

Therefore, R_T is obtained from measured Auger spectrum by deconvoluting the effects of Γ and SUO as

$$R_T = (\text{Measured Auger Spectrum}) \otimes \Gamma^{-1} \otimes SUO^{-1} , \tag{3}$$

where -1 on the shoulder expresses the deconvolution procedure. This is one of distinctive features in our experiment. Separate measurements of R_{opt}, EL and BS are not necessary and reliable deduction of the R_T function is possible without them. Sources used in the experiment had the thickness of two-molecular layers and activities of ∿ 8 mCi and ∿ 120 μCi for ^{3}H and ^{109}Cd, respectively. A background due to the evaporated ^{3}H molecule was completely suppressed by cooling the sources down to $-35°$C. No increase of background was observed during the measurement for more than one month. Our background (BG) is mainly due to cosmic rays and natural radioactivity in metals around the β-detector. We could subtract these BG from the measured ^{3}H β-spectrum by using the separate BG run without the ^{3}H source. When the source is cooled down, however, the energy resolution was deteriorated due to gas adsorption on the source surface. The resultant resolution including all the effects mentioned above, except Γ, was

$$\Delta E = 10 - 16 \text{ eV} \tag{4}$$

depending upon the detector position on the focal plane. Thickness of the adsorbed gas layer was measured to be about a half of mono-layer acid.   Both ^{3}H and ^{109}Cd sources were set in a vaccum chamber interchangeably and then their spectra were measured alternately.   Performance of our system described above was already reported in more detail in ref.3).

Several types of measurements as described below were carried out with the electrostatic way by changing the acceleration voltage (V_{acc}) and keeping the magnetic field strength constant.   In a general case, V_{acc} was set at $\sim$ 2.8 kV to reduce BG onto the ^{3}H β-spectrum.   This method is the same as ITEP [4) used.   Afterward it was found that this approach was unnecessary because of the complete suppression of ^{3}H evaporation mentioned above.

(a)   The ^{3}H β-spectrum was measured in the energy range from 17.4 keV to 19.2 keV by a scanning method with the energy step of 2.4 eV and the total number of 119 steps.   With this method, the spectrum measured by the individual detector cell was overlapped with these measured by neighbouring 6 cells.   Relative detection efficiency of the individual cell was determined so as to have the same total counts in the spectrum at the same energy range measured by the neighbouring cell.

(b)   The same ^{3}H β-spectrum was measured with V_{acc} $\sim$ 3.4 kV, but with the energy step of 41 eV and the total number of 22 steps.   These data were used to confirm correctness of the correction for the energy-dependent transmission variation through the examination of constancy of the end-point energy deduced from the data of the individual cell, and also used to check the relative detection efficiency of the individual cell.

(c)   Auger spectra were measured alternately with ^{3}H β-spectrum with the similar manner as in (a).   These spectra were used to deduce the total responce function R_T.   Peak positions of Auger lines were used for monitoring the long-term stability of the system and for calibrating the absolute energy of the ^{3}H β-spectrum.   Their shapes were also carefully measured to monitor the source quality, because the increase of the gas adsorption on the surface might change the EL contribution to the responce function.   Since variation of the Auger line shape was not observed, stable source quality was confirmed throughout the experiment.

(d)   The BG spectrum was measured by closing a gate valve in front of the sources.   Increase was not observed during the measurement. The background count rate was 0.005/sec/cell which was consistent with that of ^{3}H β-spectrum in the energy region higher than the end-point.

(e)   Energy dispersion at the detector position was investigated by measuring the shifts of Auger lines with V_{acc}.   13.5 eV/mm was obtained.

5 K events in the energy range of 100 eV below the end-point were accumulated for the ^{3}H source.   We applied several corrections for such as dead time of the data acquisition system (<5%), for the variation of the energy dependent transmission (<2%) and for the long-term drift of the system (<5eV).   The last one was due to the mechanical shrink of the source supporting rod by the temperature difference between the inner and outer parts of the source cooling device, while the source temperature was kept constant.   This drift was corrected by using the monitored Auger lines.   After the BG rate was subtracted, the spectrum for each cell was summed up.   Resultant ^{3}H β-spectrum is shown in Fig.2 in the form of Kurie plot.

The SUO spectrum was investigated by another separate experiment as follows.   Auger spectra were measured for two different kinds of ^{109}Cd sources without cooling, where one source had the effective thickness of 1 molecular layer and the other had of 3 layers.   The spectrum in the case of no energy loss in the source and EL spectrum were obtained from the above two spectra.   Symmetrical $R_{opt} \otimes \Gamma$ was deconvoluted from the former.   Resultant spectrum was then composed of SUO and BS.   The latter contribution was inferred from the spectrum at the energy region far below the KL_1L_1 line and was known to be negligibly small.   The SUO spectrum obtained is shown in Fig.3, and its probability and its mean energy were estimated to be 22 % and 25 eV, respectively.   A relativistic Hartree-Fock calculation in the case of free ^{109}Cd atom gives 19 % and 24 eV [5], respectively and good agreement between them was obtained. Due to the covalent bond of the outermost atomic electrons of ^{109}Cd with the acid molecule, the real SUO might be more complex than that calculated. So that, the SUO spectrum determined from the experiment was chosen in the later analysis.

The total responce function R_T was deduced from Eq.(3). Energy dependence of R_T could be evaluated by using the 5 dominant Auger lines. Resultant R_T are shown in Fig.4 together with their energy resolution.

Neutrino mass deduction was carried out by the χ^2 fit with the following formula :

$$N(E) = A_o \int \{ F(E') \ p \ E_t \ [1 + \alpha_2 (E_o - E')^2] \\ \sum_j w_j (E_o - E_j - E') \sqrt{(E_o - E_j - E')^2 - m_\nu^2} \} R(E,E') \ dE' \quad , \quad (5)$$

where $F(E)$ is the Fermi function and p, E and E_t are the momentum, kinetic energy and total energy of β-rays, respectively. E_j is the final state excitation energy with the transition probability of w_j. An empirical shape correction factor α_2 which was used by Bergkvist [6], ITEP [4] and Fritschi et al. [7], was required in fitting the data. Free parameters were the normalization factor A_o, the end-point energy E_o, α_2 and the mass-squared m_ν^2. The number of used data points was 500 which covered the energy range from 17.42 keV to 18.62 keV. R_T at every data point was obtained by interpolation from those obtained from the Auger spectra at higher and lower energies.

For FSS, Arafune et al. [1] obtained the transition probabilities for the ground state (W_{oo}), the one-electron excitation ($\sum W_{on}$), the two-electron excitation and the ionization to be 59.8%, 34.0%, 4.8% and 1.4%, respectively. The transition energy distribution was calculated there only for the former two cases, but not for the latter two. We grossly evaluated the average transition energy to be $\sim$ 78 eV for the two electron excitation by calculating a weighted mean transition ener-

gy for double one-electron excitation.   The average ionization energy
was also estimated to be $\sim$ 220 eV.   For comparison, Kaplan et al.[8]
resulted in W_{oo} = 61.2% and ΣW_{on} = 28.2% in the case of valine molecule
used by ITEP [4].   Their calculated transition energy distribution
was similar with that by Arafune et al..   While for ionization including two-electron excitation, their probability was 10.6% with an average energy of 46.5 eV. [4].

Resultant parameters for the best fit are listed in Table 1 and its
Kurie plot is shown in Fig.2.   Fig.5 shows the deviation of the data
from the fit divided by σ. χ^2 distribution vs. m_ν^2 was studied by
making all other parameters free for each value of m_ν^2 and is plotted
in Fig.6.   From the statictical consideration only, the upper limit of
m_ν^2 with 95% confidence level could be set as

$$m_\nu^2 \ < \ 780 \ \ (eV)^2 \quad , \tag{6}$$

or of the neutrino mass m_ν as

$$m_\nu \ < \ 28 \ \ (eV) \quad . \tag{7}$$

To estimate the systematic error, Γ and SUO rate were changed by
the amounts of $\pm$ 2 eV and $\pm$ 20%, respectively, in the deduction stage
of the total responce function.   Above amounts were chosen to be reasonably larger than estimated uncertainties.   Resultant m_ν^2 shifted by
$^{+105}_{-\ 8}$ $(eV)^2$ and $^{+125}_{-30}$ $(eV)^2$ in the above cases, respectively.   We took
the quadratic sum of above larger shifted amounts as the systematic
error which was

$$\Delta m_\nu^2 \ = \ 163 \ (eV)^2 \quad . \tag{8}$$

It is difficult to evaluate an uncertainty of the calculated FSS.
Without this ambuguity, by adding above systematic error linearly, resultant m_ν^2 was

$$m_\nu^2 \ < \ 943 \ \ (eV)^2 \quad , \tag{9}$$

or the upper limit on the neutrino mass was set as

$$m_\nu \ < \ 31 \ \ (eV) \quad . \tag{10}$$

Resultant end-point energy was

$$E_o = 18591 \pm 5 \quad (eV) \ , \qquad\qquad (11)$$

where the systematic error of 4 eV was linearly added to the statistical error of 1 eV.   Our systematic error was mainly arose from uncertainty of knowledge of the absolute energy of M1 line of ^{169}Yb and from the source alignment error of $\pm$ 50 μm ($\pm$ 0.7 eV) through a double calibration procedure : Auger line energies were calibrated relative to the M1 line energy of ^{169}Yb and the energy of ^{3}H β–spectrum was calibrated by Auger line energies.   Absolute energy of M1 line was cited from ref. 9).   The mass difference of the ^{3}H and ^{3}He atoms was calculated using the resultant E_o.   It was

$$\Delta M(^3H-^3He) \ = \ 18608 \pm 6 \quad (eV) \quad , \qquad\qquad (12)$$

where the recoil energy was taken to be maximum of 3 eV and the difference of the electron binding energies of the acid before and after β–decay was also taken from ref. 1).

χ^2 fits were also done by assigning the FSS of ^{3}H atom and nucleus , and their results and χ^2 distributions are shown in Table 1 and Fig.6 , respectively.   ITEP-85 [4] data are listed in Table 2 for comparison. INS m_ν^2 values systematically shifted by $\sim$ –1000 $(eV)^2$ relative to ITEP 's.   Also, χ^2 fit was performed with use of the FSS of valine molecule which was read from Fig.3 in ref.4) and the fit result is listed in Table 1.   The difference between m_ν^2 = 246.6 $(eV)^2$ and 66.7 $(eV)^2$ is greatly diminished, when we take, for instance, our average energy of the two-electron excitation and ionization to be 50–70 eV , which is similar with ITEP value of 46.5 eV   and which results in m_ν^2 = –30 – +60 $(eV)^2$ with $\chi^2 \simeq$ 542.

In summary, we had successfully carried out our first round of the neutrino mass measurement with the highest energy resolution in the world.   Our distinctive approach to use the same molecule for ^{3}H and reference ^{109}Cd sources made the unambiguous measurement of the responce function possible which made our experiment highly reliable.   The resultant preliminary upper limit of the neutrino mass was set to be

$$m_\nu \ < \ 31 \quad eV \quad .$$

Final result will be reported soon after the more detailed analysis about FSS in our acid.

Improvement of statistics is now progress where the counting rate will be increased by a factor of 30.   Together with higher counting rate, the complete suppression of the background from the ^{3}H source will improve the mass sensitivity greatly.

- We should mention that the results of the analysis presented here are different from those reported at the time of the Conference due to the more detailed analysis later.

References

1) J.Arafune, N.Koga, K.Morokuma and T.Watanabe, preprint KOBE-86-02, May 1986 ; KOBE-86-03 May 1986.

2) K.B.Blodgett, J.Am.Chem.Soc. 56 (1935) 1007 ; K.B.Blodgett and I.Langmuir, Phys.Rev. 51 (1939) 964.

3) H.Kawakami, K.Nisimura, T.Ohshima, S.Shibata, Y.Shoji, I.Sugai, K.Ukai, T.Yasuda, N.Morikawa, N.Nogawa, T.Nagafuchi, F.Naito, T.Suzuki, H.Taketani, M.Iwahashi, K.Hisatake, M.Fujioka, Y.Fukushima, T.Matsuda and T.Taniguchi, Proceedings of the VIth Moriond Workshop on Massive Neutrinos in Particle and Astrophysics, Jan.25-Feb.1,1986,France, to be published.

4) S.Boris, A.Golutvin, L.Laptin, V.Lubimov, V.Nagovizin, E.Novikov, V.Nozik, V.Soloshenko, I.Tihomirov and E.Tretjakov, Phys.Lett. 159B (1985) 217.

5) K.Yokoi, private communication, to be published.

6) K.-E.Bergkvist, Nucl.Phys. B39 (1972) 317.

7) M.Fritschi, E.Holzschuh, W.Kundig, J.W.Petersen, R.E.Pixley and H.Stussi, Phys.Lett. 173B (1986) 485.

8) I.G.Kaplan, V.N.Smutny and G.V.Smelov, Phys.Lett. 112B (1982) 417 ; Sov.Phys.JETP 57 (1983) 483 ; Phys.Lett. 161B (1985) 389.

9) V.A.lubimov, E.G.Novikov, V.Z.Nozik, E.F.Tretyakov and V.S.Kosik, Phys.Lett. 94B (1980) 266.

Table 1.   Results for fits with different FSS.

Source	m_ν^2 (eV)^2	E_o (eV)	$\alpha_2 \times 10^8$ (eV)$^{-2}$	χ^2
Arachidic acid	246.6 ± 333.8	18591.0 ± 1.0	1.31 ± 0.27	550.0
atom	-131.6 ± 376.7	18586.8 ± 1.1	1.69 ± 0.28	539.7
nucleus	-894.2 ± 447.3	18574.0 ± 1.2	1.77 ± 0.30	542.4
Valine acid*	66.7 ± 333.3	18590.6 ± 1.0	1.63 ± 0.28	540.8

* FSS was read from ref.4).

Table 2.   Results of ITEP-85 [4] for different FSS

Source	m_ν^2 (eV)^2	E_o (eV)
Valine acid	1215 ± 130	18584.2 ± 1.6
atom	954 ± 95	18580 ± 1.3
nucleus	190 ± 80	18567.4 ± 1

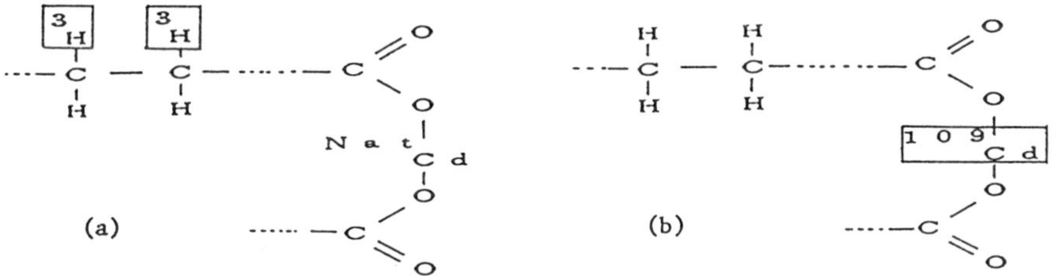

Fig.1.   Illustration of chemical structure of ^{3}H and ^{109}Cd sources.   (a) is for ^{3}H source and (b) for ^{109}Cd source. Active labelled position is indicated by ☐ .

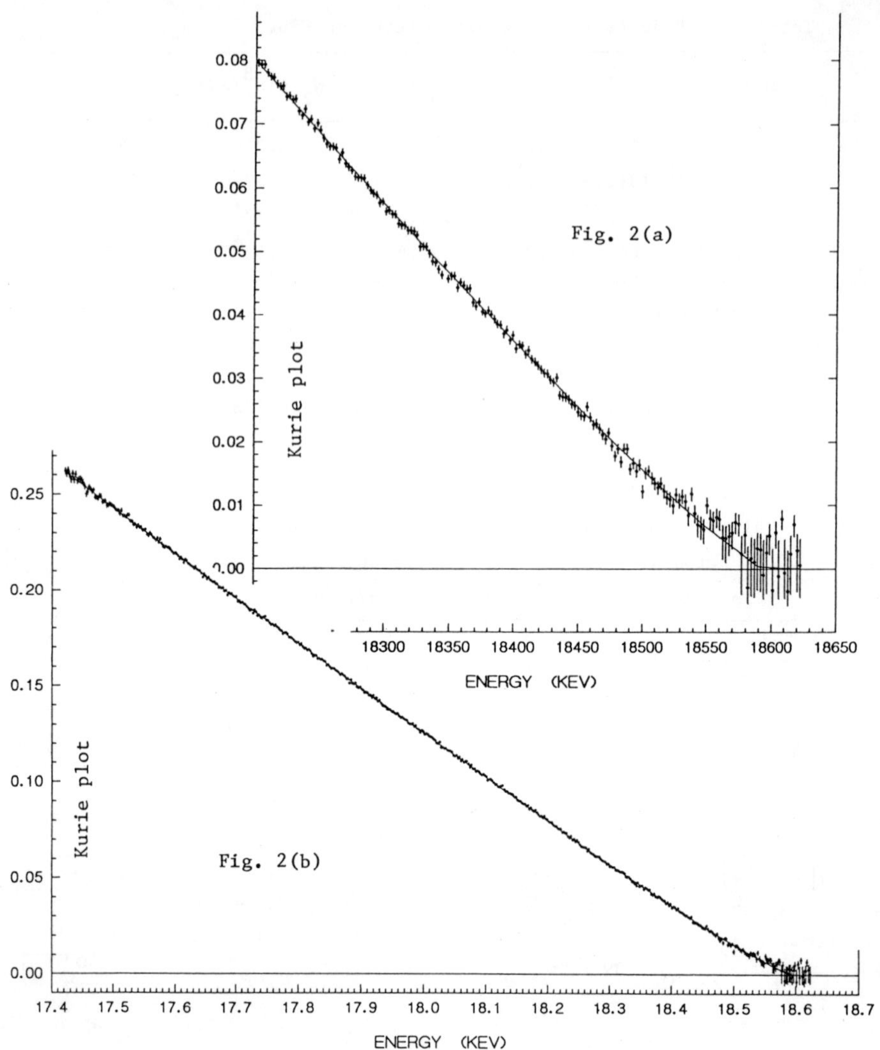

Fig. 2. Kurie plot of the observed data. Data close to the end-point are shown in Fig.2 (a) and all data are in Fig.2 (b). The line is the resultant curve at the best χ^2 fit which parameters are listed in Table 1 with "Arachidic acid".

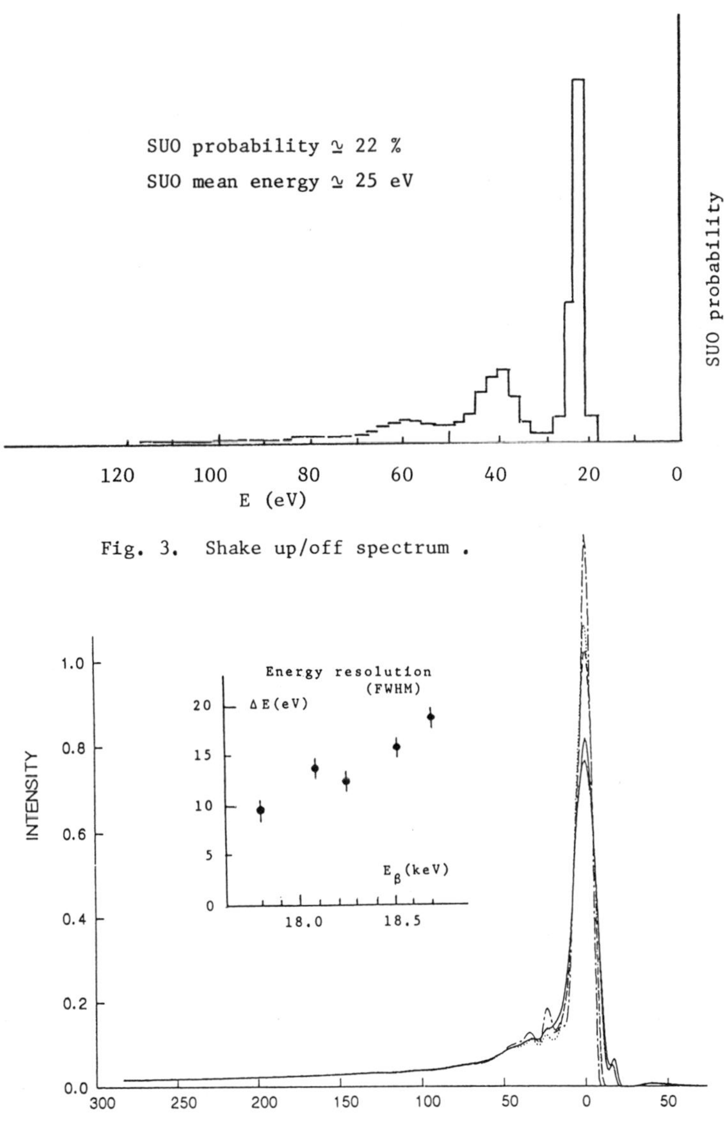

Fig. 3. Shake up/off spectrum .

Fig. 4. The total responce function, deduced from Auger spectra using Eq.(3). Sharper one in the figure corresponds to R_T at the lower energy. All R_T are normalized to have the same integrated intensity. Dependence of energy resolution on the focal plane is plotted together.

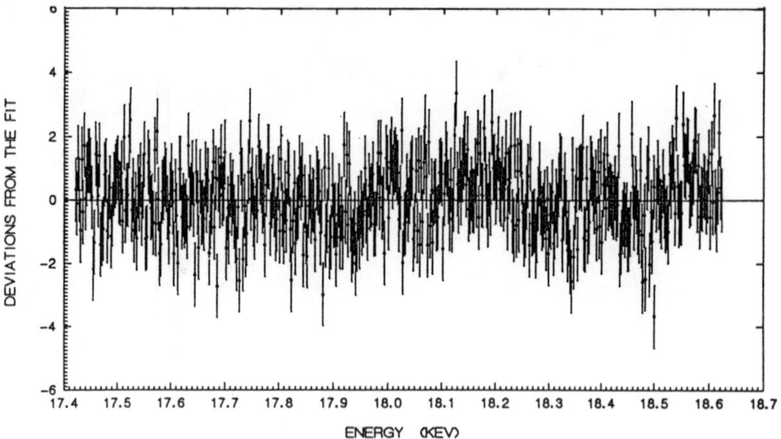

Fig. 5.   Deviation of the data from the best fit, normalized by the experimental errors σ.

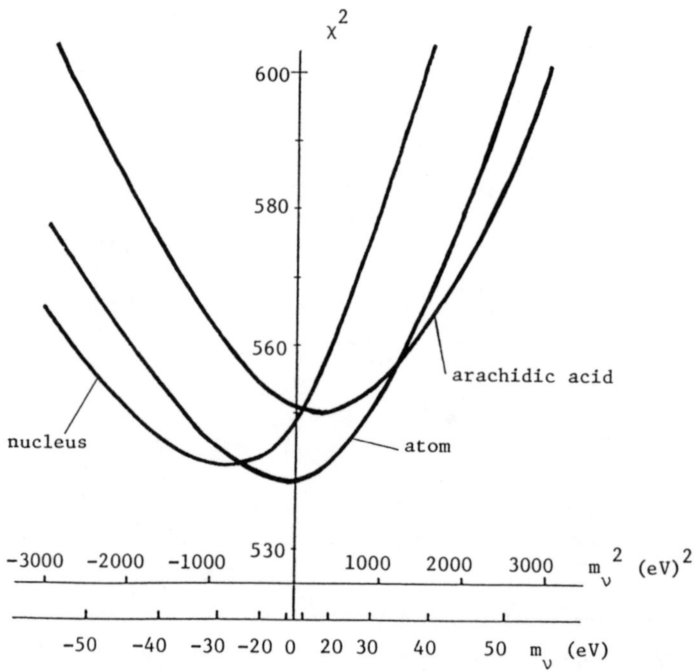

Fig. 6. χ^2 distributions vs. m_ν^2 for different FSS.

THE LOS ALAMOS EXPERIMENT ON THE BETA DECAY OF FREE
MOLECULAR TRITIUM

R.G.H. Robertson, T.J. Bowles, J.C. Browne, M.P. Maley, and
J.F. Wilkerson
Physics Division, Los Alamos National Laboratory
Los Alamos, New Mexico 87545

D.A. Knapp
Princeton University
Princeton, New Jersey 08544

J.A. Helffrich
University of California at San Diego

(Presented by R.G.H. Robertson)

ABSTRACT

The beta spectrum of the decay of free molecular
tritium has been accurately measured in order to search
for a finite $\bar{\nu}_e$ mass. The betas originating from the
decay of free tritium molecules in a differentially
pumped source region were momentum analyzed in a toroidal
beta spectrometer with 36-eV resolution. The final state
effects in molecular tritium are accurately known and the
data thus allow us to set an upper limit of 29.3 eV on
the $\bar{\nu}_e$ mass at the 95% confidence level.

Introduction

The question of a nonzero neutrino mass has received
considerable attention since Lyubimov et al.[1] in 1981 reported
evidence for an electron antineutrino mass between 14 and 46 eV, with
a best fit value of 35 eV. While the statistical evidence to support
such a claim is high, there are still concerns about possible
systematic problems in their experiment. Many of these concerns
revolve around the use of a very complex source material (tritiated
valine, an amino acid) in which the energy given up in final state
excitations of the molecule following the beta decay of a tritium
atom is comparable to the size of the neutrino mass observed. These

final state effects are difficult to calculate in a molecule as
complex as valine. In addition, ionization energy loss and
backscattering of the betas in traversing the solid source are
appreciable and must be very accurately accounted for. These
concerns have led us to develop an experiment using free molecular
tritium as the source material. The final state effects have been
accurately calculated[2,3] for the tritium molecule and the
uncertainties in these calculations are at the level of approximately
1 eV[2]. In addition, the energy loss in the source is small because
the source consists of tritium only and there is no backscattering.

<u>Method</u>

The experimental apparatus has been described in detail
elsewhere[4] and will only be briefly described here. Molecular
tritium is passed through a palladium leak, enters a 3.7-m long,
3.8-cm inner diameter aluminum tube at the center, and is pumped away
and recirculated at the ends. The tube is held at approximately 130
K to increase the source strength and is uniformly biased to
typically -8 kV. The source tube is inside a superconducting
solenoid so that betas from the decay of tritium spiral along the
field lines without scattering from the tube walls. The equilibrium
density of tritium in the source integrated along the axis was 7.8 x
10^{15} tritium atoms/cm^2. Electrons (that are not trapped in local
field minima) pass through an average thickness 2.1 times this value
as they spiral through the source gas. At one end, the betas are
reflected by a magnetic pinch and at the other end are accelerated to
ground potential. A hot filament located at the pinch emits thermal
electrons that neutralize positive ions trapped in the source,
keeping the change in source potential due to space charge buildup to
less than a volt. The betas are guided through the pumping
restriction where the tritium is differentially pumped away and then
are focused by nonadiabatic transport through a rapidly falling
magnetic field to form an image on a 1-cm diameter collimator at the
entrance to the spectrometer. The collimator defines an acceptance
radius in the source tube such that decays originating on or close to
the walls of the source tube are not viewed by the spectrometer. A
small Si detector is located at a position in front of the collimator
where it intercepts a small fraction of the betas from decays in the
source tube. Its position is set so that it does not obstruct the
view of the spectrometer and also does not see any betas originating

from decays of tritium on the source walls. This beta monitor serves
to normalize the source strength from point to point. The
spectrometer is a 5-m long, 2-m inner diameter, 72-coil toroidal beta
spectrometer similar in design to the Tretyakov instrument, but with
a number of improvements. The use of curved entrance and exit
conductors and thin (0.5 mm) conductors at the points where the betas
cross the coils results in greatly improved acceptance and
transmission. Betas from a 1.7-cm^2 area in the source tube are
transmitted with 25% efficiency through the spectrometer entrance
collimator and form a cone of 30° half angle into the spectrometer.
Betas between 19.5° and 29.5° are transmitted through the
spectrometer to a position sensitive gas proportional counter at the
focal plane of the spectrometer. The focal plane detector is 2 cm
diameter with a 2-mm-wide entrance slit. The energy resolution for
26 keV betas is 20% and the position resolution 4 mm FWHM (position
information is used to reject backgrounds outside of the slit
acceptance). The earth's magnetic field is cancelled to a level of
±10 mG in the spectrometer volume by a set of cosine coils wound
around the spectrometer, and the zero field setting is determined by
fluxgate magnetometers mounted in the spectrometer. The event rate
in the last 100 eV was typically 0.10 counts/sec.

The beta spectrum is scanned by changing the voltage applied to
the source tube so that betas of constant energy are analyzed by the
spectrometer. Accelerating the betas by several keV not only
improves the emittance of the source but also raises the energy of
betas of interest from the source well above backgrounds from betas
originating from decay elsewhere in the pumping restriction or
spectrometer. The beta monitor is biased at the same voltage as the
source tube, which results in constant-energy betas being detected by
the beta monitor.

In order to determine the overall source and spectrometer
resolution, we introduce ^{83m}Kr into the source tube in the same
manner as tritium is injected. The krypton emanates from a mixed Na-
Rb stearate[5] containing 5 mCi of ^{83}Rb, and produces a 17.835(20)-keV
K-conversion line. The intrinsic width of the line is a 2.26-eV
Lorentzian[6]. The dominant shakeup satellite is located 20 eV below
with an intensity of 8% of the main peak, as estimated by scaling the
measurements of Spears et al.[7] according to the calculations of
Carlson and Nestor[8]. The same calculations were used to assign
intensities to shakeoff satellites. The spectral distribution of

shakeoff was taken to have the 2p Levinger form[9]. Figure 1 shows the shakeoff spectrum deduced. The spectral contribution from scattering

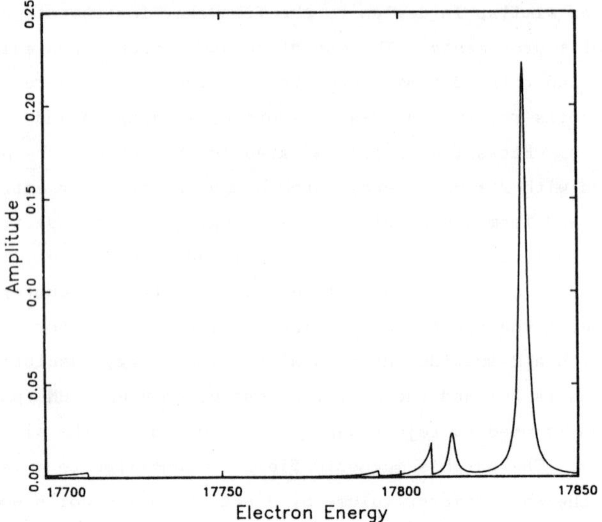

Fig. 1.   Shakeup and shakeoff spectrum of Kr.

of the conversion electrons from nitrogen molecules in the source gas (which accumulate due to the recirculation of the krypton) has been calculated using existing experimental data[10] and has been removed from the resolution function by fitting the amount of nitrogen. The fitted contributions, 10 to 20%, were proportional to measured source pressures. These measurements yield a spectrometer resolution function which has a skewed Gaussian shape with a FWHM of 52 eV for the first data set and 32 eV for the last three data sets. (The contribution of the 21-eV exit slit is not included in these figures, but is taken into account in the analysis.) The change in resolution between the data sets was due mainly to improved cancellation of residual magnetic fields from the source magnets in the region of the spectrometer. The total resolution function is obtained by correcting this instrumental contribution for scattering in the tritium gas, which has two components, both calculated by Monte Carlo methods from the known doubly differential cross sections[11] for electron scattering from H_2. Some of the electrons, 11.8%, are trapped in the source by local field minima and must multiply scatter in order to escape, and 5.2% of the untrapped electrons suffer a

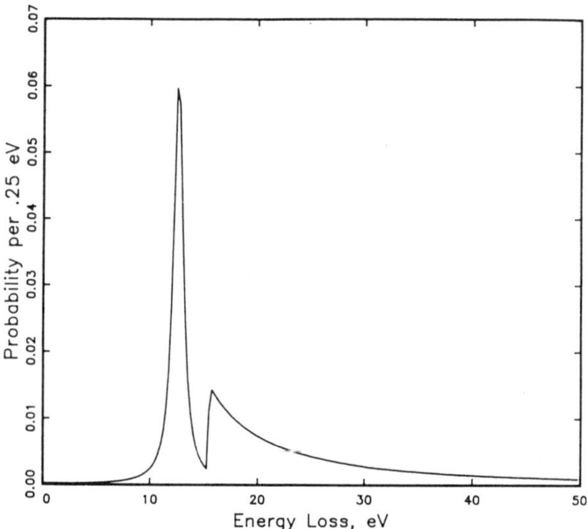

Fig. 2. Integral energy-loss spectrum for single
interactions of 18.5-keV electrons with H_2.

single scatter in the gas before being extracted (Fig. 2).

Measurements of backgrounds from the source and tritium
contamination of the spectrometer have been made and we do not
observe any backgrounds originating from the source walls or
extraction region. After operation of the source and spectrometer
with tritium for more than one month, no increase in background from
tritium contamination of the spectrometer has been observed. The
background rate in the focal plane detector has remained constant at
1 count/270 sec and is primarily due to cosmic ray muons traversing
the detector.

Three data sets were taken, each of 3-4 days duration, with
operating conditions (given in Table I) varied somewhat between runs
to check systematic effects. The first two runs were taken with the
spectrometer set to analyze 26.0-keV betas. The beta spectrum was
scanned from 16.44 to 18.94 keV in 10-eV steps. Two randomly
selected data points were taken for 600 seconds each, followed by a
200-second data run at 16.44 keV in order to check for time dependent
systematic errors. The third data set was taken in a similar manner,
except that the spectrometer was set to analyze 26.5 keV betas in
order to check for any systematic effects in varying the extraction
voltage (and therefore the extraction efficiency).   Extra data
points were taken in 5-eV steps near the endpoint in the third run.

Other data sets taken were not used because resolution measurements were not available, or the runs were incomplete.

Analysis and Results

To analyze the data, a predicted beta spectrum is generated which includes the molecular final states, Coulomb corrections, screening corrections, nuclear recoil effects, weak magnetism, and acceleration gap effects (the last three are negligible). The total resolution, including energy loss in the source, is folded with the calculated spectrum. A five-parameter fit to the amplitude (determined by the total number of events), endpoint energy, neutrino mass, background level, and a quadratic extraction efficiency term in a maximum likelihood procedure with Poisson statistics is then performed. The resulting fit (Fig. 3) is characterized by a

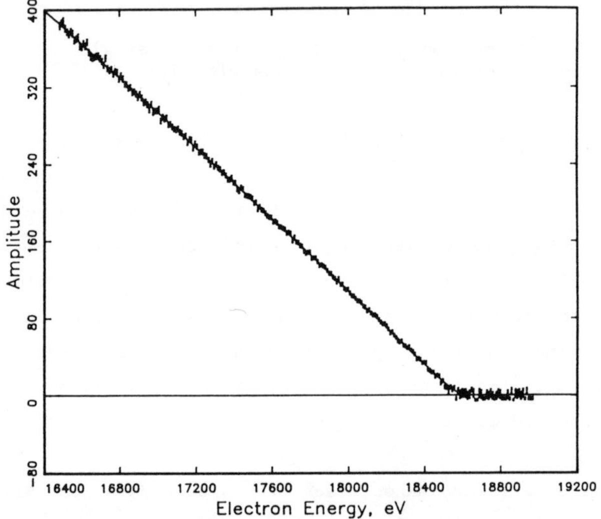

Fig. 3. Kurie plot for run 4A.

Xi-squared parameter, which is analogous to the standard chi-squared parameter:

$$\Xi^2 = 2\Sigma\{s_i - y_i - y_i\ln(s_i/y_i)\},$$

where s_i and y_i are the fit value and the measured value, respectively. (Chi-squared minimization gives a biased estimate of areas, and results in an incorrectly fitted neutrino mass.)

In Table I we summarize run parameters and fit results. The consistency between the measured endpoint energies is good, notwithstanding the large change in spectrometer resolution between runs 3 and 4A, and the change to 26.5 keV operation in run 4B. The overall uncertainty in the endpoint energy is dominated by the 20-eV uncertainty in the energy[12] of the ^{83m}Kr calibration line, however.

TABLE I. Summary of Parameters for each run and results from fitting procedure for each run. Uncertainties in Δm_ν^2 are 1σ.

	RUN 3	RUN 4-A	RUN 4-B	COMBINED
E_{SPECT} (keV)	26.0	26.0	26.5	
ΔE(FWHM,eV)	52.1 ± 1.7	32.0 ± 1.5	32.4 ± 1.3	
Skewness	.133	.153	.173	
Total Events	5,081,270	944,353	567,581	6,593,204
Counts in 100 eV	170	93	273	536
Background in 100 eV	36	28	53	117
Quadratic Term (10^{-8}/eV^2)	-1.28	-1.80	-0.64	
Number of Data Points	254	250	220	
E_0(eV)	18584.8	18585.7	18584.4	18585.0
Δm_ν^2 (statistical, eV^2)	1126	1720	688	
Δm_ν^2 (resolution, eV^2)	70	364	52	638
Δm_ν^2 (e loss, eV^2)	50	28	25	
m_ν^2 (eV^2)	-1190	1880	-63	-186

The quadratic correction term varies from run to run owing both to changes in focus coil excitation and (in run 4B) to normalization of the source intensity by interpolation between calibration points rather than by the Si detector, which had become excessively contaminated. A linear term was tried in place of the quadratic one and gave similar results but with larger variations as the fitting interval was successively truncated. Such variations were within statistics with the (fixed) quadratic term when the fitting interval below the endpoint was varied over the range 2200 to 300 eV. There was no statistical evidence that both linear and quadratic terms were required.

Statistical errors in m_ν^2 were extracted from the Ξ^2 plots (which were closely parabolic in m_ν^2). A conservatively estimated systematic error arising from imperfect knowledge of the resolution function in each run was then added linearly to the statistical error. The resolution-function uncertainties have both systematic and statistical components, but are in any case believed to be largely uncorrelated from run to run. Finally, a systematic uncertainty from the measurement of the density of the source gas and the Monte Carlo simulation of multiple scattering was added linearly to the weighted average of all runs. These were the only systematic uncertainties considered to be non-negligible.

The uncertainty in the final result is predominantly statistical (Fig. 4). An upper limit on the mass of the electron antineutrino is

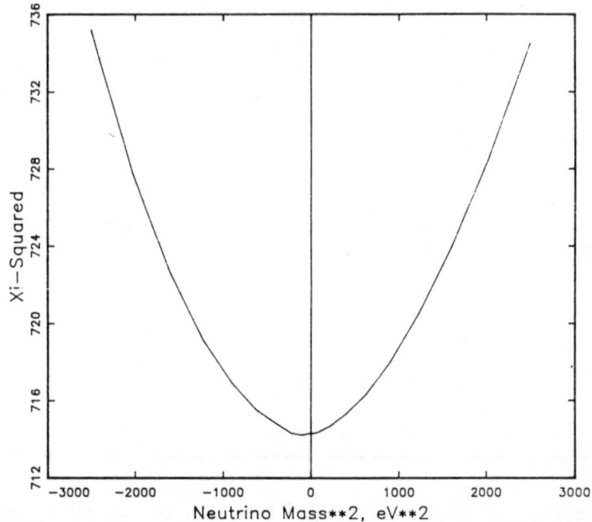

Fig. 4. Combined Xi-squared plot for the data. At the minimum, Xi-squared has the value 711 for 709 degrees of freedom.

found to be 29.3 eV at the 95% confidence level (C.L.) or 25.4 eV at the 90% C.L. It does not support the central value reported by

Lyubimov et al.[1], 30(2) eV, but neither does it exclude the lower
part of the range 17 to 40 eV. It is also in agreement with the
upper limits recently reported from solid-source experiments by
Fritschi et al.[13] and Iwahashi et al.[14]. The present result is, for
all practical purposes, model independent. Improvements to the
apparatus transmission and resolution now in progress are expected to
result in a sensitivity to neutrino mass in the vicinity of 10 eV.

We gratefully acknowledge the essential contributions of
E.Ballard, T.H.Burritt, J.L.Friar, J.D.King, D.Kleppner,
G.J.Stephenson, J.A.Wheatley, and K.Wolfsberg to the success of this
work.

References

1.   V.A.Lyubimov, E.G.Novikov, V.Z.Nozik, E.F.Tretyakov, V.S.Kosik,
     and N.F.Myasoedov, Zh. Exp. Teor. Fiz. 81, 1158 (1981); S.Boris
     et al. Phys. Lett. 159B, 217 (1985).
2.   R.L.Martin and J.S.Cohen, Phys. Lett. 110A, 95 (1985).
3.   W.Kolos et al., Phys. Rev. A31, 551 (1985); O.Fackler et al.,
     Phys. Rev. Lett. 55, 1388 (1985).
4.   See, for example, J.F.Wilkerson et al., Proc. Sixth Moriond
     Workshop on Massive Neutrinos in Particle and Astrophysics,
     Tignes, France, 1986 (to be published).
5.   K.Wolfsberg (private communication).
6.   W.Bambynek et al., Rev. Mod. Phys. 44, 716 (1972).
7.   D.P.Spears et al., Phys. Rev. A9, 1603 (1974).
8.   T.A.Carlson and C.W.Nestor, Phys. Rev. A8, 2887 (1973).
9.   J.S.Levinger, Phys. Rev. 90, 11 (1953).
10. S.M.Silverman and E.N.Lassettre, J. Chem. Phys. 42, 3420 (1965);
     T.C.Wong, J.S.Lee, H.F.Wellenstein, and R.A.Bonham, Phys. Rev.
     A12, 1846 (1975).
11. R.Ulsh, H.Wellenstein, and R.A.Bonham, J. Chem. Phys. 60, 103
     (1974); J.Geiger, Z. Phys. 181, 413 (1964).
12. S.L.Ruby et al., Phys. Lett. 36A, 321 (1971).
13. M.Fritschi et al., Phys. Lett A, 1986 (to be published).
14. H.Kawakami et al., preprint INS-Rep.-561, 1986 (unpublished).

Heavy Neutrino Searches in e^+e^- Annihilation

Takashi Maruyama

University of Wisconsin, Madison, Wisconsin, U. S. A.

ABSTRACT

Recent searches for heavy neutrino production in e^+e^- annihilation are reviewed. The results from searches by the CELLO, JADE, HRS, MARKII, and MAC groups are presented.

1. Introduction.

Fermion pair production in e^+e^- annihilation at the lowest order proceeds via virtual photon or virtual Z^0 annihilation. At PEP/PETRA energies the photon annihilation is the dominant contribution for charged fermion production. Although the Z^0 contribution is only about one per cent of the photon contribution for charged fermion production, it is the only channel for neutral fermion production since the photon annihilation term is absent. With more than five years of accumulated data, the PEP/PETRA experiments are now capable of meaningful searches for heavy neutrino production.

2. Event yields and experimental conditions.

Table 1 summarizes the experimental conditions and expected yields for the reactions $e^+e^- \to \mu^+\mu^-$ and $e^+e^- \to \nu\bar{\nu}$ at PEP and PETRA.

Table 1

	PEP	PETRA
$\sqrt{s}$ (GeV)	29	44
$\int \mathcal{L}$ dt (pb^{-1})	300	100
$\sigma_{\mu\mu}$ (pb)	103	45
N$_{\mu\mu}$ (event)	30,900	4,500
$\sigma_{\nu\bar{\nu}}$ (pb)	0.36	1.1
N$_{\nu\bar{\nu}}$ (event)	108	110

Here, $\sigma_{\nu\bar{\nu}}$ is the total cross section for massless neutrino pair production and sets the cross section scale for the heavy neutrino searches. The PETRA experiments have the advantage of a larger cross section due to the higher beam energy, while the PEP experiments have more total integrated luminosity. The PEP and PETRA experiments have comparable signals. However, the PEP experiments could have more serious backgrounds from the virtual photon annihilation processes than the PETRA experiments. Since the neutrino is a weakly interacting particle, the search is only feasible for massive neutrinos decaying inside detectors.

3. Electron-type heavy neutrino.

Heavy neutrinos are expected in a model with right handed doublets: $\left(\begin{smallmatrix} E^0 \\ e \end{smallmatrix}\right)_R$, $\left(\begin{smallmatrix} M^0 \\ \mu \end{smallmatrix}\right)_R$, $\left(\begin{smallmatrix} T^0 \\ \tau \end{smallmatrix}\right)_R$, where E^0, M^0 and T^0 are right handed neutrinos associated with e, μ and τ, respectively. The electron-type heavy neutrino can be produced in e^+e^- annihilation via W-exchange as shown in Fig. 1. The production cross section is approximately 10% of $\sigma_{\mu^+\mu^-}$. If E^0 decay proceeds through the conventional charged current interaction, the decay is similar to that of a sequential charged lepton as shown in Fig. 1. The event topology would be either i) acoplanar dileptons, ii) a single jet, or iii) acoplanar two-jet.

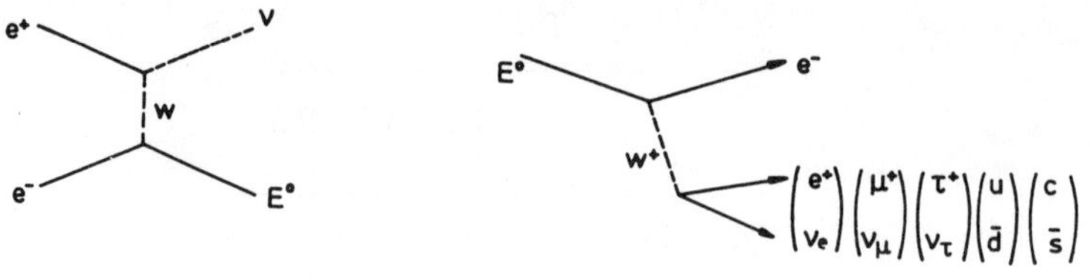

Fig. 1 *Production and decay of electron-type heavy neutrino.*

• The JADE group has searched for single jet and acoplanar two jet events at $\langle\sqrt{s}\rangle$=34.2GeV. [1] No unusual events were observed beyond expected backgrounds. E^0 production was excluded for the mass range $1 \leq m_{E^0} \leq 24.5$ GeV/c^2 (V+A), and $1 \leq m_{E^0} \leq 22.5$ GeV/c^2 (V−A), where the coupling type represents the E^0-e coupling to W.

• More recently, the CELLO group has also searched for E^0 by explicitly requiring leptons in the event. [2]   They searched for acoplanar e-μ events and single jet events associated with an electron at $\langle\sqrt{s}\rangle$=44.2GeV. No excess of such events was observed and the excluded E^0 mass ranges are $0.4 \leq m_{E^0} \leq 37.4$ GeV/c^2 (V+A) and $0.6 \leq m_{E^0} \leq 34.6$ GeV/c^2 (V−A).

4. Fourth generation heavy neutrino.

Production of a new sequential heavy charged lepton has been excluded up to a mass of 22.7 GeV/c^2. [3] Since the neutrino partner is very likely to be lighter than the associated charged lepton of such a new generation, a neutrino search may provide evidence for a new generation. If such a neutrino is massive, it may mix with other neutrinos. For example, the electron-neutrino would be a superposition of the mass eigenstates of all generations, $\nu_e = \sum_i U_{1i}\nu_i$, where the matrix U is a mass mixing matrix analogous to the Kobayashi- Maskawa matrix for quarks.

Again, if the decay proceeds through a conventional charged current interaction, it is similar to that of a sequential charged lepton as shown in Fig. 2 and the decay modes and branching ratios can be calculated in a model independent way. [4] The lifetime of a sequential heavy neutrino N is given in terms of the muon lifetime and the mass mixing matrix as:

$$\tau(N \to l^- X^+) = \left(\frac{m_\mu}{m_N}\right)^5 \frac{\tau(\mu \to e\nu\bar{\nu})B(N \to le\bar{\nu})}{|U_{lN}|^2},$$

where l is the lepton to which N primarily couples, and B is the leptonic decay branching ratio of the heavy neutrino.

4-1 Single production.

Sequential heavy neutrinos can be produced singly in association with the electron-neutrino as shown in Fig. 3. This diagram is identical to that of the E^0 production except for the mass mixing factor U_{41} at the e-N vertex. The production cross section is given by[5]

$$\sigma_{N\nu_e} = |U_{14}|^2 \frac{G_F^2 s}{6\pi}\left(1 - \frac{m_N^2}{s}\right)^2 \left(1 + \frac{m_N^2}{2s}\right).$$

The event topology would be either i) acoplanar dileptons, ii) a single jet for a small mass neutrino ($m_N \leq 10$ GeV/c^2), or iii) acoplanar two-jet for a large mass neutrino.

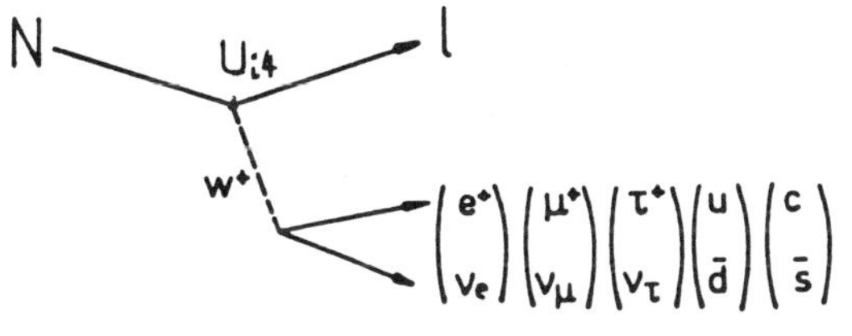

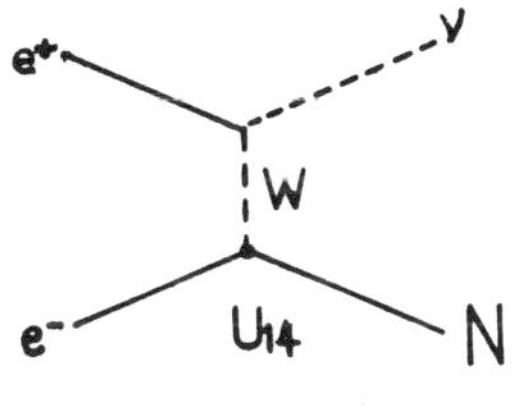

Fig. 2 *Heavy neutrino decay.*　　Fig. 3 *Single production of a heavy neutrino.*

No group has searched explicitly for heavy neutrino production using this process. However, there are good limits on single jet production. Searches for single jet production in e^+e^- annihilation have been made to test a model for UA1 "monojet" events.[6] The HRS, MARKII, and MAC groups at PEP and the CELLO and JADE groups at PETRA have made such searches.[7] No monojet production was observed beyond expected backgrounds.

Gilman and Rhie have pointed out that the upper limits obtained by the experiments can be re-interpreted to place upper limits on the mass mixing matrix $|U_{14}|^2$ in single heavy neutrino production.[5] Based on PEP monojet searches, an upper limit $|U_{14}|^2 \leq 2 \times 10^{-3}$ was obtained for the mass range $2 \leq m_N \leq 12$ GeV/c². Since the JADE group searched for acoplanar two jet events in addition to single jet events, these limits can be extended up to about 20 GeV/c².

4-2. Pair production.

Heavy neutrinos can be pair produced by virtual Z^0 decay as shown in Fig. 4(a). There is no mass mixing factor in this case. Heavy neutrinos can also be pair produced by W-exchange as shown in Fig. 4(b), where there are two mass mixing factors at the W-vertices. Considering the limits on $|U_{14}|^2$ obtained by the single jet production, this cross section is assumed neglegible compared to the Z^0 decay process. The production cross section for Fig. 4(a) is given by $\sigma_{N\bar{N}} = \sigma_{\nu\bar{\nu}}\beta(3 + \beta^2)/4$, where $\sigma_{\nu\bar{\nu}}$ is given in Table 1 and β is the velocity of heavy neutrino.

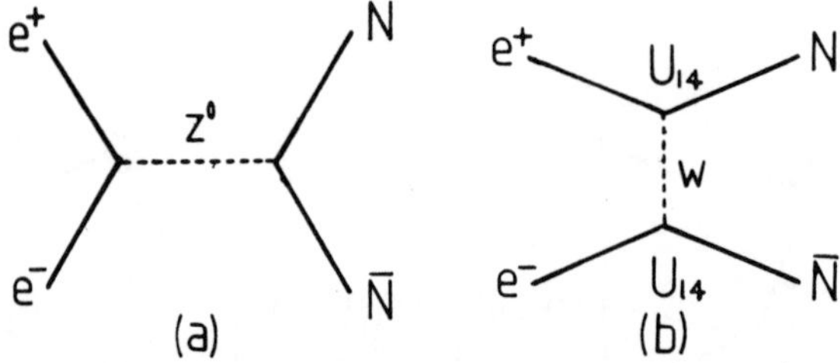

Fig. 4 *Pair production of heavy neutrino: (a) Z^0 decay, (b) W-exchange.*

Depending on the lifetime and decay mode of the heavy neutrino, six different event signatures are possible as shown in Fig. 4. Characteristics of heavy neutrino production are that there are at least two leptons and for long-lived heavy neutrino two separate decay vertices.

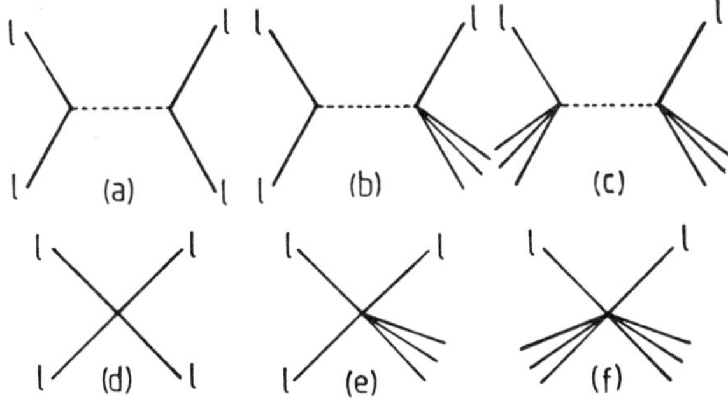

Fig. 5 *Event topologies of heavy neutrino production. (a), (b), and (c) are for long-lived N. (d), (e), and (f) are for short-lived N. (a) and (d) are for leptonic-leptonic decay. (b) and (e) are for leptonic-hadronic decay. (c) and (f) are for hadronic-hadronic decay.*

• The CELLO group has found an unusual event at $\sqrt{s}$=43.5 GeV.[8] This event has two high momentum muons (11.0 and 12.6 GeV/c), two jets, and little missing energy. More interestingly, the invariant masses of (μ_1, jet_1) and (μ_2, jet_2) are about equal, 22.2±1.6 and 19.4±1.3 GeV/c^2. Since this event was reported, many speculations have been made to explain it. Heavy neutrino pair production: $e^+e^- \to N\bar{N}$ followed by N$\to \mu$+jet decays is one of them.[9] Expected backgrounds consist of multi-hadron production with muon misidentification and the α^4 order QED process $e^+e^- \to \mu\mu q\bar{q}$. The expected backgrounds in the data sample is estimated to be on the order of 10^{-3} events. The group has continued the search for similar events with increased integrated luminosity. No new events have been found in a data sample about ten times larger than before.

• The HRS group has searched for a short-lived heavy neutrino with the event topology shown in Fig. 4(e),[10] events with an isolated electron and a minimum-ionizing charged particle in one hemisphere, and more than two charged particles in the other hemisphere. Seven candidates were observed, but these events were consistent with the expected backgrounds of 5.5±2.2 events from D-meson decay and particle misidentification. The limits on $\sigma \cdot$B depend on the heavy neutrino mass and lifetime, and range from 0.08 to 0.2pb. However, these limits are above the rate expected for a new sequential neutrino.

• The MARKII group has searched for short-lived heavy neutrinos in events with four charged particles.[11] Their search is sensitive to the event topology shown in Fig. 5(d) and to four prong final states of the types in Fig. 5(e) and (f). They have looked for events which are kinematically consistent with heavy neutrino pair-production $e^+e^- \rightarrow N\bar{N}$ followed by N$\rightarrow x^+y^-$ and N$\rightarrow u^+v^-$ decays. No events were observed with invariant masses of (x^+, y^-) and (u^+, v^-) between 0.11 and 5.4 GeV/c^2. Fig. 6 shows the upper limits on heavy neutrino production as a function of the neutrino mass, where σ_{stan} is the cross section expected from a sequential neutrino, *i.e.* $\sigma_{stan} = \sigma_{\nu\bar{\nu}}\beta(3 + \beta^2)/4$. Although the limits obtained are smaller than σ_{stan} for m$_N \leq 2$ GeV/c^2, there is a caveat. For a heavy neutrino with a mass less than 2 GeV/c^2, the expected decay distance is longer than about 1 mm if the neutrino is a conventional sequential lepton. Since the search explicitly required that the particles come from the primary interaction point, their search is only sensitive to unconventional heavy neutrinos with a short lifetime.

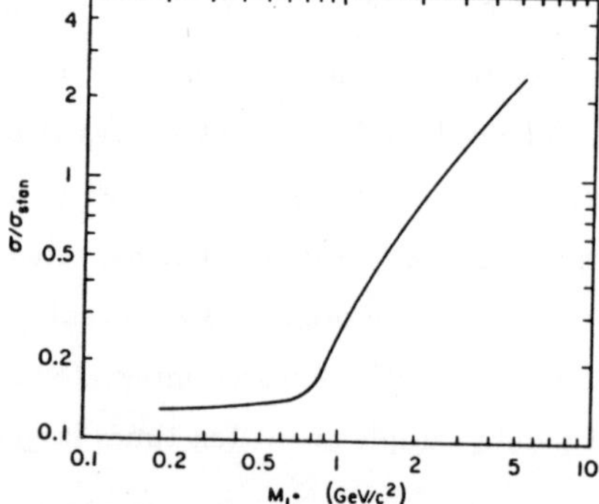

Fig. 6 *The 90% C.L. upper limits on σ/σ_{stan} obtained by the MARKII search.*

• The first significant search for heavy sequential neutrinos was made by the MARKII group.[12] They searched for a long-lived heavy neutrino by looking for finite decay distances as shown in Fig. 4(a), (b) and (c).

The basic selection criteria were

1) Charge multiplicity N$_{CHG} \geq 4$,

2) One vertex with $0.2 \leq r_1 \leq 10$cm, where r_1 is the radial distance between the vertex and the interaction point. If $r_1 \leq 3$ cm or $N_{CHG} = 4$, a second vertex with $r_2 \geq 0.2$ cm was required,

3) Tracks from identified K_s^0 and Λ particles and from beam pipe interactions were removed.

After all the cuts, three candidates were observed. However, none of these events were kinematically consistent with heavy neutrino production. The excluded region in the space of $|U_{l4}|^2$ and m_N is shown in Fig. 7. The limits were calculated separately for the heavy neutrino coupling to e, μ, and τ.

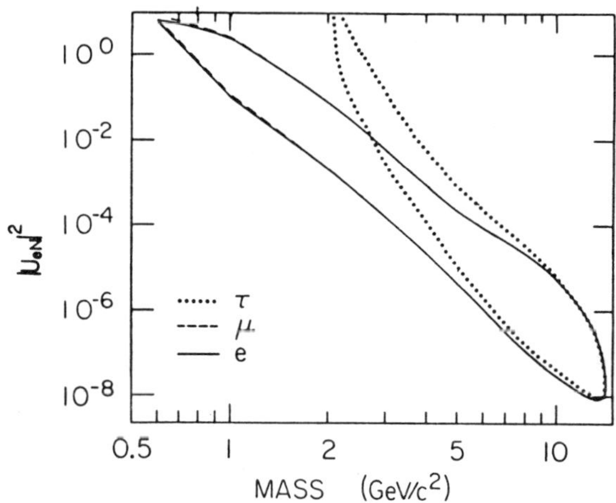

Fig. 7 *Excluded region in the space of $|U_{lN}|^2$ and m_N obtained by the MARKII search.*

• More recently the CELLO[2] and MAC groups have also searched for heavy neutrinos in hopes of extending the MARKII limits. The upper half region in Fig. 7 not excluded by the MARKII search corresponds to a short-lived heavy neutrino. To search for such short lived heavy neutrinos, a multi-lepton signature is required. Table 2 summarizes the basic selection criteria used by the CELLO and MAC experiments.

Table 2. Search criteria used by the experiments.

	CELLO	MAC
4 prong	$\cdot \Sigma$ Q = 0	$\cdot \Sigma$ Q = 0
	$\cdot \geq 2$ leptons of same type and opposite charge	$\cdot \geq 3$ leptons
	$\cdot$ at least one lepton p $\geq$ 4 GeV/c	$\cdot \Sigma$ $p^{lepton} \geq$ 5 GeV/c
	$\cdot$ kinematic fit to reject QED	$\cdot$ kinematic fit to reject QED
$\geq$ 5 prong	$\cdot \geq 2$ leptons of same type and opposite charge	$\cdot$ 2 leptons with opposite charge in one hemisphere
	$\cdot$ $p^{lepton} \geq$ 4 GeV/c	$\cdot \Sigma$ $p^{lepton} \geq$ 5 GeV/c
	$\cdot$ reject $\gamma \to e^+e^-$	$\cdot$ reject $\gamma \to e^+e^-$

The CELLO group identifies leptons with p $\geq$ 2 GeV/c. The MAC group identifies electrons with p $\geq$ 1 GeV/c and muons with p $\geq$ 1.5 GeV/c.

Dilepton events are also expected from one photon annihilation into heavy quark-pairs, $e^+e^- \to c\bar{c}/b\bar{b}$ followed by semi-leptonic decays, c/b$\to$ l +X. Since this is a serious background in the MAC search, they did not search for the particular event topology shown in Fig. 5(f). Other backgrounds expected from α^4 order QED processes are $e^+e^- \to ee\tau\tau$, $eeq\bar{q}$, $\mu\mu\tau\tau$, and $\mu\mu q\bar{q}$. These backgrounds are estimated by the Monte Carlo program of Berends, Daverveldt and Kleiss.[13] The CELLO and MAC results are shown in Tables 3 and 4.

Table 3. Results of the CELLO search.

	Candidates	Backgrounds
4 prong	1	0.74 $\pm$ 0.07
$\geq$ 5 prong *ee*	0	1.4 $\pm$ 0.3
$\mu\mu$	1	0.13 $\pm$ 0.03

Table 4. Results of the MAC search.

	Candidates	Backgrounds	α^4-QED	multi-hadron
4 prong	2	2.3 ± 0.8	1.9 $ee\tau\tau$ 0.4 $\mu\mu\tau\tau$	-
$\geq$ 5prong ee	8	6.6 ± 1.4	3.9 $eeq\bar{q}$ 0.6 $ee\tau\tau$	2.1
$\mu\mu$	3	3.0 ± 0.7	2.6 $\mu\mu q\bar{q}$ 0.1 $\mu\mu\tau\tau$	0.3
$e\mu$	1	1.2 ± 0.3	-	1.2

The candidate events are consistent with the expected backgrounds and heavy neutrino production has been excluded in the following mass regions:

	N couples to e	N couples to μ
CELLO (at 95% c.l.)	$3.1 \leq m_N \leq 18.0$ GeV/c^2	$3.2 \leq m_N \leq 17.4$ GeV/c^2
MAC (at 90% c.l.)	$1.0 \leq m_N \leq 9.4$ GeV/c^2	$1.0 \leq m_N \leq 9.6$ GeV/c^2.

These limits apply to a short-lived heavy neutrino. Since the experiments did not require that the particles come from the primary interaction point, the selection criteria are also sensitive to long-lived heavy neutrinos with decay distance less than about 20 cm. None of the candidates have finite decay vertices and thus a long-lived heavy neutrino is excluded as well. The excluded region in the space of $|U_{e4}|^2$ and m_N is shown in Fig. 8, where the dashed curve represents the region excluded by the MARKII search.

5. Conclusion.

Searches for heavy neutrinos in e^+e^- annihilation have been made at PEP and PETRA. There is no experimental evidence for the heavy neutrino production. Stringent limits on the mass mixing matrix have been obtained.

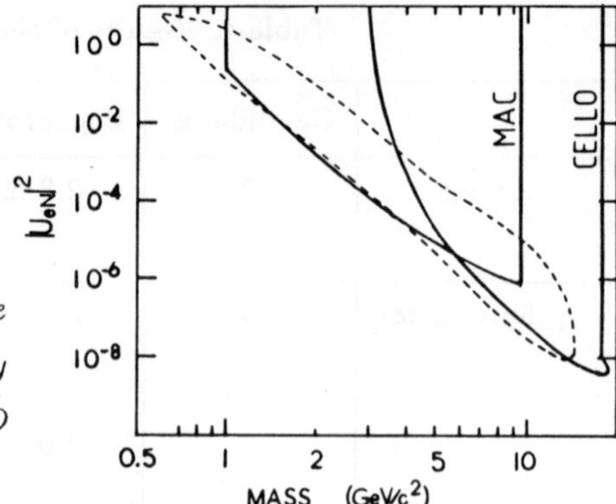

Fig. 8 *Excluded region in the space of $|U_{eN}|^2$ and m_N obtained by the MAC (90% C.L.) and CELLO (95% C.L.) searches.*

Acknowledgements

I would like to thank the conference organizers, especially Prof. T. Kitagaki, for inviting me to the conference, and R. Prepost, H. Band and T. Lavine for discussions and proofreading the manuscript.

REFERENCES

1. W. Bartel *et al.*, Phys. Lett. 123B, 353 (1983).

2. R. Aleksan, Saclay Report DPhPe 86-08 (1986, unpublished).

3. S. Komamiya, in *Proceedings of the 1985 International Symposium on Lepton and Photon Interaction at High Energies*, ed. by M. Konuma and K. Takahashi (Kyoto, 1985), p. 612

4. Y. S. Tsai, Phys. Rev. D4, 2821 (1971).

5. F. J. Gilman and S. H. Rhie, Phys. Rev. D32, 324 (1985).

6. G. Arnison *et al.*, Phys. Lett. 139B, 115 (1984).

7. H. J. Behrend *et al.*, Phys. Lett. 161B, 182 (1985); W. Bartel *et al.*, Phys. Lett. 155B, 288 (1985); C. Akerlof *et al.*, Phys. Lett. 156B, 271 (1985); G. J. Feldman *et al.*, Phys. Rev. Lett. 54, 2289 (1985); W. W. Ash *et al.*, Phys. Rev. Lett. 54, 2477 (1985).

8. H. J. Behrend *et al.*, Phys. Lett. 141B, 145 (1984).

9. J. L. Rosner, Nucl. Phys. B248, 503 (1984).

10. D. Errede *et al.*, Phys. Lett. 149B, 519 (1984).

11. M. L. Perl *et al.*, Phys. Rev. D32, 2859 (1985).

12. G. J. Feldman, SLAC-PUB-3684 (1985, unpublished).

13. F. A. Berends, P. H. Daverveldt and R. Kleiss, Nucl. Phys. B253, 441 (1985).

SEARCH FOR HEAVY NEUTRINO IN BETA RAY SPECTRA

T. Ohi, M. Nakajima, H. Tamura, T. Matsuzaki, F. Shimokoshi,
R.S. Hayano, T. Yamazaki,

Department of Physics and Meson Science Laboratory
Faculty of Science, University of Tokyo
7-3-1 Hongo, Bunkyo-ku, Tokyo 113, Japan

and O. Hashimoto

Institute for Nuclear Study, University of Tokyo
Tanashi, Tokyo 188, Japan

Heavy neutrinos can be searched for from anomalous peaks in the
two-body decay of π mesons and K mesons, as proposed by Shrock [1].
This method is sensitive to very small mixing ($\approx 10^{-6}$) in the mass
range between 5 and 300 MeV [2]. The higher mass range has been studied
by trying to observe the decay of such a hypothetical heavy neutrino
[3]. In this way very stringent upper bounds have been obtained. On
the other hand, the lower mass region can be studied from a precise
measurement of beta ray spectra. From the search for a kink typical for
the mixing of a heavy neutrino ν_i one can deduce its mass $m(\nu_i)$ and
mixing ratio $|U_{ei}|^2$. Schreckenbach et al. [4] measured β^+ and β^-
spectra from ^{64}Cu by using a magnetic spectrometer and obtained an upper
limit around 10^{-2} in the mass range 30-460 keV. Simpson [5] gave
another limit in the mass range below 10 keV from a spectrum of tritium
embedded in a Si detector. Recently Simpson [6] reported a kink in a
tritium spectrum and ascribed this to a heavy neutrino of 17 keV with a

mixing ratio of 3 %. This interpretation was, however, inconsistent
with experiments on other beta ray emitters such as ^{35}S with an end
point energy 167 keV [7-10] and of ^{63}Ni with an end point energy 65.8
keV [11].

In order to search for heavy neutrinos in a wider mass range with
a better sensitivity we made a dedicated apparatus which permits a
high-rate and high-resolution multi-channel measurement of beta rays up
to 2.8 MeV. It consists of two Si(Li) detectors, A and B, placed face
to face, sandwiching a beta ray source, as shown in Fig.1. Each
detector has size of 7 mm thickness and 40 mm sensitive diameter. The
total energy of beta rays is absorbed in this 4π set up. Since the
probability to hit either one of the detectors (no multiple hit) is 70%,

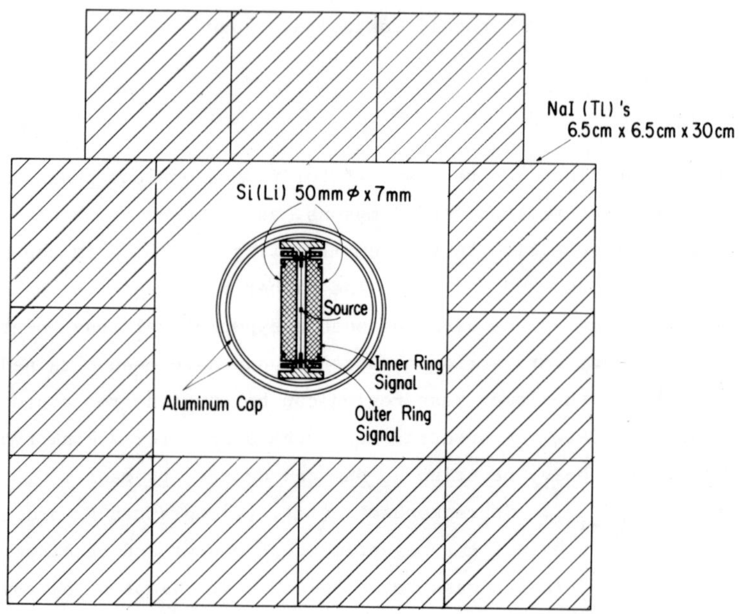

Fig.1

The present experimental set up. The detection system consists of
two large Si(Li) detectors. The beta-ray source is placed in a
narrow gap between the detector surfaces.

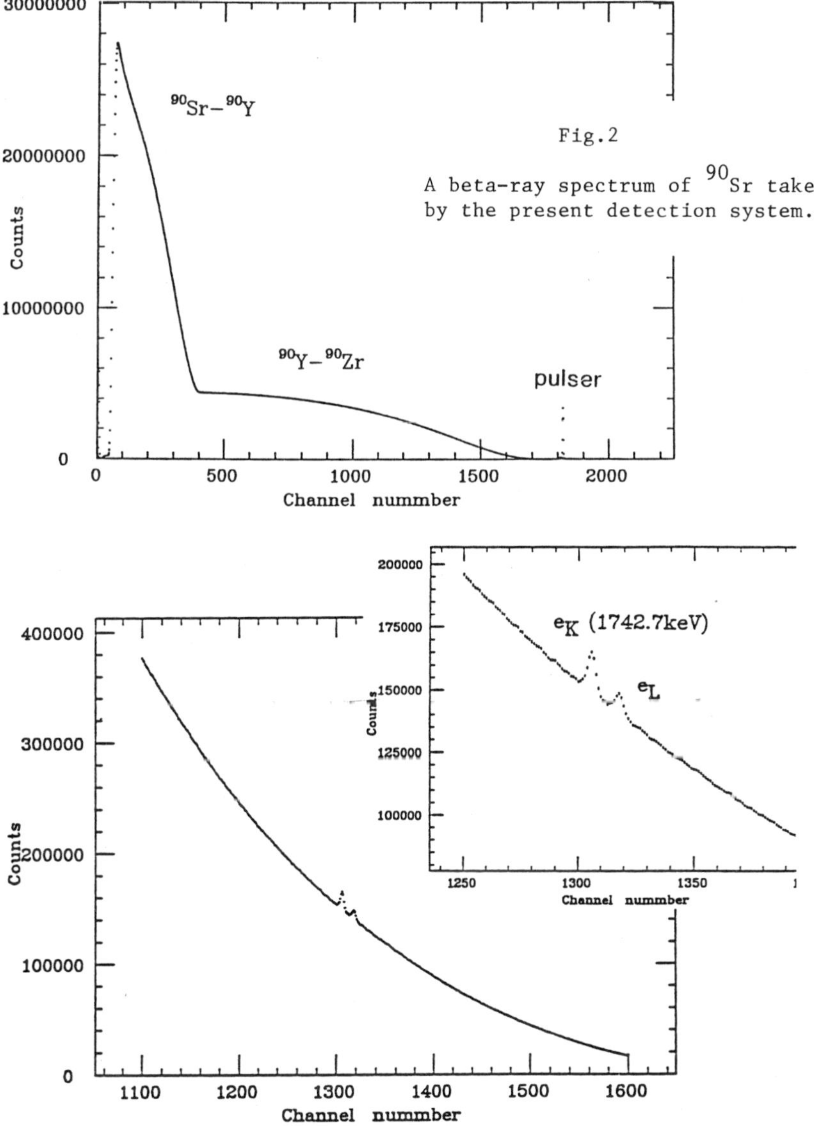

Fig.3 A portion of a beta-ray spectrum of ^{90}Sr coincident with
the opposite detector, revealing E0 conversion electron
peaks of ^{90}Zr. This logic enhances the $\beta^- \to$ E0 sequence.

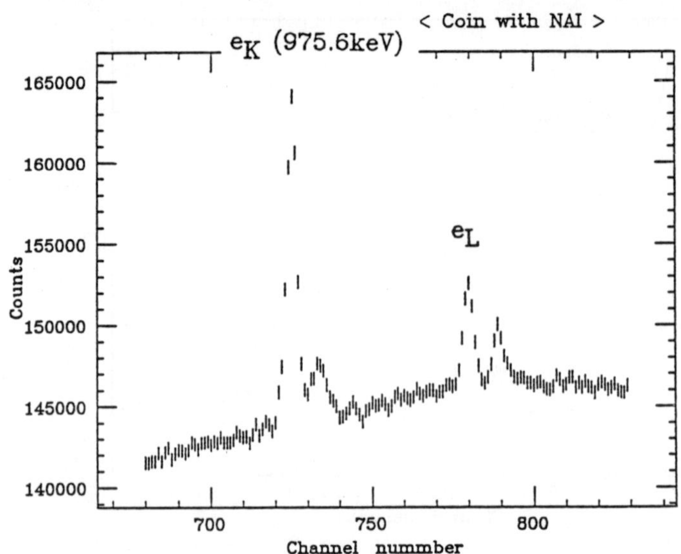

Fig.4 · A portion of a beta-ray spectrum of ^{90}Sr coincident
with the NaI(Tℓ) detectors, revealing the conversion
electron lines of the 1064 keV transition of a slightly
mixed ^{207}Bi. This logic enhances the γ-decay cascade
in the decay of ^{207}Bi.

we took only single-hit events. The resolution of the detector is
typically 2 keV, good enough to hunt for a kink. The Si(Li) detector
assembly was surrounded by NaI(Tl) detectors which served to veto
bremsstrahlung events of beta rays in the detector and also cosmic ray
background. The typical counting rate was 3 x 10^4 /sec.

Precise measurements of ^{90}Sr and ^{35}S were carried out. A
preliminary result on ^{35}S was reported already [7]. The ^{90}Sr decays to
^{90}Y with Q$_\beta$ = 546 keV and then decays to ^{90}Zr with Q$_\beta$ = 2284 keV. Fig.2
shows a spectrum of ^{90}Sr, which is composed of two components. We are
interested in the higher component. The decay of ^{90}Y populates the 0$^+$
excited state of 1761 keV (0.011%) in ^{90}Zr which proceeds to the ground
state by emitting an E0 conversion electron. This extremely small
branch was detected in the present measurement, in particular by taking
coincidence A x B, as shown in Fig.3. Thus, the K and L conversion

electron peaks were used as an internal calibration which assured that there was no drift in the pulse height in a long run. Another internal calibration was obtained from a mixture of a very small amount of ^{207}Bi. The conversion electron peaks of the 1064 keV transition were hardly

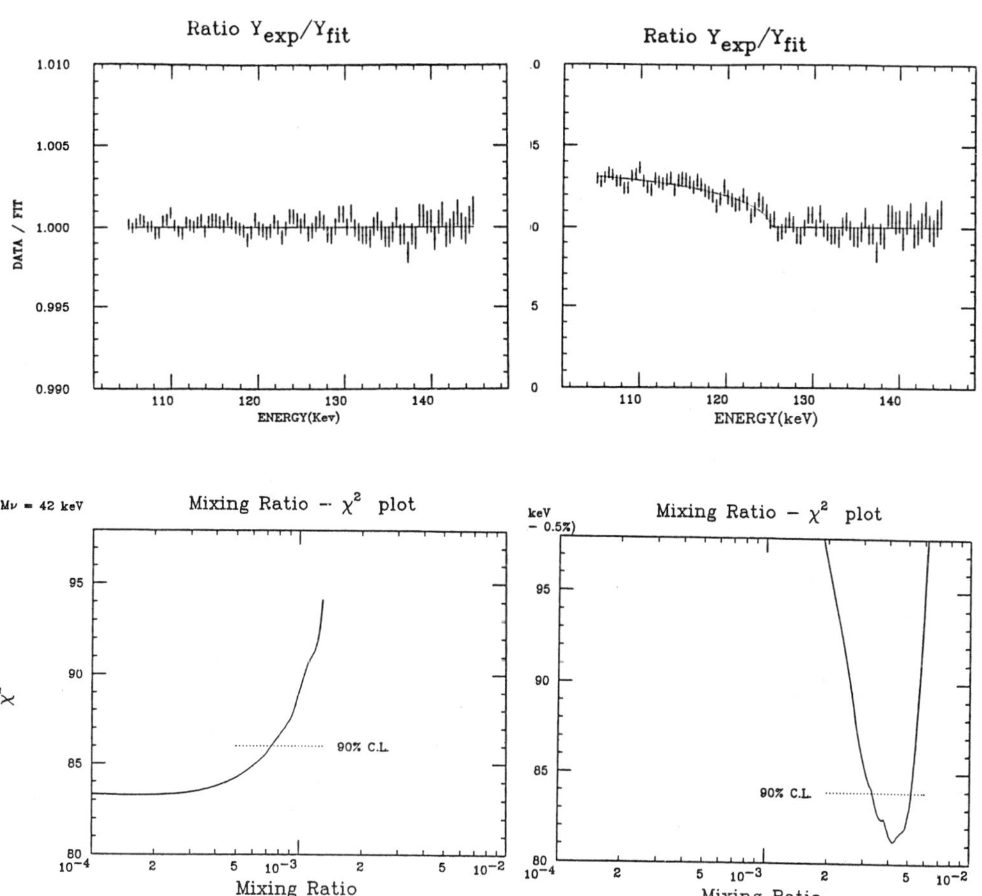

Fig.5

Experimental data of ^{35}S for the chi-squares test at $m_{\nu i}$=42 keV.

Fig.6

Fake data with $m_{\nu i}$=42 keV and B_i=0.5% for the chi-squares test.

visible in a singles spectrum, but appeared distinctly in a coincidence spectrum with the surrounding NaI(Tl) detectors, as shown in Fig.4.

Both differential and integral linearities in the ADC's are essential in the present experiment. We tested commercially available ADC's and chose the best ones. Furthermore, we employed a digital compensation system for differential non-linearity.

The spectra obtained were analysed in the following way. The ratio of the experimental spectrum to the theoretical one is not constant in general, reflecting instrumental distortion of the spectrum due to the response function of the detector system, energy losses, etc. The correction factor is, however, expected to be a smooth function of energy, thus represented by a suitable function such as polynomials, while the presence of a heavy neutrino with a mixing ratio $B_i = |U_{ei}|^2$ causes a characteristic kink which is not confused with an instrumental correction factor. So, for a given heavy neutrino mass $m_{\nu i}$, we fitted the experimental data in a region centered at $E_\beta = E_\beta^{max} - m_{\nu i}c^2$ with a test function which involves a mixing ratio B_i as a parameter. In this way we obtained a chi-squares versus B_i for each $m_{\nu i}$, which yields either a non-zero value of B_i (actually, this was not the case) or an upper limit of B_i. Fig.5 shows an experimental data of ^{35}S, with a corresponding chi-squares curve at $m_{\nu i}$ = 42 keV. This gives a 90% C.L. upper limit for B_i. On the other hand, Fig.6 shows a fake data which simulates a heavy neutrino mixing B_i = 0.5%. The chi-squares analysis gives B_i = 0.42 ± 0.08%, consistent with the input.

In this way we obtained "model independent" upper bounds for both ^{35}S and ^{90}Sr. The results are presented in Fig.7 together with other data. In this search we have observed no heavy-neutrino kink anywhere, but arrived at a stringent limit around 10^{-3} in a wide mass range. The 17-keV heavy neutrino of Simpson is definitely excluded.

We thank Drs. F. Goulding, D.A. Landis, N.W. Madden and J.T. Walton for the fabrication of the Si(Li) detector, and Dr. M. Imamura for the help in source preparation. This work was supported in part by the Grant-in-Aid of the Japanese Ministry of Education, Culture and Science.

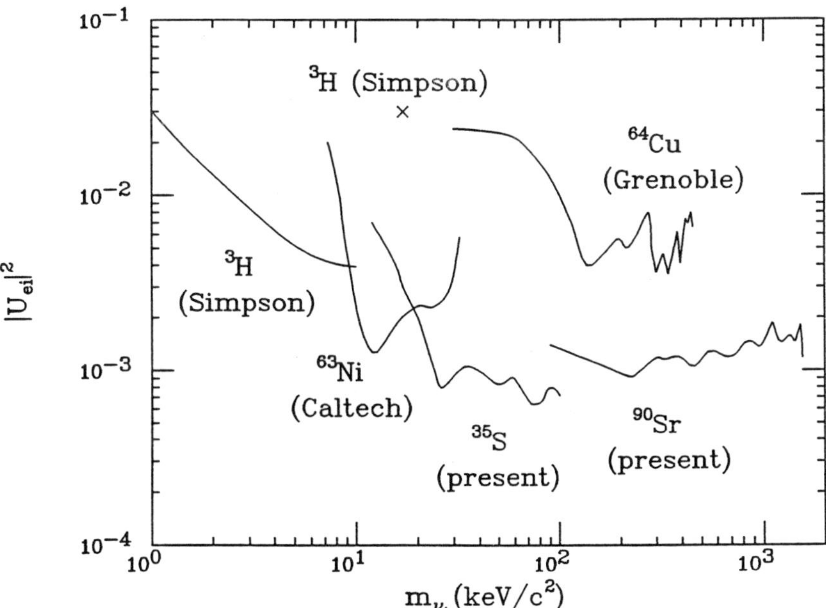

Fig.7 Upper bounds of B_i obtained in the present experiment as well as from earlier experiments.

References

[1] R. Shrock, Phys. Lett. 96B (1980) 159; Phys. Rev. D24 (1981) 1232; 1275.

[2] See, for instance, T. Yamazaki, Proc. Int. Conf. Neutrino Physics and Astrophysics, Neutrino 1984, Nordkirchen (ed. K. Kleinknecht and E.A. Paschos, World Scientific, Singapore, 1984) p.183; D.A. Bryman et al., Phys. Rev. Lett. 50 (1983) 1546.

[3] M. Gronau, Phys. Rev. D28 (1983) 2762; G. Bernardi et al., Phys. Lett. 166B (1986) 479; J. Dorenbosch et al., Phys. Lett. 166B (1986) 473.

[4] K. Shreckenbach et al., Phys. Lett. 129B (1983) 265.

[5]   J.J. Simpson, Phys. Rev. D24 (1981) 2971.

[6]   J.J. Simpson, Phys. Rev. Lett. 54 (1985) 1891.

[7]   T. Ohi et al., Phys. Lett. 160B (1985) 322.

[8]   T. Altzitzoglou et al., Phy. Rev. Lett. 55 (1985) 799.

[9]   A. Apalikov et al., ITEP-114 (1985).

[10] V.M. Datar et al., Nature 318 (1985) 547.

[11] F. Boehm, private communication (1986).

REVIEW OF DOUBLE BETA DECAY RESULTS FROM ^{76}Ge

David O. Caldwell

Physics Department, University of California[*]
Santa Barbara, California 93106, U.S.A.

ABSTRACT

Experiments using Ge detectors have obtained lifetime limits for neutrinoless double beta decay in ^{76}Ge which provide the best limits on (1) lepton number conservation,
(2) the mass of a light Majorana electron neutrino,
(3) the existence of right-handed currents for the conditions in which the "see-saw" mechanism provides neutrino
mass, (4) the mass of a heavy Majorana neutrino as a function of its coupling to the electron neutrino, and (5) the
coupling of a Majoron to the electron neutrino. Currently
six such experiments are in progress and another is in
early stages of preparation, representing work in six
countries. The status of these experiments, their results,
and the interpretation of these results in terms of quantities of interest to particle physics are presented below.

1. INTRODUCTION TO DOUBLE BETA DECAY

Since this is the first contribution on the subject of double beta
decay, it may be useful to provide a brief introduction to this extremely
sensitive nuclear probe of particle physics. Excellent extensive reviews are available.[1-4] Here we shall concentrate on those aspects of
the process which provide an understanding of how double beta decay can
yield information on lepton number conservation, light and heavy neutrino mass, right-handed currents, and Majoron-neutrino coupling.

Double beta decay is possible because the pairing energy makes a

[*]Supported in part by U.S. Department of Energy.

nucleus with an even number of neutrons and protons more tightly bound than its odd-odd neighbors. Such a decay for ^{76}Ge to ^{76}Se, by-passing ^{76}As, is shown in Fig. 1. The second-order weak decay, which will be referred to as $\beta\beta_{2\nu}$,

$$^{76}\text{Ge} \rightarrow {}^{76}\text{Se} + 2e^- + 2\bar{\nu}_e, \qquad (1)$$

must occur, although with very small probability.

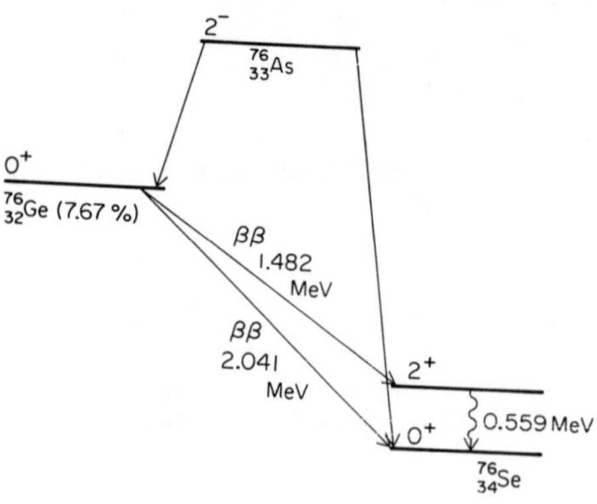

Fig. 1. Double beta decay scheme for ^{76}Ge.

A decay which would be favored by a factor $\sim 10^7$ because of the greater phase space available is

$$^{76}\text{Ge} \rightarrow {}^{76}\text{Se} + 2e^-. \qquad (2)$$

This process, labeled $\beta\beta_{0\nu}$, clearly violates lepton number and hence may not occur at all. The virtual (anti)neutrino emitted in the first neutron decay must be absorbed as a neutrino by the second neutron to produce an e^-, thus requiring the neutrino to be the same as its anti-particle, a property which defines the neutrino to be of the Majorana type.

Because of parity nonconservation, for a massless neutrino the first neutrino emitted would have helicity opposite to that required by the neutrino being absorbed. The necessary admixture of the opposite helicity can be obtained if the neutrino has a mass, m_ν, and hence the inhibition of this decay is greater the smaller that mass. If right-handed currents (RHC) exist, that mechanism could also provide the required helicity reversal, so the rate of the process increases with increasing admixture of RHC to the normal left-handed currents. A quark-level diagram as an example of RHC-induced $\beta\beta_{0\nu}$ is shown in Fig. 2. Note that with a right-handed, as well as a left-handed W boson involved, both a right-handed and a left-handed electron are emitted. In this example, the back-to-back e_L^- and e_R^- require an angular momentum of one, which is provided by p-wave emission, and the resulting total angular momentum change in the process can then be 0, 1, or 2. Thus the transition from the 0^+ ground state of ^{76}Ge can be to the 0^+ ground state or the 2^+ first excited state of ^{76}Se, as shown in Fig. 1. On the other hand, for double beta decay induced purely by neutrino mass, two left-handed W bosons are involved, producing two left-handed e^-, so their helicities cancel in the back-to-back decay, permitting s-wave emission and giving a total angular momentum of zero. Thus only the $0^+ \rightarrow 0^+$ transition is allowed, and this provides a means of distinguishing experimentally between the two mechanisms.

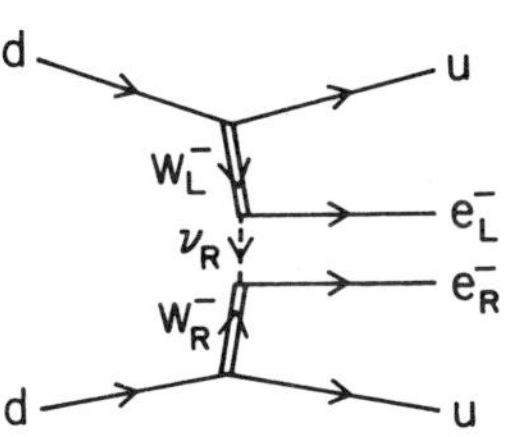

Fig. 2. An example of double beta decay induced by right-handed currents.

A special case of the non-zero neutrino mass decay could exist if there were a very heavy Majorana neutrino, M_ν, which coupled to the electron neutrino (which could then be massless) with some mixing co-efficient, U_e^L. The small admixture of M_ν would provide the helicity reversal, and hence a limit on $\beta\beta_{0\nu}$ yields also a limit on a combination of M_ν and U_e^L.

A quite distinct mechanism for inducing $\beta\beta_{0\nu}$ would be possible if a light or massless boson, such as the Majoron,[5] existed. This Goldstone boson, χ°, arising from the breaking of B-L (baryon number minus lepton number) symmetry could induce the decay,

$$^{76}\text{Ge} \rightarrow {}^{76}\text{Se} + 2e^- + \chi^\circ, \tag{3}$$

which is designated $\beta\beta_{0\nu,\chi}$. Experimentally such a three-body spectrum is quite distinguishable from the four-body spectrum of $\beta\beta_{2\nu}$ and of course from the more usual form of $\beta\beta_{0\nu}$, as is shown in Fig. 3.

All of the interesting particle physics results come from the various forms of the $\beta\beta_{0\nu}$ decay, but the values of, or limits, on these quantities are quite dependent on the nuclear matrix elements, which are difficult to calculate. While the $\beta\beta_{0\nu}$ is a short-range, high-energy process in contrast to the long-range, low energy $\beta\beta_{2\nu}$ decay, the latter provides the only check on the nuclear calculations. That the calculations give longer lifetimes than do the measurements for other nuclei has been discussed by Moe and by Pomansky at this Conference. How the resolution of this issue will affect the interpretation of the $\beta\beta_{0\nu}$ results is totally unclear at this time. However, for a nucleus as light as Ge the lifetime discrepancy is surely no worse than an order of magnitude, and if it were that bad and a similar factor applied to $\beta\beta_{0\nu}$, the quantities of interest would change by a factor of three at most. Unfortunately, $\beta\beta_{2\nu}$ is very difficult to measure in Ge, with the most stringent upper limit of 8×10^{19}y by the UCSB-LBL group[6] being

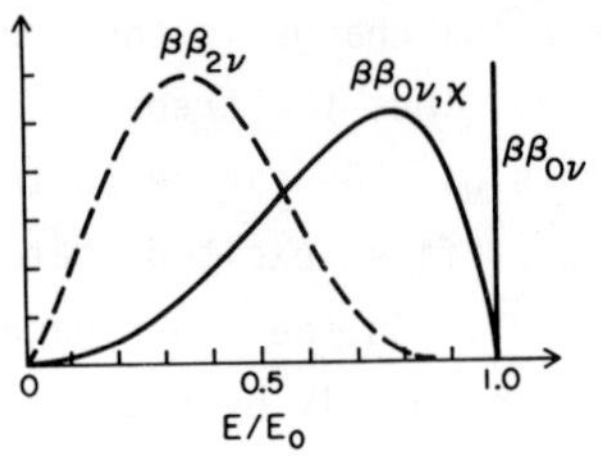

Fig. 3. The summed kinetic energies of the two electrons for $\beta\beta_{2\nu}$, $\beta\beta_{0\nu}$, and $\beta\beta_{0\nu,X}$ decay modes (arbitrary scales).

set by the systematics of the background determination and not by statistics. That limit is still shorter than any calculated value.

2.   GENERAL DESCRIPTION OF DOUBLE BETA DECAY EXPERIMENTS

Ge detectors give excellent energy resolution, so that a peak at 2.041 MeV expected for $\beta\beta_{0\nu}$ could be seen with a full width at half maximum of about 3 keV. This helps considerably in reducing background counts, as does the fact that to be a good detector the Ge has to be of unusually high purity. The zone refining and other purification steps eliminate heavier radioactive elements. The sensitivity is quite good, since 7.8% of the Ge is the double-beta-decay candidate nucleus ^{76}Ge, and the nuclear transition is believed to be favorable. An especially big advantage of Ge is that the source is intimately mixed with the detector, and hence the amount of ^{76}Ge can be made as large as can be afforded. With typical electron ranges of the order of one millimeter it is difficult to get large sources when the source and detector are separate.

With these advantages it is not surprising that Ge experiments are being done by many groups around the world. All of them have been struggling with the necessity to reduce backgrounds as much as possible. One component of the background, that due to cosmic rays, can obviously be reduced by going underground, and all experiments have done this. The depths vary greatly, as shown in Fig. 4. After the hadronic component is absorbed, the main residual effect is from muon-produced neutrons which give capture γ rays. Reducing this effect has to be balanced against the problems caused by neutrons from fission in rocks.

The main background problem is ubiquitous radioactivity. Material is selected with as little radioactivity as possible. Specially selected Pb is used as an outer shield, usually associated with some hydrogenous neutron absorber (such as borated polyethylene). Inner shields are often of multiply distilled Hg or of oxygen-free hard copper. Some experiments employ in addition an active shield of NaI. This serves not only to veto activities from the outside, but also to suppress Compton scattered γ's.

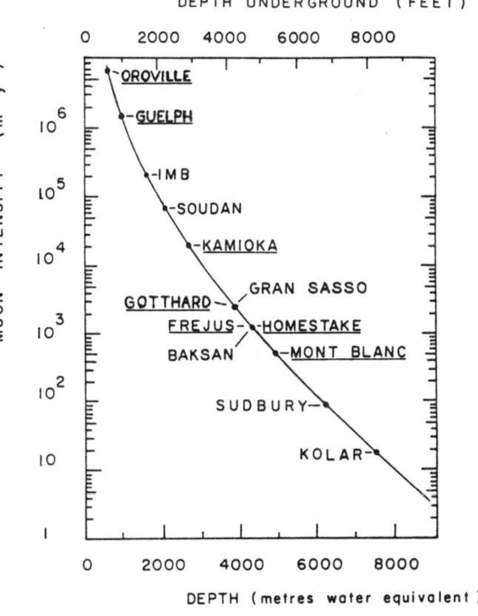

Fig. 4. Depth of underground laboratories, with those underlined having double beta decay experiments.

A γ, usually from material inside the NaI, can Compton scatter in the Ge and leave the requisite 2 MeV, but if it then passes into the NaI that event is eliminated. It is thus essential to have as little material inside the NaI as possible and to have that of as low a Z as can be managed so the Compton γ's are not absorbed before reaching the NaI. The NaI is not as free from radioactivity as the other materials used, and while it is mainly self-vetoing, experiments using NaI tend to have larger and more numerous photopeaks. However, these are at discrete energies, and the bothersome Compton continuum is so well suppressed that these activities do not matter for the $\beta\beta_{0\nu}$ decay. Experiments with NaI try to use the NaI to detect the 559 keV deexcitation γ ray in coincidence with the 1.482 MeV 2e⁻ energy in the Ge from the $0^+ \to 2^+$ transition and hence can set better limits on this $\beta\beta_{0\nu}$ decay.

The background reduction has been so effective that many experiments are now actually limited by activities in the Ge itself, such as ^{68}Ge. Some of these have been induced by cosmic rays when the material

was above ground, but others are very likely due to the unfortunate re-
cycling of this valuable substance. Old detectors go back into the
melt and contaminate the new product, albeit at a level that other experiments could not detect.

3.   STATUS OF SPECIFIC EXPERIMENTS

With this general description as a base, we now proceed to the
specific experiments. In each case the lifetime limits obtained for the
$0^+ \to 0^+$ (and in several cases the $0^+ \to 2^+$) $\beta\beta_{0\nu}$ transitions will be given,
along with background levels and resolutions achieved, counting times,
quantity of Ge, and any special features.

The Zaragoza-Bordeaux-Strasbourg experiment[7] in the Frejus tunnel
is shown schematically in Fig. 5. This experiment uses four Ge detectors of about 417 cm^3 active volume inside 19 hexagonal NaI detectors.
Five of the latter have not yet been replaced by new scintillators and
are quite radioactive, so the background level (40 counts/keV·y·kg) near
2 MeV is still high, but when the NaI is used in coincidence for the
$0^+ \to 2^+$ transition the background is only 3 counts/keV·y·kg. Good resolutions have been achieved: 3.1 keV at 2 MeV and 2.6 keV at 1.5 MeV
for the Ge and 11% at 0.6 MeV for the NaI. After 2252 hours of counting,
lifetime limits of 2.2×10^{22}y $(0^+ \to 0^+)$ and 1.8×10^{22}y $(0^+ \to 2^+)$ were obtained.

The Cal Tech group used a 90 cm^3 detector of 3 keV resolution inside a plastic
scintillator anticoincidence at sea level for
3820 hours and got a lifetime limit of
1.9×10^{22}y.[8] Collaborating now with SIN and
Neuchatel, they moved the detector to the St.
Gotthard tunnel where the background was reduced by a factor of 16 to 4 counts/keV·y·kg.
In addition to using the detector to test the
radioactivity level for components used in
the construction of a system of eight Ge detectors of 140 cm^3 each, it was run for 6728
hours to get a $0^+ \to 0^+$ lifetime limit of
6.2×10^{22}y.[9] The spectrum above 2 MeV for

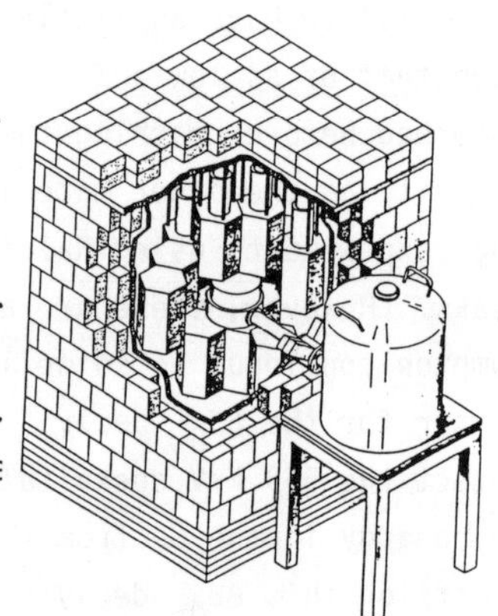

Fig. 5.   Schematic view
of the Zaragoza-Bordeaux-
Strasbourg apparatus.

this run is shown in Fig. 6. The installation of the larger system is now proceeding.

The Osaka experiment[10] was operated first above ground and then moved into the Kamioka mine. A 164 cm^3 detector was used inside a NaI shield, as shown in Fig. 7. Resolutions were 2.7 keV ($0^+\to0^+$) and 2.4 keV ($0^+\to2^+$). With backgrounds of 6 c/keV·y·kg ($0^+\to0^+$) and a coincidence background of 3 c/keV·y·kg ($0^+\to2^+$), an 8621-hour run produced limits of 7.4×10^{22}y ($0^+\to0^+$)

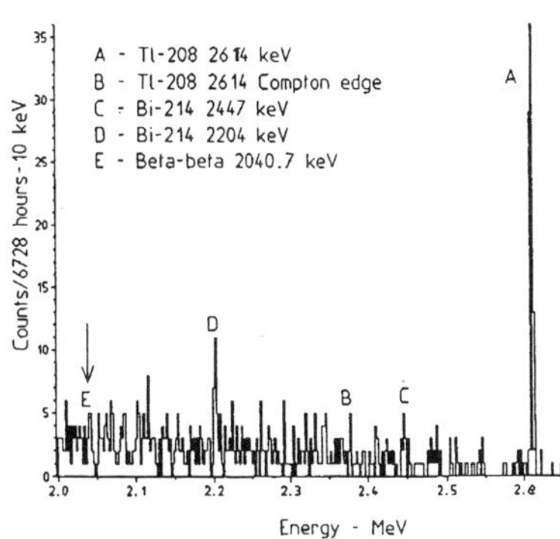

Fig. 6. The spectrum above 2 MeV for the Caltech-SIN-Neuchatel experiment.

and 6×10^{22}y ($0^+\to2^+$). At present the Ge has been replaced by a ^{100}Mo double beta decay experiment using Si detectors, as has been described at this Conference by Ejiri.

The Pacific Northwest Laboratory, University of South Carolina group first did an above-ground experiment[11] which gave a limit of 1.3×10^{22}y in 4054 hours using a 125 cm^3 detector of 3.4 keV resolution inside a NaI anticoincidence and a complex bulk shield. A detector of similar volume and 3.7 keV resolution but much lower intrinsic radioactivity was placed inside Cu and Pb shields in the

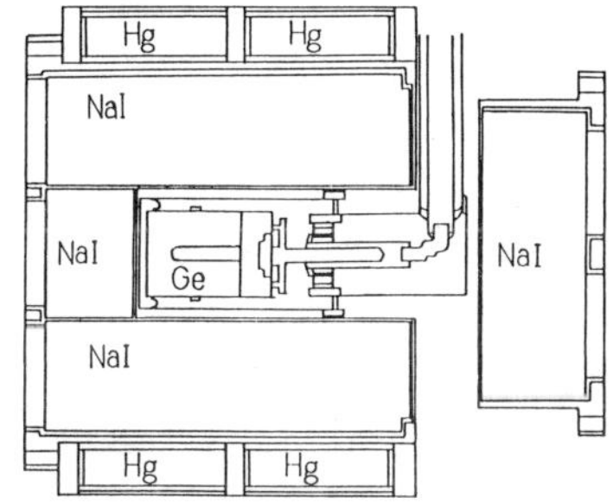

Fig. 7. Cross-sectional view of the Osaka experiment.

Homestake mine and a background of 2.4 c/keV·y·kg achieved.[12] By counting for about 8000 hours they obtained a lifetime limit of 1.4×10^{23}y. They are still trying to reduce backgrounds while continuing to count and to prepare a larger (720 cm^3 and eventually twice that volume) detector system.

The Guelph-Aptec-Queens group used the largest detector then developed of 194 cm^3, which was shielded by Hg, Pb, and a plastic scintillator anticoincidence and placed in a salt mine near Windsor, Ontario. With a background of about 34 c/keV·y·kg at 2 MeV and a resolution of 3 keV they obtained in 2363 hours lifetime limits of 3.2×10^{22}y ($0^+ \rightarrow 0^+$) and 1.6×10^{22}y ($0^+ \rightarrow 2^+$).[13] Now with three detectors of similar size and a much improved 2 MeV background of about 2 c/keV·y·kg, a run of 3200 hours has given a $0^+ \rightarrow 0^+$ limit of 1.6×10^{23}y. The data in the vicinity where the $0^+ \rightarrow 0^+$ line (2.041 MeV) should appear is shown in Fig. 8.

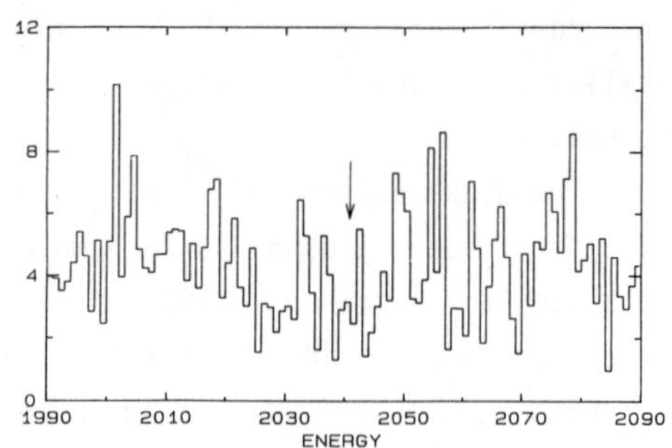

Fig. 8. Data from the Guelph-Aptec-Queens experiment in the region of the 2.041 MeV $\beta\beta_{0\nu}$ line (arrow).

The first Ge experiment[14] was performed by the Milan group. For their more recent results[15] they have used two quite different Ge detectors in the Mt. Blanc tunnel. The first has been operated for 21,000 hours and has a useful volume of 116 cm^3, a resolution of 3 keV, and a background of 24 c/keV·y·kg, and achieved a $0^+ \rightarrow 0^+$ limit of 1.2×10^{23}y. A larger detector (138 cm^3) with about the same resolution but a much improved background (4 c/keV·y·kg) has been operating in the same place for 10,000 hours, and a lifetime of 1.3×10^{23}y has been obtained. If these data are combined, a $0^+ \rightarrow 0^+$ lifetime limit of 1.8×10^{23}y is reached. Since the group is turning to other techniques, the first detector is no longer operating in the Mt. Blanc tunnel, but the second is still collecting data.

The last of these active Ge experiments is that of the University of California Santa Barbara, Lawrence Berkeley Laboratory collaboration. They set out from the beginning to build a multidetector array completely enclosed in a 15-cm-thick assembly of ten NaI counters. However, the cryostats were constructed to house units of two Ge detectors each, and the first experiment was done with just one such unit, which was operated above ground. Using 336 cm^3 of Ge of 4 keV combined resolution with a background of 11 c/keV·y·kg, they obtained a $0^+ \rightarrow 0^+$ limit of 5×10^{22}y in 1618 hours.[16] The experiment was then moved into a powerhouse built into a mountain at the Oroville, California dam and two more detectors were installed for a total of 658 cm^3. The combined resolution was improved to 3.7 keV, and the average background (which dropped by a factor of two in six months) was reduced to 2.2 c/keV·y·kg. In 3550 hours of counting, a $0^+ \to 0^+$ limit of 2.5×10^{23}y was obtained.[6] Using 3404 hours of data, a limit of 5×10^{22}y was achieved for the $0^+ \to 2^+$ transition. With 2456 hours of data the already mentioned $\beta\beta_{2\nu}$ 90% C.L. limit of 8×10^{19}y was reached, and a 90% C.L. limit of 6×10^{20}y was obtained for the $\beta\beta_{0\nu,\chi}$ decay.

The UCSB-LBL experiment has since been improved by adding two more detectors for a total of about 990 cm^3. These were operated for about 5000 hours, and for a small part of that time the total complement of 8 detectors (1.3 liters) was functioning. The resolution has been improved to 3.4 keV and the time-averaged background to 1.7 c/keV·y·kg. For 7.8×10^6 hour·cm^3 (about 47,000 detector hours) the $0^+ \to 0^+$ limit has become 4.1×10^{23}y and the $0^+ \to 2^+$ limit, 1.3×10^{23}y. The spectra in these two energy regions are shown in Fig. 9 and 10. Counting continues, and it is hoped to make some obvious improvements in the detectors to improve backgrounds.

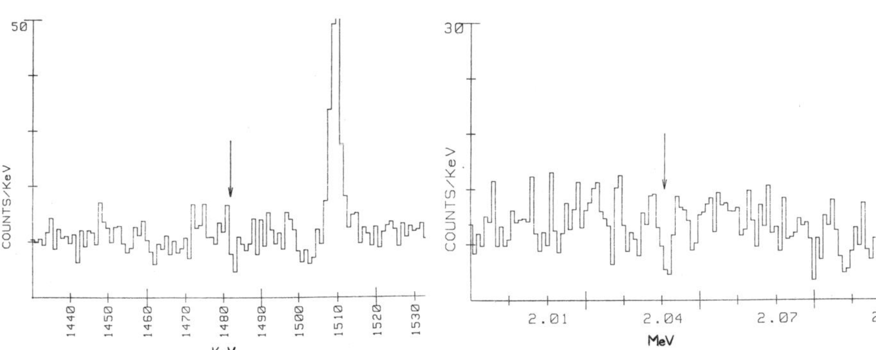

Fig. 9. Ge spectrum with NaI coincidence in the region of the expected $0^+ \to 2^+$ peak (arrow) for the UCSB-LBL experiment.

Fig. 10. Ge spectrum with NaI anticoincidence in the region of the expected $0^+ \to 0^+$ peak (arrow) for the UCSB-LBL experiment.

A summary of recent results from each of the experiments is given in the table below for the $\beta\beta_{0\nu}$ $0^+ \to 0^+$ transition. Also shown is the most recent amount of Ge used and background levels obtained. Note that the lifetime limits are half-lives generally given for 1σ (68% C.L.). Most are obtained by taking the square root of the number of background counts in a broad region in the vicinity of the expected peak. Some are done by a maximum likelihood method which will give a higher limit if there is a dip where a peak is expected. Thus the data in Figs. 9 and 10 would give much higher limits if this procedure were used.

TABLE I: Recent $\beta\beta_{0\nu}$ Results for the $0^+ \to 0^+$ Transition in ^{76}Ge

Group	Background	Ge	$T_{1/2}$ Limit
	c/keV·y·kg	kg	10^{23}y (1σ)
Zaragoza-Bordeaux-Strasbourg	40	2.2	0.2
Caltech-SIN-Neuchatel	4	0.48	0.6
Osaka	6	0.88	0.7
Pacific Northwest-South Carolina	2.4	0.67	1.4
Guelph-Aptec-Queens	∿2	3.0	1.6
Milan	24,4	1.4	1.8
UCSB-LBL	1.7	7.1	4.1

While the limits have improved appreciably in recent years, further improvements will come slowly. Quantities of Ge will not undergo very large increase, backgrounds will improve but not by large factors, and the limit improves only as the square root of the counting time. The one way there could be a significant change in the prospects for Ge experiments is if the Moscow-Leningrad-Heidelberg collaboration succeeds in fully utilizing the large quantity of enriched ^{76}Ge they and they alone are able to obtain. The promised 15 kg of ^{76}Ge would be worth several hundred million dollars in the West. If that could all be used effectively, they would have a factor 20 advantage over the UCSB-LBL experiment. There is normally a large loss of material in making the crystal and a lesser loss in making detectors from the crystal. Also, separated isotopes tend to have radioactive contamination. However, if these problems can be overcome, the next big step for Ge experiments should be taken by this group.

4.   INTERPRETATION OF THE $\beta\beta_{0\nu}$ RESULTS

The limit of 6×10^{20}y for the $\beta\beta_{0\nu,\chi}$ decay in the UCSB-LBL experiment has been evaluated by Doi, Kotani, and Takasugi[3] to give 90% C.L. limits for the coupling of the Majoron to the electron neutrino. The result, however, depends on the matrix elements used, with values varying from $<1.1\times10^{-3}$ for the Los Alamos[2] matrix elements to $<4.1\times10^{-4}$, using those of Heidelberg.[17] A somewhat lower limit[3] from ^{130}Te depends on additional assumptions, but the present more direct results is more stringent than those from ^{48}Ca[18] and ^{150}Nd.[19]

The variation in these two results indicates only a part of the problem regarding interpreting the lifetime limits in terms of interesting particle physics quantities. Already in the first section of this paper the discrepancy between measured and calculated $\beta\beta_{2\nu}$ lifetimes was noted, and this casts doubt on the $\beta\beta_{0\nu}$ calculations. Conservatively one might scale up the results given below by ~3, but it is quite likely that no such factor is needed.

An especially important caveat applies to deducing a neutrino mass limit from the data. If there is a mixture of neutrinos, these can have opposite CP eigenstates, producing a cancellation which would make the effective neutrino mass for double beta decay, $<m_\nu>$, much less than it might be for, say, single beta decay.[20]

To understand the right-handed-current parameters it is useful to employ a simplified Hamiltonian for $\beta\beta_{0\nu}$ decay:

$$H_W = \frac{-G_F\cos\theta_c}{\sqrt{2}}\ [j_L^\mu(J_{L\mu}^\dagger + n_{LR}J_{R\mu}^\dagger) + j_R^\mu(n_{RL}J_{L\mu}^\dagger + n_{RR}J_{R\mu}^\dagger)] + h.c. \qquad (4)$$

Here the $j_{L(R)}^\mu$ are the components of the leptonic current, with L and R standing for left- and right-handedness, while the J's are similar components of the hadronic current, and G_F is the weak coupling constant measured in μ decay, with θ_c the Cabibbo angle. Note that RHC parameters n_{LR}, etc. are defined so that the first subscript refers to the leptonic current and the second to the hadronic current. Another notation used is $\kappa = n_{LR}$, $\eta = n_{RL}$, and $\lambda = n_{RR}$.

For the table below, two more caveats must be stated. First, the

$0^+ \rightarrow 0^+$ lifetime limit used, 4×10^{23} y, from the UCSB-LBL group is only at the 68% C.L., so that value has a significant uncertainty. Second, the parameters $\langle m_\nu \rangle$, η_{RL}, and η_{RR} are treated as being independent, with only one at a time taken to be non-zero. In this way the sensitivity of each parameter to the lifetime limit is clearly shown. If there were a positive effect, it is likely that more than one of these would be involved. In particular, Kayser, Petcov, and Rosen[21] have shown that for any gauge theory if RHC exist, then $m_\nu \neq 0$.

TABLE II: Neutrino Mass and RHC Parameter Limits for $T_{1/2} > 4 \times 10^{23}$ y

Parameter	Los Alamos[2]	Tübingen[22]	Heidelberg[17]
$\langle m_\nu \rangle$	2.0	1.5	0.8
$\langle \eta_{RR} \rangle$	3.3×10^{-6}	2.9×10^{-6}	--
$\langle \eta_{RL} \rangle$	4.9×10^{-6}	3.5×10^{-8}	

The differing values for the same parameter give an idea of the variations in the matrix elements used, with the exception of $\langle \eta_{RL} \rangle$. The more recent Tübingen-Jülich calculation includes the effects of using a relativistic p-wave electron wave function and of including nucleon recoil. These impressively small values for the RHC parameters are many orders of magnitude more stringent[3] than those from other experiments for the case of neutrino mass generation by the see-saw mechanism,[23] in which all right-handed Majorana neutrinos are heavier than all left-handed leptons.

The $\beta\beta_{0\nu}$ experiments can also set limits on these heavy Majorana neutrinos, as discussed in the first section. A fourth-generation neutrino could be either a Dirac or a Majorana particle, but extra neutrinos, such as required in the low-energy limit of string theories or in left-right symmetric models, are most frequently Majorana particles. The limits obtained for left-right symmetric theories involve knowledge of the ratio of left-handed to right-handed W boson mass and are given elsewhere.[24] Here the case in which only left-handed W bosons are involved is considered. The Haxton-Stephenson[2] calculation is used to give the straight-line limit seen in Fig. 11, in which the square of the mixing coefficient, U_e^L (coupling of the heavy neutrino to the light or massless electron neutrino), is plotted against the heavy neutrino mass, M_ν. Limits from other experiments are taken from the compilation of Gilman and Rhie.[25] Values below and to the right of the lines are allowed. While the limit from $\beta\beta_{0\nu}$ is much more stringent than those from other experiments, with one small exception, the other limits apply to Dirac neutrinos as well.

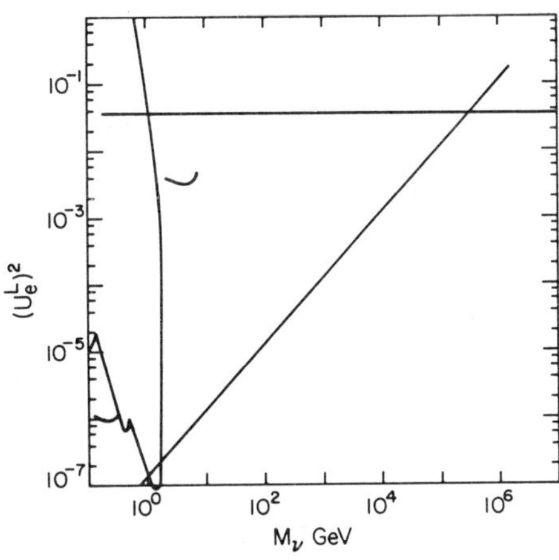

Fig. 11. Limit on the mass of a heavy Majorana neutrino as a function of its probability for mixing with an electron neutrino.

While the limitations imposed by $\beta\beta_{0\nu}$ on light Majorana neutrino mass are relatively well known, those on right-handed currents are much less known. However, the limitation $\beta\beta_{0\nu}$ places on the mass and coupling of heavy Majorana neutrinos seems not to be known even to many experts, despite this not being a new subject.[1] It is necessary to call this to the attention particularly of theorists for two very different reasons. First, the calculations which went into Fig. 11 can be improved considerably. For instance, finite nuclear size effects are very important and have been taken into account quite approximately. Second, the M_ν vs. U_e^L limits can be very useful in restricting models, and even reviews of this subject ignore the $\beta\beta_{0\nu}$ contribution. It may well be that the limitations $\beta\beta_{0\nu}$ imposes on M_ν may turn out to be more important than those it imposes on m_ν.

So far I have covered rather predictable topics, but I wish to close with something less expected. The UCSB-LBL experiment is located outside a small town in the foothills of the Sierra mountains called Oroville, and the people of that town have shown a lot of interest in the experiment. For example, one sees bumper stickers on cars saying, "I brake for neutrinos." The Chamber of Commerce even sponsored a contest for school children to draw their idea of what a neutrino looks

like. This was won by an eight-year old, Jim Barnes. For a conference
devoted to neutrinos this is an appropriate place to reveal for the
very first time a picture of a neutrino:

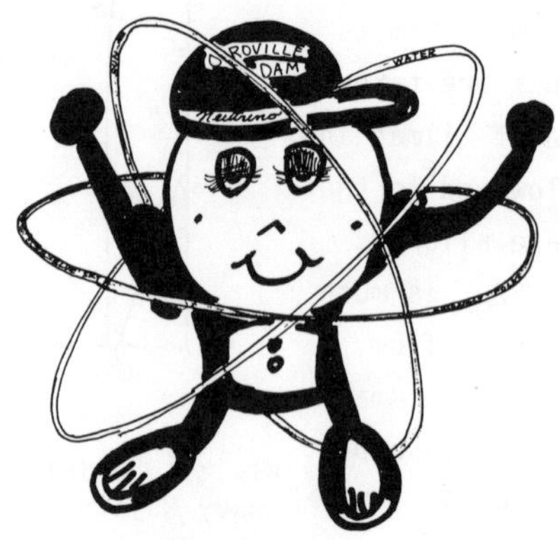

Its mass has not yet been measured!

REFERENCES

1. H. Primakoff and S.P. Rosen, Ann. Rev. Nucl. Part. Sci. $\underline{31}$, 145
    (1981).

2. W.C. Haxton and G.J. Stephenson, Jr., Progr. in Part. and Nuc.
    Phys. $\underline{12}$, 409 (1984).

3. M. Doi, T. Kotani, and E. Takasugi, Prog. of Theo. Phys. Supp.
    No. 83 (1985).

4. J.D. Vergados, Phys. Rep. $\underline{133}$, No. 1 and 2 (1986).

5. Y. Chikashige, R.N. Mohapatra, and R.D. Peccei, Phys. Lett. $\underline{98B}$,
    265 (1981); G.B. Gelmini and M. Roncadelli, Phys. Lett. $\underline{99B}$,
    411 (1981).

6. D.O. Caldwell, R.M. Eisberg, D.M. Grumm, D.L. Hale, M.S. Wither-
    ell, F.S. Goulding, D.A. Landis, N.W. Madden, D.F. Malone,
    R.H. Pehl, and A.R. Smith, Phys. Rev. $\underline{33D}$, 2737 (1986).

7. F. Leccia, Ph. Hubert, D. Dassie, P. Mennrath, M.M. Villard,
    A. Morales, J. Morales, and R. Nunez-Lagos, Nuovo Cimento $\underline{78A}$,
    50 (1983); Nuovo Cimento $\underline{85A}$, 19 (1985).

8. A. Forster, H. Kwon, J.K. Markey, F. Boehm, and H.E. Henrikson, Phys. Lett. <u>138B</u>, 301 (1984).

9. P. Fisher, Caltech preprint (1986).

10. H. Ejiri, N. Takahashi, T. Shibata, Y. Nagai, K. Okada, N. Kamikubota, T. Watanabe, T. Irie, Y. Itoh, and T. Nakamura, Nuc. Phys. <u>A448</u>, 271 (1986).

11. F.T. Avignone, III, R.L. Brodzinski, D.P. Brown, J.C. Evans, Jr., W.K. Hensley, J.H. Reeves, and N.A. Wogman, Phys. Rev. Lett. <u>50</u>, 721 (1983).

12. F.T. Avignone, III, R.L. Brodzinski, D.P. Brown, J.C. Evans, Jr., W.K. Hensley, H.S. Miley, J.H. Reeves, and N.A. Wigman, Phys. Rev. Lett. <u>54</u>, 2309 (1985).

13. J.J. Simpson, P. Jagam, J.L. Campbell, H.L. Malm, and B.C. Robertson, Phys. Rev. Lett. <u>53</u>, 141 (1984).

14. E. Fiorini, A. Pullia, G. Bertolini, F. Cappelani, and G. Restelli, Phys. Lett. <u>25B</u>, 602 (1967) and Nuovo Cimento <u>13A</u>, 747 (1973).

15. E. Bellotti, O. Cremonesi, E. Fiorini, G. Ligouri, A. Pullia, P. Sverzellati, and L. Zanotti, Phys. Lett. <u>146B</u>, 450 (1984).

16. D.O. Caldwell, R.M. Eisberg, D.M. Grumm, D.L. Hale, M.S. Witherell, F.S. Goulding, D.A. Landis, N.W. Madden, D.F. Malone, R.H. Pehl, and A.R. Smith, Phys. Rev. Lett. <u>54</u>, 281 (1985).

17. K. Grotz and H.V. Klapdor, Phys. Lett. <u>153B</u>, 1 (1985).

18. J.D. Vergados, Phys. Lett. <u>109B</u>, 96 (1982); erratum, Phys. Lett. <u>113B</u>, 513 (1982).

19. A.A. Klimenko, A.A. Pomansky, and A.A. Smolnikov, Proc. "Neutrino-84" Conf., Dortmund, W. Germany, 161 (1984).

20. M. Doi, T. Kotani, H. Nishiura, K. Okuda, and E. Takasugi, Phys. Lett. <u>102B</u>, 323 (1981); L. Wolfenstein, Phys. Lett. <u>107B</u>, 77 (1981) and Nuc. Phys. <u>B186</u>, 147 (1981); B. Kayser and A.S. Goldhaber, Phys. Rev. <u>D28</u>, 2341 (1983); M. Doi, T. Kotani, H. Nishiura, and E. Takasugi, Prog. Theor. Phys. <u>69</u>, 602 (1983); B. Kayser, Phys. Rev. <u>D30</u>, 1023 (1984); S.M. Bilenky, N. Nedelcleva, and S.T. Petcov, Nuc. Phys. <u>B247</u>, 61 (1984).

21. B. Kayser, S.T. Petcov, and S.P. Rosen, to be published.

22. T. Tomoda, A. Faessler, K.W. Schmidt, and F. Grümmer, Nuc. Phys. <u>B153</u>, 1 (1985).

23. M. Gell-Mann, P. Raymond, and R. Slansky in "Supergravity", ed. by P. van Niewenhulzen and D. Freedman (North-Holland, Amsterdam, 1979), p. 317; T. Yanagida, Proc. Workshop on Unified Theory and Baryon Number in the Universe, ed. by Sawada and Sugamoto (KEK, 1979).

24. R.N. Mohapatra, University of Maryland preprint, 1986.

25. F.J. Gilman and S.H. Rhie, Phys. Rev. $\underline{D32}$, 324 (1985).

SEARCH FOR DOUBLE BETA DECAY IN ^{82}Se

S. R. Elliott, A. A. Hahn, and M. K. Moe
University of California, Irvine, CA 92717

ABSTRACT

A time projection chamber with a selenium double beta decay source as the central electrode has yielded a lower limit of 1.0 x 10^{20} years at 68% confidence for the two-neutrino half life of ^{82}Se. This limit is consistent with the results of geochemical measurements, and disagrees with the shorter half life predictions of nuclear theory. For the neutrinoless mode we find a corresponding lower limit of 1.0 x 10^{22} years, also at 68% confidence.

INTRODUCTION

A double beta decay search is being carried out with a time projection chamber (TPC) containing 14 grams of selenium enriched to 97% in isotope 82. The chamber is 80 cm across with its 20 cm depth bisected into two 10 cm drift regions by a central electrode containing the selenium source. The TPC is immersed in a 700 G magnetic field, and is surrounded by 15 cm of lead and a 4π cosmic ray veto. It operates with a filling of helium and propane at one atmosphere absolute pressure, and is situated in an on-campus laboratory.

The energies and opening angles of the two beta particles are derived from the parameters of helices fit to the tracks. Conversion electrons from ^{207}Bi and ^{208}Tl sources serve as calibration lines. Further details have been reported elsewhere.[1,2]

RESULTS

The sum energy spectrum for 88 double beta decay candidate events accumulated in 5426 live hours is shown in Fig. 1a. A threshold at 1.1 MeV excludes background from beta decay with internal conversion in ^{214}Pb. This contaminant is deposited on the surface of the double beta decay source by decaying ^{222}Rn in the TPC gas. Tests have shown that radon, in turn, comes from the Be-Cu wire used for the grid, cathode, and field wires of the chamber.

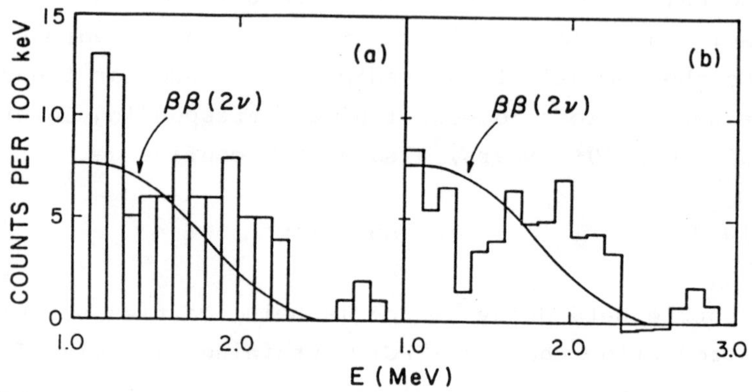

Fig. 1(a) The electron sum energy spectrum of the 88 double beta decay candidates found in 5426 live hours. The theoretical $\beta\beta(2\nu)$ spectrum is normalized to the geochemical rate of $(1.26 \pm 0.04) \times 10^{20}$ yr. reported by Kirsten.
(b) An attempt at background subtraction. Removing identified background affects mostly the low energy end of the spectrum, although the absolute amount to remove is uncertain within a factor of 2.

Also eliminated from the spectrum are ^{214}Bi events, similar in origin to ^{214}Pb, but of higher energy. In this case the alpha particle from the polonium daughter tags the false double beta events.

The little cluster of 4 counts at the high energy end
of the spectrum deserves some attention since it is close
to the neutrinoless double beta decay ($\beta\beta(0\nu)$ position at
the 3.0 MeV endpoint. The individual electron energies,
their uncertainties, and whether the tracks appeared on
the same side or on opposite sides of the source are given
in Table I. (Opposite-side events are easier to find and
generally easier to measure accurately.)

Table I. The 4 events just below the endpoint.

E_1	E_2	σ_1	σ_2	E_{sum}	Same side	Opposite sides
1.06	1.61	.06	.16	2.67		X
.77	2.00	.02	.13	2.77		X
.59	2.12	.01	.07	2.71		X
1.36	1.44	.04	.18	2.80	X	

A 9% shift to the right would center the cluster at
the ($\beta\beta(0\nu)$) energy. The probability of a calibration
error of this magnitude can be estimated with the aid of
the ^{207}Bi spectrum in Fig. 2. Although the resolution
allows some error in individual events, the spectral peaks
are within 2% of their respective conversion lines. It is
unlikely that all 4 counts in question belong at 3.0
MeV.

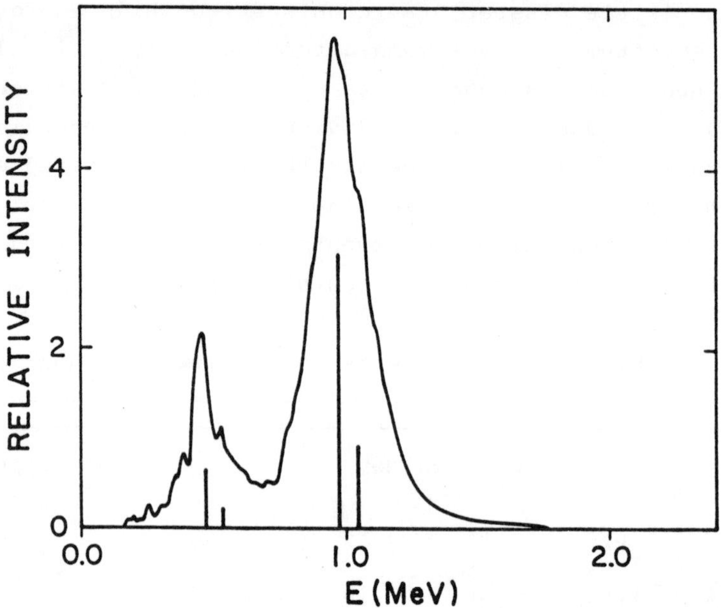

Fig. 2. A Gaussian idiogram of 730 electrons from a 1 nCi
bismuth-207 conversion source placed inside the TPC.

Superimposed on the data of Fig.1a is the theoretical
spectrum[3] for the two neutrino mode ($\beta\beta(2\nu)$) normalized to
the geochemical rate reported by Kirsten.[4] The histogram
is not a good fit to this curve, the most probable explanation being residual background. Certain background processes are known to be present from manifestations of
associated activity. For example, the thorium decay chain
includes a rapid beta-alpha sequence from the decay of
^{212}Bi and its 0.3 µs daughter. These events have a
distinctive appearance in the TPC, and their rate can be
measured. With the thorium series activity thus
determined, the background contribution from the series'
most troublesome member, ^{208}Tl, can be calculated to be
between 5% and 9% of the 88 candidates.

The energy spectrum of the ^{208}Tl background is known

from measurements following the injection of a sizable dose of the parent ^{220}Rn (thorium series) into the TPC. No permanent contamination results, because the radon daughters all decay away in a few days. Following the radon injection, conversion lines from ^{208}Tl decay are very conspicuous in the spectrum of single electrons belonging to false double beta decay pairs (Fig. 3.) The absence of conversion lines in the corresponding singles spectrum of the 88 double beta candidates confirms the lack of strong ^{208}Tl background.

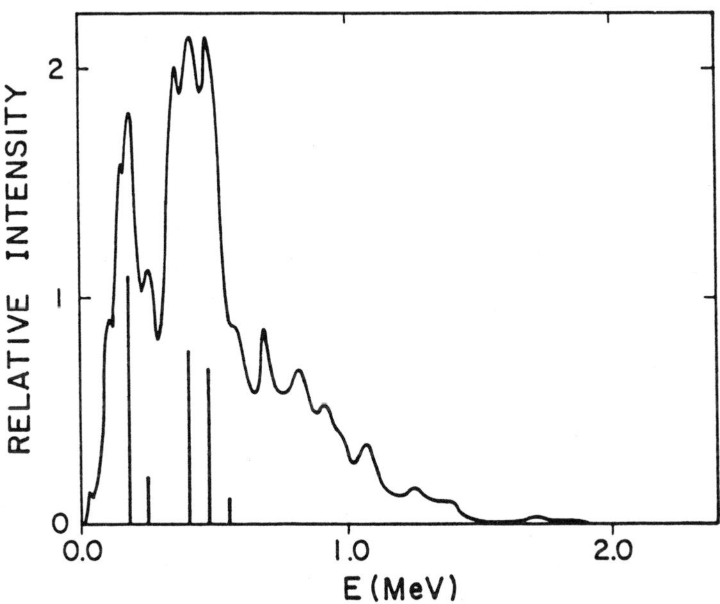

Fig. 3. A Gaussian idiogram of single electrons that are members of false double beta decay pairs resulting from the injection of radon-220. The conspicuous thallium-208 conversion lines show that internal conversion is the dominant mechanism for background production by this contaminant. These lines do not appear in the singles spectrum of the double beta decay candidates.

Similarly, an injection of ^{222}Rn (uranium series) produced the conversion line spectrum of ^{214}Bi (Fig. 4.) These, too, are absent in the candidate spectrum.

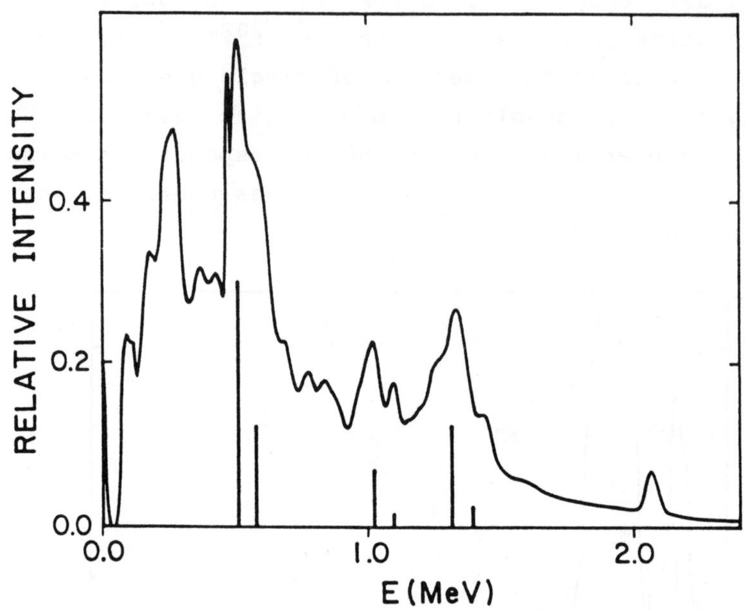

Fig.4. Similar to Fig.3, but showing bismuth-214 conversion lines following injection of radon-222.

Background spectra from Möller scattering and double Compton scattering can also be estimated. The input for these calculations is the spectrum of single electrons (not members of pairs) originating at the double beta decay source. The single electron spectrum and a tabulation of estimated intensities of all identified backgrounds can be found in reference 5.

Our estimate of the absolute magnitude of total identified background is uncertain within a factor of 2, but the <u>shape</u> of the background spectrum is better known. Qualitatively, a subtraction of identified

background pulls down the low energy end of the spectrum
of Fig. 1a with the result suggested by Fig. 1b.

CONCLUSIONS

The geochemical half life for ^{82}Se is in close
agreement with the area under the histogram of Fig. 1b,
and suggests that double beta decay is responsible for
most of these events. However, there is little
independent evidence from our data confirming double beta
decay to be the cause. The spectrum appears to be a bit
harder than the superimposed theoretical $\beta\beta(2\nu)$ curve, and
the opening angle distribution[5] deviates from the expected
form. Of course, it is possible that the double beta
decay phenomenon involves more than is predicted by
$\beta\beta(2\nu)$ theory, but a likely alternative at this point is
an incomplete understanding of background, combined with
the effects of counting statistics.

While the experiment continues to run, a new TPC is
being constructed without the radioactive wire and other
materials suspected of causing background. Depending on
the outcome we may also consider relocating the experiment
in an underground laboratory.

Meanwhile, a half life limit for ^{82}Se can be found by
subtracting the estimated background (26 counts) from the
88 candidates, and assigning the remainder to double beta
decay. The result is $T_{\frac{1}{2}}(2\nu) > 1.0 \times 10^{20}$ yr at 68% c.l.
The sum energy for a $\beta\beta(0\nu)$ ground-state transition in
^{82}Se is 3.0 MeV. The probability that a neutrinoless
event would be detected in a 300 keV window centered on
3.0 MeV is estimated to be 0.21 ± 0.02. The absence of
counts in this window during 7106 hours yields $T_{\frac{1}{2}}(0\nu) >$
1.0×10^{22} yr at 68% c.l., or about 3 times the previous
published limit[6] for ^{82}Se.

REFERENCES

1. M.K. Moe, A.A. Hahn, and H.E. Brown, in <u>The Time Projection Chamber</u>, edited by J.A. Macdonald, AIP Conference Proceedings No. 108 (American Institute of Physics, New York, 1984), p. 37.

2. M.K. Moe, A.A. Hahn, and S.R. Elliott, in <u>Proceedings of the Conference on Neutrino Mass and Low Energy Weak Interactions, Telemark, Wisconsin, 1984</u>, edited by Vernon Barger and David Cline (World Scientific, Singapore, 1985).

3. H. Primakoff and S.P. Rosen, Rep. Prog. Phys. <u>22</u>, 121 (1959).

4. T. Kirsten, Private communication, June, 1986.

5. S.R. Elliott, A.A. Hahn, and M.K. Moe, Phys. Rev. Lett. <u>56</u>, 2582 (1986).

6. B.T. Cleveland, et. al., Phys. Rev. Lett. <u>35</u>, 757 (1975).

Lepton Number Non-conservation Studied by $\beta \cdot \gamma$
Spectroscopic Method on ^{100}Mo and ^{76}Ge Double Beta Decays

H. Ejiri, N. Kamikubota, Y. Nagai*, T. Nakamura,
K. Okada, T. Shibata,
T. Shima, N. Takahashi and T. Watanabe
Dept. Phys. & Lab. Nucl. Studies, Osaka Univ.
Toyonaka, Osaka 560, Japan

The neutrinoless double beta decay ($0\nu\beta\beta$) provids a
very sensitive probe for the lepton number non-conservation.
The $0\nu\beta\beta$ excludes the Dirac particle of the neutrino, and
gives an evidence for the Majorana neutrino, and a finite
mass $<m_\nu>$ and/or a finite right-handed weak current. General aspects of the $0\nu\beta\beta$ have been already discussed by previous speakers: Details of the $0\nu\beta\beta$ theories and experiments may be found in recent reviews [1,2,3] and references
therein.

This report gives our recent results on the $0\nu\beta\beta$ from
^{76}Ge to both the 0^+ ground and 2^+ first excited states in
^{76}Se and the present status of the $0\nu\beta\beta$ from ^{100}Mo. Some
of them have preliminarily been reported [3,4,5].

The weak interaction Hamiltonian is written as [1]

$$H = \frac{G}{\sqrt{2}}(j_L J_L + \eta j_R J_L + \lambda j_R J_R + \kappa j_L J_R) \tag{1}$$

where j_i and J_i are lepton and hadron currents, respectively,
with i = L and R standing for the left-handed and right-
handed ones. Then the $0\nu\beta\beta$ transition rate $t^{0\nu} \equiv \ln2/T^{0\nu}_{1/2}$ is

$$t^{0\nu} = GM^2_{\beta\beta} \left[|<m_\nu>|^2 / m_e^2 + C_\lambda <\lambda>^2 + C_\eta <\eta>^2 \right] \tag{2}$$

* Present address, Dept. Phys. Tokyo Inst. Technology,
  Megro, Tokyo 152.

where G is the phase factor, $M_{\beta\beta}$ is the nuclear matrix element, $\langle m_\nu \rangle = \Sigma m_j U_{ej}^2$, $\langle \lambda \rangle = \lambda \Sigma U_{ej} V_{ej}$, and $\langle \eta \rangle = \eta \Sigma U_{ej} V_{ej}$. The $0\nu\beta\beta$ shows up as a sharp peak among huge continuum background in the electron sum energy $E_\beta + E_{\beta'}$ spectrum. Then the fluctuation of the background counts (N_B) gives the lower limits on the peak yield (N_t) which can be identified, namely $N_t > \sqrt{N_B}$. Then

$$0.76 \; t^{0\nu} \; N_o t \; k/ \sqrt{N_{BG} \cdot \Delta E \cdot t} > 1 \qquad (3)$$

where N_o is the number of source nuclei, t is the measurement time in units of year, ΔE is the energy resolution, k is the detection efficiency, N_{BG} is the background count rate per keV per year. The coefficient 0.76 is the probability that the peak falls within the energy window ΔE. Thus $\langle m_\nu \rangle$, $\langle \lambda \rangle$, and $\langle \eta \rangle$ are written in terms of the nuclear sensitivity S_N and the detection sensitivity S_D as

$$\{ |\langle m_\nu \rangle| /m_e, \; C_\lambda \langle \lambda \rangle, \; C_\eta \langle \eta \rangle \} = 1/S t^{1/4} \qquad (4a)$$

$$|\langle m_\nu \rangle|^2/m_e^2, \; \ldots\ldots = t^{0\nu}/S_N, \; t^{0\nu} > 1/(S_D \sqrt{t}) \qquad (4b)$$

$$S = \sqrt{S_N \cdot S_D}, \; S_N = GM_{\beta\beta}^2, \; S_D = 0.76 \; N_o k/\sqrt{N_{BG} \cdot \Delta E} \qquad (4c)$$

S is the overall sensitivity. Since G is nearly proportional to the $\beta\beta$ Q value $(Q_{\beta\beta})^5$, a nucleus with large $Q_{\beta\beta}$ and $M_{\beta\beta}$ has a large S_N. A detector with large N_o and k and small N_{BG} and ΔE has a large S_D.

It is important to perform an experimental measurement with a large sensitivity $S = (S_N S_D)^{1/2}$. Thus one has to measure $\beta\beta$ decays from nuclei with large S_N by means of a detector with a large S_D. For the present study we have selected ^{100}Mo and ^{76}Ge for the $0\nu\beta\beta$ experiment. The ^{100}Mo has a large $Q_{\beta\beta}$ value and accordingly a large S_N as shown in Fig. 1. Natural Ge contains 7.8% of ^{76}Ge. Thus one can

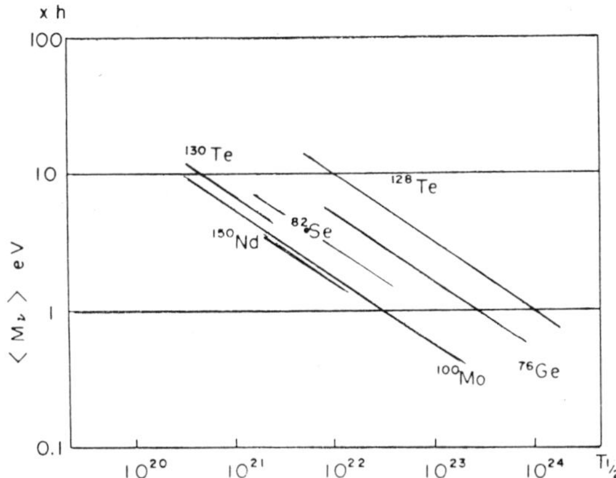

Fig. 1 Lower limits on half-lives for $0\nu\beta\beta$ decays and upper limits on the Majorana neutrino mass $\langle m_\nu \rangle$. Scaling factor $h = M_{\beta\beta}(K.G)/M_{\beta\beta}$, where $M_{\beta\beta}(K.G)$ is given in ref. 11.

use a Ge crystal for both a $\beta\beta$ ^{76}Ge source and a high resolution (small ΔE) high efficiency (large k) detector.

A dector system ELEGANTS (ELEctron GAmma-ray NeuTrino Spectrometer) has been developed as shown in Fig. 2. For ^{76}Gc $\beta\beta$ the 171cc Ge serves as both the $\beta\beta$ detector and the ^{76}Ge source [6]. $N_o(^{76}$Gc) is $5.7 \cdot 10^{23}$. It is surrounded by a 4π geometry 6-NaI crystals. Recording all energy and time signal from the Ge and 6-NaI in a list mode, being followed by the online-offline analysis, is very useful to identify the $0\nu\beta\beta$ $0^+ \to 0^+$(no γ ray) and $0\nu\beta\beta$ $0^+ \to 2^+$ ($2^+ \to 0^+$ γ ray) events among huge electron events followed by many γ-rays (including Compton ones) [3,5,7]. Thus such β-γ spectroscopic device of the ELEGANTS gives high detector sensitivities S_D because of the low N_{BG} for both the $0^+ \to 0^+$ and $0^+ \to 2^+$ $\beta\beta$. Advantages of using the Ge detector are high resolution (small ΔE) and high efficiencies (k = 1 for $0^+ \to 0^+$ and k = 0.34 for the $0^+ \to 2^+$).

In case of ^{100}Mo a mosaic type source detector system

consisting of 11 layers of Si(Li) detectors with 10 ^{100}Mo foils with 50mg/cm^2 sandwiched between the Si detectors with 1500mm^2 × 4mm is used in place of the Ge detector, as shown in Fig. 3. The total ^{100}Mo nuclei is $N_o = 0.5 \cdot 10^{23}$. Major advantages of using ^{100}Mo are firstly the large S_N because of the large $Q_{\beta\beta}(Q_{\beta\beta}(0^+) = 3.03$ MeV) in comparison with the $Q_{\beta\beta}(0^+) = 2.0407$ MeV for ^{76}Ge, and secondary the large S_D (small N_{BG}) because of requirement that two β-rays emitted into opposite directions are detected in coincidence by the adjacent two Si detectors. The energy resolution of the Si detector cooled down to Liq. N_2 temperature is as good as 5 keV and the overall one of around $\Delta E \approx 100$ keV is entirely due to the source thickness. The efficiency k, however, is small. k is evaluated for the 0νββ process due to $<m_\nu>$ by a Monte Carlo calculation.

The electron energy spectra were measured at the underground laboratory located in the 1000m deep (2700m we) Kamioka mine. Run 1 for ^{76}Ge was carried out for 1600hrs with the mercury shield around the NaI. Run 2 was done for

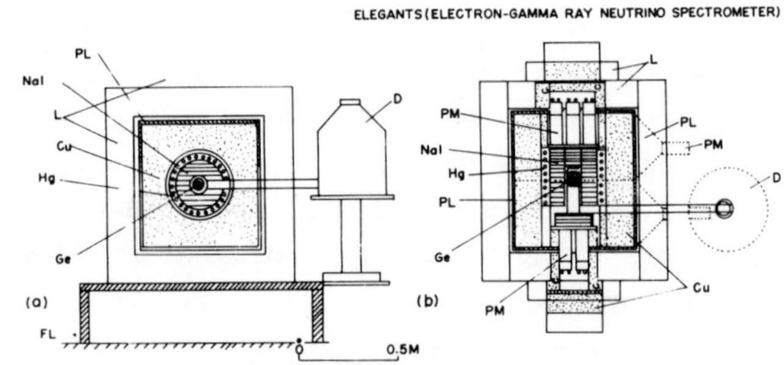

Fig. 2　Left:　Side view of ELEGANTS.　Right:　Top view.　Ge:　171cc active volume intrinsic Ge. NaI:　10"ϕ·12" NaI segmented into 5 and 8"ϕ× 3" NaI.　Cu:　15cm thick OFHC.　Pb:　15cm thick lead.　D: Liq. N_2 dewar.

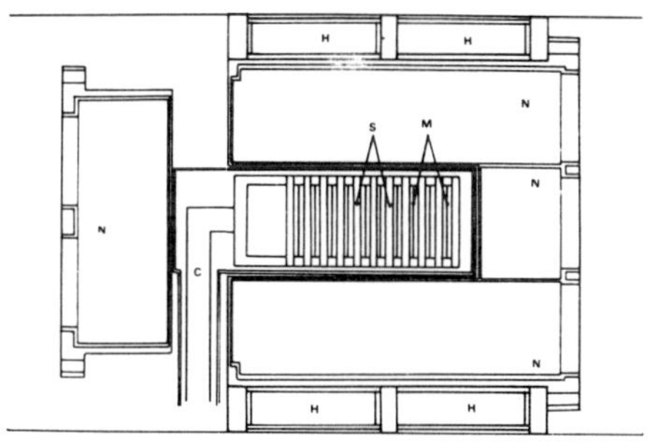

Fig. 3   Si(Li) detectors and ^{100}Mo foils set in
place of the Ge in Fig. 2.

7021 hrs without the mercury shield to avoid the 2044 keV
line from ^{194}Hg contained in the shield. The energy spectra
measured for the Ge detector (Run 2) are shown in Fig. 4. No
appreciable $0\nu\beta\beta$ peaks at $Q_{\beta\beta}(0^+)$ and $Q_{\beta\beta}(2^+)$ are found.

The sensitivities and half-life limits are evaluated
as follows. For the ^{76}Ge $0^+ \to 0^+$ transition $N_{BG} = 5/\text{keV·y}$,
$\Delta E = 3$ keV, $k = 1$ and $S_D(0^+) = 1.1\cdot10^{23}y^{1/2}$, while for the
^{76}Ge $0^+ \to 2^+$ transition $N_{BG} = 2/\text{keV·y}$, $\Delta E = 2.5$ keV, $k = 0.35$,
and $S_D(2^+) - 0.6\cdot10^{23}y^{1/2}$. The value for $S_D(0^+)$ are in the
same order of magnitude as other sensitive detectors [8,9,
10], and the $S_D(2^+)$ is the best. The half-life limits are
$T^{0\nu}_{1/2}(0^+) \gtrsim 7.4\cdot10^{22}y$ and $T^{0\nu}_{1/2}(2^+) \gtrsim 6.0\cdot10^{22}y$, both on 68%
CL. The $0\nu\beta\beta$ process accompanied by the Majoron and the
$2\nu\beta\beta$ process are examined at $E_\beta + E_{\beta'} = 0.9\ Q_{\beta\beta}(0^+)$ and 0.5
$Q_{\beta\beta}(0^+)$, respectively. The lower limits on the half lives
are $T^{0\nu M}_{1/2} \gtrsim 2.5\cdot10^{20}y$ and $T^{2\nu}_{1/2} \gtrsim 0.2\cdot10^{20}y$.

The nuclear sensitivity for ^{76}Ge is rather small, S_N
for $0^+ \to 0^+$ is $7.0\cdot10^{-13}$ by using the $M_{\beta\beta}$ in ref. 11. Then
the overall sensitivity is $S = \sqrt{S_N\cdot S_D} = 3\cdot10^5$. The present
limit on $T^{0\nu}_{1/2}(0^+)$ gives $\langle m_\nu\rangle \lesssim 1.9$ eV on the 68% CL. This
is in accord with the value $m_e/S \approx 1.8$ eV. The upper limit
on the Majoron coupling is $\langle g_B\rangle \lesssim 6.4\cdot10^{-4}$ by using the $M_{\beta\beta}$

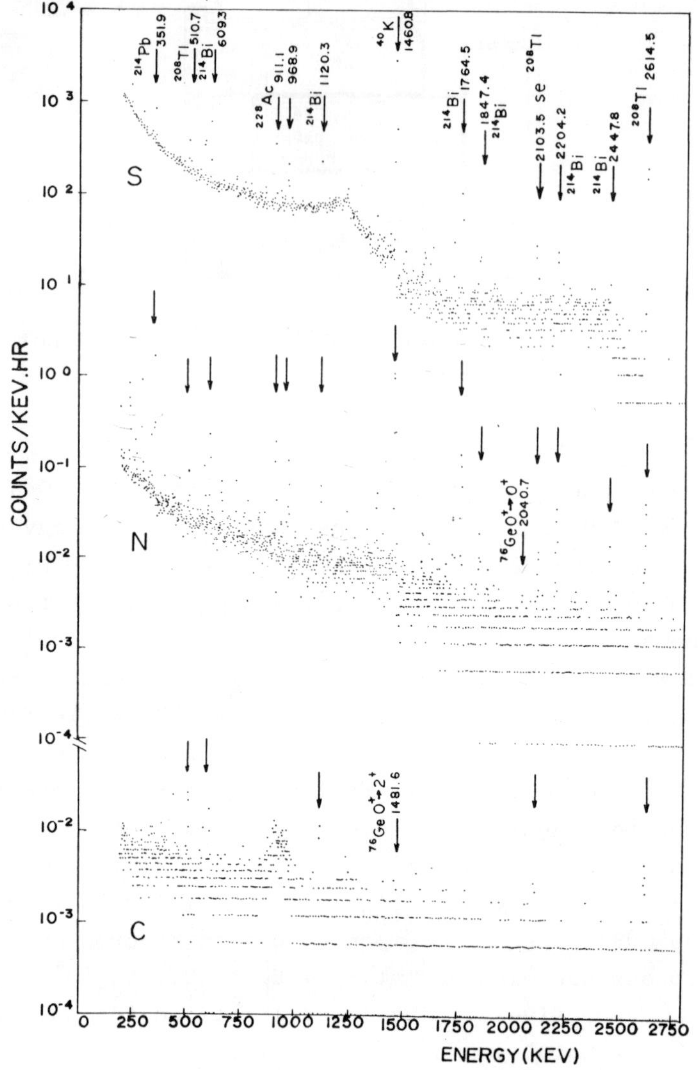

Fig. 4 Energy spectra measured by the Ge detector at
Kamioka (run 2): S - singles without any shields;
N - $0^+ \rightarrow 0^+$ followed by no signals from any of the
NaI segments; C - $0^+ \rightarrow 2^+$ followed by 559.1 keV
γ-signals from one of the NaI segments. N and C
are with shields.

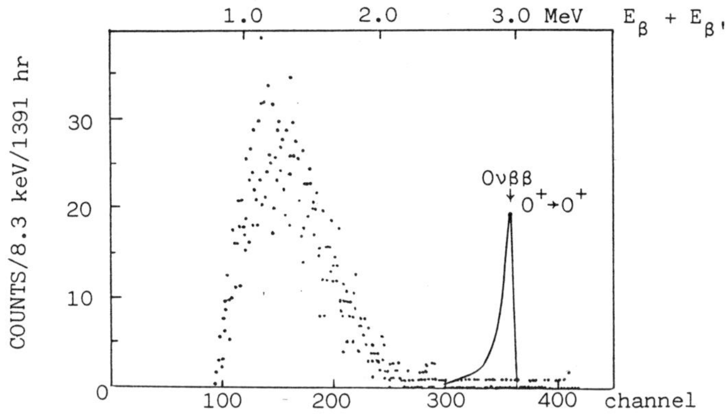

Fig. 5   Energy spectrum of $E_\beta + E_{\beta'}$ for 6 pairs of the adjacent two Si detectors with ^{100}Mo. The solid line is the Monte Carlo calculation.

[11]. The nuclear parameters in ref. 12 lead to $<m_\nu> \lesssim 4.8$ eV, $<\eta> \lesssim 1.1 \cdot 10^{-5}$, $<\lambda> \lesssim 0.8 \cdot 10^{-5}$, and $<g_B> \lesssim 1.6 \cdot 10^{-3}$, while those in ref. 13 are $<m_\nu> \lesssim 3.6$ eV, $<\eta> \lesssim 8.7 \cdot 10^{-8}$ and $<\lambda> \lesssim 5.7 \cdot 10^{-6}$. The recent limit of $T_{1/2}^{0\nu}(0^+) \gtrsim 2.5 \cdot 10^{23}$y [10] leads to smaller limits on these quantities by a factor 2.

The sum energy spectrum for ^{100}Mo was obtained by adding signals from two adjacent Si-detectors. The result of the preliminary run for 1391hrs is shown, together with the Monte Carlo calculation for the $0\nu\beta\beta$ (0^+), in Fig. 5. Here the result obtained by only the 6 sets of the Si·Mo units out of ten is shown because of α activity contaminants in other sets. The background rate is indeed very small, but no appreiable $0\nu\beta\beta$ peak appears. The sensitivity for the 6 operating Si-detectors is obtained as $\tilde{S}_D = 5 \cdot 10^{20} \text{y}^{1/2}$ from $\tilde{N}_{BG} = 0.13/\text{keV} \cdot \text{y}$, $\Delta E = 200$ keV, $N_o = 0.28 \cdot 10^{23}$ and $k = 0.11$. It is estimated to be $S_D = 1.7 \cdot 10^{21}$ for the present 11 Si-detector system with $N_o = 0.5 \cdot 10^{23}$ and $k = 0.22$. The half-life limit for $0\nu\beta\beta$ process is $\tilde{T}_{1/2}^{0\nu}(0^+) \gtrsim 1.7 \sim 1.0 \cdot 10^{20}$y, depending of the way of analysis. The half-life limit measurable for the one year run with the 11 Si-detector is

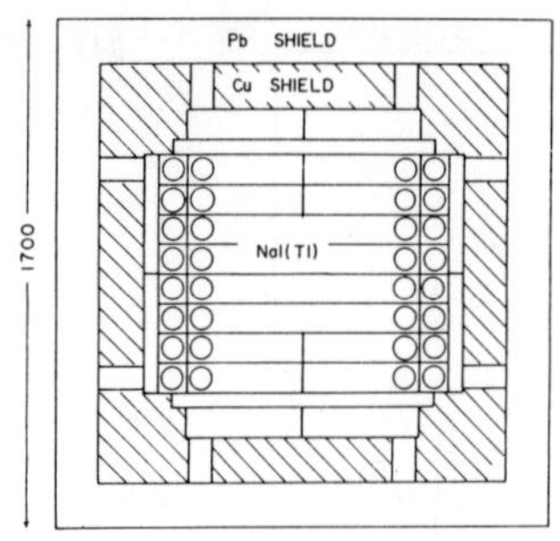

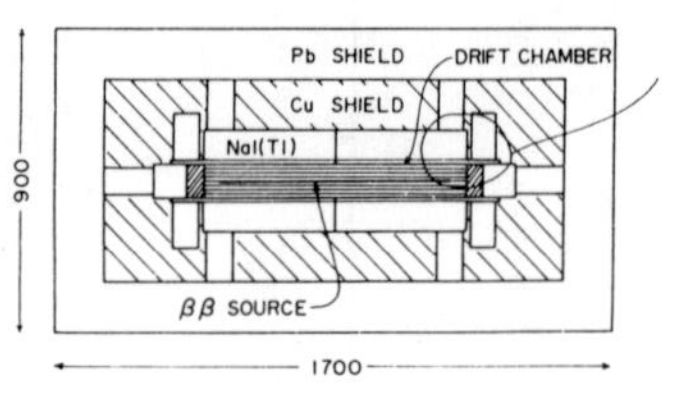

Fig. 6   Schematic picture of ELEGANT V.   Left: the top of view.   Right: the side view.

$T^{0\nu}_{1/2} \simeq S_D \sim 1.7 \cdot 10^{21}$ y.   The lower limit for the $0\nu\beta\beta$ with the Majoron is $\tilde{T}^{0\nu M}_{1/2} \gtrsim 7 \cdot 10^{18}$ y on the 68% CL.

The ^{100}Mo has a large nuclear sensitivity of $S_N = 55 \cdot 10^{-13}$ by using the $M_{\beta\beta}$ in ref. 10, and the overall sensitivity is $S = \sqrt{S_N \cdot S_D} = 1.0 \cdot 10^5$.   The present half-life limit gives $\langle \tilde{m}_\nu \rangle \lesssim 14 \sim 18$ eV on the 68% CL, and the $\langle m_\nu \rangle$ measurable with the 11 Si detector is $\langle m_\nu \rangle / m_e \sim S^{-1} \sim 10 \cdot 10^{-6}$ ($\langle m_\nu \rangle$ measurable: ~ 5 eV).   The upper limit on the Majoron coupling is $\langle \tilde{g}_B \rangle \lesssim 10^{-3}$.   Because of low background rate one year run with the 11 Si-detector array may elucidate the $\langle g_B \rangle$ in the order of $3 \cdot 10^{-4}$ and $T^{2\nu}_{1/2}$ in the order of $10^{17} - 10^{18}$ y.

A detector system "ELEGANT V" with drift chambers surrounded by NaI scintillators is now under progress.   The schematic drawing of ELEGANT V is shown in Fig. 6.   The two β-ray tracks and the vertex are measured by the drift chamber in order to select the true $\beta\beta$ event.   It is used to search for $0\nu\beta\beta$, $0\nu\beta\beta$ with Majoron and $2\nu\beta\beta$ processes of ^{100}Mo and ^{150}Nd with fairly large S_N.   Evaluated sensitivities are $S_N = 75 \sim 55 \cdot 10^{-13}$, $S_D = 5 \cdot 10^{23}$ and $S \simeq 20 \cdot 10^5$.   Thus this can access to $\langle m_\nu \rangle$ in the order of 0.3 eV.

References

* A major part of the present paper has been presented in
the International Symposium on Weak and Electromagnetic
Interaction in Nuclei, Heidelberg, July, 1986 (Proc.
Symposium edited by Klapdor).

1. M. Doi, T. Kotani and E. Takasugi: Prog. Theor. Phys.
Supp. No. 83 (1985).
2. S.P. Rosen: Proc. Int. Conf. Neutrino '86 (Univ.
Hawaii, 1982).
3. H. Ejiri: Proc. Int. Workshop Grand Unification/ICOBAN,
Toyama, (World Scientific, Singapore 1986).
4. H. Ejiri et al.: Proc. Int. Symp. Nucl. Spectroscopy &
Nucl. Interactions (Osaka, World Scientific, Singapore,
1984) p. 284.
5. H. Ejiri et al.: Nucl. Phys. $A448$ (1986) 271,
N. Kamikubota et al.: Nucl. Instr. Methods $A245$ 379
(1986).
6. E. Fiorini et al.: Nuovo Cim. $13A$ 747 (1973).
7. F. Leccia et al.: Nuovo Cim. $78A$ 50 (1983).
8. E. Bellotti et al.: Phys. Lett. $121B$ 72 (1983).
9. F.T. Avignone et al.: Phys. Rev. Lett. 54 2309 (1985).
10. D.O. Caldwell et al.: Phys. Rev. Lett. 54 281 (1985),
Preprint 1986.
11. K. Grotz and H.V. Klapdor: Phys. Lett. $157B$ 242 (1985).
12. W.C. Haxton et al.: Phys. Rev. $D25$ 2360 (1982).
13. T. Tomoda et al.: Phys. Lett. $157B$ 4 (1985).

AN INVESTIGATION OF THE DOUBLE BETA DECAY OF ^{136}Xe

[1]A.S.Barabash, V.V.Kuzminov, V.M.Lobashev,
V.M.Novikov, B.M.Ovchinnikov, A.A.Pomansky

Institute for Nuclear Research of the USSR
Academy of Sciences, 60-th October Anniversary Prospect, 7a, Moscow, USSR

[1]Institute for Theoretical and Experimental
Physics, Moscow, USSR

Abstract. Lower limits were obtained for the
half-life of the ^{136}Xe double beta decay,
using pulse ionization chamber detector. The
experiment was performed at 850 m.w.e. underground in the low-background laboratory. The
observed count rates are consistent with a
lower limit of 1.8×10^{19} y for the half-life
of the $\beta\beta(2\nu)$ decay, 1.2×10^{21} y for the
$\beta\beta(0\nu)$, 0^+-0^+ decay and 4.9×10^{20} y for the
$\beta\beta(0\nu)$, 0^+-2^+ decay.

The usual theory of beta decay, assuming lepton conservation, predicts double beta decay of a nucleus with
emission of two neutrinos ($\beta\beta(2\nu)$). If leptons are not
conserved the process can go without the emission of neut-
rinos ($\beta\beta(0\nu)$). There are many isotopes for which $\beta\beta$
decay can occur. There have been many direct-detection
experiments performed but no direct observation of the $\beta\beta$
decay with or without neutrinos has been reported.

^{136}Xe appears to be one of the most favorable nuclei
to study. It has high decay energy (2.48 MeV). Moreover,
a recent theoretical prediction / 1 / gives a relatively
small half-life of 2.3×10^{19} y for $\beta\beta(2\nu)$ of ^{136}Xe.

First experimental studies of $\beta\beta$ (0ν) of ^{136}Xe have been made recently. These measurements gave a $\beta\beta$ (0ν) half life limits of 2.36×10^{21} y (with use of a liquid scintillator detector filled with enriched xenon, / 2 /) and 5.5×10^{19} y / 3 / obtained with use of ionization chamber filled with the natural xenon.

In the present experiment a pulse ionization chamber was used. The chamber filled with both the natural (8.87% of ^{136}Xe) and enriched (93% of ^{136}Xe) xenon. To obtain the high drift velocity of electrons 0.8% H_2 is added to xenon. The chamber has an active volume of 3.14 1. The working pressure is 25 atm. The energy resolution is 3.8% at 2.5 MeV. The detector is installed at the Baksan Neutrino Observatory at the depth of 850 m.w.e. and is shielded by copper and lead of 10 cm and 15 cm thickness, respectively.

The present experiment consisted of two runs. In the first one the detector has been operated for 243 h live time with the enriched xenon. Then, in the second run, the detector has been operated for 120 h live time with the natural xenon.

Fig.1 shows background spectra obtained in the low background shielding. The background due to presence of ^{85}Kr (E_β^{max} = 0.672 MeV) in the natural xenon is clearly seen in the energy region of 0-0.7 MeV.

In the neutrinoless double beta decay one would expect to see peaks at 2.48 MeV and 1.66 MeV for 0^+-0^+ and 0^+-2^+ transition respectively. No peak appears in the region of 0^+-0^+ and 0^+-2^+ neutrinoless double beta decay (Fig.1). Corresponding 68% c.l. lower limits on half lives are

$$T_{1/2}(\beta\beta (0\nu)) \geq 1.2\times10^{21} \text{ y} \quad \text{for } 0^+\text{-}0^+,$$

$$T_{1/2}(\beta\beta (0\nu)) \geq 4.9\times10^{20} \text{ y} \quad \text{for } 0^+\text{-}2^+$$

Also, the possibility to carry out a difference experiment allowed us to set the limit for the half life of $\beta\beta$ (2ν):

$$T_{1/2}(\beta\beta (2\nu)) \geq 1.8\times10^{19} \text{ y} \text{ , 68\% c.l.}$$

We wish to thank N.Likhovid and B.Pritichenko for their assistance.

REFERENCES

1. P.Vogel, P.Fisher, Phys. Rev. C, $\underline{32}$ (1985) 1362
2. I.Barabanov et al, Pisma v JETF, $\underline{43}$, N4 (1986) 166
3. A.Barabash et al, Low Radioactivities'85 (in press)

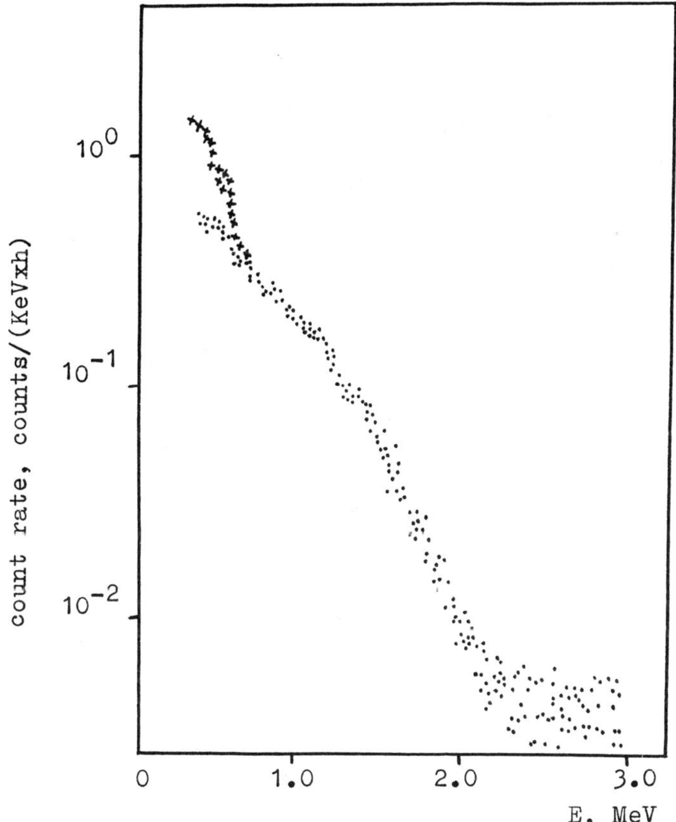

Fig1 ^{136}Xe (.) and Xe^{nat} (x) backgro-
und spectra obtained with use of ionization chamber (Xe+0.8%H$_2$) in low back-
round shielding

Are massive neutrinos Majorana particles?

Tsuneyuki Kotani

Institute of Physics, College of General Education
Osaka University
Toyonaka, Osaka 560, Japan

ABSTRACT

One Dirac field can be constructed from two Majorana
fields theoretically. The origin of the different
properties between the Dirac and Majorana neutrinos is
explained. The theory of the neutrinoless double beta
decay is briefly reviewed, and the data analysis and the
ambiguities due to the nuclear matrix elements are
discussed. Consistencies are considered for various experimental bounds from the neutrinoless double beta
decay, the tritium beta decay and the neutrino
oscillations. A consistent solution for these bounds is
shown for the SO(10) model with three generations.

§ 1. Introduction

Among several assumptions in the minimum standard model of
electroweak interaction, two followings will be discussed;

(1) Are all neutrinos massless, $m_\nu = 0$? If $m_\nu \neq 0$, are neutrinos
Dirac or Majorana particles?

(2) Is the gauge group only $SU(2)_L \times U(1)$, i.e., the V-A interaction?

At present, there is no definite result for the finite mass of
neutrino, except the experiment on the tritium β decay by the Moscow
group. However, there are two indirect evidences indicating the
finite neutrino mass; the Mikheyev-Smirnov-Wolfenstein enhancement of
oscillation as a possible solution to the solar-neutrino problem and
the dark matter problem in the astrophysics.

If neutrinos are massive, we can say that the Majorana neutrino
is theoretically more fundamental than the Dirac neutrino, because one
massive Dirac field consists of two Majorana fields (§ 2 and § 3). As

114

it is well known, the neutrinoless double beta decay (the $(\beta\beta)_{0\nu}$ mode) is one of experiments presently accessible for testing directly whether neutrinos are Dirac or Majorana. It has been shown that the observation of this mode implies at least the electron neutrino to be a massive Majorana particle.[1] [See § A.1 of Ref.(2).]

Until now, there is no direct observation of this $(\beta\beta)_{0\nu}$ mode. But the present experimental limits for this mode give some upper bounds on the neutrino mass, though it is a sum of virtual neutrino masses weighted by the neutrino mixing matrix elements squared. Also, some information on the right-handed weak interaction can be derived. These limits correlate with some uncertainties of the theoretical es- timates of nuclear matrix elements.(§ 4 and § 5).

Although various experimental informations on the neutrino mass are not definite, it is useful to check whether they are consistent each other within the present experimental limits (§ 6).

In this paper, theoretical formulae are presented on the $SU(2)_L \times SU(2)_R \times U(1)$ model with 3 generations.

§ 2. Relation between Dirac and Majorana neutrinos

It will be shown that one Dirac neutrino field consistes of a pair of mass degenerate Majorana neutrino fields. This property has been shown by many authors implicitly[3], and by Schechter and Valle[4] and Doi et al.[5] explicitly.

Let us show it by starting from the Lagrangian density for a classical Dirac field $\psi(x,t)$,

$$\mathcal{L} = \overline{\psi}\,(i\gamma^{\circ}\,\partial_{t} + i\vec{\gamma}\cdot\vec{\nabla} - m)\psi . \tag{2.1}$$

Hereafter, the Weyl representation of γ matrices is employed.

[Case I] If ψ is expressed in the two component form as

$$\psi = \begin{pmatrix} \phi \\ \chi \end{pmatrix} , \tag{2.2}$$

then the Lagrangian becomes

$$\mathcal{L} = \phi^{\dagger} O_{+}\phi + \chi^{\dagger} O_{-}\chi - \phi^{\dagger} m\chi - \chi^{\dagger} m\phi , \tag{2.3}$$

where

$$O_\pm = i(\partial_t \pm \vec{\sigma}\cdot\vec{\nabla}). \qquad (2.4)$$

Note that the mass term is given by the mixing between χ and ϕ^+ or χ^+ and ϕ (the Dirac type mass term). Two operators O_+ and O_- mean that ϕ and χ represent neutrinos with the positive (h = +1/2) and negative (h = -1/2) helicities, respectively, as classical fields, if m = 0. Thus, the original Dirac field with m $\neq$ 0 has 4 freedoms, i.e., two spin states for each ϕ or χ component.

[Case II] There is another decomposition of ψ to express it as a linear combination of two independent fields η and ξ , each of which is a two component field;

$$\psi = [N_1 + i N_2]/\sqrt{2}, \qquad (2.5)$$

where four component forms N_1 and N_2 are defined as

$$N_1 = \begin{pmatrix} -i\sigma_2\eta^* \\ \eta \end{pmatrix} \quad \text{and} \quad N_2 = \begin{pmatrix} \xi \\ i\sigma_2\xi^* \end{pmatrix}. \qquad (2.6)$$

In this case, the original Lagrangian Eq.(2.1) is separated into two independent parts as

$$\mathcal{L} = \mathcal{L}_L(\eta) + \mathcal{L}_R(\xi) , \qquad (2.7)$$

where

$$\mathcal{L}_L(\eta) = \eta^+ O_- \eta - (m/2)(\eta^T i\sigma_2 \eta - \eta^+ i\sigma_2 \eta^*), \qquad (2.8)$$

$$\mathcal{L}_R(\xi) = \xi^+ O_+ \xi + (m/2)(\xi^T i\sigma_2 \xi - \xi^+ i\sigma_2 \xi^*). \qquad (2.9)$$

Note that the field N_j in Eq.(2.6) satisfies the Majorana (self-conjugate) condition $N_j^c \equiv C \bar{N}_j^T = N_j$ where the superscript T means to take the transpose and C is the charge conjugation matrix.   Therefore, $\mathcal{L}_L$ and $\mathcal{L}_R$ are Lagrangian densities for the left- and right-handed Majorana neutrino fields in the two component form, respectively. Two components of η (or ξ) represent two spin states of the Majorana field (two freedoms), while the four component form N_1 (or N_2) should be understood as a convention to express physical quantities in the relativistic form including the γ matrix. Also it is worthwhile to note that it is necessary to have a factor i (or -i) in Eq.(2.5) in order to express ψ as the superposition of two Majorana fields. This factor i means that η and ξ have opposite signs under a CP transformation if ψ is required to have a definite CP value. Finally, note that the Majorana type mass term $\eta^T i\sigma_2 \eta$ in Eq.(2.8) does not include the Hermitian conjugate of η in contrast to the Dirac type mass term $(\phi^\dagger \chi)$ in Eq.(2.3). This is the reason why the particle and antiparticle creation operators can not be distinguished in the case of the quantized Majorana field η . In the classical field, it appears as the mixing between the positive and negative frequency parts in the momentum representation of η . [For detail, see § 2.4 of Ref.(2).]

The lepton number can not be assigned to the Majorana neutrino (η and ξ), while does to the Dirac field (ψ). Where does this difference come from? In the gauge theory, the conservations of the lepton number and electric charge are expresed as the freedom of the phase transformation. In the case of the Majorana field, there is no phase freedom because of the Majorana condition, $N_j^C = N_j$. However, one Dirac field can be constructed from two Majorana fields related to N_1 and N_2 by an orthogonal transformation; namely

$$\psi_\alpha = [N_1(\alpha) + i N_2(\alpha)]/\sqrt{2}, \tag{2.10}$$

where

$$\begin{pmatrix} N_1(\alpha) \\ N_2(\alpha) \end{pmatrix} = \begin{pmatrix} \cos\alpha & -\sin\alpha \\ \sin\alpha & \cos\alpha \end{pmatrix} \begin{pmatrix} N_1 \\ N_2 \end{pmatrix}. \tag{2.11}$$

Then, this ψ_α is related to ψ in Eq.(2.5) by $\psi_\alpha = \psi \exp(i\alpha)$. This is the origin of the phase freedom for the Dirac field. It means that all charged fermions should be described by the appropriate Dirac fields in order to ensure the conservation of charge.

A Majorana neutrino can not have nonzero electric and magnetic moments, independently of whether massless or massive. However, a massive Dirac neutrino can have a magnetic moment and if the time reversal is violated, has an electric moment, too. Why is this difference? For the Majorana neutrino, the above statement is derived from the following relation,

$$\int d^4x \bar{N}_j \, i\sigma_{\mu\nu}(a+b\gamma_5)\partial^\nu N_k = -\int d^4x \bar{N}_k \, i\sigma_{\mu\nu}(a+b\gamma_5)\partial^\nu N_j . \tag{2.12}$$

It is clear that the Majorana neutrino has no moment, but the transition moments between different Majorana neutrinos are allowed. On the other hand, the massive Dirac neutrino gets the magnetic moment which comesfrom the transition moment between N_1 and N_2;

$$\int d^4x\,\bar{\psi}\,i\sigma_{\mu\nu}\,\partial^\nu\psi = i\int d^4x\,\bar{N}_1\,i\sigma_{\mu\nu}\,\partial^\nu N_2 \ . \tag{2.13}$$

Thus, it can be concluded that one Dirac neutrino field can be constructed from two mass degenerate Majorana neutrino fields, which should have opposite signs under the CP transformation. Therefore, we can say that the Majorana description for neutrinos is more fundamental than the Dirac description theoretically.

§ 3. Mass matrix for the Majorana neutrino system

Let us consider a system of 6 Majorana neutrinos in the $SU(2)_L \times SU(2)_R \times U(1)$ model with 3 generations. Three left-handed and right-handed original Majorana neutrinos are denoted by ν_L^o and $\nu_R^{o'}$ in the four component form, i.e., $\nu_L^{oT} = (\nu_e^{oT}, \nu_\mu^{oT}, \nu_\tau^{oT})_L$. The Lorentz invariant mass term is expressed as

$$\mathcal{L}_M = -(1/2)(\ \overline{(\nu_L^o)^C},\ \overline{\nu_R^{o'}}\ )\ \mathcal{M}_\nu \begin{pmatrix} \nu_L^o \\ (\nu_R^o)^C \end{pmatrix} + \text{h.c.}, \tag{3.1}$$

where

$$\mathcal{M}_\nu = \begin{pmatrix} \mathcal{M}_L & \mathcal{M}_D^{\ T} \\ \mathcal{M}_D & \mathcal{M}_R \end{pmatrix}. \tag{3.2}$$

The Majorana type mass matrices, $\mathcal{M}_L$ and $\mathcal{M}_R$, are 3×3 symmetric ones, and so is true for $\mathcal{M}_\nu$. Therefore, $\mathcal{M}_\nu$ is diagonalized by a unitary matrix U_ν (transformation matrix for neutrinos) as

$$U_\nu^T\,\mathcal{M}_\nu U_\nu = D_\nu = \begin{pmatrix} m_1 & & & \\ & m_2 & & \\ & & \ddots & \\ & & & m_6 \end{pmatrix}, \tag{3.3}$$

where $m_j > 0$. We assume $m_1 < m_2 < m_3$ and $m_4 < m_5 < m_6$. Three original neutrinos are related to 6 Majorana mass eigenstate neutrinos N_j as

$$\begin{pmatrix} \nu_L^o \\ (\nu_R^{o'})^c \end{pmatrix} = U_\nu N_L \ , \qquad U_\nu = \begin{pmatrix} U_{\nu I} & U_{\nu II} \\ V_{\nu I}^* & V_{\nu II}^* \end{pmatrix} , \tag{3.4}$$

where $N^T = (N_I^T , N_{II}^T)$. These N_I and N_{II} are assumed to consist of 3 Majorana neutrinos corresponding mainly to the left- and right-handed ones, respectively. The weak eigenstate neutrinos are defined as

$$\nu_L = U N_L = U_\ell^\dagger \nu_L^o \quad \text{and} \quad \nu_R' = V N_R = V_\ell^\dagger \nu_R^{o'} , \tag{3.5}$$

where the neutrino mixing matrix is defined as

$$\begin{pmatrix} U \\ V^* \end{pmatrix} = \begin{pmatrix} U_\ell^\dagger U_{\nu I} & U_\ell^\dagger U_{\nu II} \\ V_\ell^T V_{\nu I}^* & V_\ell^T V_{\nu II}^* \end{pmatrix} , \tag{3.6}$$

and U_ℓ and V_ℓ are the transformation matrices for left- and right-handed changed leptons, respectively.

Let us choose two Majorana neutrinos among 6 N_j's, say N_j and N_k. If they are in a mass degenerate state $(m_j = m_k)$ and have the opposite CP values, then they become either the ordinary Dirac neutrino for $k=j+3$, or the Konopinski-Mahmoud type Dirac one for $k \neq j+3$, as shown in Table 1. In order to form such a Dirac neutrino, there should be some special symmetry for the neutrino mass matrix to ensure both the mass equality $m_j = m_k$ and the relation between the neutrino mixing matrix elements

$$U_{\ell j} = \pm i U_{\ell k} \quad \text{or} \quad V_{\ell j} = \pm i V_{\ell k}, \tag{3.7}$$

where the latter relation is necessary to get the superposition in Eq.(2.5). All other elements of U and V in the j-th and k-th columns should be zoro, because if not so, we have $m_j \neq m_k$ (the PD neutrino).

Suppose that the above special symmetry for the mass degeneracy of neutrinos is not the symmetry for the weak interaction. Then the mass degeneracy is relaxed by the higher order perturbations of weak interaction. Thus neutrinos which form a Dirac neutrino at tree level split into two Majorana neutrinos with a tiny mass difference. A pair of these Majorana neutrinos are called the pseudo-Dirac (PD) neutrino, as shown in Table 1. [For detail, see § 2.5 of Ref.(2).]

Name of ν		Masses	CP values	Combination
Ordinary Dirac		$m_j = m_k$	opposite	$k = j+3$
Konopinski-Mahmoud type Dirac				$k \neq j+3$
Two Majorana neutrinos	Pseudo Dirac	$0 < \left| \dfrac{m_j - m_k}{m_j} \right| \ll 1$		any combination
		others		

Table 1

§ 4. Theory of neutrinoless double beta decay

Within the framework of gauge theory, the leptonic part of the amplitude for the $(\beta\beta)_{0\nu}$ mode is given as the second order perturbation of the leptonic weak interaction;

$$\mathcal{L}_w = [\, g_L \, j_L^\rho \, \overline{W_{L\rho}} + g_R \, j_R^\rho \, \overline{W_{R\rho}} \,] \, / \, 2\sqrt{2} + \text{h.c.}, \tag{4.1}$$

where the left-handed leptonic current (V-A) is

$$j_L^\rho = \bar{e} \, \gamma^\rho \, (1-\gamma_5) \, \nu_{eL} \quad \text{with} \quad \nu_{eL} = \Sigma_j \, U_{ej} \, N_{jL} \,, \tag{4.2}$$

and the right-handed current (V+A) is

$$j_R^\rho = \bar{e} \, \gamma^\rho \, (1+\gamma_5) \, \nu'_{eR} \quad \text{with} \quad \nu'_{eR} = \Sigma_j \, V_{ej} \, N_{jR} \,. \tag{4.3}$$

In this paper the $(\beta\beta)_{0\nu}$ mode mediated by Higgs bosons and the contributions from the mirror leptons are not condidered. [See § 5.1 and Appendix(A.2) in Ref.(2).]

The decay amplitude is divided into two parts by the lepton vertices due to the helicity matching of the propagating neutrino. If both verticies are V-A(or V+A), then the neutrino mass m_j part of the virtual neutrino propagator contributes (the m_ν part), as shown in

(V − A) − part	(V + A) − part	
m_ν − part	ω − term ($\omega = q^0$)	$\vec{q}$ − term
e_1^- (S) e_2^- (S) m_j W_L^- W_L^-	e_1^- (S) e_2^- (S) ω W_L^- W_R^-	e_1^- (S) e_2^- (P) $\vec{q}$ W_L^- W_R^-
Both e^- : S-wave ($j=1/2$)	Both e^- : S-wave ($j=1/2$)	one e^- : S-wave ($j=1/2$) other e^- : P-wave ($j=3/2$)
nuclear transition: $0^+ \to J^+$		
$0^+ \to 0^+$ 0^+ $\beta\beta$ — 2^+ — 0^+	$0^+ \to 0^+$ 0^+ $\beta\beta$ — 2^+ — 0^+	$0^+ \to 0^+$ $0^+ \to 2^+$ 0^+ $\beta\beta$ — 2^+ γ — 0^+
$\left(\dfrac{m_j}{m_e}\right) \sim 10^{-5}$	$\left(\dfrac{\langle\omega\rangle}{m_e}\right) \sim \left(\dfrac{\langle\vec{q}\rangle}{m_e}\right) \sim \dfrac{1}{R m_e} \sim 80$	
$\langle m_\nu \rangle = \left| \Sigma'_j m_j U_{ej}^2 \right|$	$\langle \lambda \rangle = \lambda \left| \Sigma' U_{ej} V_{ej} \dfrac{g_V'}{g_V} \right|$	$\langle \eta \rangle = \left| \eta \Sigma'_j U_{ej} V_{ej} \right|$

Table 2

Table 2. Since the electron neutino mixing matrix U_{ej} appears at each vertex, this m_ν part is proportional to the effective neutrino mass which is given as

$$< m_\nu > = \left| \Sigma'_j \, m_j \, U^2_{ej} \right| \, ,$$

(4.4)

where the primed sum means to extend over only the light neutrinos $(m_j \lesssim 10 \text{ MeV})$. These virtual neutrino propagations give the Coulomb type potentials.

If one vertex is V–A and another is V+A, then the decay ampritude should be proportional to 4 momentum part q^μ of the neutrino propagator and also includes parameters characteristic to the V+A part, λ and η , which are defined approximately as

$$\lambda \simeq (M_{WL} / M_{WR})^2 \quad \text{and} \quad \eta \simeq - \tan \zeta \, ,$$

(4.5)

if $M_{WR} >> M_{WL}$, where ζ is the mixing angle between the left gauge boson W_L with the mass M_{WL} $(\sim M_1)$ and the right gange boson W_R with M_{WR} $(\sim M_2)$. [See Eqs. (A.2.7) and (A.2.8) of Ref.(2).] If neutrino mixing matrices U_{ej} and V_{ej} are taken into account at verticies V–A and V+A as defined in Eqs.(4.2) and (4.3), then the effective right-handed parameters become

$$< \lambda > = \lambda \left| \Sigma'_j \, U_{ej} \, V_{ej} \, (g'_V / g_V) \right| \, ,$$

(4.6)

and

$$< \eta > = \left| \eta \, \Sigma'_j \, U_{ej} \, V_{ej} \right| \, ,$$

(4.7)

where $g_V = \cos \theta_C$ and $g'_V = \cos \theta'_C \exp(i\alpha)$. Here θ_C and θ'_C are the Cabibbo–Kobayashi–Maskawa mixing angles for the left- and right-handed d and s quarks, and α is the CP violating phase due to both the mixing of right-handed quarks and the mixing between W_L and W_R. The energy ω $(= q^0)$- and momentum $\vec{q}$-terms of virtual neutrino give different features. The important point is that $\vec{q}$ plays a parity odd operator between initial and final nuclei, while ω (and m_j) does not. Since the $(\beta \beta)_{0\nu}$ mode takes place only as the $0^+ \to J^+$ transition practically and the parity is conserved in the non–relativistic treatment of nucleons, there should be some additional parity odd operators, which come from either one P-wave electron with $j=3/2$ or the nucleon recoil term with S-wave electrons. Although the latter nucleon recoil term is not listed in Table 2, it may give the largest contribution.[7] [See table 3.3 of Ref.(2).]

Finally, let us compare contributions from the m_ν - and (V+A)-parts. The amplitude proportional to the neutrino mass m_j includes a factor $(m_j/m_e) \sim 10^{-5}$, while the latter $(<\omega>/m_e) \sim (<\vec{q}>/m_e) \sim (1/R\, m_e) \sim 80$, where R is the nuclear radius. At a glance, it is hopeless to measure the m_ν-part in comparison with the (V+A)-part by the factor $10^{-5}/80$. However, the reality is not. Let us examine it for various cases:

[Case D] If all neutrinos are Dirac, then there should be pairs of two Majorana neutrinos with the same mass $m_j = m_k$ and the relation in Eq.(3.7) should be satisfied. Therefore, there is no $(\beta\beta)_{0\nu}$ mode, because we have exactly

$$< m_\nu > \, = \, < \lambda > \, = \, < \eta > \, = \, 0 \, . \tag{4.8}$$

[Case M-I] If all neutrino masses are less than 10 MeV, we have

$$< m_\nu > \neq 0 \quad \text{but} \quad < \lambda > \, \simeq \, 0, \quad < \eta > \, \simeq \, 0 \, , \tag{4.9}$$

because of the unitarity condition, $\Sigma_j U_{ej} V_{ej} = 0$. The deviation from the zero value of $< \lambda >$ and $< \eta >$ is the order of the ratio of m_j to the average virtual neutrino energy, $m_j/<\omega> \sim 10^{-7}$.

[case M-II] Both light $(m_{N_I} \lesssim O(\text{MeV}))$ and heavy $(m_{N_{II}} \gtrsim O(\text{GeV}))$ Majorana neutrinos coexist. This is the case where the see-saw mechanism is effective. Since the mixing matrix elements between left- and right-handed neutrinos are small, i.e., $(U_{\nu I})_{ej} >> (V_{\nu I})_{ej}$ and $(U_{\nu II})_{ej} << (V_{\nu II})_{ej}$, there is a small deviation from the unitarity condition, say $\bar{v}_e = |\Sigma_j U_{ej} V_{ej}|$. Thus we have

$$< m_\nu > \neq 0, \quad < \lambda > = \lambda \, | \, \bar{v}_e \, (g'_V/g_V) | \quad \text{and} \quad < \eta > = | \eta \bar{v}_e | \, . \tag{4.10}$$

In conclusion, it is not unnatural that contributions from both the m_ν - and (V+A)-parts are of the same order. [For details, see Tables 3.3 and 10.1 of Ref.(2).]

In order to obtain the experimental result on $< m_\nu >$ independently of $<\lambda>$ and $<\eta>$, the $0^+ \to 2^+$ transiton in the $(\beta\beta)_{0\nu}$ mode is useful, because it offers only the information on $<\lambda>$ and $<\eta>$, as shown in Table 2. Though the nuclear matrix elements for this transition may be smaller than those for the $0^+ \to 0^+$ transition,[8] the observation of the $(\beta\beta)_{0\nu}(0^+ \to 2^+)$ transition will give us some useful information. [For details, see Tables 3.5 and 7.6 of Ref.(2).]

§ 5. Data analysis of the double beta decay

Let us consider the $(\beta\beta)_{0\nu}$ mode in the 2n-mechanism, where two neutrons in the same neucleus decays successively without neutrinos in the final state. The half-life formula for the $(\beta\beta)_{0\nu}$ $(0^+ \to 0^+)$ transition is

$$[\, T_{0\nu}\, (0^+ \to 0^+)]^{-1} = |\, M_{GT}^{(0\nu)}\, |^2\, C_1 \tag{5.1}$$

$$\left\{ \left(\frac{<m_\nu>}{m_e}\right)^2 + [\, C_2' <\lambda> \cos\psi_1 + C_3' <\eta> \cos\psi_2]\left(\frac{<m_\nu>}{m_e}\right) \right.$$

$$\left. + C_4' <\lambda>^2 + C_5' <\eta>^2 + C_6' <\lambda><\eta> \cos(\psi_1 - \psi_2) \right\},$$

where ψ_1 and ψ_2 are the CP violating phases and the double Gamow-Teller nuclear matrix element is defined as

$$M_{GT}^{(0\nu)} = -\Sigma_a <0^+ \|\, \Sigma_n \tau_n \vec{\sigma}_n \,\| N_a><N_a \|\, \Sigma_m \tau_m \vec{\sigma}_m \,\| 0^+> h_+(r_{nm}, E_a). \tag{5.2}$$

Here h_+ is a potential due to the neutrino propagation and depends on both the energy (E_a) of intermediate neucleus N_a and the distance (r_{nm}) between two decaying n-th and m-th neutrons. Rigorously speaking, h_+ includes the mass (m_j) of virtual neutrino, but this m_j dependence is neglected by assuming $m_j < 10$ MeV $< (<\omega>)$, $<\omega>$ being the average energy of virtual neutrino. Its neglection requires the primed sum in Eq.(4.4). Six coeffecients C_j consist of both nine known constants G_{0k} and seven ratios of other nuclear matrix elements to $M_{GT}^{(0\nu)}$, which should be determined by the theory of nuclear structure. [For details, see Table 3.4 and Eq.(3.5.15) of Ref.(2).]

In Table 3, by using estimates of various nuclear matrix elements

	$M_{GT}^{(0\nu)}$	$\langle m_\nu \rangle$ (eV)	$\langle\lambda\rangle$	$\langle\eta\rangle$
Grotz-Klapdor[6]	10.40	<0.99		
Tomoda et. al[7]	5.70	<1.9 (<2.2)	$<3.7\cdot10^{-6}$ $(<3.9\cdot10^{-6})$	$<4.5\cdot10^{-8}$ $(<5.2\cdot10^{-8})$
Haxton-Stephenson[8]	4.18	<2.5	$<4.2\cdot10^{-6}$	$<6.3\cdot10^{-6}$
Scaled from Haxton-Stephenson[10]	1.59	<6.7 (<7.5)	$<1.2\cdot10^{-5}$ $(<1.2\cdot10^{-5})$	$<1.2\cdot10^{-6}$ $(<1.4\cdot10^{-6})$

Table 3

by Grotz-Klapdor,[6] Tomoda et al.[7] and Haxton-Stephenson,[8] the upper limits of three parameters are given for the ^{76}Ge data, $T_{0\nu}(0^+ \to 0^+) < 2.5\cdot10^{23}$ yr measured by the Santa Barbara-LBL group.[9] Thus, we may say

$$\langle m_\nu \rangle < (1 \sim 2.5)\ \text{eV} \qquad \text{for} \quad \langle\lambda\rangle = \langle\eta\rangle = 0 . \tag{5.3}$$

However, can we accept this limit on the face value? There are some discrepancies between experimental values and theoretical estimates for the neutrino emitting $\beta\beta$ decay, the $(\beta\beta)_{2\nu}$ mode, as mentioned below. In order to be as conservative as possible, Doi et al[10] have used the scaled value for $M_{GT}^{(0\nu)}$ and the estimates by Haxton-Stephenson for the ratios of other nuclear matrix elements to $M_{GT}^{(0\nu)}$. The scale factor for ^{76}Ge is assumed to be the same as for ^{82}Se, by which the data on the $(\beta\beta)_{2\nu}$ mode of ^{82}Se are reproduced, because there is no precise data on the $(\beta\beta)_{2\nu}$ mode of ^{76}Ge. The use of the same scale factor, of course, is not supported theoretically, but offers some idea on the ambiguities. Also note that the theoretical estimates of $\xi = |M_{GT}^{(2\nu)} / M_{GT}^{(0\nu)}|$ for ^{76}Ge and ^{82}Se show the similar values within 10%.[6] Another problem is that the half-life formula in Eq.(5.1) is a function of three unknown parameters, $\langle m_\nu \rangle$, $\langle\lambda\rangle$ and $\langle\eta\rangle$. The upper limit of $T_{1/2}$ defines the allowed region for them within an ellipsoid. A little larger bounds have been found as shown in the parenthesis of Table 3. Because of these ambiguities, the conservative bound may be

$$\langle m_\nu \rangle < (1 \sim 7.5)\ \text{eV} \qquad \text{for} \quad \langle\lambda\rangle \neq 0 \text{ and } \langle\eta\rangle \neq 0 . \tag{5.4}$$

	^{76}Ge	^{82}Se	^{130}Te	^{150}Nd
$T_{2\nu}$ exp.data	$>8\cdot10^{19}$ (SB-LBL)	$\dfrac{(1.30\pm0.05)\cdot10^{20}}{\text{(Heidelberg)}}$ $\dfrac{(1.0\pm0.4)\cdot10^{20}}{\text{(Missouri)}}$ $>1.0\cdot10^{20}$ (Irvine)	$\dfrac{>1.5\cdot10^{21}\text{yr}}{\text{(Heidelberg)}}$ $\dfrac{(7\pm2)\cdot10^{20}}{\text{(Missouri)}}$	$>1.3\cdot10^{19}$ (Moscow)
$\left\lvert \dfrac{M_{GT}^{(2\nu)}}{\mu_0} \right\rvert_{exp.}$	<0.309	$\underline{0.042}$ $\underline{0.048}$ <0.048	$\underline{<0.012}$ $\underline{0.017}$	0.025
Theoretical Estimate				
Klaptor et al.[6]	0.191	0.124	0.042	0.132
Tomoda et al.[7]	0.188			
Haxton et al.[8]	0.139	0.0951	0.114	
Vogel-Fisher[15]	0.066	0.082	0.090	0.064
Vogel-Zirnbauer[15]	0.035~ 0.123	0.019~ 0.065	0.029~ 0.069	

Table 4

SB-LBL: (9)   Heidelberg: (11)   Missouri: (12)
Irvine: (13)   Moscow: (14)

	Experiment (Geochemical)		Theory	
$\dfrac{T_{2\nu}[^{130}T_e]}{T_{2\nu}[^{82}Se]}$	12.5±0.9 [Heidelberg[11]]	7±2 [Missouri[12]]	8.0 [Grotz-Klapdor[6]] 0.70 [Vogel-Fisher[15]] 0.65 [Haxton-Stephenson[8]] 0.4~0.8 [Vogel-Zirnbauer[15]]	
$\dfrac{G_{GT}(^{82}Se)}{G_{GT}(^{130}Te)}$			$\dfrac{4.35\cdot10^{-18}}{4.80\cdot10^{-18}}= 0.906$	

Table 5

In order to see such ambiguities for theoretical estimates of nuclear magtrix elements, let us consider the $(\beta\beta)_{2\nu}$ mode. The half-life formula for this mode is

$$[\,T_{2\nu}\,]^{-1} \simeq G_{GT}\,|\,M_{GT}^{(2\nu)}/\mu_0\,|^2\;, \tag{5.5}$$

where G_{GT} is a known constant. The experimental and theoretical values of $|\,M_{GT}^{(2\nu)}/\mu_0\,|$ are listed in Table 4, where the underline means the data obtained by the geochemical methode. The result of direct measurement on ^{82}Se by the Irvine group is consistent with the smallest theoretical estimates by Vogel and Zirnbauer.[15] However, the ratio of half-lives measured by the geochemical method still disagrees with theoretical estimates, as shown in Table 5. These discrepancies in Tables 4 and 5 require to refine the theoretical estimates of $|\,M_{GT}^{(2\nu)}/\mu_0\,|$. Also other possible mechanisms such as pion exchange [Fig. 3.1 of Ref.(2)] may have to be considered.

There are other geochemical data on the ratio of half-lives of ^{130}Te to ^{128}Te,

$$R_T^{-1} = \frac{T_{1/2}(^{130}\mathrm{Te})}{T_{1/2}(^{128}\mathrm{Te})} = \frac{T_{2\nu}(^{130}\mathrm{Te})}{T_{2\nu}(^{128}\mathrm{Te})}\;\frac{[\,1+(\Gamma_3/\Gamma_{2\nu})+(\Gamma_4/\Gamma_{2\nu})\,]_{128}\mathrm{Te}}{[\,1+(\Gamma_3/\Gamma_{2\nu})+(\Gamma_4/\Gamma_{2\nu})\,]_{130}\mathrm{Te}}\;, \tag{5.6}$$

where $\Gamma_{2\nu}$, Γ_3 and Γ_4 are transition probabilities for $(\beta\beta)_{2\nu}$, $(\beta\beta)_{0\nu}$ (3 body decay) and the process $(A,\,Z{-}2)\rightarrow(A,\,Z)+2e^-+M^0$ (Majoron emmision), respectively. The maximum kinetic energy release (T) for ^{128}Te is 1.7 m_e, while $T = 4.957\ m_e$ for ^{130}Te. The large T value for ^{130}Te means that $(\Gamma_3/\Gamma_{2\nu})$ and $(\Gamma_4/\Gamma_{2\nu})$ can be neglected in comparison with unity, because of the small values of $\langle m_\nu\rangle/m_e$, $\langle\lambda\rangle$ and $\langle\eta\rangle$. The ratio of $T_{2\nu}(^{130}\mathrm{Te})/T_{2\nu}(^{128}\mathrm{Te})$ is given roughly by $G_{GT}(^{128}\mathrm{Te})/G_{GT}(^{130}\mathrm{Te})\sim 1.78\cdot 10^{-4}$. This is because the ratio of $(M_{GT}^{(2\nu)}/\mu_0)$ for ^{128}Te and ^{130}Te is expected to be unity, as Pontecorvo pointed out,[16] and theoretical estimates of this ratio are 1.05,[8] 1.1[6] and (1.3$\sim$1.6).[15] Two experimental data have been reported;

$$(R_T)^{-1} = 5 \cdot 10^{-4} \qquad \text{by the Missouri group,[12]}$$
$$(R_T)^{-1} = (1.03 \pm 1.13) \cdot 10^{-4} \qquad \text{by the Heidelberg group.[17]} \qquad (5.7)$$

The Missouri data show the existence of the $(\beta\beta)_{0\nu}$ mode, while the Heidelberg data may not, because the largest theoretical prediction for $T_{2\nu}(^{130}\text{Te})/T_{2\nu}(^{128}\text{Te})$ is $4.6 \cdot 10^{-4}$ until now. By using the various ratios of nuclear matrix elements estimated by Haxton and Stephenson,[8] we find

$$\langle m_\nu \rangle \lesssim 8.6 \text{ eV (12 eV)} \qquad \text{from Missouri's data,}$$
$$\langle m_\nu \rangle < 5.7 \text{ eV (7.9 eV)} \qquad \text{from Heidelberg's data.} \qquad (5.8)$$

For the Heidelberg data, theoretical values are assumed to be within two standard deviation from the experimental central values.

The upper bounds for the effective right-handed parameters, $\langle \lambda \rangle$ and $\langle \eta \rangle$, are also obtained from the $(\beta\beta)_{0\nu}$ mode, as shown in Table 3. The limit on $\langle \eta \rangle$ obtained by Doi et al.[2] is by one order smaller than $\langle \lambda \rangle$. Its origin comes from both one electron P-wave with $j=3/2$ and the nucleon recoil term, while these effects have no contribution for the $\langle \lambda \rangle$ case. The tighter bound on $\langle \eta \rangle$ is obtained by Tomoda et al..[7] This sizable difference comes from the nucleon recoil term, which contains the neutrino potential h_R with a singular part proportional to $\delta(\vec{r}_{nm})$. Doi et al.[18] assumed that the relative distance between two decaying neutrons can not vanish, because of the hard core. Therefore, the singular term $\delta(\vec{r}_{nm})$ did not contribute. However, Tomoda et al. assumed the extended nucleon picture and thus the singular term gave the large contributions for the $\langle \eta \rangle$ parts.[19] The geochemical data in Eq.(5.7) also give the similar results.[19]

Before closing this section, it is worthwhile to mention other mechanisms in addition to the 2n-mechanism. Primakoff and Rosen proposed the N^*-mechanism shown in Fig. 1. Recently, Fazely and Liu pointed out other mechanism shown in Fig. 2, which is referred to as the N^*-T mechanism.[20] The latter have a contribution to the $(\beta\beta)_{0\nu}$ $(0^+ \rightarrow 0^+)$ trantion, as shown in Table 6. Watanabe and Toki[21] have concluded that this contribution is not large in compariosn with those from the 2n-mechanism. It is desirable to investigate these mechanisms from the view point of the relativistic quark model.

Mechanisms	m_ν – part			V + A part		
	2n	N*	N*–T	2n	N*	N*–T
$0^+ \to 0^+$	O	X	O	O	X	O
$0^+ \to 2^+$	X	X	X	O	O	O

Table 6

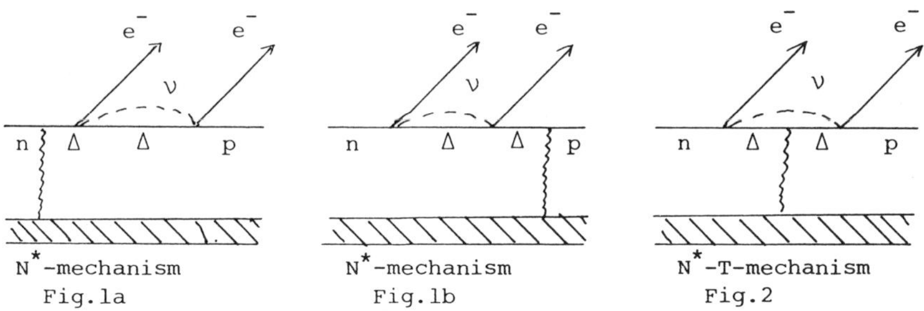

§ 6. Consistency among various experimental bounds

Data on the electron anti-neutrino mass from the tritium beta decay have been reported by four groups at this conference. Three groups, Zurich,[22] Los Alamos[23] and INS,[24] obtained the upper limits 18 eV, 29 eV and 31 eV, respectively. On the other hand, the ITEP group reported the finite value, 17 eV $< m_{\overline{\nu}_e} <$ 40 eV, again.[25]

If the ITEP result is taken seriously and ν_{eL} does not mix with other neutrinos, then this result contradicts with the bounds on $\langle m_\nu \rangle$ from the $(\beta\beta)_{0\nu}$ mode given in Eqs.(5.4) and (5.8). Several scenarios have been proposed to solve it: (a) The simplest one is to assume all neutrinos to be Dirac, i.e. $\langle m_\nu \rangle = 0$ in Eq.(4.8). (b) Two neutrinos form a pseudo-Dirac neutrino, namely, the destructive interference between almost degenerate neutrinos.[5] For two generation case, we have

$$< m_\nu > = \left| m_1 \cos^2\theta - m_2 \sin^2\theta \right| \quad \text{with } m_1 \simeq m_2 \,. \tag{6.1}$$

(c) The light (~ 30 eV) and heavy (~ 100 MeV) neutrinos interfere destructively.[26]   The propagation of virtual heavy neutrino gives the Yukawa type potential, but for such heavy neutrino, the extended nucleon model and its recoil has to be taken into account, so that the potential has a A-dependence, A being the mass number. Data on ^{82}Se and ^{128}Te require to suppress $\langle m_\nu \rangle$ for different A values. It means that the fine tuning of parameters should be taken into account.

Let us consider whether a realistic model can be found for the scenario (b), namely the SO(10) model with 3 generations. Of course, the neutino mixing in this model should be consistent with the data on neutrino oscillations. Nishiura, Kim and Kim[27] extended the model by Stech[28] by including the Majorana type mass matrix. This model is interesting because mass matrices for quarks and leptons are derived under the fairly general assumptions; namely

the up quark sector: $\qquad \mathcal{M}_u = S,$ (6.2)

the down quark sector: $\qquad \mathcal{M}_d = \alpha S + S' + A,$ (6.3)

the charged lepton sector: $\mathcal{M}_e = [\ \alpha S - 3S' + \delta A\ ] / r_1,$ (6.4)

the neutrino sector in Eq.(3.2):

$$\mathcal{M}_D = S / r_2,$$

$$\mathcal{M}_L = \beta S' / r_3,$$

$$\mathcal{M}_R = \gamma S' / r_4,$$ (6.5)

where 3×3 matrices S, S' and A are derived from matrices for Yukawa couplings multiplied by the non-zero vacuum expectation values $\langle \phi_{10} \rangle$, $\langle \phi_{126} \rangle$ and $\langle \phi_{120} \rangle$. Here ϕ_j stands for 10, 126 and 120 dimensional Higgs bosons, so that S and S' are symmetric matrices, while A is an antisymmetric one. The parameters α, β, γ and δ are ratios of various $\langle \phi_j \rangle$ and assumed to be real, for example,

$$\alpha = \langle \phi_{10}\,(\bar{5})\rangle \,/\, \langle \phi_{10}\,(5)\rangle,$$
$$\beta = \langle \phi_{126}\,(15)\rangle \,/\, \langle \phi_{126}\,(\overline{45})\rangle,$$
$$\gamma = \langle \phi_{126}\,(1)\rangle \,/\, \langle \phi_{126}\,(\overline{45})\rangle, \tag{6.6}$$

where numbers in parenthesis indicate the representation under SU(5). By assumption, $|\alpha|$ should be much smaller than 1, and S, S' and A are hermitian matrices. The parameters r_j come from the different renormalization effects for the lepton and quark masses when scale change is taken from the unification energy ($\sim 10^{15}$ GeV) down to the energy of the order of masses of quarks and charged leptons (~ 1 GeV). Numerically, $r_1 \simeq 3$ and all r_j are of order of unity.

The sea-saw mechanism which is considered as one of possibilities to explain the smallness of neutrino mass is expressed as

$$(\mathcal{M}_\nu)_{\text{light}} \simeq \mathcal{M}_L - \mathcal{M}_D\,(\mathcal{M}_R)^{-1}\,\mathcal{M}_D, \tag{6.7}$$

and

$$(\mathcal{M}_\nu)_{\text{heavy}} \simeq \mathcal{M}_R, \tag{6.8}$$

corresponding to mass eigenstate Majorana neutrinos N_{I} (light) and N_{II} (heavy) in Eq.(3.4), respectively. The energy scale of $\langle \phi_{126}\,(1)\rangle$ in γ of Eq.(6.6) is around the SU(5) smmetry breaking ($\sim 10^{15}$ GeV), while $\langle \phi_{126}\,(45)\rangle$ is of order of a few GeV, so that

$$\|\mathcal{M}_L\|,\ \|\mathcal{M}_D\| \ll \|\mathcal{M}_R\|, \tag{6.9}$$

where $\|A\|$ means the absolute value of matrix A, that is $\|A\| = (\Sigma_{jk}\,|A_{jk}|^2)^{1/2}$. If $\|\mathcal{M}_D\| \sim \|\mathcal{M}_u\|$ is assumed, the contribution from the second term in the light neutrino $(\mathcal{M}_\nu)_{\text{light}}$ is of order 10^{-3} eV. On the other hand, β in Eq.(6.6) is generally considered to be small, but essentially no restriction on it. Therefore, if $m_{\overline{\nu_e}}$ is around 30 eV as suggested by the ITEP group, then the left-handed Majorana type mass term $\mathcal{M}_L$ in Eq.(6.7) should be responsible to give the light neutrino mass. In summary, the light neutrinos (N_{I}) consist of the left-handed neutrino mainly, and their mixing to the right-handed neutrino is of order $\|\mathcal{M}_D\| / \|\mathcal{M}_R\| \sim 10^{-13}$. Therefore, the mixing between the left- and right-handed neutrinos can be neglected.

Now, masses of quarks (u, c, d, s, b) and charged leptons

(e, μ, τ) and the experimentally determined values of quark mixing (Kobayashi–Maskawa matrix) are taken into account as input data. Thus, the parameters α, δ and the top quark mass (m_t) are determined. Since β and r_3 are unknown, only the ratios of neutrino masses are obtained. Under the constraint $(\langle m_\nu \rangle / m_{\overline{\nu e}}) < 1/4$, Takasugi, Taruya and I[29] found one type of solution: The neutrino mixing matrix is

$$\begin{pmatrix} \nu_e \\ \nu_\mu \\ \nu_\tau \end{pmatrix}_L = \begin{pmatrix} 0.762 & i\,0.647 & 0.035\ e^{i247^\circ} \\ 0.040 & i\,0.039\ e^{i35^\circ} & 0.998\ e^{i21^\circ} \\ 0.647 & i\,0.761 & 0.044\ e^{i161^\circ} \end{pmatrix} \begin{pmatrix} N_1 \\ N_2 \\ N_3 \end{pmatrix}_L , \quad (6.10)$$

and the top quark mass is

$$m_t \simeq (\,110 \sim 140\,)\ \text{GeV}. \tag{6.11}$$

If $\Delta m^2 = |\,m_1^2 - m_2^2\,|$ is large $(> 10^{-3}\ \text{eV}^2)$, the neutrino oscillation between ν_{eL} and $\nu_{\tau L}$ is large from Eq.(6.10). In order to prevent it, the constraint such as $m_1 \simeq m_2$ is required. It means that N_1 and N_2 form one pseudo Dirac neutrino, because of the i factor for N_2 in the mixing matrix. The mass of N_3 is given by $(m_3/m_1) \sim (10 \sim 50)$. Since $\Delta m^2 > 20\ \text{eV}^2$, we found that the neutrino oscillations are

$$\begin{aligned} P(\nu_e \to \nu_\mu) &= 0.00237, \\ P(\nu_e \to \nu_\tau) &= 4.7 \cdot 10^{-6}, \\ P(\nu_\mu \to \nu_\tau) &= 0.0039. \end{aligned} \tag{6.12}$$

which are consistent with the present experimental limits.

In this model, the mass matrices for the light and heavy neutrinos have the same structure as given by $\mathcal{M}_L$ and $\mathcal{M}_R$ in Eq.(6.5). Therefore, contributions from two heavy neutrinos cancel each other in our model. Caldwell showed the order of $|U_{ej}^2|$ as a function of heavy neutrino mass $(> 1\ \text{GeV})$ at this conference.[9] He considers only one generation. This assumption seems not to be realistic, because mixings among generations exist in GUTs generally and all U_{ej}^2 are not positive. Also his use of the dipole form factor at each nucleon weak vertex and the non-relativistic treatment of nucleon seems to be questionable for such heavy neutrino ($> 10^{4\sim 8}$ GeV).[8]

§ 7 Summary of present status

If the $(\beta\beta)_{0\nu}$ mode is observed, the electron neutrino should be the massive Majorana neutrino. This mode has not yet been confirmed.

The Missouri data on the ratio of half-lives of ^{128}Te to ^{130}Te in Eq.(5.7) may indicate the existence of this mode. However, there are some uncertainties. First, their result is not consistent with the Heidelberg data.[17] Second, discrepancies between geochemical data and theoretical estimates have not yet been settled, as shown in Tables 4 and 5. Therefore, the ratio of the $(\beta\beta)_{2\nu}$ mode for ^{128}Te to ^{130}Te is not definite. The Irvine group will soon observe the $(\beta\beta)_{2\nu}$ mode of ^{82}Se by the visible method and give us some useful information.[13]

While, the very precise experimental data on ^{76}Ge has been reported by the Santa Barbara-Berkeley group.[9] In order to draw the definite information on $\langle m_\nu \rangle$ from them, it is required to know the absoulte values of the nuclear matrix elements, as shown in Table 3. I hope that the progress of the nuclear theory will solve these complicated circumstances in near future.

References

1) J. Schechter and J.W.F. Valle, Phys. Rev. D25 (1982), 2951.
   J.F. Nieves, Phys. Lett. 147B (1984), 375.
   E. Takasugi, Phys. Lett. 149B (1986), 372.
2) M. Doi, T. Kotani and E. Takasugi, Prog. Theor. Phys. Supplement 83 (1985), 1.
3) K.M. Case, Phys. Rev. 107 (1957), 307.
   R.N. Mohapatra and G. Senjanovic, Phys. Rev. D23 (1981), 165.
   T.P. Cheng and Ling-Fong Li, Phys. Rev. D22 (1980), 2860.
4) J. Schechter and J.W.F. Valle, Phys. Rev. D22 (1980), 2227.
5) M. Doi, M. Kenmoku, T. Kotani, H. Nishiura and E. Takasugi, Prog. Theor. Phys. 70 (1983), 1331.
6) K. Grotz and H.V. Klapdor, Phys. Lett. 153B (1985), 1; 142B (1984), 323.
   H.V. Klapdor, Proc. Int. Symp. on Nuclear Beta Decays and Neutrino, ed. by Kotani, Ejiri and Takasugi, Osaka, June, 1986 (World Scientific, Singapore, 1986).
7) T. Tomoda, A. Faessler, K.W. Schmid and F. Grummer, Nuclear Phys. A452 (1986), 591.

8) W.C. Haxton and G.J. Stephenson, Jr., Progress in Particle and Nuclear Physics $\underline{12}$ (1984), 409.

9) D.O. Caldwell et al., Phys. Rev. $\underline{D33}$ (1986), 2737.
    D.O. Caldwell et al., Proceedings of this conference.

10) M. Doi, T. Kotani, H. Nishiura and E. Takasugi, Prog. Theor. Phys. $\underline{70}$ (1983), 1353.

11) T. Kirsten, E. Hensser, D. Kaether, J. Oehm, E. Pernicka and H. Richter, Proc. Int. Symp. on Nuclear Beta Decays and Neutrino, ed. by Kotani, Ejiri and Takasugi, Osaka, June, 1986 (World Scientific, Singapore, 1986).

12) O.K. Manuel, Proc. Int. Symp. on Nuclear Beta Decays and Neutrino, ed. by Kotani, Ejiri and Takasugi, Osaka, June, 1986 (World Scientific, Singapore, 1986).
    J.F. Richardson, O.K. Manuel, B. Sinha and R.I. Thorpe, Nucl. Phys. $\underline{A453}$ (1986), 26.
    W.J. Lin, O.K. Manuel, L.L. Oliver and R.I. Thorpe, Nucl. Phys. (1986) in Press.

13) M.K. Moe, Proceedings of this conference.

14) A.A. Klimenko, A.A. Pomansky and A.A. Smolnikov, Proc. "Neutrino 84", ed. by Kleinknecht and Paschos, June, 1984 (World Scientific, Singapore, 1984), P.161.

15) P. Vogel and P. Fisher, Phys. Rev. $\underline{C32}$ (1985), 1362.
    P. Vogel, Proc. Int. Symp. on Nuclear Beta Decays and Neutrino, ed. by Kotani, Ejiri and Takasugi, Osaka, June, 1986 (World Scientific, Singapore, 1986).

16) B. Pontecorvo, Phys. Lett. $\underline{26B}$ (1968), 630.

17) T. Kirsten, H. Richter and E. Fessberger, Phys. Rev. Lett. $\underline{50}$ (1983), 474; Z. Phys. $\underline{C16}$ (1983), 189.

18) M. Doi, T. Kotani and E. Takasugi, Phys. Lett. $\underline{158B}$ (1985), 164.

19) E. Takasugi, Proc. of '86 Massive Neutrino in Astrophysics and in Particle Physics, ed. by Tran Thanh Van, Tigne, France, January, 1986 (Editions Frontieres, France, 1986), p.589.

20) A. Fazely and L.C. Liu, Phys. Rev. Lett. $\underline{57}$ (1986), 968.

21) M. Watanabe and H. Toki, Proc. Int. Symp. on Nuclear Beta Decays and Neutrino, ed. by Kotani, Ejiri and Takasugi, Osaka, June, 1986 (World Scientific, Singapore, 1986).

22) E. Holzschuh, Proceedings of this conference.

23) T. Ohshima, Proceedings of this conference.

24) R.G.H. Robertson, Proceedings of this conference.

25) V. Lubimov, Proceedings of this conference.

26) A. Halprin, S.T. Petcov and S.P. Rosen, Phys. Lett. $\underline{125B}$ (1983), 335.
    J.D. Vergados, Phys. Rev. $\underline{D28}$ (1983), 2887.
    S.P. Rosen, Proceedings of the Fourth Moriond Workshop, ed. by Tran Thanh Van (Editions Frontieres, France, 1984), p.425.
    P. Langacker, B. Sathiapalan and G. Steigman, Pensylvania preprint UPR-0285T.

27) H. Nishiura, C.W. Kim and J. Kim, Phys. Rev. $\underline{D31}$ (1985), 2288.

28) B. Stech, Phys. Lett. $\underline{130B}$ (1983), 189.

29) T. Kotani, E. Takasugi and K. Taruya, in preparation.

NEUTRINO OSCILLATIONS AT REACTORS

F. Boehm

California Institute of Technology
Pasadena, CA   91125

ABSTRACT

The results from neutrino oscillation experiments
at nuclear reactors are reviewed with focus on
the experiment at the Gösgen reactor.

INTRODUCTION

The objective for carrying out neutrino oscillation
experiments is the search for neutrino mass and neutrino
mixing. The results of oscillation experiments are usually
given in terms of a two-parameter model for neutrino
oscillations[1] and are characterized by the mass parameter
Δm^2 and the mixing strength $\sin^2\theta$. Reactor based
experiments explore the $\bar{\nu}_e$ disappearance channel $\bar{\nu}_e \rightarrow$ X.
The probability of finding a $\bar{\nu}_e$ state is given by

$$P_{\bar{\nu}_e} = 1 - \sin^2 2\theta \, \sin^2[1.27 \frac{\Delta m^2(eV^2)L(m)}{E_\nu(MeV)}] \qquad (1)$$

The sensitivity for oscillations depends on the argument of
the second $\sin^2$ term in Eq.(1). For a given Δm^2 the
sensitivity depends on the value of L/E_ν. Figure 1 serves
to illustrate the regions of L/E_ν for the reactor

experiments reviewed here and also shows the corresponding regions of L/E_ν associated with other oscillation experiments.

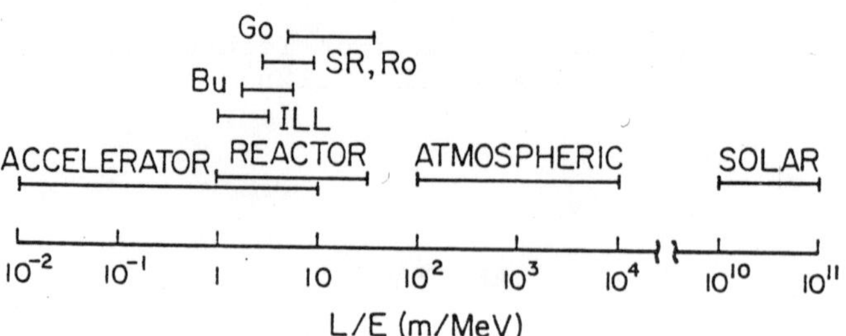

Fig. 1   Illustration of the region of L/E_ν accessible for various experiments.

Power reactors are prolific sources of $\bar{\nu}_e$ with energies of $0 < E_\nu$ (MeV) < 10. The neutrino source strength of a 2800 MW(th) reactor is about 5×10^{20} s^{-1}. The $\bar{\nu}_e$ are detected by the reaction $\bar{\nu}_e + p \rightarrow e^+ + n$. Correlated position and neutron events constitute a valid neutrino signal.

Table I provides an overview of current reactor experiments. The reactor power, the neutrino detecting scheme, the distance between reactor core and detector, and the number of events accummulated in each experiment are listed.

THE GOSGEN EXPERIMENTS

Three experiments have been carried out between 1981 and 1985 at the Gösgen power reactor in Switzerland at distances of L = 37.8m, 45.9m, and 64.7m. The work was a

TABLE I: REACTOR EXPERIMENTS

Reactor(MW)	Detector	L	#.Counts
ILL(57) *Caltech* *ISN Grenoble* *TU Munich*	377 l Liq.Sc. $+ {}^3$He $\varepsilon = 0.20,\ 0.17$	8.8 m	0.5×10^4
Gösgen(2800) *Caltech* *SIN* *TU Munich*	$\Delta t = 250\ \mu s$ $0.8 < E_{e^+} < 5.6$ MeV	37.8 m 45.9 m 64.7 m	1.1×10^4 1.1×10^4 0.9×10^4
Bugey(2800) *ISN Grenoble* *Annecy*	321 l Liq.Sc. $+ {}^3$He $\varepsilon = 0.26$ $\Delta t = 200\ \mu s$ $1.5 < E_{e^+} < 6.5$ MeV	13.6 m 18.3 m	4.0×10^4 2.3×10^4
Savannah River(2300) *Irvine*	300 l Liq.Sc. $+$ Gd $\varepsilon \approx 0.5$ $\Delta t = 15\ \mu s$ $1.0 < E_{e^+} < 9$ MeV	18.2 m 23.7 m	3.8×10^4 1.9×10^4
Rovno(1400) *Moscow*	240 l Liq.Sc. $+$ Gd 136 kg Polyeth. $+ {}^3$He $\varepsilon = 0.29,\ 0.52$	18.5 m 25.0 m	

References: ILL Ref. 2, Gösgen Ref. 3-6, Bugey Ref. 7,
Savannah River Ref. 8, Rovno Ref. 9.

collaboration between Caltech, SIN, and TU Munich. It has
been communicated in Refs. 3-5 and is reviewed in Ref. 6.
The detector consisted of an array of liquid scintillation
counters to detect the positrons, sandwiched between ^{3}He
multiwire proportional chambers for detection of the
neutrons. The detector system and its veto counter and
shielding are depicted in Fig. 2.

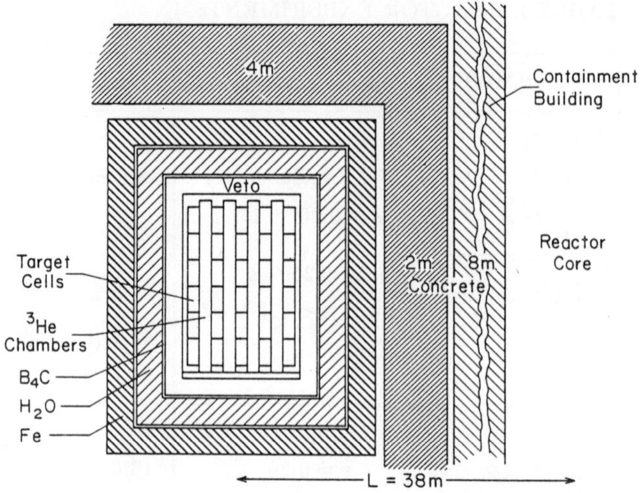

Fig. 2.   Neutrino detector at Gösgen.   The central neutrino detector unit consists of 30 liquid scintillator cells, arranged in five planes, for positron detection, and four ^{3}He filled wire chambers for neutron detection.

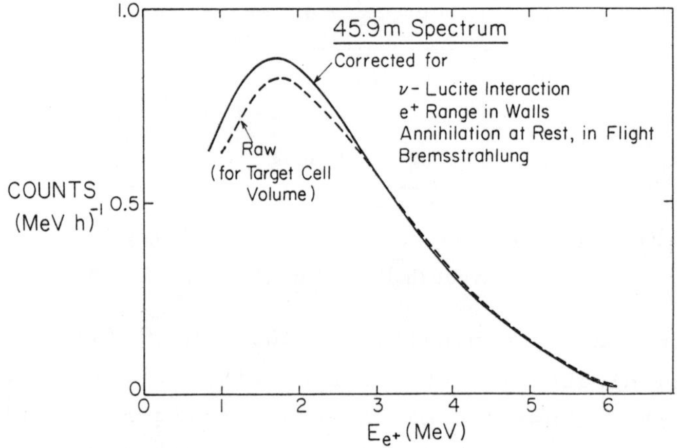

Fig. 3   Illustration of corrections   to raw   positron spectrum.

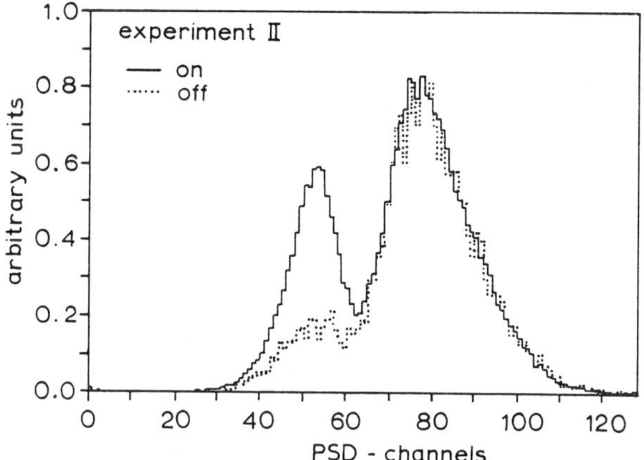

Fig. 4 Example of a PSD-spectrum obtained in experiment II for reactor-on data (solid line) and reactor-off data (dashed line). The channel number is proportional to the decay time of the light pulse associated with the recoiling particle. For reactor-on data the peak on the left is enhanced due to neutrino induced positrons, while the neutron peak on the right remains unchanged.

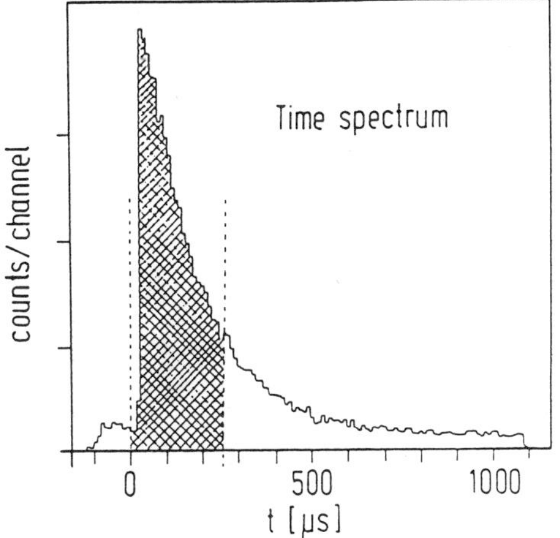

Fig. 5 Distribution of time intervals between neutrino induced events in a scintillator cell and a wire chamber. The shaded area corresponds to the employed time window of 250 µs.

In order to obtain the correct positron spectra a detailed knowledge of the scintillation detector response was necessary. Using sources as well as Monte Carlo simulations the response of the scintillator as a function of positron energy was derived taking into account the neutrino interactions in the scintillation liquid and in the Lucite walls of the detector cells. Positron finite ranges, annihilation at rest and in flight and bremstrahlung were also taken into account. Figure 3 illustrates the effect of the mentioned corrections in the spectrum.

Pulse shape discrimination served to reduce the background rate associated with neutrons from cosmic rays as shown in Fig. 4. There was no reactor associated neutron or gamma ray background, as could be verified from the data for reactor on and reactor off. A neutron pulse in the ^{3}He counter triggers the time-to-amplitude conversion reading out preceding events in the scintillation counter. A time spectrum is shown in Fig.5. In order to compare the spectra obtained at the 3 positions it was necessary to know the relative reactor spectrum for each position. Although each experiment was conducted over nearly a full fuel cycle of the reactor, there were small differences in fuel composition that needed to be taken into account.

Figure 6 shows the fuel composition of the Gosgen core as a function of time, toghether with the running cycles for each experiment. From these differences in fuel composition the relative spectra shown in Fig.7 were obtained. The differences in the spectra amount to a few % and their uncertainties thus are negligible.

We now come to the data analysis. The experimental positron spectrum is compared to a calculated spectrum which is given by an expression (see Ref.5,6) that

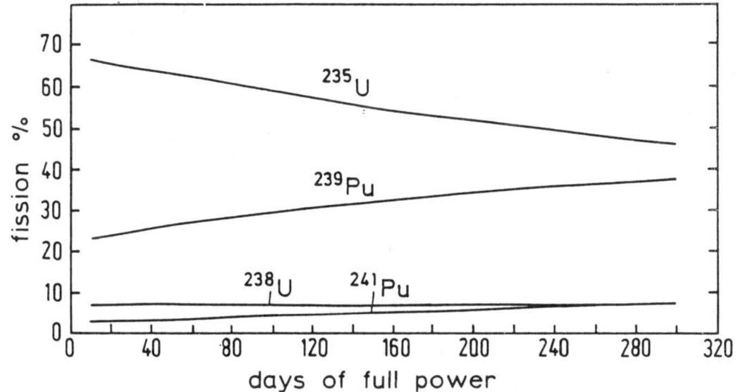

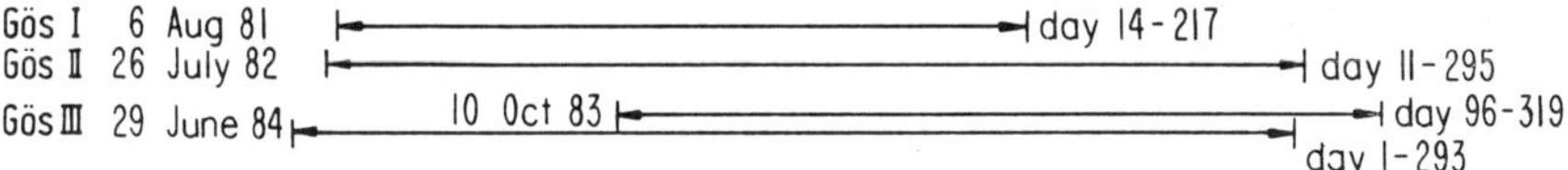

Fig. 6   Relative contributions to the number of fissions from the four relevant isotopes as a function of days of reactor at full power. The data taking periods of experiments I to III are indicated.

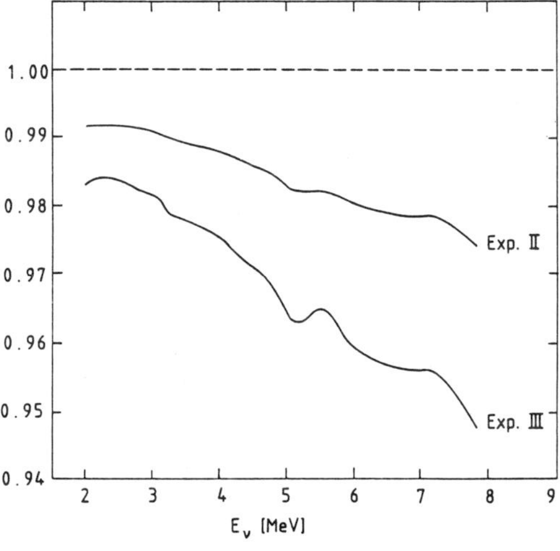

Fig. 7 Relative changes of the reactor antineutrino yields of experiments II and III, as compared to the yield of experiment I, as a function of the neutrino energy. These differences are caused by slight changes in the reactor fuel compositions for the individual measuring periods.

contains the product of the detector efficiency ε, a coefficient η that accounts for the small difference in fuel composition, the cross section, the neutrino spectrum $S(E_\nu)$ and the oscillation function $P(E_\nu L \Delta m^2 \theta)$, all this integrated over the energy resolution function of the detector and the finite solid angle.

Two analyses were performed and are briefly described below.

<u>Analysis A</u>: This analysis is independent of the source neutrino spectrum. The neutrino spectrum $S(E_\nu)$ is parametrized as

$$S_A(E_\nu) = e^{(A_0 + A_1 E + A_2 E^2)} \tag{2}$$

and a χ^2 is calculated for the difference between the experimental yield and the expected yield obtained in the manner described above, summed over all the data bins and positions. The χ^2 was minimized for a fixed set of parameters Δm^2 and $\sin^2 2\theta$, by varying the coefficients A in Eq. (2) and three normalization coefficients, one for each position. For no-oscillations it was found that $\chi^2(0,0)$ = 41.1/45. A maximum likelyhood test was used to obtain the exclusion plot shown in Fig.8. There is no evidence for oscillations and the parameter region excluded at 90% c.l. is to the right of the curves. Figure 9 shows the experimental positron spectra and compares them with the spectra derived from Eq. (2).

<u>Analysis B</u> is based on neutrino spectra $S_B(E_\nu)$ obtained from measured on-line electron spectra of the fission targets ^{235}U and ^{239}Pu and from calculated spectra for fission of ^{238}U and ^{241}Pu. These spectra are also shown in Fig. 9 and agree quite well with the experimental spectra. All three spectra are displayed as a function of L/E_ν in Fig. 10.

Exclusion plots obtained by similar procedures are shown in
Fig. 11.

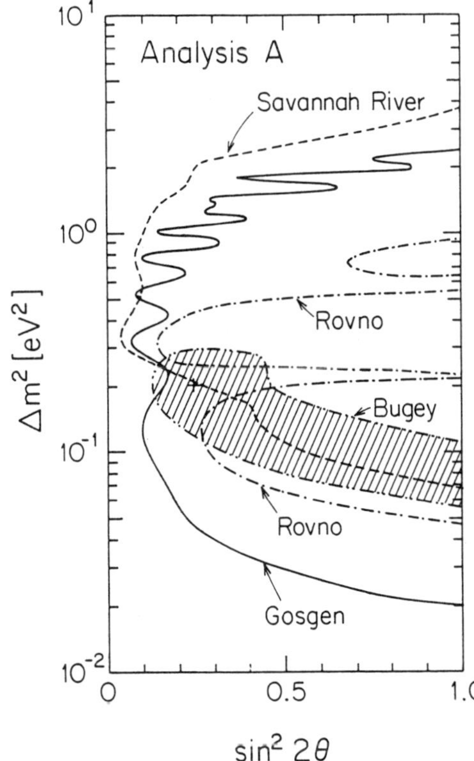

Fig. 8 Exclusion plots for oscillation parameters Δm^2 from 3-position experiments at Gosgen, 2-position experiments at Bugey, Rovno, and Savannah River. In the Bugey experiment the shaded area is allowed, in all other experiments the area to the right of the curve is forbidden. The Rovno results are based on the integral yield only.

The recent experiments at Savannah River have now
also provided data at two positions. The detector, a Gd
loaded liquid scintillation counter detected both the
positrons and the neutrons. While the results will be
presented by H. Sobel in the following lecture, we indicate
for reason of comparison in Fig. 8 the excluded region as
obtained in these experiments.

Another recent experiment to be reported in this
session by A. Pomansky is the work by Afonin et al.[9] at the
Rovno power reactor mesuring the integral neutrino yield.

For comparison we have drawn the exclusion plots obtained for a single position (18.5 m) and for two positions (18.5 m and 25 m) in Figs. 8 and 11, respectively.

The exclusion plots in Fig. 8 also include the work at the Bugey reactor published two years ago by Cavaignac et al.[7]. The Bugey results are indicative for oscillations with allowed parameters within the shaded area, with a most likely value of (Δm^2 = 0.2 eV^2, $\sin^2 2\theta$ = 0.25). These results do not agree at a high confidence level with the Gosgen data and the disagreement is presently not understood.

Conclusion

Results for the integral ratios of the experimental spectra at various positions are shown in Table II. These ratios are consistent with 1.0 (no oscillations) except for the Bugey data.

TABLE II: RESULTS FROM REACTOR EXPERIMENTS

Ratios of Integral Yields

	Expt./Calc.	Expt(2)/Expt(1)	Evidence for Osc.
ILL	0.955±0.110		No
Gö 1	1.018±0.065	(2)/(1) = 1.027±0.034	No
2	1.047±0.065	(3)/(1) = 0.958±0.053	No
3	0.975±0.076		
Bu 1	≈ 1.0	0.907±0.025	Yes
2	≈ 0.9		
SR 1		0.963±(0.013st.)	No
2			
Rov 1	0.997±0.06	0.986±0.037±0.029	No
2			

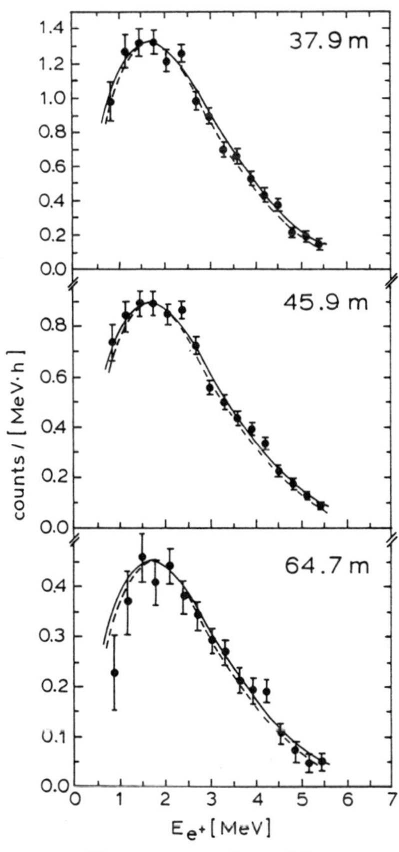

Fig. 9 Measured and predicted positron yields for experiments I to III. The solid lines represent the predicted positron yields derived by using the data from experiments I to III (Analysis A). The dashed lines represent the predicted positron yields derived by using the spectrum $S_B(E_\nu)$ based on independent β-spectroscopic data.

To summarize the present state of oscillation experiments we show in Fig. 12 several curves indicating the result of high energy experiments and reactor experiments, as well as the expected area in the parameter space from the solar neutrino flux from ^{8}B interpreted in terms of matter oscillations[10].

As to the future of reactor experiments it appears that essential progress to improve the sensitivity to Δm^2 can only be made at high cost. To improve the sensitivity by a factor of 10 it will be necessary to build a 100 times larger and also more efficient detector to be installed at 10 times the present largest distance, i.e. at 650 m. Such a 40,000 liter low background detector is a formidable undertaking. Its sensitivity, however, would still be insufficient to observe possible effects from matter oscillations, as indicated in Fig. 12.

This work is supported in part by the U.S. Department of Energy.

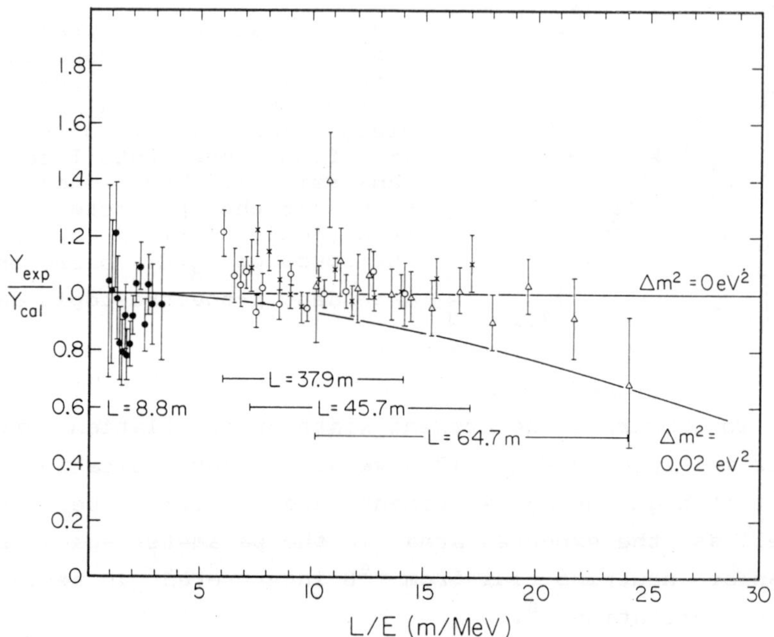

Fig. 10   Display of positron spectra from the ILL experiment and the three Gosgen experiments vs L/E$_\gamma$. The positron yield divided by the expected yield for no oscillation is plotted.

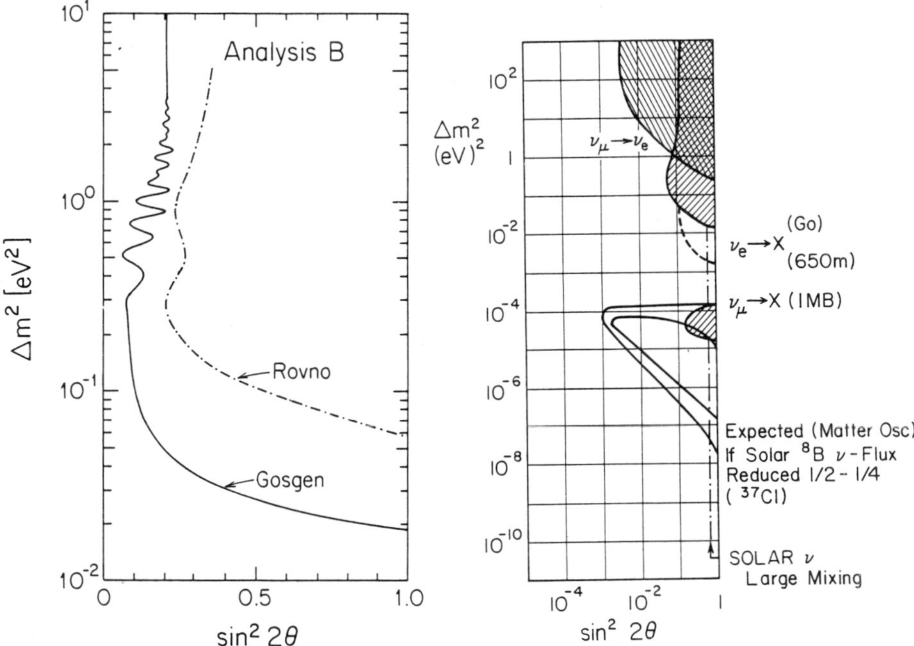

Fig. 11 Exclusion plots based on predicted neutrino spectrum (Gosgen) and predicted integral yield (Rovno).

Fig. 12 Limits for neutrino oscillations from various experiments. The dotted curve below the Gosgen line is for a hypothetical experiment at 650 m.

References

1. F. Boehm and P. Vogel, Ann. Rev. Nucl. Part. Sci. <u>34</u>, 125 (1984).
2. H. Kwon et al., Phys. Rev. <u>D24</u>, 1097 (1981)
3. J.-L. Vuilleumier et al., Phys. Lett. <u>114B</u>, 298 (1982)
4. K. Gabathuler et al., Phys. Lett. <u>138B</u>, 449 (1984)
5. V. Zacek et al., Phys. Lett. <u>164B</u>, 193 (1985)
6. G. Zacek et al., Phys. Rev. D., to be published
7. J.-F. Cavaignac et al., Phys. Lett. <u>148B</u>, 387 (1984)
8. H. Sobel, this Conference
9. A. Afonin et al., JETP Lett. <u>42</u>, 285 (1985) and
   A. Pomansky, this Conference
10. S. Mikheyev, this Conference.

THE UC IRVINE MOBILE NEUTRINO-OSCILLATION EXPERIMENT

N. Baumann[2], Z.D. Greenwood[1], H. Gurr[3], W.R. Kropp[1]
M. Mandelkern[1], L. Price[1], F. Reines[1], and H.W. Sobel[1]

(1) University of California, Irvine, CA
(2) The Savannah River Plant, Aiken, S.C.
(3) The University of South Carolina, Aiken, S.C.

ABSTRACT

Preliminary results are reported from the Irvine Neutrino Oscillation
Experiment. This is an electron-antineutrino disappearance
experiment at a reactor, and is sensitive to mass squared differences
in the range $\sim .02$ to $\sim 3.$ eV^2.

1. INTRODUCTION

This report gives preliminary results from the University of
California, Irvine (UCI) Neutrino Oscillation Experiment. We have
constructed and installed a detector in our Savannah River Plant
(SRP) Neutrino Laboratory. The detector is designed to be movable,
and can, with relative ease, be positioned at any distance from the
reactor core between about 18 and 50 meters. The neutrino flux at
the closest position is approximately $10^{13}/\text{cm}^2/\text{sec}$.

A single detector, measuring the neutrino spectra from one
reactor, at various distances from the core, avoids many of the
problems associated with reactor and detector systematics which are
possible in single-position experiments.

2. THE REACTOR AS A NEUTRINO SOURCE

Power reactors, used in most other (reactor) neutrino
oscillation experiments, produce neutrinos from the fission of
significant fractions of each of the isotopes ^{235}U, ^{238}U, and
^{239}Pu. The neutrino spectrum of each component is different,

yielding a source spectrum which is fuel-mix dependent, and therefore
time dependent. At the SRP production reactor, ^{239}Pu produces less
than 8% of the fissions, and ^{238}U less than 4%. As a result, for
neutrino energies between 2 and 8 MeV, the difference from a pure
^{235}U neutrino spectrum is less than 1.5%.

3. A DISAPPEARANCE EXPERIMENT

The detector is designed to observe neutrinos via the inverse-
beta reaction:

$$\bar{\nu}_e + p \rightarrow e^+ + n$$

Since the neutrino energy is less than 10 MeV, the neutron receives
negligible kinetic energy, and the positron kinetic energy is:

$$E_{e+} = E_{\bar{\nu}_e} - (M_n - M_p + m_e) = E_\nu - 1.8 \text{ MeV}$$

A measurement of the positron spectrum together with the known
interaction cross section yields the neutrino spectrum.

If the $\bar{\nu}_e$ were to change into another type, the new neutrino
would be below its lepton production threshold, and thus would be
unobservable. A reactor experiment is, therefore, a disappearance
experiment: a deficiency in the $\bar{\nu}_e$ signal is expected if oscillations
occur.

4. THE DETECTOR

A schematic of the detector is shown in Figure 1. The "target"
is 300 liters of NE313, a xylene-based scintillator loaded with 0.5%
gadolinium. This scintillator provides pulse-shape information,
distinguishing between lightly and heavily ionizing particles
(positrons and protons). The gadolinium provides fast, high-
efficiency neutron detection. The target is surrounded by 1100
liters of a mineral oil based scintillator, divided into two volumes;
we call these the "blanket" detectors. These in turn are surrounded
by two inches of low background lead, a three-inch thick plastic
scintillator anticoincidence, and finally, eight inches of additional
lead.

An electron antineutrino from the reactor can interact with a proton in the target volume. The resultant positron stops and promptly annihilates. Each 0.5 MeV annihilation gamma ray has a high probability of depositing its energy in the liquid scintillator. The neutron, with kinetic energy of tens of KeV, quickly thermalizes and is captured with a characteristic time of 10 µs, by a Gd nucleus. The resultant ~ 8 MeV of gamma rays are detected with high efficiency. The signature of an inverse beta event is then Fig. 1.:

a. A pulse in the target with an electron-like pulse shape and energy between 1.8 and 8 MeV, and,

b. A second pulse within ~ 15 µs for which the sum of the target and blanket energies lies between 4 and 10 MeV.

With the efficiencies and thresholds discussed below, we have an event rate of ~ 400 per day at 18.2 meters. At 50 meters the expected rate would be ~ 54 per day.

5. DETECTOR CHARACTERISTICS

a. <u>Energy Resolution</u>: We have measured the energy resolution of the Target and Blanket detectors using various sources. The results can be expressed as follows:

$$\text{and} \qquad \begin{aligned} \sigma_{target} &= 0.12 \ \sqrt{E(MeV)} \\ \sigma_{blanket} &= 0.20 \ \sqrt{E(MeV)} \end{aligned}$$

b. <u>Detector Efficiencies</u>: Monte Carlo simulations of the detector properties have been made and give the positron and neutron detection efficiencies. Integrated over the expected energy spectrum, the positron detection efficiency, with a threshold energy of 1.8 MeV is found to be ~ 0.7. The delayed neutron efficiency with a gamma threshold of 4 MeV is found to also be ~ 0.7. Preliminary measurements of these quantitites verify the results of the Monte Carlo.

c. <u>Neutron Capture Time</u>: The neutron capture time also enters into the overall detection efficiency. This is measured with various neutron sources and is found to be ~ 10 µs, in agreement with the Monte Carlo, and with direct calculations.

6. ANALYSIS METHODS

We treat our data in two fashions:

a. The experimental results are compared to combined theoretical[1] and experimental[2] determinations of the reactor neutrino spectrum. (This technique suffers from the possibility that any observed deficiencies could be due to inaccuracies in the predicted spectra, or to systematic uncertainties in detector parameters.) This analysis will be described in a future report.

b. The spectra measured at different distances from the reactor core are compared, giving conclusions relatively free of detector systematics.

7. PRELIMINARY RESULTS

We have taken data at two distances from the reactor core, 18.2 and 23.8 meters:

a. <u>18.2 Meter Data</u>: These data are shown in Figure 2. The data represent an equivalent of about 38,000 neutrino interactions in the target. The abscissa is the observed energy deposited as the prompt pulse, and represents the sum of the positron kinetic energy and the fraction of the resultant annihilation gamma-ray energy contained in the target region. A Monte Carlo calculation gives 0.75 MeV for this additional energy. This energy boost is helpful in that it allows us to be sensitive to a lower positron energy for a given discriminator level.

b. <u>23.8 Meter Data</u>: Figure 2 also shows the corresponding spectrum for the ~ 39,000 events collected thus far at the 23.8 meter position.

The integrated rates, greater than 1.8 MeV (observed energy), corrected for power variations and distance give:

$$\frac{\text{position 1}}{\text{position 2}} = 1.040 \pm 0.012 \text{ (statistics)} \pm 0.020 \text{ (systematic)}.$$

A ratio of unity is expected in the absence of oscillations.

Figure 3 shows the ratio of the data for the two positions, as a function of observed energy. The data has been normalized for power and distance. Superimposed on this data is the predicted ratio for this experiment using the values of $\sin^2 2\theta$ and Δm^2, favored by the results of the Bugey experiment[3].

In the absence of oscillations, this ratio is expected to equal 1.0. The comparison of our data to this null hypothesis yields a $\chi^2 = 32$ for 24 degrees of freedom. A comparison of our data with the Bugey-favored parameters yields $\chi^2 = 20$. We conclude that the current data does not significantly favor either hypothesis.

For each pair of values $(\Delta m^2, \sin^2 2\theta)$ in the range $.01 \leq \Delta m^2 \leq 10.0$ eV^2 and $.01 \leq \sin^2 2\theta \leq 1.0$ we calculate a value of χ^2 and the probability of exceeding this χ^2 value. This probability is expressed in terms of a number of standard deviations (σ) and shown as a three dimentional plot in Figure 4. These results can also be presented as an exclusion region in the parameter space of Δm^2 and $\sin^2 2\theta$. The results of this analysis are shown in Figure 5. The region to the right of the UCI curve is excluded at the 95% confidence level (2 sigma).

Also shown in Figure 5 are the regions of parameter space allowed by the Bugey experiment, and the region excluded by the Goesgen experiment[4]. Note that this latter group expresses their result as a 90% confidence limit, while the Bugey allowed region has a 2 sigma (95%) significance. Using this technique, neither UCI nor Goesgen has as yet either fully ruled out or confirmed the Bugey experimental results.

8.   THE FUTURE

We finished taking data at 23.8 meters in April 1986, and have returned to the 18.2 meter position to complete this two position comparison. The next sequence will involve the original 18.2 meter position and a third location currently expected to be at about 52 meters.

This work is supported in part by the U.S. Department of Energy. We thank the E. I. Dupont de Nemours Company, operators of the Savannah River Plant, for their continued support and hospitality.

REFERENCES

1. Davis, B.R., et al., Phys. Rev. C, **19**, 2259 (1979)
    Vogel, P., et al., Phys. Rev. C, **24**, 1543 (1981)
2. Schreckenbach, K., et al., Phys. Lett., **99B**, 251 (1981); von
    Feilitzsch, F., et al, Phys. Lett., **118B**, 162, (1982).
3. Cavaignac, J.F., et al, Phys. Lett., **148B**, 387 (1984).
4. Kwon, H., et al., Phys. Rev. D, **24**, 1097 (1981); Zacek, V. et
    al., Phys. Lett., **164B**, 193 (1985).

FIGURES

Figure 1. Schematic of the detector. The operation of the detector
is indicated by the superimposed sketch of a neutrino interacting in
the target.

Figure 2. The observed-energy spectrum at the 18.2 and 23.8 meter
positiono.

Figure 3. Plot of the ratio of the rates at the two positions,
versus energy. The rates are normalized for reactor power and
distance. Also shown are the expected ratios for the null hypothesis
(ratio equal unity) and the Bugey hypothesis.

Figure 4. Preliminary exclusion plot for the UCI experiment. Also
shown are results from the Goesgen and Bugey experiments. Regions to
the right of the Goesgen and the UCI curves are excluded at the
confidence levels indicated. The Bugey curve shows a region of
parameters allowed by that experiment at 2 sigma.

Neutrino Oscillation Detector
$\bar{\nu}_e$
e+
NE 313 300 ℓ
1100 ℓ MINERAL OIL SCINTILLATOR
ANTICOINCIDENCE 3" PLASTIC SCINT.
SHIELDING 2"Pb + 8"Pb

Fig. 1

100.0
10.0
1.0
0.1
EVENTS/DAY/0.2 MeV
18.2 m
23.8 m
0 1 2 3 4 5 6 7 8 9 10
OBSERVED ENERGY (MeV)

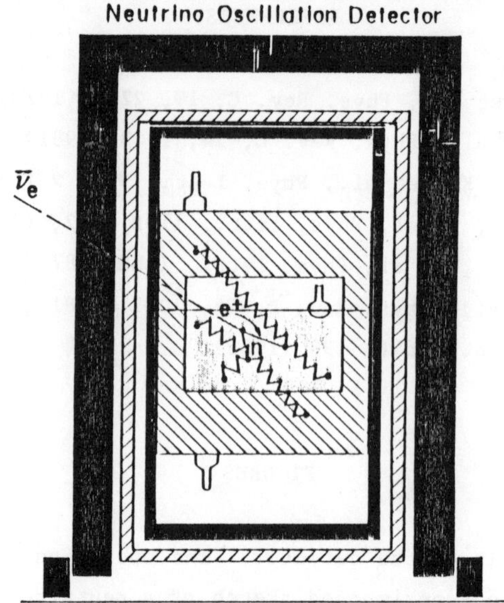

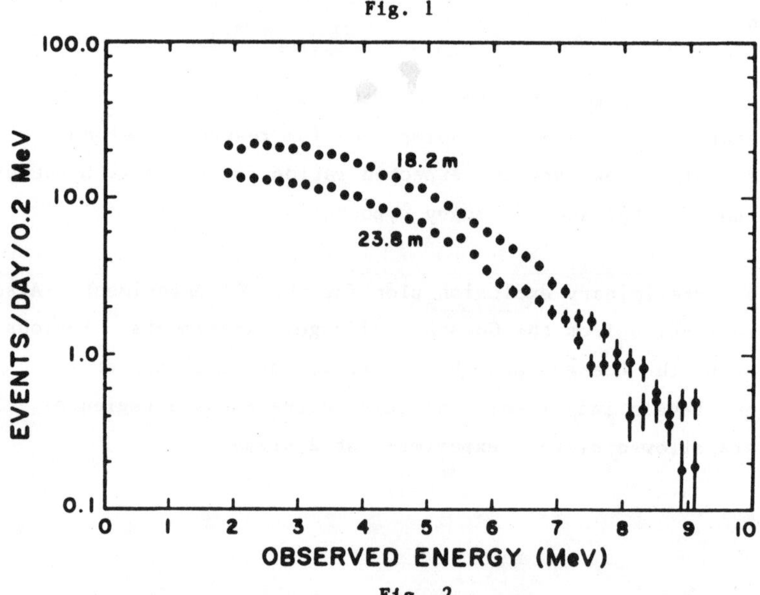

Fig. 2

1.50
1.00
0.50
0.00
POSITION 1 / POSITION 2
1.8 MeV
SOFTWARE THRESHOLD
sin² 2θ = 0.24
ΔM² = 0.2
0 1 2 3 4 5 6 7 8 9 10
OBSERVED ENERGY (MeV)
σ
10.0
1.0
0.2
ΔM²
0.01
0.01
0.01 0.1 0.24 1.0
Sin² 2θ

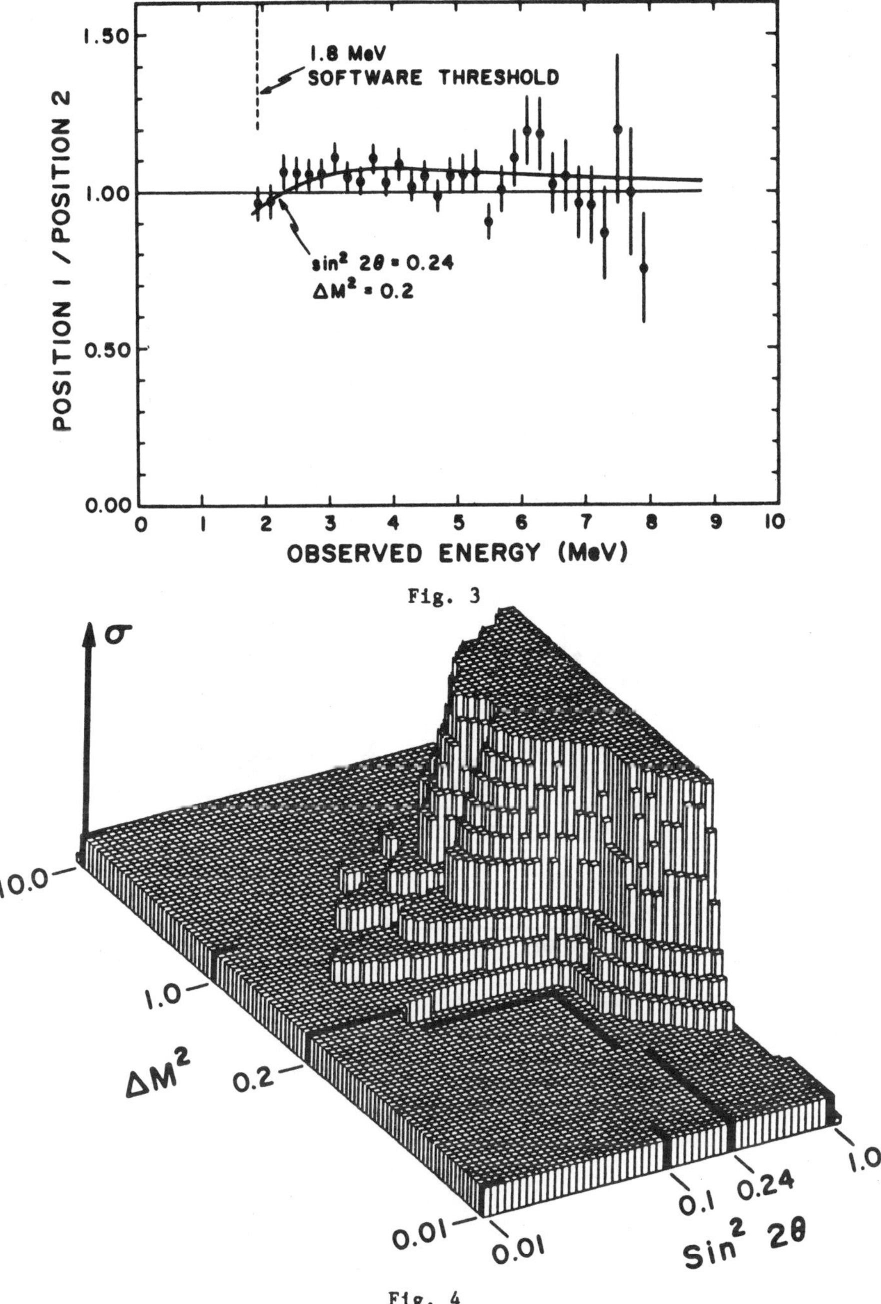

Fig. 3

Fig. 4

10.0
(Preliminary)$^{6/86}$
(SRP)
(18.2, 23.8)
2σ
1.0
Bugey
(13.6, 18.3)
2σ
0.2
0.1
GOESGEN
(37.9, 45.9, 64.7)
90% C.L.
0.01
0.01
0.1
1.0
ΔM² (eV)²
Sin² 2θ

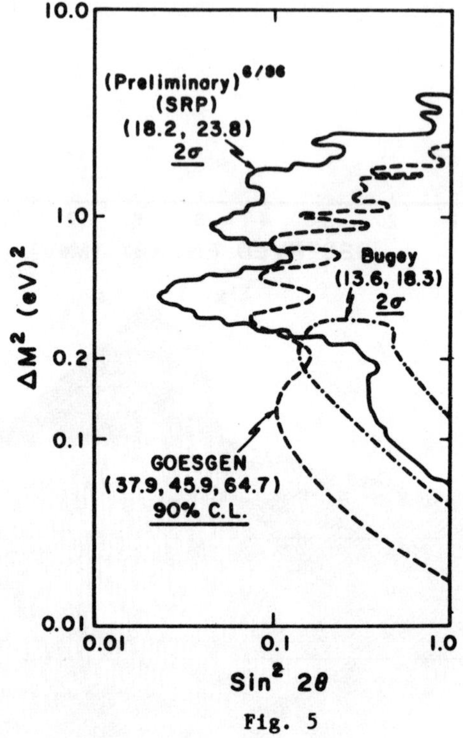

Fig. 5

NEUTRINO OSCILLATION EXPERIMENTS AT ACCELERATORS

Wonyong Lee

Columbia University, New York, NY 10027

Although at the present time there is no clear experimental evidence for neutrino mass, there is no inherent reason for neutrinos to be massless. If neutrinos have masses and if flavor changing currents exist, then neutrinos could oscillate between different flavors. More precisely, neutrino oscillations will occur if the following two conditions are satisfied. 1) The weak eigenstates must not be the same as the mass eigenstates. The weak eigenstates can then be written as linear mixtures of the mass eigenstates, in the following manner:

$$\nu_\alpha = \Sigma u_{\alpha j}\nu_j$$

where $\alpha = e,\mu,\tau,\ \cdots$

$j = 1,2,3,$

$|\nu_\alpha\rangle$ are the weak eigenstates

$|\nu_j\rangle$ are the mass eigenstates.

2) At least one of the mass eigenstates must have a non-zero mass with

$$\Delta m_{ij}{}^2 = \left|m\nu_i - m\nu_j\right|^2 \neq 0 \ \ .$$

For the case of two-component mixing, we write

$$\begin{pmatrix} \nu_\alpha \\ \nu_\beta \end{pmatrix} = \begin{pmatrix} \cos\theta & \sin\theta \\ -\sin\theta & \cos\theta \end{pmatrix} \begin{pmatrix} \nu_1 \\ \nu_2 \end{pmatrix}$$

where θ is a mixing angle.

From these equations, one can then derive the following probability distributions for the transitions of a weak eigenstate $|\nu_\alpha\rangle$:

$$\text{i) } P(\nu_\alpha \rightarrow \nu_\alpha) = 1 - \sin^2 2\theta \, \sin^2\left(\frac{1.27\ \Delta m^2 L}{E_\nu}\right) \tag{1}$$

$$\text{ii) } P(\nu_\alpha \rightarrow \nu_\beta) = \sin^2 2\theta \, \sin^2\left(\frac{1.27 \Delta m^2 L}{E_\nu}\right) \quad . \tag{2}$$

Here L is the distance between the ν source to the detector in meters, E_ν the energy of the neutrino in MeV, and Δm^2 in eV^2.

Two types of oscillation experiments, inclusive (or disappearance) and exclusive (or appearance) experiments are possible corresponding to Eqs. (1) or (2), respectively. In the exclusive experiment, one looks for appearance of ν_β in ν_α beam. In the inclusive experiment, one looks for the change in the number of ν_α as a function of L and E_ν.

Inclusive Oscillation Experiments

Equation (1) can be written as

$$P(\nu_\alpha \rightarrow \nu_\alpha) = 1 - \sum_{\alpha \neq \beta} P(\nu_\alpha \rightarrow \nu_\beta) \quad \ldots \tag{3}$$

As one can see from Eq. (3), inclusive experiments search for all possible channels of oscillations. In principle, this type of experiment can be carried out with one detector at a fixed position while varying the neutrino energy. However, measurement and comparison of neutrino interaction rates at two distances make it possible to estimate the systematic errors reliably. Systematic errors come from the uncertainty in the ν-flux at the source and from the changing detector response to various reactions.

Searches for inclusive muon neutrino oscillations have been performed by the CCFR[1] at Fermilab, and CDHS[2] and CHARM[3] at CERN. In the CCFR experiment, two detectors were located at 715m and 1115m from the neutrino source. Data were taken with dichromatic beam with neutrino energy between 50 and 200 GeV.

The oscillation measurement was performed by comparing the number of charge current events in the two detectors as a function of neutrino energy (Fig. 1). A comparison of the data with the hypothesis of no oscillation gave an acceptable χ^2 and, therefore, was consistent with a null signal. The 90% C.L. limits are shown in Fig. 2 and we summarize the data in Table 1. Both CERN experiments used a low energy bare target beam from the CERN PS with two detectors at 140m and 880m from the source. The experiments reported a null signal and we summarize the data in Table 1. In Fig. 2, we show the best limits by the CERN experiments. As one can see in Fig. 2, there is room for improvement in the ν_μ disappearance experiment in $\Delta m^2 \sim 0.1 \sim 10$ eV^2 region.

Exclusive Oscillation $(\nu_\mu \to \nu_e)$ Experiments

Appearance experiments have been performed at LAMPF, FERMILAB, BNL and CERN. In all experiments, a search was made for an anomalous yield of ν_e in a relatively pure beam of ν_μ. If no statistically significant number of ν_e were observed well above various sources of background, then the limits were set on $\nu_\mu \to \nu_e$ oscillation. The backgrounds are due to the small amount of ν_e in the beam as well as ordinary neutrino reactions faking ν_e.

The currently available results are summarized in Table 3 and the best limits are shown in Fig. 3. Two component mixing is assumed and the limits are at the 90% C.L. The references to published works given in Table 2 can be found in a review talk by M. Shaevitz.[4]

New Results. 1. The BEBC PS Experiment.[5]

The BEBC PS experiment was carried out with a horn focused neutrino beam ($<E_\nu> \sim 1.5$ GeV) from the CERN PS. The decay tunnel was 45m long and the BEBC bubble chamber was located 825m from the source. 470 ν_μ CC events and 4 ν_e events, with an estimated background of 3 ν_e CC events, were observed. The limits on oscillation parameters are given in Table 2 and new limits are shown in Fig. 4.

    2. <u>PS 191 Experiment</u>.[6]

The PS 191 experiment was carried out with a low energy bare target beam from the CERN PS with a fine grain calorimeter at 130m from the neutrino source. The detector is a fine sampling (0.17 x_o sample) calorimeter using flash chambers (5 x 5 mm^2) and 3mm iron sheets as the absorber. The neutrino beam consists of ν_μ with an energy spectrum peaking at 600 MeV and a small mixture of 0.7 $\pm$ 0.2% of ν_e at somewhat higher energies.

The total number of ν_μ detected is estimated to be 608 events. A total of 57 events with one shower plus one or two associated tracks were found. The shower energy was required to be larger than 400 MeV. The same topology for the case of two showers (presumably π^o) gave a total of 23 events.

In order to estimate π^o contamination in the single shower sample, the gamma conversion point from the vertex was studied for two samples. In Figs. 5a and 5b, the distribution of the distance between the vertex and the beginning of the closet shower for two and one shower events, respectively, as shown. The solid line is data and the dashed curve is calculated by a Monte Carlo simulation. For Fig. 5a the Monte Carlo is normalized for the total number of events. In Fig. 5b the Monte Carlo is normalized only for events beyond the first bin.

    The results are:

        Total number of electron candidates   42

        π^o contamination in the sample   7 $\pm$ 3

        ν_e beam contamination   12 $\pm$ 3

There are an excess of 23 $\pm$ 8 events.

The authors speculate that this could be due to neutrino oscillations, but wish to wait until the BNL run is completed before making any definite conclusions.

The data presented in the paper is not sufficient to judge if there is a significant signal for neutrino oscillations. For example, I would like to see the energy and angular distributions of electron and π^0 candidates. Still, we can say a few words about the data. We go back and look at Fig. 5b. Although the conversion distance distribution between the data and the Monte Carlo simulation is statistically consistent, we can observe clear excess of π^0 in the first bin. This may be due to π^0 converting away from the vertex with some overlapping tracks making it appear as if it was coming from the vertex. Remember that the detector has only one stereo view. If this interpretation is correct, then the number of π^0 in single shower events is underestimated and the statistical significance of the results is much reduced.

Exclusive Oscillation $(\nu_\mu - \nu_\tau)$ Experiments

The limits in the exclusive experiments $\nu_\mu \rightarrow \nu_e$ can also be used to set a limit on the oscillations $\nu_\mu \rightarrow \nu_\tau$ if the energy of the neutrino is high enough to allow the reaction $\nu_\tau + N \rightarrow \tau + \text{anything}$ to proceed. These results are summarized in Table 3 and the best limits are shown in Fig. 6. The most sensitive search was performed by the E531 hybrid emulsion spectrometer using the Fermilab wide band horn beam. The experiment used an emulsion target, 700m from the neutrino source and search for τ-lepton with finite decay length.

Future Experiments

E776 BNL (Columbia/U of Illinois/Johns Hopkins)

This group has accumulated $\sim 3 \times 10^{19}$ proton-on-target with a horn focused dichromatic beam ($E_\nu \sim 1.4$ GeV) and also $\sim 3 \times 10^{19}$ protons-on-target with a horn focused wide band beam ($<E_\nu> \sim 1.5$ GeV). The detector is at 1000m from the neutrino source.

LAMPF E645 (Argonne/Caltech/Los Alamos/LSU/Ohio State)

This group is currently taking data with 800 MeV on a beam dump. The expected sensitivity of both experiments is $\Delta m^2 \sim 0.1$ eV^2 and $\sin^2 2\theta \sim 10^{-3}$.

<u>Conclusions</u>

In the past, a few experiments have reported tantalizing hints for neutrino oscillations. However, at present there is no clear experimental evidence for the oscillation. As experimental techniques are improved, and new opportunities open up, there will be more experiments to study this very important area of physics.

References

1) I.E. Stockdale et al, Phys. Rev. Lett. <u>52</u>, 1384 (1984).
   I.E. Stockdale et al, Z. Phys. C<u>27</u>, 53 (1985).
2) F. Dydak et al, Phys. Lett. <u>134B</u>, 34 (1984).
3) F. Bergsma et al, Phys. Lett. <u>142B</u>, 103 (1984).
4) M. Shaevitz, Rencontre de Moriond (1986):
   Massive Neutrino Workshop.
5) C. Angelini et al, Phys. Lett. (to be publ.).
6) G. Bernardini et al, Phys. Lett. "   "   "   .

Table 1. $\nu_\mu \rightarrow \nu_\mu$

Expt.	Accel.	ℓ/E	Lower Limit Δm^2	Best Limit $\sin^2 2\alpha$
CDHS	PS	.8 - .03	.3 eV^2	6×10^{-2}
CHARM	PS	.8 - .03	.3 eV^2	2×10^{-1}
CCFR	FNAL	.03- .003	15 eV^2	2×10^{-2}

Table 2. $\nu_\mu \rightarrow \nu_e$

Expt.	Accel.	ℓ/E	Lower Limit Δm^2	Best Limit $\sin^2 2\alpha$
BNL/Col	FNAL	.16	.6 eV^2	6×10^{-3}
FNAL/Mich II	FNAL	.04	3 eV^2	8×10^{-3}
Gargamelle	PS	.05	.7 eV^2	2×10^{-3}
Nemethy	LAMPF	.3	.9 eV^2	2×10^{-1}
Gargamelle	SPS	.03	2 eV^2	1.2×10^{-2}
BEBC	SPS	.03	1.7 eV^2	1.0×10^{-2}
BFH	FNAL	.03	2.3 eV^2	1.6×10^{-2}
E734	BNL	.15	.43 eV^2	3.4×10^{-3}
LAMPF/UCI	LAMPF	.2	.37 eV^2	$.8 \times 10^{-2}$
BEBC	PS	.84	.08 eV^2	1×10^{-2}
Vannucci	PS	.15	5 eV^2	3×10^{-2}

Table 3. $\nu_\mu \rightarrow \nu_\tau$

Expt.	Accel.	ℓ/E	Lower Limit Δm^2	Best Limit $\sin^2 2\alpha$
BNL/Col	FNAL	.04	3 eV^2	6×10^{-2}
FNAL/Mich II	FNAL	.04	2.2 eV^2	4.4×10^{-2}
Emulsion	FNAL	.02	3 eV^2	1.3×10^{-2}
Gargamelle	SPS	.03	5 eV^2	8×10^{-2}
BEBC	SPS	.03	6 eV^2	5×10^{-2}
BFH	FNAL	.04	7.4 eV^2	8.8×10^{-2}

W. Y. Lee

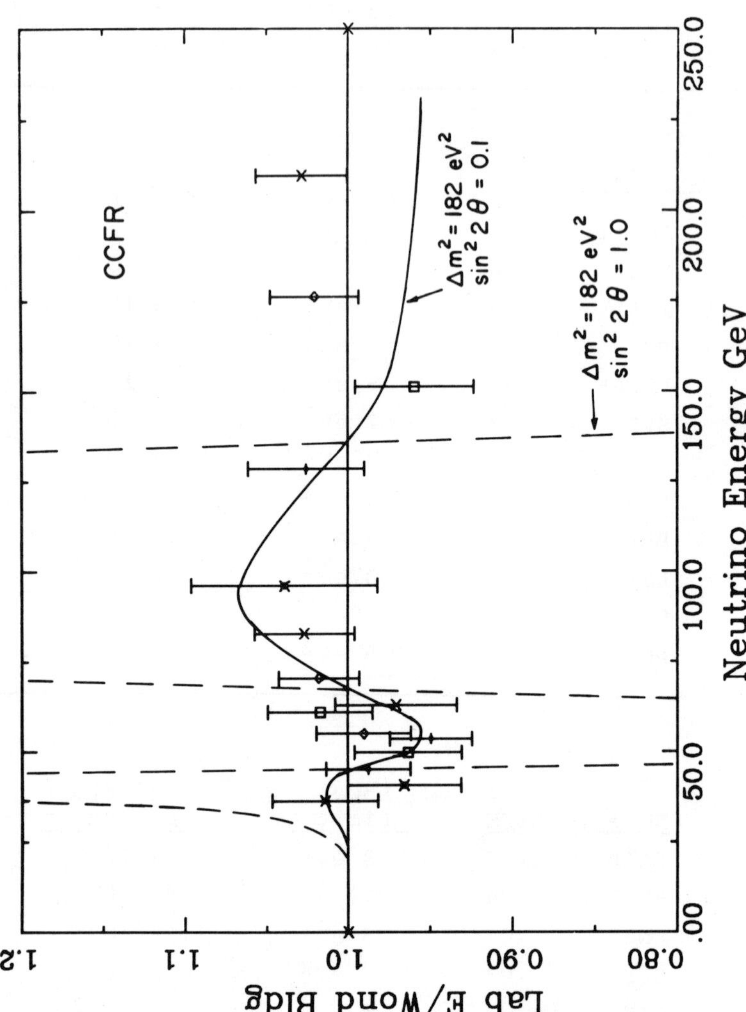

Fig. 1. Ratio of events in two detectors from the CCFR experiment.
A curve (solid) for a typical Δm^2 value and 10% mixing is shown.
The curve for maximal mixing (dashed) at the same Δm^2 is largely
off-scale.

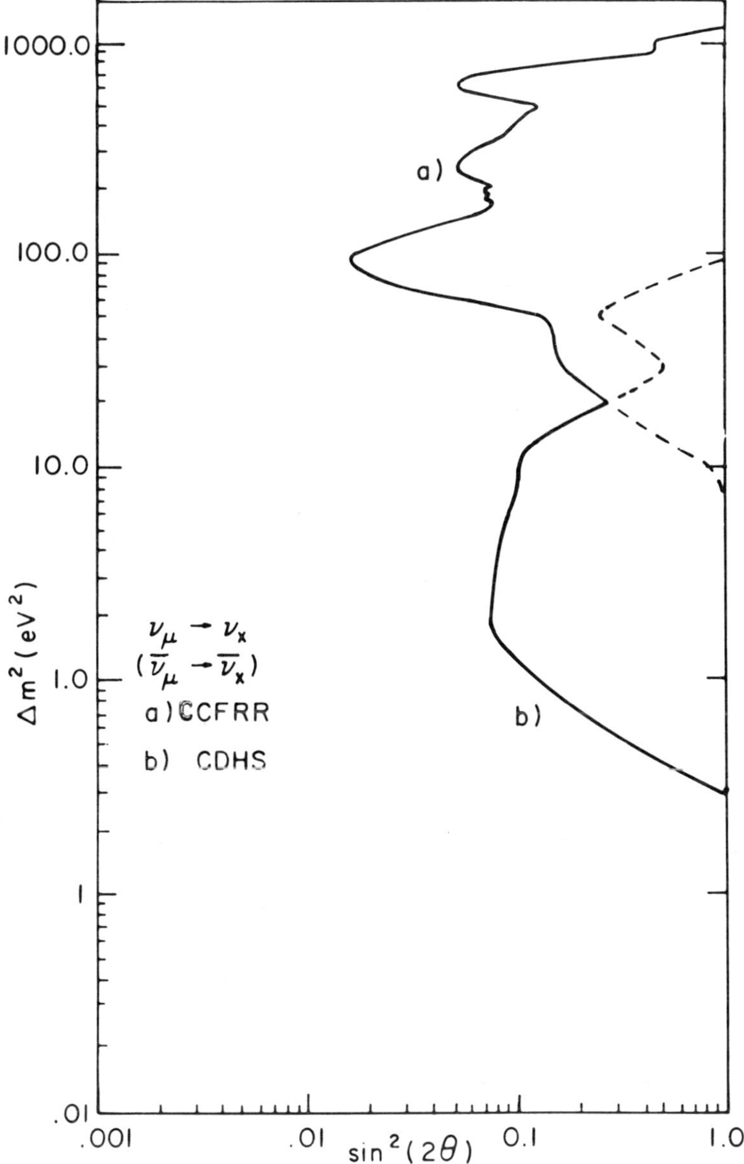

<u>Fig. 2.</u> Inclusive oscillation limits for $\nu_\mu \rightarrow \nu_x$.

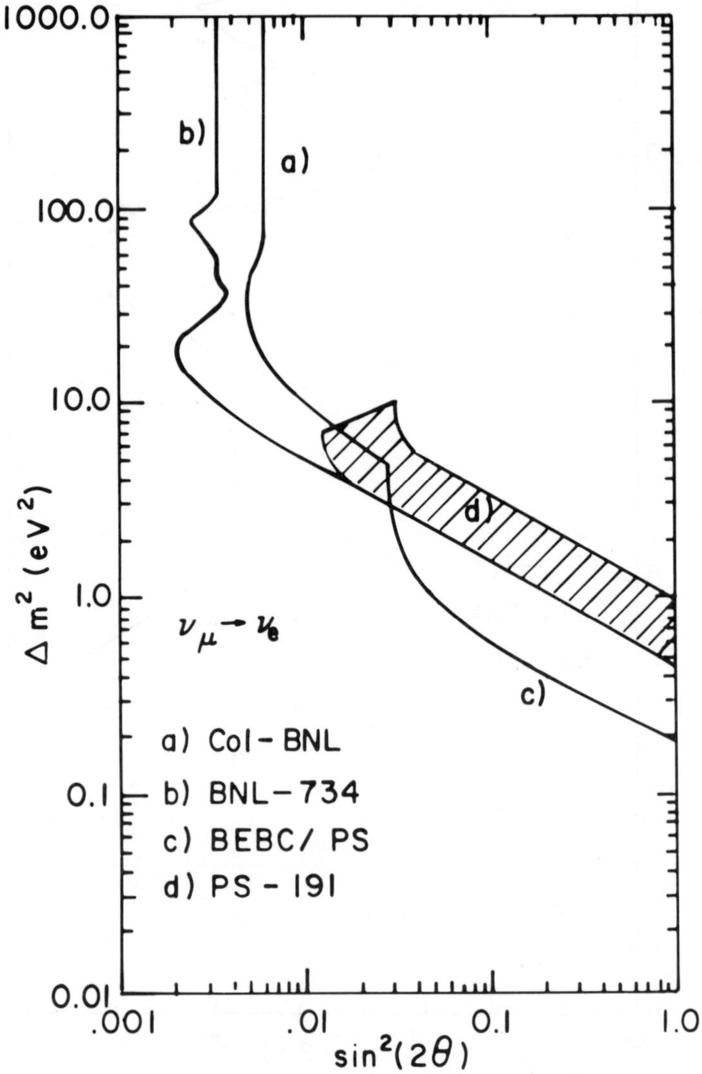

<u>Fig. 3.</u> Exclusive oscillation limits for $\nu_\mu \rightarrow \nu_e$.

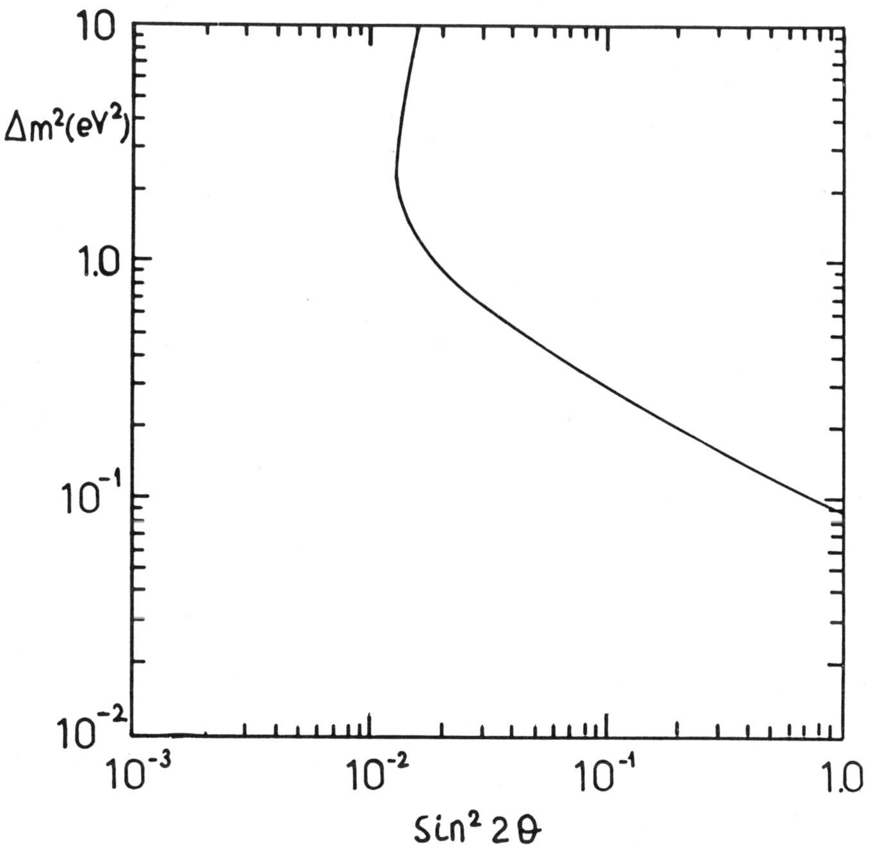

<u>Fig. 4.</u> New limits on the $\nu_\mu \to \nu_e$ oscillation parameters by Ref. 5.

W. Y. Lee

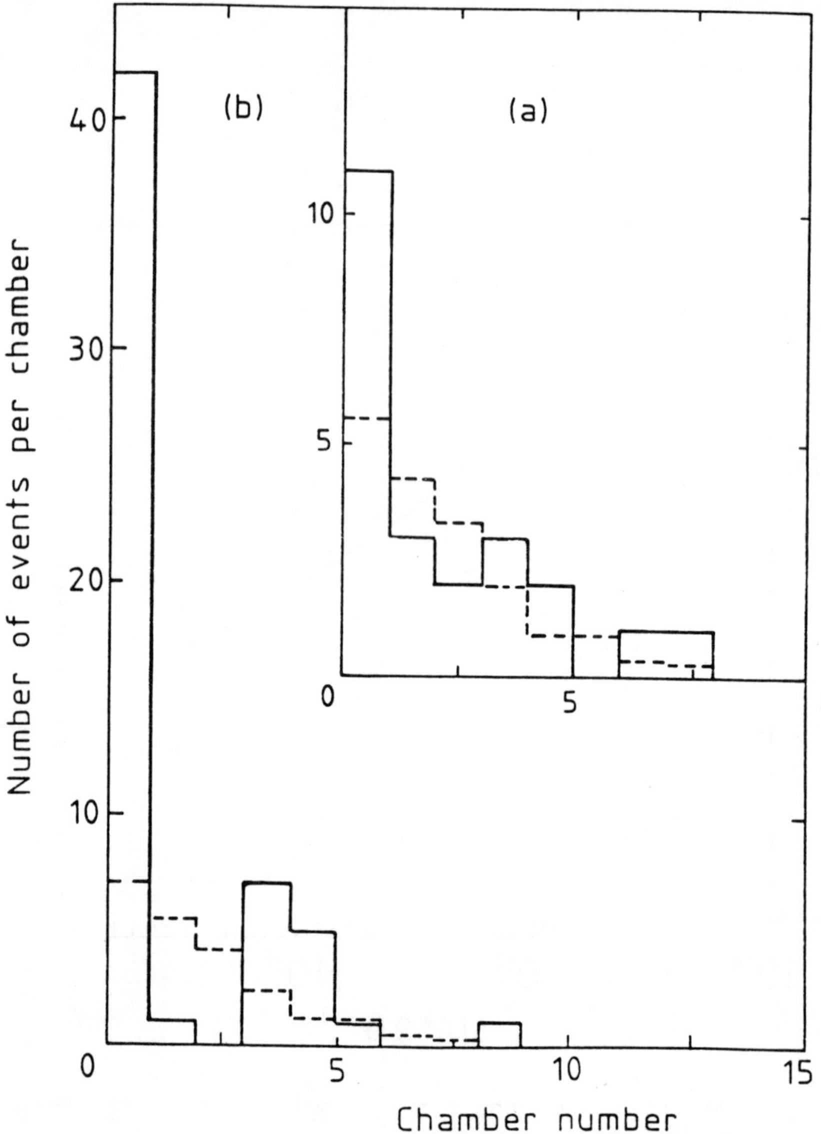

Fig. 5. a) Distribution of the distance between vertex and beginning of the closest shower for the two-shower events. Each bin represents one chamber, i.e. two planes. b) Distribution of the distance between vertex and beginning of the shower for the one-shower events.

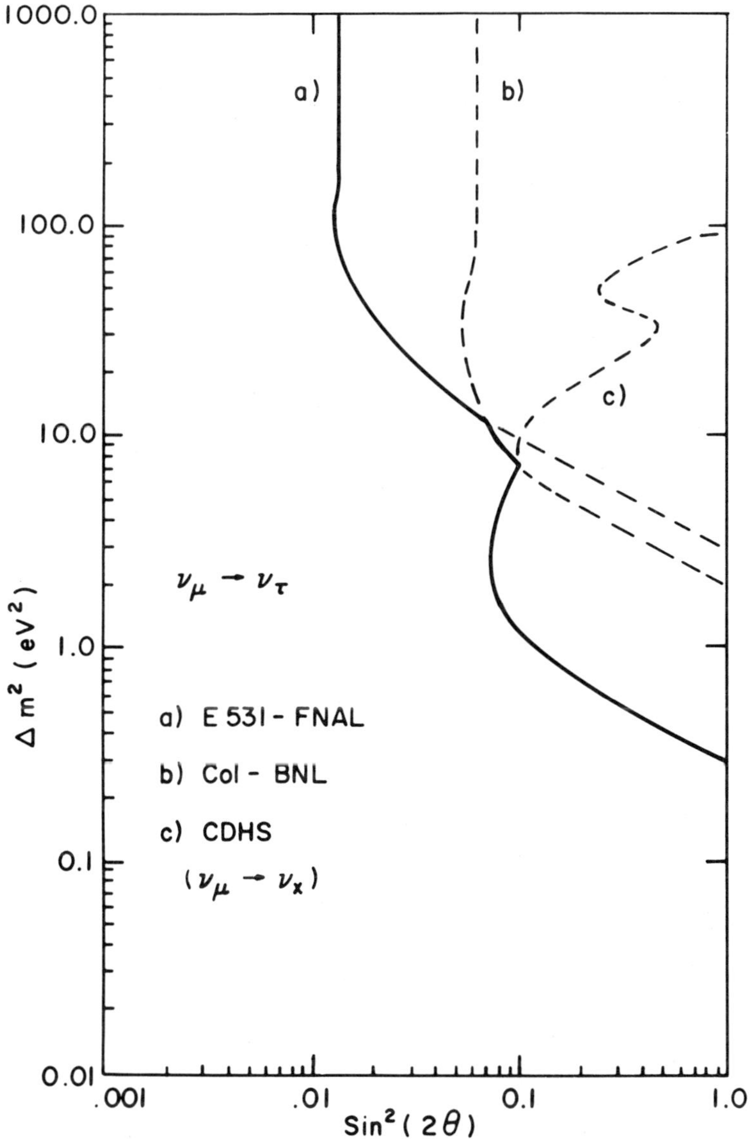

Fig. 6. Exclusive oscillation limits for $\nu_\mu \rightarrow \nu_\tau$.

NEW EXPERIMENTAL LIMITS ON $\nu_\mu \to \nu_e$ OSCILLATIONS

Milla BALDO-CEOLIN

Dipartimento di Fisica and Sezione INFN

Padova, Italy

(Padova-Pisa Athens-Wisconsin Collaboration) [1]

Abstract

The results of an experiment which searched for ν_e events arising from oscillations of a low energy ν_μ beam are presented. The BEBC heavy liquid bubble chamber, placed at a distance of 825 m from an external proton target at the CERN PS, was used as a neutrino detector. 470 ν_μ CC events and 4 ν_e events, with an estimated background of 3 ν_e CC events, have been observed. The resulting limits on the oscillation parameters, at the 90% confidence level, are: $\delta m^2 \leq 0.09$ eV^2 (for maximal mixing) and $\sin^2 2\vartheta \leq 0.013$ for $\delta m^2 = 2.2$ eV^2.

1. Introduction

As an introduction, I will simply recall that the study of the ν oscillations is the most sensitive method to search for very small neutrino masses [2]. In the simplified hypothesis of mixing between ν_μ and ν_e only, the oscillation probability is:

$$P(\nu_\mu \to \nu_e) = \sin^2(2\,\vartheta)\,\sin^2(1.27\,\delta\,m^2\,L/E) \tag{1}$$

where $\delta m^2 = |m_1^2 - m_2^2|$, in (eV)^2 and ϑ is the $\nu_\mu - \nu_e$ mixing angle. δm^2 and $\sin^2(2\,\vartheta)$ are the parameters that determine the oscillation probability. L is the neutrino propagation length in meters and E the neutrino energy in MeV. From Eq. (1), the sensitivity to small δm^2 values increases in proportion to L/E. Moreover, in order to obtain a precise measurement of the transition probability, a clear, unambiguous identification of the ν_e events is required as well as a small ν_e background which must be quantitatively well known. A good neutrino energy determination is also essential.

2. Experimental conditions

The present experiment was performed at the CERN PS, with the intent of exploring the small δm^2 region ($\cong 0.1$ eV^2).

The low energy neutrino beam was generated by 19.2 GeV/c protons. The ν_μ flux was enhanced by focussing with a pulsed magnetic horn π^+'s and K^+'s. The horn was followed by a 45 m decay tunnel and a concrete-iron shielding. The average ν_μ energy was 1.5 GeV.

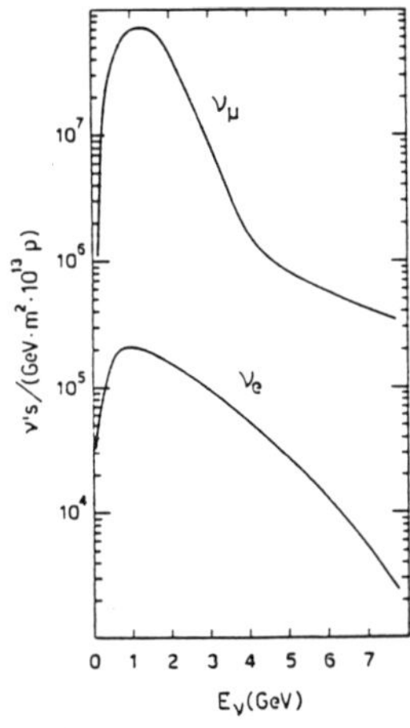

Fig. 1

The background ν_e flux was estimated from existing measurements[3] of the π and K production rates at this energy and from the known geometry of the beam, and found to be 0.4 % of the ν_μ flux[4].

The energy dependence of this background is shown in Fig. 1, together with the ν_μ spectrum.

The BEBC bubble chamber was used as a neutrino detector. It was located 825 m from the target so that L/E resulted $2.6 \leq L/E \leq 0.16$. The chamber, filled with a 73% Ne/H$_2$ molar mixture ($\rho=0.68$ g/cm^3, $X_0=43$ cm), with a high magnetic field (3.5 T) and the Internal Picket Fence - IPF (a cylindrically-symmetric set of proportional wire chambers placed outside the containment vessel of the chamber), provided an extremely clean and clear electron and muon identification, as well as good energy determination.

A total of 794,000 pictures, corresponding to 0.9 10^{19} protons on target were obtained.

3. Candidate event collection

All the pictures were visually scanned for nuclear events induced by neutrals, with the exception of bare one-prong events, and for isolated gammas or electrons having momentum larger than 120 MeV/c. Approximately 60% of the pictures were scanned twice and, in addition, a special scan was performed on the frames flagged by a track that gave a hit in the IPF during the 2.2 μsec beam-gate. The average scanning efficiency was $\varepsilon = 0.90$.

The identification of the ν charged current (CC) events rests primarily on the identification of the μ^- or e^- produced in the interaction. Candidate muons were considered the negative, non-interacting tracks that left the chamber after a track length of at least 20 cm, or those that came to rest with or without a visible decay and which have a range compatible with that of a muon. To allow a good π/μ discrimination and to ensure the background to be negligible, muons stopping without a visible decay were required to have a momentum greater than 150 MeV/c.

In the case of electrons the short radiation length enhanced the probability of the characteristic electromagnetic processes which were used for their identification. To be accepted as an electron a track was required to satisfy at least two of the following conditions:

a) have an e^+e^- pair ($P_{e^+e^-}$ >20 MeV/c), or a Compton electron (P_e->20 MeV/c) pointing tangentially to the track

b) have a δ-ray, carrying at least 20% of the initial energy

c) have an e^+e^- pair lying on the track (trident)

d) spiral and stop with a range longer than that of a muon.

The probability, for an electron track, to produce at least two of the above signatures, was evaluated as a function of energy by two independent Montecarlo programs. For instance, a 1 GeV/c electron produces an average of 6 signatures and has an identification probability of 95%. The possibility of misidentifying a π or μ track as an electron is completely negligible.

4. Event selection

To be accepted as ν_μ or ν_e CC events the candidates had to fulfill the following kinematical conditions, chosen in order to reduce the cosmic ray background: (a) $(E_{vis}-P_{vis})< 1$ GeV, (b) $P_{//}>200$ MeV/c, where $E_{vis} = \Sigma_i E_i$, $P_{vis} = \Sigma_i P_i$, E_i and P_i are energy and momentum of the ith track and $P_{//}$ is the total longitudinal momentum. Furthermore, P_{vis} in the events produced by neutrino interactions points to the ν_μ beam direction, while for cosmic ray interactions it is expected to be distributed over the whole solid angle. Thus the selected events had to satisfy the further conditions $\cos\vartheta>0.8$, $\mathrm{sen}\vartheta<0.8$, where ϑ is the ν_μ beam direction. According to Montecarlo studies, these cuts give an acceptance of 93.5% for genuine CC events.

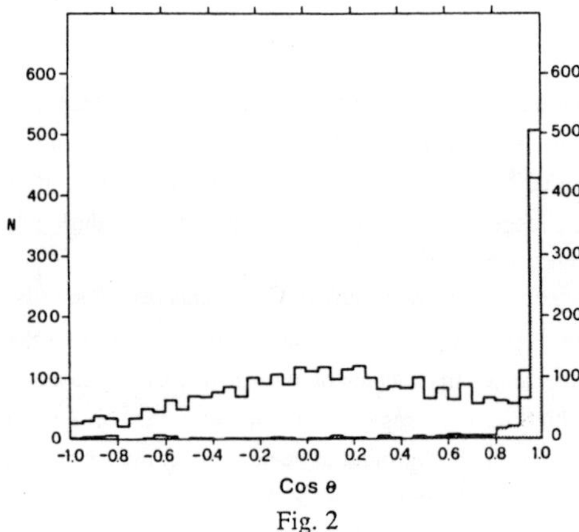

Fig. 2

In Fig. 2 the $\cos\vartheta$ distribution is shown for all the events with at least one negative track, the dotted area represents the selected events. It can be seen that the requirement: $\cos\vartheta>0.8$ optimizes the signal to background ratio.

These criteria led to the selection of 470 ν_μ CC events; in addition, 4 events with an electron were found, and they were in agreement with all the above selection criteria.

In Table I are given the numbers of ν_μ CC events, according to the μ^- track type, weighted for scanning efficiency. The corresponding Montecarlo predictions are also given in the same table. The agreement is satisfactory.

The background – due to cosmic rays, neutrino N.C. or neutron interactions – in the sample of 470 ν_μ CC candidates was determined by several methods and corresponds to 24 ± 5 events.

Energies and topologies of ν_e events are given in Table II.

The expected number of events mimicking ν_e interactions is entirely negligible.

Table I

ν_μ CC events	μ leaving with IPF hit	μ leaving without IPF hit	μ decay+stop
Experiment	194	207	120
Montecarlo	193	193	135

Table II

Electron CC events (Roll/Frame)	event energy GeV	electron energy GeV	$\cos\vartheta$
1158/850	2.5	1.50	0.98
1280/428	5.9	3.40	0.99
1287/1241	2.8	2.00	0.99
1306/1047	1.4	0.86	0.99

5. Event energy determination

Then, in order to better determine the ν interaction energy the variable:

$$\eta = \sum_{i=1}^{N} (E^i - c\, P^i_{//})\ \text{GeV} \tag{2}$$

was used; here N is the number of visible particles in the final state, $p^i_{//}$ is the longitudinal momentum of the ith particle and E^i its energy or, if identified as a baryon, its kinetic energy. Neglecting Fermi momentum, η would be zero for ν interactions if all the tracks were visible and their masses correctly assigned. The loss of neutrons would smear the η distribution and slightly shift its average to positive values, the misidentification of protons as pions wouls shift η to higher positive values. Thus the ambiguity between π^+ and p was solved for each event by matching the total visible energy and longitudinal momentum $P_{//}$. The resulting η distribution is shown in Fig. 3a. The average energy lost in gammas and neutrons was estimated as a function of the event energy and found to be ~5%. In the final analysis discussed below, a correction factor for this loss was applied in a proper, energy dependent, way . The energy spectrum of the ν_μ events is shown in Fig. 3b; the μ^- momentum spectrum is shown in Fig. 3c. In all figures the solid lines are obtained by a Montecarlo simulation normalized to the total number of events. The agreement corroborates both particle identification and the scanning and measuring procedure.

6. Conclusions and analysis

Montecarlo studies indicate that from 470 ν_μ CC events a background of 3 ν_e CC events is expected, which is compatible with what is observed. It is to be noted that the energies of the ν_e events are somewhat higher than those of the ν_μ events, as expected from the ν_e contamination in the beam.

Thus, the conclusion is that in the present data there is no evidence for $\nu_\mu \rightarrow \nu_e$ oscillations.

To set limits on the oscillations parameters, a two neutrino ($\nu_\mu \rightarrow \nu_e$) oscillation scheme was assumed. The likelihood ratio $L = L / L_0$ was computed as a function of the parameters, using the function:

$$L\,(\delta m^2, \sin^2 2\vartheta) = \prod_{i=1}^{N} P_i^{n_i}\,(1-P_i)^{1-n_i}$$

where $n_i = 1(0)$ if event i is an electron (muon) event, and P_i is given by:

$$P_i = \frac{[q_i + \alpha_i(1-q_i)]\zeta_i}{[1-\alpha_i(1-q_i)]/(1-f) + [q_i + \alpha_i(1-q_i)]\zeta_i} \tag{3}$$

with $\alpha_i = \sin^2 2\vartheta \, \sin^2(1.27 \times L \times \delta m^2/(E_i))$, where E_i is the visible energy (in MeV) of event i, corrected for the undetected fraction of energy carried away by neutrals, q_i is the expected ratio of background ν_e to ν_μ in the beam at the energy E; ζ_i is the detection efficiency for electrons relative to that of muons, including the ratio of ν_μ to ν_e cross sections at the energy E_i, and f is the fraction of background events in the muon sample. L_0 is the likelihood function corresponding to the hypothesis of no oscillations.

The resulting confidence region is shown by the solid line in Fig. 4. The lowest value of the oscillation parameters that are excluded at the 90% confidence level by the present experiment are: $\delta m^2 = 0.09$ eV^2 (for $\sin^2 2\vartheta = 1$) and $\sin^2 2\vartheta = 0.013$ (for $\delta m^2 = 2.2$ eV^2).

The result on $\delta m^2 (\sin^2 2\vartheta)$ is essentially unchanged when the parameters E_i, q_i, ζ_i and f in eq. (3) are allowed to vary within the uncertainties assigned to them.

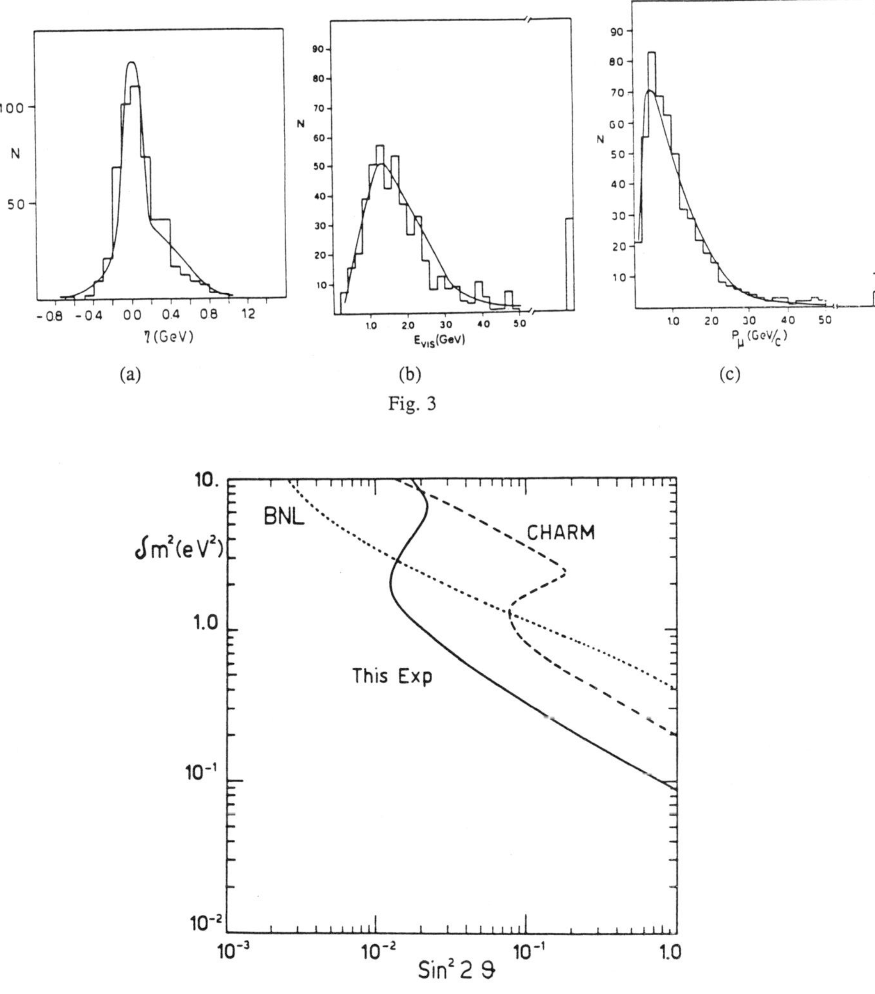

Fig. 3

Fig. 4

A comparison of the present data with the most recent results is made in Fig. 4. It can be seen that the results obtained in the present experiment are by far the most stringent limits on the transition $\nu_\mu \to \nu_e$ in the small δm^2 region[4]. This is mainly due to the ability of the detector in providing a clear and unambiguous electron identification, and to its good energy resolution.

References and footnotes

[1] A. Apostolakis, P. Ioannou, S. Katsanevas, C. Kourkoumelis, J. Koutentakis, P. Pramantiotis, L.K. Resvanis, M. Vassiliou, Physics Department, University of Athens, Athens, Greece; M. Baldo-Ceolin, F. Bobisut, E. Calimani, S. Ciampolillo, H. Huzita, M. Loreti, G. Miari, G. Puglierin, Dipartimento di Fisica dell'Università di Padova, Padova, Italy; C. Angelini, A. Baldini, L. Bertanza, B. Saitta, Dipartimento di Fisica dell'Università di Pisa, Pisa, Italy; U. Camerini, W.J. Fry, R. Loveless, M. Procario, D.D. Reeder, Physics Department, University of Wisconsin, Madison, Wisconsin, USA.

[2] B. Pontecorvo, JETP 34, 247 (1958); JETP 7, 172 (1958); Zh. Eksp. Teor. Fiz. 53, 1717 (1967) [JETP 26, 984 (1968)].

S.M. Bilenky and B. Pontecorvo, Physics Reports 41, 225 (1978).

The possibility of neutrino oscillation among flavours had independently been suggested in 1962 by Z. Maki, M. Nakazawa and S. Sakata; Prog. Theor. Physics 28, 870 (1962).

For recent reviews of the theoretical aspects of lepton non-conservation see G. Costa and F. Zwirner, preprint DFPD 10/85, April 1985; to appear on Rivista del Nuovo Cimento, and Vergados, Physics Reports 133, 1 (1986).

[3] J.V. Allaby et al., CERN REPORT 70-12 (1970).

H. Grote, R. Hagedorn and J. Ranft, Atlas of Particle Production Spectra, CERN (1970).

[4] C. Visser, NUBEAM: Neutrino Beam Simulator, Hydra Application Library, CERN (1979).

[5] L.A. Ahrens et al., Physical Review D 31, 2732 (1985).

F. Bergsma et al., Physics Letters B 142, 103 (1984).

For a recent review of neutrino oscillation experiments see V. Flaminio and B. Saitta, "Neutrino Oscillation Experiments" INFN PI/AE 85/6, to be published on the Rivista del Nuovo Cimento.

NEUTRINO OSCILLATIONS IN MATTER

S.P.Mikheyev and A.Yu.Smirnov

Institute for Nuclear Research

the USSR Academy of Sciences

Moscow, USSR

ABSTRACT

Properties of neutrino mixing in matter and ν-oscillations for different regimes (different density distributions) are considered. Main results conserning the applications of matter effects to ν-fluxes in the Sun, in cores and envelopes of collapsing stars and in the Earth are summed.

1. INTRODUCTION

The picture of vacuum neutrino oscillations[1] is modified in matter[2,3,17]. For a wide regions of densities and energies this modification is due to elastic neutrino scattering on electrons and nuclei only. It can be described in terms of refraction indexes[2,17] or effective neutrino masses which depend on density of matter[5].

Matter effects have a resonant character (resonance on density of matter or energy of neutrinos[4]). The picture of ν-oscillations in matter with varying density differs essentially from that in vacuum or in matter with constant density. In particular, near to complete transformation of initial type neutrinos into another ones in a wide energy region is possible[4]. Two things are needed for this: adiabatic (slow) changing of density and neutrino crossing of resonance (layer with $\rho = \rho_{res}$)[4].

Resonance matter effects are essential for neutrino astrophysics and geophysics. In particular, they give new insight into solar neutrino puzzle[4].

Recently many aspects of neutrino oscillations in matter have been elaborated. Among them:

i) The theory of effects itself. Properties of ν-oscillations in adiabatic regime have been considered in details[4,6-12]. Divergence of neutrino wave packets in matter is studied[7,9,10,13]. Some results beyond adiabatic approximation have been obtained[7,8,13].

ii) Applications. Neutrino oscillations in the Sun[4,5,8,10-14], in cores and envelopes of collapsing stars[4-7], in the Earth[7,13-15], in the early Universe[7] have been considered.

iii) Consequencies for neutrino mass spectrum and theory of particles. Overlapping of Δm^2 vs. $\sin^2 2\theta$ region of strong matter effect with "see-saw" predictions region is remarked[5,16].

It should be noted, that the discussed ν-oscillations effects are similar to many phenomena from other fields, in particular, to coherent regeneration of K-mesons[2], to some effects with polarized light in active mediums, to spin precession in alternating magnetic field[15] and to others. The following analogy gives very clear insight into considered oscillation effects. The mixed neutrinos, for example ν_e, ν_μ, are equivalent to the system of two weakly coupled oscillators (for definiteness - pendulums). Oscillations of one pendulum corresponds to propagation of ν_e, oscillations another one - to propagation ν_μ. Generation of ν_e is the exitation of "ν_e-pendulum". Coupling between pendulums results in exchange of oscillations between them. It is this periodic exchange corresponds to ν-oscillations. Matter changes the frequencies of pendulums. Moreover this changing is different for "ν_e" and "ν_μ". Coincidence of frequencies corresponds to resonance in oscillations: the exchange of exitations may be complete.

2. NEUTRINO MIXING IN MATTER.

2.1 Neutrino eigenstates and mixing.

Consider two neutrinos mixing (for definiteness ν_e and ν_μ):

$$\vec{V}_\alpha = \hat{S}\,\vec{V} \qquad (1)$$

Here $\vec{V}_\alpha = (\nu_e, \nu_\mu)$, $\vec{V} = (\nu_1, \nu_2)$ are the states with definite masses m_1 and m_2, $\hat{S}$ is the unitary matrix with $(\hat{S}_o)_{e1} = \cos\theta$, $(\hat{S}_o)_{e2} = \sin\theta$, ... θ is mixing angle in vacuum.

The evolution equations for $\vec{V}_\alpha$ in matter are[2]

$$i\frac{d}{dt}\vec{V}_\alpha = \hat{M}\vec{V}_\alpha \qquad (2)$$

Here $\hat{M}$ is evolution matrix

$$\hat{M} = \begin{vmatrix} M_e & \overline{M} \\ \overline{M} & M_\mu \end{vmatrix} \tag{3}$$

Diagonal elements in (3) contain the terms are proportional to the density of matter (see appendix A, some relations and definitions from the previous papers are summed in appendix A). Note that the matter effect is diagonal in flavour; nondiagonal element $\overline{M}$ is determined by vacuum parameters only.

Mixing angle θ_m and neutrino eigenstates in matter ν_{im} are generalizations of θ and ν_i. Eigenstates ν_{im} are determined as the states which diagonalize evolution matrix $\hat{M}$ in (2):

$$\vec{\nu}_\alpha = \hat{S}_m \vec{\nu}_m \tag{4}$$

$$\hat{S}_m^{-1} \hat{M} \hat{S}_m = \hat{M}^d \tag{5}$$

Here

$$\hat{S}_m = \begin{pmatrix} C_m & S_m \\ -S_m & C_m \end{pmatrix}, \quad C_m = \cos\theta_m, \quad S_m = \sin\theta_m, \quad \vec{\nu}_m = (\nu_{1m}, \nu_{2m}), \quad M^d$$

is diagonal matrix $M^d = \mathrm{diag}(M_1^d, M_2^d)$. The angle θ_m in matrix $\hat{S}_m$, which relates (according to (4,5)) the eigenstates of weak interaction and neutrino eigenstates is called the mixing angle in matter. Using (5) one has

$$\sin 2\theta_m = 2\cdot\overline{M}\cdot(M^2 + 4\overline{M}^2)^{-1/2} \tag{6}$$

and

$$M_{1,2}^d = \frac{1}{2}(M_e + M_\mu \pm (M^2 + 4\overline{M}^2)^{1/2}) \tag{7}$$

In neutrino oscillations ν_{im} and θ_m play the same role as ν_i and θ to do for vacuum case. The relations between them are straitforward: at $\rho \to 0$, $\theta_m \to \theta$, $\nu_{im} \to \nu_i$, $M_i^d \to m_i^2/2k$

Both θ_m and $M_{1,2}^d$ depend on density. In connection with this two properties of mixing are important: i) resonance dependence of $\sin^2 2\theta_m$ ii) flavour changing of neutrino eigenstates.

2.2 Resonance in a system of mixing neutrinos.

$\sin^2 2\theta_m$ depends on ρ and E; moreover the dependence of $\sin^2 2\theta_m$ on ρ or E has a resonance character[4]. At

$$L_V = L_o \cdot \cos 2\theta \tag{8}$$

$\sin^2 2\theta_m$ reaches the maximum: $\sin^2 2\theta_m = 1$. In (8) $L_V = 4\pi E/\Delta m^2$ is oscillation length in vacuum; L_o is the eigenlength in matter $(L_o \sim (G_F \rho)^{-1})$. The equality (8) is called the resonance condition. The density (ener-

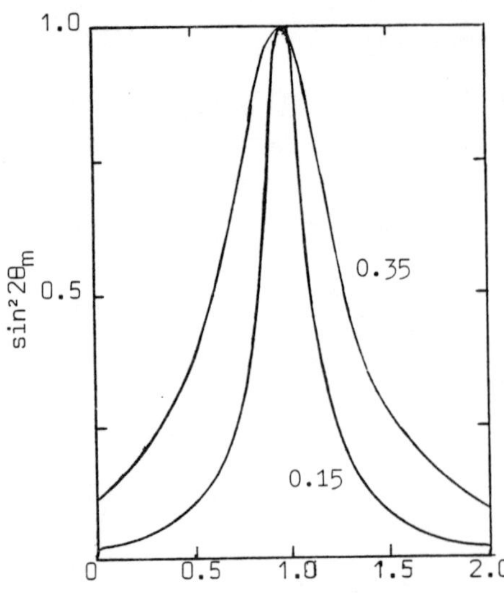

Fig.1 Resonance dependencies of $\sin^2 2\theta_m$ on L_V/L_O for different $\sin 2\theta$ (figures on the curves)

gy) which satisfies (8) is called the resonance density (energy):

$$\rho_{res} = \frac{m_N \Delta m^2 \cdot \cos 2\theta}{2 \sqrt{2} \, G_F \cdot E} \qquad (9)$$

The half width of resonance is

$$\Delta \rho_{res} = \rho_{res} \cdot \tan 2\theta \qquad (10)$$

or $\Delta E_{res} = E_{res} \cdot \tan 2\theta$. In resonance matter mixing is maximal: $\theta_m = 45°$. The existance of resonance is quite clear. As it have been pointed in sect.1 mixed neutrinos are system of weakly coupled oscillators. Matter changes their eigenfrequecies. When these frequencies are equal ("resonance") the oscillations of one oscillator can be passed to another one completely. The hardness of coupling and consequently, the width of resonance, is proportional to vacuum mixing. Note also that the considered neutrino resonance is equivalent to nuclear magnetic resonance. According to (8) for small θ the L_V - parameter of system is equal to L_O - parameter of external medium.

In terms of ρ and ρ_{res}, $\sin^2 2\theta_m$ can be rewritten as

$$\sin^2 2\theta_m = \tan^2 2\theta / ((1 - \rho/\rho_{res})^2 + \tan^2 2\theta) \qquad (11)$$

Resonant density determines the scales of matter phenomena. The manifestations of resonance depend on character of density changing.

2.3 Flavour changing of neutrino eigenstates[4,7,10].

On the defenition mixing angle θ_m determines flavours (ν_e, ν_μ-content) of neutrino eigenstates:

$$\vec{\nu}_m = \hat{S}_m^{-1} \vec{\nu}_\alpha \qquad (12)$$

θ_m and consequently flavours depend on the density of matter. When ρ diminishes from $\rho \gg \rho_{res}$ to zero, the angle θ_m varies from $\theta_m = \pi/2$ to θ. At small θ basis $\vec{\nu}_{1m}$, $\vec{\nu}_{2m}$ rotates with respect to $\vec{\nu}_e$, $\vec{\nu}_\mu$ on the an-

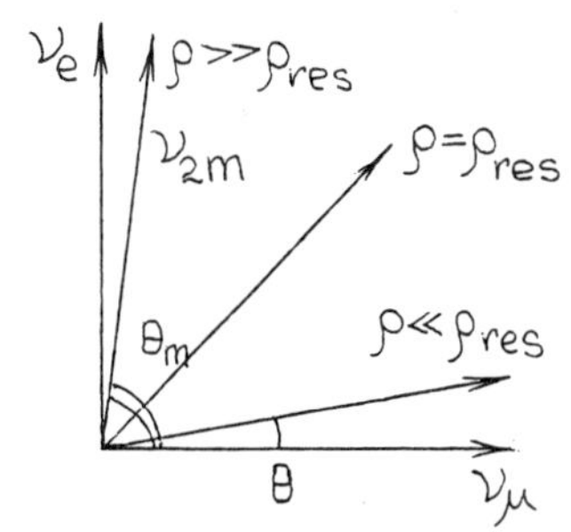

Fig.2 The dependence of flavour of neutrino eigenstate in matter on density.

gle $\pi/2$ (Fig.2). Flavours of ν_{im} change almost completely. If, for example, $\nu_{1m}(\rho \gg \rho_{res}) \simeq \nu_\mu$, then $\nu_{1m}(\rho \ll \rho_{res}) \simeq \nu_e$. It is this property lies in the basis of discussed neutrino transformations.

We deal with eigenstates immediately, but the same property can be established in terms of eigenvalues[5] (7). Without mixing ($\overline{M}=0$): $M_1^d = M_e$, $M_2^d = M_\mu$ for any ρ. At $\rho = \rho_{res}$ M=0 $M_1^d = M_2^d$. With mixing: $M_1^d(\rho \gg \rho_{res}) \simeq M_\mu(\rho)$ (if $M_1^d(\rho \ll \rho_{res}) \simeq M_e$) and, inversely, $M_2^d(\rho \gg \rho_{res}) \simeq M_e(\rho)$. Mixing rejects level crossing*.

3. NEUTRINO OSCILLATIONS IN DIFFERENT REGIMES.

3.1 Oscillations in matter. General relations.

Consider dynamics of ν-oscillations. In vacuum neutrino oscillates around ν_i the states with definite masses. In matter ν oscillates around ν_{im}. In analogy with vacuum case we can represent neutrino state as

$$(13) \quad |\nu(t)\rangle = \cos\theta_a(L)\, e^{i\varphi_1(t)}|\nu_{1m}(L)\rangle + \sin\theta_a(t)\, e^{i\varphi_2(t)}|\nu_{2m}(t)\rangle$$

Here angle $\theta_a(t)$ determines the admixtures of ν_{im} in neutrino states at a moment t. The monotonic changing of phase difference $\varphi = \varphi_1 - \varphi_2$ with time results in oscillations of flavours in ν-beam. Indeed, taking into account that $\langle \nu_e | \nu_{1m}(t)\rangle = \cos\theta_m$ and $\langle \nu_e | \nu_{2m}(t)\rangle = \sin\theta_m$ one can write the amplitude of probability to find ν_e in $\nu(t)$ (13):

$$\langle \nu_e | \nu(t)\rangle = \cos\theta_a \cdot e^{i\varphi_1}\cos\theta_m + \sin\theta_a \cdot e^{i\varphi_2(t)}\sin\theta_m \qquad (14)$$

For probability $P(t) = |\langle \nu_e|\nu(t)\rangle|^2$ it follows from (14)

$$P(t) = \overline{P}(t) + \frac{1}{2}A_p(t)\cos\varphi(t) \qquad (15)$$

with the average value

$$\overline{P}(t) = \cos^2\theta_a \cos^2\theta_m + \sin^2\theta_a \sin^2\theta_m \qquad (16)$$

the depth of oscillations

$$A_p(t) = \sin 2\theta_a \sin 2\theta_m \qquad (17)$$

---

*) Such description have been remarked by N.Cabibbo aloo[18].

S. Mikheyev

and local period

$$T = 2\pi/\dot{\varphi} \qquad (18)$$

θ_m is known function of density and the problem turn to finding of admixture angle θ_a and phase φ.

3.2 Evolution equations for neutrino eigenstates in matter.

The products

$$\Psi_{1m} = \cos\theta_a \, e^{i\varphi_1} \; , \qquad \Psi_{2m} = \sin\theta_a \, e^{i\varphi_2} \qquad (19)$$

in (13) are in fact the wave functions of neutrino eigenstates ν_{1m} and ν_{2m} normalized so that $|\Psi_{1m}|^2 + |\Psi_{2m}|^2 = 1$. The equation for ν_{im} can be found from system (2). Substituting in (2) $\vec{V}_\alpha = \hat{S}_m \vec{\Psi}_m$, $\vec{\Psi}_m = (\Psi_{1m}, \Psi_{2m})$ we have[8-11]

$$i\frac{d}{dt}\begin{pmatrix} \Psi_{1m} \\ \Psi_{2m} \end{pmatrix} = \begin{pmatrix} M_1^d & -i\dot{\theta}_m \\ i\dot{\theta}_m & M_2^d \end{pmatrix}\begin{pmatrix} \Psi_{1m} \\ \Psi_{2m} \end{pmatrix} \qquad (20)$$

where

$$\dot{\theta}_m = \frac{1}{2\rho_{res}} \, \dot{\rho} \, \frac{\tan^2 2\theta}{(1-\rho/\rho_{res})^2 + \tan^2 2\theta} \qquad (21)$$

Note that nondiagonal elements in (20) describing transitions $\nu_{1m} \leftrightarrow \nu_{2m}$ are proportional to $\dot{\theta}_m \propto \dot{\rho}$. From (20) and (19) equations follows for admixture angle θ_a and phase difference $\varphi = \varphi_1 - \varphi_2$:

$$\dot{\theta}_a = \dot{\theta}_m \cos\varphi \qquad (22)$$

$$\dot{\varphi} = \dot{\varphi}^d - 2\dot{\theta}_m \sin\varphi \cdot \cot 2\theta_a \qquad (23)$$

where $\dot{\varphi}^d = M_1^d - M_2^d$.

If the eigenstate of weak interaction is generated, for example ν_e, then according to (4): $\nu(t=0) = \nu_e = \cos\theta_m^0 \, \nu_{1m}(0) + \sin\theta_m^0 \, \nu_{2m}(0)$, where θ_m^0 is mixing angle at beginning, $\theta_m^0 = \theta_m(\rho_0)$. Comparing this equality with (13), we obtain initial conditions for (22), (23):

$$\theta_a(0) = \theta_m^0 \qquad\qquad \varphi(0) = 0 \qquad (24)$$

Consider now the properties of ν-oscillations in matter depending on density distributions.

3.3 Constant density.

$\dot{\theta}_m = 0$. ν_{im} diagonalize system (20). There is no transition $\nu_{1m} \leftrightarrow \nu_{2m}$ ν_{im} evolute independingly: $|\nu_{im}(t)\rangle = |\nu_{im}\rangle \exp(-iM_i^d t)$. In the other words: i) the admixtures of ν_{im} do not change $\dot{\theta}_a = 0$ (see (22)) ii) θ_m is a constant, flavours of ν_{im} conserve. This means that the pictures

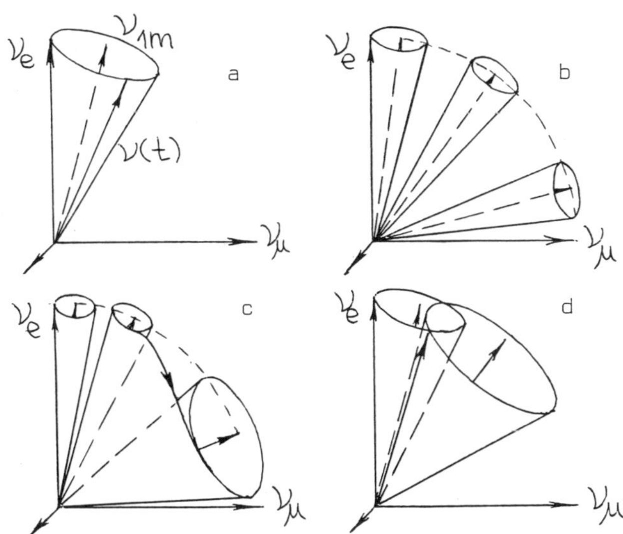

Fig.3 Geometrical representation of ν-oscillations. a)vacuum or matter with constant density, b)adiabatic regime, c)fast density changing (violation of adiabatic condition in resonant layer), d)jump of density.

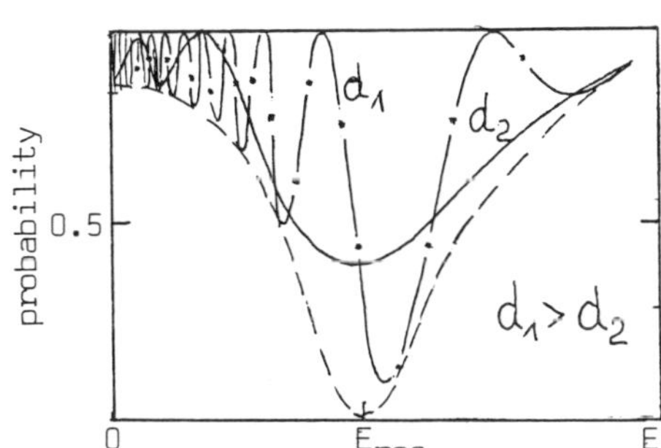

Fig.4 The dependence of probability of transition $\nu_e - \nu_e$ on E in layer with constant density and size d.

of ν-oscillations in matter with constant density and in vacuum are the same. The values of $\overline{P}$, A_p, T are changed only. If ν_e is generated, then $\theta_a=\theta_m=\theta_m^o$, $\dot{\mathcal{Y}} = \dot{\mathcal{Y}}^d = M_1^d-M_2^d$ and from (15-18) Wolfenstein's results follow: $\overline{P}=1- \frac{1}{2}\sin^2 2\theta_m$, $A_p=\sin^2 2\theta_m$, $T=L_m=2\pi/|M_1^d-M_2^d| =2\pi/\sqrt{M^2 +4\overline{M}^2}$. (see Fig.3)

In matter with $\rho=\rho_{res}$ the depth of oscillations is maximal, $A_p=1$. If neutrinos with

continuous energy spectrum are generated then in interval $E = E_{res} \pm \Delta E_{res}$ ($E_{res}=E_{res}(\rho)$) resonant amplification of ν-oscillations occures[4] (amplification of depth of oscillations). At the exit from the layer with the thickness d the probability to find the initial type of neutrinos - P(E) is the oscillating function of energy with frequency depending on d (Fig.4). Moreover the skirtting curve is the resonant one: $\sin^2 2\theta_m(E)$.

3.4 Slow density changing. Adiabatic regime.

If the density changes slowly the adiabatic condition is fulfilled[4,5,8,11]:

$$|\dot{\theta}_m|^2 \ll |M_1^d - M_2^d|^2 = |\dot{\varphi}^d|^2 = 4\pi^2/L_m^2 \qquad (26)$$

Note, $\dot{\theta}_m \sim \dot{\rho}$. Under this condition nondiagonal elements in evolution equations for neutrino eigenstates (20) can be neglected. The deviation of exact solution for probability from adiabatic one is of order $|\dot{\theta}_m/\dot{\varphi}|^2$.

In geometrical representation[7,9,10] (see appendix B) $\dot{\theta}_m$ is the angular velocity of vector $\vec{\nu}_{1m}$ rotation respect to $\vec{\nu}_e$, $\vec{\nu}_\mu$; $\dot{\varphi}^d$ is the angular velocity of vector ν rotation around $\vec{\nu}_{1m}$. Adiabatic condition means that $\vec{\nu}_{1m}$ should rotate much slowly than $\vec{\nu}$ does: $\dot{\theta}_m \ll \dot{\varphi}^d$.

Inequality (26) should be satisfied for any moment t, but in resonance layer $(\rho = \rho_{res})$ it becomes the most crucial. According to (21) $\dot{\theta}_m$ is amplified by resonance factor. On the other hand in resonance $L_m = = L_V/\sin 2\theta$ is maximal. For $\rho = \rho_{res}$ the adiabatic condition is

$$|2\Delta r_{res}|^2 \gg L_m^2/4\pi^2$$

where $\Delta r_{res} = (d\rho/dr)^{-1} \cdot \Delta\rho_{res}$ is spacial width of resonance layer. This inequality is satisfied if at least one oscillation length may get in resonance layer: $2\cdot\Delta r_{res} \geqslant L_m^{res}$ [4,6,7].

Under adiabatic condition transitions $\nu_{1m} \leftrightarrow \nu_{2m}$ are negligible; ν_{im} evolute independingly as in vacuum, but in contrast with vacuum flavours of ν_{im} are not constant. Shortly, the adiabatic approximation is $\dot{\theta}_m/M^d \simeq 0$, but $\theta_m = \theta_m(\rho) \neq$ constant. In adiabatic regime: i) admixtures of ν_{im} in $\nu(t)$ do not change and equal to those in initial moment. From (22) one has: $\dot{\theta}_a \simeq 0$, $\theta_a = \theta_a^0$. If initial condition is $\nu(0) = \nu_e$, then according to (24):

$$\theta_a = \theta_a^0 = \theta_m^0 \qquad (27)$$

That is the admixtures of ν_{im} are determined by mixing angle in matter at moment t=0. Vector $\nu(t)$ follows the eigenstate ν_{1m} (fig.3b). ii) Flavours of ν_{im} are changed with density. iii) At θ_m eqs.(23,24) give $\dot{\varphi} = \dot{\varphi}^d$ and $\varphi = \varphi^d = \int (M_1^d - M_2^d) dt$.

Substituting (27) in (16-18), we get parameters of ν-oscillations adiabatic regime:

$$\bar{P} = \cos^2\theta_m^0 \cos^2\theta_m + \sin^2\theta_m^0 \sin^2\theta_m$$
$$A_p = \sin^2 2\theta_m^0 \sin^2 2\theta_m; \quad T = 2\pi/|M_1^d - M_2^d| \qquad (28)$$

The essential property of this solution is universality[7]. $\overline{P}$ and A_p are functions of one variable - mixing angle in matter θ_m and one parameter - the value of θ_m in initial moment θ_m^0. θ_m, θ_m^0 and consequently $\overline{P}$ and A_p are determined by the densities at the beginning and at given moment. They do not depend on density distribution $\rho(r)$. $\overline{P}$ and A_p do not depend on phase of oscillations also. All dependences of $\overline{P}$ and A_p on E, m^2, $\sin^2 2\theta$, t are in θ_m and θ_m^0.

It is convinient to represent the universality of $\overline{P}$ and A_p in terms of dimensionless variable $n = (\rho - \rho_{res})/\Delta\rho_{res}$, which is related with density of matter directly[6,7]. According to (11) $\sin 2\theta_m = (n^2+1)^{-1/2}$ and from (28) we have[6,7,11]

$$\overline{P} = \frac{1}{2}(1 + n_0 \cdot n/\sqrt{(n_0^2+1)\cdot(n^2+1)}) \tag{29}$$

$$A_p = 1/\sqrt{(n_0^2+1)(n^2+1)} \tag{30}$$

Here $n_0 = n(\rho_0)$ (see fig.5).

Let us stress that the transformations in ν-beam are due to flavour changing of ν-eigenstates. The smaller vacuum mixing and the greater

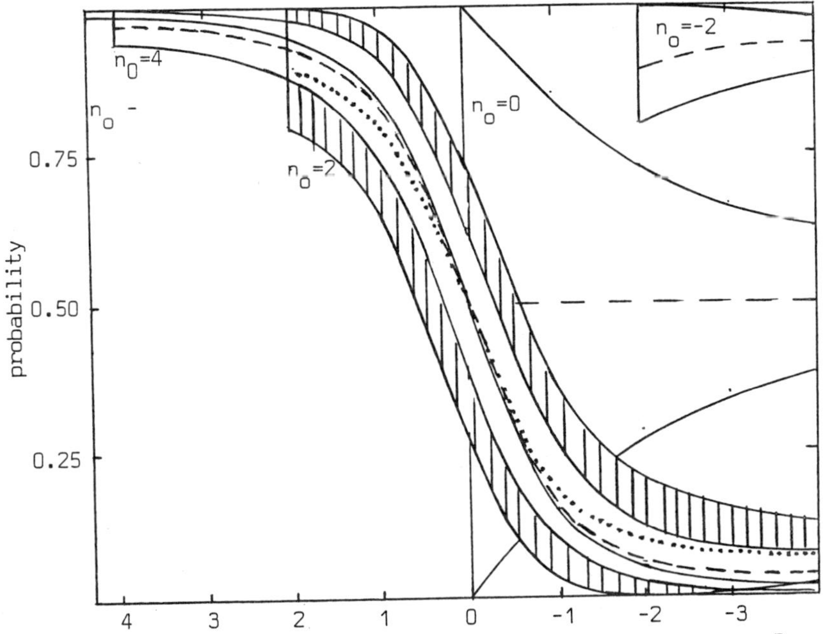

Fig.5 The dependences of average probability $\overline{P}$ and oscillation depth on n for different initial conditions(n_0) in adiabatic regime.

difference between initial and final densities are, the stronger changing of ν_{im} flavours and the more complete transition of one type neutrino into another one take place.

The depth of oscillations is maximal in resonance and it decreases when neutrinos move away from ρ_{res}. When n_o rises A_p diminishes: $A_p \simeq \frac{1}{p\, n_o}$ In the limit $n_o \rightarrow \infty$ A_p go to zero and $\bar{P}$ converges to asymptotic dependence: $P_{as} = (1 + n \cdot (n^2+1)^{-1/2})/2$. The propagation of neutrinos takes form of oscillationless transformation of one type neutrino into another one[6,7]. In this case $\nu(t)$ coincides practically with one of neutrino eigenstates. Note that in matter with varying density the depth of oscillations (17) do not determined uniqually by mixing angle θ_m. In particular for resonant layer, where the mixing is maximal ($\theta_m = 45°$) the depth may be negligible ("oscillationless transformation"). A_p equals to unit if neutrinos are generated at resonant density ($n_o = 0$) and within resonance layer only.

Resonant character of matter effects manifestes itself in the following. $\rho = \rho_{res}$ or $n = 0$ is the special point. For any values n_o: $\bar{P}(n_o, 0) = \frac{1}{2}$ the depth of oscillation is maximal here $A_p(n_o, 0) > A_p(n_o, n \neq 0)$; A_p may reach absolute maximum only in resonance $A_p(0,0) = 1$. It is in resonance layer the most strong changing take place in ν-state: $\Delta \bar{P} = \bar{P}(n_o, 1) - \bar{P}(n_o, -1) = n_o/(2(n_o^2+1))^{1/2}$; when n_o increases from 1 to $n_o \gg 1$, $\bar{P}$ rises from 1/2 to $1/\sqrt{2}$.

The result of -propagation in matter is determined by initial and final values of n (by ρ_o and ρ_f): $\bar{P} = \bar{P}(n_o, n_f)$, $A_p = A_p(n_o, n_f)$. It does not depend on size of the object and density distribution. If $\rho_f = 0$, then $n_f = -1/\tan 2\theta$ and from (29),(30) one has :

$$\bar{P} = \frac{1}{2}(1 - n_o \cos 2\theta/(n_o^2+1)^{1/2}), \quad A_p = \sin 2\theta/(n_o^2+1)^{1/2} \tag{31}$$

For $n_o \rightarrow \infty$, this gives $\bar{P} = \sin^2\theta$[4]. Note, in contrast with vacuum, the smaller mixing angle θ is the more complete transformation of initial type neutrino into another one can be reached.

3.5 Fast density changing.

If the adiabatic condition is wrong the transition $\nu_{1m} \leftrightarrow \nu_{2m}$ becomes essential. Now the admixtures of ν_{im} in a given ν-state do not conserve , $\dot{\theta}_a \neq 0$. Both θ_a and, consequently, A_p, $\bar{P}$ depend on phase of oscillations (see(22)). Period of oscillations $\dot{\varphi}$ distinguishes from

$M_1^d - M_2^d$.

Integrating (22) over t one gets

$$\theta_a = \theta_a^0 + \int_0^t dt' \, \dot{\theta}_m \cos\varphi \qquad (32)$$

For two limit cases solutions of system (22), (23) are rather simple.

i) θ_m varies slowly, so that the second term in (23) is small. Changing of is due to $\mathcal{V}$-rotation around $\mathcal{V}_{1m}$ mainly, $\dot{\varphi} - \dot{\varphi}^d$, $\varphi \simeq \int_0^t dt' M^d$. Introducing in (32) integration over φ, one has

$$\theta_a = \theta_a^0 + \int_0^\varphi d\varphi' \, \dot{\theta}_m \cos\varphi' / M^d \qquad (33)$$

This describes, in fact, first correction to the adiabatic solution[8].

If $\dot{\theta}_m / M^d = \text{const}$[13] it follows from (33): $\theta_a = \theta_a^0 + \dot{\theta}_m \sin\varphi / M^d$.

ii) θ_m varies rapidly, so that the first term in (23) can be neglected. This is quite possible situation for resonant layer where $\dot{\theta}_m$ is amplified (see (21)). Phase changes mainly due to rotation of $\vec{\mathcal{V}}_{im}$ in $\mathcal{V}_e$, $\mathcal{V}_\mu$-space. The neglection of $\dot{\varphi}^d$ in (23) corresponds to the jump of density and, consequently, to the jump of θ_m ($\theta_m \to \theta_m + \theta_m$) in some moment t_c. Solution (22),(23) coincides with that obtained graphically[7]. Vector $\vec{\mathcal{V}}$ is at the rest, when $\mathcal{V}_{1m}$ changes its position. After jump the rotation angle θ_a and phase φ are determined by $\vec{\mathcal{V}}(t_c)$ and new position of $\vec{\mathcal{V}}_{1m}$ (see fig.3d). Solutions are very simple if in the moment of jump φ_c is zero or $\pi/2$. From (22,23,32) one has: $\theta_a = \theta_a^0 + \theta_m$ and $\Delta\varphi = 0$[7].

4. DIVERGENCE OF NEUTRINOS WAVE PACKETS.

The considered theory corresponds to the complete overlap of neutrinos wave packets. Really $\mathcal{V}_{im}$ have different group velocities and on the spacial scales, where strong matter effects develope the divergence of wave packets becomes essential[6,7,9,10].

4.1 Group velocities of neutrino eigenstates.

In matter with a constant density the group velocities v_i are

$$v_i = dE_i/dk - \frac{d}{dk}(k + M_i^d(k)) = 1 + dM_i^d/dk$$

Substituting here M_i^d from (7) one has

$$\frac{1}{2}(v_{1m} + v_{2m}) = 1 - \overline{m}^2 / 2k^2, \qquad \overline{m}^2 = (m_1^2 + m_2^2)/2$$

(vacuum result) and

$$v_m = v_{1m} - v_{2m} = v_v \cos2\theta \, (-n + \tan2\theta)/(n^2 + 1)^{1/2} \qquad (34)$$

where $v_v = -\Delta m^2 / 2k^2$ is difference of velocities in vacuum (fig.6).

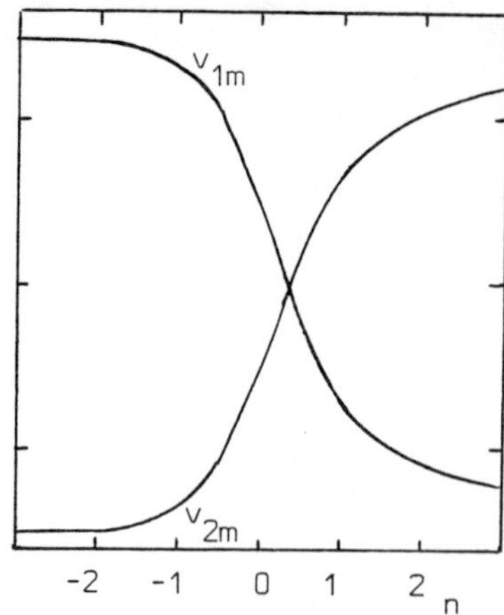

Fig.6 The dependencies of group velosities of ν_{im} on n.

When ρ approaches ρ_{res}, $|\Delta\nu_m|$ diminishes. In resonance: $\Delta\nu_m = \Delta\nu_v \sin^2 2\theta$. At $\rho' = \rho_{res}/\cos^2 2\theta$ $\Delta\nu_m$ equals to zero and changes the sign. ρ' turns out to be the resonant density for $\nu_1 - \nu_2$ oscillations. For small θ and $\rho \gg \rho_{res}$ one has $\Delta\nu_m \simeq -\Delta\nu_v$. Note that in resonance channel for any values of ρ matter suppresses divergence of wave packets: $|\Delta\nu_m| < |\Delta\nu_v|$.

4.1 Effects of wave packets divergence.

In matter with a constant or slowly changing density (adiabatic regime) the picture of divergence is rather simple. ν_{1m} and ν_{2m} evolute independingly and D – the divergence of their packets on distance L is $D(L) = \int_0^L \Delta\nu_m(\rho(r))dr$. The divergence effect can be neglected if $|D| \ll \tau$, where τ is the length of wave packet.

Note that the divergences in matter with $\rho > \rho'$ and $\rho < \rho'$ have opposite signs. If neutrinos cross the resonant layer, then divergence effects before resonance and after it compensate for each other. The restoration of packets overlap is possible.

Obviously the probability to find ν_e summed over wave packets ($P_s = P_1 + P_2$) in the case of complete or partial divergence is equal to average probability $\overline{P}$, when the packets overlap: $P_s = \overline{P}$[9,10]. The depth of oscillations is proportional to the size of overlapping part of packets. Taking the packets in form $\exp(-t/\tau)$, we obtain $A_p = A_p^0 \exp(-|D|/\tau)$, where A_p^0 is the depth without divergence.

5. APPLICATIONS OF MATTER EFFECTS:

5.1 General conditions[7].

In order to strong observable matter effects take place the follwing conditions should be satisfied.

i) The thickness of matter should be suffesiently large: $d \gtrsim L_0\rho$. For $\nu_e \leftrightarrow \nu_\mu$ oscillations, one has $L_0^{\nu}\rho \approx m_N G_F \simeq 3.5 \cdot 10^9 g/cm^2$, where m_N is nucleon mass, G_F is Fermi constant.

ii) Matter should be asymmetric respect to oscillating ν-components.

iii) Fluxes of oscillating components should be different. Otherwise the initial ν_e´s, for example, transform into ν_μ´s and initial ν_μ´s transform into ν_e´s equally.

iv) Strong matter effects ($\bar{P}_f < P_V^{max} = 1/2$ for two neutrinos mixing) take place, when neutrinos cross resonant layer. For given object (given distribution $\rho(r)$, ρ_{max}, ρ_{min}) resonance condition is fulfilled in definite intervals $E/\Delta m^2$, $\sin^2 2\theta$. Minimal value $(E/\Delta m^2)_{min}$ is determined by ρ_{max}. Upper edge of $E/\Delta m^2$ - interval for strong matter effects is determined by ρ_{min}, or by the size of the object, or by adiabatic condition.

In contrast with vacuum case, in matter practically complete transformation of one type neutrino into another one may take place for large interval of neutrino energies.

v) Resonance condition fixed sign of the products $\Delta f(0) \cdot \Delta m^2$. Because of $f_\nu = -f_{\bar{\nu}}$ in matter with given composition resonance effects may take place for neutrinos or antineutrinos only.

The listed conditions may be fulfilled in the Sun, in cores and envelopes of collapsing stars, in the Earth, in early Universe. Summarize main consequencies.

<u>5.2 The Sun.</u>

i) Strong matter effects take place for $\Delta m^2 = 10^{-4} \div 10^{-8} eV^2$ and $\sin^2 2\theta \gtrsim 10^{-3.5}$ [4,8,10-14] (see fig.7). Large part of this region corresponds to ν-oscillations in adiabatic regime. On its lower edge adiabatic condition is violated.

ii) The region on Δm^2 vs. $\sin^2 2\theta$ - plot exists in which for standard solar model (2 ÷ 4)-fold suppresion of ν-caipture rate in Cl-Ar - experiment is predicted [4,5,13,14]. The explanation of Davis result can be achived for mixing angle $\sin^2 2\theta \gtrsim 10^{-3}$.

iii) In these Δm^2, $\sin^2 2\theta$ - regions predictions for Ga-Ge - experiment varies from standard one (~ 123 SNU) to 20-fold suppressed in comparision with the standard [13,14]. Because of known astrophysical fac-

Fig.7 The m^2 versus $\sin^2 2\theta$ regions of strong matter effects ($\overline{P} \leqslant 1/2$) for the Sun, cores and envelopes of collapsing stars, for the Earth. Shaded region is experimental restrictions.

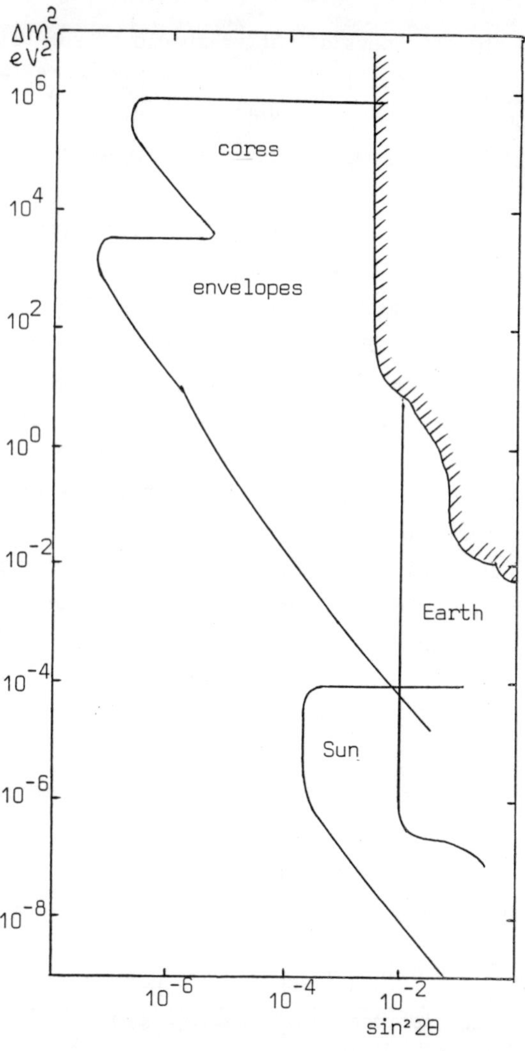

tors may diminish Q_{Ga-Ge} by factor two only, the observation in future Ga-Ge experiments Q (60-70) SNU will be very important indication on the existance of strong matter effects.

iv) Distorsions of energy spectrum of ^{8}B- and pp-neutrinos are predicted[13,14]. This enables to distinguish any astrophysical reasons of ν-deficiency from suppresion factor due to ν-oscillations in matter.

v) Data comparison from Cl-Ar and Ga-Ge experiments as well as direct measurement of boron (and may be pp) neutrinos spectrum enable one to establish if strong matter effect takes place in the Sun and if so to find Δm^2 and $\sin^2 2\theta$ almost unambiguously[13].

vi) The Δm^2 vs. $\sin^2 2\theta$ region of strong matter effect in the Sun overlap the region of "see-saw" predictions with large mass of order of Grand unification scale[5,16,17].

5.3 Cores and envelopes of collapsing stars[6,7].

i) The region of strong matter effect is $\sin^2 2\theta \geqslant 10^{-7}$, $\Delta m^2 = (10^{-5} - 10^6)$

for $\nu_e - \nu_\mu$ and $(10^{-5}-10^{7})eV^2$ for $\nu_e - \nu_s$ (ν_s - sterile state).
The most part of this region corresponds to oscillationless transformation in adiabatic regime. Divergence of wave packets is essential.
The suppression of initial neutrino flux as strong as $\sin^2 2\theta = 10^{-6}$ can be obtained.

ii) Observable effects depend on mixing channel for which resonance condition is fulfilled.

a) $\nu_e - \nu_x$: disappearance of ν_e peak from neutronization is expected.

b) $\nu_e - \nu_\mu (\nu_\tau)$: ν_e and $\nu_\mu (\nu_\tau)$ exchange their energy spectra. Effect can be observable if $F_{\nu_e}(E,t) \neq F_{\nu_\mu}(E,t)$.

c) $\bar{\nu}_e - \bar{\nu}_\mu (\bar{\nu}_\tau)$: $\bar{\nu}_e$ and $\bar{\nu}_\mu (\bar{\nu}_\tau)$ exchange their spectra.

d) $\nu_e - \nu_s$ or $\bar{\nu}_e - \bar{\nu}_s$: practically complete disappearance of ν_e or $\bar{\nu}_e$-fluxes may occure. It is important to search for ν-bursts by ν_e and $\bar{\nu}_e$-detectors simultaneously.

5.4 The Earth.

i) The region of strong matter effect is $E/\Delta m^2 \sim (10^{3}-10^{4})Gev/ev^2$ and $\sin^2 2\theta \gtrsim 10^{-2}$. Zenith angle should be sufficiantly small: $\cos 0.8$. For larger Ψ the thickness of matter becomes small.

ii) Situation corresponds roughly to several layers with different constant densities. Strong effect is due to that about one half of resonance oscillations length can be placed in such layers.

iii) The dependence of suppression factor on $E/\Delta m^2$ and reflects the density distribution within the Earth[7,15]. In this context the viewing of the Earth by accelerator ν-beam have been considered[15].

iv) Oscillations in matter of the Earth may results in distorsion of atmospheric neutrinos spectrum[7]. The scanning of this spectrum from $E_\nu \sim 0.5GeV$ to $E_\nu \sim 10^{5}Gev$ may give in principle the information on Δm^2 in regions $10^{-1}-10^{-4}eV^2$.

v) Effect in the Earth may have character of day-night modulation of solar neutrinos flux. This modulation results in regeneration of initial type of neutrinos ($\nu_e \to \nu_x$ inside the Sun and $\nu_x \to \nu_e$ inside the Earth). For radiochemical experiments the effect is averaged and small $(< 10\%)^{[7,13]}$. It can be seen in direct detection of ν_e-scattering[7].

vi) Neutrino fluxes from collapsing stars measured by installations, placed in different parts of the Earth may be different. This enables in principle to fix the direction on collapsing star[7].

ACKNOWLEDGE

We are grateful to V.S.Berezinsky, V.A.Kuzmin, V.A.Matveyev, A.Messiah, V.A.Rubakov, M.E.Shaposchnikov, A.N.Tavhelidze, L.Wolfenstein and G.T.Zatsepin for many fruitful discussions and remarks.

Appendix A. Relations and definitions.

The elements of matrix $\hat{M}$ obey the following relations:

$$M_e + M_\mu = (m_1^2 + m_2^2)/2k + \Sigma_e + \Sigma_\mu$$
$$M = M_e - M_\mu = \Delta m^2 \cdot \cos 2\theta/2k + \Sigma_e - \Sigma_\mu \qquad (A.1)$$
$$2\overline{M} = -\sin 2\theta \cdot \Delta m^2/2k$$

where $\Delta m^2 = m_1^2 - m_2^2$, k is the momentum of neutrino,

$$\Sigma_\alpha = \sum_i f_\alpha^i(0) \cdot n_i/k \qquad (A.2)$$

Here $f_\alpha^i(0)$ is the amplitude of forward ν_α scattering on i-component of matter, n_i is the concetration of i-component. The eigenlength in matter is equal to

$$L_o = 2\pi/(\Sigma_e - \Sigma_\mu) = 2\pi (\sum_i \Delta f^i(0) \, n_i/k)^{-1} \qquad (A.3)$$

In terms of L_v and L_o M can be rewritten as $M = \frac{2\pi}{L_o}(\cos 2\theta - L_v/L_o)$.

Appendix B. Graphical representation of ν-oscillations[7,9,10]

Any neutrino state may be written as

$$\nu(t) = \cos\theta_a | \nu_{1m}\rangle + \sin\theta_a | \nu_{2m}\rangle \exp(-i\varphi(t)) \qquad (B.1)$$

so that the coefficient in front of $| \nu_{1m}\rangle$ is real. Introduce basis $\{\nu_m\} = \{\nu_{1m}, \nu_{2m}^{Re}, \nu_{2m}^{Im}\}$, where $\nu_{1m}, \nu_{2m}^{Re}, \nu_{2m}^{Im}$ corresponds to real $(\nu_{1m}, \nu_{2m}^{Re})$ and imagine parts of ν_{1m}, ν_{2m} - wave functions. Evolution of ν-state (B.1) is equivalent to the rotation of unit vector $\vec{\nu}$ around $\vec{\nu}_{1m}$, the angle of rotation (angle between $\vec{\nu}$ and $\vec{\nu}_{1m}$) is θ_a, the angular velocity is $\dot\varphi(t)$. Projection of $\vec{\nu}(t)$ on some axis ν_x gives the amplitude of probability to find ν_x in ν-state. ν_e and ν_μ are described by vectors $\vec{\nu}_e = \{\cos\theta_m, \sin\theta_m, i\sin\theta_m\}$, $\vec{\nu}_\mu = \{-\sin\theta_m, \cos\theta_m, i\cos\theta_m\}$ in basis $\{\nu_m\}$, where θ_m is mixing angle in matter.

REFERENCES

1. Pontecorvo B.M., ZhETF, 33, 549 (1958)

   Bilenky S.M. and Pontecorvo B.M., Phys. Rep., C41, 225 (1978)
2. Wolfenstein L. Phys. Rev., D17, 2369 (1978); ibid D20, 2634 (1979)

3. Barger V., Whisnant K., Pakvasa S. and Phillips R.J.N. Phys. Rev., D22, 2718 (1980)

4. Mikheyev S.P. and Smirnov A.Yu. Talk given at 10th International Workshop on Weak Interactions, Savonlinna, Finland June 16-22, 1985; Yadernaja Fizika (Sov. J.), 42, 1441 (1985); Nuovo Cimento, C9, 17 (1986)

5. Bethe H., Phys. Rev. Lett., 56, 1305 (1986)

6. Mikheyev S.P. and Smirnov A.Yu. ZhETF (Sov. J.), 91, 7 (1986)

7. Mikheyev S.P. and Smirnov A.Yu., Talk given at VIth Moriond Workshop on "Massive Neutrinos in Particle Physics and Astrophysics", Tignes, France, Jan.25 - Feb.1 (1986), to be published

8. Messiah A., Talk given at VIth Moriond Workshop (see ref.7)

9. Mikheyev S.P. and Smirnov A.Yu. Talk given at International Seminar "Quarks-86", Tbilisi, USSR, April 15-17, 1986, to be published

10. Mikheyev S.P. and Smirnov A.Yu. Talk given at 7th WOGU/ICOBAN'86, April 16-18, 1986, Toyama, Japan, to be published

11. Barger V., Whisnant K. and Phillips R.J.N. Preprint Univ. of Wisconsin, MAD/PH/280, April, 1986

12. Haxton W.C. Preprint Univ. of Washington, Seattle, May, 1986

13. Hampel W., Preprint Max-Planck-Inst., Heidelberg (Jan. 1986); Bouchoz J., Cribier M., Rich J., Spiro M., Vignaund D., Hampel W. Talk given at VIth Moriond Workshop (see ref.7); Preprint Saclay, DPhPE 86-10, May, 1986, to be published in Z. Phys. C24; Bouchez J. Talk given at this conference.

14. Rosen S.P. and Gelb J.M. Talk given at VIth Moriond Workshop (see ref.7); Rosen S.P. Talk given at this conference

15. Chechin V.A., Ermilova V.K. and Tsarev V.A., Preprint 45, Lebedev Physical Institute, 1986

16. Langacker P., private communication

17. Wolfenstein L., Talk given at this conference

18. Cabibbo N. Summary talk at 10th International Workshop on Weak Interactions, Savonlinna, Finland, June 16-22, 1986

MATTER EFFECTS FOR SOLAR NEUTRINO OSCILLATIONS

J. BOUCHEZ

DPhPE, CEN-Saclay.

Abstract

A Monte Carlo analysis of matter oscillations for solar neutrinos is presented. Consequences for future experiments are emphasized. One shows in particular how the Gallium experiment could lead to a determination of Δm^2, $\sin^2 2\theta$.

I - Introduction

For a long time, the puzzling result of the Davis experiment [1] finding only 1/3 of the expected flux for solar neutrinos, if taken at face value, had only two possible explanations :

- either the standard solar model is wrong, at least for ^{8}B neutrinos to which this experiment is sensitive,

- or there are oscillations between three kinds of neutrinos with maximal mixing, an unsatisfactory ad-hoc condition.

Recently, a more exciting alternative has appeared : analyzing the work of Wolfenstein [2] who showed that oscillation patterns for neutrinos are different in vacuum and in matter, Mikheyev and Smirnov [3] showed thar this difference could in some cases lead to a dramatic reduction of the ν_e flux observed on earth, even for small mixing angles in vacuum. This is known as the MSW effect.

We have investigated this effect in detail, using Monte Carlo simulations. The main result is that the future Gallium experiment will help to solve the solar neutrino puzzle and maybe allow to determine the mixing parameters Δm^2 and $\sin^2 2\theta$.

We shall first recall the formalism of the matter oscillations using a geometrical approach and then present the main results of our simulations. More details can be found elsewhere [4].

II - Formalism

A - A well-known case

Let's first consider the case of a spin 1/2 particle, with its spin oriented along $\vec{s}$ and lying in a magnetic field $\vec{H}$. Energy eigenstates have their spins aligned along $\vec{H}$ with eigenvalues $m \pm \mu H$.

The evolution equation for a given state reads

$$\frac{d\vec{s}}{dt} = 2\mu \ \vec{s} \wedge \vec{H} \ . \tag{1}$$

If the magnetic field is constant, this equation simply states that $\vec{s}$ rotates around $\vec{H}$ with the Larmor frequency.

If H varies in modulus and direction with time, the equation (1) is no longer solvable. Let's consider the case where $\vec{H}$ moves in the x, z plane. Equation (1) can be rewritten in time dependent space coordinates X, Y, Z chosen so that $\vec{H}$ stays along Z at all times and $Y \equiv y$, as :

$$\frac{d}{dt} s_X = -\lambda \ s_Y - \frac{d\psi}{dt} \ s_Z$$

$$\frac{d}{dt} s_Y = \lambda \ s_X \tag{2}$$

$$\frac{d}{dt} s_Z = - \frac{d\psi}{dt} \ s_X$$

$\lambda = 2\mu H$ is the difference of energy eigenvalues,

J. Bouchez

ψ is the angle between $\vec{Oz}$ (fixed axis) and $\vec{H}$.

Equations (2) show that, if $d\psi/dt$ is sufficiently small compared to λ, the spin $\vec{s}$ still rotates with time dependent frequency around $\vec{H}$: this corresponds to the so called adiabatic approximation, where spins follow the magnetic field in its movement; this is how one rotates spins in polarized targets using a slowly rotating magnetic field.

If on the contrary H rotates very quickly ($d\psi/dt \gg \lambda$) equations (2) just say that $\vec{s}$ rotates around Y with angular speed $- d\psi/dt$, which simply means that $\vec{s}$ doesn't move in xyz frame. In this sudden change approximation, the spin of the particle is insensitive to very rapid changes of the magnetic field : this is used to keep the polarization of beams when passing depolarizing accelerator resonances very quickly.

In intermediate cases, one has to resort to numerical computations. However, if H is simply rotating with constant speed ω, equations (2) are analytically solvable : $\vec{s}$ rotates around the axis of coordinates

$$X = 0 , \quad Y = \omega , \quad Z = \lambda$$

which means in xyz frame that the spin precession is accompanied by a nutation whose amplitude is proportional to ω, as shown graphically on figure 1.

One should note that in this geometric representation, the probability of finding $\vec{s}$ in a state $\vec{s}'$ is given by $2 \, \vec{s}.\vec{s}' + 1/2$.

B - The 2-neutrino case in matter

There is a complete analogy with the spin 1/2 particle.

Any superposition of $|\nu_e\rangle$ and $|\nu_\mu\rangle$ can be written (up to a global phase)

$$|\nu\rangle = \cos\theta \, |\nu_e\rangle - \sin\theta \, e^{i\varphi} \, |\nu_\mu\rangle$$

and mapped onto a sphere at a point M of coordinates

$$x = 1/2 \, \sin2\theta \, \cos\varphi$$

$$y = 1/2 \; \sin 2\theta \; \sin\varphi$$

$$z = 1/2 \; \cos 2\theta$$

$\vec{OM}$ is the equivalent of $\vec{s}$ and the probability of finding a ν_e in that state is $z + 1/2$.

In a medium of given electronic density, propagation eigenstates ν_{1m} and ν_{2m} with a difference λ in energy eigenvalues define on the sphere a direction $\vec{\lambda}$ which is the equivalent of $\vec{H}$, with $2\mu H = \lambda$.

$\vec{\lambda}$ components are :

$$\lambda_x = - \frac{\Delta m^2}{2E_\nu} \; \sin 2\theta_v$$

$$\lambda_y = 0 \tag{3}$$

$$\lambda_z = - \frac{\Delta m^2}{2E_\nu} \; \cos 2\theta_v + G \sqrt{2} \, N_e$$

$$\Delta m^2 = m_2^2 - m_1^2 \; .$$

m_1, m_2 are the masses of the vacuum eigenstates closest to ν_e and ν_μ respectively. θ_v is the usual mixing angle in vacuum. N_e is the electronic density and G is the Fermi constant.

When N_e goes from infinity to zero, $\vec{\lambda}$ goes from $\vec{\nu}_e$ to either $\vec{\nu}_1$ (the closest to $\vec{\nu}_e$) if $m_1 > m_2$ or to $\vec{\nu}_2$ (the closest to $\vec{\nu}_\mu$) if $m_1 < m_2$.

The MSW effect is simply what happens in the last case when the above-mentionned adiabatic condition holds : a neutrino ν_e born in a medium of very high electronic density, so that $\vec{\lambda}$ is nearly along $\vec{\nu}_e$, follows $\vec{\lambda}$ with decreasing N_e and ends up along $\vec{\nu}_2$ at the outcome of the sun. It propagates to earth staying a $\vec{\nu}_2$ (since this is a propagation eigenstate in vacuum) and gives rise to a measured ν_e flux $\sin^2\theta_v$ times the original flux. The MSW effect thus corresponds to an oscillationless driving·of ν_e's towards ν_2's through the sun ; for this to happen, three conditions are necessary :

1) $m_1 < m_2$ (the electronic neutrino mean mass is higher than the muonic one)

2) the adiabatic approximation $d\psi/dt \ll \lambda$ must hold.

Noting from (3) that $d\psi/dt$ is maximal near $\lambda_z = 0$ and taking $N_e \sim N_c$ $e^{-L/Lo}$ with $N_c \sim 5.4\ 10^{25}$ cm^{-3} being the electron density at the center of the sun and $L_o = 75000$ km, we get

$$\Delta m^2 (eV^2)\ x\ \frac{\sin^2 2\theta_v}{\cos 2\theta_v}\ \gg\ 6\ 10^{-9}\ E_\nu\ (MeV) \tag{4}$$

3) the electronic density at the center of the sun (where neutrinos are born) must be big enough so that λ can have its full excursion from $\sim \nu_e$ to ν_2. This implies

$$\Delta m^2 (eV^2)\ \ll\ 1.5\ 10^{-5}\ E_\nu\ (MeV) \tag{5}$$

Relations (4) and (5) define a triangle in the Log-Log plot $\sin^2 2\theta_v$, Δm^2 for a given neutrino energy. This triangle moves up when the neutrino energy increases (see figure 2).

Inside this triangle, the expected ν_e flux will be $\sin^2 \theta_v$ times the original flux, whereas outside, the original flux will be conserved for very small Δm^2 or multiplied by $1 - 1/2 \sin^2 2\theta_v$ for high Δm^2. Numerical simulations are necessary to describe quantitatively the transition regions.

III – Monte Carlo simulations

A – Ingredients and method

In the following, θ will stand for θ_v , the mixing angle in vacuum.

For each value of Δm^2 and $\sin^2 2\theta$, a few hundred neutrinos have been generated according to the detectable momentum spectrum (the low cut-off being experiment dependent) and were created in the sun at locations whose distribution follows the standard solar model [5].Inside the sun, evolution equations were integrated numerically by propagating each neutrino by steps of length δx. On each step, we took into account the first order correction due to the density variation over the step. The convergence of the method was checked by decreasing the step size. Between sun and earth, the analytical propagation was used. Earth effects were also taken into account. Actually, matter effects in the earth may lead to a partial regeneration of ν_e when the MSW effect is at work in the sun, and so have to be considered.

We also tried a wave packet treatment for each neutrino rather than plane wave approximations to check coherence problems and found no difference in the results. This can be expected at least when the MSW effect is at work since in that case, the neutrino has an oscillationless propagation.

B – Results

We first show in figure 3 the propagation of a 500 keV neutrino inside the sun. As an illustration, the figure is drawn for $\sin^2 2\theta = 0.1$ and $\Delta m^2 = 10^{-7}$ eV^2. A perfect drivig is seen up to the "resonant" density ($\lambda_z = 0$ in equation 3) for which the adiabatic condition (4) doesn't fully hold. So the neutrino cannot follow exactly the eigenstate, and this explains the little oscillations on the second part. However, averaging over one oscillation length, one finds the MSW predicted result $P(\nu_e) = \sin^2 \theta = 0.025$.

Figure 4 shows in the same conditions the probability of finding a ν_e on earth when the neutrino momentum is varied. This corresponds to a vertical cut of the triangle shown in figure 2 (E_e being varied rather than Δm^2). The transition due to the density condition (around p=10 keV/c) is sharper than the end of the adiabatic regime (between 1 and 10 MeV/c). Between 10 keV/c and 1 MeV/c, the MSW effect fully works.

Figure 5 shows the same result when earth effects such as experienced at midnight on Winter Solstice (for a detector at Gran Sasso or at a similar latitude) are taken into account. A dramatic regeneration of ν_e's inside the MSW band can be seen. We would like to emphasize here that such day/night effects would be a clear signature of matter oscillations, independent of any solar model.

We then come to the simulation of experiments such as the Chlorine and Gallium experiments which measure an integrated flux above a given threshold.

Figure 6 shows the predicted measured flux for the Gallium experiment (threshold = 233 keV) if $\sin^2 2\theta = 0.4$. One sees that the MSW effect works significantly on a wide band of Δm^2.($P(\nu_e < 0.5$ for Δm^2 between 3 10^{-9} and 8 10^{-6} eV^2) The three curves in this area show the mean earth effect integrated over time, inbetween daylight measurements (lower curve) and midnight measurements (upper curve). Even though less dramatic than in figure 5, since one has here integrated over the momentum spectrum, one sees that these effects can stay quite sizeable and have to be taken into account in the analysis of the future results. Outside the Δm^2 band where the MSW suppression is effective, earth effects on the contrary are insignificant.

Figure 7 shows the sensitivity domain of the future Gallium experiment. The shaded area corresponds to the zone where the measured flux is smaller than 0.7 times the expected one (assuming no oscillation). One sees that if the MSW effect operates, one reaches sensitivities in $\sin^2 2\theta$ down to 6 10^{-4}, as opposed to only 0.6 if no MSW suppression occurs. The analog domain for the Chlorine experiment (threshold = 814 keV) is shown for comparison. Due to a higher threshold, the sensitivity area is shifted upwards in Δm^2.

Figure 8 is certainly the most interesting. It shows the domain in Δm^2, $\sin^2 2\theta$ selected if one interprets the experimental result of Davis (2 +/- 0.3 SNU instead of 5.8 predicted [1]) as a manifestation of the

MSW effect. This domain is a triangular band. The vertical band corresponds to $\sin^2\theta = 1/3$ and lies in a zone where the MSW suppression is fully effective over the whole momentum spectrum. The horizontal band corresponds to the transition zone given by the relation (5), when the central solar electron density is of the order of the resonant density ($\lambda_z =0.$ in equation 3) for the mean momentum. The oblique band corresponds to the transition region where the adiabatic condition (5) is no longer perfectly fulfilled, at least for part of the spectrum. Inside this triangular band are shown the corresponding predictions for the Gallium experiment, if we have matter oscillations with $m_1 < m_2$. In the absence of oscillations, one expects 123 SNU [6]. Below $\Delta m^2 = 10^{-6}$ eV^2, the MSW suppression is fully effective for all momenta above 233 keV/c, so that the prediction is 123 x $\sin^2\theta$. For higher Δm^2, the MSW suppression becomes only partly effective, which explains the higher predicted values. The predicted fluxes range between 20 and 110 SNU depending upon $\sin^2 2\theta$ and Δm^2 values. On the contrary, if no MSW effect takes place, the measured flux will necessarily be above 60 SNU (in the 2 neutrino case), value obtained for maximal mixing. So a measured value below 60 SNU (or 40 SNU if one goes to the 3-neutrino case) would unambiguously prove the presence of neutrino oscillations in matter giving rise to the MSW effect and would, together with the Chlorine result, determine Δm^2 and $\sin^2 2\theta$ quasi unambiguously. Between 60 and 110 SNU, one could either interpret the result within the MSW scheme or suppose that something is wrong in the solar model for ^{8}B neutrinos and that oscillations with $m_1 > m_2$ occur for pp and ^{7}Be neutrinos (which dominate in Gallium). One way out would then be to measure the momentum spectrum of the neutrinos. This could be done in an Indium or heavy water experiment. Shape distorsion of the spectrum would prove the presence of the MSW effect and give more indications on Δm^2 and $\sin^2 2\theta$ values. This is illustrated in figure 9 . As already mentionned, flux variations between day and night in any experiment detecting in real time the solar neutrinos would prove the presence of neutrino oscillations.

In conclusion, new generation experiments will certainly greatly help in solving the longstanding solar neutrino puzzle.

Acknowledgments.

This work was done in collaboration with M.Cribier,J.Rich,M.Spiro, D.Vignaud from Saclay and W.Hampel from Heidelberg.

REFERENCES

[1]   J.K. Rowley, B.T. Cleveland, R. Davis,
         "Solar neutrinos and neutrino astronomy", Homestake (1984),
         AIP Conf. Proc. n°126, p.1.

      J.N. Bahcall, B.T. Cleveland, R. Davis, J.K. Rowley,
         Astrophysical Journal Letters $\underline{292}$, L79 (1985).

[2]   L. Wolfenstein, Phys. Rev. $\underline{D17}$, 2369 (1978).

[3]   S.P. Mikheyev and A.Yu. Smirnov,
         Yad. Fiz. $\underline{42}$, 1441 (1985)
         Sov. J. Nucl. Phys. $\underline{42}$, 913 (1985)
         Nuovo Cimento $\underline{9C}$, 17 (1986).

[4]   J. Bouchez, M. Cribier, J. Rich, M. Spiro, D. Vignaud, W. Hampel,
         Z. Phys. C - Particles and Fields, in press.

[5]   J.N. Bahcall et.al.,
         Rev. Mod. Phys. $\underline{54}$, 567 (1982).

[6]   W. Hampel, to be published
      D. Krofcheck et al., Phys. rev. Lett. $\underline{55}$, 1051 (1985).

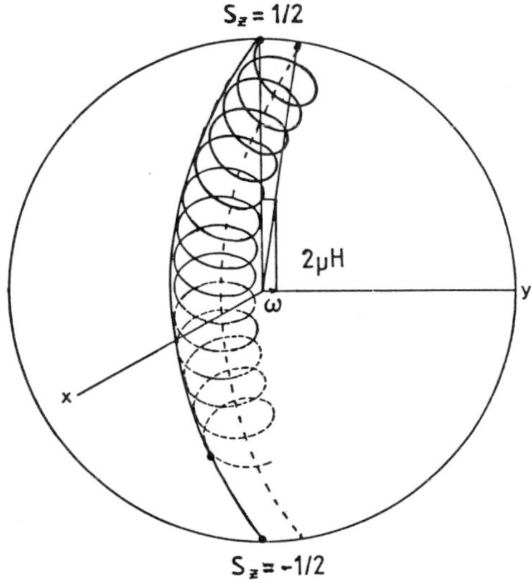

Fig. 1- Spin driving by magnetic field rotating around y with constant angular speed ω. The spin and the magnetic field are originally aligned along z.

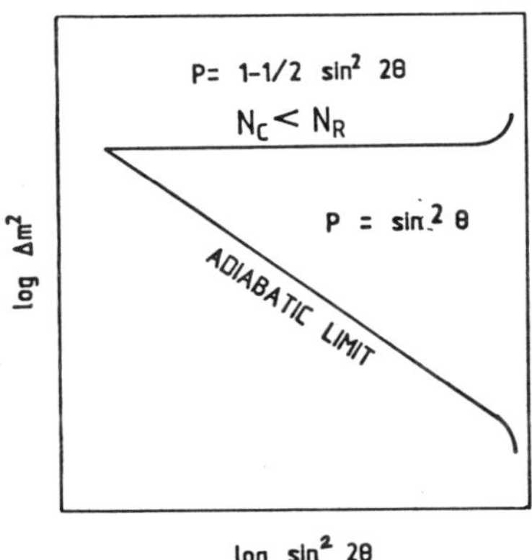

Fig.2 - Schematic display of the area where the MSW effect occurs for a neutrino of given energy, provided that $m_1 < m_2$.

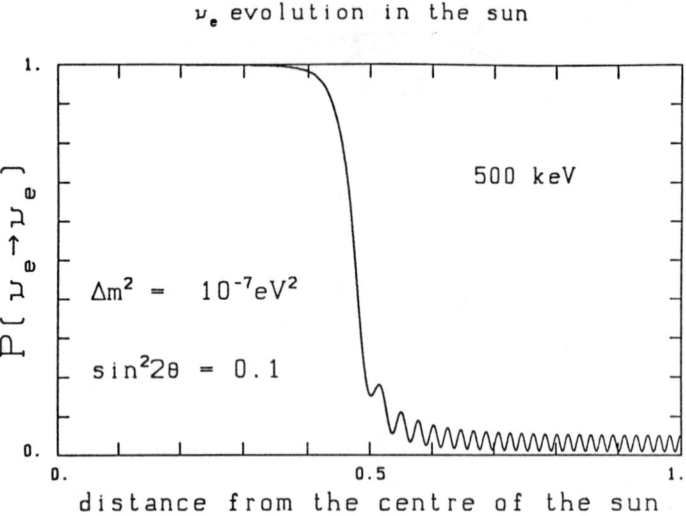

Fig. 3

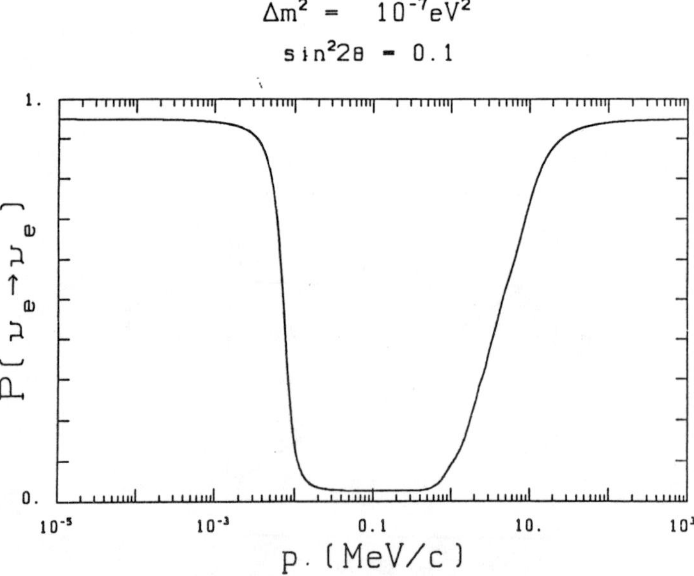

Fig. 4 – $P(\nu_e)$ at the outcome of the sun as a function of the neutrino momentum.

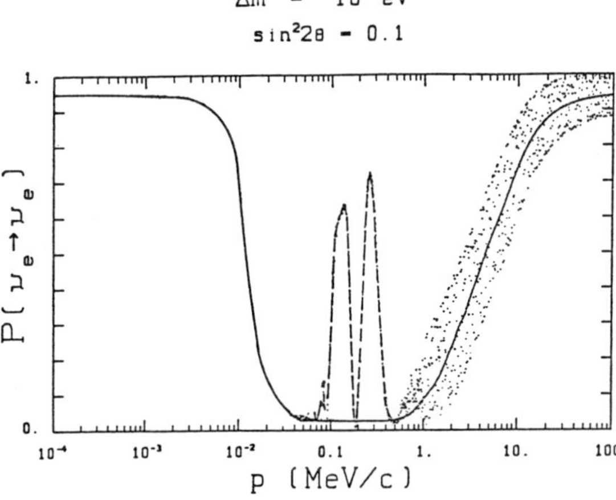

Fig.5 — Same as 4 : the dashed curve shows the earth effect on
winter solstice for Gran Sasso latitude.

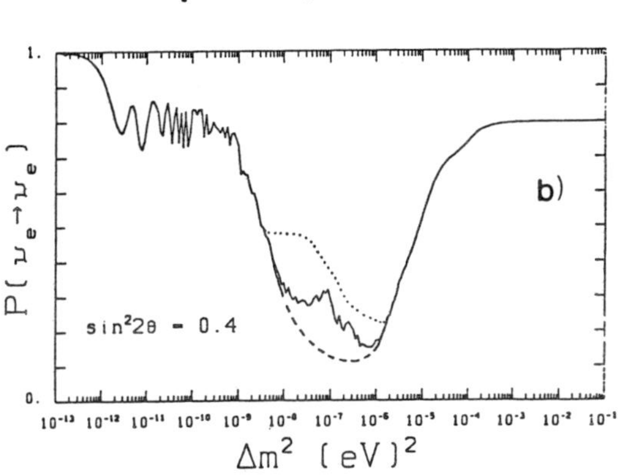

Fig.6 — Full curve : mean earth effects in the Gallium experiment
Dashed curve : daylight flux
Dotted curve : midnight flux.

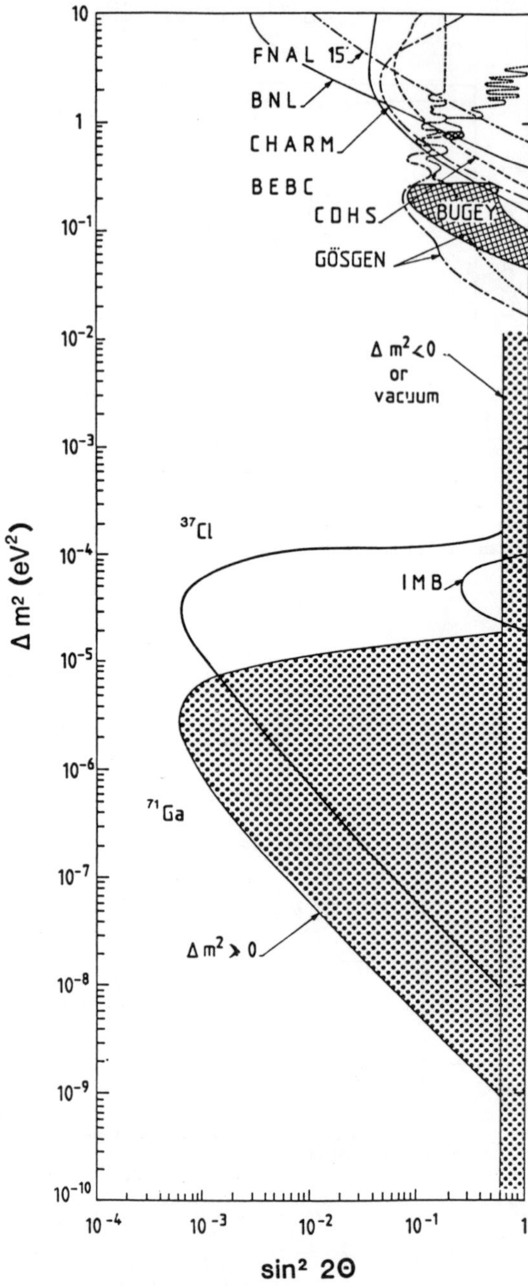

Fig.7 – Sensitivity area of the Gallium experiment in case of MSW effect ($\Delta m^2 > 0$) and in its absence ($\Delta m^2 < 0$).

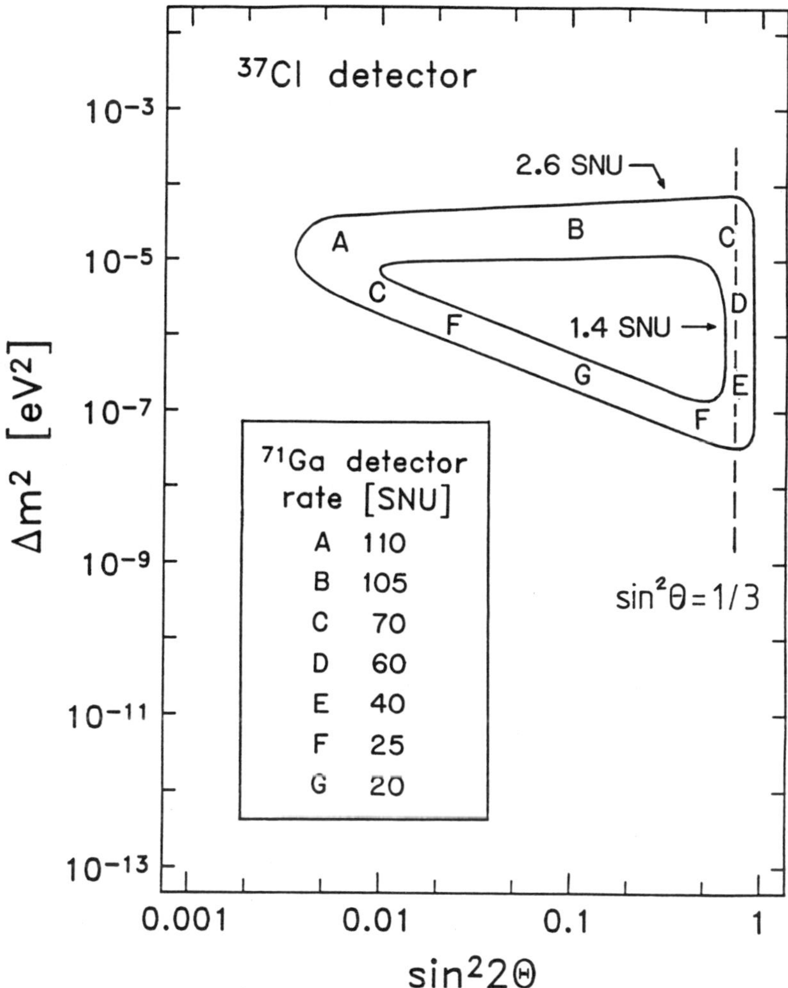

Fig.8 – Interpretation of the Chlorine result in MSW scheme and corresponding predictions for Gallium.

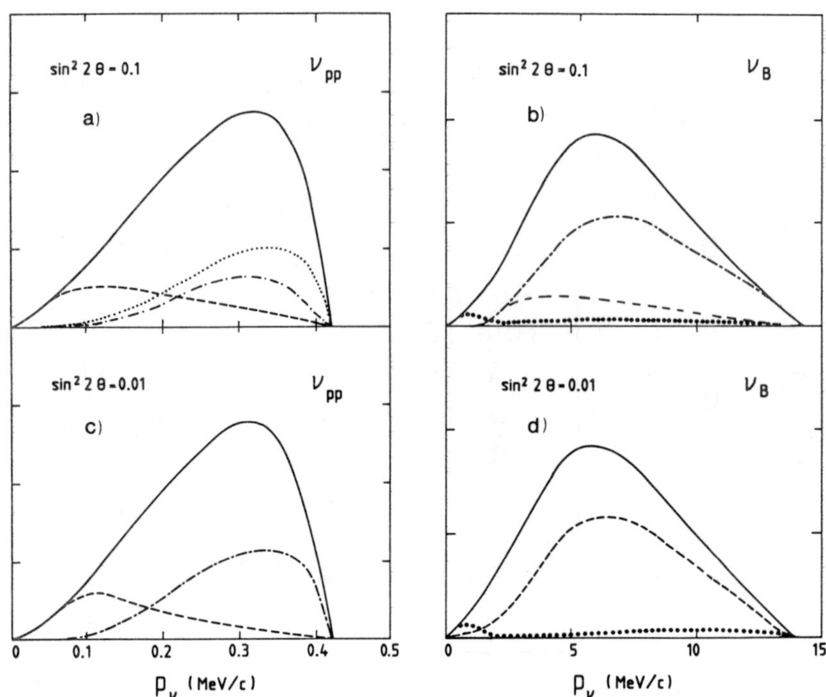

Fig.9 - Spectrum distorsion due to matter oscillations if $m_1 < m_2$ for pp and B neutrinos.

Full curve : no oscillation   Dotted curve : $\Delta m^2 = 10^{-8}$ eV^2
Dashed-dotted curve : $\Delta m^2 = 10^{-7}$ eV^2
Dashed curve : $\Delta m^2 = 10^{-6}$ eV^2
Full circles : $\Delta m^2 = 10^{-5}$ eV^2 .

The MSW Effect and Solar Neutrino Experiments*

S. P. Rosen

T-Division, Los Alamos National Laboratory

Los Alamos, New Mexico 87545, USA

ABSTRACT

We describe the MSW solutions to the ^{37}Cl solar neutrino experiment, and their implications for the ^{71}Ga experiment. Measurement of the spectrum of electron-type neutrinos arriving at earth is emphasized.

The MSW Effect and Solar Neutrino Experiments

S. P. Rosen

T-Division

Los Alamos National Laboratory, Los Alamos, New Mexico USA

In his excellent opening remarks, our chairman[1] discussed the prejudices held by particle physicists regarding neutrino masses and mixings. In this talk, I would like to adopt a prejudice too, namely that the Mikheyev-Smirnov-Wolfenstein effect is so elegant that it has to be true. This statement is highly colored by emotion, but I would like to use it as the basis for what I hope is a rational discussion of the implications of MSW for future solar neutrino experiments.

The two preceding speakers[2] in this session have discussed the more theoretical aspects of the MSW effect, and so I will confine myself to a few remarks on this aspect of the phenomenon before moving on to the more phenomenological aspects.

As is now well-known, the condition for maximal mixing induced by matter oscillations is

$$L = L_0 \cos 2\theta \tag{1}$$

where θ is the *in vacuo* mixing angle and L the corresponding oscillation length

$$L = 2.5 \frac{(E/\text{MeV})}{(\Delta m^2/\text{eV}^2)} \text{ meters} \quad . \tag{2}$$

The matter-associated "scattering length" L_0 is given by

$$L_0 = \frac{2.5 \times 10^7}{\sqrt{2}\, \rho_e} \text{ meters} \tag{3}$$

where ρ_e is the density of electrons measured in units of Avogadro's Number. The density of the sun varies with radial position, the variation being approximately exponential:

$$\rho_e \approx \rho_0\, e^{-R/R_c}$$

$$\rho_0 \approx 117, \quad R_c \approx \frac{1}{10}\, (R_{SUN}) \approx 7 \times 10^7 \text{m.} \tag{4}$$

and so for each $E/\Delta m^2$ in a range from roughly 10^4 to 10^9, there will always be one radius at which the enhancement condition is satisfied.

The impact of this enhancement upon the nature of the neutrino arriving at earth depends upon a comparison between the radial size of the enhancement region in the sun ΔR_0 and the effective neutrino oscillation length L_m at enhancement. If the radial size is much greater than L_m, then the oscillation of neutrino flavors will be in the adiabatic regime. The electron neutrino born in the central region of the sun will be a particular linear combination of eigenstates of the effective mass matrix; as it moves through the enhancement region, it remains essentially the same combination of eigenstates, but the meaning of the eigenstates in terms of neutrino flavors slowly changes.[3] At the center of the sun the heavier of the two eigenstates is mainly an electron-type neutrino, but as it moves adiabatically through the sun, it changes slowly and emerges as a predominantly muon-type neutrino. This can therefore lead to a large suppression of electron-type neutrinos at earth.

If the radial size ΔR_0 is much smaller than L_m, then the oscillation occurs within the sudden approximation. As the neutrino crosses the enhancement region, it suddenly becomes a combination of new eigenstates and can make a sudden transition from one flavor to another. This mechanism can also lead to large flavor changes at earth, but it is not as efficient as the adiabatic one.

To make the comparison between ΔR_0 and L_m, we use the relations

$$\sin^2 2\theta_m = \frac{\sin^2 2\theta}{\{(L/L_0 - \cos 2\theta)^2 + \sin^2 2\theta\}} \tag{5}$$

$$L_m = L / \{(L/L_0 - \cos 2\theta)^2 + \sin^2 2\theta\}^{1/2} \tag{6}$$

for the effective mixing angle and oscillation length in matter. It is clear from eq. (5) that the width of the enhancement region as a function of (L/L_0) is proportional to $\sin 2\theta$. To convert this to a radial distance, we presume that ρ_e is a function of radius R (see eq. (4)), and we then find that in the enhancement region

$$\Delta R_0 = \frac{\tan 2\theta}{h_0}$$

$$h_0 = \left| \frac{1}{\rho} \frac{d\rho}{dR} \right|_{\text{at enhancement point}} \quad . \tag{7}$$

For the exponential dependence in eq. (4)

$$\Delta R_0 = \tan 2\theta \times 7 \times 10^7 \, \text{m} \quad . \tag{8}$$

At the point of enhancement, the effective oscillation length is

$$(L_m)_0 = L / \sin 2\theta \tag{9}$$

and so the critical ratio of ΔR_0 to $(L_m)_0$ is given by

$$r = \frac{2\Delta R_0}{(L_m)_0} = \frac{\Delta m^2}{E} \frac{\sin 2\theta \ \tan 2\theta}{1.25 \ h_0} \tag{10}$$

where E is measured in MeV, $\Delta m^2 \equiv (m_2^2 - m_1^2)$ is in $(eV)^2$ and h_0 is measured in inverse meters. When r is much greater than unity, the MSW effect occurs in the adiabatic regime, and when r is much smaller than unity, it is in the sudden approximation.

Now let us turn to the phenomenology of MSW. Gelb and I[4] have taken the position that the suppression of electron neutrinos by a factor between 2 and 4 as seen in the ^{37}Cl experiment of R. Davis and coworkers is due to the MSW effect and we have calculated the two-flavor parameters Δm^2 and $\sin^2 2\theta$ which lead to such a suppression, especially in the regime of small mixing angles. For each set of parameters we then calculate the predicted capture rate for solar neutrinos in the ^{71}Ga experiment, and the probability that a neutrino of a given energy from ^{8}B decay remains an electron neutrino when it arrives at earth.

Our results indicate that there are two classes of solution for the ^{37}Cl experiment in the limit of small mixing angles (see Fig. 1). In the first class, recently discussed by Bethe,[5] Δm^2 is approximately 10^{-4} for all mixing angles, and in the second there is a relation

$$(\Delta m^2) \ (2\theta)^2 \approx 10^{-7.5} \tag{11}$$

between the mass difference and the mixing angle. The first class solutions correspond to the adiabatic approximation and the second to the sudden approximation.

These two classes differ in the predictions they yield for the ^{71}Ga experiment and in the nature of the probability distribution for electron neutrinos arriving at earth. The first class solutions leave the standard model predictions for the gallium experiment virtually unchanged, while the second class can lead to large suppressions in the gallium capture rate, especially for slightly larger mixing angles (see Table 1).

This difference is readily understood when we look at the probability distributions for electron-neutrinos arriving at earth (see Figure 2). For the first class of solutions, all neutrinos with energies less than some value in the neighborhood of 5-7 MeV remain as electron-type neutrinos, while all neutrinos with energies above this value are converted to muon-type neutrinos. For the second class of solutions the case is somewhat reversed: neutrinos of all energies tend to be converted to muon-type neutrinos, but the conversion is much more pronounced for low energy neutrinos than it is for the higher energy ones. Since the gallium experiment is dominated by the low energy neutrinos, the impact upon it of the second class of solutions will be much more severe than that of the first class.

From this analysis it is clear that the next solar neutrino experiment beyond gallium should have the capability of measuring the actual spectrum of electron neutrinos arriving at earth, rather than an integral over the spectrum, as is the case for ^{37}Cl and ^{71}Ga. If nothing else, such a measurement can confirm the MSW effect; but, more importantly, it can be used to resolve any ambiguities that may arise once the result of the gallium experiment is known.

For example, should the gallium experiment yield a result consistent with the standard model, we shall have to choose between the MSW effect and some modification of the temperature of the solar core as the correct explanation of the ^{37}Cl experiment. In the former case, high energy neutrinos from ^{8}B will be converted to muon-type neutrinos, whereas in the latter the overall normalization, but not the spectral shape of the neutrinos, will be changed. Thus a spectral measurement of sufficient accuracy would resolve the ambiguity.

Another possible ambiguity might arise if the result of the gallium experiment is roughly one third of the standard model expectation. In this case we will have to choose betwen MSW with small mixing angles and relatively large Δm^2 ($\approx 10^{-6,7}$) on the one hand, and large mixing angles with small Δm^2 ($\approx 10^{-10}$) on the other. In the latter case we expect a uniform suppression of essentially all neutrinos, whereas in the former the low energy neutrinos are much more suppressed than the latter. Thus again a spectral measurement would enable us to decide which alternative is correct.

For these reasons we believe that the next major thrust in solar neutrino physics should be to obtain spectral information.

Similar calculations have been carried out by the Gallex collaboration[6] and have been reported at this conference by Dr. Bouchez[2]. Barger, Phillips and Whisnant[7] have also carried out a numerical analysis of the adiabatic regime, while Haxton[8], and more recently Parke[9], have shown that an exponential function of r (eq.

(10)) gives a very good fit to the calculations in the "sudden" regime.

Gelb and I are extending our work in two directions. We are collaborating with John Bahcall to make a more refined calculation of capture rates in the two-flavor model, and we are beginning to explore the case of three flavors. The latter case has many more parameters, and so we are exploring models that reduce the number of free parameters to as few as possible.

<u>References</u>

1.   L. Wolfenstein, These Proceedings.

2.   S. P. Mikheyev and J. Bouchez, These Proceedings.

3.   For discussions of the adiabatic approximation see for example S. P. Mikeyev and A. Y. Smirnov, Proceedings of ICOBAN '86, Toyama City, Japan 16–18 April 1986; and A. Messiah, Proceedings of the Moriond Workshop, 25 January – 1 February 1986.

4.   S. P. Rosen and J. M. Gelb, Los Alamos Report LA–UR–86–804 and Phys. Rev. D (15 August, 1986; to be published).

5.   H. A. Bethe, Phys. Rev. Lett. <u>56</u>, 1305 (1986).

6.   J. Bouchez, M. Gribier, J. Rich, D. Vignaud, and W. Hampel, Saclay Report DPhPe 86–10 (May, 1986).

7.   V. Barger, R. J. N. Phillips and K. Whisnant, Wisconsin Report MAD/Ph/280 (April, 1986) and Phys. Rev. <u>D</u> (to be published).

8.   W. C. Haxton, Phys. Rev. D. (to be published).

9.   S. J. Parke, Fermilab Preprint (1986).

<u>Figure Captions</u>

Figure 1.   The two solutions for the ^{37}Cl experiment based upon MSW enhancement.

Figure 2.   Probability $P(\nu_e \to \nu_e : E)$ as a function of energy for ^{8}B neutrinos to remain electron-type neutrinos on arrival at earth. The stars correspond to the first class of solutions and the circles to the second class; the value of $\sin^2 2\theta$ is the same for the two curves in each figure.

TABLE I: Predictions for the ^{71}Ga experiment for parameters (circled) which yield a 1/3 reduction in the ^{37} experiment.

$$\sin^2\theta$$

	$10^{-3.0}$	$10^{-2.5}$	$10^{-2.0}$	$10^{-1.5}$	$10^{-1.0}$	$10^{-0.5}$
1.1E–4	100	100	100	100	100	100
1.0E–4	100	100	100	100	100	100
9.5E–5	100	100	100	100	100	100
5.8E–5	100	100	100	100	100	100
5.0E–5	100	100	100	100	100	100
1.7E–5	100	100	100	100	100	95
3.6E–6	65	60	60	60	50	45
1.1E–6	45	15	10	10	20	25
3.5E–7	70	40	10	5	5	20
1.0E–7	85	70	40	10	5	20

SOLUTIONS TO 37CL EXPERIMENT

(8B PLUS 7BE ON 37CL)

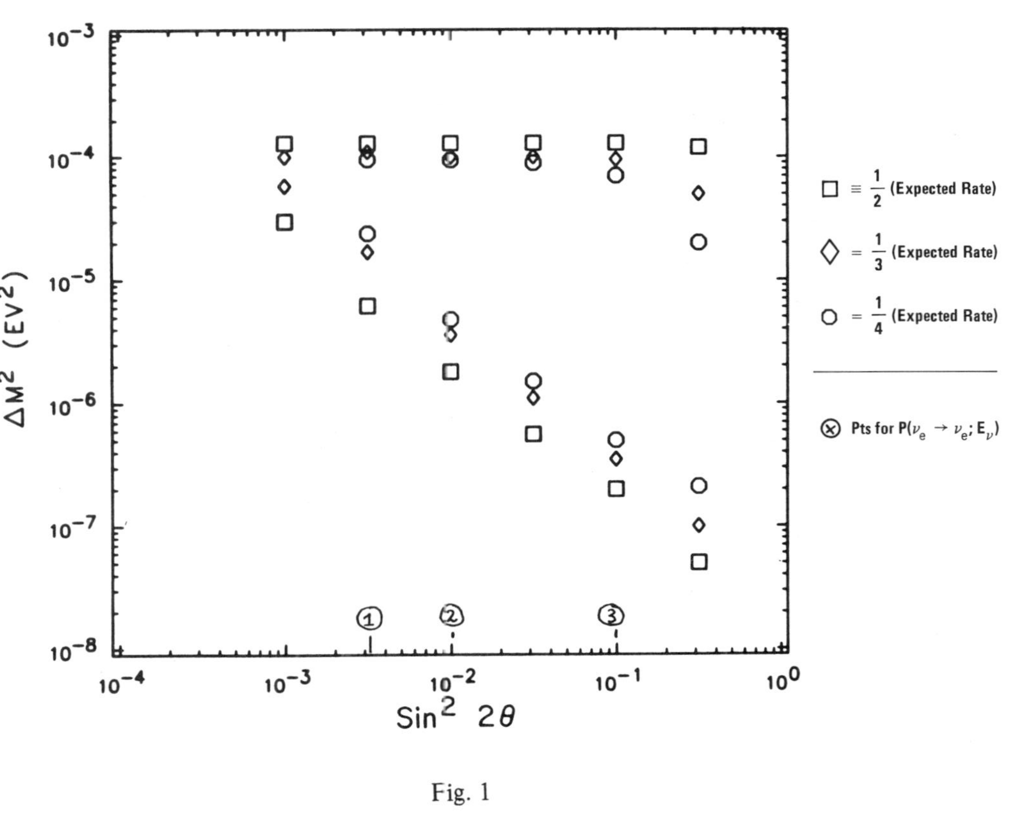

Fig. 1

PROBABILITY AT EARTH
$10^{-2.5}$
$0 = 1.7 \times 10^{-5}$
$* = 1.1 \times 10^{-4}$
PROBABILITY AT EARTH
$10^{-1.0}$
$0 = 3.5 \times 10^{-7}$
$* = 9.5 \times 10^{-5}$
PROBABILITY AT EARTH
$\sin^2 2\theta = 10^{-2.0}$
$0 = 3.6 \times 10^{-6} (eV)^2$
$* = 1.0 \times 10^{-4}$
E (MEV)

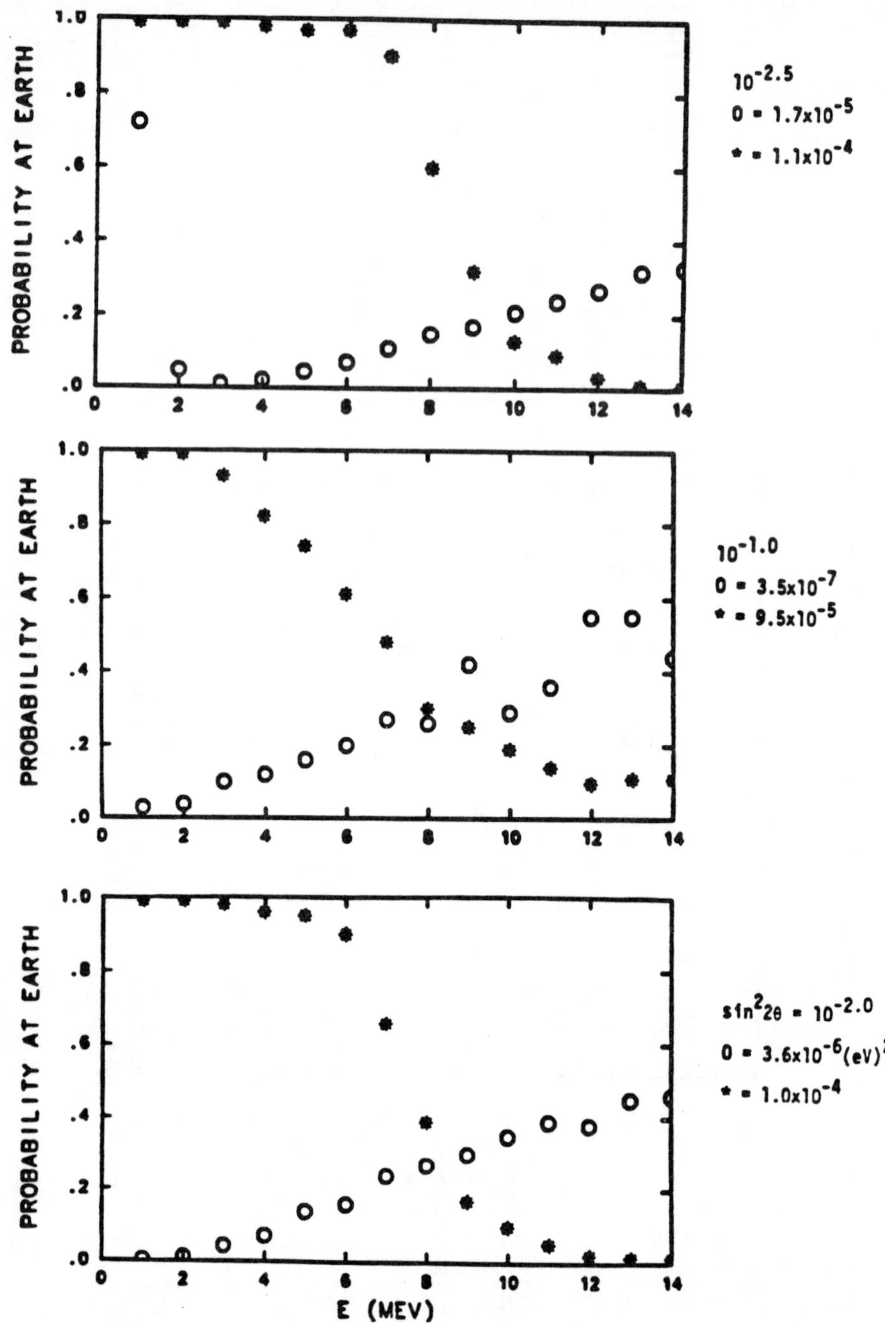

Fig. 2

EVIDENCE OF GRAVITY-MODE COUPLING
AND ITS RELEVANCE TO SOLAR NEUTRINO PRODUCTION RATES

Henry A. Hill

Department of Physics and Arizona Research Laboratories

University of Arizona, Tucson, AZ 85719, U.S.A.

ABSTRACT

Four internal gravity mode multiplets of the Sun which have $m = 0$ eigenfrequencies near 350 μHz and are of intermediate degree ($\ell \approx 30$) have been identified with differential radius observations. The frequency patterns of the multiplet fine structure for these four multiplets were found to consist of sections linear in m, and the phases of the oscillations were found to have a linear dependence on m. These two properties are interpreted as the manifestation of phase-locking of the internal gravity modes. It is observed that the presence of phase-locking indicates temperature perturbations in the core that may be sufficiently large to be relevant to the solar neutrino paradox. Also, the presence of several sets of phase-locked modes indicates that there may be periodic modulation of the solar neutrino flux with a typical period of ≈ 2 y.

I. INTRODUCTION

The first evidence of the excitation of solar gravity modes which may have relatively large temperature perturbations in the core was obtained in the 1970s by Hill and Caudell (1979). They found evidence which indicated the excitation of a number of the gravity modes of oscillation with frequencies near 250 μHz and 370 μHz and $20 \lesssim \ell \lesssim 40$, where ℓ is the degree of the angular dependent part of the eigenfunction given by Y_ℓ^m, the spherical harmonic. The implications of those findings were quite important for several areas of solar physics: if the values of ℓ were of this magnitude, then the implied

221

amplitudes of the associated oscillations were sufficiently large in the Sun's core to affect its properties in a significant way.

One of the more important effects of the presence of such intermediate-degree gravity modes is to lower the effective central temperature of the Sun and thus to alter the production rates of solar neutrinos. This possibility was noted in the 1970s by Hill (cf. Barabanov et al. 1978 and Thomsen 1978). The reduced core temperature would be, in part, a consequence of the additional energy transport mechanism furnished by the large-amplitude gravity modes.

The lack of a detailed knowledge of the internal structure of the Sun and the potential problem presented by the complexity of the boundary conditions (Hill and Logan 1984) make it difficult to infer with some confidence the amplitudes of these oscillations in the core from knowledge of their properties at the solar surface. Thus, the work of the 1970s left us with this impasse: we needed to know in particular the equilibrium properties of the core and the boundary conditions at the surface to calculate the properties of the large-amplitude oscillations, and if the oscillation amplitudes were indeed large, we needed to know their properties to infer the equilibrium properties of the core.

If the amplitudes of oscillation are in fact sufficiently large to alter the effective temperatures in the core, then second- and possibly higher-order processes are important. If higher-order processes are important, it is possible that the oscillations with large amplitudes in the core are no longer independent of each other, i.e., they are coupled. Coupling may be manifested in different ways. One of the generally recognized ways is that an oscillation may couple to excite harmonics of its own frequency. A second is that two or more modes may be phase locked. In this case, the modes of a rotationally split multiplet may have their frequencies and phases nearly uniformly spaced (amplitudes $\propto \exp i[2\pi(\nu_{n,\ell} + m\nu'_{n,\ell})t + m\phi_{n,\ell}]$, where m is the angular order of Y_ℓ^m, n is the radial order of the mode, and $\nu_{n,\ell}$, $\nu'_{n,\ell}$, and $\phi_{n,\ell}$ are constants). It should be noted that the set of phase-locked modes need not include the entire set of $(2\ell+1)$ modes of a multiplet.

The demonstration of the two primary observable features of mode coupling of the second class, i.e., uniform spacing in phase and frequency, would side-step the impasse described above and indicate that the amplitudes of the particular g-modes are sufficiently large to have higher-order processes important in the core of the Sun. We present in the following sections evidence of the second class of mode coupling for a set of gravity modes which in particular may include the modes identified by Hill and Caudell (1979), and we consider the implications of mode coupling for the production rates of solar neutrinos. The mode-coupling evidence is obtained from the analysis of observations made in 1979, while the mode identifications of Hill and Caudell (1979) were obtained from observations made in 1973.

II. OBSERVATIONS

The solar observations examined here for evidence of resolved low-order, intermediate-degree gravity modes with eigenfrequencies between $\approx$ 250 μHz and $\approx$ 400 μHz were made at SCLERA[1] in 1979 by Bos (1982). These observations have received considerable attention (Bos and Hill 1983) and have proven quite valuable in studying properties of normal modes of oscillation (Hill, Bos, and Goode 1982; Hill 1984, 1985a,b; Hill, Tash, and Padin 1986).

The observations span a 42-day period from June 1–July 12, 1979, and were obtained on 18 days for an average of 9.2 hours per day. For discussions of the relative strengths of the signals and noise in this data, reference is made to Bos and Hill (1983), Hill (1984, 1985b), and Hill, Tash, and Padin (1986). One of the most important results of these previous works is that the power density due to noise is typically 1% of that due to solar oscillations, and with this signal-

---

[1] SCLERA is an acronym for the Santa Catalina Laboratory for Experimental Relativity by Astrometry, a facility operated jointly by the University of Arizona and Wesleyan University.

to-noise ratio, individual modes of oscillation can be resolved in the power spectrum with a frequency resolution of 0.28 µHz. The analyses of the observations leading to power spectra for frequency information and to Fourier transforms for phase information of the oscillations are the same as those used by Hill (1984).

III. MULTIPLET IDENTIFICATION AND CLASSIFICATION

The criteria used in the identification of intermediate-degree g-mode multiplets were quite similar to those used by Hill (1984) in the study of low-order, low-degree acoustic modes. The principal differences have to do with the starting values of $\nu'_{n,\ell}$ used in the search for multiplets and the method used to determine the m = 0 mode and the degree of the multiplet. For this work, the values of $\nu'_{n,\ell}$ obtained by Hill, Bos, and Goode (1982) and Hill (1985b) were used as the initial values assumed in a search for evidence of multiplets. The difference with regard to the assignment of m and ℓ arose because of the limited number of multiplets included in the current analysis, the complexity of the g-mode spectrum, and the reduced effectiveness of D_{ij} in the study of intermediate-degree g-modes. The D_{ij} is a functional of m (see Hill 1984; Hill, Alexander, and Caudell 1985). Its reduced effectiveness in this study is due in part to the closeness of $\nu'_{n,\ell}$ to 4/d (see Hill, Bos, and Goode 1982; Hill 1985b) and in part to the effect of atmospheric seeing on the D_{ij} for $m \geq 20$ (see Sec. IIIa of Hill, Alexander, and Caudell 1985).

The technique used to identify the m = 0 mode and to determine the value of ℓ for a given multiplet is based in this analysis on the properties of $\overline{Y^m_\ell}$, where $\overline{Y^m_\ell}$ is the average of Y^m_ℓ over the detector geometry used by Bos (1982) (see Bos and Hill 1983). In particular, the $\overline{Y^m_\ell}$ have zeros which introduce 180° phase shifts in the plots of the phases of the modes versus m for a phase-locked section of a multiplet. The relationship between ℓ and the values of m, m_0, for which $\overline{Y^m_\ell} = 0$ are shown in Table 1. The tentative assignments of n, ℓ, and $\nu_{n,\ell}$ for the multiplets for which mode-coupling evidence was found are given in Table 2. The standard solar model of Saio (1982) was used in the assignments of n.

TABLE 1	TABLE 2
Values of m, m_0, for $\overline{Y_\ell^m} = 0$	Phase-Locked Gravity Mode Multiplets (tentative assignments)

ℓ	m_0	Classification	$\nu_{n,\ell}$ (μHz)
26	9	$g_{14,28}$	341.81
28	13	$g_{15,30}$	341.43
30	17	$g_{13,30}$	≈ 350
32	21	or $g_{15,34}$	
34	25	($\ell \approx 32$)	≈ 350

IV. EVIDENCE OF MODE COUPLING

Evidence of the two features of mode coupling discussed in the Introduction has been found in the investigation of four multiplets. The frequency patterns of the multiplet fine structure were found to consist of sections linear in m, and the phases of the oscillations were found to have a linear dependence on m. We present in Figure 1 the phase-versus-m diagram for one of the multiplets, which is tentatively classified as $g_{15,30}$.

We see from Figure 1 that the phase diagram for the $g_{15,30}$ multiplet consists of three groups of phase-locked modes. The π shifts in slope at m = ±17 are identified as the phase shifts due to the zeros of $\overline{Y_\ell^m}$ discussed in Section III. It is the separation in m for these two π phase shifts that was used to determine the value of ℓ as outlined in Section III.

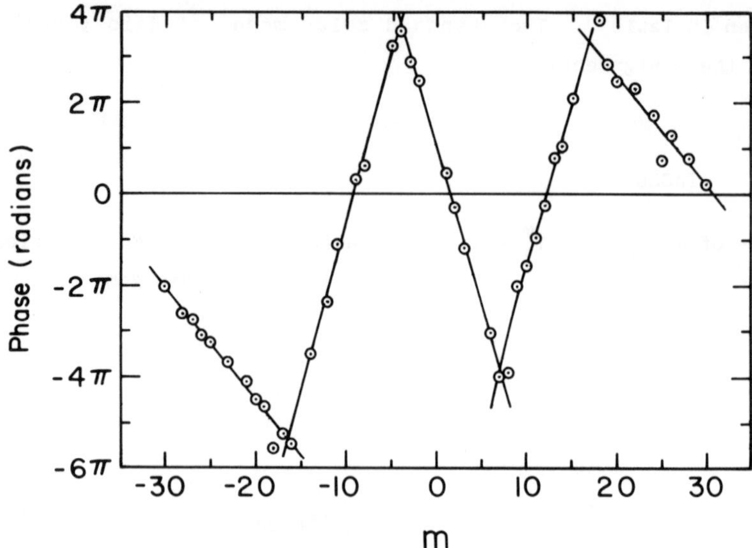

Figure 1.   The phases of the modes belonging to the $g_{15,30}$ multiplet plotted as a function of m.

The results shown in Figure 1 represent highly statistically significant support for the phase-locking hypothesis. Similar results were obtained for the other three multiplets indicated in Table 2. In addition, highly statistically significant independent results in support of the phase-locking hypothesis were obtained in the examination of $\nu_{n,\ell,m}$ as a function of m for each of the four multiplets. It is also noted that the subsets of the members of a multiplet determined to be phase locked by examination of phase versus m coincided with those found on examination of frequency $\nu_{n,\ell,m}$ versus m. This last result represents an important self-consistency test.

V.   RELEVANCE TO SOLAR NEUTRINO PRODUCTION RATES

The evidence of phase locking of a single set of gravity modes leads to a modification of the neutrino production rates predicted by a standard solar model with no solar oscillations present. The evidence of two or more different sets of phase-locked gravity modes should lead to a periodic modulation in time of the neutrino production rates. The periods of modulation are determined by the sets of properties of the phase-locked gravity modes, while the amplitudes of modulation are determined by both the properties of the modes and the details of the neutrino production. Two facts -- that a modulation is a natural consequence of phase locking and that the periods can be derived from solar seismological studies -- place the mode-locking hypothesis in a unique position relative to the many other suggestions that have been offered for resolving the solar neutrino paradox: a periodic modulation of the neutrino production rates is not a natural consequence of most if not all of these other suggestions, and the relevance of the phase-locking hypothesis can be tested. Such a test can be made by looking for a correlation between the predicted periodic behavior of the neutrino production rates and their observed fluctuations.

Effective Temperature: Temperature Eigenfunction and Energy Transport

The presence of the relatively large temperature eigenfunctions of phase-locked gravity modes for $0.05 \leq r/R \leq 0.45$ and the associated energy transport of the modes (Thomsen 1978) are both expected to alter the neutrino production rates. These two mechanisms lead to a modification of the effective temperature profile defined for a given process in the Sun. It is anticipated in particular that the former will be the more important process for the modification of the production rate of the neutrino from the decay of ^{8}B.

Periodic Modulations of Neutrino Production Rates

For a set of phase-locked modes belonging to the multiplet $g_{n,\ell}$, there exists a rotating coordinate system with a rotational speed $\Omega = -2\pi\nu'_{n,\ell}$ where each mode of the set has the same frequency of oscillation. In this rotating coordinate system, there will be one or more regions where the temperature eigenfunctions for the members of the set of modes are all in phase. These regions of constructive interference thus all rotate at $-2\pi\nu'_{n,\ell}$, the same rate as the rotating coordinate system. For a second set of phase-locked modes, an identical set of properties exists except for one: the respective rotating coordinate systems will in general have slightly different rotation rates because the respective $\nu'_{n,\ell}$ are slightly different.

It is the slightly different rotation rates of the respective regions of constructive interference that are of interest here. Periodically, the regions of constructive interference belonging to two different sets of phase-locked modes will pass through each other. Because of the importance of the higher-order processes indicated by phase locking, this passage of two different phase-locked regions of oscillations will lead to enhanced neutrino production rates. The periodicity of the enhanced neutrino production is obtained directly from the observed $\nu'_{n,\ell}$ for the phase-locked modes. The results of the analysis in Section IV indicate a typical period for this periodicity of ≈ 2 y.

VI. DISCUSSION

The findings of Hill and Caudell (1979) with regard to the detection of excited intermediate-degree gravity modes presented an immediate theoretical problem: the treatment of excitation and radiative damping of gravity modes by Dziembowski and Pamjatnykh (1978) indicated that all gravity modes with $\ell \geq 6$ were damped. This problem motivated the work of Christensen-Dalsgaard, Dziembowski, and Gough (1980), in which attempts were made to match the eigenfrequency spectra of solar models having very shallow convection zones with the observational results of Hill and Caudell (1979). The problem remains, although the theoretical restriction on ℓ is relaxed when the contribution to the excitation of the gravity modes by the ^{3}He(^{3}He,2^1H)^{4}He reaction is properly included in the Dziembowski and Pamjatnykh (1978) analysis. The new results presented here in Section III, which are based on a different set of observations, corroborate the findings of Hill and Caudell (1979) at a statistically significant level. When combined, these two independent sets of findings with regard to the excitation and detection of intermediate-degree gravity modes further establish the reality of the differences between observations and the predictions based on the standard treatment of excitation and radiative damping of gravity modes.

Although the analysis of the 1979 observations for the properties of the intermediate-degree gravity modes is not yet completed, the statistical significance of the evidence for the detection of phase-locked modes is sufficiently high to warrant certain preliminary observations. First, the presence of the phase-locked gravity modes may be important to the resolution of the solar neutrino paradox. More importantly, should the mode-locking hypothesis be correct in part, it can be subjected to a direct observational test, an opportunity which places this hypothesis in a unique position with respect to the other suggestions for the resolution of the paradox. This test, based on the observed presence of several sets of phase-locked gravity modes which should lead to a periodicity in the neutrino production rates, will consist of looking for predicted periodicities in, for example, the neutrino capture rates recorded at the Homestake Gold Mine (Rowley, Cleveland, and Davis 1985). The predicted periodicities would be based on the 1979 solar seismological observations.

It should be noted that a positive result from the test described above would have an impact on several areas of research. First, of course, such a result would confirm the presence of phase-locked gravity modes. Second, a positive result would provide the opportunity to infer in a quantitative manner the net contribution of the phase-locked modes to the solar neutrino paradox relative to other possible mechanisms such as neutrino oscillations. Third, and most important, the confirmation of the mode-locking hypothesis would demonstrate the value of solar seismology for future studies of solar neutrinos through the phase-sensitive detection of modulation in the neutrino production rates.

ACKNOWLEDGMENTS

We wish to acknowledge the support and encouragement of Academician G. T. Zatsepin for this work. Also acknowledged are the calculations of the $\overline{Y_\ell^m}$ by C. Barry. This work was supported in part by the Department of Energy, the Air Force Office of Scientific Research, and the Astronomy Division of the National Science Foundation.

REFERENCES

Barabanov, I. R., Egorov, A. I., Gavrin, V. N., Kopysov, Yu. S., and Zatsepin, G. T. 1978, in Neutrino '77 (Academy of Sciences of the USSR, Moscow), p. 20.

Bos, R. J. 1982, Ph.D. thesis, University of Arizona.

Bos, R. J., and Hill, H. A. 1983, Solar Phys., 82, 89.

Christensen-Dalsgaard, J., Dziembowski, W., and Gough, D. 1980, in Nonradial and Nonlinear Stellar Pulsation, ed. H. A. Hill and W. Dziembowski (Springer, Berlin), p. 313.

Dziembowski, W., and Pamjatnykh, A. A. 1978, in Pleins feux sur la physique solaire (CNRS, Paris), p. 135.

Hill, H. A. 1984, Ap. J., submitted.

_________. 1985a, Ap. J., 290, 765.

_________. 1985b, Ap. J., submitted.

Hill, H. A., Alexander, N., and Caudell, T. P. 1985, SCLERA Monograph Series in Astrophysics, no. 1.

Hill, H. A., and Caudell, T. P. 1979, MNRAS, 186, 327.

Hill, H. A., and Logan, J. D. 1984, Ap. J., 285, 386.

Hill, H. A., Tash, J., and Padin, C. 1986, Ap. J., 304, 560.

Rowley, J. K., Cleveland, B. T., and Davis, R., Jr. 1985, in Solar Neutrinos and Neutrino Astronomy (AIP Conf. Proc. No. 126), M. L. Cherry, K. Lande, and W. A. Fowler, eds. (American Institute of Physics, New York), p. 1.

Saio, H. 1982, private communication.

Thomsen, D. E. 1978, Science News, 113, 253.

MASSIVE NEUTRINOS AND COSMOLOGY

GARY STEIGMAN

Bartol Research Foundation, University of Delaware

Newark, DE 19716, USA

ABSTRACT

Massive relic neutrinos could be candidates for the dark matter in the Universe. Cosmological data may be used to constrain the masses, lifetimes and number of flavors of neutrinos. The assets and liabilities of light (tens of eV) and heavy ($\geq$ GeV) stable and unstable neutrinos are reviewed as are the limits on the number of neutrino flavors. Although very light ($< 10^{-2}$ eV) neutrinos may provide a solution to the solar neutrino problem, heavier, stable neutrinos are seriously ill. Big bang nucleosynthesis is consistent with three flavors of light ($\leq$ MeV) neutrinos; no more than one extra, fully coupled (to Z_L^0), light neutrino is allowed.

INTRODUCTION

It is abundantly clear that, at present, there is little compelling evidence in favor of significant ($\geq$ eV) neutrino masses. A discussion of the role of massive neutrinos in cosmology may very well be [23] "Much Ado About Nothing". In considerations of the connections between astrophysics and elementary particle physics, the best strategy is to use cosmology to derive constraints on models of high energy physics - to employ the Universe as the laboratory of last resort to probe scales inaccessible at terrestrial accelerators. This approach has led to limits to the masses, lifetimes, couplings and numbers of flavors of neutrinos which, in many cases, are more constraining than those derived from conventional experiments. This subject has been reviewed many times in many places[1]. It is not the goal of the present talk to challenge any of the previous conclusions. It is the more modest goal of this talk to review those results with some attention paid to placing them in their proper prospective. This may be of some value since the Universe - as a laboratory for particle physics - provides only indirect evidence which, if it is to be properly exploited, must be placed in its correct context.

DARK MATTERS

There are many dark matter problems. On scales ranging from smaller than galaxies to larger than clusters of galaxies, most of the Universe is "dark" - not shining in the visible part of the spectrum[2]. For example, roughly half of the total (dynamically measured) mass in the disk of the Galaxy in the vicinity of the Sun is unobserved. It is, however, most unlikely that this local dark mass is due to massive neutrinos or other exotic dark matter candidates. The reason is that, to have settled in a thin disk, the dark matter must have been dissipative. Ordinary, nucleonic matter is the best candidate for the local dark matter. That at least <u>some</u> of the dark matter is nucleonic is well worth remembering.

On the scale of entire galaxies, flat rotation curves provide evidence for unseen halo matter[2]. Since this matter is "observed" only by its dynamical effect, its nature is unknown; on scales of galactic halos and beyond, the dark matter may, or may not, be nucleonic in origin. What is clear is that on such large scales, most of the mass is dark; this is the "missing light problem". The missing light problem ($M_{TOT} \gg M_{LUM}$) exists on scales of galactic halos, small groups of galaxies, rich clusters of galaxies and, perhaps, beyond.

It is convenient to measure the mass density in units of the critical mass density, ρ_c ; low density universes with $\rho \leq \rho_c$ will expand forever, high density universes ($\rho > \rho_c$) will recollapse (throughout the discussion here it is assumed that the cosmological constant vanishes). In terms of the present value of the Hubble parameter (H_o) and Newton's constant (G), the critical density is,

$$\rho_c = 3H_o^2 /8\pi G = 1.05 \times 10^4 h_o^2 \ (eVcm^{-3}), \tag{1}$$

where

$$H_o = 100h_o(kms^{-1}Mpc^{-1}) = 50h_{1/2}(kms^{-1}Mpc^{-1)}). \tag{2}$$

Large systematic uncertainities dominate current determinations of H_o ; it is suggested[4] that $1/2 \lesssim h_o \lesssim 1$ (or, $1 \lesssim h_{1/2} \lesssim 2$). The ratio of the density ρ to ρ_c, Ω, is an important cosmological parameter; the present total density of the Universe is,

$$\rho_o = 1.05 \times 10^4 \ \Omega_o h_o^2 \ (eVcm^{-3}) = 2.6 \times 10^3 \Omega_o h_{1/2}^2 (eVcm^{-3}). \tag{3}$$

In terms of Ω_0 and H_0, the present age of the Universe (assumed to be dominated by pressureless matter) is,

$$t_0(\text{Gyr}) = 9.78 \; h_0^{-1} \; f(\Omega_0). \qquad (4)$$

For $\Omega_0 > 0$, $f < 1$; for $\Omega_0 = 1$, $f = 2/3$ so that in this case: $h_{1/2} = 13 \; \text{Gyr}/t_0$.

If the inner, luminous parts of galaxies were the "typical" mass in the Universe, then by adding up the contributions from "Galaxies" (i.e.: their inner, luminous parts)[2], $\Omega_{\text{GAL}} \approx 0.01$. The contribution of the dark mass on the larger scales of halos, groups and clusters suggests, however, a higher average universal mass density. These dynamically determined mass estimates suggest, $\Omega_{\text{DYN}} \approx 0.2 \pm 0.1$. There is, at present, no direct observational support for a higher value of Ω_0. It is, of course, well known that the inflationary scenario[5] does predict (for a vanishing cosmological constant) that $\Omega_0 = 1$.

Since the density (ρ_0) is proportional to the combination $\Omega_0 h_{1/2}^2$ (see eq. 3), it is of interest to bound this combination from above. For $\Omega_0 \leq 1$,

$$\Omega_0 h_{1/2}^2 \; \lesssim \; (13\text{Gyr}/t_0)^2, \qquad (5)$$

where, for $t_0 \geq 13(10)\text{Gyr}$, $\Omega_0 h_{1/2}^2 \lesssim 1 \; (1.7)$.

How much of the dark matter could be nucleonic in origin (eg: dead or failed stars, large planets, etc.)? To answer this question we appeal to the results of big bang nucleosynthesis[6]. Consistency of the predictions of primordial nucleosynthesis with the observed abundances of the light elements D, ^{3}He and ^{7}Li requires that the present ratio of nucleons to photons lie in the narrow range[6],

$$\eta_{10} \equiv 10^{10} \; N/\gamma = 5 \pm 2. \tag{6}$$

If the present temperature of the microwave background radiation is 2.7K, there are $\approx$ 400 relic photons in every cubic centimeter so that,

$$\Omega_N h_{1/2}^2 \approx 0.014 \eta_{10} \approx 0.07 \pm 0.03. \tag{7}$$

If the Hubble parameter lies in the anticipated range ($1 \lesssim h_{1/2} \lesssim 2$), then $0.01 \lesssim \Omega_N \lesssim 0.1$. It is not without significance that the lower bound on Ω_N inferred from big bang nucleosynthesis is comparable to the density inferrred from the luminous parts of galaxies: $\Omega_N \gtrsim \Omega_{GAL}$. Also noteworthy is that the upper bound on Ω_N is comparable to the observational lower bound on Ω_{DYN}; perhaps most or, even, all the dynamically inferred mass is nucleonic.

LIGHT NEUTRINOS (tens of eV)

Light neutrinos are produced copiously in the hot, dense epochs during the early evolution of the Universe. For the "standard" weak interactions, neutrinos with $m_\nu \lesssim$ MeV are still relativistic when they decouple at temperatures of a few MeV. The ratio of such light neutrinos, with g_ν helicity states, to photons (with $g_\gamma = 2$) is,

$$n_\nu/n_\gamma = 3/4 \; (T_\nu/T_\gamma)^3 (g_\nu/2). \tag{8}$$

In (8), T_ν (T_γ) is the neutrino (photon) temperature. When $e^\pm$ pairs annihilate the photon gas is heated with respect to the (already decoupled) neutrino gas so that, at present, $(T_\nu/T_\gamma)^3 = 4/11$. The present mass density in (all) light neutrinos may then be written as,

$$\rho_\nu = \Sigma m_\nu n_\nu \approx 10^4 \; \Omega_\nu h_0^2 \; (eVcm^{-3}), \tag{9}$$

where the sum is over all neutrino flavors. Combining (8) and (9) and adopting $T_{\gamma 0}$ = 2.7K. it follows that,

$$M_\nu = \Sigma (g_\nu/2)m_\nu \simeq 24 \, \Omega_\nu h_{1/2}^2 \ (eV).\tag{10}$$

Could the Universe be Neutrino Dominated ? To answer this question, compare Ω_N (eq. 7) with Ω_ν (eq. 10),

$$\Omega_\nu/\Omega_N \simeq 3M_\nu/n_{10}.\tag{11}$$

Clearly, if M_ν exceeds $n_{10}/3 \simeq 1\text{-}3$ (eV), the Universe is Neutrino Dominated (provided, of course, that the light but massive neutrinos are stable).

How heavy could such light neutrinos be? Comparing (10) with (5) we find that stable, light neutrinos have the sum of their masses restricted to.

$$M_\nu(eV) \lesssim 24 \, (13 \ Gyr/t_0)^2 \lesssim 24(41),\tag{12}$$

provided that $t_0 \gtrsim 13$ (10) Gyr (indeed, since $\Omega_\nu \lesssim 1 - \Omega_N$ and $\Omega_N h_{1/2}^2 \gtrsim 0.04$, $M_\nu \lesssim 23$ (40) eV).
It is often - carelessly - claimed that the limit to M_ν is $\sim$ 100eV. If, indeed, $\Omega_0 \lesssim 1$ and, if $M_\nu \simeq 100$eV, the Universe would be extraordinarily young: $t_0 \simeq 6.5$ Gyr. Equation (12) provides a more realistic - and reliable - upper limit to the sum of the masses of all light (stable) neutrinos.

Light but massive, stable neutrinos are candidates for "hot" dark matter. A light neutrino dominated Universe scenario has several assets[7]. Firstly, neutrinos do <u>exist</u>! And they are <u>dark</u>! Compared to a nucleon dominated scenario, galaxy formation is "improved" in the sense that the perturbations in the decoupled neutrinos have more time to be amplified compared to the perturbations in the nucleons which can't grow until recombination. Finally, such light neutrinos can only cluster on large scales - where most of the dark mass appears to be. Unfortunately, a light

neutrino dominated Universe scenario has some very serious
liabilities. For example, free streaming neutrinos can mix regions
of high and low density, damping initial perturbations on very large
scales[8].

$$\lambda_{FS} \approx 12(100 \text{ eV}/m_\nu) \text{ Mpc} \gtrsim 30 - 50 \text{ Mpc}. \qquad (13)$$

Since the surviving perturbations would only exist on very large
scales, large peculiar velocities are predicted[9].

$$\langle v_p^2 \rangle^{1/2} \approx 350 h_{1/2}^{-1} (1+z_{GF})(\text{kms}^{-1}), \qquad (14)$$

where z_{GF} is the redshift of the epoch of galaxy formation. Since
quasars exist at $z \gtrsim 3.$ (14) suggests that $(v_p^2)^{1/2}$ should
exceed ~700 - 1400 kms^{-1}.

Furthermore, since perturbations on small scales are erased in
this scenario, small scale structure (galaxies) should result from
the collapse and fragmentation of larger scale inhomogeneities. In
such a picture, galaxies form tomorrow[9]!

Rumors of the demise of the light neutrino scenario may,
however, be exaggerated. A recent survey of the large scale
Universe[10] reveals a bubble-like structure with enormous voids of
size ~ $50 h_{1/2}^{-1}$ Mpc. If some other "seeds" can be found for the
origin of galaxies (cosmic strings?), perhaps relic light neutrinos
could be responsible for the observed <u>large</u> scale structure of the
Universe.

HEAVY NEUTRINOS ($\gtrsim$ GeV)

Heavy neutrinos remain in equilibrium in the early Universe
after they have become non-relativistic ($T < m_\nu$). As the
Universe cools, the abundance of such neutrinos decreases
exponentially ($\propto e^{-x}$, $x = m_\nu/T$). Eventually, the heavy neutrino
gas becomes so dilute that equilibrium can no longer be
maintained[11] (this occurs at T_*, when the $\nu\bar{\nu}$ interaction rate
Γ_* exceeds the universal expansion rate H_* by a factor $x_* =$

$m_\nu/T_\star$; for massive, weakly interacting particles, $x_\star \simeq 20$).
Below $T_\star$ annihilations continue to reduce the abundance of heavy
neutrinos until the density becomes too low for any further
annihilations to occur. The final, "frozen-out" abundance of
neutrinos is smaller than the abundance at $T_\star$ by, roughly, a
factor of $x_\star$ ($N_f/N_\star \simeq x_\star^{-1}$). The surviving abundance of
heavy relic neutrinos has been calculated by many authors[12].
Gaisser et al.[13] find that $\Omega_\nu h_{1/2}^2 \geq 1$ for $m_\nu \leq 3.6\,(6.3)$ GeV
for heavy, relic Dirac (Majorana) neutrinos. The density varies
with the neutrino mass roughly as $\Omega_\nu \propto m_\nu^{-2}$; for $m_\nu \geq 17(42)$
Gev for Dirac (Majorana) neutrinos, the density of relic, heavy
neutrinos is less than the density of nucleons ($\Omega_\nu \lesssim \Omega_N$).

A heavy neutrino scenario has several assets. Galaxy formation
is improved compared to the nucleon dominated and/or light neutrino
dominated scenarios since perturbations in the heavy neutrinos are
undamped (the free-streaming scale for non-relativistic, heavy
neutrinos is very small: $\lambda_{FS} \ll \lambda_{GAL}$) and, they may begin
growing as soon as the Universe becomes matter dominated. Structure
will form on small scales and, small peculiar velocities are
predicted.

As popular as such cold dark matter scenarios have become, there
are serious liabilities for heavy neutrino scenarios. In contrast
to possibly massive but light neutrinos which are known to exist
(ν_e, ν_μ, ν_τ), there are no experimental candidates for heavy
($\geq$GeV) neutrinos. Furthermore, such heavy neutrinos should be
unstable with very short lifetimes. For example[13], unless the
ν_x- ordinary lepton mixing is incredibly suppressed, $\nu_x \to \ell\pi$
decays should be very rapid. Furthermore, a heavy neutrino scenario
shares in liabilities common to all cold dark matter scenarios.
Namely, what you see is what you get. Since cold dark matter will
cluster on small scales, the average mass density should be that
provided by the dynamical determinations: $\Omega_0 = \Omega_{DYN} \lesssim 0.3$;
$\Omega_0 \neq 1$. That is, unless you're biased! To avoid this conclusion, it
must be assumed that the mass associated with the luminous material
(galaxies) does not provide a fair sample of the average mass in the

Universe. In any case, it is very difficult to reproduce the observed[10] large scale structure of the Universe in either biased or normal cold dark matter scenarios.

A heavy neutrino - cold dark matter - scenario has yet another liability. Such heavy neutrinos would be the dark mass of the Galactic halo and would occasionally be captured by the Sun, accumulate in the solar core and annihilate[13]. The high energy prompt ordinary (e,μ,τ) neutrinos from the annihilation of heavy Dirac or Majorana neutrinos would be detectable in deep underground experiments[13]. For Dirac neutrinos with $m_\nu \gtrsim 3$ GeV, Gaisser et al.[13] find for the predicted event rate in an underground detector R ~ 7-20 events per kiloton year, and for Majorana neutrinos R ~ 5-6 events per kiloton-year; present limits are R $\lesssim$ 4 events per kiloton-year. Rumors of the demise of heavy neutrino scenarios may not be premature.

UNSTABLE NEUTRINOS

Massive neutrinos should be unstable. Many authors have considered the astrophysical and cosmological constraints on the lifetimes of massive neutrinos; for very recent summaries see reference 14. The problems associated with both the light neutrino (hot dark matter) scenario and the heavy neutrino (cold dark matter) scenario have led to consideration[15] of a hybrid scenario in which heavy neutrinos live long enough to form the small scale structure (galaxies) and then they decay, their relativistic (or, high velocity) decay products distributing themselves smoothly throughout the Universe (such a smoothly distributed component of the total mass density would have not been included in the dynamical estimates of the universal mass density).

Even such "designer" scenarios have serious liabilities. A Universe dominated by relativistic matter is very young.

$$t_0 \text{ (Gyr)} = 19.6 h_{1/2}^{-1} (1 + \Omega_0^{1/2})^{-1}. \tag{15}$$

If, indeed, $\Omega_{o_1} = 1$, then $t_o = \approx 9.8\ h_{1/2}^{-1}$ Gyr so that H_o < 50 kms^{-1}Mpc^{-1} is required if the present Universe is at least 10 billion years old. Furthermore, since structure is determined by the cold dark matter (before it decays), it is difficult to reproduce the observed[10] large scale structure. Finally, the absence of observed perturbations in the cosmic background radiation[16] leads to severe constraints on the parameters of this scenario[17]. Vittorio and Silk[17] find that this scheme may barely survive provided that the Hubble parameter is in the narrow range ~40-50 kms^{-1}Mpc^{-1} ($0.8 \lesssim h_{1/2} \lesssim 1$) and that the redshift of the decay epoch is in the range: $2 \lesssim z_D \lesssim 4$. These parameters correspond to a range of (heavy) neutrino masses: $m_{\nu H} \approx 40\text{-}110$ eV ($m_{\nu L} \ll m_{\nu H}$) and lifetimes: $t_D \approx 2 - 1/2$ Gyr. This scenario would seem to be as sick as the others discussed earlier.

MISCELLANEOUS

The Mikheyev-Smirnov-Wolfenstein mechanism[18] of neutrino oscillations in matter has been discussed extensively at this meeting. The requirement of small Δm^2 ($\lesssim 10^{-4}$eV^2) is a natural result of the see-saw mechanism[19]. However, such small masses ($m_{MAX} \lesssim 10^{-2}$eV) would be of no cosmological consequence.

Light (<< MeV) neutrinos play a significant cosmological role, regulating the universal expansion rate during primordial nucleosynthesis[20]. The standard model is consistent with the observed abundances of D, ^{3}He, ^{4}He and ^{7}Li for $N_\nu = 3$ (three flavors of light neutrinos). The data lead to the upper limit[20] $N_\nu \lesssim 4.0$; at most, on extra flavor of fully coupled (to the Z_L^0), light neutrino flavor is permitted. Some models based on the superstring suggest 3 extra flavors of light neutrinos; such models are severely constrained by big bang nucleosynthesis[20].

The ITEP experiment[21] on tritium beta decay suggest a nonzero e-neutrino mass in excess of ~20 eV. However, the absence of evidence for neutrinoless double beta decay suggests - for ν_e an unmixed Majorana neutrino - an upper limit of ~2eV[22]. Taken at face value, these results imply that ν_e is either a Dirac neutrino or, a mixture of Majorana mass eigenstates which cancel in the double beta decay amplitude. Langacker et al[23] considered the latter possibility and employed a variety of laboratory and cosmological constraints to (almost) rule it out. The loopholes are that the experimental results could be reconciled if the ν_e mixes with a heavy Majorana neutrino, ν_H, which has fast invisible decays (Majoran/familon models) or annihilations or, possibly if $M_{\nu H} \approx 40$ MeV or $m_{\nu H} \approx 2$ GeV. The alternate possibility for reconciling the ITEP result with the limits from neutrinoless double beta decay is that the electron neutrino is a Dirac particle. If, however, all three neutrinos were Dirac particles, then the effective number of neutrino flavors, N_ν, which contributes to the primordial abundance of ^{4}He[20] should be doubled: $N_\nu = 6$. This is inconsistent with the upper limit, $N_\nu \lesssim 4$, from current data[20]. Note that even if $m_{\nu\tau}$ exceeds several MeV, the contribution of ν_e and ν_μ - if they are Dirac - would just saturate the current upper limit.

CONCLUSIONS

There is no evidence from astrophysics or cosmology for a nonzero neutrino mass. Nor is it clear that a massive neutrino would be a blessing for astrophysics or cosmology. The MSW mechanism[18] does provide an elegant solution to the solar neutrino puzzle. But, is it the correct solution? If it is, then the tiny neutrino masses required ($\lesssim 10^{-2}$ eV) play no cosmological role. If, however, the ITEP results[21] were correct and there is a neutrino whose mass is ~ 20-40 eV, some ingredients must be missing in our understanding of the formation and evolution of structure in the Universe. Very heavy ($\gtrsim$ GeV) neutrinos may supply the halo dark mass. It is, however, hard to understand how such massive neutrinos would be stable on cosmological timescales. Furthermore, these heavy WIMPs would be captured by the Sun and annihilate in the solar core producing a detectable flux of high energy neutrinos[13]. The apparent absence of such a signal argues against such heavy neutrinos.

"Much Ado About Nothing"[23] is, perhaps, a not overly pessimistic view of the present status of massive neutrinos in astrophysics and cosmology.

ACKNOWLEDGMENTS

In preparing this talk I have profitted from valuable discussions with many colleagues whom I thank collectively. I am especially indebted to T. K. Gaisser and P. Langacker. This work is supported at Bartol by the DOE (DE-ACO2-78ER-05007).

REFERENCES

1) M. Turner, Proceedings Neutrino '81, Vol. I, 95 (1981); G. Steigman, Nucl. Phys. B252, 73 (1985); G. Steigman and M. S. Turner, Nucl. Phys. B253, 375 (1985).

2) S. Faber and J. S. Gallagher, Ann. Rev. Astron. Astrophys. 17, 135 (1979).

3) M. Aaronson and J. Mould, Ap. J. 265, 1 (1983); R. Buta and G. DeVaucouleurs, Ap. J. 266, 1 (1983).

4) A. Sandage and G. Tammann, Nature 307, 326 (1984); D. Branch et al, Ap. J. 270, 123 (1983).

5) A. Guth, Phys. Rev. D23, 347 (1981).

6) J. Yang, M. S. Turner, G. Steigman, D. N. Schramm and K. A. Olive, Ap. J. 281, 493 (1984); A. M. Boesgaard and G. Steigman, Ann. Rev. Astron. Astrophysics 23, 319 (1985).

7) R. Cowsik and J. McClelland, Phys. Rev. Lett. 29, 669 (1972); A. S. Szalay and G. Marx, Astron. Astrophysics. 49, 437 (1976); D. N. Schramm and G. Steigman, Ap. J. 243, 1 (1981).

8) J. R. Bond, G. Efstathiou and J. Silk, Phys. Rev. Lett. 45, (1980); I. Wasserman, Ap. J. 248, 1 (1981); F. R. Klinkhamer and C. A. Norman, Ap. J. (Lett.) 243, L1 (1981).

9) S. White, C. Frenk and M. Davis, Ap. J. (Lett.) 274, L1 (1983).

10) V. DeLapparent, M. Geller and J. Huchra, Ap. J (Lett.) 302, L1 (1986).

11) Ya. B. Zeldovich, Adv. Astron. Astrophys. $\underline{3}$, 241 (1965); H. Y. Chiu, Phys. Rev. Lett. $\underline{17}$, 712 (1966); G. Steigman, Ann. Rev. Nucl. Part. Sci. $\underline{29}$, 313 (1979).

12) D. A. Dicus, E. W. Kolb and V. L. Teplitz, Phys. Rev. Lett. $\underline{39}$, 168 (1977); S. Miyama and K. Sato, Prog. Theor. Phys. $\underline{60}$, 1703 (1978); E. W. Kolb and K. A. Olive, Phys. Rev. $\underline{D33}$, 1202 (1986); G. L. Kane and I. Kani, UM Preprint TH 85-20 (1986).

13) J. Silk, K. A. Olive and M. Srednicki, Phys. Rev. Lett. $\underline{55}$, 257 (1985); L. M. Krauss, M. Srednicki and F. Wilczek, Phys, Rev. $\underline{D33}$, 2079 (1986); T. K. Gaisser, G. Steigman and S. Tilav, Bartol preprint BA-86-42 (Phys. Rev. D. submitted June 1986).

14) M. Kawasaki and K. Sato, Phys. Lett. $\underline{B169}$, 280 (1986); M. Takahara and K. Sato, Preprint UTAP-39 (1986).

15) M. Turner, G. Steigman and L. Krauss, Phys. Rev. Lett. $\underline{52}$, 2090 (1984); S. Bludman and Y. Hoffman, Phys. Rev. Lett. $\underline{52}$, 2087 (1984).

16) J. M. Uson and D. T. Wilkinson, Nature $\underline{312}$, 427 (1984).

17) N. Vittorio and J. Silk, Ap. J. (Lett.) $\underline{285}$, L39 (1984), Y. Suto, H. Kodama and K. Sato, MNRAS 218, 637 (1986).

18) S. P. Mikheyev and A. Yu. Smirnov, INR Preprint, Moscow (1985); L. Wolfenstein, Phys. Rev. $\underline{D17}$, 2369 (1978).

19) M. Gell-Mann, P. Ramond and R. Slansky, in Supergravity, eds. P. van Nieuwenhuisen and D. Freedman, North Holland p. 315 (1979); T. Yanagida, Prog. Theor. Phys. $\underline{B135}$, 66 (1978); P. Langacker, S. T. Petcov, G. Steigman and S. Toshev, CERN Preprint TH 4421/86 (1986).

20) G. Steigman, D. N. Schramm and J. E. Gunn, Phys. Lett. $\underline{B66}$, 202 (1977), G. Steigman, K. A. Olive, D. N. Schramm and M. S. Turner, BA-85-30, Phys. Lett. B. (In Press) (1986).

21) V. A. Lubimov, these Proceedings (1986); S. Boris et al, Phys. Lett. $\underline{B159}$, 217 (1985).

22) D. Caldwell, these Proceedings (1986); T. Kirsten et al., Phys. Rev. Lett. $\underline{50}$, 474 (1983); K. Grotz and H. V. Klapdor, Phys. Lett. $\underline{B153}$, 1 (1985).

23) W. Shakespeare, London (1600).

SEARCH FOR A SIGNAL FROM CYCNUS X-3 IN UNDERGROUND EXPERIMENTS.

E. Bellotti

Dipartimento di Fisica dell'Università and I.N.F.N. - Milano

1) Introduction

Cygnus X-3 is a very peculiar celestial object, discovered in 1966 by Giacconi and collaborators [1] trough observation of its X-ray emission. After its discovery, Cygnus X-3 was recognized to be a strong source of radiowaves, infrared and X-rays; on the contrary observations of MeV photons gave contradictory results [2] . Finally it has been observed at very high energy by Cerenkov light detectors and at ultra high energy by means of EAS detectors [2, 3] .

Position of this source (1950) is:declination 40° 47; right ascension 20^h 30^m (307,6°);distance $\sim$12 Kpc.

Cygnus X-3 is a strongly variable source of radiowaves; sporadic out bursts can reach an intensity 1000 times larger than the quiescent emission intensity. In the KeV region, a sinusoidal component with a period of $\sim$4.8 h is superimposed to a steady component, the corresponding ephemeris, as determined by Van der Klis and Bonnet-Bidaud [4] are:

$$P = .199683 \ d \quad and \quad \dot{p} = 1.18 \ 10^{-9}$$

In the TeV region, emission seems to be concentrate at two phases: around 0.2 and around 0.6.

The intensity is very large over all the energy spectrum, however Cygnus X-3 is not visible, being shadowed by cosmic dust.

In the TeV region ,i.e. the interesting region for underground experiments, the intensity roughly follows the law I$\sim$ 10^{-8} E^{-1} (E in GeV), up to 10^{16} eV.

From the above recalled observations at low energy, Cygnus X-3 is believed to be a binary system, with a compact object (a neutron star ?) orbiting around a more massive object. Radio bursts are explained by accretions of infalling matter from the massive companion.

Observations at high energy with surface and underground detectors pose many questions like:

- are there long term intensity variations [5] ?
- are the emitted particle really photons, as expected from the fact that direction and time information is retained [6] ?
- is it possible to concile above ground and underground measurements[7].

2) Underground measurements

In this section,observations from some of the underground detectors presently in operation are summarized; Table I reports some of the relevant features of these detectors which were designed to search for nucleon instability . Two of them (Soudan I and NUSEX) have reported positive results.

TABLE I - UNDERGROUND DETECTORS

EXPERIMENT.	LOCATION	DEPTH(mwe)	PERIOD	ANGULAR RES.
HPW wat. C	40.6° N 111° W	1450	Oct.'83-July'84	3° - 7°
SUDAN I tr. cal.	48° N 92° W	1800	Sept.'81-Nov'83 (0.96 d)	1.4°(±1.5°)
KAMIOKA wat. C	36.42° N 137.31° E	2400	July'83-Sept.'84	2.7°
FREJUS tr. cal.	45.1° N 6.7° E	4000	feb.'84-Jan.'86 (1/2 of data after June '85)	1°
NUSEX tr. cal.	45.9° N 6.9° E	4600	June '82...	1°

a)   Soudan I [8-9]

Results obtained by the Soudan I collaboration can be summarized as follows:

- a flux $\langle\phi\rangle$ of $\sim 7 \cdot 10^{-11}$ μ's/ (cm^2 x s) was observed in an angular window centered 3° off Cygnus X-3 in the phase interval 0.65-0.9 ($\langle\phi\rangle$ means a flux averaged over a period).

- an excess of multimuons events correlated in time was observed from a region of 30°x30° around Cygnus position at phase 0.7.

- a flux $\langle\phi\rangle$ = 2.5 10^{-11} μ's / (cm^2x s) in the phase interval 0.65 -0.9 in a direction pointing backward to Cygnus X-3 was detected.

b)   NUSEX [10-11]

A clear signal in the phase interval 0.7-0.8 was observed in data collected before Jan. 1985 (fig. 1), in a cone of half-aperture of 5° centered in direction pointing backward to Cygnus X-3. Not yet an explanation exists for the angular spread around the Cygnus X-3 direction (fig. 2); in fact this angular distribution is much broader than expected from angular resolution and indetermination in position of the detector, multiple scattering in the rock etc. Data collected duing '85 and part of '86 don't show any signal, in agreement with the results obtained by other underground detectors . Fig. 3 shows the time dependence of the signal. Finally the hypothesis that these events are due to neutrino interactions seems very unlikely [11] .

c)   HPW [12]

No excess was observed in data collected by this water Cerenkov detector. The maximum flux allowed by data (at 90% C.L.) is 5 10^{-11}, 3.5 10^{-11}, 2.6 10^{-11}, 1.5 10^{-11} μ's/ (cm^2x s), at slant depth greater than 1450, 2200, 2500 and 3000 m.w.e. respectively,and in the phase interval 0.7-0.8. This result is somewhat in contradiction with the Soudan I result.

d)   Kamiokande [13]

The maximum flux allowed by data is $\langle\phi\rangle \sim 5.4 \; 10^{-12} \, \mu$'s $/ \; (cm^2 \times s)$ regardless the phase; in the 0.7-0.8 phase interval, maximum fluxes (at 90% C.L.) are: $2.2 \; 10^{-12}$, $2.6 \; 10^{-12}$, $2.0 \; 10^{-12}$, $1.7 \; 10^{-12} \, \mu$'s $/ \; (cm^2 \times s)$ at depth larger than 2400, 3400, 4400, 5000 m.w.e.

<u>Frejus</u> [14]

Frejus detector is a digital tracking calorimeter, with very good angular resolution, located at a depth very similar to the M.nt Blanc one. Data have been collected partly in '84, when the detector was not yet completely assembled, and partly during '85 and '86.

Data can be summarized as follows:

in an angular window of 2° of half-aperture, 70 events were observed the corresponding maximum flux is:

all phases $0.8 \; 10^{-12} \, \mu$'s/ $(cm \times s)$

phase bin 0.7-0.8 $0.12 \; 10^{-12}$ "   "   "

Is a cone of 5° of half-aperture

all phases $1.5 \; 10^{-12} \, \mu$'s/ $(cm \times s)$

phase bin   0.7-0.8 $0.12 \; 10^{-12}$ "   "   "

Finally, it must mentioned that data collected in the Baksan detector[15], which is very shallow, don't show any excess at any phase, in the muon flux from Cygnus X-3 direction.

<u>Comments and future perspectives</u>

From the above summary of underground data, it is clear that a contradiction exists between Soudan I and NUSEX data one side, and HPW, Kamiokande and Frejus on the other side. However, different depths, acceptance and period of observation of detectors, leave a not large indeed possibility that the positive result observed by Soudan I and

NUSEX is a real physical effect . This possibility will be verified by future detectors being installed like Soudan II or in preparation , like LVD, MACRO and ICARUS, which will be installed in the Gran Sasso Laboratory.

R E F E R E N C E

1) R. Giacconi et al. - Astrophys J. <u>148</u>, L119 (1967)

2) For a review up to 1985,
   B.M. Vladimirskii et al. - Sov. Phys. Usp. <u>28</u>, 153 (1985)

3) M.Van der Klis and J.M. Bonnet-Bidaud - Astron. and Astrophys. <u>95</u>,
   L5   (1981)

4) For a critical discussion of high energy data, see also
   G. Chardin - Preliminary results on atmospheric neutrinos and Cyg
   X-3 in the Frejus detector
   in Neutrinos and the Present-day Universe. Ed. by Thierry-Montmerle
   and M. Spiro.

5) M.F. Cawley et al. - 19th Int. Conf. Cosmic Ray Phys
   La Jolla, 1,131 (1985)

6) Y. Kifune et al. Proc. - 19th Int. Conf. Cosmic Ray Phy. La Jolla
   1,67 (1985)
   M. Samorsky and W. Stamm - Ap. J. (Letter) <u>268</u>, L17 (1983)
   and Proc. 18th Int. Conf. Cosmic Ray Phys. 11,284 (1983)

7) see f.i. A. Castellino et al. - Lettere al N.C. <u>44</u>, 491 (1985)

8) J. Bartelt et al. - Phys. Rev. <u>D32</u>, 1630 (1985)

9) M. Marshak et al. - Phys. Lett. <u>54</u>, 2079 (1985)
   M. Marshak et al. - Phys. Lett. <u>55</u>, 1965 (1985)

10) G. Battistoni et al. - Phys. Lett. <u>155B</u>, 465 (1985)

11) B. D'Ettorre-Piazzoli. - Proc. of the 1st Symposium on underground
    physics - Saint Vincent (Italy), April 1985

12) W. Worstell et al. - A search for high energy muons from Cygnus X-3
    Harvard - Purdue - Wisconsin preprint (1985)

13) Y. Oyama et al. - Phys. Rev. Lett. <u>56</u>, 991 (1986)

14) Ch. Berger et al. - Phys. Lett. <u>B118</u>, 118 (1986)

15) Alexenko et al. - Proc. of the 19th Int. Conf. Cosmic
    Ray Phys. - La Jolla 1985

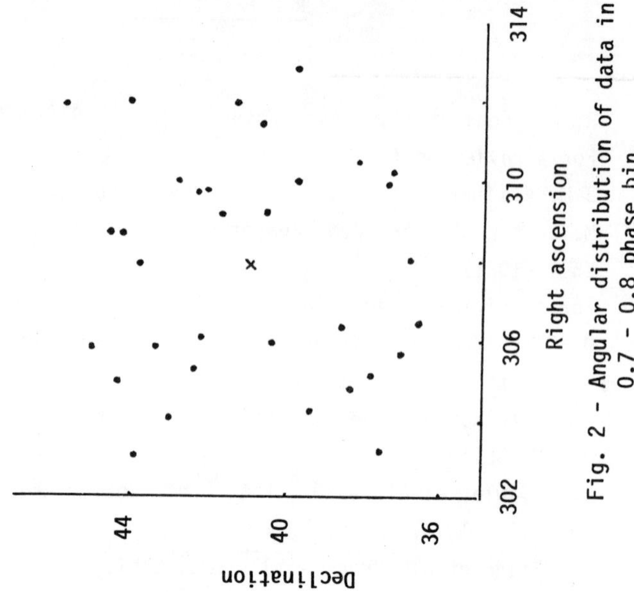

Fig. 2 - Angular distribution of data in 0.7 - 0.8 phase bin.

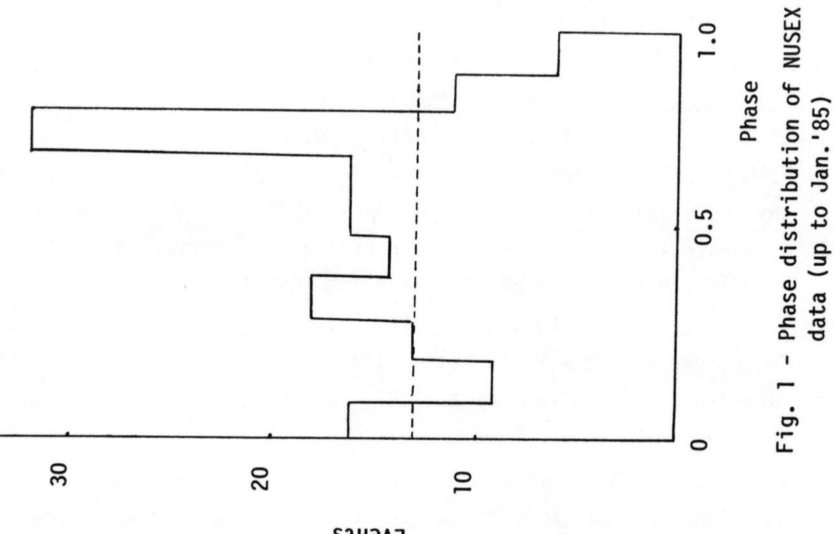

Fig. 1 - Phase distribution of NUSEX data (up to Jan. '85)

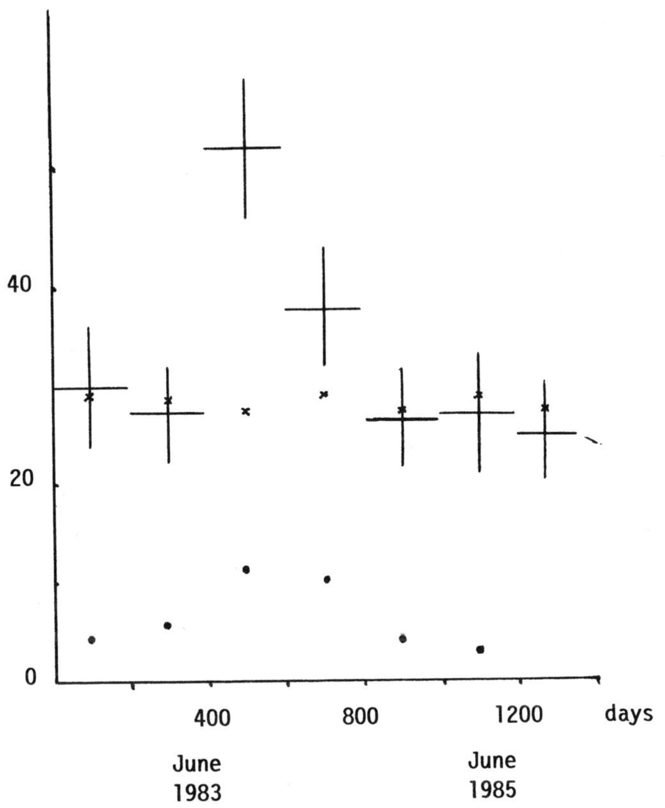

Fig. 3 - NUSEX data - Time distribution of:
  + all phases on source events
  × 0.7 - 0.8 phase on source events
  • all phases off source events

Cosmological Constraints on the lifetime of neutrinos

Katsuhiko Sato

Department of Physics, Faculty of Science
The university of Tokyo, Tokyo 113, Japan

Cosmological effects of the radiative decay of heavy neutrinos(1eV $<$ $m_\nu <$ 1GeV) are reviewed and current constraints on the lifetime from cosmology are shown. It is also shown that by combining the cosmological and experimental constraints, the possibility that a tau neutrino has the mass greater than 50 eV is almost ruled out, although there exists a narrow region which has not yet been ruled out near the mass $\sim$ 50MeV and the lifetime $\sim$ 1000 sec.

1. Introduction

Recently it has been suggested that neutrinos have non-vanishing mass from both sides of experimental and theoretical investigations[1].

If neutrinos have finite masses, they play important roles in cosmology[2].

In the very early stage of the universe, every type of neutrinos is in thermal equilibrium with matter. In the cause of the expansion of the universe, however, interactions between neutrinos and the matter are switched off when the time scale of weak interaction becomes longer than the time scale of cosmic expansion. The number ratio of heavy neutrinos to photons at present universe is approximately given by[3]-[8]

$$n_\nu /n_\gamma = 3/11 \ \mathrm{Min}\{1, \ (10 \ \mathrm{MeV}/m_\nu)^{2.75}\} \ .$$

From the condition that mass density of heavy neutrinos in the universe should be smaller than the observational limit of the energy density of the universe, ρ_0;

$$2n_\nu \, m_\nu \; < \; \rho_0,$$

we can conclude that neutrinos in the mass range, 50 eV $\sim$ 3 GeV should decay or are ruled out if they are stable[3]-[5].

In the present talk, I review the cosmological effects of Neutrino decay and show the current constraints on the lifetime assuming radiative decay.

2. Distortion of the Spectrum of Microwave Background Radiation

If neutrinos decay in the very early stage of the universe, photons emitted by the decay are quickly thermalized and no remnant can survive. But the decaying time is later than the freezing time of Compton scattering and Bremsstrahlung, t_f, the photons are superimposed on the cosmic background radiation and the spectrum of background radiation is deformed. Sato and Sato[9] discussed the decay of a weekly interacting massive particle in the big bang universe and gave a constraint on its lifetime from the upper limit of the cosmic background radiation. This constraint was extended to the heavy neutrinos by Sato and Kobayashi[5] and also Dicus Kolb and Teplitz[6], Miyama and Sato[8], and Gunn et al[10]. In these papers, authors obtained the constraints on the proper lifetime τ_0 from the condition that effective lifetime (τ_0 x Lorentz factor) should be smaller than the freezing time t_f. But it is no easy to estimate t_f analytically, for example, Sato and Kobayashi[5] estimated $t_f = 10^{10}$ sec, but, Gunn et al[10] considered $t_f = 10^4$ sec. More quantitative investigation was carried out by Danese and Zotti[11]. However, in order to investigate the distortion precisely, it is necessary to calculate the evolution of the radiation spectrum numerically. Silk and Stebins[12] first performed this numerical calculation, but only in the region where the mass of neutrinos is smaller than their momentum. In addition, they only considered the high baryon density universe($\Omega_b \sim 1$) and neglected the double Compton process. in the present universe, however, it is presumed that $\Omega_b \ll 1$ and double Compton scattering play an important role in thermalization. Recently Kawasaki and Sato[13] investigated the time evolution of the distorted spectrum more quantitatively by computing the Kampaneets equation and derived a precise constraint on the lifetime, which is considerably different from that of previous works.

In Fig.1, examples of the final spectrum of photons for neutrinos of various masses are shown. This figure suggests that neutrinos with the lifetime 10^7 sec and the mass 20 eV and also 10 keV should be ruled out from the observations[14], which are also shown in this figure. The constraints on the lifetime in the mass range 10 eV to 10^5 eV are shown in Fig. 2 for baryon/photon ratio, $\eta = 10^{-8}$, 10^{-9} and 10^{-10}. Because it is believed that the value of η is $(3-10) \times 10^{-10}$, we can conclude that the lifetime should be

$$\tau_0 < 2 \times 10^6 \, [1 + (1000 \; eV/m_\nu)^2]^{-1/2} \; sec.$$

3. Constraint From Primordial Nucleosynthesis

If the massive neutrinos decay and disappear before the era of nucleosynthesis ($T \sim 100$ keV), no effects can be expected on the abundance of elements. On the contrary, if they decay after the nucleosynthesis has begun, abundance of elements produced by primordial nucleosynthesis are significantly modified by the two mechanism. As is well known, the abundance of elements produced by the primordial nucleosynthesis are controlled by two parameters. One is the expansion rate of the universe at this period. If the massive neutrinos exists, the expansion speed of the universe becomes faster than that of the case without massive neutrinos, because the energy density of the former case is larger than the later case. The speed–up of the expansion of the universe makes the the number ratio of neutron/proton ratio increase, at the first step, because the more rapid expansion of the universe makes the freezing time of the weak interaction shift to the earlier time with the higher n/p ratio. As the result of the increase of neutrons, ^{4}He is more abundantly synthesized.

The other is the baryon/photons ratio at the period of the nucleosynthesis, η_{NS}. Under the standard model without decaying neutrinos, this value is essentially the same one of the present universe, $\eta_0(\sim 3\times 10^{-10})$. However under the presence of the unstable neutrinos, the entropy of the universe is greatly increased by the decay. This suggests that the baryon/photon ratio at the nucleosynthesis, η_{NS}, was very large compared with the present value η_0. As is well known, if the value of η_{NS} is very large, ^{4}He is overproduced and D cannot be synthesized sufficiently. Therefore, from the conditions that ^{4}He is not overproduced, $Y(^4He) < 0.26$, and enough abundant D is synthesized, $D/H > 10^{-5}$, we can impose the constraints on the lifetime and mass of heavy neutrinos. This constraint was first obtained by Sato and Kobayashi[5] analytically and successively more precise constraint was calculated numerically by Miyama and Sato[8] and Dicus et al [6]. Recently, Terasawa, Kawasaki and Sato[15] calculated this constraint using newly revised nuclear reaction rates. The result is shown in Fig. 7 as the curve No. 2, which is, however, essentially the same as previous works.

4. Photo–Destruction of Light Elements by The Decay of Heavy Neutrinos

When the mass of neutrinos m_ν is very heavy(> 10 MeV), high energy photons are produced by the decay. If they decay soon after the primordial nucleosynthesis, light elements which were synthesized by primordial nucleosynthesis are destroyed by this photons. Even if neutrinos decay into pairs of electrons, $\nu \to e^- + e^+ + \nu'$, (which may be faster than the mode $\nu \to \gamma + \nu'$ when m_ν is greater than $2m_e$), high energy photons are created, because emitted high energy electrons and positrons scatter with background photons (inverse Compton scattering). Lindley[16] showed that the photofission of light elements imposes the stringent constraints on its lifetime, because the destruction rate depends on the lifetime and its number density in the universe. The later neutrinos decay, the more light elements are destroyed, because the thermalization process of high energy photons proceeds more slowly. Thus the upper limit of the lifetime can be obtained by requiring that the light elements should not destroyed so much[16)-18)].

In order to obtain a reliable constraint, we must know the spectrum of high energy photons and its time evolution precisely. Recently, Kawasaki, Terasawa and Sato[19)] calculated the evolution by computing Boltzmann equation for high energy photons and electrons simultaneously, and obtained the constraint on the lifetime. In Fig. 3, the upper limit of the lifetime which are obtained by the destruction of ^{4}He, ^{3}He and ^{2}H are shown. Obviously, the destruction of ^{2}H impose the most stringent limit, because the threshold energy of the photofission reaction for ^{2}H $+ \gamma \rightarrow p + n$, 2.225 MeV, is the lowest one.

In the work Kawasaki et al[19)]., however, recombination of ^{2}H from the fragments, n and p, was not taken into account. Recently, Terasawa, Kawasaki and Sato[15)] calculated the evolution of the abundance of light elements by computing the nuclear reaction network in which photofission processes also included. An example is shown in Fig. 4. As seen in this figure, deuterium which was synthesized at $10^{2.5}$ sec is destroyed almost completely at $10^{4.8}$ sec and a small amount of tritium is synthesized from the deuterium and neutrons, which are the fragment of deuterium. The result shows that although the abundance is modified a little, the constraint on the lifetime is not altered even if full reaction network of nucleosynthesis is included.

5.Cosmic Gamma-Ray Background Radiation and Supernova Neutrinos

As is well known, Most of the energy released by the gravitational collapse of stars is emitted as neutrino energy($\sim 3 \times 10^{53}$ erg). If heavy neutrinos emitted from supernova cores decay outside stars, photons produced by decays form the cosmic gamma-ray background radiation. Even if neutrinos decay into electron pairs, $\nu \to e^- + e^+ + \nu'$, high energy photons are created by the process of pair annihilation $e^+ + e^-$(interstellar electrons) $\to 2\gamma$. From the condition that this gamma-ray flux should be smaller than the observed background flux[20], a constraint on the lifetime is obtained. Cowsik[21], Miyama and Sato[8], and Falk and Schramm[22] investigated effects of heavy neutrinos on the cosmic background radiation and imposed constraints on the lifetime and mass from supernova explosions. Although they ruled out very wide range of the mass and lifetime, no constraints are imposed in the mass range higher than 10 MeV. Recently Takahara and Sato[23] investigated in the mass range, $m_\nu > 10$ MeV by using the recent results of the numerical simulation of gravitational collapse[24], and obtained a constraint on the mass and lifetime, which is shown in Fig. 5. This constraint is the most stringent one in the mass range $m_\nu < 65$ MeV.

6. Discussion and Summary

Until now, many experiments to search heavy neutrinos have been carried out. In particular, from the precise measurement of the decays of π and K mesons, and from the beam dump experiments, stringent constraints on the lepton mixing parameters have been obtained, see, for example, reviews[25]. In Fig. 6, limits on the mixing parameter U_{ei}^2 which are obtained recent experiments are shown as a function of the neutrino mass m_ν in the range $m_\nu < 150$ MeV[26]–[29].

In Fig. 7, constraints obtained from the cosmological discussions are summarized, and also constraints from the experiments are superimposed. The line **a** indicates the upper limit for the tau neutrino mass, which was obtained recently from the tau lepton decay[30].

This figure shows that even if tau neutrinos have mass greater than 1

MeV, its mass and lifetime should be in a small region near $m_\nu \sim 50$ MeV and $\tau \sim 1000$ sec. A few years ago, Sarkar and Cooper[18] discussed that unstable tau neutrinos heavier than 1 MeV are completely excluded by a cosmological argument concerning the deuterium destruction , together with experimental results. However precise calculation of deuterium destruction by Kawasaki et al[19] shows that their destruction rate was overestimate. At present, therefore, there exists a narrow region which has not yet been ruled out. It is, however, likely that this region is completely disappeared by the more stringent experiments near future.

Finally we mention that our cosmological constraints can be applied to WIMPs(Weakly Interacting Massive Particles), which are now predicted by some particle theories, such as supersymmetric theories, and now are actively investigated as possible candidates of dark matter in the universe. In particular, for particles which were thermal equilibrium in the early universe, present results can be applied by trivial modification.

Acknowledgments

The author thanks M. Kawasaki, M. Takahara, and N. Terasawa for variable discussion. Present paper is based on the collaborations with them. He also thanks J. M. Levy and R. Prieels for showing me new experimental results and stimulating discussion.   This work was supported in part by the Grant-in-Aid for Science Research Fund of the Ministry of Education, Science and Culture No. 60302024.

References

[1]   L. Wolfenstein, in this volume.

[2]   G. Steigman, in this volume.

[3]   P. Hut, Phys. Lett. **69B** (1977) 85.

[4]   B.W. Lee and S. Weinberg, Phys. Rev. Lett. **39** (1977) 165.

[5]   K. Sato and M. Kobayashi, Prog. Theor. Phys. **58** (1977) 1775.

[6]   D.A. Dicus, E.W. Kolb and V.L. Teplitz, Phys. Rev. Lett. **39** (1977) 168 ; Astrophys. J. **221** (1978) 327.

[7]   M.I.Vysotsky,et al, Sov. Phys. JETP lett. **26** (1977) 188.

[8]   S. Miyama and K. Sato, Prog. Theor. Phys. **60** (1978) 1703.

[9]   K.Sato and H. Sato, Prog. Theor. Phys. **54** (1975) 912.

[10] J.E. Gunn et al., Astrophys. J. **223** (1978) 1015.

[11] L. Danese and G. de Zotti, Astron. Astrophys. **107** (1982) 39.

[12] J. Silk and A. Stebbins, Astrophys. J. **269** (1983) 1.

[13] M. Kawasaki and K. Sato, Phys. Lett.**169B** (1986) 280.

[14] G.F. Smoot et al., Phys. Rev. Lett. **51** (1983) 1099.
       J.B. Peterson,et al, Phys. Rev. Lett. **55** (1985) 332.

[15] N. Terasawa, M. Kawasaki and K. Sato, preprint Sept.1986.

[16] D. Lindley, Mon. Not. R. Astron. Soc. **188** 15P.

[17] D. Lindley, Astrophys. J. **294** (1985) 1.
       L.M. Krauss, Phys. Rev. Lett. 53 (1984) 1976.

[18] S. Sarkar and A.M. Cooper, Phys. Lett. **148B** (1983) 347.

[19] M. Kawasaki, N. Terasawa and K. Sato, Phys. Lett. in press.

[20] N. Bartel et al., Nature **318** (1985) 25.
       J.I.Trombka, et al, Astrophys. J. **212** (1977) 925.
       G.E.Fichtel, et al, Astrophys. J. **222** (1978) 833.

[21] R. Cowsik, Phys. Rev. Lett. **39** (1977) 784.

[22] S.W. Falk and D.N. Schramm, Phys. Lett. **79B** (1978) 511.

[23] M. Takahara and K. Sato, Phys. Lett. **174B** (1986) 373.

[24] J.R.Wilson,R. Mayle, S. E.Woosley and T. A. Weaver,
       Ann. New York Acad. Sci. in press.
       W. Hillebrandt, Proc. of NATO-ASI on High Energy Phenomena
       around Collapsed Stars (Cargese, September, 1985), in press.

[25] T. Yamazaki, Prog. in Particle and Nucl. Phys. **13** (1985) 489.
       A. M. Cooper-Sarkar et al. preprint CERN/EP85-104 (July 1985)

[26] D.A. Bryman et al., Phys. Rev. Lett. **50** (1983) 1546. (TRIUMF)

[27] F. Bergsma et al., Phys. Lett. **128B** (1983) 361..(CHARM)

[28] G. Bernaldi et al, **166B**(1986), 479.(CERN PS-191)

[29] N. De Leener-Rosier et al., Preprint May 1986 (Louvain,
       Lousanne and Zurich group at SIN)

[30] H. Albrecht et al., Phys. Lett. **163B** (1985) 404.

Figure Captions

Fig. 1: The final spectrum of photons for radiative decay of massive neutrinos with the mass indicated on each curve. In this example τ_0 $=10^7$sec and $n_i=10^{-8}$. The symbol denotes the observational data[14].

Fig. 2: Constraints on the lifetime and the mass for $n_i=10^{-8}$, 10^{-9} and 10^{-10}. The solid lines are lifetime limits for massive neutrinos and the dashed lines are for weakly interacting massive particles which decoupled when the effective number of species is ~ 50.

Fig. 3: Constraints on the lifetime of heavy neutrinos from the destruction of ^{4}He, ^{3}He and ^{2}H for the decay mode $\nu \rightarrow e^- + e^+ + \nu$'.

Fig. 4: An example of the time evolution of elements with unstable neutrinos. In this case, it is assumed that m_ν = 1 GeV and τ $=10^5$sec.

Fig. 5: Constraints on mass and lifetime of heavy neutrinos. imposed by the observation of the γ-ray background radiation by supernova explosions. The dashed curves denoted by x 10^{-1} (or x10) show the constraints imposed when the γ-ray flux from supernovae is decreased (or increased) by factor 10.

Fig. 6: Upper limits of U_{ei}^2 obtained from recent experiments[26]–[29] in the mass range m_ν < 150 MeV.

Fig. 7: Summary of the constraints on the lifetime of heavy neutrinos: **1**: distortion of the spectrum of cosmic microwave radiation, **2**: primordial nucleosynthesis, **3**: destruction of deuterium[17],[19], **4**: gamma-ray background radiation due to the decay of supernova neutrinos. Constraints from experiments are also shown. **a**: Albrecht et al[30], **b**: CHARM[27], **c**: TRIUMF[26], **d**: SIN[29].

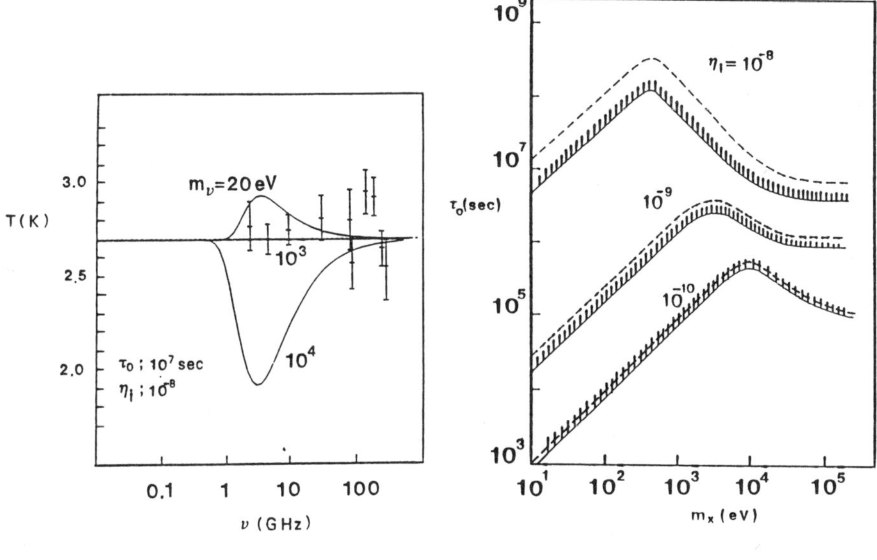

Fig. 1

Fig. 2

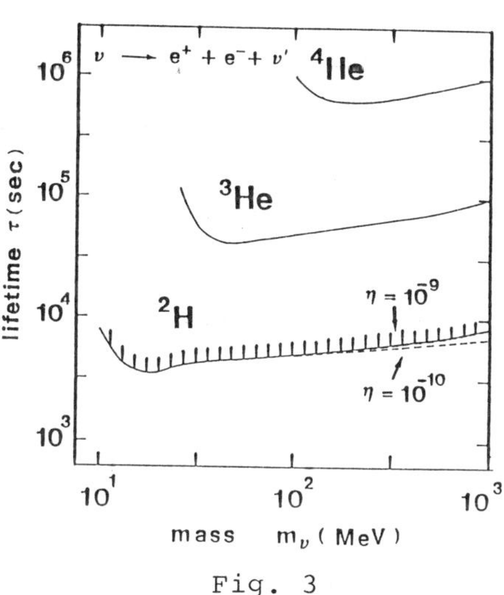

Fig. 3

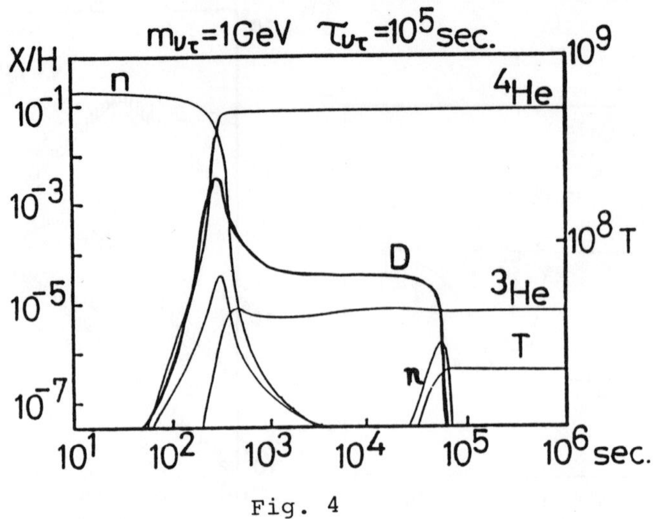

Fig. 4

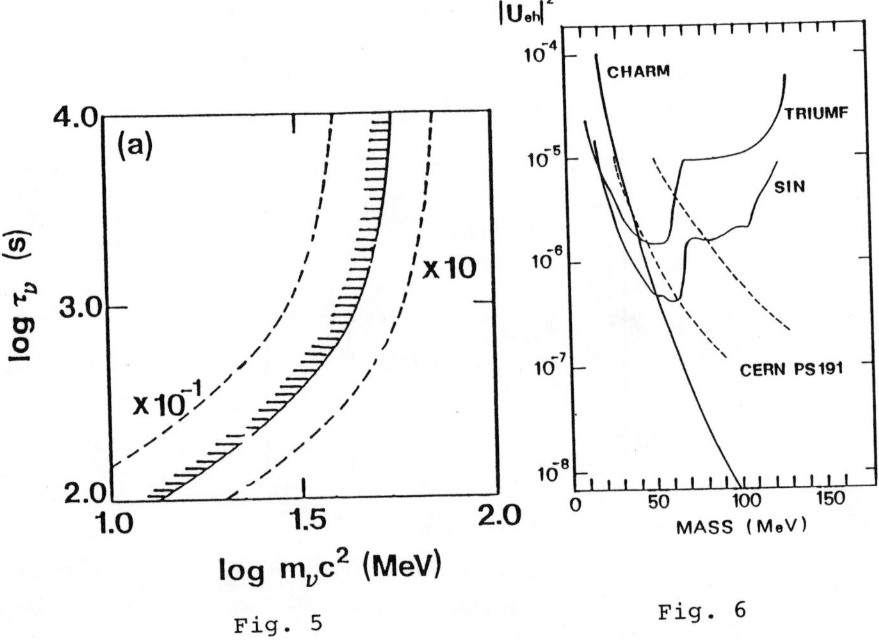

Fig. 5   Fig. 6

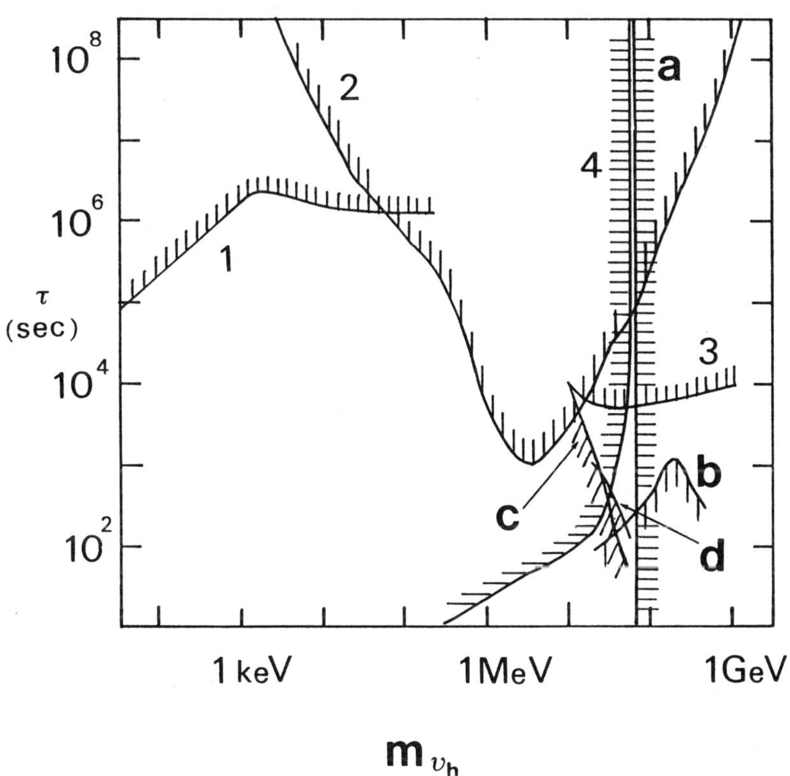

Fig. 7

ON THE POSSIBLE ASTROPHYSICAL EVIDENCE

FOR A RADIATIVELY DECAYING NEUTRINO

J. Maalampi
Research Institute for Theoretical Physics
University of Helsinki, Helsinki, Finland

K. Mursula
NORDITA, Copenhagen, Denmark

M. Roos
Department of High Energy Physics
University of Helsinki, Helsinki, Finland

ABSTRACT

We consider the possibility that the step-like structure observed in the
cosmic background radiation at $\lambda = 1667 \pm 10$ Å would indicate radiative decays
of massive cosmic neutrinos. In the standard electroweak theory the corresponding radiative lifetime is several orders of magnitude longer than observed,
due to the purely left-handed structure of the charged weak currents. We point
out that in models which include also right-handed currents, the lifetime can
naturally be short enough to account for the astrophysical observations.

The cosmic ultraviolet background radiation may contain information on the decay of a big bang relic massive neutrino[1-4]. For a neutrino ν_2 with mass m_2 in the eV region, the radiative decay into a lighter neutrino ν_1

$$\nu_2 \rightarrow \nu_1 + \gamma , \tag{1}$$

would manifest itself as a monochromatic line at E_0 in the far UV spectrum. Decays during the distant past would redshift this line, producing a step-like increase at $E \leq E_0$.

A first hint of such a step reported[5] an increase in the intensity of the diffuse high-latitude spectrum from 260 ± 40 (in photons/s·cm^2·sr·Å units) in the range $1300 - 1525$ Å, to ~ 600 in the range $1680 - 1800$ Å. However, since different detectors were used in the two wavelength bands, systematic errors could have caused the step.

Subsequently, rocket observations[6] with three UV broad band photometers covering the bandpasses $1450 - 1780$ Å (peak sensitivity at 1590 Å), $1610 - 1950$ Å (peak 1720 Å), and $1700 - 2420$ Å (peak 2125 Å) indicated extragalactic diffuse radiation intensities of 550 ± 150, 900 ± 150, and $\lesssim 1300$, respectively.

Recently, a compilation[7] of exposures at high galactic latitudes with the International Ultraviolet Explorer claims a step-like signal of 5σ significance at $\lambda_0 = (1667 \pm 10)$ Å in the otherwise flat continuum. This corresponds to a photon energy $E_0 = 7.4$ eV or, due to the relation $m_1^2 = m_2 \, (m_2 - 2E_0)$, a neutrino mass of $m_2 = 14.9$ eV, assuming $m_1 = 0$ eV. Note that photons of this energy propagate freely through the intergalactic medium as well as through the interstellar medium and are consistent with the existence of neutral hydrogen clouds in galactic halos[8].

The intensity change ΔI_0 at the step is[7] 454 ± 274, or within errors the same as in the previous observations[5,6]. One then deduces the lifetime of a hypothetical ν_2 from[1,2]

$$\tau_{obs} = \frac{c \, n(\nu_2)}{4\pi \, \Delta I_0 \, H_0 \, \lambda_0} , \tag{2}$$

where $n(\nu_2)$ is the number density of ν_2's in the Universe today. Assuming $n(\nu_2) \simeq 100$ cm^{-3} for the decaying neutrinos, one finds

$$2 \cdot 10^{15} \text{ yr} \lesssim \tau \lesssim 16 \cdot 10^{15} \text{ yr}. \tag{3}$$

Note that this result does not contradict the lower bound, $\tau \geq 3 \cdot 10^{15}$ yr, derived[2,3] from other astrophysical data. It is remarkable that the upper limit is several orders of magnitude smaller than the lifetime one would expect[9] within the standard $SU(2)_L \times U(1)_Y$ theory of electroweak interactions.

The effective matrix element for the decay (1) compatible with gauge invariance is of the general form[10]

$$\mathcal{M} = \epsilon^\mu \, q^\nu \, \bar{v}_1(p_1)\sigma_{\mu\nu}(a + b\gamma_5)v_2(p_2), \tag{4}$$

where ϵ^μ and $q^\nu = (p_2-p_1)^\nu$ are the polarization vector and the momentum of the photon, respectively. From (4) one derives

$$\frac{1}{\tau} = \Gamma(\nu_2 \to \nu_1\gamma) = \frac{1}{8\pi}\left(\frac{m_2^2 - m_1^2}{m_2}\right)^3 (|a|^2+|b|^2) = E_0^3\,(|a|^2+|b|^2)/\pi \tag{5}$$

The decay (1) in the standard model proceeds (in the unitary gauge) through the two one-loop diagrams shown in Fig. 1.

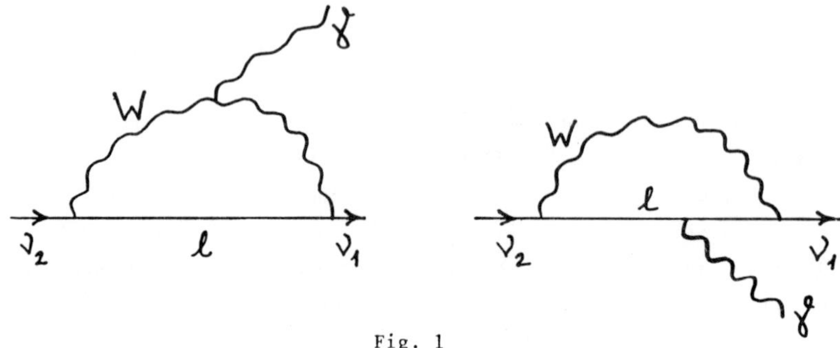

Fig. 1

A full standard model calculation[9] with $m_2 = 15$ eV and $m_1 = 0$ gives

$$\tau_{\text{Stand.}}^{-1} = \Gamma(\nu_2 \to \nu_1\gamma) = \frac{K\alpha G_F^2}{128\pi^4}\, m_2^5 \, \left|\sum_\ell U_{1\ell}U_{2\ell}F(r_\ell)\right|^2 \simeq K(3\cdot10^{23}\,\text{yr})^{-1}\left|\sum_\ell U_{1\ell}U_{2\ell}F(r_\ell)\right|^2 \tag{6}$$

where $K = 1$ (or 2) for Dirac (or Majorana) neutrinos, U is the leptonic mixing matrix and $F(r_\ell) = O(1)$ is a smooth function of $r_\ell = m_\ell/M_W$. With three light charged leptons ($r_\ell \ll 1$), the leading term in the sum is cancelled due to the unitarity of U, and the rate is thus additionally suppressed by the factor r_ℓ^2. Nevertheless, even if this GIM cancellation were avoided by adding new heavy generations[9] or by making the unitarity sum incomplete (e.g. with an extra singlet ν), the natural lower bound is $\tau \gtrsim 10^{23}$ yr for a 15 eV neutrino in the $SU(2)_L \otimes U(1)_Y$ gauge theory.

The essential feature of Eq. (6) is that the scale is (apart from G_F) set solely by the mass of the neutrino as a consequence of the pure left-handed structure of $W^\pm$-lepton interactions in the standard model. Indeed, if the two $W\ell\nu_i$ vertices in the decay amplitude (see Fig. 1) are both of the same chirality, only the momentum part of the internal charged lepton propagator $(\not{p}_\ell - m_\ell)^{-1}$ contributes to the matrix element. All terms of the gauge invariant form (4) are then proportional to the neutrino mass, thus $\Gamma \sim m_2^5$. However, if the $W\ell\nu$

vertex would contain a right-handed piece, also the mass part of the lepton propagator would contribute to the amplitude.

In the first model based on the $SU(2)_L \times SU(2)_R \times U(1)$ gauge group, a RH neutrino $\nu_R(1,2,-1)$ can decay radiatively to an LH neutrino $\nu_L(2,1.-1)$, provided the vector bosons $W_R^\pm$ and $W_L^\pm$ mix (see Fig. 2). The lighter of the

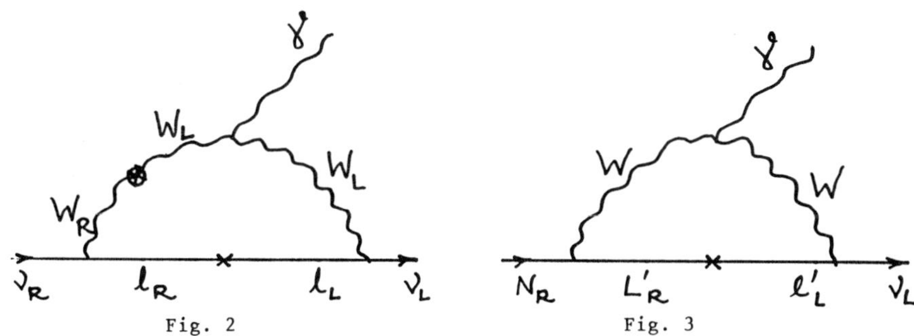

Fig. 2   Fig. 3

two vector boson mass eigenstates, $W_1 \equiv \cos\zeta W_L + \sin\zeta W_R$, couples to leptons according to

$$\mathcal{L} = -W_1^\mu \frac{g}{\sqrt{2}} (\cos\zeta \, \bar{\nu}_L \gamma \, \ell_L + \sin\zeta \, \bar{\nu}_R \gamma_\mu \, \ell_R) + h.c., \tag{7}$$

where ζ is the bosonic mixing angle. We neglect here the possible intergenerational mixing and the mixing between ν_L and ν_R, which are inessential to our argument. The contribution of the diagrams mediated by the heavier vector boson W_2 can be neglected since $M_{W_2} \geq 20 \, M_{W_1} \simeq 1.6$ TeV from the $K_L - K_S$ mass difference and other low-energy constraints[14]. The decay width for $\nu_R \to \nu_L \gamma$ (or $\nu_L \to \nu_R \gamma$) is given by

$$\tau_{LR}^{-1} = \Gamma(\nu_R \to \nu_L \gamma) \simeq (4 \cdot 10^{13} \, yr)^{-1} \left(\frac{E_0}{7.4 \, eV}\right)^3 \left(\frac{m_\ell}{MeV}\right)^2 \sin^2 2\zeta \tag{8}$$

independently of the neutrino masses for fixed E_0.

The lifetime one deduces from astrophysics depends on the present cosmological density $n(\nu_R)$ of the ν_R. This may be smaller than the density of the standard LH neutrinos, $n(\nu_L) \simeq 100 \, cm^{-3}$, since the more weakly interacting RH neutrinos decouple from the thermal equilibrium earlier[15]. If they decoupled after the muons were annihilated they would today be as abundant as the ν_L. This corresponds to the situation when $M_{Z_R} \lesssim 3$ TeV. In this case we obtain, by comparing Eqs. (3) and (8), the condition

$$0.05 \lesssim \frac{m_\ell}{MeV} \sin 2\zeta \lesssim 0.14 . \tag{9}$$

If ν_R belongs to the electron family, Eq. (9) requires $\zeta \gtrsim 0.05$. This possibility is clearly excluded since the experimental upper bound from the K_L - K_S mass difference is $\zeta_{exp} \lesssim 0.004$[14] On the other hand, the ν_R could still be a member of the μ or τ families: then Eq. (9) predicts $\zeta = (2-7) \cdot 10^{-4}$ or $\zeta = (1-4) \cdot 10^{-5}$, respectively.

If the ν_R density $n(\nu_R)$ is less than the standard neutrino density $n(\nu_L)$, the value of τ_{obs} in Eq. (3) is reduced by a factor $n(\nu_R)/n(\nu_L)$ and correspondingly larger values of the angle ζ are required.

In the mirror model[12] the situation is favourable in two respects. Firstly, the cosmological density of mirror neutrinos N_R is never suppressed since their neutral current interactions have the same strength as those of the standard LH neutrinos. Secondly, charged mirror leptons, L, must be heavy, $20 \lesssim m_L \lesssim 250$ GeV, and therefore the radiative width increases. The relevant part of the weak Lagrangian is given by[16]

$$\mathcal{L} = -W^\mu \frac{g}{\sqrt{2}} (-\sin\theta_\ell \, \bar{\nu}_L \gamma^\mu L_L + \cos\theta_\ell \bar{N}_R \gamma^\mu L_R) + \text{h.c.}, \tag{10}$$

where θ_ℓ is the ℓL mixing angle parametrizing the mixing between the weak eigenstates of the charged lepton ℓ and its mirror partner L. One typical diagram contributing to the decay $N_R \to \nu_L \gamma$ is shown in Fig. 3. The width is, independently of the neutrino masses for fixed E_0, given by

$$\tau_{Mirror}^{-1} = \Gamma(N_R \to \nu_L \gamma) \simeq K(2 \cdot 10^5 \text{ yr})^{-1} \left(\frac{E_0}{7.4 \text{ eV}}\right) \left(\frac{m_L}{20 \text{ GeV}}\right) \sin^2 2\theta_\ell. \tag{11}$$

Comparing this to τ_{obs} of Eq. (3), we find for $20 \le m_L \le 250$ GeV

$$2 \cdot 10^{-7} \lesssim \theta_\ell \lesssim 1 \cdot 10^{-5} \tag{12}$$

which is far below the present experimental upper limit[16], $\theta_\ell \lesssim 0.1$. Let us note that values of the mixing angle in this range are suggested by some specific mirror models[17].

We conclude that the standard electroweak model cannot account for the short radiative lifetime, $\tau_{obs} \lesssim 10^{16}$ yr, of a massive neutrino, as suggested by astrophysical observations[5-7]. This is due to the purely LH structure of the weak interactions which makes the radiative width $\Gamma_{rad}(\nu_2 \to \nu_1 \gamma)$ proportional to $E_0^3 m_2^2$, where $E_0 = 7.4$ eV is the observed photon energy and m_2 is the mass of ν_2. Accordingly, in the standard model the radiative lifetime for neutrinos in the allowed range ($m_i \lesssim 50$ eV) is always larger than 10^{22} yr.

In non-standard electroweak models which encompass also RH currents, Γ_{rad} is proportional to $E_0^3 m_\ell^2$, where ℓ is a charged lepton. Consequently, the lifetime is shortened by the enormous factor $(m_\ell/m_2)^2$ compared to the standard model.

This enables RH models to explain the astrophysical observation[18]. The predictions for the model parameters are given in Eqs. (9) and (12), respectively. The implied small mixing angles would make it very hard to see the effects of RH interactions of known particles in laboratory experiments. However, N_R would contribute to the total width of Z^0 by an equal amount as the standard neutrino.

We note that ν_R or N_R will together with ν_{eL}, $\nu_{\mu L}$ and $\nu_{\tau L}$ saturate the famous cosmological bound of at most 4 light neutrinos, derived from the present helium abundance in the Universe. If this limit is taken seriously, the RH or mirror neutrinos of the other families should be much heavier or much more unstable than the one discussed here.

Although the astrophysical information used is controversial[5], or scanty[6], or preliminary[7], there is hope for very precise data in the near future, when the Berkeley Extreme Ultraviolet/Far Ultraviolet Shuttle telescope will be launched on the Space Shuttle[19].

It is a pleasure to acknowledge the useful and stimulating discussion we have had (K.M. and M.R.) with D. Fargion, Rome, and (M.R.) with S. Bowyer and C. Martin, Berkeley. J.M. has been supported by Bundesministerium für Forschung und Technologie, Bonn, F.R. Germany.

REFERENCES

1. R. Cowsik, Phys. Rev. Lett. **39**, 784 (1977);
   D. Dicus, E. Kolb and V. Teplitz, Astrophys. J. **221**, 327 (1978);
   A. de Rujula and S. Glashow, Phys. Rev. Lett. **45**, 942 (1980).
2. F.W. Stecker, Phys. Rev. Lett. **45**, 1460 (1980);
   F.W. Stecker, and R.W. Brown, Astrophys. J. **257**, 1 (1982).
3. R. Kimble, S. Bowyer and P. Jacobsen, Phys. Rev. Lett. **46**, 80 (1981).
4. D. Fargion and M. Roos, Phys. Lett. **147B**, 34 (1984).
5. M. Maucherat-Joubert, P. Cruvellier and J.M. Deharveng, Astron. Astrophys. **70**, 467 (1978); R.C. Anderson et al., Astrophys. J. **234**, 415 (1979).
6. P. Jakobsen et al., Astron. Astrophys. **139**, 481 (1984).
7. G. Auriemma et al., INFN-Rome preprint **447** (1985, unpublished).
8. A. Melott and D. Sciama, Phys. Rev. Lett. **46**, 1369 (1981);
   Y. Rephaeli and A. Szalay, Phys. Lett. **106B**, 73 (1981).
9. P.B. Pal and L. Wolfenstein, Phys. Rev. **D25**, 766 (1982), and references quoted therein.
10. W. Marciano and A. Sanda, Phys. Lett. **67B**, 303 (1977);
    S. Petcov, Sov. Nucl. Phys. **25**, 340 (1977); E **25**, 698 (1977).
11. J.C. Pati and A. Salam, Phys. Rev. **D10**, 275 (1974);
    R.N. Mohapatra and J.C. Pati, Phys. Rev. **D11**, 566 (1975); ibid. 2558;
    G. Senjanovic and R.N. Mohapatra, Phys. Rev. **D12**, 1502 (1975).
12. See e.g., J.C. Pati and A. Salam, Phys. Lett. **58B**, 333 (1975);
    J. Maalampi and K. Enqvist, Phys. Lett. **97B**, 217 (1980);
    F. Wilczek and A. Zee, Phys. Rev. **D25**, 553 (1982).
13. K. Enqvist, K. Mursula and M. Roos, Nucl. Phys. **B226**, 121 (1983).
14. See e.g., L. Wolfenstein, Phys. Rev. **D29**, 2130 (1984).
15. G. Steigman, K.A. Olive and D.N. Schramm, Phys. Rev. Lett. **43**, 239 (1979).
16. J. Maalampi, K. Mursula and M. Roos, Nucl. Phys. **B207**, 333 (1982).
17. K. Enqvist and J. Maalampi, Z. Phys. **C21**, 345 (1984).
18. See also J. Maalampi, K. Mursula and M. Roos, Phys. Rev. Lett. **56** (1986).
19. S. Bowyer, private communications.

THE BAKSAN EXPERIMENT ON THE SEARCH FOR STELLAR
COLLAPSE AND SOLAR FLARE NEUTRINOS

Alexeyev E.N., Alexeyeva L.N., Chudakov A.E., Krivosheina I.V.
Institute for Nuclear Research, Academy of Sciences
of the USSR, Moscow, USSR

Abstract. The results on the search for stellar collapse
events during June, 30, 1980 - May, 14, 1986, using the
Baksan underground scintillation telescope, are summarized.
These data have been used to look for solar flare neutrinos.
There is no correlation between the Baksan and the Homestake solar neutrino detector data. No events interpreted
as due to stellar neutrino burst or solar flare-produced
neutrinos have been obtained. Some of the Baksan observational results are presented.

Introduction. The experiment is made to search for rare
event of the expected antineutrino burst, accompanying the
stellar collapse and cooling of newborn hot neutron star.

The theory after more than two decades of its development leads to a conclusion that neutrinos are emitted in a
burst of $\sim$1-20 seconds, with average energies between 8 and
20 Mev and the total radiated energy of $\sim 10^{53}$ergs /1,2/.
The total neutrino flux of all species will bring the information on the collapse dynamics and the main physical processes: possible core oscillations and shock waves, cooling
of new born pulsar or the black hole birth. From various
observational data the mean time between collapses in our
Galaxy is estimated of 10 - 30 years /3,4/.

The Baksan experiment on the search for stellar collapse
neutrinos started in autumn, 1979. The continuous Galaxy observation began in June, 1980.

The another subject of the experiment is the solar flare-
produced neutrinos. The main sources of neutrinos in solar
flares should be π-μ-e decays and radionuclei β^+ decays
due to interactions of particles accelerated to high energies /5,6/.

The duration of generation and interactions of the particles trapped in flare region is less than $\sim$ 1000 second.

This estimate results from x-ray and γ-ray observations /5/. The correlation between Davis' experiment results /7/and cosmic ray intensity have been analysed in detail by Bazilevskaya G.A. et al./8/. Suggesting the solar flares were responsible for the increase of the ^{37}Ar production rate in runs 19 and 27 of solar neutrino detector, the authors predicted that the large solar flare of October 7-12,1981 might be detected by the chlorine experiment. This prediction for run 71 (September, 2- October,28,1981) happened to be. Also the last highest run 86 correlated in time with the great solar flare of August 5, 1984 /9/. As the measurements of nuclear line intensities from flares can give the information on both the spectrum and total number of accelerated particles /6/, one can calculate the neutrino flux and its contribution to the ^{37}Ar production in chlorine detector. The spectra of electron and muon neutrinos produced from π-μ-e decays in solar flares were calculated by G.A.Kovaltsov /10/. So made estimates of flare neutrino fluxes from flare of August,4,1972 give a negligible contribution to the ^{37}Ar production in Davis's detector (run 27).

The contribution form β^+ decay neutrinos has been estimated by R.E.Lingenfelter et al /6/, using direct measurements of electron-positron annihilation line, and was found to be negligible (ten orders of magnitude less than Davis's rate). Thus there is no hope to explain the correlation of Davis's data with solar flares using a conventional flare model. A "hidden" neutrino source, though very unprobable, could do the job, a source in which not only the majority of accelerated protons, but also majority of produced gammas, absorbed in solar atmosphere.

Obviously, this situation makes worthwhile to seek for confirmation from others neutrino detectors, if they operated in the same periods. Any way we would like to try a pure experimental approach.

<u>Baksan scintillation telescope.</u> The Baksan scintilla-

tion telescope is located at a depth of 850 m.w.e. at the North Caucasus. The telescope consists of 3156 standard detectors arranged in eight planes, forming a closed parallelepiped with two horizontal planes inside it (fig.1.). As the telescope construction materials a special low background concrete has been used.

Each of 3156 detectors is an aluminium tank ($70 \times 70 \times 30$ cm^3), containing liquid scintillator, which is viewed by one 15 cm photomultiplier tube. The total telescope target mass is 330 tons. The detailed description of the telescope and the detection technique have been published elsewhere /11/.

The detection of electron and muon neutrinos in the scintillation telescope (target $C_n H_{2n+2}, n \sim 9$) is realized mainly via neutrino-target interactions:

$$\left.\begin{aligned} \nu_e + {}^{12}C &\rightarrow {}^{12}N + e^- \\ \tilde{\nu}_e + {}^{12}C &\rightarrow {}^{12}B + e^+ \\ \tilde{\nu}_e + p &\rightarrow n + e^+ \end{aligned}\right\} (1) \qquad \left.\begin{aligned} \nu_\mu + {}^{12}C &\rightarrow {}^{12}N + \mu^- \\ &\quad\ \hookrightarrow e^- + \tilde{\nu}_e + \nu_\mu \\ \tilde{\nu}_\mu + {}^{12}C &\rightarrow {}^{12}B + \mu^+ \\ &\quad\ \hookrightarrow e^+ + \nu_e + \tilde{\nu}_\mu \\ \tilde{\nu}_\mu + p &\rightarrow n + \mu^+ \\ &\quad\ \hookrightarrow e^+ + \nu_e + \tilde{\nu}_\mu \end{aligned}\right\} (2)$$

The threshold energies (12,5 MeV-discriminators included) are:

$$E_{th} = \begin{cases} 12,5 \text{ MeV for electron antineutrinos} \\ 30 \text{ MeV for electron neutrinos} \\ 120 \text{ MeV for muon neutrinos} \end{cases}$$

<u>Electron neutrino</u> interactions are recorded, using so called "single detector mode", assuming the energy of electron or positron resulted in reactions (1) as relatively small. The trigger system selects an event, if only one detector from all 3156 gives a signal. Such selection mode is similar to the anticoincidence scheme, when each detector is protected by all the rest from cosmic ray muons (the main source of background).

The total information (detector coordinates, amplitude,

time is stored in on-line computer. The detectors of three
internal planes (1200 detectors, 130 tons of scintillator)
were chosen as a fiducial mass of the telescope because of
their low background in the "single detector mode" /11/.

__Muon neutrinos__ interacted due to reactions (2) in the
scintillator or in the construction concrete layers (fig.
1) can be identified by succession of two signals: stopped
muon signal and then muon decay electron signal. In this
particular case three internal horizontal planes serve as
fiducial mass, five external planes serve as anticoincidence shield.

The main source of background in this case are cosmic
ray muons passed nondetected through the external plane detectors and then stopped and decayed in the fiducial scintillators.

A. Stellar collapse neutrinos

__Analysis of the experimental data.__ To search for the stellar neutrino bursts, the Baksan scintillation telescope is
most sensitive to the electron antineutrinos via the char-
ged-current reaction on protons (reaction (1): $\tilde{\nu}_e + p \rightarrow n + e^+$.
The information is analysed by sliding time window method.
The window of 20 seconds slides along the time scale, being opened by each signal, and the number of signals is
counted in every such time interval.

We mark a succession of signals in time window of 20
seconds by the "event". The $\tilde{\nu}_e$ -burst search experiment
started in autumn, 1979 with the background counting rate
in fiducial telescope region of $\sim 0.43 \text{ s}^{-1}$.

At present the background rate is $\sim 0.014 \text{ sec}^{-1}$. The
general decrease of the background is due to gradual elimination of the contribution from "noisy" photomultiplier
tubes (second main source of background).

The expected rates for background of $\sim 0.014 \text{ sec}^{-1}$
are shown in Table 1.

Table 1. The expected rates of background events

Number of pulses in event	3	4	5	6	7	8
Rate of events	36 day^{-1}	3 day^{-1}	7 month^{-1}	5 yr^{-1}	1 per 4.5 yr	1 per 112yr

The minimal number of signals in an event to be analysed is equal to 4 at present time.

Results. The results of the first six years of observation are shown in fig.3-5. The live time up to May 14, 1986 is 4.74 years, or 80 per cent of total observational time. The 20 per cent of total time was lost due to apparatus "dead" time, repair days and interruptions.

Fig.3 shows all events with number of signals K = 5, 6... (black points, curve 1) detected during the period of 1980-1985 yrs. All of them were apparently of background origin. Since the averaged background counting rate in that time was relatively high (~ 0.03 s^{-1}), mainly due to photomultiplier tube noises, we applied discrimination of events with signals majority of which was made by "noisy" detectors.

These detectors were found out by weekly monitors. Results after such statistical analysis are shown in the same figure curve 2. The distribution of remaining events well described by effective background rate of ~ 0.0215 sec^{-1}, which was essentially constant in spite of the season changes of the total counting rate.

Fig.5a shows the relation "number of signals versus event duration", calculated for the effective background rate. The figures to the right of the curve indicate the amount of the selected events.

There were no events with probability of appearance much less than that expected for background. As the total counting rate has been decreased and the individual counting rates of detectors have been considerably equa-

lized, since July 1985 the discrimination of events has
not been used any more.

Fig.4 and fig.5b show the spectra analogous to, respectively, fig.3 and fig. 5a, but obtained since July 1985.
Discussion. In the live time of 4.74 years no candidates
for the stellar neutrino burst have been found in our experiment. Corresponding lower limit on the mean time between collapses is:

$$\Delta T > 2 \text{ years} \qquad\qquad (90 \text{ \% C.L.})$$

The maximum number of signals in the selected events was
K = 7 (fig.5a). If we require a 90% probability of recording a real event, its mean value should be $K \geqslant 11$. This
corresponds to the neutrino fluence $J_{\widetilde{\nu}_e}$ of: $J_{\widetilde{\nu}_e} \cdot \sigma_{\widetilde{\nu}_e p} \cdot \eta \geqslant 11$,

$J_{\widetilde{\nu}_e} \geqslant 1.85 \cdot 10^{11} \, \widetilde{\nu}_e / cm^2$, which is a characteristic sensitivity of the Baksan scintillation telescope (here N_p-
number of protons in target, $\sigma_{\widetilde{\nu}_e p}$ - cross-section of $\widetilde{\nu}_e p$
interactions, η -detection efficiency).

We can compair this fluence with that predicted by different
model calculations. Fig.8 shows the Baksan telescope ran-
ge versus the total energy of the $\widetilde{\nu}_e$ emission. Horizontal lines indicate the total energy of emission released
in the models of "hot"/1/ and "cold" /2/ collapse. Thus,
we derive the Baksan telescope ranges and corresponding
fraction of stars in the Galaxy /3/ to be 10 Kpc and 0.53
for "cold" collapse model and 15 Kpc and 0.84 for "hot"
one.

The main characteristics of the Baksan scintillation
telescope are shown in Table 2.

B. Solar flare-produced neutrinos

We utilized the experimental data of the previous research programme to look for the effect from the great solar flares of October 7-12,1981 and August 5,1984, which,
as supposed, affected Davis' solar neutrino detector.
The responses of Baksan and Homestake detectors versus
neutrino energy are shown in fig.2. The calculated spect-

ra of flare neutrinos were taken from /10/, the cross - sections of neutrino-carbon interactions from /14/, neutrino-chlorine interactions- from /15,16/. One can conclude that the flare neutrino effect with neutrino energies of $E_{\nu_e, \nu_\mu} \sim 100$-150 MeV is much stronger in Baksan telescope. We took $E_{\nu_e} \sim$ 100 MeV and $E_{\nu_\mu} \sim 150$ MeV to estimate the effects in both Baksan and chlorine detectors. These energies correspond to the effective values from /10/.

Only part of the collapse fiducial scintillators is used to detect the flare-produced $\nu_e, \widetilde{\nu}_e$, (reactions (1), namely ~ 90 per cent of three internal horizontal plane detectors. This part of the telescope has the lowest background counting rate in "single detector mode".

The fiducial mass for solar $\nu_e \widetilde{\nu}_e$ detection corresponds to ~ 110 tons ($5,3.10^{30}$ ^{12}C nuclei). The background rate is 0.01 sec^{-1}.As to muon neutrino detection (reactions (2)) the fiducial mass corresponds to ~ 130 tons ($5,6.$ $.10^{30}$ ^{12}C nuclei), the background rate is ~ 1.4 day^{-1}.

In both cases the data were analysed, using time windows of 900 seconds, maximum expected solar flare duration. <u>Results.</u>We examine at first the telescope data obtained in the period of the great solar flare of October 7-12, 1981. In the chlorine detector the measured ^{37}Ar production rate was equal to 1.21 $\pm$ 0.37 day^{-1}, while the average one was 0.40 $\pm$ 0.04 day^{-1}. The ^{37}Ar atom excess produced, possibly, by flare neutrinos was 56 $\pm$ 30 /7/. Hence, we can find the additional neutrino fluence:

$$\mathcal{J}_{\nu_e} = \frac{56 \pm 30}{2.2 \cdot 10^{30} \cdot 5 \cdot 10^{-40}} = (5.1 \pm 2.7) \cdot 10^{10} \quad \nu_e / cm^2$$

where $2,2.10^{30}$- number of ^{37}Cl atoms in the chlorine detector /7/, $5.10^{-40} cm^2$-cross-section of $^{37}Cl (\nu,e)^{37}Ar$ reaction by neutrino E_{ν_e} = 100 MeV /16/.

In the fiducial mass for solar $\nu_e \widetilde{\nu}_e$ detection of the Baksan telescope the total detector counts thus expected from the electron neutrinos would be:

$$N = (5,1 \pm 2,7) \cdot 10^{10} \cdot 1,3 \cdot 10^{-39} \cdot 5,3 \cdot 10^{30} \cdot 0,5 = 176 \pm 93$$

where $1,3 \cdot 10^{-39}$ cm^2 - cross-section of ^{12}C $(\nu_e , e)^{12}$N reaction by neutrino $E_{\nu_e} = 100$ MeV,

0,5 - detection efficiency.

Fig.6 shows counting rate per 900 sec during the period of October 6-13,1981, including both flares of October 7 and October 12. The pointer indicates the flare local time. No excess of counts were observed. The average number of counts was 9.

The baksan upper limit (90% C.L.) on solar neutrinos ($E_{\nu_e} = 100$ Mev) made in flare of October 7-12,1981 versus the flare duration is shown in Fig . 7, curve 1. The darkened horizontal region represents the fluence needed to account for measured excess of ^{37}Ar atoms in run 71. Fig.7 shows, that 100 MeV ν_e-hypothesis is strongly disapproved, if we assume the neutrino flare duration of $\lesssim 10^5$ sec ($\lesssim 1$ day).

The similar situation was in the case of another great solar flare of August,5,1984. No excess was observed in the Baksan telescope, though the measured ^{37}Ar production rate in respective run 86 in the Homestake solar neutrino detector was ~ 1.26 atom ^{37}Ar.day^{-1}/9/. This would imply a flare induced production of ~ 75 atoms of ^{37}Ar, when corrected for decay.

There were no positive effects in the Baksan telescope in the periods of these two flares, while the measured excess summarized over the flares of October 12, 1981 and August,1984 was estimated as $(56 + 75)=131$ atoms ^{37}Ar in Davis's detector. So,the summarized for two flares Baksan excess of detector counts was no more than 6 counts at the 90 % C.L. This implies a total flare-produced neutrino fluence of no more than $(\mathcal{I}_{\nu_e})_\Sigma \lesssim 1,7 \cdot 10^9$ ν_e/cm^2 ($E_{\nu_e} \approx 100$ MeV). We can also find the upper limit on solar flare neutrino fluence with $E_{\nu_e} \approx 30$ MeV. The energy of ~ 30 MeV is the mean value of rest μ^+-decay ν_e-spectrum . The average cross-section for $^{12}C\,(\nu_e,e)\,^{12}N$ reaction by these neutrinos was taken $1,6.10^{-41}$ cm^2, then the upper limit is: $\left(\Im_{\nu_e}\right)_\Sigma \lesssim 1,4.10^{11}$ ν_e/cm^2 at the 90% C.L. ($E_{\nu_e} \approx 30$ MeV). In the chlorine detector the ^{37}Ar productions due to these fluences would be only 43 atoms for $E_{\nu_e} \sim 30$ Mev or 2 atoms for $E_{\nu_e} \sim 100$ MeV. We now turn to a result of flare muon neutrino search in the Baksan telescope. Table 3 shows the observed and expected amounts of events with different number of μe decays for the period of September "-October 28,1981, the duration of the run 71 of chlorine detector, including the solar flare of October 7-12,1981.

Table 3. The events detected in the period of September 2, 1981-October,28,1981.

Number of μe decays in a event	1	2	3	3
Amount of expected background events	78	0.6	-	-
Detected events	67	1	-	-

All detected events were apparently due to the background. Because of the low background rate we obtain relatively good upper limit of muon neutrino fluency, even suggesting extremely long duration of neutrino flare, say 57 days for run 71. In this case $\Im_{\nu_\mu} < 0.8.10^9$ ν_μ/cm^2 (90% c.l.) The last extreme value is smaller than the above-stated upper limit on electron neutrinos with energies of ~ 100 MeV and flare duration of 900 seconds $\left(\left(\Im_{\nu_e}\right)_\Sigma \lesssim 1.7.10^9 \right.$ ν/cm^2). Suggesting the fluxes of ν_μ and ν_e in $\gtrsim 100$ MeV region to be of the same order of magnitude , this means that conventional neutrino energy spectra /10/ can be excluded in the discussion on solar flare-Homestake data correlations, independently on assumed flare duration.

Also the well known great flares of June 3,5 and 15,1982
were analysed too. These flares were not detected by Davis's detector, but were observed by high-energy γ -ray
emission ($E_{max} \sim$ 30-80 MeV) /12,13/. All these flares did
not give any visible effects in Baksan scintillation telescope.

<u>Conclusion</u>.The analysis of the Baksan experimental data
obtained in the periods of several great solar flares, especially the flares of October 12,1981 and August 5,1984,
in comparison with Homestake solar experiment data gives
no evidence of neutrino fluxes from the Sun, which could
be responsible for observed excess in Davis' detector. The
suggestion of $E_{\nu_e} \gtrsim$ 30 MeV is exclused, if assumed neutrino
flare duration is $\lesssim 10^3$ seconds, the conventional $\nu_e \, \nu_\mu$
mixture of $E_\nu \gtrsim$ 100 MeV is excluded independently on assumed duration. If the hypothesis of the flare neutrino-
produced excess of ^{37}Ar atoms in runs 71 and 86 $_w$ere correct we should suggest the neutrino energies to be less
than 30 MeV and the neutrino origin, probably, from ther-
monuclear reactions.

<u>Acknowledgements.</u> The authors would like to thank Bazi-
levskaya G.A. and Stozhkov Yu.I.for making some data on
solar flares available for us.

<u>References</u>

1. Imshennik V.S. and Nadyozhin D.K., Itogi Nauki i Tekh-
niki,ser.Astronomiya, v.21,p.63, 1982.
2. Burrows A., AIP Conf.Proc. No.126, p.283, New York,1985
3. Bahcall J.N., Piran T., Astrophys.J.Lett., v.267,
p.L77, 1983.
4. Alexeyeva L.N. Proc.Workshop INR "Particles and Cosmology", USSR, Baksanvalley, part. 1, p.32, 1984.
5. Kurzhevsky B.M., Uspekhi Fizicheskikh nauk, v.137,
p.237, 1982.
6. Lingenfelter R.E. et al., AIP Conf. Proc. No.126, p.
121, New York,1985.

7. Rowley J.K., Cleveland B.T., Davis R., AIP Conf. Proc. No.126, p.1, New York, 1985.

8. Bazilevskaya G.A. et al., J.Nuclear Phys.(USSR), v.39, p.856.

9. Davis R.,private communication.

10. Kovaltsov G.A., Izv.Akad.Nauk SSSR,ser.fiz.,v.45,p. 1151,1981.

11.Alexeyev E.N. et al., Proc.17th ICRC, Paris, France, 1981, v.7,p.110.

12. Rieger E. et al ., Proc.18th ICRC, Bangalore,India, 1983, v.10,p.338

13. Mc.Donald F.B., Van Hollebeke M.A.I., Astrophys.J. Lett.,v.290,p.L67,1985.

14. Bugaev E.V. et al., Nucl.Phys.,v.A324,p.350,1979.

15. Bahcall J.N., Rev.Mod.Phys.,v.50,p.881,1979.

16. Domogatsky G.V., Eramzhyan R.A., Izv.AN SSSR, ser. fiz.,v.41,p.1969,1977.

Table2.The main characteristics of the Baksan telescope for the collapse neutrino search programme

Depth (m.w.e.)	850
Target	C_nH_{2n+2}, $n \sim 9$
Total mass (tons)	330
Fiducial mass (tons)	130
Energy threshold (MeV)	12,5
Background counting rate (sec^{-1})	$\sim$0.014
Detection efficiency with account for the detector geometry and energy threshold	$0.5^{x)}$
Average dead time (%)	$\sim$7
Time window (sec)	20
Minimal number of signals in a time window to be analysed	4
Expected number of signals due to $\bar{\nu}_e$ detection	$11-24^{xx)}$
Detectable with probability of 90% fluence ($\bar{\nu}_e$ /cm^2)	$1.85 \cdot 10^{11}$

x) The spectrum of electron antineutrinos was taken with kT= 4 MeV, α = 0.04 /1/.

xx) The star is in the Galaxy centre.

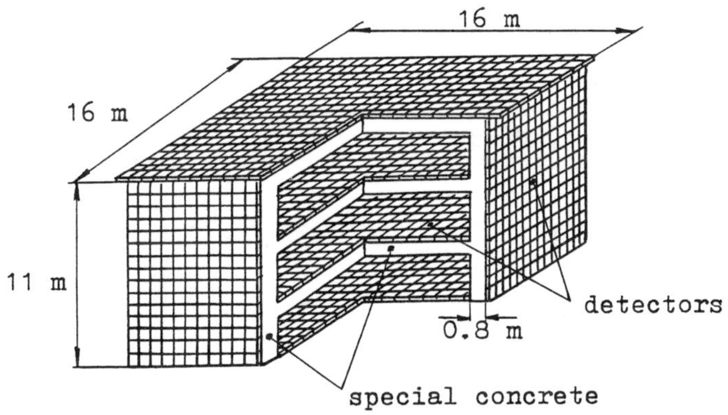

Fig.1 Baksan underground scintillation telescope

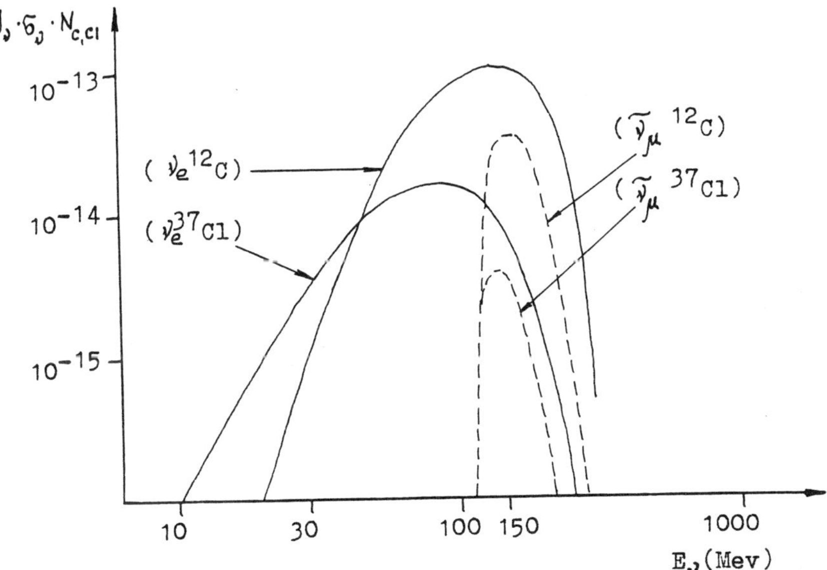

Fig.2 The responses of the Baksan and the Homestake detectors to the flare-produced neutrinos versus neutrino energy. $\mathcal{I}_\nu$ - spectra of ν_e and $_{37}\widetilde{\nu}_\mu/10/$, $_{12}\sigma_\nu$- respective cross-sections, $N_{C,Cl}$- amount of ^{37}Cl and ^{12}C atoms.

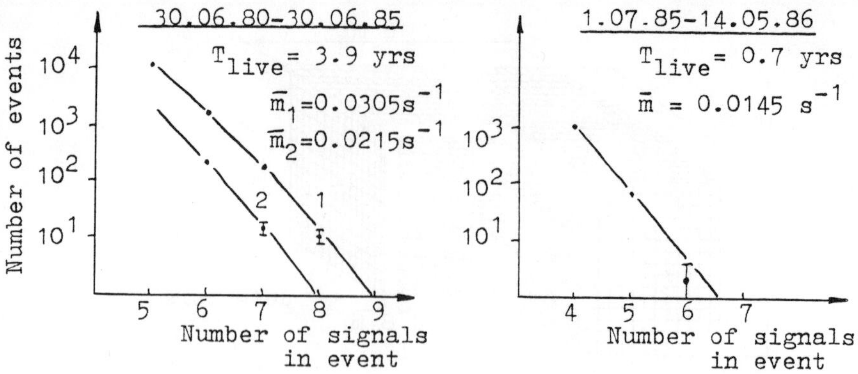

Fig.3 The spectrum of events detected in the period of June 30, 1980–June,30,1985. Curve 1-all events, curve 2 – selected events.

Fig.4 The spectrum of all events detected in the period of July,1,1985–May,14, 1986.

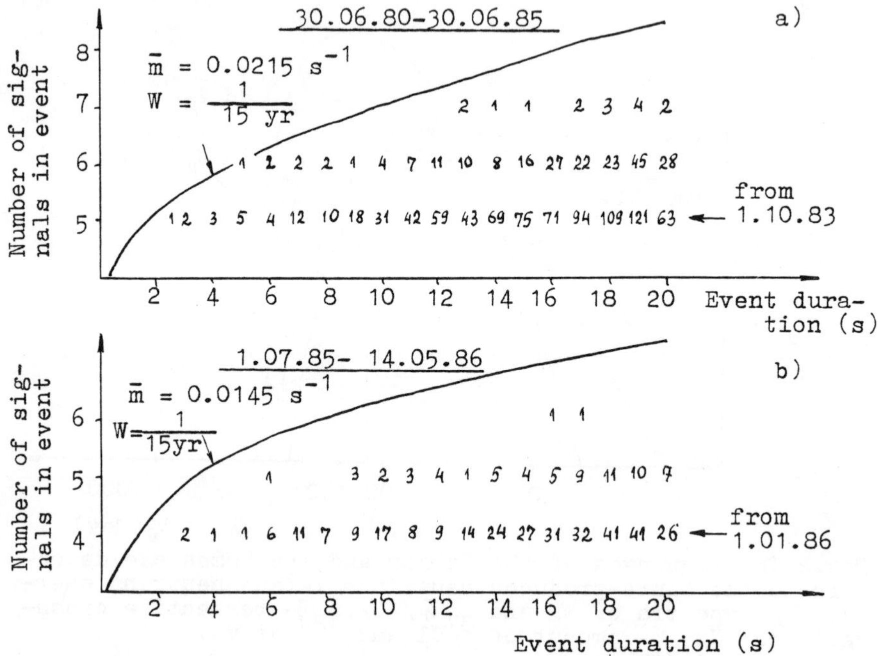

Fig.5 Relation "number of signals – event duration" for a)selected events (30.06.80–30.06.85) and b)all events (1.07.85–14.05.86).

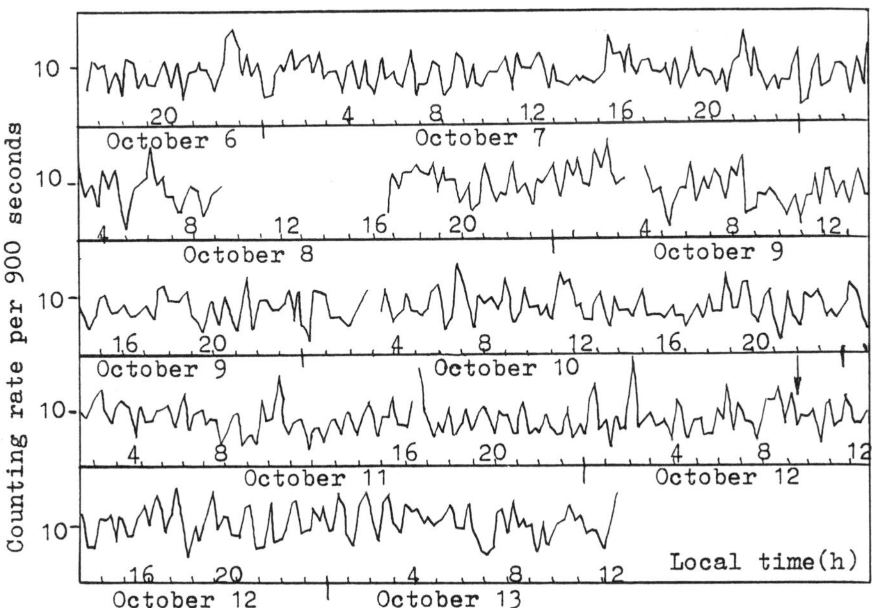

Fig.6 Counting rates per 900 sec in the period of October
6-13, 1981. The pointer indicates the solar flare local time.

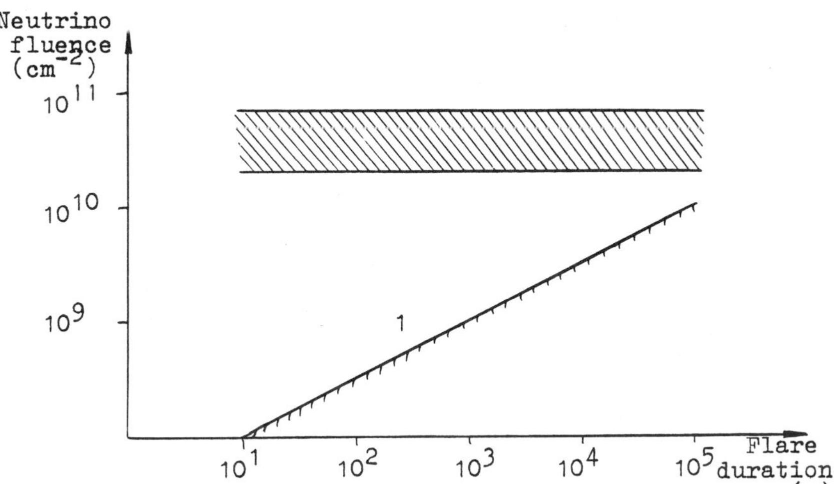

Fig.7 The sensitivity of the Baksan telescope to the flare-
produced neutrino fluence versus flare duration.
Horizontal lines indicates the fluence needed to account for
apparent excess of ^{34}Ar production in the Homestake detector.

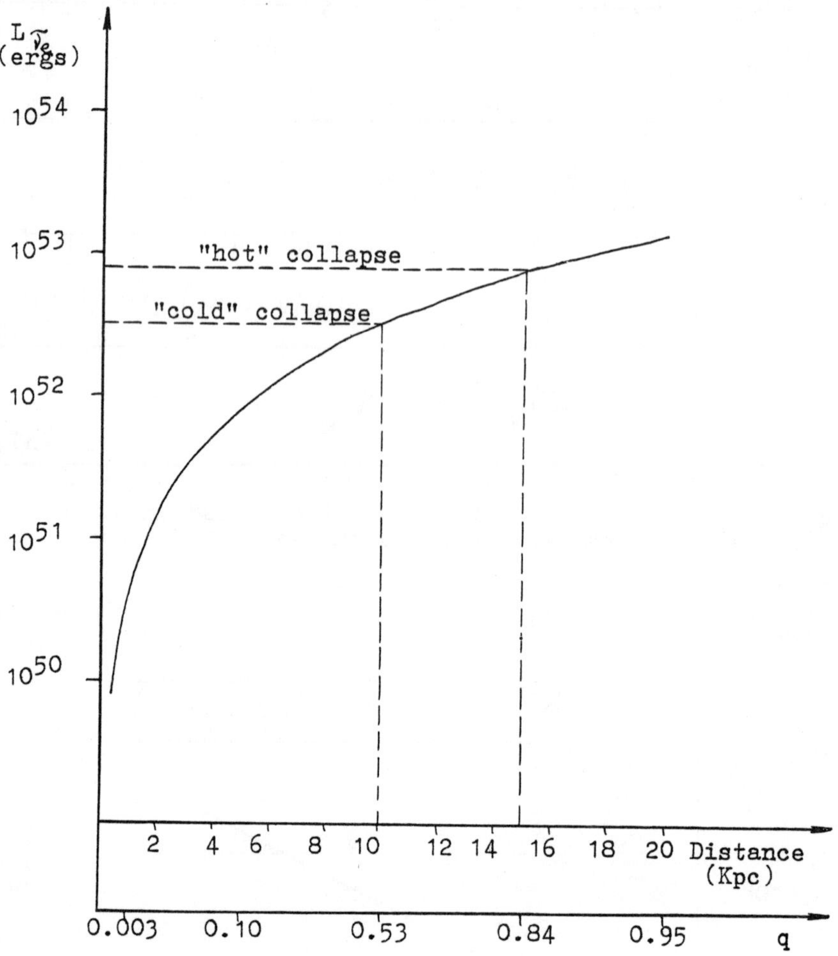

Fig. 8 The Baksan telescope sensitivity on the total antineutrino energy versus the distance to the collapsing star. (Detection probability - 90%).
q - fraction of Galaxy sampled /5/.

NEUTRINO EXPERIMENTS AT LSD AND ASD INSTALLATIONS

V.L.Dadykin, F.F.Khalchukov, V.B.Korchagin, P.V.Korchagin, E.V.Korol-
kova, A.S.Malgin, V.G.Ryassny, O.G.Ryazhskaya, V.P.Talochkin, V.F.Ya-
kushev, G.T.Zatsepin.
.Institute for Nuclear Research, Academy of Sciences, Moscow, USSR

M.Aglietta, G.Badino, G.F.Bologna, C.Castagnoli, A.Castellina, W.Ful-
gione, P.Galeotti, O.Saavedra, G.C.Trinchero, P.Vallania, S.Vernetto.
Istituto di Cosmogeofisica del CNR, Torino, Italy
Istituto di Fisica Generale dell'Università, Torino, Italy
(presented by O.G.Ryazhskaya)

Abstract

A 90 tons Liquid Scintillation Detector (LSD) is fully running since
october 1984 in the Mont Blanc underground laboratory, at the depth
of 5200 hg/cm^2 of standard rock. Being very well shielded with iron
slabs against the local radioactivity, the experiment is extremely
sensitive to detect low energy particles, at a threshold of 5 MeV,
and has a very good signature to detect electron antineutrinos
through capture reactions by protons. We report here the results of
measurements showing the possibility to detect solar neutrinos, and
discuss also the possibility to search for low energy neutrinos
correlated with the solar activity, in particular during large solar
flares. Using 6 internal scintillation counters, containing 7.2 tons
of liquid scintillator, we have observed 2 events during 75 days of
running time. From our data an upper limit of 5 10^4 cm^{-2} s^{-1} at the 90%
c.l. for the atmospheric $\bar{\nu}_e$-flux from μ-e decay was estimated in the

energy range 12-60 MeV.

The Artyomovsk Scintillation Detector (ASD), which is a single 105 tons module, operates since 1978 in a salt mine at the depth of 570 hg/cm^2 of standard rock. During more than 4 years of the ASD running time, and 250 days of the LSD running time, no candidate of a $\bar{\nu}_e$-burst from collapsing stars has been recorded. The simultaneous operating of both detectors during 1985-86 does not show any burst candidate too.

The LSD experiment

The LSD experiment, described in detail elsewhere[1,2], consists of 72 scintillation counters (1.5 m^3 each) containing 90 tons of liquid scintillator[3]. The scintillator is watched by 3 FEU-49B photomultipliers (15 cm diameter) on the top of each counter.

Each counter is inserted in a Fe container (2 cm thick), and the whole detector is also shielded with Fe slabs against the local radioactivity background: 15 cm thick on the floor, and 2 cm thick on the walls and on the ceiling of the laboratory. The total iron mass is 200 tons. The shielding allows us to adjust the energy threshold of the counters to rather low levels: E_{th} = 5 MeV for the 16 internal scintillation counters, and E_{th} = 7 MeV for the 56 external ones.

The main purpose of the experiment is to detect bursts of electron antineutrinos from collapsing stars through the capture reactions:

$$\bar{\nu} + p = n + e^+$$
$$\llcorner\!\!\longrightarrow n + p = d + \gamma$$

$$\tau_{capture} = 170 \ \mu sec; \ E_\gamma = 2.2 \ MeV$$

The most important feature of LSD is its capability to detect both products of an $\bar{\nu}_e$ interaction with protons, namely positrons and gammas from (n,p) capture in a delayed coincidence. For this aim, a 500 μs gate ($\sim$ 3 times the average delay) is opened by pulses with energy $E \geqslant E_{th}$; pulses with $E \geqslant 0.7$ MeV are recorded during the gate duration. The efficiency to detect gammas from neutron capture is on the average $\varepsilon \approx 0.7$, depending on the position of the neutrino interaction inside the counter.

The total background counting rate over the energy threshold is $N(\geqslant E_{th}) = 6.2 \ 10^{-3} s^{-1}$; the counting rate above the 0.7 MeV energy threshold is in the range $10^2 - 10^3 \ s^{-1}$ depending on the position of the scintillation counter. Indeed, the inner and well shielded counters have a very low background counting rate: on the average 0.05 pulses during the gate duration. Background pulses in the gate are detected in a delayed coincidence with trigger pulses opening the gate, at the rate $N_c = 2.8 \ 10^{-4} \ s^{-1}$.

The background analysis

We discuss here the background features in LSD connected with the items of the experiment we are reporting in this paper, namely: 1) search for neutrino bursts from collapsing stars, 2) measurements of the electron antineutrino atmospheric flux from muon decay, and 3) possibility to detect solar neutrinos with massive scintillation detectors.

Fig.1 shows the background energy spectra of 3 scintillation counters in different positions in the Mont Blanc laboratory: the internal counter n.15, the lateral one n.27, and the counter n.72 at the top corner of the detector. Fig.1 clearly shows the background attenuation of an internal counter in comparison with an external one.

As regards the correlated pairs of pulses, the first one above E_{th} and the delayed second one above the 0.7 MeV threshold, our data show that 14% of the total counting rate of the internal counters on the ground layer is due to such events, while no correlated event has been detected by the internal counters of the middle layer during 75 days of analyzed data. Since the detection efficiency of prompt gammas from uranium decay is $\varepsilon = 10.5\%$ [4], and that of neutrons is $\varepsilon = 70\%$, we can conclude that the effective uranium contamination in LSD (averaged over the scintillator, phototubes glass, steel, etc..) is $\leq 1.8 \cdot 10^{-10}$ g/g at the 90 % c.l.

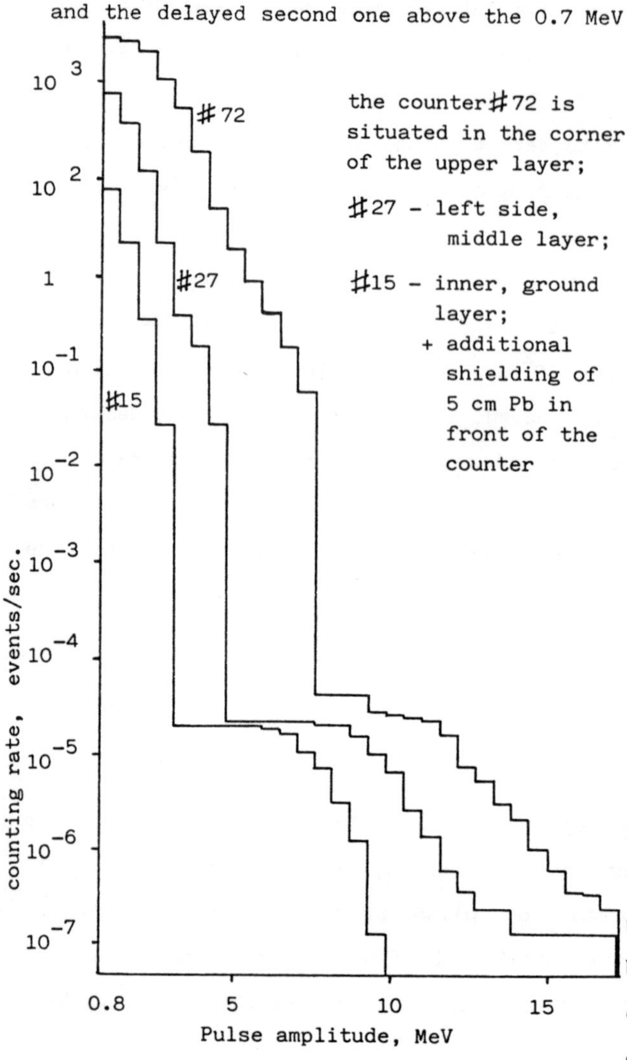

the counter #72 is situated in the corner of the upper layer;

#27 - left side, middle layer;

#15 - inner, ground layer;

+ additional shielding of 5 cm Pb in front of the counter

Fig. 1 - Integral pulse amplitude spectra for LSD counters

Search for $\bar{\nu}_e$-bursts from collapsing stars

Theoretical models predict[5,6,7] that a gravitational stellar collapse is accompanied by a powerful neutrino burst, which carries out almost all the binding energy of the collapsed object (10^{53} - 10^{54} erg). The temporal structure of the neutrino burst is spread over several seconds, giving a clear detectable effect[8] in massive underground detectors for galactic collapses. The $\bar{\nu}_e$ energy spectrum is predicted to be in the range 1 - 60 MeV, with an average value of 10 - 13 MeV. Thus a neutrino detector should work at the lowest possible energy threshold, not only to increase the number of interactions, but also to reconstruct the energy spectrum and establish the peak value of the emission from the measured positron distribution.

The rate of gravitational stellar collapses in our Galaxy is still rather uncertain, but the expected value from the pulsar birth-rate function and from the rate of optical supernavae ranges from one every few years to every some dozens years. Obviously, an experiment searching for neutrino bursts must be alive for a long time and must have an extremely low background imitation rate. However, whatever low the latter is, one cannot avoid an unexpected noise of electric, or other, origin; thus it is necessary to perform independent measurements in different laboratories, leading to the idea of a "neutrino network".

Both the Mont Blanc LSD and the Artyomovsk Scintillation Detector (ASD) are well suited to search for $\bar{\nu}_e$ - bursts from collapsing stars. These detectors have a low energy threshold for the positron detection (also for ASD this value is 5 MeV), and the radioactive background has a counting rate low enough for neutron detection in a delayed coincidence (the efficiency for ASD is 82 %). Both installations will produce a sufficiently high statistics of $\bar{\nu}_e$ interactions for a galactic collapse: the expected values are 36 for LSD and 47 for ASD for a collapsing stellar core of mass 2 M_o at the distance 10 kpc. The background counting rate is extremely small for LSD (as discussed previously) and sufficiently small for ASD (0.18 s^{-1} in the energy range of interest), taking into account also the delayed signal from the neutron capture. In ASD the pulse energy spectrum is practically the same of the positron one, while in LSD the energy distribution is slighly broader than the positron one, but with the position of the peak unchanged[9]. From these properties, it follows that ASD is sensitive to at least 90% and LSD to almost 100%[10] of the gravitational collapses in our Galaxy, if they have the characteristics suggested by current models of collapsing stars.

The ASD searches for $\bar{\nu}_e$-bursts over time intervals Δt = 20 s width, starting from any pulse with E $\geqslant$ 5 MeV; each of these pulses opens the 500 μs wide gate (10 μs after the trigger), and pulses in the gate with E $\geqslant$ 0.8 MeV are counted. The first selection criterium for a burst candidate is to record at least 15 pulses during the time Δt (this limits the sensitivity of the detector to $\overline{N}_e + \geqslant 7$). Then, the number N_{ac} of "accompaning" pulses in the gate is counted.

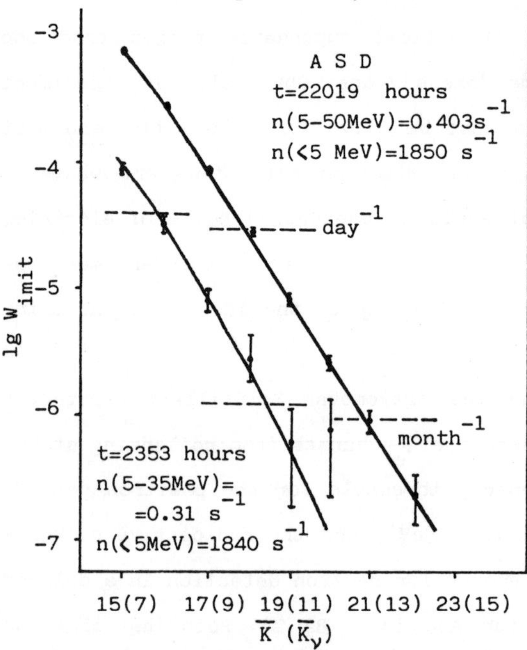

Fig. 2 – Probability to imitate bursts at ASD. K-the multiplicity of pulses; $\overline{K}_\nu$- the most probable estimate for the number of $\overline{\nu}_e$-interactions.

Fig.2 shows the rate of possible $\bar{\nu}_e$-burst candidates in comparison with the expected one; from this fig. it follows that the backgro-

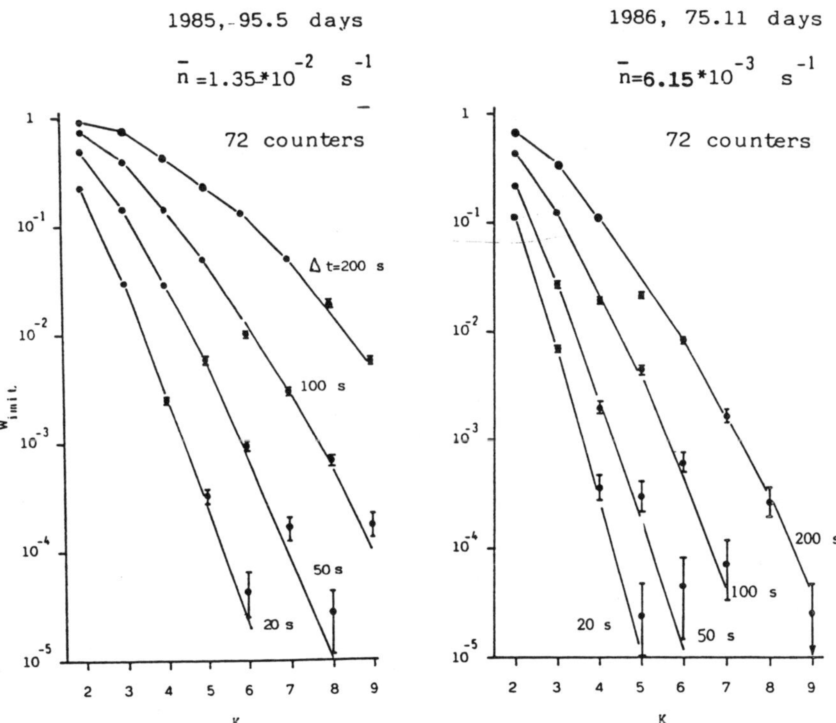

Fig. 3 — The probability to imitate $\tilde{\nu}_e$-bursts at LSD, for different intervals Δt.

und in ASD is randomly distributed and that there are no unusual events in our set of data. From the lifetime of ASD one can thus derive the limit on the rate of galactic stellar collapses:

$$r \leqslant 0.83 \text{ year}^{-1}, \quad \text{at 90\% c.l.}$$

taking into account the probability to detect a stellar collapse in ASD[11]. Additional data, not reported in Fig.2, do not indicate other candidates, so the quoted limit should be further increased. Fig.3 shows the LSD data for different time intervals, for 2 different running conditions, since from december 1985 the iron shield was improved. Since no candidate was recorded, we obtained the lower limit to the rate of galactic stellar collapses:

$$r \lesssim 3.7 \text{ year}^{-1}, \quad \text{at 90\% c.l.}$$

Also in this case not all existing data are reported, because the analysis of time intervals wider than 20 s is still in progress. This is in fact very important since there are theoretical expectations[12] for a prolonged neutrino burst from a rotating stellar collapse. Finally, no coincidence from the overlapping running times of ASD and LSD during 1985 and 1986 has been observed.

Atmospheric low-energy $\bar{\nu}_e$ search in LSD

The analysis of 75 days of LSD running time with the improved shielding does not show any signal in the (12-60) MeV energy range from the internal scintillation counters, whose total mass is 15 tons. Therefore, we can conclude that the atmospheric $\bar{\nu}_e$ flux from μ -e decay is

$$j \lesssim 5 \; 10^4 \text{ cm}^{-2}\text{s}^{-1} \quad \text{at 90\% c.l.}$$

This limit does not contradict the estimates made by Gaisser and Stanev[13], from which one expects $\lesssim 1$ event in 15 tons of scintillator during 75 days.

Possibility to detect ^{8}B solar neutrinos in LSD

The reaction to be considered to detect solar neutrinos in the scintillator is electron-neutrino scattering. Fig.4 shows the response of our detector, computed with a Monte Carlo simulation, to the solar neutrino flux. The total flux has been normalized to the Brookhaven's data of 2.2 SNU and to the energy distributions computed by Bachall et al.[14]. The spread in the pulse amplitude spectrum of

Fig.4 is due to light collection conditions in the detector.

During 75 days of analysed data, 2 events have been recorded with E⟩ 12 MeV in the 6 inner scintillation counters (total mass 7.2 tons) where the background counting rate is lower; from the distribution reported in Fig.4, one expects 0.4 such events. We can conclude that the low background in the internal counters seems to allow detection of solar neutrinos with LSD, after further shielding the whole detector to have an additional attenuation in the background counting rate by a factor of 5 to 10.

In this case, the total mass of LSD (90 tons) would be enough to give a significant limit on the solar neutrino flux. A very massive underground scintillation detector, such as the Large Volume Detector (LVD)[15] to be installed in the Gran Sasso laboratory, with 1 kton of internal mass, will have a very good statistics and, probably,

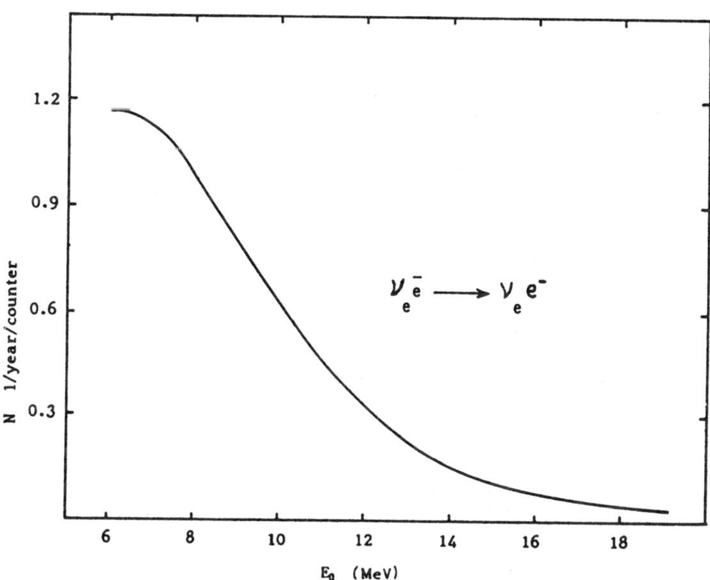

Fig. 4 – Response of the LSD counter to solar neutrino interactions (integral spectrum)

will measure also the variations of the solar neutrino flux on time scales of order of a few days.

Possible correlation between solar activity and solar neutrinos

A possible correlation between the ^{37}A production rate and powerful solar flares was suggested by Basilevskaya et al.[16]. We have looked for such a correlation by using the LSD data recorded in the period 16th to 24th July, 1985, since two large solar proton flares were observed on the 17th and 20th that month; three other proton flares observed during the period 25th to 29th April 1985 were not considered because of a stop over of LSD in that period.

Fig.5 shows the LSD counting rates above the energy thresholds 7, 10, and 13 MeV in the period of interest. For each energy range the number of LSD scintillation counters taken into account for the analysis was choosen in order to have the same normalization: a

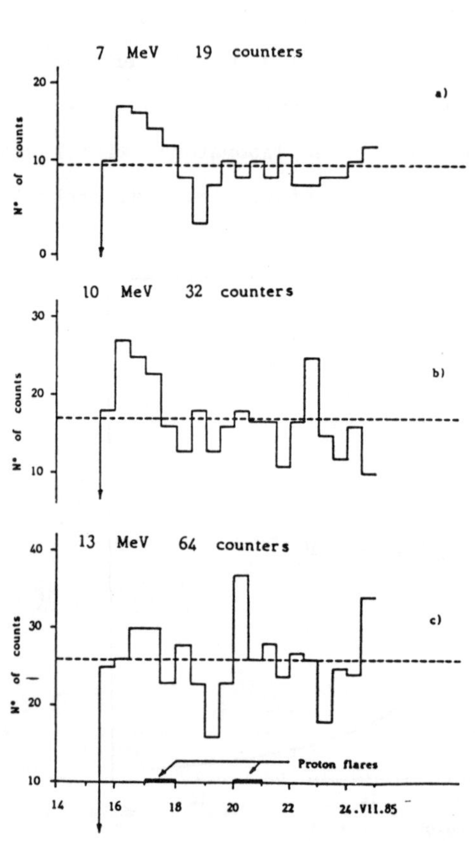

Fig. 5 - The numbers of counts in LSD during 15-25 July 1985. Dashed lines - average values of background count. rates

total background counting rate for each set of data of $\leqslant 5 \; 10^{-4} \; \text{s}^{-1}$;
thus groups of 19, 32, and 64 counters were considered for the
analysis of the 3 sets of data.

At July 15th the LSD started running after a 4 day interruption. One
can see from Fig.5 that, during the period 16-17 July, the counting
rates of the selected groups of counters reasonably exceeded the
background level. Furthermore, it should be noticed that the counter
groups with different energy threshold seem to behave in a practically independent way, because of the steep decrease of the background
with energy. Assuming a statistical independence among the counter
groups, the probability to obtain a simultaneous increment in the
counting rates, as observed for the various groups of counters, is $\approx$
$3 \; 10^{-4}$, which corresponds to an expected event rate of $4 \; 10^{-3}$ during
that running time. We regard this single event to be hardly interpreted in definite terms, but anyway showing the capability of LSD to
search for solar neutrino bursts associated with large solar flares.

Conclusions

No candidate for an $\bar{\nu}_e$-burst from collapsing stars has been observed
during more than 4 years of ASD and 250 days of LSD lifetime. The
data collected by the LSD installation were used to obtain the upper
limit on the flux of atmospheric neutrinos, and to examine the
possibility to detect solar neutrinos and the correlation between
their flux and the solar activity.

O. Ryazhskaya

References

1. G.Badino et al., Nuovo Cimento C, 1984,7, 573.

2. M.Aglietta et al., Nuovo Cimento C, 1986, in press

3. A.V. Voevodsky, V.L. Dadykin, O.G. Ryazhskaya, Pribori i Tekhnika Experimenta, 1970,1, 85

4. M.Aglietta et al., Proc. 19th Int. Cosmic Ray Conf. 1985, 8, 108.

5. Ya.B. Zeldovitch, O.Kh. Gusseinov, Pisma Zh. ETF, 1965, 1, 4; Doclady Ac.Sci.USSR, 1965,162, 791.

6. V.S. Imshennik, D.K. Nadyozhin, in: Uspekhi Nauki i Jekhniki, vol. 21 (Astronomia), Moscow: VINITI, 1984, pag. 75.

7. A.Burrows, Ap. J. 1984,283, 848.

8. G.V. Domogatsky, G.T. Zatsepin, Proc. 9th Int. Conf. on Cosmic Rays, 1965,2, 1030.

9. F.F.Khalchukov et al., Proc. 19th Int. Cosmic Ray Conf., 1985,8, 140, Preprint of INR USSR Ac. Sci., 1985, P - 0388.

10. G.Badino et al. Neutrino '84, Proc. 11th Int. Conf. on Neutrino Phys. and Astrophys., Dortmund 1984, 556.

11. V.I. Beresnev et al.; Proc. 16th Int. Cosmic Ray Conf., 1979,10, 293.

12. V.S. Imshennik, D.K. Nadyozhin, Pisma AZh (Sov. AJ Lett.), 1977, 3, 353.

13. T.K. Gaisser, T.Stanev, Proc. 19th Int. Cosmic Ray Conf., 1985, 8, 156.

14. J.N. Bahcall et al., Astroph. J., 1974,184, 1.

15. C.Alberini et al., to be published in Nuovo Cimento C, 1986.

16. G.A. Bazilevskaya, Yu.I. Stozhkov, T.N. Charakhchyan, Sov. J.ETP Lett., 1982,35, 341.

ν Results from IMB

J. M. LoSecco, R. M. Bionta, G. Blewitt, C. B. Bratton, D. Casper,
R. Claus, B. Cortez, S. Errede, G. Foster, W. Gajewski, K. S. Ganezer,
M. Goldhaber, T. J. Haines, T. W. Jones, D. Kielczewska, W. R. Kropp,
J. G. Learned, E. Lehmann, H. S. Park, F. Reines, J. Schultz, S. Seidel,
E. Shumard, D. Sinclair, H. W. Sobel, J. L. Stone, L. Sulak, R. Svoboda,
J. C. Van der Velde, C. Wuest

University of California, Irvine, CA 92717
University of Michigan, Ann Arbor, MI 48109
Brookhaven National Laboratory, Upton, NY 11973
California Institute of Technology, Pasadena, CA 91125
Cleveland State University, Cleveland, OH 44115
University of Hawaii, Honolulu, HI 96822
University of Notre Dame, Notre Dame, IN 46556
University College, London WC1E 8BT, United Kingdom
Warsaw University, Warsaw PL-00-681, Poland

Abstract

Neutrinos are studied with the deep underground 3300 metric ton IMB detector. Properties such as neutrino lifetime and oscillations are examined taking advantage of the 10^7 meter flight path of some of these atmospheric neutrinos. Matter oscillation effects occuring from propagation through the earth are investigated. Bounds on possible point sources are set. General properties of the data are in good agreement with expectations for atmospheric neutrinos. Geomagnetic effects in the data are noted.

J. LoSecco

1 Data Sample

The IMB detector is a deep underground imaging water Čerenkov detector with a fiducial mass of 3,300 metric tons. The data discussed here has been collected in the period September 1982 to June 1984. 417 days of livetime in this period have given us 401 contained events in the detector[1]. The efficiency for finding multitrack proton decay like events is in excess of 95%. The efficiency for finding single visible track neutrino interactions is about 85%.

The detector measures the time of the event, the vertex, the visible energy of the event, the event direction, the number of tracks (at wide angles) and the time of a muon decay if present.

310 of the events have the topology of single track neutrino interactions and are used to study neutrinos.

Several additional data samples have been collected but are not discussed in the present work.

In general the interactions have energy and direction distributions as expected from neutrinos from atmospheric sources. The data sample is unique in that the neutrinos originate from natural sources. They travel long distances ($\sim 10^7$ meters) to the detector. And they travel through the earth. The energy of the parent neutrinos is low, averaging about 920 MeV. The source is a mixture of ν_e and ν_μ. It also has a substantial $\bar{\nu}$ component. Though the lower cross section significantly reduces the $\bar{\nu}$ interaction rate. In the IMB detector ν_μ interactions are distinguished by the presence of a muon decay. The efficiency for this signal is not high but it produces a clean ν_μ sample.

2 Neutrino Lifetime

Upward going neutrinos have a time of flight in excess of 1/30 of a second. By comparing the event rate at different angles we can study neutrinos traveling different distances. By fitting the observed flux to an exponential decay we can get a mode independent bound on the lifetime[2]. The data for all upward going events is plotted in figure 1.

Since there is independent evidence on the stability of the ν_e we can interpret our result as a bound on ν_μ

$$\frac{p\tau}{m} > .1 \text{ seconds } (90\% \text{ confidence limit})$$

At present the bound is statistics limited. Comparable results are achieved with and without taking into account the small geomagnetic modulation of the flux.

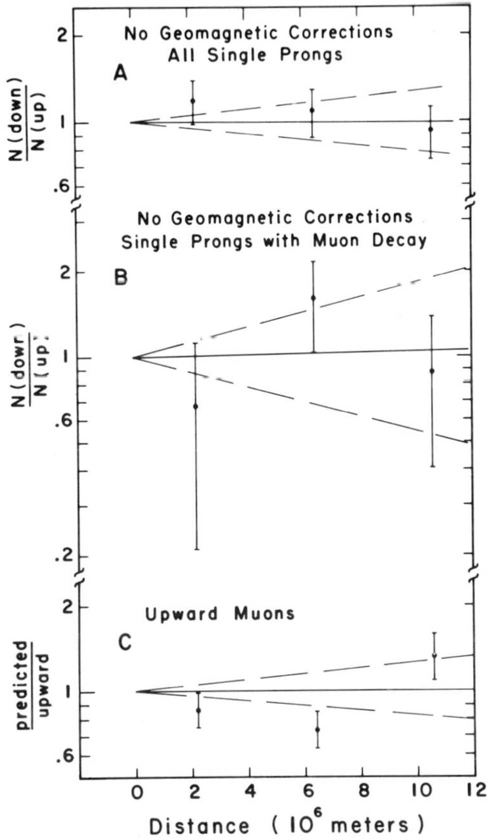

Figure 1: Neutrino rate as a function of distance

3 Matter Oscillations

Wolfenstein[3] has pointed out that electron neutrinos traversing matter see a different index of refraction due to the charged current contribution to the forward scattering amplitude. This has consequences for neutrino oscillations since it provides a modification to the velocity differences associated with Δm^2. In particular the two effects can work together to greatly amplify neutrino oscillations with very small mixing angles. A number of observations are possible with atmospheric neutrinos[4].

The effect is density dependent. The earth's density varies from 2.8 to 12 gm/cm^3. The neutrino spectrum runs from 100 MeV to > 1 GeV. The enhancement should occur for small angles and:

$$2.1 \times 10^{-5} eV^2 < \Delta m^2 < 8.9 \times 10^{-4} eV^2$$

The mean density of the earth is 5.5 gm/cm^3 and so large effects could be found for:

$$4.1 \times 10^{-5} eV^2 < \Delta m^2 < 4.1 \times 10^{-4} eV^2$$

Recent work studying this effect in the sun indicates[5] that the value $\Delta m^2 = 1.1 \times 10^{-4} eV^2$ could explain the anomalously low value of the solar neutrino flux observed on earth. It is clear that atmospheric neutrinos can provide an independent test of this hypothesis. ν_μ to ν_e oscillations are easily studied at the energies of several hundred MeV since the transition of ν_μ into ν_e would produce a significant event rate increase due to the lower threshold for ν_e interactions and the higher fraction of ν_μ present. $\nu_e \rightarrow \nu_\tau$ oscillations would be evident by a decrease in the ν_e event rate. A direct comparison between downward going, non oscillated neutrinos, and upward going, potentially oscillated ones can be performed. Such a comparison removes many possible systematic errors.

IMB has presented the limit[6]

$$sin\theta_v(\nu_\mu \leftrightarrow \nu_e) < .14 \ (90\% \ \text{confidence limit})$$

4 Vacuum Oscillations

Since the majority of neutrinos observed in underground experiments are ν_μ, an experiment that looks for an E/L dependence in the rate is sensitive to ν_μ oscillations into a sterile or non interacting state such as ν_τ. Such a search has been carried out[7] and yields a lower limit of $\Delta m^2 < 2.2 \times 10^{-5} eV^2$ at maximum mixing. This limit has been expanded to larger Δm^2 by reducing L by selecting neutrino events coming from away from the vertical. Path lengths of $< 10^6$ meters can be used. The excluded region is shown in figure 2. Due to limited statistics the result does not probe very small mixing angles.

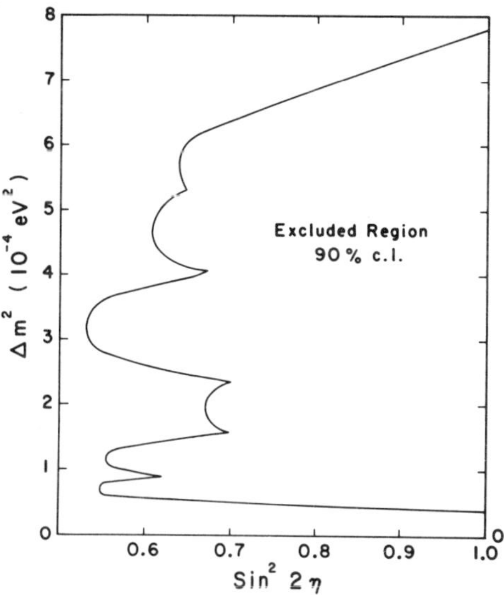

Figure 2: Vacuum Oscillation Limits

5   Point Sources of Neutrinos

Proton decay detectors have accumulated very long exposures in very low background environments. The contained signals come from either nucleon decay or neutrino interactions. Aside from the general, cosmic ray induced neutrino background, there are many potential point sources of neutrinos.

Stars producing energetic γ's may also produce ν's from charged pion decay. A number of authors have suggested that dark matter may get trapped in the earth or sun and produce a neutrino flux due to the annihilation process. Each of these potential sources requires a search in a different reference frame. Equitorial coordinates were used for stellar searches; local coordinates were used for earth based sources and a modified ecliptic coordinate system was used to look at the sun.

Both searches for local and stellar sources found directions with 38 events within a 30° half angle cone[8]. This is 1.8 times the average value but since the

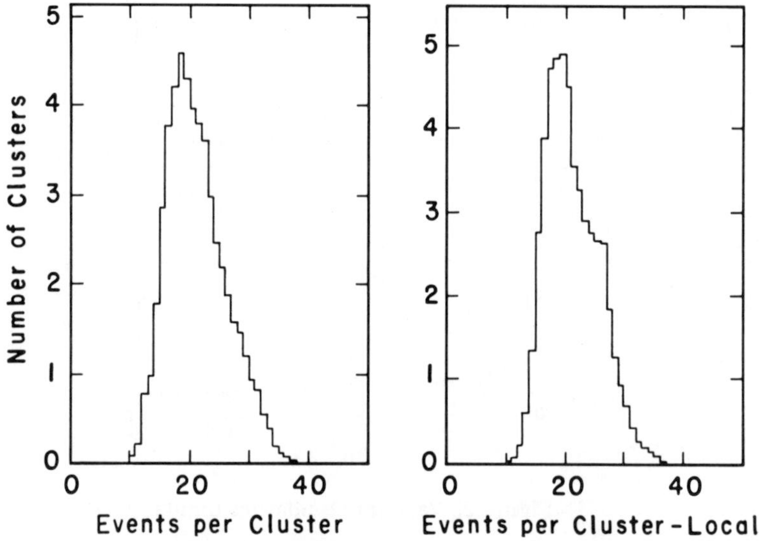

Figure 3: Number of clusters of events

search was a whole sky survey the result is not significant. Figure 3 shows the number of clusters found as a function of the number of events in the cluster for both the stellar search and the terrestrial search.

In the direction of the sun 25 events were found when 20.8 were expected. Supersymmetric particle annihilation is expected to provide excess flux at high energies ($E_\nu > 1$ GeV). 11 events were found with $E_\nu > 1$ GeV. Only 6.03 $\pm$ .64 were expected. These results do not probe the neutrino flux at high energies because the detection efficiency drops rapidly to zero above 2 GeV since tracks can exit the detector.

6   Geomagnetic Effects

In general the neutrino flux may be treated as isotropic. In actual fact magnetic effects permit a greater cosmic ray flux near the magnetic poles of the earth. A greater primary flux will produce a greater secondary flux of neutrinos. This effect

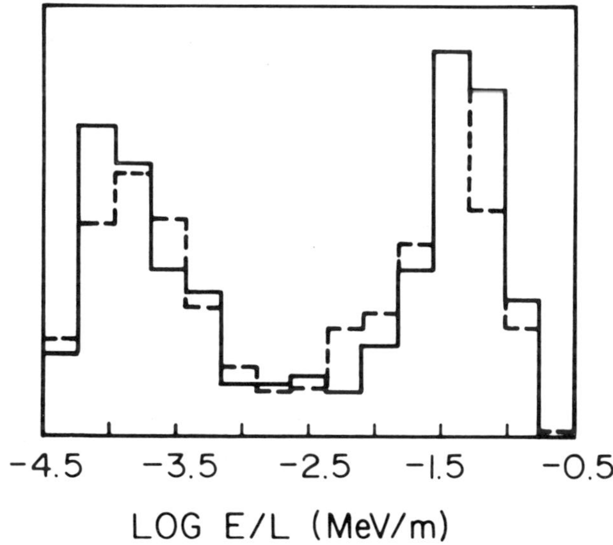

Figure 4: E/L distribution. Data compared to predictions

is most significant at low energies. It is important since the downward component of neutrinos is created locally within 500 km of the detector. Comparisons of data from all directions gives good agreement with expectations[9]. Figure 4, for example, compares the observed E/L ratio, solid line, to that expected.

Geomagnetic corrections give some improvement in our neutrino lifetime limit and are essential in understanding our result on matter oscillations. In both cases statistical errors dominate the result. In the search for a low energy excess of upward events one must compare the upward sample with about 71% of the downward sample. The raw data[6], for example, has a ratio of up/down = .72 ± .22.

7 Conclusions

Non-accelerator physics is a unique tool for studying neutrinos. Natural point sources can be searched for. The very long baseline and low energy provides a sensitive neutrino oscillation test. Since half of the beam traverses matter, interference effects can be studied.

As larger detectors are built statistics will rise and a better understanding of geomagnetic effects will become important. Better angular resolution will greatly enhance the ability to search for point sources. At present angular resolution is limited at low energies because much of the hadronic part of the event goes undetected. Future detectors will have full event reconstruction and hence much better angular resolution.

8 References

1. J. LoSecco, *et al.*, Phys. Rev. Lett. <u>54</u>, 2299 (1985).
   G. Blewitt, *et al.*, Phys. Rev. Lett. <u>55</u>, 2114 (1985).

2. J. LoSecco, *et al.*, "Limits on the Neutrino Lifetime" UND-PDK-86-6 (to be

published).

3. L. Wolfenstein, Phys. Rev. D17, 2369 (1978).

4. J. LoSecco, Comment on "Possible Explanation of the Solar Neutrino Problem" UND-PDK-86-5 (to be published).

5. H. Bethe, Phys. Rev. Lett. 56, 1305 (1986).

6. J. LoSecco, *et al.*, "Investigation of Matter Enhanced Neutrino Oscillations Relevant to the Solar Neutrino Problem", UND-PDK-86-7.

7. J. LoSecco, *et al.*, reference 1 above.

8. J. LoSecco, *et al.*, "Neutrino Physics with a Massive Underground Detector", Proceedings of the XXIth Rencontre de Moriond, Workshop on Massive Neutrinos (Tignes, France, 1986) (to be published) J. Tran Than Van, editor.

9. T. Stanev, "Muons and Neutrinos", MAD/PH/270 (to be published in the Proceedings of the XIXth International Conference on Cosmic Rays, La Jolla, California, 1985).

   J. LoSecco, *et al.*, Proceedings of the XIXth International Conference on Cosmic Rays, (La Jolla, California, 1985) Volume 8, pages 116-119.

The Search for Solar ^{8}B Neutrinos at KAMIOKANDE II[*]

A. Suzuki

University of Tokyo

for the KAMIOKANDE-II collaboration[**]

Abstract

Solar ^{8}B neutrinos are searched for in the low energy contained events with energy > 9 MeV by the upgraded KAMIOKA nucleon decay detector (KAMIOKANDE II). No definite signature as yet of solar neutrinos has been observed in the angular correlation plot with respect to the sun. From 48.5 days effective run time, the flux upper limits at 90 % C.L. are obtained to be 4.1 x 10^6, 6.0 x 10^6 and 10.4 x 10^6 (1/cm^2/sec), corresponding to the different energy calibration errors, -10 %, 0 % and +10 %, respectively. These results are to compared with the value predicted by the standard solar model, 4.0 x 10^6 (1/cm^2/sec), and indicate that the KAMIOKANDE II is now beginning to explore the real time directional observation of solar ^{8}B neutrinos.

Introduction

The discrepancy in solar neutrino flux between the results of Chlorine experiment (2.1 ± 0.3 SNU)[1] and the standard solar model prediction (5.8 ± 2.2 SNU)[2] is the so-called missing solar neutrinos. It casts serious doubt on our understanding of nuclear astrophysics and elementary particle physics. The recent theoretical conjecture of neutrino oscillation in matter sheds light on this question and stresses the renewed importance of the solar ^{8}B neutrino measurement[3].

The KAMIOKANDE II is the upgrade of the KAMIOKA nucleon decay detector and incorporates a 4 π water

Cherenkov anti-counter and multi-hit timing electronics, which not only improve the sensitivity to nucleon decay, but also lower the energy threshold thereby allowing it to search for solar neutrinos. Measurement is done by detecting (ν_e, e^-) elastic scattering in the water in real time, and this allows the observation of the energy spectrum and the angular distribution of ν_e. In this paper we report the first preliminary results of the solar neutrino experiment at KAMIOKANDE II.

KAMIOKANDE II detector

The KAMIOKA nucleon decay detector is a 3000 ton imaging water Cherenkov detector and was described elsewhere[4]. Fig. 1 shows the schematic outline of the KAMIOKANDE II detector. 1071 20" photomultipliers (948 for the inner part and 123 for the anti-part) are mounted on the detector walls. The mean thickness of the top, barrel and bottom anti-layers is 1.5 m. These active water layers shield against entering photons and neutrons in addition to vetoing entering cosmic ray muons. The electronics system is also improved with multi-hit time and charge measurements. The detailed description of the KAMIOKANDE II detector is given in ref. 5. The new discriminators (at threshold of 0.4 photoelectron) and the timing electronics allow a simple trigger scheme based on the number of hit PMT's. The current data taking is carried out with $N_{hit} \geq$ 20 (9 MeV for electrons) within 100 nanosecond coincidence time. The total trigger rate is less than 1 Hz. Timing information gives a remarkable improvement of the vertex reconstruction especially for low energy events. The accuracy (1 standard deviation) for 10 MeV electrons is found to be 1.7 m for position and 38° for direction. Grading up the water purification system is one of the major items to improve the detection of low energy events because backgrounds come from radioactive impurities in the water.

Data analysis and results

The KAMIOKANDE II data taking has been continuing since December, 1985. 48.5 days effective run time data are used for the present analysis. The average trigger rate is 0.8 Hz under the trigger condition of $N_{hit} \geq 20$, of which 0.3 Hz is due to cosmic ray muons. The expected event rate from solar ^{8}B neutrino is 0.33 events /day. Therefore the utmost care in background reduction is essential and is described in the following.

The first step of reduction processes includes:

1) The total number of photoelectrons of the event is less than 100, $PE_{tot.} \leq 100$ ($\sim$ 25 MeV)
2) No event signal in the anti-counter.
3) The time elapse from the last event, ΔT, is more than 100 microsecond to eliminate muon decay electrons.
4) Rejection of PMT noise events, $PE_{max.}/PE_{tot.} \leq 0.25 \sim 0.125$, depending on $PE_{tot.}$

The second step is to reduce the environmental backgrounds by applying the fiducial volume cut. Fig. 2 is the vertex point distribution of the reconstructed events with $N_{hit} \geq 25$ in R-Z plane, where R and Z stand for the radial and height coordinates in the cylindrical detector. The concentration of events is seen near the edges (top, barrel and bottom) of the detector. The Z-distribution of events with R $\leq$ 5.2 m is shown in Fig. 3. To reduce background contamination, the fiducial volume is set in the region of -3.98 m $\leq$ Z $\leq$ 4.0 m and R $\leq$ 5.2 m which corresponds to a total fiducial volume of 680 m^3.

One of the major background sources after the first and the second reduction processes, is due to the interactions of high energy cosmic ray muons in the detector. In these reactions, a muon breaks up ^{16}O and produces various radioactive nuclei which have relatively long life times and energetic beta rays. Also the radioactive products from the interactions of the secondary pions with ^{16}O will contribute. The third step of event reduction is to reject these beta rays. To identify a specific isotope, the time interval (ΔT) distribution between the preceding event and the low energy one is taken. Fig. 4-a shows the time distribution of events with $N_{hit} \geq 20$ and $\Delta T < 100$ milisecond. A clear decay structure is observed. A similar plot for $0.1 \leq \Delta T \leq 9$ second and $N_{hit} \geq 25$ is shown in Fig. 4-b, for those events with the total pulse height of the preceding event $PE \geq 20000$. These curves are well fitted by decay constants of $t_{1/2} = 15.9 \pm 2.3$ milisecond and $t_{1/2} = 0.61 \pm 0.17$ second, respectively. Fig. 5 shows the Cherenkov pattern of a typical preceding event in the expanded view of the detector, where a circle means a PMT and its area is proportional to the pulse height it produced. This Cherenkov pattern clearly shows the contribution of a large interaction. The total pulse height of the preceding events is usually much larger than that of a simple through-going muon as shown in Fig. 6. It is apparent that the low energy events are strongly correlated with the muon induced nuclear interactions. Energy distributions of two decay groups, $\Delta T \leq 30$ milisecond and $0.1 \leq \Delta T \leq 2$ second, are plotted in Fig. 7-a and Fig. 7-b under the additional event selection; i.e., the total pulse height of the preceding event ≥ 20000 and ≥ 40000 photoelectrons, respectively. In these figures, the distributions are well reproduced with the beta-ray spectra of $^{12}N + ^{12}B$ and ^{8}B, respectively. Based on the above discussion, the criteria to eliminate the muon induced backgrounds are determined by using the time interval and the total pulse height of preceding event.

Fig. 8 shows the integral event number distribution as a function of N_{hit} of the events surviving up to the first, the second and the third reduction processes. The expected solar ^{8}B neutrino event distribution passed through the same reduction processes is also shown in Fig. 8. It can be seen that the S/N ratio is about 1/10 around $N_{hit} \sim 25$.

Finally, the directional correlation to the sun is used for the last reduction step. Fig. 9 is the angle distribution of the remained events with respect to the sun. At present the angular resolution is obtained to be 38° which includes both the reconstruction error and the coulomb scattering effect. No significant enhancement is yet seen in the region of $\cos\theta \geq 0.79$ (1 standard deviation limit) region. The distribution is consistent with a flat one. In Fig. 9, expected angle distributions from the standard solar model are superimposed by dashed curves which correspond to the different energy calibration errors, +10 % (lower one) and -10 % (upper one). The present error (ΔE), ±10 % is estimated from the energy spectra of muon decay electrons and beta-rays of ^{12}N + ^{12}B and ^{8}B (Figs. 7-a and 7-b). Flux limits of 90 % C.L. for the present 48.5 days effective exposure are derived to be 4.1×10^6, 6.0×10^6 and 10.4×10^6 (1/cm^2/sec) for $\Delta E = -10$ %, 0 % and +10 %, while the standard solar model prediction[2] is 4.0×10^6 (1/cm^2/sec).

Conclusions

The low energy electrons with energy down to 9 MeV are measured in the upgraded KAMIOKANDE II detector. The observed results indicate that the KAMIOKANDE II has the unique potential to detect solar ^{8}B neutrinos. The most of the remained low energy events after all the reduction processes are considered to have come from the distribution

tail of the environmental backgrounds and the radioactive impurities in the detector water. In order to improve the quality of the data, and to further lower the detection threshold we are working on:
  a) Establishing the energy calibration around 10 MeV region by using the standard γ sources and
  b) Lowering the energy threshold down to 6 MeV by installing a new ion-exchanger in the water purification system to eliminate the remaining radioactivities.

References

*   This little paper is based on the updated analysis after the conference.

**   The Kamiokande II Collaboration includes:
     B. G. Cortez, CaliforniaInstitute of Technology; K. Takahashi,KEK; K. Miyano, U. of Niigata; E.W. Beier, L. Feldscher, S.B. Kim, A. K. Mann, F. M. Newcomer, R. Van Berg, W. P. Zhang, U. of Pennsylvania; and K. Hirata, T. Kajita, T. Kifune, M. Koshiba, M. Nakahata, Y. Oyama, N. Sato, T. Suda, A. Suzuki, M. Takita, Y. Totsuka, Univ. of Tokyo

[1] J. K. Rowley, B. T. Cleveland and R. Davis, in Solar Neutrinos and Neutrino Astronomy, Homestake (1984), AIP Conf. Proc. No. 126, p1.

[2] J. N. Bahcall, B. T. Cleveland, R. Davis and J. K. Rowley, Astrophysical Journal, 292 (1985) L79.

[3] L. Wolfenstein, Phys. Rev. D17 (1978) 2369.
    S. P. Mikheyev and A. Yu. Smirnov, talk given at the Int. Workshop on Weak Interactions and Neutrinos, Savaonlinna (Finland), June (1985).
    H. A. Bethe, Phys. Rev. Lett. 56 (1986) 1305.

[4] K. Arisaka et al., J. Phys. Soc. Jpn. 54 (1985) 3213.

[5] E. W. Beier, talk given at the 7th Workshop on Grand Unification/ICOBAN '86, Toyama (Japan), April (1986).

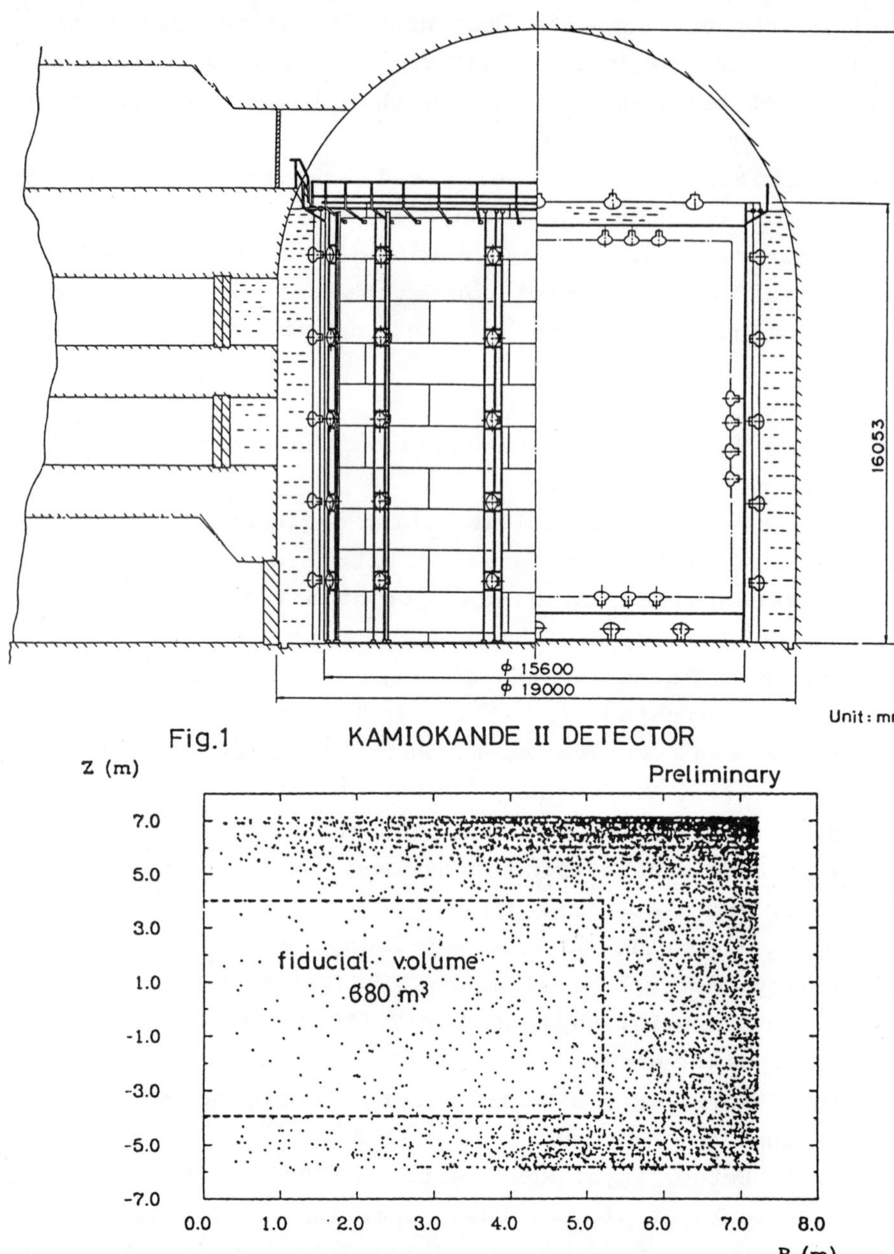

Fig.1   KAMIOKANDE II DETECTOR

Fig. 2   VERTEX POSITION (NHIT>=25) Distribution

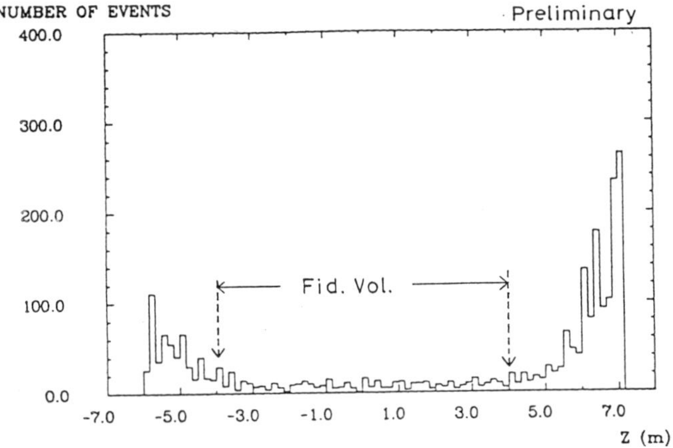

Fig.3   Z DISTRIBUTION (NHIT >= 25)   (R≤ 5.2 m)

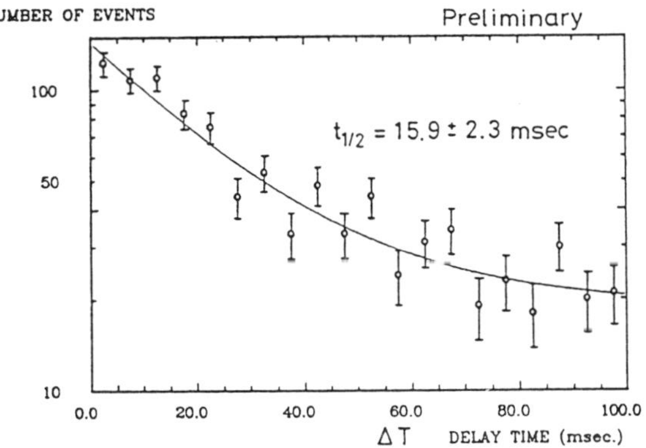

Fig. 4-a   Time Interval Distribution

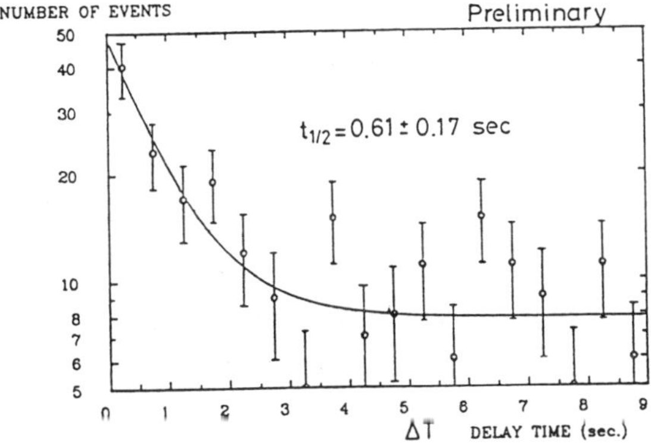

Fig. 4-b   Time Interval Distribution

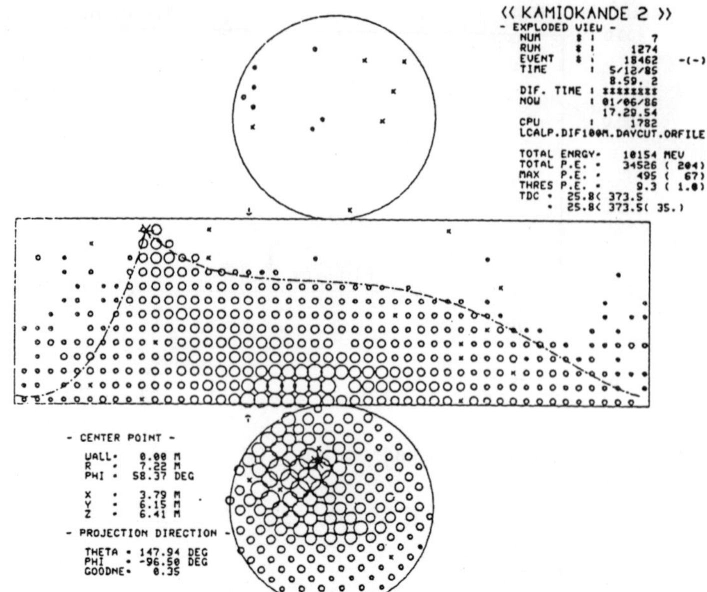

Fig. 5   Typical Pattern of Preceding Event

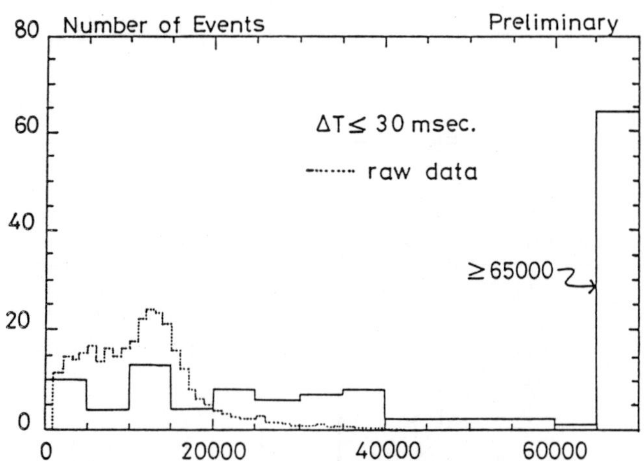

Fig.6   Pulse Height of Preceding Events (P.E.)

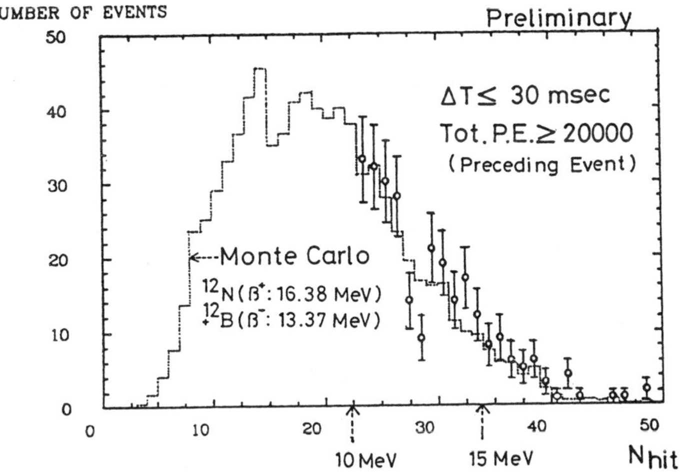

Fig. 7-a   No of Hit PMT Distribution

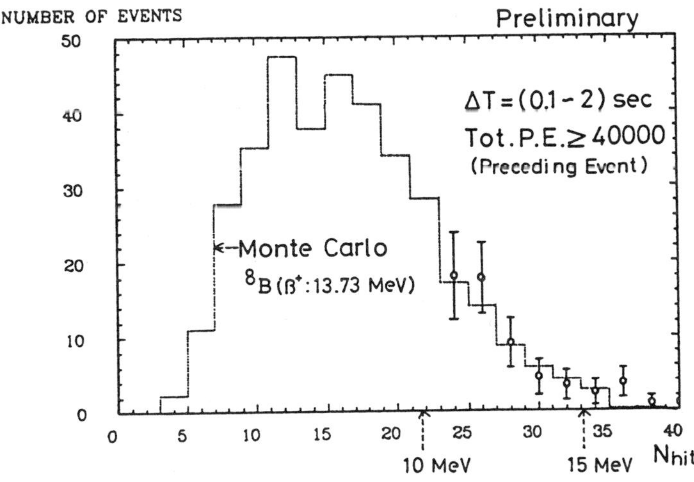

Fig. 7-b   No of Hit PMT Distribution

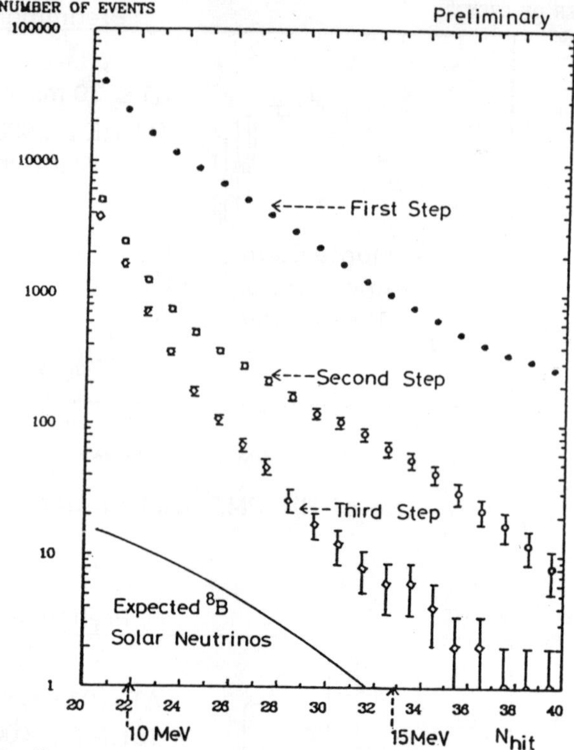

Fig.8   INTEGRATED NHIT DISTRIBUTION

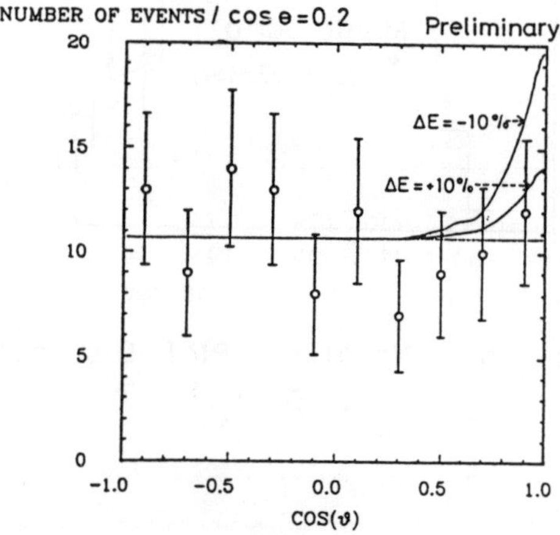

Fig. 9   DIRECTION FROM THE SUN

PROGRESS REPORT ON THE "GALLEX" SOLAR NEUTRINO PROJECT

Till Kirsten[*]

Max-Planck-Institut für Kernphysik
Heidelberg
P.O. Box 103 980
D-6900 Heidelberg, F.R. Germany

ABSTRACT

The "Gallex"-experiment aims at the radiochemical detection of solar pp-neutrinos via the reaction $^{71}Ga(\nu_e,e^-)^{71}Ge$. The project is now in its implementation phase. A status report of the major aspects of the experiment is given. Acquisition of the full 30 tons of Gallium will last until 9/1989. This marks the beginning of the actual measurements in the Gran Sasso European Underground Laboratory.

Introduction

The Gallex collaboration intends to measure for the first time the pp-neutrino flux from the sun by using a low threshold radiochemical detector based on the reaction $^{71}Ga(\nu_e,e^-)^{71}Ge$. The goals of this experiment are:
- the first experimental verification that nuclear fusion is the principal solar energy source,
- the search for neutrino flavor oscillations down to neutrino mass differences as small as $\Delta m_i^2 \approx 10^{-12}$ eV^2,
- to test the standard solar model[1] (SSM) in view of the solar neutrino problem (^8B neutrino deficit in the chlorine solar neutrino experiment of Davis[2].

---

[*] GALLEX-COLLABORATION: MPI Heidelberg: W. Hampel, G. Heusser, J. Kiko, T. Kirsten, A. Lenzing, E. Pernicka, B. Povh, S. Richter, K. Schneider, M. Schneller, H. Völk, R. Wink; KFK Karlsruhe: R. v. Ammon, K. Ebert, E. Henrich; TU München: R.L. Mößbauer; CEN Saclay: M. Cribier, M. Dumont (Grenoble), P. Pichard, J. Rich, M. Spiro, D. Vignaud; Nice University: G. Berthomieu, E. Schatzman; Milano University/INFN: E. Bellotti, O. Cremonesi, E. Fiorini, C. Liguori, S. Ragazzi, L. Zanotti; Roma University/INFN: S. d'Angelo, R. Bernabei, L. Pauluzi, R. Santonico; WIS Rehovot: I. Dostrovsky.

Non-standard solar models have the potential to drastically affect the fluxes of ^{8}B- and ^{7}Be-neutrinos. In fact, often they are tailored to reproduce the low neutrino capture rate of the chlorine detector. On the other hand, as long as the sun presently is producing its energy by nuclear fusion, the pp-neutrino flux is hardly altered at all by particularities of solar models but it is fixed by the solar luminosity. If the result of a pp-sensitive detector (Ga-detector) falls short of the SSM-expectation for the pp-neutrino production rate, this would be an unambiguous proof for neutrino flavor oscillations (or decay) and, consequently for non-zero restmass of at least one neutrino mass eigenstate.

Among the potential pp-sensitive neutrino experiments ($E_{threshold}$ <420 keV), the radiochemical gallium experiment ($E_{threshold}$=233 keV)[3] is the only one which has been demonstrated to be technically feasible at present. Its feasibility was the major result of the BNL/MPI pilot experiment with 1.3 tons of Ga[4]. Due to funding problems, this collaboration dissolved in 1983. A new collaboration, GALLEX was formed in 1984 (see title page for its members). Major funding, e.g. for the required 30 t of Ga and for the underground laboratory installations, was assured in 1985. Now, the GALLEX project has undergone the crucial transition from hope to reality. Status and plans are described in the following.

Outline of the Experiment

Natural Ga contains 39.6% ^{71}Ga and 60.4% ^{69}Ga. The latter is useless for pp-neutrino detection but it can play a role in monitoring interfering (p,n) reactions. The SSM prediction for the total capture rate is approximately 20 times higher for ^{71}Ga than for ^{37}Cl. As an immediate consequence, shielding requirements to reduce interfering (p,n) reactions are less severe for Ga than for the Cl-detector. Table 1 summarizes the expected production rates from the various neutrino sources (a) for the SSM and (b) for a model consistent with the low ^{8}B-neutrino flux as indicated by the results of the Cl-experiment (2.0+.3) SNU vs. 6.1 SNU SSM-prediction. Excited state contributions are based on GT strengths deduced from Indiana (p,n) experiments[6]. The (almost) star model independent contributions of pp+pep neutrinos to the total production rate are 60% (SSM) and 78% (Consistent model) respectively.

The high costs of Ga force us to choose the smallest target size which is still sufficient to produce a +10% result within approx. three years of measurements. Monte Carlo calculations based on

Neutrino source	Standard Solar Model			"Consistent" Model		
	g.s.	exc.s.	total	g.s.	exc.s.	total
pp	70.3	-	70.3± .6	71.5	-	71.5± .6
pep	2.5	.7	3.2± .2	2.5	.7	3.2± .2
^{7}B	28.5	2.1	30.6+1.3	14.2	1.0	15.2± .7
^{8}B	1.2	11.5	12.7+2.0	.3	3.0	3.3± .5
^{13}N	2.5	.2	2.7± .1	.28	.02	.3± .01
^{15}O	3.4	.6	4.0± .2	.68	.12	.8± .04
Total	108.4	15.1	123.5+4.4	89.5	4.8	94.3+2.1

Table 1: ^{71}Ge production rates from ^{71}Ga for ground state and excited state contributions for the SSM and for a "Consistent" model[5]. Unit is 1 SNU=1 capture per 10^{36} target atoms per second.

achieved experimental capabilities lead to a required minimum quantity of 30 tons of natural Ga, corresponding to 1.1 neutrino captures per day for the SSM production rate.

The concept of the experiment is

(i)   to expose the target in a chemical form suitable for Ge-extraction of about one half life of ^{71}Ge (11.13 d) in an environment which is sufficiently low of disturbing natural radiations,

(ii) to extract the Ge,

(iii) to detect the extracted ^{71}Ge via its EC decay (back to ^{71}Ga) in extremely background-free counting devices,

(iv) to accumulate sufficient statistics by repeated runs (2-4 years, $\sim$ 25 runs per year),

(v)   to demonstrate the concerted performance of all experimental components and processes by means of a strong artificially produced low energy neutrino source inserted into the target tank.

Ga Target and Ge Extraction

Like the pilot experiment, the 30 t experiment will be performed with highly concentrated ($\sim$ 8 m) $GaCl_3$ solution. This chemical form is chosen to benefit from the clear-cut kinetics of Ge-extraction from an aqueous solution. As soon as a Ge atom is formed within the $GaCl_3$ solution, it will transform into germanium chloride, $GeCl_4$. This compound is highly volatile and can be swept out from the solution by a circulating stream of air or helium. Due to the generous dimensions and the ideal access logistics of the Gran Sasso Lab (see below), the total solution (~ 55 m^3) can now be contained in one single large tank (Fig. 1) in which it can be swept in situ. An identical second tank will be installed both, as a safety provision in case of a tank leak and to provide the option to perform Ge-extraction in a mode in which

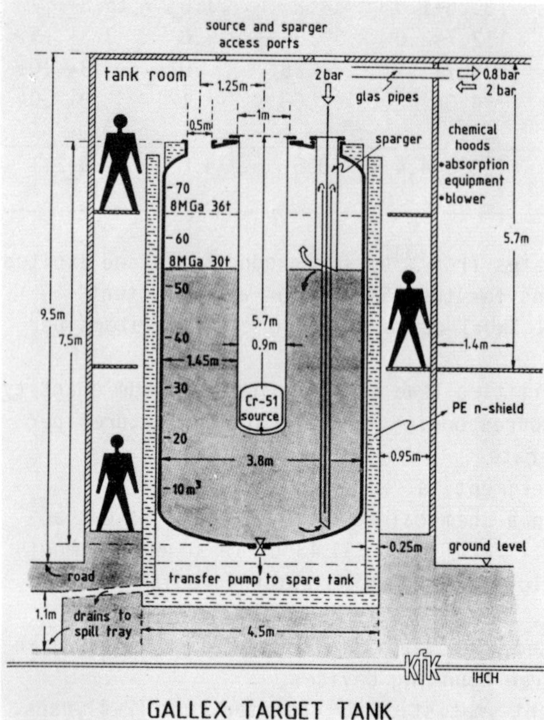

Figure 1:
Sketch of the GALLEX target tank. For acid resistance it is teflon-lined. The insert to accommodate the calibration source will be narrower as shown.

the liquid is physically transferred from tank 1, through the extraction column, into tank 2. At least for the latter mode, there will be no upscaling problems (relative to the pilot experiment). After sweeping, the GeCl$_4$ is taken out by gas scrubbers, extracted with CCl$_4$, and back extracted with water. Now that all the Ge is confined to a small volume, it is reduced to gaseous germane (GeH$_4$). After gaschromatographic purification, the germane is admixed to the counting gas of proportional counters described below. These procedures have been successfully tested in more than 30 test runs during the pilot experiment at BNL both, with and without inactive carrier added (.1-3 mg Ge). The main purpose of the carrier is to determine the extraction yield in every individual run. The overall recovery, including conversion into GeH$_4$, was $\sim 98\% \pm 2\%$ after ~ 15 hours of extraction time (pilot experiment). Meanwhile, apparatuses for $GeCl_4$-GeH_4 conversion, GC-purification, and counter filling stations have been set up at MPI. In gaschromatographic purification, the earlier problem of complete separation between GeH_4 and CO_2 has been solved (Fig. 2).

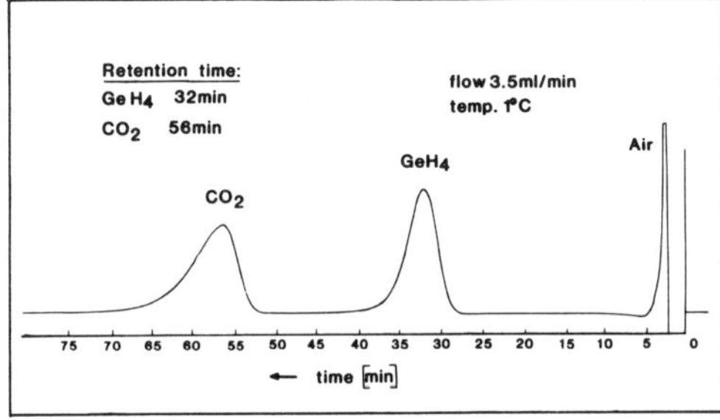

Figure 2:
GeH_4-CO_2 separation after optimalization of operating conditions in the GC-purification procedure.

Underground Laboratory and Side Reactions

The critical side reaction for the production of ^{71}Ge in $GaCl_3$ solution is ^{71}Ga(p,n)^{71}Ge ($E_{threshold}$=1.02 MeV). The protons may be generated as secondaries by (α,p), (n,p), and cosmic ray muon interactions. This necessitates shielding in an underground laboratory, a radiochemically clean target solution and provisions against fast neutrons from the surroundings (tank walls, rock walls).

The GALLEX experiment has been approved for installation in hall A of the Gran Sasso Underground Physics Laboratory. This impressive facility provided by INFN[7] consists of three enormous cavities (approx. 100x18x18 m^3 each) which are readily excavated next to a highway under the Gran Sasso massif in the Abruzzese mountains near L'Aquila (Fig. 3). Presently, utilities are being installed, inauguration of the facility is scheduled for 1987. The GALLEX experiment will be accommodated in two large buildings: the GALLEX main building (12x10x10 m^3) and the Low-Level counting lab (10x10x6 m^3), with very convenient access from the highway. The muon flux measured with a hodoscope detector is $\sim 11 \mu/m^2$, d in accordance with the estimated effective shielding depth of 3500 m.w.e. The depth dependence of ^{71}Ge production in $GaCl_3$-solution by cosmic ray muons has been derived from measurements and calculations on the same effect for the Cl-detector and from the measured cross section ratio (^{71}Ge from $GaCl_3$)/(^{37}Ar from C_2Cl_4) for 225 GeV muons[8]. Accordingly, the ^{71}Ge

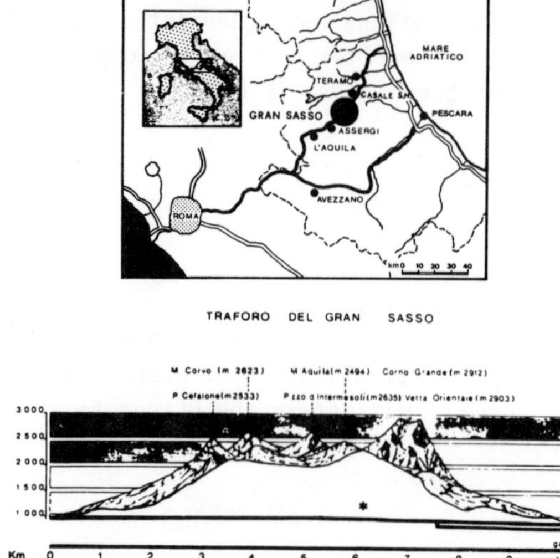

Figure 3:
Situation of the Gran Sasso European Underground Physics Laboratory, the site of the GALLEX experiment (star).

production rate in 30 tons of Ga at the GALLEX site is $\lesssim$.01 atoms/day or $\sim 1\%$ of the SSM ν_0 induced production rate.

The U/Th concentrations in the dolomitic wall rocks of the cavities are favourably low, neutron activation analyses of various specimens gave results always well below 1 ppm. In a quick look experiment, the integral fast neutron flux in the cavity was found to be $<10^{-6}$ n/cm^2, s. With the projected 25 cm PE-shield surrounding the tank, this leads to ^{71}Ge production rates below 1% of the SSM ν-signal according to yield determinations with a Pu-Be n-source in the pilot tank.

Concerning contaminants in the $GaCl_3$ solution itself, it was established in the course of the pilot experiment that the tolerable concentrations are ≤ 2 ppb Th, ≤ 30 ppb U, and $\leq.5$ pCi ^{226}Ra/kg solution if each of them is to contribute $\leq 1\%$ of the neutrino-induced SSM production rate. It has also been shown that multiton quantities of $GaCl_3$ solutions meeting these specifications can be industrially produced, corresponding specifications are part of the Ga-acquisition contract. For assurance, we routinely analyze relevant test solutions and chemicals for U, Th by neutron activation and for ^{226}Ra by measuring swept out ^{222}Rn using proportional counters.

Low Level Counting

^{71}Ge decays by K (88%), L(10%), and M(2%) electron capture. It is best detected in proportional counters using 1-2 atm GeH$_4$-Xe mixtures as counting gas. The energy deposition from Auger electrons and X-rays leads to the L-peak at 1.2 keV and to the K-peak at 10.4 keV. All provisions typical of low level counting have to be applied to achieve the extremely low background rates required: miniaturization (Fig. 4), use of ultrapure materials for counter construction, anti-

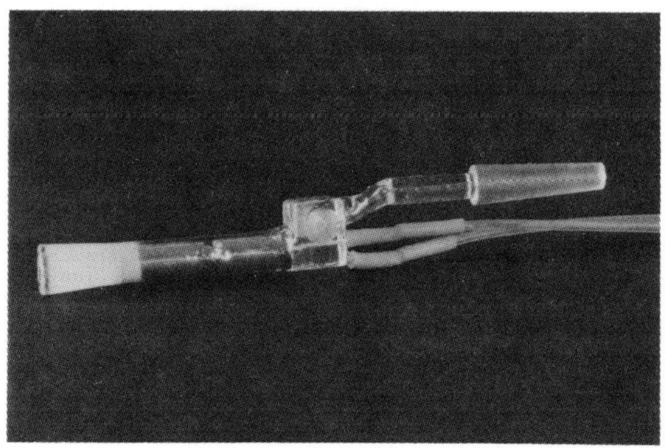

Fig. 4: Miniaturized proportional counter made from suprasile quartz and directly evaporated Al-cathode. The active volume (to the left) is .5cc only. The tungsten anode wire is directly sealed into the quartz ends. Right hand parts serve for gas filling. Wire connections to the preamp are low capacity (<1pF). The dead volume is $\lesssim$5%.

coincidence shield with NaI and plastic scintillation detectors (Fig. 5), heavy passive shielding with Pb and Fe. Still, the remaining background in the L- and K-peak windows is of the order of 1 cpd. It is mainly caused by (a) beta particles from natural radioactivities of the construction materials; (b) Compton electrons caused be external gamma rays; (c) electronic noise pulses. Fortunately in most cases, all these events produce pulse shapes distinct from those of ^{71}Ge decays (Fig. 6). Our computer-controlled counting system (Fig. 8) is therefore designed to record the whole shape of each PC-pulse by means of a fast transient digitizer. All counting data including the pulse shapes are finally stored on magnetic tapes. Up to 8 counters can be operated simultaneously in the well of the NaI pair spectrometer

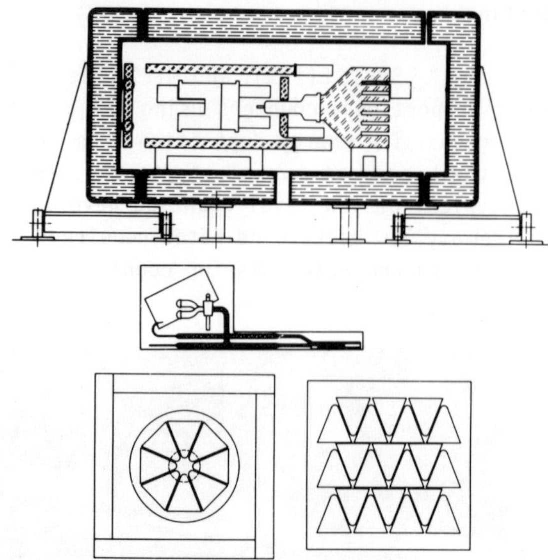

Fig. 5:   Scheme of the low level counting arrangement. A Pb-
filled steel tank with movable side ends (top) houses a
well type NaI crystal (interior left), a plastic scintillator shield (above and below) and a plastic scintillator block (right). The NaI-well accommodates up to
8 counters encapsulated, together with the preamp, in a
Cu-box (middle and bottom left). The scintillator block
can accommodate 21 additional counters for operation
without NaI crystal (top and bottom right). Plastic
scintillators operate in anticoincidence, NaI either in
coincidence or anticoincidence. Multipliers are
specially selected low potassium mps.

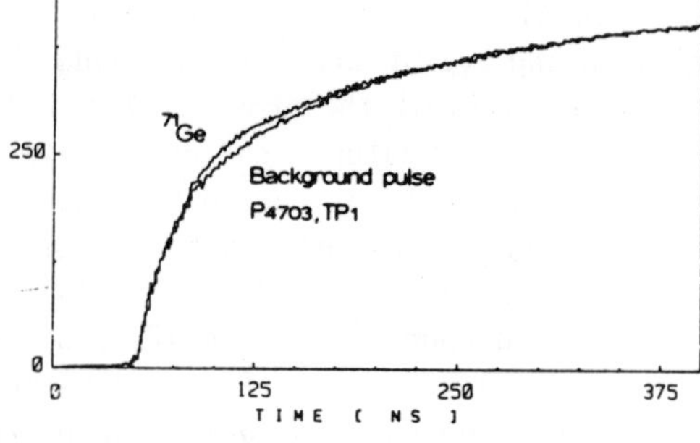

Figure 6:
A ^{71}Ge decay pulse and a
background pulse indis-
tinguishable by rise time
only but distinguishable
by the G*I criterium.

Apart from its anticoincidence function, it serves, in the anticoincidence mode, to identify β^+ decays of ^{69}Ge($T_{1/2}$=1.6 d). In this way, one can monitor side reactions in the tank via ^{69}(p,n)^{69}Ge.

Figure 7 shows the rejection efficiency between ^{71}Ge decays and Compton background simulated by a ^{60}Co source. The pulse form para-

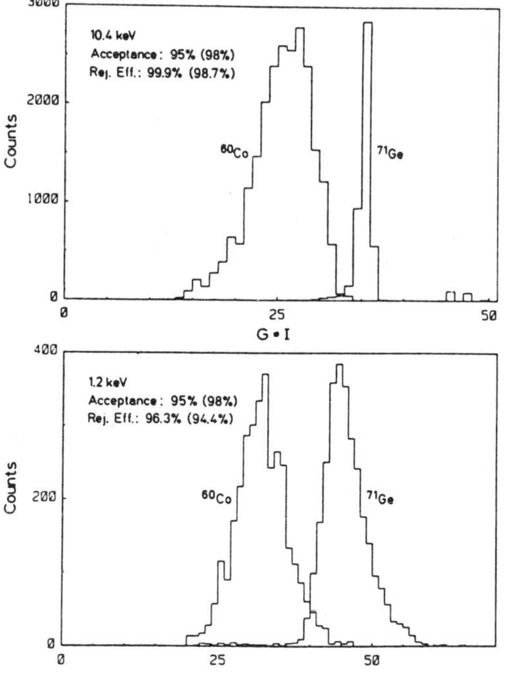

Figure 7:
Rejection efficiencies of Compton background as simulated by a ^{60}Co source in ^{71}Ge counting.

meter G*I has proved to be particularly discriminating. It is defined as

$$G*I = 100 \left[S(T)/S(t_{max}) \right] \left[\frac{1}{T} \int_0^T S(t)\,dt/S(t_{max}) \right]$$

where $S(t)$ is the pulse height at time t after pulse onset, t_{max} is the time at which $S(t)$ is maximal, and T is an empirically selected parameter, here T=160 ns. Overall efficiencies are 70% or better (2/5 L, 3/5 K). The background rates achieved are 0.2 cpd (L-peak), 0.08 cpd (K-peak) for many counters and 0.08 cpd (L-peak), 0.04 cpd and better (K-peak) for a few particularly good counters. A further improvement may result once the low level spectrometer set-up is transferred into the Gran Sasso Underground Laboratory. Presently, it is operated in the weakly shielded MPI Heidelberg Low-Level laboratory ($\sim$15 m.w.e. only).

The presently achieved background rates allow to measure 94 SNU (consistent model) during 2 years of operation of the 30 t Ga detector to $\pm$12%.

Calibration Source

We consider it as indispensable to demonstrate the concerted performance of all experimental components by means of a strong artificially produced low energy neutrino source. The most favourable choice is the use of ^{51}Cr produced by neutron activation of ^{50}Cr (4.35% natural abundance). ^{51}Cr decays be electron capture ($T_{1/2}$=27,7 d) into ^{51}V. It emits monoenergetic neutrinos of 746 keV (90%) and 426 keV (10%). A source strength of 600-800 kCi is necessary in order to perform a sensible experiment. The feasibility of a 600 kCi source has recently been demonstrated in a test irradiation at the 35 MW Siloe reactor at Grenoble[9]. The test was performed with 12.4 kg Cr powder for 21 days at an undisturbed n-flux of $\sim$9x10^{13} n/cm^{2}s. Accurate calibration of the source strength via 320 keV gamma rays (10% of all decays) is facilitated by the use of Cr-powder since it can be homogenized through mixing after irradiation. Among other things, the test irradiation has demonstrated the solution of such practical problems as containments, cooling during irradiation, mixing, and other hot cell operations, definition of shielding and transport requirements. The necessary shielding is entirely determined by the activated impurities. The selected Cr used in the test irradiation was sufficiently pure, the highest remaining γ-activities came from ^{110m}Ag and from ^{124}Sb. The actual source would consist of 120 kg natural Cr powder in a sealed container of dimensions r$\sim$10 cm, h$\sim$1 m; surrounded by a Pb or W-shield of 10 cm thickness. One source could be used during $\sim$60 days for 4-6 Ge-extractions. The first extraction would yield $\sim$3 times more ^{71}Ge produced by the source as by the sun (SSM). With one 600 kCi source, the obtainable accuracy would be $\sim$20%. The goal of a $\pm$10% result could be achieved with altogether 4 sources or by using Cr enriched in ^{50}Cr.

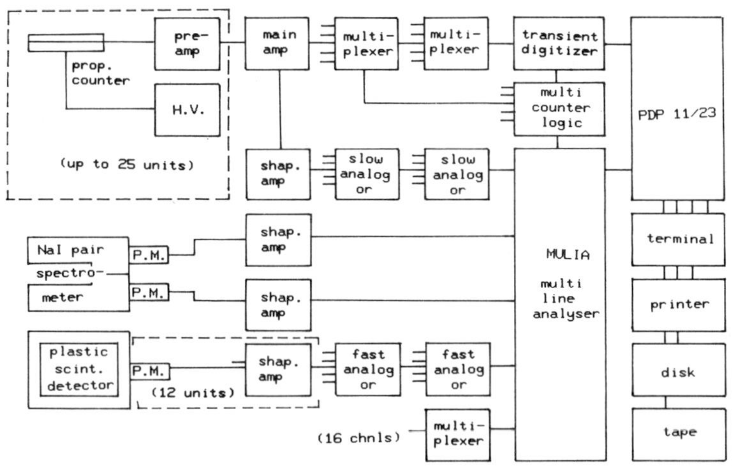

Figure 8: Block diagram of the measuring and
data acquisition system.

Time Schedule and Required Accuracy

The time schedule for the GALLEX project is shown in Figure 9. It is largely determined by the Ga-delivery schedule. Our contracts with the supplier define October 1989 as the date when the full 30 t. of Ga will be available to us. The experiment installations in the Gran Sasso lab should be completed in early 1988 when the first Ga will be delivered. Test runs of the full scale equipment will be performed in early 1989 using partial Ga quantities already delivered. The actual measurements will start in 1990. After two years of running (50 extractions), we would acquire a result with ±12% statistical error (consistent model) or ±11% (SSM). However, we must consider the possibility of a signal far below 94 or 123.5 SNU if neutrino flavor oscillations occur (Fig. 10). For maximal mixing of three neutrino flavor oscillations the production rate would be as low as $\sim$ 40 SNU. With the MSW matter oscillation effect[10,11] even lower production rates are possible. Clearly, an accurate measurement of such low rates would call for much longer measurement times. However, the principal question to be solved is the distinction between deviations from the SSM and neutrino oscillations as the cause of the solar neutrino problem. Provided that there is not a combination of both effects, the Ga experiment can distinguish these two possibilities if the absolute

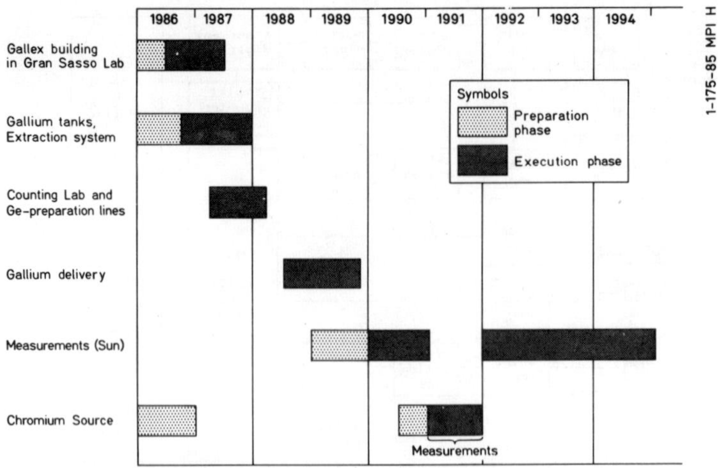

Figure 9: Projected time schedule for the GALLEX project

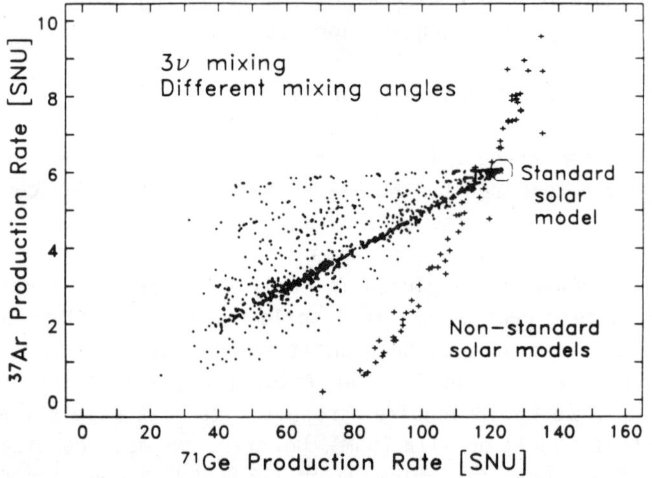

Figure 10: (from [12]) Expected production rates for the Ga- and Cl-
detectors. Crosses apply to all kinds of published non-
standard solar models, dots result from Monte Carlo varia-
tions of the neutrino mixing parameters Δm^2_{ij} and Θ_{ij} for
vacuum oscillations. The measured ^{37}Ar production rate is
2.0+.3 SNU. Without neutrino mixing, ^{71}Ge production rates
below 80 SNU are not possible. Not applicable for $m^2 \sim 10^{-5}$
eV^2 (see text).

uncertainty is <15 SNU (see Fig. 10). To approach 2 σ-significance, we are presently planning to measure for a period of 4 years (100 runs) plus one year for the ^{51}Cr source calibration runs.

A final remark concerns the MSW effect. Matter oscillations make the Ga experiment sensitive to neutrino oscillations even for small mixing angles in the range of $10^{-9}< \Delta m^2 < 10^{-5}$ eV^2. The price paid for this is an ambiguity, however, in a very narrow mass range at $\sim 10^{-5}$ eV^2. Here it is possible that the Cl-result is due to neutrino oscillations, yet the Ga-experiment still yields a rate around ~ 90 SNU.

References

1 J. Bahcall, W. Huebner, S. Lubov, P. Parker, R. Ulrich, Rev.Mod.Phys.54, 767 (1982).
2 J.K. Rowley, B.T. Cleveland, R. Davis, AIP Conf.Proc. 126, 1 (1985).
3 W. Hampel, R. Schlotz, Proc. AMCO-7 (Darmstadt) (Ed. E. Klepper, GSI Darmstadt) 89 (1984).
4 W. Hampel, AIP Conf.Proc. 126, 162 (1985).
5 W. Hampel, The Cross Section for Capture of Solar Neutrinos in ^{71}Ga. MPI Heidelberg Report, Feb. 1986, to be published.
6 D. Krofcheck, E.Sugarbaker, J. Rapaport, Phys.Rev.Lett. 55, 105 (1985).
7 A. Zichichi, ICOMAN Proc., Frascati 1983 (Ed. E. Bellotti), p.3.
8 J. Bahcall, B. Cleveland, R. Davis, I. Dostrovsky, J. Evans, W. Frati, G. Friedlander, K. Lande, J. Rowley, R. Stoenner, J. Weneser, Phys.Rev.Lett. 40, 1351 (1978).
9 M. Cribier, B. Pichard, J. Rich, M. Spiro, D. Vignaud, A. Besson, A. Bevilacqua, F. Caperan, G. Dupont, J. Gorry, W. Hampel, T. Kirsten, Production of a High Intensity 746 keV Neutrino Source for the Calibration of Solar Neutrino Detectors. Subm. to Nucl.Instr.Meth. (1986).
10 S. Mikheyev, A. Smirnov, Jadern.Fiz. 42, 1441 (1985).
11 J. Bouchez, M. Cribier, J. Rich, M. Spiro, D. Vignaud, W. Hampel, Matter Effects for Solar Neutrino Oscillations. Subm. to Z.Physik (1986).
12 W. Hampel, The Impact of Neutrino Oscillations on the Cl and Ga Solar Neutrino Experiments. MPI Heidelberg Report, Jan. 1986.

<u>TOTAL CROSS SECTIONS FOR CHARGED CURRENT NEUTRINO</u>
<u>INTERACTIONS IN Fe, Ne AND H$_2$</u>

H. Wachsmuth

CERN, Geneva, Switzerland

ABSTRACT

Total cross sections for charged current neutrino and anti-neutrino
interactions in iron, neon and hydrogen are presented. Details and
methods of the three CERN neutrino experiments are briefly described.

1. INTRODUCTION

The cross section results presented in this paper come from three
experiments in CERN neutrino beams: the first is a new measurement of
σ_{tot} in Fe by the CDHS Collaboration [1] using event samples
collected from 1982 to 1984 in the Narrow Band Beam (NBB) at three
momenta (100, 160, 200 GeV/c) and an improved neutrino flux measurement.
The second is an update of σ_{tot} in Ne obtained by the BEBC Collaboration WA47 [2] from data collected in the NBB from 1977 to 1980. The
third is a new measurement of σ_{tot} on protons by the BEBC (H$_2$)
Collaboration WA21 [3] using the Wide Band Beam (WBB) and a fourth ν
experiment, the Ne/H$_2$ experiment in the same WBB, done by the BEBC (Ne)
Collaboration WA59 to link $\sigma(\nu p)$ to the isoscalar cross section $\sigma(\nu I)$.

2. <u>LAYOUT OF THE NEUTRINO EXPERIMENTS AT CERN</u>

Fig. 1 shows the layout of the CERN West Area Neutrino Facility
from the proton extraction point to the neutrino detectors. The
enlarged part shows details of the focusing hall where the WBB and the
NBB are installed side-by-side.

In the WBB pions and kaons are produced by 400 GeV protons in a 2-3
absorption lengths long Be target of 3 mm diameter. Mesons of a broad

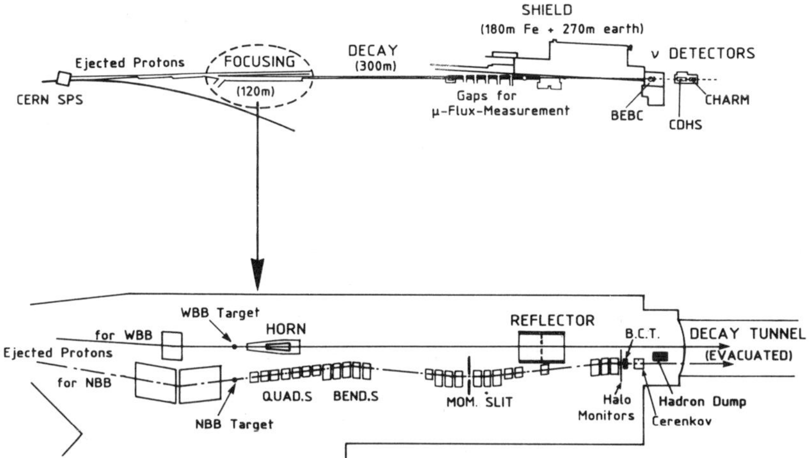

Fig. 1 Layout of CERN West Area Neutrino Facility

momentum range are charge selected and focused by pulsed rotation symmetric magnetic fields – called horn and reflector. They yield a neutrino spectrum with peak at ~ 20 GeV and mean $\nu(\bar{\nu})$ event energy of ~ 55 (~ 38) GeV above 10 GeV.

In the NBB pions and kaons emerging from a 50 cm long transmission target are momentum and charge selected and focused by a system of quadrupoles and bending magnets. The resulting momentum byte is ~ 6% (in 1984 ~ 8%). The 400 (1984: 450) GeV proton beam points horizontally and vertically 11 mr away from the neutrino detectors in order to reduce WBB background flux. Mean $\nu(\bar{\nu})$ event energies above 20 GeV are 84 (72) GeV from 160 GeV parents.

Also indicated in fig. 1 are the muon flux measurement gaps in the steel part of the shielding after the 300 m long decay pipe. These gaps house the Neutrino Flux Monitoring (NFM) system [4] used to measure the neutrino flux [5]: an array of some 80 fully depleted silicon diode Solid State Detectors (SSD's) which record the muon flux at radii of of 0, 15, 30 and 45 cm at five depths in the shielding corresponding to minimum muon momenta ranging from ~ 10 to ~ 180 GeV. These SSDs were calibrated relative to each other using moveable sets of SSDs. The

absolute calibration (to ~ ± 3%) is done by counting tracks in photo-
graphic emulsions exposed during a few pulses and separating muon from
delta electron tracks by their different angular distributions [6].

In the NBB the neutrino flux can be determined also from the flux
of pions and kaons measured in the Beam Current Transformers (BCTs);
their location is indicated in fig. 1. The BCTs are toroids around the
beam pipe consisting of high permeability alloy foils and a pickup coil
[7]. They are calibrated electromagnetically to a few 10^{-3}. A
differential Cerenkov detector downstream of the BCT analyses the
composition of the hadron beam.

Downstream of the shielding are (were) the neutrino detectors BEBC,
CDHS and CHARM in ~ 820, 890 and 910 m distance from the target.

3. THE ISOSCALAR CROSS SECTION

3.1 From the CDHS Experiments in Fe

The data come from three largely independent NBB exposures with
parent momenta 100 and 200 GeV/c in 1982/1983 and 160 GeV/c in 1984
using the upgraded CDHS detector.

3.1.1 Neutrino flux determination

(a) From the muon flux

In order to relate the muon flux recorded by the SSDs to the
neutrino flux traversing the neutrino detector the so-called trapped
muons have to be subtracted. These are muons from π and K decays
upstream of the decay tunnel which are trapped in the hadron beam line
and transported along with the hadrons. In 1984 their amount could be
measured by moving the dump shown in fig. 1 in front of the decay tunnel
into the beam line. The muon flux distribution observed with the dump
in place must be corrected for the effects of multiple scattering and
energy loss in the 1.5 m steel of the dump. The trapped muons were
found [8] to contribute 18 ± 1% (in gap 2 at r = 15 cm) to ~ 30% (in

gap 4) of the total μ flux in ν running (in $\bar{\nu}$ running $\sim$ 13% less due to the smaller K/π ratio). Fig. 2 shows the muon flux distribution after this correction. The neutrino flux corresponding to this muon flux has an accuracy of $\pm$ 3.6% including a 2% uncertainty in the parent beam momentum. In the 100 and 200 GeV beams the accuracy is 5-6% due to larger uncertainties in calibration and beam conditions.

(b) <u>From the pion and kaon flux</u>

The ratio of π^+/π^- fluxes extracted from the BCT recordings, R_{BCT}, differed from the one extracted from the muon fluxes (SSD recordings), R_{SSD}, by ΔR = 9 $\pm$ 2, 4.5 $\pm$ 2 and 22 $\pm$ 2.5% in the 100, 160 and 200 GeV beams, respectively. The errors are mainly due to uncertainties in particle ratios (p/π, K/π,...) as determined in the Cerenkov detector. Since R_{SSD} is known to $\pm$ 1%, the observed differences must be due to additional charge induced in the BCT. The smaller difference in the 160 GeV beam was achieved by extending the beam vacuum through the BCT (eliminating delta ray effects) and correct grounding connections in the 1984 arrangement.

Changing quadrupole focusing strengths and thus increasing the amount of beam halo which could be measured in the 1984 set-up (with ionisation chambers serving as halo monitors, see fig. 1) resulted in a linear dependence of ΔR on the halo, with opposite sign in a nearly identical second BCT (see fig. 3). When extrapolated to zero halo, both BCTs yield the same R_{BCT} and - within the 2% scale error - equal to R_{SSD}. From these studies it was concluded that the sum of the BCT signals in positive and negative beam had to be corrected by ΔR. In the 160 GeV experiment the amount of ΔR was shared between negative and positive polarity in the ratio 1.2 : 1, since in standard run conditions in the negative beam 20% more halo per beam hadron was recorded than in the positive beam. For the 100 and 200 GeV experiments where no halo monitors were in the beam line a 1:1 sharing was applied, supported by the fact that a change of BCT grounding conditions had about equal effects on the BCT response in both polarities.

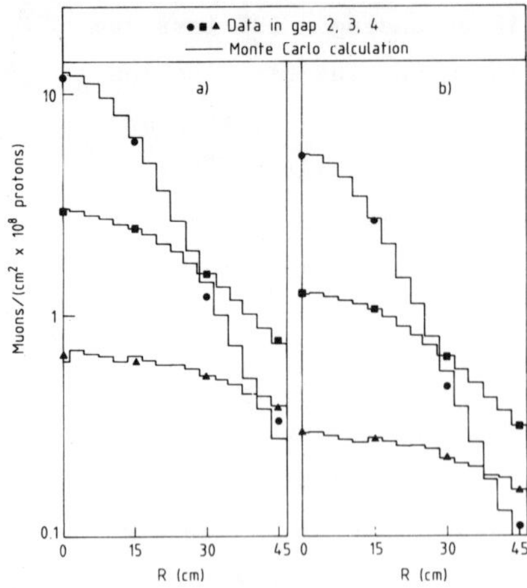

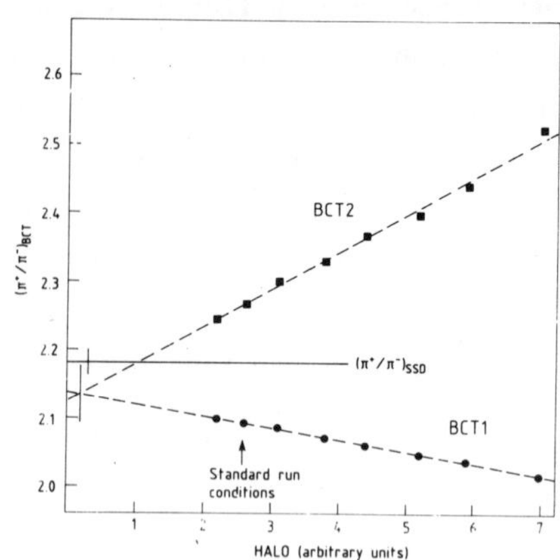

Fig. 2 Radial muon flux distr. in the 160 GeV (a) pos., (b) neg. beam after correction for trapped μ's. Calculation normalised to data at R = 30 cm.

Fig. 3 The π^+/π^- ratio extracted from the BCTs compared to the one extracted from the μ^+/μ^- ratio (SSDs) as a function of beam halo.

The total error on the BCT derived antineutrino flux due to uncertainties in calibration, K^-/π^- ratio and halo correction (varying slightly with time) is 2.3, 3 and 3.3% in the 160, 100 and 200 GeV beam, respectively. Neutrino fluxes derived from the π and K fluxes (BCT) agree with those derived from the μ fluxes (SSD) to 1–2%, well within the respective calibration uncertainties.

3.1.2 <u>The event samples</u>

Neutrino interactions were recorded with the upgraded CDHS detector [9]. It consisted of toroidally magnetized iron plates sandwiched with planes of scintillators. They were grouped in 21 modules of three types: modules 1–10 with 2.5 cm, modules 11–15 with 5 cm, and modules 16–21 with 15 cm thick iron plates. The total iron thickness of each module was 50 cm for modules 1–10 and 75 cm for the others. Each module had a diameter of 3.75 m. In between the modules drift chambers were inserted for muon tracking (96 ± 1% reconstruction efficiency). The momentum of the muon was measured with an average resolution of ± 9%.

The hadronic energy E_h is measured by calorimetry with a resolution of $\sigma/E_h \sim 0.58/\sqrt{E_h}$ for modules 1-10 and $\sim 0.70/\sqrt{E_h}$ for modules 11-15.

The apparatus was triggered whenever the energy deposition in three modules was larger than one third of the energy deposited by a minimum ionizing particle traversing the complete module. For muons of momenta larger than 5 GeV, the trigger efficiency was measured to be better than 99.9% using events which were selected by an independent trigger based on hadronic energy only. The dead-time varied between 5 and 25%, depending on the beam intensity and the length of the extraction; it was measured to better than 10%.

Events were selected for the analysis when:

- at least five drift chambers were available for muon tracking;

- the muon had more than 7 GeV momentum (160 GeV and 200 GeV data) or penetrated more than 3 m of iron (100 GeV data);

- the vertex was inside the fiducial volume, corresponding to 460 ± 5 t (100 and 200 GeV data) or 280 ± 3 t (160 GeV data). The size of the event samples are given in table 1.

<u>TABLE 1</u> Event statistics

E_{beam} (GeV)	100		200		160	
	ν	$\bar{\nu}$	ν	$\bar{\nu}$	ν	$\bar{\nu}$
Protons on tgt. (10^{18})	0.21	0.48	0.48	0.42	5.50	0.60
Events after cuts	10,000	9,000	11,000	2,500	200,000	9,000

3.1.3 The cross sections

The event samples were corrected for the p_μ cut (1% uncertainty) and for the other experimental inefficiencies. The flux measurements via BCT and SSD were averaged, weighted according to their systematic uncertainties. Table 2 shows the resulting total cross sections. The cross section ratios were obtained using the SSDs only.

TABLE 2 Average cross section slopes separately and combined for the three beam energies. The first error is statistical, the second systematic. The systematic error of the weighted average contains the common 2% overall scale error in quadrature.

E_{beam} (GeV)	σ^{ν}/E (10^{-38} cm^2/GeV)	$\sigma^{\bar{\nu}}/E$ (10^{-38} cm^2/GeV)	$\sigma^{\bar{\nu}}/\sigma^{\nu}$
100	.691±.007±.021	.330±.004±.009	.478±.008±.007
160	.707±.002±.018	.333±.004±.007	.471±.006±.007
200	.708±.007±.024	.333±.007±.010	.470±.011±.007
Weighted average	.702±.002±.019	.332±.003±.009	.473±.005±.007

The $\nu(\bar{\nu})$ cross sections corrected for the neutron excess in Fe using measured cross sections on neutrons and protons [10,11] are 2.5% smaller (2.3% larger).

Since in a NBB ν experiment ν_{π} and ν_{K} induced events are separated spatially in the detector, cross sections at six energies can be determined from the three exposures (fig. 4).

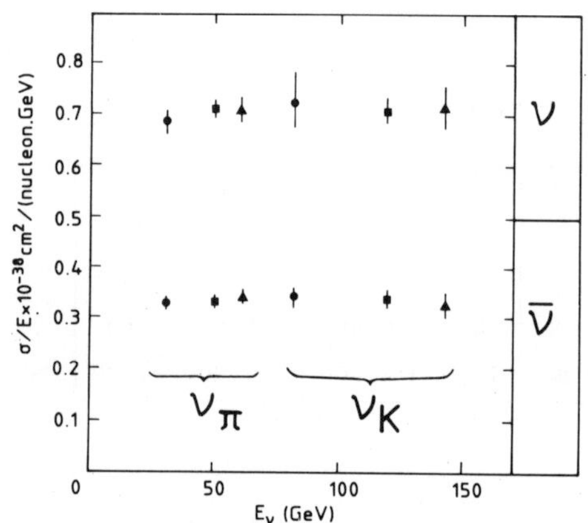

Fig. 4 Average ν_{π} and ν_K cross sections from 100 (●), 160 (■), and 200 (▲) GeV data. Statistical and systematic errors are added in quadrature. Not included is a 2% error common to all three exposures.

They were used to determine limits for a possible energy dependence of σ/E. A straight line fit yields 2 ± 5% and –1 ± 3% rise of σ/E between 50 and 150 GeV for ν and $\bar{\nu}$, respectively.

3.2 <u>From the BEBC NBB Experiment in Ne/H$_2$ (an update)</u>

In this experiment, WA47, done with BEBC filled with a Ne/H$_2$ mixture and exposed to the ν's of a 200 GeV NBB during 1977–1980, some 3000 ν and 1000 $\bar{\nu}$ events were used to determine σ_{tot} [2]. The ν flux was measured via the associated muon flux. These cross sections were recently revised [3] because two corrections had been underestimated as was found in later experiments: (a) The event samples of the WA47 analysis were underestimated due to loss of events with low prong topologies. This was found by comparing rates of events with one or two tracks with those with many tracks energy bin by energy bin in WA47 and in another BEBC ν experiment, WA59. In the WA59 experiment, using also Ne/H$_2$ but a WBB, special scans were made for such low prong events. Correcting for this additional scanning inefficiency increases the $\nu(\bar{\nu})$ event samples by 1.8 (7.5)%. (b) From the trapped muon flux measurements done in the CDHS experiment (see above) and applied to the 1980 set-up it was found that in the WA47 analysis the corresponding correction was underestimated such that the $\nu(\bar{\nu})$ flux has to be decreased by 6 (4)%. The updated cross section values are shown in table 3 together with the above CDHS and other results with similar precision (better than 6%).

<u>TABLE 3</u> Average slopes of isoscalar cross sections between 20 and 200 GeV from experiments quoting $\leq$ 6% measurement error (units: 10^{-38} cm^2/GeV).

Experiment	σ^ν/E	$\sigma^{\bar{\nu}}/E$	$\sigma^{\bar{\nu}}/\sigma^\nu$
CDHS	.685±.002±.019	.340±.003±.009	.496±.005±.007
BEBC–WA47	.723±.013±.036	.351±.010±.016	.485±.017±.011
CHARM [11]	.604±.032	.301±.018	.498±.019
CCFRR [12]$^{(a)}$	.668±.003±.024	.343±.003±.020	.513 ± (< .035)

(a) With the same correction for non-isoscalarity as applied for CDHS: −2.5% for ν, + 2.3% for $\bar{\nu}$.

4. <u>THE NEUTRINO PROTON CROSS SECTION</u>

The νp cross section was determined from an experiment by the
WA21 Collaboration with BEBC filled with hydrogen and exposed to the
WBB. It is difficult to achieve in a WBB a cross section precision
better than $\pm$ 8%: The absolute SSD calibration is less certain than
in a NBB because of the broader angular distribution of the muons in
the shield (δ–μ separation in the emulsions!). There is an error of
5–6% due to uncertainties in critical assumptions in the conversion of
μ to ν flux (shield density, μ–momentum range relation, multiple
scattering, particle production, ...). The event sample has to be
corrected for scan inefficiency, energy measurement resolution
(smearing), etc. ($\sim$ 4% error).

Neglecting nuclear ("EMC like") effects a more precise $\sigma(\nu p)$ was
obtained by relating it to the isoscalar cross section $\sigma(\nu I)$ (now known
to $\pm$ 3%, table 3) via a ν event sample N_{Ne} taken in Ne by the WA59
Collaboration in the same BEBC volume under the same WBB flux
conditions as the νH_2 sample (N_H):

$$\sigma(\nu p)/\sigma(\nu I) = (N_H/N_{Ne}) \cdot (\rho_{Ne}/\rho_H) \cdot (\phi_{\mu Ne}/\phi_{\mu H}) \cdot f_{is}$$

The two exposures were done consecutively (WA21 in March 1980 and
WA59 in April 1980). The event samples (3184 νH_2, 3223 $\bar{\nu} H_2$, 7577 νHe,
11857 $\bar{\nu} Ne$, all corrected) were completely reanalysed for this comparison
such that systematic errors are only due to differences in the corrections. The largest uncertainty lies in the difference of the correction
for event loss due to scan inefficiency (1.2% in ν, 2.3% in $\bar{\nu}$) and
smearing (1% in ν and $\bar{\nu}$). The ratio of the BEBC liquid densities
ρ_{Ne}/ρ_H was measured to be 11.3 $\pm$ 0.1. The ratio of the ν fluxes is
given by the ratio of muon fluxes $\phi_{\mu Ne}/\phi_{\mu H}$ selected for equal beam
condition and integrated over the two experimental periods and was
0.140 $\pm$ 0.002 for the ν and 0.459 $\pm$ 0.005 for the $\bar{\nu}$ samples. The proton
excess in the 75 mole% Ne/H_2 mixture is corrected for by f_{is} = 0.992
(ν) and 1.009 ($\bar{\nu}$).

The resulting ratios agree with those derived in other experiments [10,11,14,15] and are $0.657 \pm 0.015 \pm 0.014$ for ν, $1.419 \pm 0.030 \pm 0.042$ for $\bar{\nu}$. Multiplying them with $\sigma(\nu I)$ from the CDHS experiment (see table 3) yields[*]:

$$\sigma(\nu p) = 0.451 \pm 0.011 \pm 0.016$$

$$\sigma(\bar{\nu}p) = 0.451 \pm 0.011 \pm 0.016$$

$$\frac{\sigma(\bar{\nu}p)}{\sigma(\nu p)} = 0.482 \pm 0.017 \pm 0.019$$

The cross section ratio can also be obtained from the H_2 samples alone. The $\nu/\bar{\nu}$ flux ratio is 0.854 ± 0.017 and the ratio of Monte-Carlo calculated event rates per muon flux in the positive and negative beam is 1.220 ± 0.024. The event rate ratio $N(\bar{\nu}H_2)/N(\nu H_2)$ of $1.012 \pm 0.028 \pm 0.034$ corresponds to

$$\frac{\sigma(\bar{\nu}p)}{\sigma(\nu p)} = 1.055 \pm 0.030 \pm 0.044$$

5. CONCLUSION AND DISCUSSION

A new measurement of the isoscalar cross section for charged current ν interactions in the CDHS detector has been done at three NBB momenta. The limits for a possible energy dependence of σ/E are 5% (3%) for $50 < E_{\nu(\bar{\nu})} < 150$ GeV. The absolute values are 13% (ν) and 11% ($\bar{\nu}$) higher than obtained previously [16]. The difference is most likely due to the beam halo effects in the BCT which had not been taken into account.

The isoscalar cross sections from the BEBC NBB 200 GeV experiment have been updated: based on newly gained information corrections to the ν flux ("trapped muons") and to the event samples ("low-prong events") were re-evaluated leading to 10% (ν) and 13% ($\bar{\nu}$) larger cross sections.

---

[*] In ref. [3] for $\sigma(\nu I)$ the updated result of the BEBC-WA47 experiment was used.

The neutrino proton cross section has been determined with 5% precision from a BEBC-H$_2$ experiment via $\sigma(\nu p)/\sigma(\nu I)$ using a BEBC-Ne experiment and $\sigma(\nu I)$ from the CDHS experiment. This analysis neglects EMC or other nuclear effects leading to a possible $\sigma(\nu I)$ dependence on the mass number.

<u>REFERENCES</u>

[1]   A. Blondel et al., paper in preparation (Zeitschr. für Phys. C);

      M.W. Krasny, Proc. Intern. Europhysics Conf. on HEP, Bari (1985) 555.

[2]   P. Bosetti et al., Phys. Lett. 110B (1982) 167, updated in [3].

[3]   M. Aderholz et al., Phys. Lett. 173B (1986) 211;

      P. Shotton, W. Venus and H. Wachsmuth, CERN/EP/NBU 85-2.

[4]   G. Cavallari et al., IEEE Trans. on Nucl. Sci., NS-25 (1978) 600;

      H.M. Heijne, CERN Yellow report, CERN 83-06.

[5]   D. Bloess et al., Nucl. Instr. & Meth. 91 (1971) 605.

[6]   C. Penfold and H. Wachsmuth, CERN/EP/NBU 81-4;

      I. Abt and R.Jongejans, Nucl. Instr. & Meth. A235 (1985) 85.

[7]   K. Unser, CERN/ISR-OP/81-14.

[8]   H. Wachsmuth and J. Wotschack, CERN/EP/NBU 85-1.

[9]   M. Holder et al., Nucl. Instr. and Meth. 148 (1978) 235;

      H. Abramowicz et al., Nucl. Instr. & Meth. 180 (1981) 429.

[10] J. Hanlon et al., Phys. Rev. Lett. 45 (1980) 1817.

[11] D. Allasia et al., Nucl. Phys. B239 (1984) 301.

[12] M. Jonker et al., Phys. Lett. 99B (1981) 265, erratum Phys.
      Lett. 100B (1981) 520.

[13] R. Blair et al., Phys. Rev.Lett. 51 (1983) 343.

[14] M.A. Parker et al., Nucl. Phys. B232 (1984) 1.

[15] H. Abramowicz et al., Zeitschr. für Phys. C25 (1984) 29.

[16] J.G.H. de Groot et al., Zeitschr. für Phys. C1 (1979) 143.

RESULTS ON CHARGED CURRENT STRUCTURE FUNCTIONS

CDHSW Collaboration

Presented by P. Perez

Département de Physique des Particules Elémentaires,

CEN-Saclay 91191 Gif sur Yvette Cedex, France.

ABSTRACT

Neutrino and antineutrino differential cross sections on iron have been measured with the CDHS detector exposed to the CERN Wide Band Beam. Nucleon structure functions are derived and compared to QCD predictions.

Neutrino beam

Data were collected in 1983 using the CERN Wide Band (anti)neutrino Beam. The energy distribution of (anti)neutrinos can roughly be described by 2 exponentials corresponding to π or K decays (fig. 1). Such a beam provides high statistics and its detailed shape is determined in 2 dimensions (energy versus radius to apparatus axis) using observed events.

Detector

The improved CDHS detector consists of 10 calorimeter modules with 2.5 cm of iron sampling, followed by 5 modules with 5 cm sampling and 6 modules with 15 cm sampling. The energy resolution for hadronic showers is $\Delta E/E = (.02^2 + .52^2/E)^{1/2}$, $.70/\sqrt{E}$ and $1.35/\sqrt{E}$ (GeV) in the respective sections. The calorimeter is magnetized (1.6 kG) and interspersed with drift chambers giving a resolution of $\Delta p/p \sim 9\%$ on muon momenta. For a muon momentum of 10' GeV/c, the average angular resolution is 10 mr and 2 mr for 100 GeV/c.

Cuts and Statistics

The vertex of each event is required to be at a distance relative to the apparatus axis bigger than 35 cm and smaller than 1.6 m in order that hadronic showers be well contained in the sensitive part of the calorimeter. After selecting events with a vertex found in modules 2 to 9, where the resolution on E_{had} is best, the fiducial mass is 247 tons. To ensure a good track reconstruction, the muon must leave a signal in at least 5 consecutive drift chambers. The momentum of the muon must be greater than 5 GeV/c for a high reconstruction efficiency. Events below E_{had} = 5 GeV, where the resolution is poor, are also rejected. The total energy range for this analysis is 20 < E < 212 GeV.

The resulting statistics after all cuts is shown in Table 1. Such high statistics, corresponding to ~ 5 times more events before cuts, enabled a precise study of systematic effects.

Differential cross-sections

The differential cross sections are computed according to :

$$\frac{1}{E} \frac{d^2\sigma}{dxdy} = \frac{N(x, y, E)}{N(E)} \frac{\sigma tot(E)}{E} \frac{1}{\Delta x \, \Delta y} \, ,$$

where the numbers of events, N, have been corrected for acceptance and smearing. Cells, in the (x, y, E) volume, for which this correction is larger than 30% are rejected. The value of the total cross section is taken from our latest measurements in Narrow Band Beam exposures [1] : $\sigma_{tot}^{\nu}/E = 0.702 \; 10^{-38}$ cm^2, $\sigma_{tot}^{\bar{\nu}}/E = 0.332 \; 10^{-38}$ cm^2.

The above formula ensures a normalization such that σ_{tot}/E is the same for each E bin.

A series of possible effects have been studied to evaluate systematic errors. They include
   - A shift of $\pm$ 500 MeV as well as a scale error of $\pm$ 2.5% and a tilt of $\pm$ 2.5% over 65 GeV, on E_{had}.
   - An error on the scale of the muon momentum of $\pm$ 2%.
   - An under / overestimate of the smearing of the measured kinematical quantities (P_μ, E_{had}, θ_μ) by $\pm$ 20%.
   And a tilt of the computed beam energy spectrum of 20% over 200 GeV.

Resulting differential cross sections are used as input to the simulation program which computes the acceptance and smearing corrections. This iterative procedure converges after 2 steps. Figure 2 shows the y distributions in different x bins for a total energy bin centered at 111 GeV. The inner error bar is statistical and the total error includes systematic errors added in quadrature.

In order to derive structure functions, a correction for bin centering is applied which may be as big as 5% for x = 0.015. Electroweak radiative corrections are performed using a program from D. Bardin. The propagator term of the W is taken into account with M_W = 82 GeV/c^2 (effect of 6% at Q^2 = 200 GeV^2).

x F$_3$

The difference of neutrino and antineutrino cross sections gives the structure function xF_3 on iron. In order to extract the nucleon structure function a correction has to be made for the non isoscalarity of iron. It is achieved with the formula :

$$xF_3 = \frac{\pi}{G^2 mE} \frac{[\frac{d^2 \sigma^{\nu F_e}}{dxdy} - \frac{d^2 \sigma^{\bar{\nu} F_e}}{dxdy}]}{[1 - (1-y)^2 + \delta (1 + (1-y)^2 - 2 (mxy/Q)^2)]}$$

where $\delta = \dfrac{N - Z}{N + Z} \dfrac{1 - d_v/u_v}{1 + d_v/u_v}$ is the isoscalar correction factor in which valence quark distributions appear as a ratio : d_v/u_v. We ascribe the world average [2] d_v/u_v = (0.55 $\pm$ 0.1) (1-x). For a given (x,Q^2) point the value of xF_3 is obtained after averaging over the available y range.

On top of the above quoted systematic errors on differential cross sections which have been propagated to structure functions, one of the most important systematic errors is due to the total cross section to which we assign a 1% scale error and a 7% tilt over 160 GeV total energy.

It should be noticed that, for the first time the statistical errors on xF_3 [fig. 3] are comparable to the systematical errors.

F$_2$ and 2xF$_1$

The sum of ν and $\bar{\nu}$ cross sections :

$$F = (\pi /G^2 mE) < (\frac{d^2 \sigma^{\nu N}}{dxdy} + \frac{d^2 \sigma^{\bar{\nu}N}}{dxdy}) / (1 + (1-y)^2) >_y \text{ is a combination}$$

of F_2 and $2xF_1$:

$F(x, Q^2) = a 2xF_1 + (1 - a) F_2$ with $a = < y^2 / (1 + (1-y)^2) >_y$.
F_2 or $2xF_1$ are computed from the measured combination F, given a prejudice for $R(x,Q^2) = \sigma_L / \sigma_T$:

$$F_2(x,Q^2) = F(x,Q^2) / [\, a\,(1 + (2\,mx/Q)^2)/(1 + R(x,Q^2)) + (1 - a)\,]$$

$$2xF_1(x,Q^2) = F(x,Q^2) / [\, a + (1 - a)(1 + R(x,Q^2)) / (1 + (2\,mx/Q)^2)\,]$$

We have parametrized our measurement of R obtained via the combination $d^2\sigma^\nu/dxdy - (1-y)^2\, d^2\sigma^{\bar\nu}/dxdy$ [3] as $R(x,Q^2) = 1.5\,(1-x)^4/\mathrm{Ln}(Q^2/0.2^2)$, which is not really a parametrization of what QCD predicts.

In our case, F is closer to F_2 than to $2xF_1$ in most of our kinematical range.

The corrections for the non isoscalarity of iron and for the strange sea are taken care of via :

$$F = (\pi/G^2 mE)\; \frac{[\,(1-b)\,\dfrac{d^2\sigma^{\nu F_e}}{dxdy} + (1+b)\,\dfrac{d^2\sigma^{\bar\nu F_e}}{dxdy}\,]}{[\,1 + (1-y)^2 - 2\,(mxy/Q)^2 + C_{sc}\,[\,1 - (1-y)^2\,]\,]}$$

with $b = (\delta - C_{sc})(1 - (1-y)^2) / [\,1 - (1-y)^2 + \delta\,[\,1 + (1-y)^2 - 2\,(mxy/Q)^2\,]\,]$,

and the strange sea correction factor $C_{sc} = x(s-c)/\bar{q} = 0.1 \pm 0.1$ (from our measurement of the rate of $\mu^-\mu^+$ at $\nu < 50$ GeV, not corrected for $s \to c$ suppression).

In this case

$$a = [\, y^2 + C_{sc}(1 - (1-y)^2)\,] / [\, 1 + (1-y)^2 - 2\,(mxy/Q)^2 + C_{sc}(1 - (1-y)^2)\,].$$

We may compare this result (fig. 4) to the previous CDHS publication [4] if we use the same ingredients : the old value of the total cross section, R was assumed constant equal to 0.1, and $C_{sc} = 0.2$. Although the binning was done directly in (x,y,Q^2) while we have to do it in (x,y,E) so that a given Q^2 point does not correspond to the same kinematical domain, especially at low x, most of the points are in agreement to better than 10%.

QCD analysis of xF_3

The s (or d) to c transition is suppressed below the charm threshold. Although this has little effect on xF_3 compared to F_2, a correction is applied using the "slow rescaling" model [5].

We have compared different QCD "programs" which either evolve structure functions from a starting value of Q_0^2, after the Altarelli-Parisi equations, or perform a Mellin inversion.

The resulting value of Λ has been studied by varying cuts on x, Q^2, W^2, the starting value Q_0^2 and the shape of xF_3 at this Q_0^2, and by introducing target mass effects.

As seen on figure 5, showing an example of a second order fit with $Q^2 > 5$ and $W^2 > 11$ GeV^2, fits are not satisfactory in the sense that the QCD prediction chooses the extreme slopes in Q^2 allowed by the error bars and, in our range of Q^2, Λ always increases with the Q^2 cut.

None of the systematic effects studied could explain to this level the different slopes.

It seems that the low Q^2, low W^2 region cannot be described by these fits together with the higher Q^2 regime.

More work is going on but, at the time of this conference, we have to present a very conservative range : $50 < \Lambda_{\overline{ms}} < 300$ MeV.

Table 1

	ν	$\bar{\nu}$
20 < E < 212	577868	368091
100 < E < 212	164541	41355

References

[1] H. Wachsmuth, these proceedings.

[2] J. Feltesse, proceedings of the Int. Europhysics Conf. on High En. Physics, Bari 1985, L.Nitti ed., p977.

[3] P. Buchholz, proceedings of the Int. Europhysics Conf. on High En. Physics, Bari 1985, L. Nitti ed., p557.

[4] H. Abramowicz et al., Z. Phys. C.17 (1983) 283.

[5] R.M. Barnett, Phys. Rev. D14 (1976) 70.

arbitrary unit
10 3
10 2
10
1
___ WBB neutrino energy spectrum
_ _ WBB antineutrino
20 40 60 80 100 120 140 160 180 200
E_ν

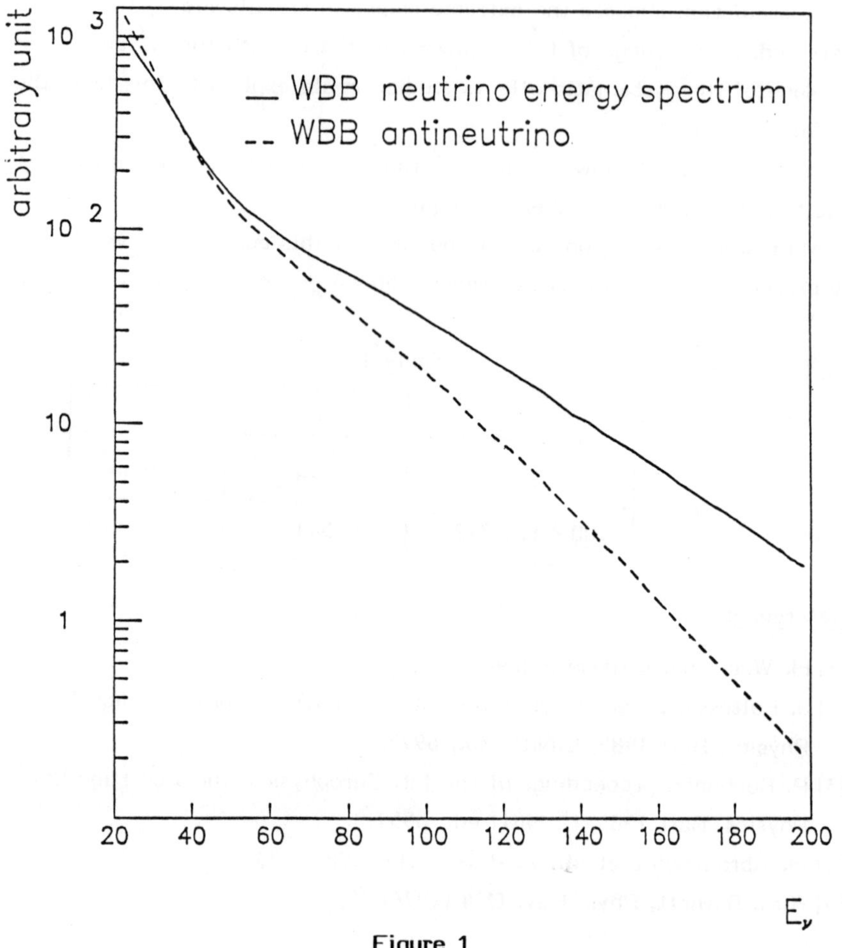

Figure 1

dσ/Edxdy
O neutrino
□ antineutrino
E = 111
X=0.015
X=0.045
X=0.080
X=0.125
X=0.175
X=0.225
X=0.275
X=0.350
X=0.450
X=0.550
X=0.650
Y

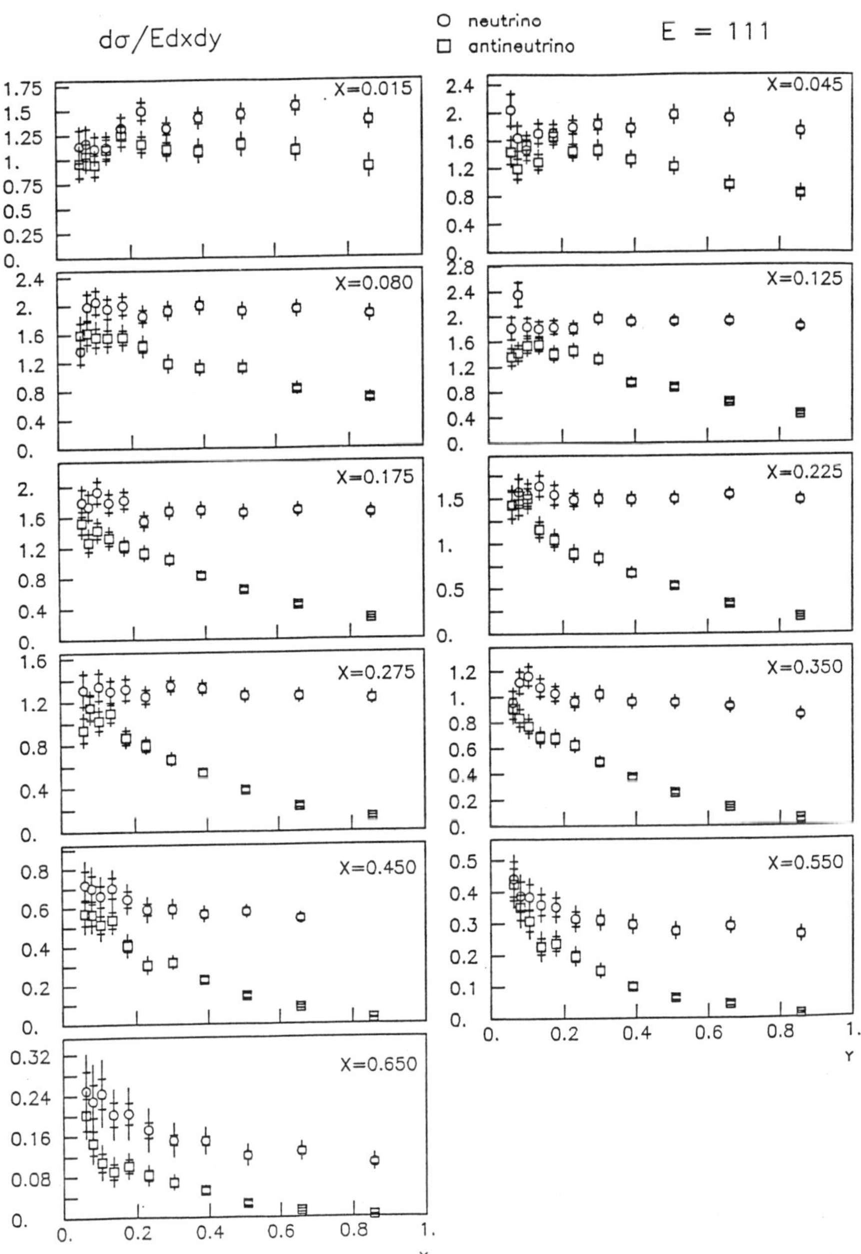

Figure 2

CDHSW 86
xF₃
X=0.015
X=0.045
X=0.080
X=0.125
X=0.175
CDHSW 86
xF₃
X=0.225
X=0.275
X=0.350
X=0.450
X=0.550
X=0.650
Q² (GeV²/c²)
Q² (GeV²/c²)

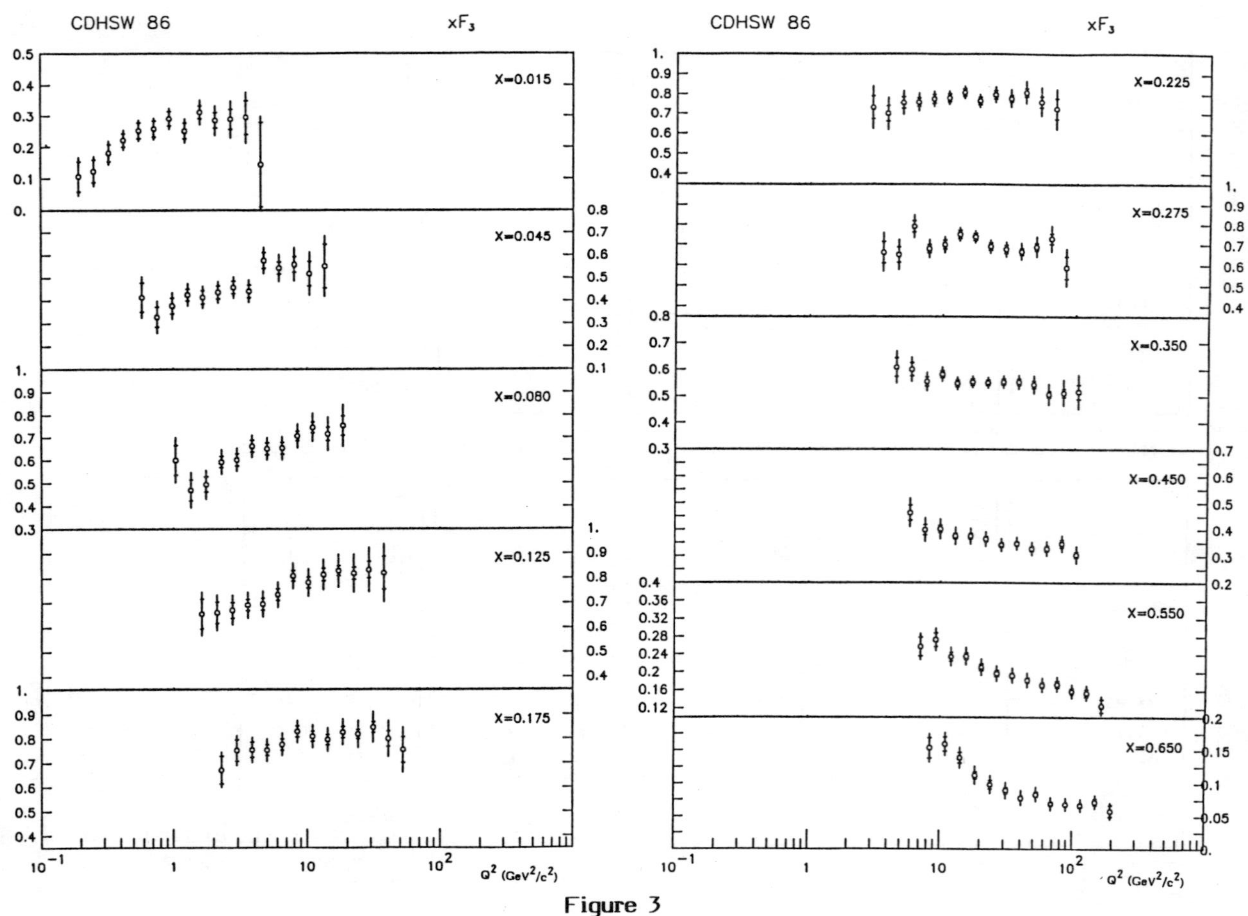

Figure 3

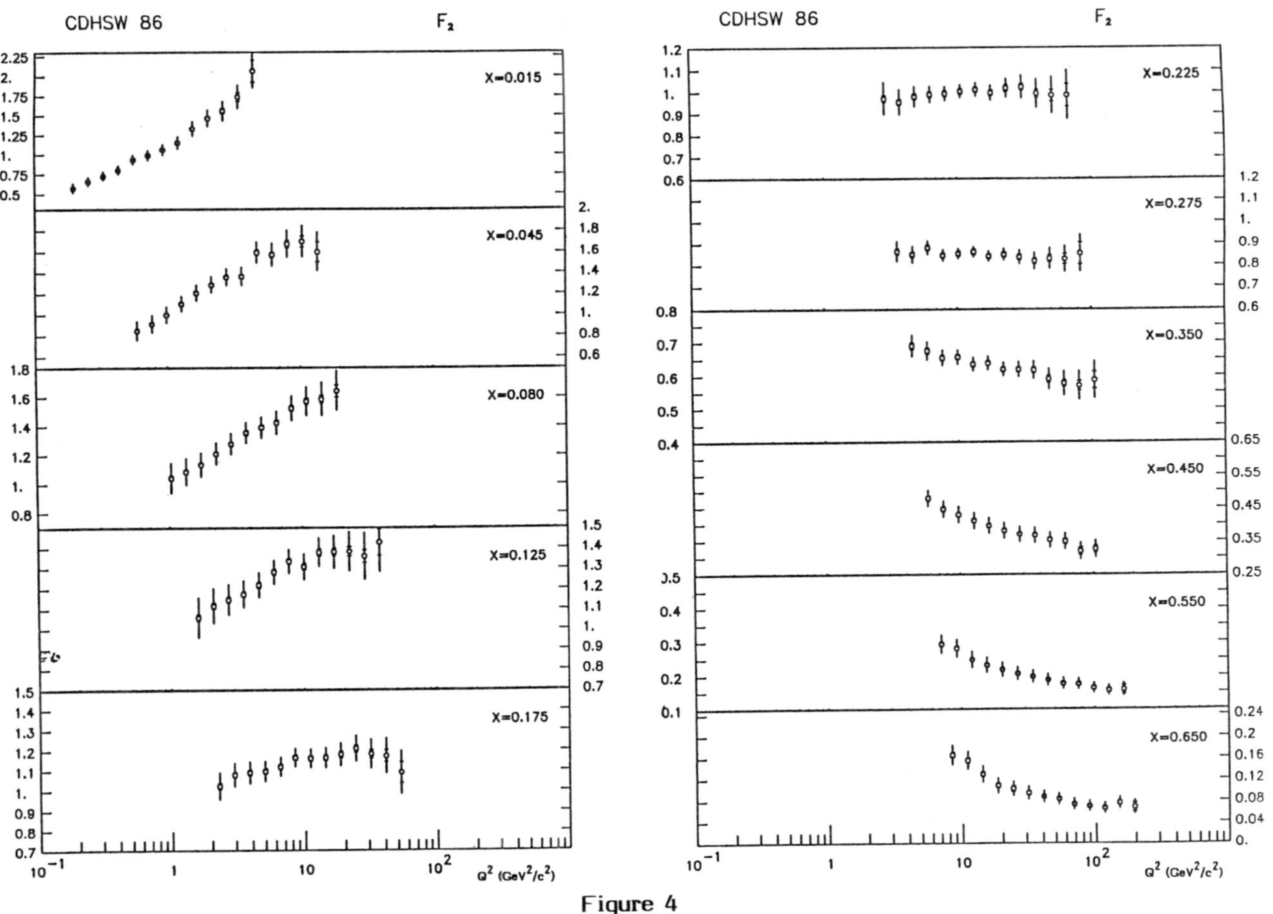

Figure 4

xF₃
X=0.015 (x150)
X=0.045 (x50)
X=0.080 (x25)
X=0.125 (x12.5)
X=0.175 (x8)
X=0.225 (x4)
X=0.275 (x2.5)
X=0.35 (x1.5)
X=0.45 (x1)
X=0.55 (x.5)
X=0.65 (x.25)
10
1
10⁻¹
10⁻²
10⁻¹
1
10
10²
Q² (GeV²/c²)

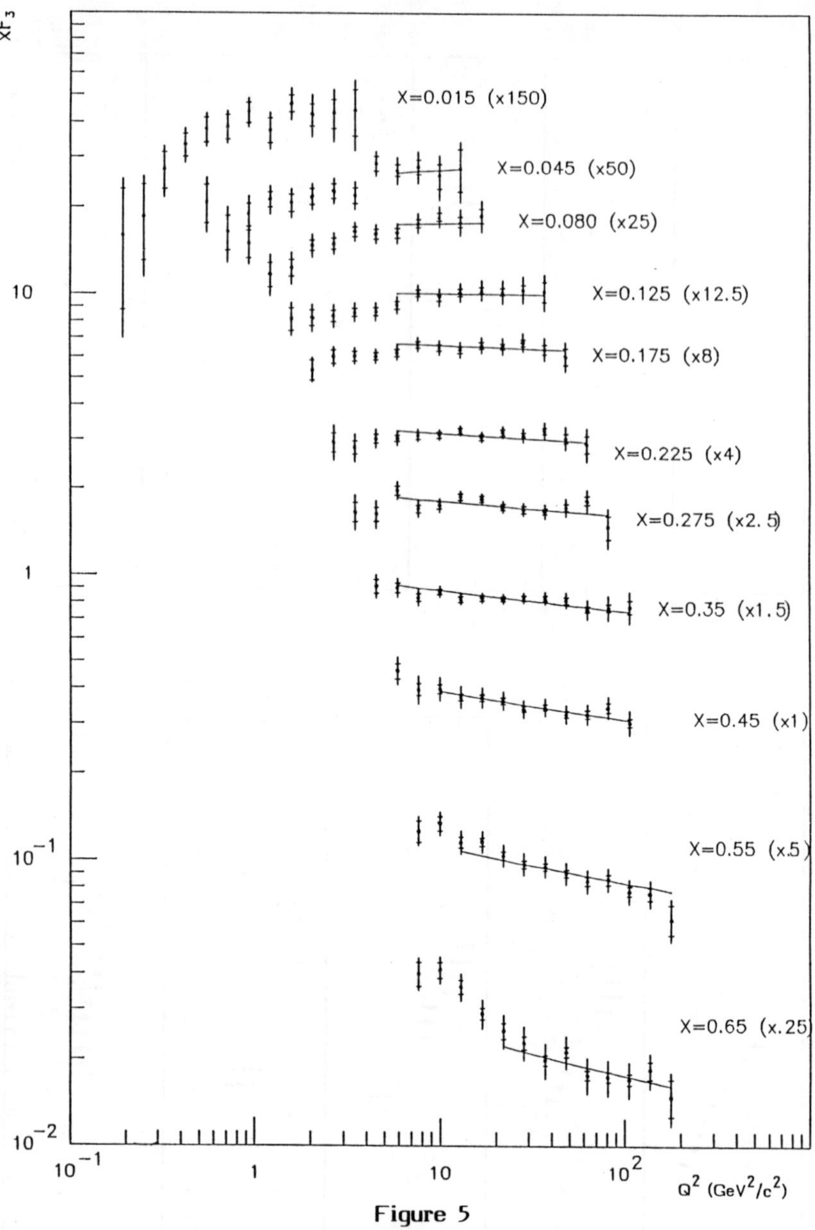

Figure 5

CHARGED CURRENT NEUTRINO SCATTERING RESULTS FROM THE CCFR COLLABORATION

P. Auchincloss, R.E. Blair, C. Haber, W.C. Leung,
S.R. Mishra, E. Oltman, M. Ruiz,
F.J. Sciulli, M.H. Shaevitz, W.H. Smith, R. Zhu
Columbia University, NY, NY 10027

K. Lang, A. Bodek, F. Borcherding,
N. Giokaris, I.E. Stockdale
University of Rochester, Rochester, NY 14627

D. Garfinkle, F.S. Merritt, M. Oreglia, P. Reutens
Enrico Fermi Institute and Physics Department
University of Chicago, Chicago, IL 60637

H.E. Fisk, Q. Kerns, B. Jin, D. Levinthal, T. Kondo,
W. Marsh, P.A. Rapidis, H.B. White, D. Yovanovitch
Fermi National Accelerator Laboratory, Batavia, IL 60510

Talk presented by R. E. Blair

ABSTRACT

Cross sections from dichromatic neutrino beam running are presented. A new method is used to normalize the high energy data (100-300GeV), and is described. This method will be applicable to wide band neutrino data where accurate flux monitoring is impossible. It is compared to conventional flux measurements for earlier data and agrees within expected uncertainties.

At present most of the charged current measurements made by the high statistics neutrino experiments are in reasonably good agreement. This has not always been true.[1] As we finish the analysis of our last dichromatic beam neutrino data and begin analysis on our recent wide band Tevatron data, we have begun to explore new methods of normalizing charged current events. We feel that these methods may be less prone to the systematic uncertainties that limit the accuracy of the more conventional ones. The data described here are from two runs of the dichromatic neutrino beam at Fermilab. The cross section results from an earlier run, E616, have been published previously;[2]

we present here for the first time results from the later run (E701). For a brief review of the beamline and detector see the talk by Wesley Smith in these proceedings.

One of the most difficult things to measure in neutrino experiments is the neutrino flux. Errors in this quantity directly affect the value of the cross sections and the scale of the structure functions. A relative error between neutrino and antineutrino flux results in incorrect values for the structure functions which depend on the sum and difference of the differential cross sections for ν and $\bar{\nu}$. The importance of this may be seen in the recent new measurements reported by the CDHSW group.[3] Their later data, with improvements in beam monitoring, yielded closer agreement with the CCFRR cross sections and structure functions than did their previously published data. Their new structure functions differ by as much as 32% at low x from their old published values.[1] Much of this difference is explained by the absence of any correction for suppression of the strange sea due to charm threshold in their earlier analysis. About a third of this difference is due to improvements in the flux measurement. As indicated in figure 1, using a wide range of reasonable assumptions should produce very little difference in the extracted structure functions (F_2 has been used here since it is much more sensitive than xF_3 to such effects). Provided all physics effects are included in the analysis, one of the largest sources of uncertainty is that due to the flux measurement.

The methods used to measure neutrino flux and the uncertainties associated with these methods are outlined in detail elsewhere[2]. In brief, the normalization is obtained by measurements of the total charged particle flux in the decay path, plus a separate determination of the beam fractions of π's and K's. This information is folded with the decay kinematics to evaluate the neutrino flux at the detector. The total charged particle measurements require a device that can measure the intensity of beams of 10^{11} charged particles in a one millisecond spill. The devices chosen by our group are ionization chambers and an RF cavity. The particle fraction measurements require a device that can function in similar beam conditions and yield kaon fractions, which can be as low as

1.3% of the total beam, to an accuracy of about 5%. An integrating Cherenkov counter is used for this. Light is measured at fixed Cherenkov emission angle from a helium radiator. During each secondary beam setting a curve of light output versus radiator pressure is taken, and from this curve particle fractions can be determined.

A check on the fluxes determined in this way is to look only at events in a region of x and Q^2 which is sampled by all beam settings and decay particle types. A comparison can then be made of the consistency of the measured numbers of events for each setting and type. This method was applied to data taken in experiment E616 and yields agreement at a level consistent with the errors on the measurement of fluxes (see figure 2). This method works provided the structure functions ($2xF_1(x,Q^2)$, $xF_3(x,Q^2)$ and $F_2(x,Q^2)$) depend only on x and Q^2 and that they are adequately modeled in the limited region of x and Q^2 involved. Given these assumptions, the number of events observed for a particular beam energy setting and decay particle type (π or K) is given by:

$$N^{\nu(\bar{\nu})}{}_i = f_i \int \int \int \Phi_i(E) \frac{d\sigma^{\nu(\bar{\nu})}}{dxdy} dxdydE$$

$$\frac{d\sigma^{\nu(\bar{\nu})}}{dxdy} = \frac{G_F{}^2 mE}{2\pi}[(2xF_1{}^{\nu(\bar{\nu})} \pm xF_3{}^{\nu(\bar{\nu})}) + (2xF_1{}^{\nu(\bar{\nu})} \mp xF_3{}^{\nu(\bar{\nu})})(1-y)^2$$

$$+2(F_2{}^{\nu(\bar{\nu})} - 2xF_1{}^{\nu(\bar{\nu})})(1-y)]$$

$N^{\nu(\bar{\nu})}{}_i$ = acceptance corrected number of events measured (the index i indicates the beam momentum and secondary decay type, the type is determined by the event energies)

E = neutrino (antineutrino energy), $y = \frac{\nu}{E}$, $x = \frac{Q^2}{2m\nu}$

$\Phi_i(E)$ = the number of target nucleons per unit area times the neutrino (antineutrino) flux per unit energy.

f_i = the flux adjustment factor (the index i is as above)

We have parameterized the structure functions using a Buras-Gaemers form[5], with the flux factors, f_i, and parameters of the model determined iteratively.

Another check on this technique is to look at the number of events at low y ($y=\frac{\nu}{E_\nu}$). Since $\frac{1}{E_\nu} \frac{d\sigma}{dy}|_{y=0}$ is approximately constant with energy (i.e., it represents physics at W^2 below charm threshold), deviations from beam setting to beam setting of this measured value can aid in interpreting the reliability of the flux evaluation. (See figure 3 for example y distributions from E616.) One difficulty with this method is that the very low y region (y<.007) includes nonscaling coherent processes. We see pileup of events in this region which is within a factor of two of what is expected from the quasielastic contribution. The *excess* cross section at low y is constant as a function of energy, rather than rising linearly as the rest of the differential cross section does. The energy dependence and magnitude indicate that this *excess* at low y may be due to coherent processes like quasielastic scattering and scattering off of nuclear fragments. In order to eliminte this non-scaling contribution, we include events only above y of about .01 and fix the level of the very low y cross section at the level measured between .01 and .025. The y distributions after resolution, acceptance and radiative corrections were fit between 0 and 0.8 using the form:

$$\frac{1}{E_\nu}\frac{d\sigma}{dy} = (\frac{1}{E_\nu}\frac{d\sigma}{dy}|_{y=0})((1-\alpha)+\alpha(1-y)^2)$$

α and $(\frac{1}{E_\nu}\frac{d\sigma}{dy}|_{y=0})$ are fit parameters

Deviations of the y=0 intercept from the mean taken over all beam settings and decay particle types indicates the reliability of the fluxes (see figure 2). The effects on the fitted y intercept of charm production, scaling violations and non zero $R = \frac{(F_2-2xF_1)}{2xF_1}$ have been estimated (see figure 4). For these calculations we have used $\kappa_S = .5$, $m_c = 1.5 GeV/c^2$, slow rescaling, the Buras-Gaemers fits mentioned above and $R = \frac{1.5(1-x)^4}{\ln(Q^2/.04(GeV/c)^2)}$ [3]. Their effects in combination introduces very little energy dependence (<1%) so no correction was applied. The effects of charm production and scaling violations each contribute at the 5% level. We have yet to estimate the uncertainty in these calculations. It is encouraging that the level of the corrections taken individually is comparable to the magnitude of the error on the experimentally determined kaon flux; as a consequence, this technique may yield results good enough to rival or exceed the more conventional methods. Since accurate flux measurements are precluded in wide band neutrino beams this method should prove particularly valuable for such running. The enhanced statistics available will also help improve the accuracy of this method.

As an application of this *y intercept* method we have used it to normalize the high energy points (from kaon decay neutrinos) in calculating a cross section from our latest dichromatic run (E701)[6]. In order to set the overall scale we have used the conventional flux determinations for the low energy points (E<100GeV from π decay neutrinos). The flux monitor calibrations were determined independently from those of the earlier run (E616). The pion fraction of the beam is more accurately determined than the kaon fraction because the pions constitute more of the beam. We believe that the pion fraction has been determined to an accuracy of 1.5%. The calibration of the RF cavity and ion chambers, which measure the total charged particle flux is less accurate than in the earlier run, and we estimate the error on this to be 5%. Other sources of uncertainty in the level of the cross section include: beam energy determination (2%), beam angular dispersion (2%) and detector livetime correction (1.5%). The total uncertainty is 5.7%.

The cross section over energy determined from the pion decay neutrinos for E701 is $\frac{\sigma^\nu}{E} = .660 \pm .005(stat.) + .039(syst.) \times 10^{-38} cm^2/GeV$ and for pion decay antineutrinos it is $\frac{\sigma^{\bar\nu}}{E} = .314 \pm .006 \pm .018 \times 10^{-38} cm^2/GeV$. The energy dependence using the y intercept method to normalize the high energy (kaon decay) neutrinos is exhibited in figure 5. Figure 5 also shows the previously published data from E616 after adjustment by the average of the flux factors from the two methods described above (*y intercept* and x and Q^2 overlap). There is good agreement for the E616 data between our original estimates of systematic errors and the adjustment factors. This agreement may be quantified by computing a χ^2 using the figure 5 values as the *true* values and the original unadjusted measurements as differing from them due to the quoted systematic point to point flux uncertainties. This yields $\chi^2 = 27.3$ for 24 degrees of freedom. For neutrinos alone it is 20. for 14 degrees of freedom, and antineutrinos alone give 7.3 for 10 degrees of freedom.

R. Blair

Some work still remains to be done in investigating how sensitive the y intercept method is to the details of the assumptions used in estimating corrections. The technique is promising and should be useful for our high statistics wide band data sample.

356

REFERENCES

[1] F. Sciulli, **Proceedings of the 1985 International Symposium
on Lepton and Photon Interactions**, Kyoto, 7 (1985).

[2] R. Blair, et al., *Phys. Rev. Lett.* 51:5 (1983);
R. Blair, et al., *Nucl. Inst. Meth.* **226**, 281 (1984).

[3] M. W. Krasny et al., **Proc. Int. Europhysics Conf. on High-Energy Physics, Bari,**
555 (1985); P. Buchholz et al., Ibid., 557 (1985) ;P. Perez et al., submission to this conference.

[4] D. B. MacFarlane, et al., *Z. Phys.* C25, 1 (1984).

[5] A. Buras and K. Gaemers, *Nucl. Phys.* **B132**, 249 (1978).

[6] I. E. Stockdale, et al., *Phys. Rev. Lett.* **52**:16, 1384 (1984);
I. E. Stockdale, et al., *Z. Phys.* *C27*, 53 (1985);
C. Haber, " A Search for Inclusive Oscillations of Muon Neutrinos",Ph. D. thesis,
Columbia University (1985);
I. E. Stockdale, "A Search for Neutrino Oscillations with Large Values of Δm^{2}",
Ph. D. thesis, University of Rochester (1984).

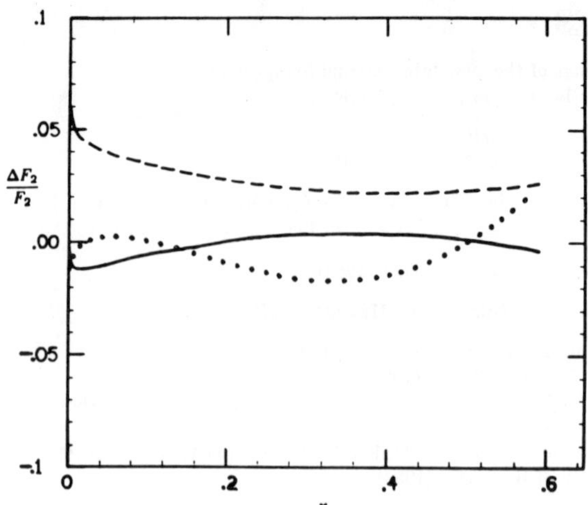

Figure 1. The difference in the measured value of F_2 induced by various assumptions in the analysis (all at $Q^2 = 10 GeV^2$). The dashed curve is $\frac{F_2(m_c=2.0GeV)-F_2(m_c=1.0GeV)}{F_2(m_c=1.5GeV)}$ where $\kappa_S = .5$ and slow rescaling are assumed. The solid curve is $\frac{F_2(\kappa_s=.55)-F_2(\kappa_s=.32)}{F_2(\kappa_s=.4)}$ where $\kappa_S = \frac{2S(x)}{(\overline{U}(x)+\overline{D}(x))}$. The dotted curve is $\frac{F_2(R=R_{QCD})-F_2(R=.1)}{F_2(R=R_{QCD})}$.

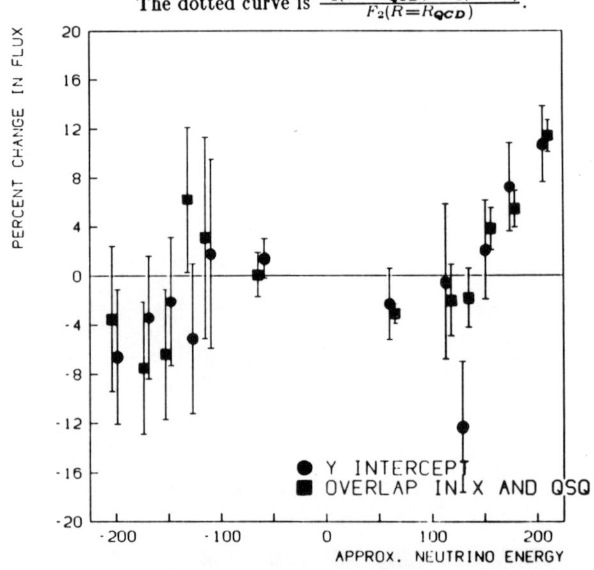

Figure 2. The flux adjustment factors determined from fits to the overlap region of x and Q^2 and similar factors determined by comparing the y intercept fit to a particular subset of data. Each point is plotted at approximately the energy of the average neutrino from that beam setting, and type (π's are plotted as one point).Negative values of energy represent antineutrino settings and positive,neutrino. The errors plotted are only statistical.

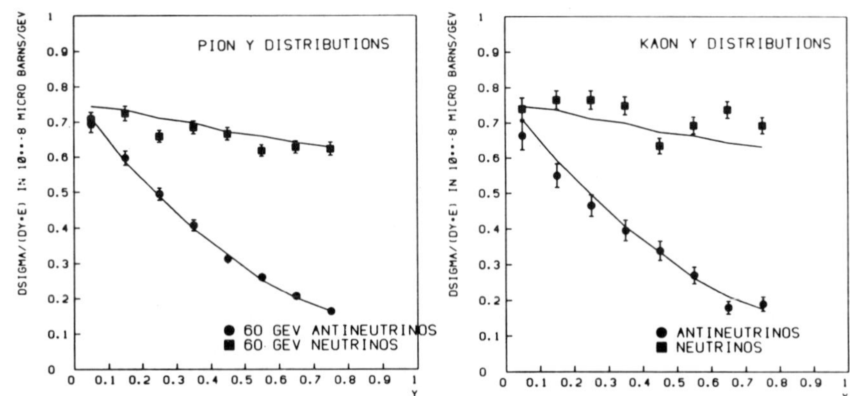

Figure 3. Measured y distributions averaged over π and K decay neutrinos.
The solid lines represent the expected values including
all known effects (i.e. charm production, R and scaling violations).
Note that the level differences between solid curves and the data are
consistent with the flux adjustment factors described above.

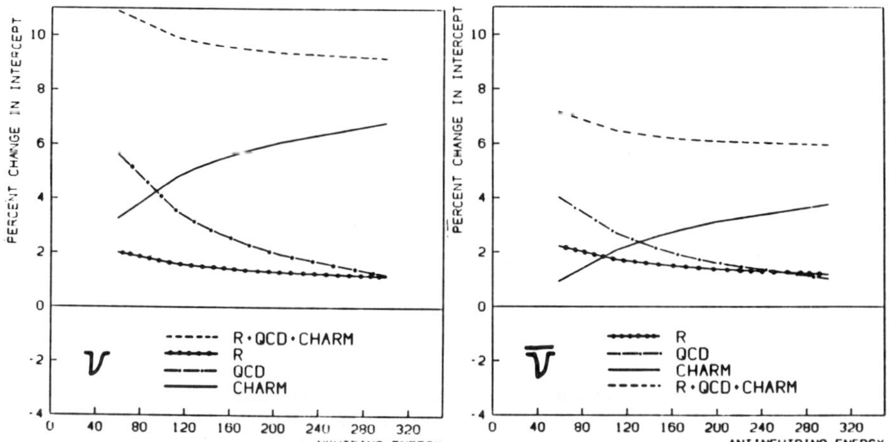

Figure 4. The expected energy dependence of the y intercepts
due to Q^2 dependence of the structure functions (labeled QCD),
charm production and non zero R.

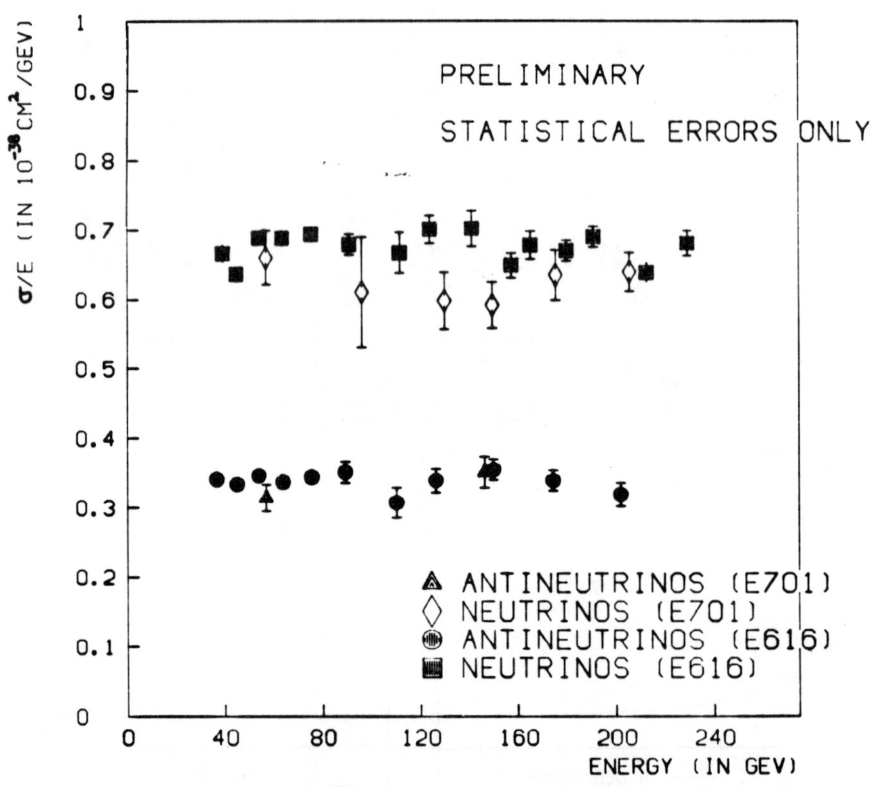

Figure 5. Cross section slopes versus energy for neutrinos and antineutrinos from two different dichromatic runs. The data refered to as E616 is the same as that published in reference 1, except for the adjustment of the data using flux factors as described in the text.

THE NON-SINGLET VALENCE QUARK DISTRIBUTION FROM NEUTRINO-DEUTERIUM DEEP INELASTIC SCATTERING

J.B. Cole, S. Kunori, G.A. Snow,

C.Y. Chang, D. Son,[1] P.H. Steinberg, and D. Zieminska [2]

University of Maryland College Park, Maryland 20742 USA

R.A. Burnstein, J. Hanlon,[3] and H.A. Rubin

Illinois Institute of Technology Chicago, Illinois 60616 USA

T. Kitagaki, S. Tanaka, H. Yuta, K. Abe, K. Hasegawa, A. Yamaguchi,

K. Tamai, Y. Otani, H. Hayano, H. Sagawa, and Y. Yanokura

Tohoku University Sendai 980 Japan

T. Kafka, W.A. Mann, A. Napier, and J. Schneps

Tufts University Medford, Massachusetts 02155 USA

Abstract

The deep inelastic scattering reaction, $\nu_\mu N \to \mu^- X$, in the 15 ft. deuterium bubble chamber at Fermilab, has been analyzed under the assumption of isospin invariance to extract $x(u_V - d_V)$ for the proton, where $xu_V(x)$ and $xd_V(x)$ are the valence up- and down-quark momentum distributions, respectively. The results are compared with other data and with different theoretical fits to data.

1 INTRODUCTION

Lepton-nucleon deep inelastic scattering experiments are the source of many important insights into the quark structure of matter that now form the basis for the current theory of strong interactions, quantum chromodynamics (QCD). But QCD cannot yet predict structure functions or quark momentum distributions in the even simplest hadronic systems because of the importance of non-perturbative effects.

As models for quarks in a nucleon become more detailed and attempts are made to understand nuclear effects and to define the range of applicability of perturbative QCD, it is increasingly important to accurately test theory against experimental data.

Perhaps the simplest quark momentum distribution is that of the difference

---

[1]Present address: Kyungpook National University Taegu 635 Korea

[2]Present address: Indiana University Bloomington,Indiana 47405 USA

[3]Present address: Fermi National Accelerator Laboratory Batavia, Illinois 60510 USA

between the fractional momentum distributions of the valence u and d quarks in the proton, $xu_V(x) - xd_V(x)$, where x is the fraction of the proton's momentum carried by the quark. This quantity, often called the non-(iso)singlet momentum distribution, has the advantage that it is decoupled from the singlet and gluon structure functions in the Altarelli-Parisi equations for Q^2 evolution [1]; furthermore it can be extracted from scattering data independently of any assumptions about the behaviour of $R = (F_2 - 2xF_1)/2xF_1$, which measures the violation of the Callen-Gross relation.

We have studied the charged current $\nu_\mu p$ and $\nu_\mu n$ interactions in the 15 foot deuterium-filled bubble chamber at Fermilab and extracted the non-singlet nucleon quark momentum distribution.

2   THE PARTON MODEL

In the quark-parton model, neutrino-nucleon charged current cross sections can be written

$$\frac{d^2\sigma^{\nu p}}{dxdy} = \frac{2xG_F^2ME_\nu}{\pi}[(d+s) + (1-y)^2(\bar{u} + \bar{c}) + \frac{1}{2}(1-y)q_L^{\nu p}] \tag{1}$$

$$\frac{d^2\sigma^{\nu n}}{dxdy} = \frac{2xG_F^2ME_\nu}{\pi}[(u+s) + (1-y)^2(\bar{d} + \bar{c}) + \frac{1}{2}(1-y)q_L^{\nu n}], \tag{2}$$

where each quark symbol stands for a function of x and Q^2. We assume isospin invariance so that $u^n = d^p \equiv d$, $d^n = u^p \equiv u$, $s^n = s^p \equiv s$, $c^n = c^p \equiv c$; and q_L is the longitudinal quark distribution defined by $F_L = F_2 - 2xF_1 \equiv xq_L$. Subtracting (1) from (2) we obtain

$$\frac{d^2\sigma^{\nu n - \nu p}}{dxdy} = \frac{2xG_F^2ME_\nu}{\pi}[(u_V - d_V) + \frac{1}{2}(1-y)(q_L^{\nu n} - q_L^{\nu p})], \tag{3}$$

where $d^2\sigma^{\nu n - \nu p}/dxdy \equiv d^2\sigma^{\nu n}/dxdy - d^2\sigma^{\nu p}/dxdy$. We assume that $u_S = d_S = \bar{u} = \bar{d}$ (see ref. [2] for a discussion of possible differences - expected to be small) where the subscripts V and S denote the valence and ocean components of the quark distribution, respectively. The second term in (3) is relating to the violation of the Callen-Gross relation. We examine the first order QCD correction to q_L given by Buras[3] and find that the second term, $(1/2)(1-y)(q_L^{\nu n} - q_L^{\nu p}) < 0.02(u_V - d_V)$ is negligible compared to $(u_V - d_V)$ in the x-range for which we present data.

We are thus left with the simple result

$$\frac{d^2\sigma^{\nu n - \nu p}}{dxdy} = \frac{G_F^2ME_\nu}{\pi}2x(u_V - d_V). \tag{4}$$

This is the starting point for our extraction of $x(u_V - d_V)$ from the experimental data.

3 EXPERIMENTAL PROCEDURE

The 15 foot bubble chamber at Fermilab was exposed to a wideband neutrino beam produced by 350 GeV protons. The analysis presented here was based on 320,000 frames, corresponding to 4.76×10^{18} protons on target. We describe the experimental procedure briefly in this section and some details will be found in [4,5].

The film was scanned at least twice and measured. The neutrino charged current events were selected by a kinematical method based on the P_T behavior of scattered muons. The neutrino energy was estimated by the method of Heilman [6]. We accepted events with 10 GeV$< E_\nu < 200$ GeV in a fiducial volume 16.7 m^3. The Monte Carlo calculation indicates that the selected carged current sample includes 4.6% contamination mostly from the neutral current events. In order to obtain deep inelastic events, we imposed conditions, $W > 2$ GeV and $Q^2 > 2$ GeV^2. The kinematic region was further restricted with the cut $y < 0.9$.

Finally we obtained 9794 νD events, of which 6002 are even-pronged and 3792 are odd-pronged events. In the prong count we excluded identified protons with momentum ≤ 340 MeV as spectator protons.

From charge conservation, a νn (νp) event should be even (odd) pronged excluding spectator neucleon in the prong count. The rescattring process in deuterium changed the νn event from even-pronged to odd-pronged but did not change oddness of the prong count for νp event. Each bin in the x distributions for even pronged and odd pronged events was corrected for the rescattering to get those for νn and νp interactions using rescattring fraction, $f_{r_s} = 0.094 \pm 0.035$ [7].

In addtion to the rescattering correction we made following corrections to data. (i) prong dependent correction for event loss in the scan-measure procedure, (ii) Monte Carlo correction for experimental distortion due to mainly our method of charged current event selection and neutrino energy estimation, (iii) correction for 2% hydrogen contamination in deuterium.

In order to get absolute cross section, we used the neutrino flux determined from the quasielastic reaction rate, $\nu_\mu n \to \mu^- p$, in the same experimental run [4].

4   RESULTS AND CONCLUSIONS

Fig.1 displays the $x(u_V - d_V)$ distribution for Q^2=11 GeV^2. We adjusted our results to the bin centers and to a common Q^2-value by multiplying the $x(u_V - d_V)|_{x_i,\bar{Q}_i^2}$ with the adjustment factors

$$a_i(Q^2) = \frac{x(u_V - d_V)_{\text{fit}}|_{x_i,Q^2}}{x(u_V - d_V)_{\text{fit}}|_{x_i,\bar{Q}_i^2}},$$

where $x(u_V - d_V)_{\text{fit}}$ is a theoretical or phenomenological fit to some data set. We choose the one due to Glück, Reya, and Hoffmann [8], as it fits our data quite well. Our adjustments average 3% or less for the first five x-bins, and are always less than the the statistical uncertainty. The figure plots $x(u_V - d_V)$ for both our standard value of the rescattering fraction (f_{r_s}), and its upper and lower limits. Notice that the uncertainty in f_{r_s} dominates the statistical uncertainty in most of the x range. There is also a systematic error in the neutrino flux determination, but since the total νD cross section that we determine from this flux is very close to the world average[5], we do not include it with the much larger systematic error in f_{r_s}.

We have compared with data from the CDHS [9], WA25 (BEBC) [10], EMC [11], and SLAC [12] collaborations and with the fits due to Glück, Reya, and Hoffmann (GRH) [8], Duke and Owens (DO) [13], and Abbot, Atwood, and Barnett (AAB) [14] in Figs.2 and 3. Since AAB is an excellent description of most SLAC data we use it instead of the actual data in our comparisons.

None of the above experiments determines $x(u_V - d_V)$ in the same manner as we do. They all derive the non-singlet structure function by subtracting two singlet structure functions, obtained in different experimental runs. CDHS compares νp with $\bar{\nu} p$ interactions, while EMC studies the the $\mu^+ p$ and $\mu^+ D$ reactions, and SLAC the eD and ep reactions. WA25 analyze the νD and $\bar{\nu} D$ interactions in a bubble chamber and determine the structure functions F_2 and F_3 separately for νn and νp events, whence they deduce $x(u_V - d_V)$. As can be seen in Fig.2b our results agree well with WA25 results. It should be noted that the extraction procedure in the two experiments used somewhat different input parameters: they take the value of f_{r_s} to be 0.12, whereas we use 0.094, but their average νN cross section is 8% lower than ours. These two differences happen to approximately cancel in the final result. In the low x region our results are higher than SLAC (AAB) and EMC. Much closer agreement with these data could be realized if f_{r_s} were near 0.06.

DO, and GRH, are more or less "traditional" QCD-type fits to deep inelastic scattering data, while AAB incorporate "higher-twist" $(1/Q^2)$ terms to account for all of the Q^2-evolution. Our results seem to be reasonably well described by the GRH fit. DO fit EMC and SLAC data causing the "oscillation" in the resulting fit (Fig.3), which arises from the different rates at which $x(u_V + d_V)$ and xd_V go to zero.

Fig.4 shows $(u_V - d_V)$ computed from our data compared with GRH. An integral of data points on the 0.05-1.00 x-range gives $0.65 \pm 0.04^{+0.13}_{-0.10}$, where the first uncertainty is statistical, and the second reflects the systematic uncertainty in f_{rs}. While the integral for GRH's curve gives 0.663. Our data agrees quite well with GRH.

To summarize, we have extracted the non-singlet nucleon quark momentum distribution from νD bubble chamber data. Our results are free of the relative systematic errors that are implicit in subtracting cross sections from different experimental runs, but they contain a systematic uncertainty due to rescattering.

We are indebted to the Fermilab accelerator and neutrino groups, to our scanning staffs, and to Professor R. Engelmann and the Stony Brook group who originally participated in this collaboration. This research was supported by the United States Department of Energy, the National Science Foundation, and by the University of Maryland General Research Board and assisted by the Computer Sciences Center.

References

[1] M. Diemoz, F. Ferroni, and E. Longo, Phys. Rep. **130**, 293 (1986).
G. Altarelli, Phys. Rep. **81** (1), 1 (1982).

[2] D. A. Ross, C. T. Sachrajda, Nuc. Phys. **B149**, 497 (1979).

[3] A. J. Buras, Rev. Mod. Phys. **52**, 199 (1980). *We apply eqn. 8.87.*

[4] T. Kitagaki, et al.,Phys. Rev., **D28**, 436 (1983).

[5] T. Kitagaki, et al. Phys. Rev. Lett, **49**, 98 (1982).

[6] H. G. Heilmann, Bonn Internal Report No. WA21-int 1, 1978 (unpublished).
J. Blietschau et al., Phys. Lett. **87B**, 281 (1979).

[7] J. Hanlon, et al., Phys. Rev. Lett **45**, 1817 (1980). *Source for $\sigma^{\nu n}/\sigma^{nup}$ data.*
J. Hanlon et al., Proc. of Neutrino '79, International Conf. on Neutrinos, Weak Interactions and Cosmology, Bergen, Norway, 1979; A. Haatuft, C. Jarlskog, eds. (Fisik Institute, Bergen, 1980); Vol. II, p.286.

[8] M. Glück, E. Reya, E. Hoffmann, Z. Physik **C13**, 119 (1982).

[9] H. Abramowicz, et al., Z. Physik **C25**, 29 (1984).

[10] D. Allasia et al. Phys. Lett. **135B**, 231 (1984). *Source for the data of Fig.3b. Other data appear in*
D. Allasia et al., Z. Physik **C28**, 321 (1985).

[11] J. J. Aubert, et al., Phys. Lett. **123B**, 123 (1983). *Source for data in Fig.3c.*

[12] C.G. Wohl et al., Rev. Mod. Phys. **56**, S61 (1984).
A. Bodek, et al. Phys. Rev. **D23**, 1070 (1981).
A. Bodek, et al. Phys. Rev. **D20**, 1471 (1979).

[13] D. W. Duke, J. F. Owens, Phys. Rev. **D30**, 49 (1984).

[14] L. F. Abbot, W. B. Atwood, R. M. Barnett, Phys. Rev. **D22**, 582 (1980). *We have multiplied eqn. 3.18 by 3 to correct an apparent error or misprint.*

FIGURE CAPTIONS

Fig.1. $x(u_V - d_V)$ at $Q^2 = 11$ GeV^2. $+$ rescattering coefficient, $f_{rs} = 0.094$ (standard value) - error bars are statistical - and its extreme values, $\triangle$, $f_{rs} = 0.129$, and ∇, $f_{rs} = 0.059$.

Fig.2 $x(u_V - d_V)$ distributions. (a) $+$ This Experiment (at $Q^2 = 15$ GeV^2), $\diamondsuit$ CDHS($Q^2 = 15$ GeV^2). (b) $+$ This Experiment (variable Q^2) $\diamondsuit$ BEBC (variable Q^2) (c) $+$ This Experiment ($Q^2 = 15.0$), $\diamondsuit$ EMC (variable Q^2).

Fig.3. $x(u_V - d_V)$ at $Q^2 = 11$ GeV^2. This Experiment, $+$; different fits, solid lines. GRH, AAB, and, DO set 2. AAB represents the best fit to SLAC data.

Fig.4. $u_V - d_V$ at $Q^2 = 11$ GeV^2. $+$ This Experiment; Curve, GRH-fit.

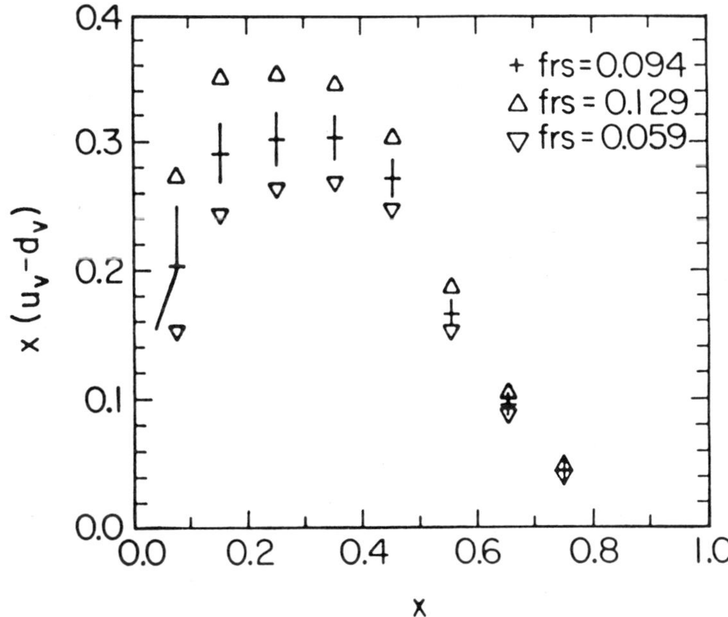

Fig.1.

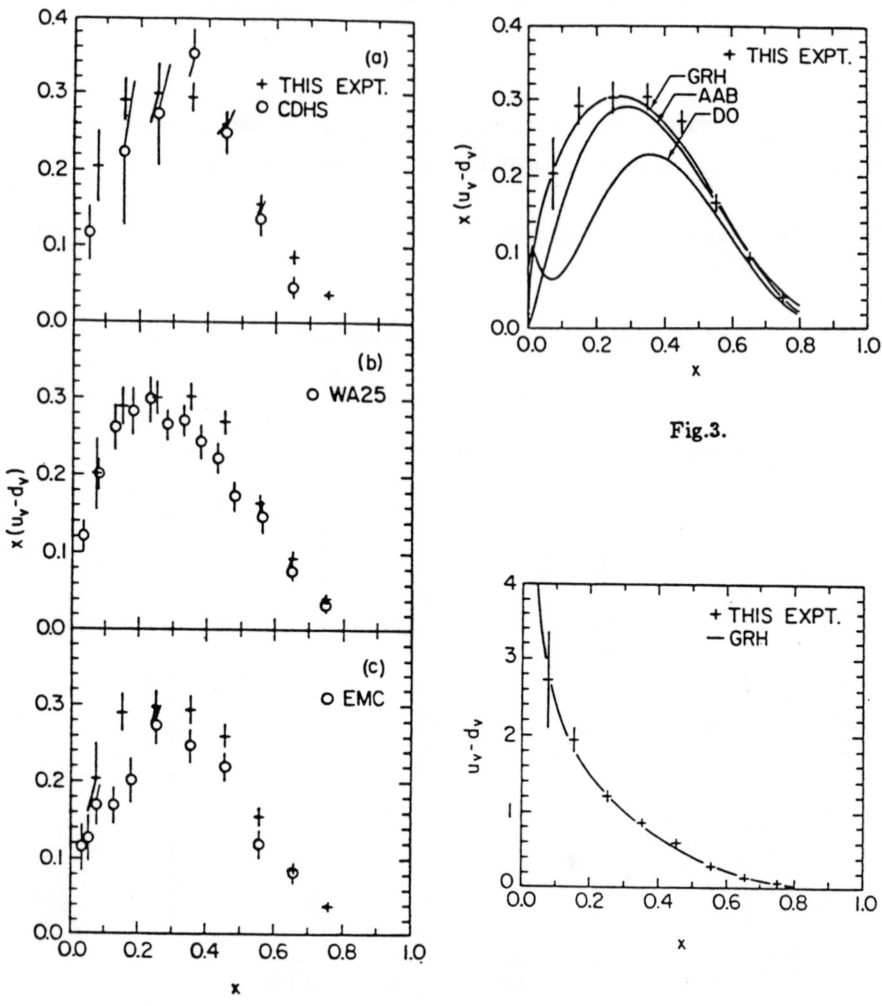

Fig.3.

Fig.4.

Fig.2

STRUCTURE FUNCTIONS OF NUCLEONS IN NUCLEI
AS MEASURED IN ELECTRON AND MUON EXPERIMENTS

Arie Bodek

Dept. of Physics & Astronomy

University of Rochester

Rochester, NY 14627 USA

ABSTRACT

A review of present experimental data on the structure functions of bound nucleons is presented. Results from electron, muon and neutrino deep inelastic scattering experiments, results from Drell Yan production of dimuons in hadronic collisions, and results from photoproduction experiments on nuclei are presented. Emphasis is placed on recent muon and electron scattering experiments.

INTRODUCTION

Presently, data on the structure functions of bound nucleons exist from muon, electron and neutrino experiments. In this review for the Neutrino 86 conference at Sendai, I will emphasize on electron and muon results. A review of results from neutrino experiments is given by Kitagaki in these proceedings. Comparisons between the various results is complicated by the fact that various experiments have different values of momentum transfers (Q^2), there are various systematic errors, and neutrino experiments suffer from poor statistics. The primary muon experiments are the EMC collaboration[1a,b] and the BCDMS[2] collaboration, and the recently formed New Muon Collaboration, NMC[3]. Some lower precision data are also available from an early muon experiment at the CCM Facility at Fermilab[4]. The most recent electron scattering data from nuclear targets come from the American University-Rochester-SLAC-Bonn collaboration[5] (SLAC experiment E139), Lower precision data from the Rochester-SLAC-MIT analysis of old empty target data are available for Iron and Aluminum[6]. Early low x and low Q^2 electron scattering data in the shadowing region are available from a SLAC experiment

using the 20 GeV/c spectrometer (Stein, ref. 7). More precise studies
of the nuclear dependence of the structure functions W1 and W2
separately are expected soon from SLAC experiment[8] E140.

The discovery that the structure functions of a nucleon bound in
a heavy nucleus are different from the structure functions of
nucleons bound in a deuteron has resulted in much theoretical
activity during the last year. These experimental results have been
interpreted as evidence for multiquark bags, or other manifestations
of the quark degrees of freedom in nuclei, the swelling of the
nucleon inside the nucleus, and QCD rescaling models. The more
conventional nuclear physics approach involves the effect of Fermi
motion and binding in conjunction with additional pions inside
nuclei. All the models rely heavily on data for support and
guidance. Therefore, quantitative consistency between the various
experiments is important for test of such models. Some experiments
supported one set of models, while other experiments indicated
contrary effects. During the last few months, with the emergence of
new data[1,2] from the EMC and BCDMS collaborations, it appears that
there is some convergence of results, especially in the controversial
small X region.

Detailed studies of the A dependence have been performed[5] by SLAC
experiment E139. The effect is seen in He, Be, C, Al, Ca, Fe, Ag,
and Au and appears to be increasing logarithmically with atomic
number A (see figure 1). Figure 2a shows the ratio σ_A/σ_D versus A
for two representative values of X. The data may be equally well
described by a two parameter fit of the form $\sigma_A/\sigma_D=CA^{\alpha}$ or
$\sigma_A/\sigma_D=a[1+b\rho(A)]$ where $\rho(A)$ is a measure[5] of the average nuclear
density. Values of α and b from fits to the data are shown in Figure
2b.

The effect appears to be increasing as A^{α} (i.e. logarithmically
with A) or linearly with the nuclear density (which is approximately
proportional to (lnA). If the effect is proportional to the nuclear
density then one expects the effect in Deuterium to be much smaller
than the effect in Iron[5] (in the ratio of 0.024 to 0.117). A test of
the nuclear effects in Deuterium has been done[9] by comparing electron

scattering data on Hydrogen and Deuterium to neutrino and
antineutrino data on Hydrogen (from which u and d quark distribution
can be extracted). This comparison (figure 3) indicates that aside
from Fermi motion, nuclear effects in the deuteron are small.

A comparison of all muon electron and neutrino experiments indicates that the fall at large X is confirmed in all experiments. The
electron data of E139 is the only data precise enough for a detailed
study of nuclear dependence. The effect at low X is controversial.
The early high Q^2 EMC data for Fe indicated a rise at small X. These
data had large systematic errors, as well as a normalization error of
±7%. The low Q^2 SLAC E139 data do not show this rise. There are new
data from both the EMC and BCDMS collaboration. These new data have
lower systematic errors, because the data for various targets were
either taken simultaneously (BCDMS) or interspersed frequently. The
high Q^2 muon data for Iron and copper nuclei are shown in figures 4a
(old EMC), 4b(new EMC) and 4c(new BCDMS). The additional normalization errors are ±7% for the old EMC and probably about a factor of
2 to 3 smaller for the new data. However, the new data is still
preliminary. The lowest X points have a Q^2 of about 10 (GeV/c2).
The lower Q^2 electron scattering data for iron and copper are shown
in Figures 5a (E139) and 5b (Stein et al.). The normalization are
about 1.5% for the E139 data and 4.2% for the Stein data. Note that
the Stein data were taken at small angles and the E139 data were
taken at larger angles.

The conclusions on Fe/C and Cu/D are:

(a) The large rise in the early EMC muon data can be attributed to
large systematic errors.

(b) The new data for Fe/D may indicate a small rise (about 5%)
between x=0.1 and x=0.2.

(c) Indications are that there is no rise at low at very small X.
More data will become available because the new muon experiments will
concentrate on this region.

(d) The 5% difference between E139, Stein, and the new preliminary
BCDMS data could be (i) systematic errors, (ii) a Q^2 dependence or
(iii) a small (0.05) difference in R = σ_L/σ_T between iron and

deuterium. This last possibility will be answered within a few
months by results from SLAC experiment E140.

The comparison of the new BCDMS data on N_2/D_2 ratio with the
Carbon/D_2 ratio measured at SLAC and the C/D ratio from EMC indicates
that there is no rise in the ratio at small X both at SLAC and at SPC
energies for these light nuclei. However, a small rise of about 3%
at high Q^2 cannot be ruled out. These data are shown in Figures 6a
and 6b.

Neutrinos offer a unique tool in establishing the origin of the
distortion of the structure functions of bound nuclei. For example,
the systematic errors are different because the radiative corrections
in neutrino experiments are different from those in electron and muon
experiments. Neutrino experiments yield information on the valence
quarks and sea antiquarks separately. This information is important
if any low X enhancement, is interpreted as an increase in the quark
antiquark sea in a heavy nucleus, or as pions in nuclei.

The neutrino experiments include a comparison of neutrino and
antineutrino data on Hydrogen and Iron done by the CDHS collaboration[10] and a comparison of Hydrogen and Neon data by the BEBC TST
group[11c]. A comparison of neutrino and antineutrino Ne and D data
has been done by the BEBC WA25/WA59 group[11b] and also by groups using
Fermilab 15'BC data[11d]. Hydrogen data from BEBC were also compared
to Iron structure function data[11e]. The basic conclusions of the
neutrino experiments are:

(1) Either the statistics are too poor to see the EMC effect at
large X, or the effect is confirmed by only a few standard
deviations. However, new data from the Fermilab 15' bubble
chamber[11f] on the Fe/D ratio (preliminary data with smaller errors)
show the effect at large X (see talk by Kitagaki in ν86 proceedings).

(2) At small X none of the experiments see any evidence for a
large increase in the antiquark fraction in a heavy nucleus.
However, the errors are large and the experiments are not at quite as
large a Q^2 value as the EMC experiment.

The CDHS data[10] for F_2^{Fe}/F_2^{H} is shown in Figure 7. At large X, the
data is consistent with a falling ratio but the errors are large.

At small X, there is no indication of the large rise observed in the old EMC muon data. The CDHS ratio is less than 1.0 in the first bin (X<0.1). The extracted antiquark distributions for Iron and Hydrogen indicate that they are very similar. Large increases of order factor of 1.5 to 2.0 in the antiquark content of nucleons in Iron have been proposed as an explanation of the low X rise in the old EMC data. Such an increase was not consistent with the CDHS data. The CDHS data is consistent with the new muon results which do not show a large rise. The CDHS data yield:

$$(\bar{u}+\bar{d}+2\bar{s})_{\nu N}/2(\bar{d}+\bar{s})_{\nu p} = 1.10 \pm 0.11 \pm 0.07 \quad .$$

Note however that the mean Q^2 range of the CHDS data is given by the expression $Q^2 = 66x (GeV/c)^2$, which indicates a mean Q^2 of only 3.3 $(GeV/c)^2$ for the first X bin. This is considerably lower than $Q^2 > 9$ $(GeV/c)^2$ for the low X EMC data. Currently planned experiments with better data on dimuon (Drell Yan) production on iron nuclei may answer this antiquark question (Fermilab experiment E772).

The WA25/WA59 antineutrino data[11b] for Neon and Deuterium are shown in Figure 8; and other neutrino data are shown in Figures 9a,b. Here also, no rise is observed at small X or at large y. The solid line is the expected curve if the the old EMC rise at small X were interpreted as evidence for a sea increase. The antineutrino data is particularly sensitive to the antiquark contents. The new muon results are in better agreement, since they indicate that there may be only a small rise for Fe, and show no evidence of any significant rise for N2 or Carbon.

The EMC effect has finally been seen in dimuon production in hadronic collisions (Drell Yan). Experiment NA10 at CERN has observed the EMC effect in the valence quarks of W nuclei. The fall at large X can be seen in Figure 10. However, the present data is unable to determine whether there is an antiquark increase at small X.

And finally, recent data on the photoproduction[12] of ψ meson from the Fermilab tagged photon spectrometer, indicate that the cross section per nucleon in iron is less than the hydrogen cross section. The Fermilab experiment obtains a ratio of 0.72 ± 0.07, as compared to earlier EMC results which yielded a ratio of 1.45 ± 0.12 ± 0.22. Within the photon-gluon fusion model, the EMC data implied that the gluon distributions in nuclei were enhanced at small X. The Fermilab data implies the opposite, and show evidence for shadowing. The new Fermilab data[12] is in agreement with lower energy photoproduction data from SLAC.

New experiments and results will help resolve some of the above questions.

(1) The preliminary[2] results of BCDMS on D and Fe are being finalized. Final analysis of systematic errors are necessary for definite conclusions.

(2) The more recent[1] EMC data with D,C, and Cu at low X are also being finalized. Only 50% of the data are were shown and the very low x data is still being studied.

(3) SLAC experiment[8] E140 (American Univ.-Rochester-SLAC-Caltech-Fermilab-Livermore-U.Mass-Tel Aviv Collaboration) has taken data at the end of 1985. This experiment measured $R = \sigma_L/\sigma_T$ for H, D, Fe and Au nuclei. Results are expected in September 86.

(4) Final analysis of the Fermilab 15' bubble chamber Fe wall[11f] data (preliminary results were shown by Kitagki in ν86).

In the future we will have two muon experiments:

(5) The E665 Tevatron muon collaboration will use a 650 GeV muon beam in the spring of '87.

(6) The New Muon Collaboration (NMC) proposal has been approved at CERN. Additional information on antiquark contents of nuclei will come from hadronic production of dimuons (Drell Yan).

(7) CERN Drell-Yan experiment NA10, and Fermilab experiment E772.

<u>References</u>

1. a)J. J. Aubert et al, Phys. Lett <u>123B</u>, 275 (1983)(Old EMC); b)New 1986 data, P.R. Norton, to be published in proceedings of the Int. Conf. HEP, Berkeley July 1986; see also talk by Taylor, Lake Louise Conference, May 1986.

2. BCDMS Collaboration(NA4), G. Bari et al. Phys. Lett. <u>163B</u>, 28 (1985)(Fe/D and N2/D), new 1986 Fe/D results, R. Voss to be

published in Proceedings of the Int. Conf. HEP, Berkeley July
1986.

3. New Muon Collaboration (NMC) see talk by Taylor (ref. 1); CERN 85-18 SPSC/p210 Feb 1985 D. Allasia et al, K. Rith, D. V. Harrach Spokesmen.

4. Muon Scattering, Chicago Cyclotron Magnet (CCM), (Fermilab E98);
    M. S.Goodman et al, Phys. Rev. Lett. $\underline{47}$, 293 (1981) (shadowing).

5. Electron Scattering, American U.-Rochester-SLAC-Bonn (E139) (A
    dep.); R. G. Arnold et al, Phys. Rev. Lett. $\underline{52}$, 727 (1984) and
    SLAC-PUB-3257.

6. a) A. Bodek et al, Phys. Rev. Lett. $\underline{50}$, 1431 (1983) (Fe/D);
    b) A. Bodek et al, Phys. Rev. Lett. $\underline{51}$, 534 (1983) (Al/D).

7. S. Stein et al, Phys. Rev. $\underline{D12}$, 1884 (1975)(Electron scattering).

8. SLAC E140, A. Bodek and S. Rock spokesmen. See S. Dasu at al.
    Univ. of Rochester preprint UR 958 (presented by S. Rock) to be
    published in Proceedings of the Int. Conf. HEP, Berkeley July
    1986.

9. A. Bodek and A. Simon, Z. Phys. $\underline{C29}$, 231 (1985).

10. Neutrino Scattering, CDHS (CERN WA1) H. Abramowicz et al, Z. Phys.
    C. $\underline{25}$, 29 (1984) (Fe/H comparison).

11. EMC Effect, Neutrino Experiments
    a) CDHS (see ref. 10)
    b) WA25/WA59 neutrino and antineutrino Neon/D data, A. M. Cooper
    et al, talk at Leipzig Conference, July 1984 and talk at 'Neutrino
    84', Dortmund, June 1984; A. M. Cooper et al, Phys. Lett. $\underline{141B}$,
    133 (1984)
    c) BEBC TST neutrino and antineutrino Ne and H data; M. A. Parker
    et al, Nucl. Phys. $\underline{B232}$, 1 (1984)
    d) Fermilab 15' BC $\bar{\nu}$ Ne/D data; A. E. Asratyan et al,
    ITEP-83-110
    e) BEBC neutrino and antineutrino H data; H. Grassler et al, paper
    C-206, Lepton-Photon Conference, Cornell 1983; see also F. Dydak,
    same conference
    f)Fe(wall)/D from Fermilab 15' BC, talk by Kitagaki, Neutrino 86
    conference, Sendai, June 1986.

A. Bodek

12. Drell Yan production on W nuclei, CERN NA10, talk by K.
    Freudenreich, to be published in proceedings of the Int. Conf.
    HEP, Berkeley July 1986
13. Photoproduction data, talk by N. Sokolov, to be published in
    proceedings of the Int. Conf. HEP, Berkeley July 1986.

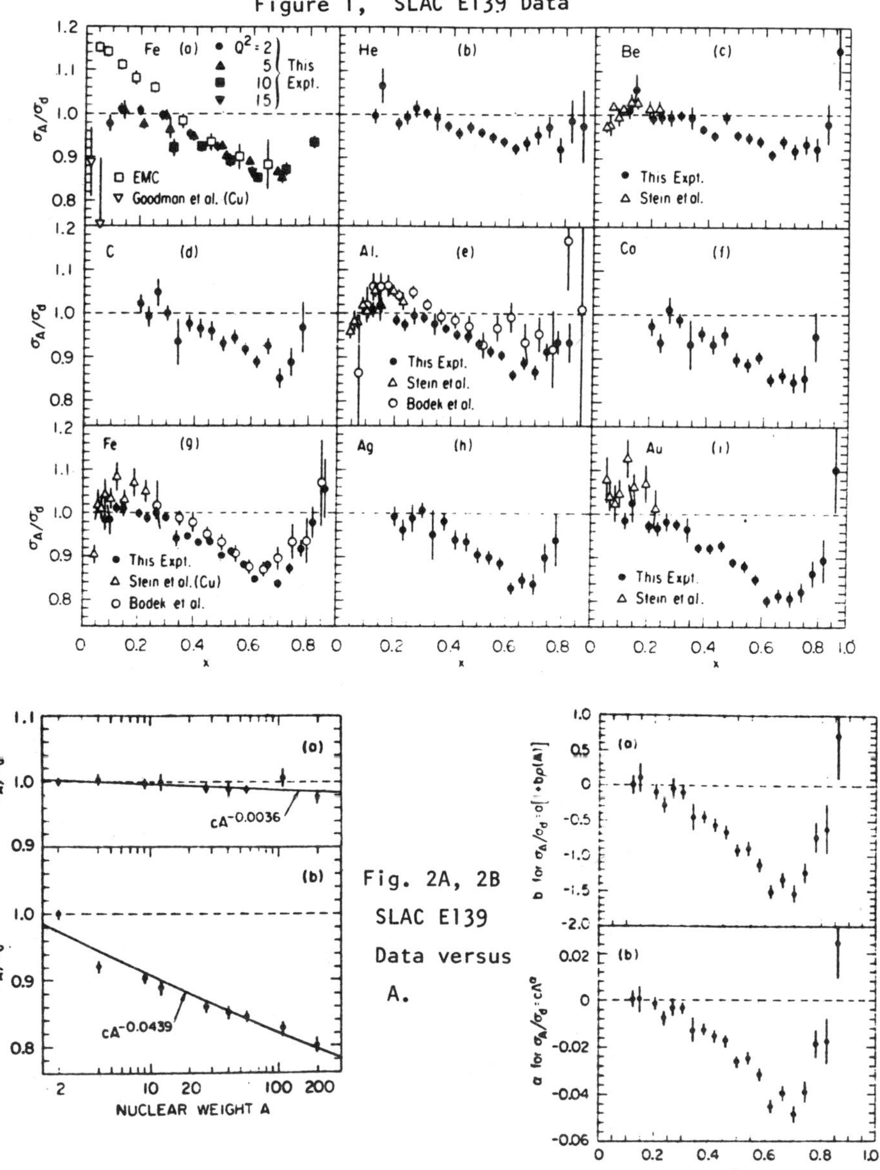

Fig. 2A, 2B SLAC E139 Data versus A.

A. Bodek

Figure 3: A test of
Nuclear effects in Deut.
F_2^{ed}/F_2^{ep} from Bodek et al.
• d/u (CDHS)
○ d/u (BEBC)
× d/u (BEBC-TST)
from $\bar{\nu}p/\nu p$
Fermi motion (Atwood-West)
Fermi motion (Frankfurt-Strickman)
$F_2^{ed}/(F_2^{ep}+F_2^{en})_{Free}$
ξ

$F_2^A(x)/F_2^N(x)$
● EMC Fe/D,
Fig. 4(a)
Old EMC
Fe/D
Normalization
error +-7%
Bjorken x

$F_2^A(x)/F_2^N(x)$
● BCDMS Fe/D, (prel.)
Fig. 4(b)
1986 BCDMS
preliminary Fe/D
normalization error +-2%
Bjorken x

EMC PRELIMINARY
COPPER : DEUTERIUM
Fig 4(c)
$F_2^N(Cu)/F_2^N(D_2)$
1986 EMC
Data.
x

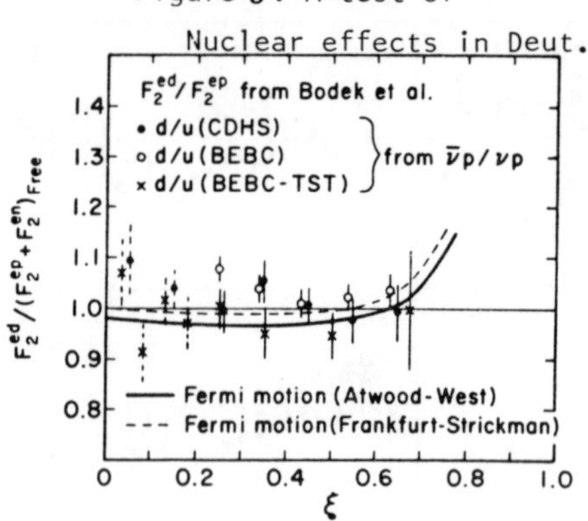

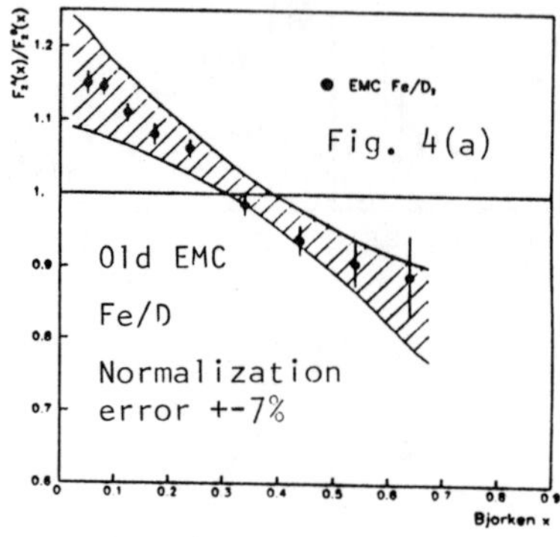

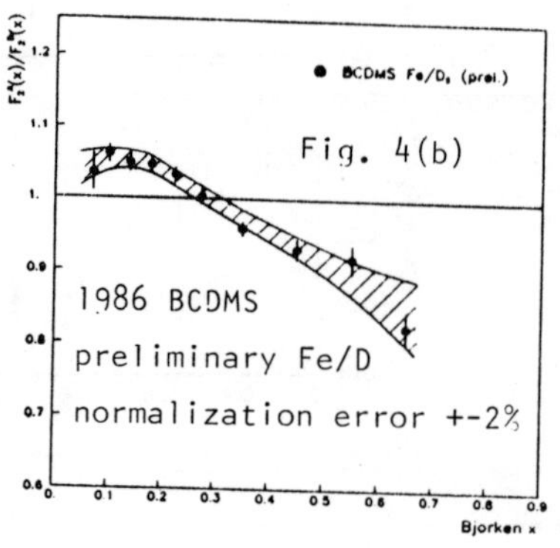

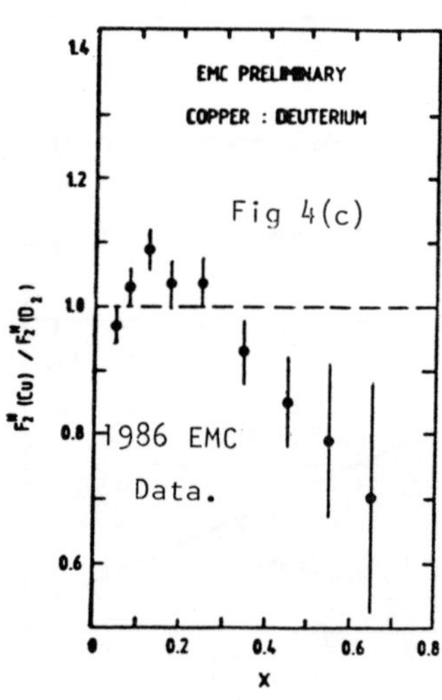

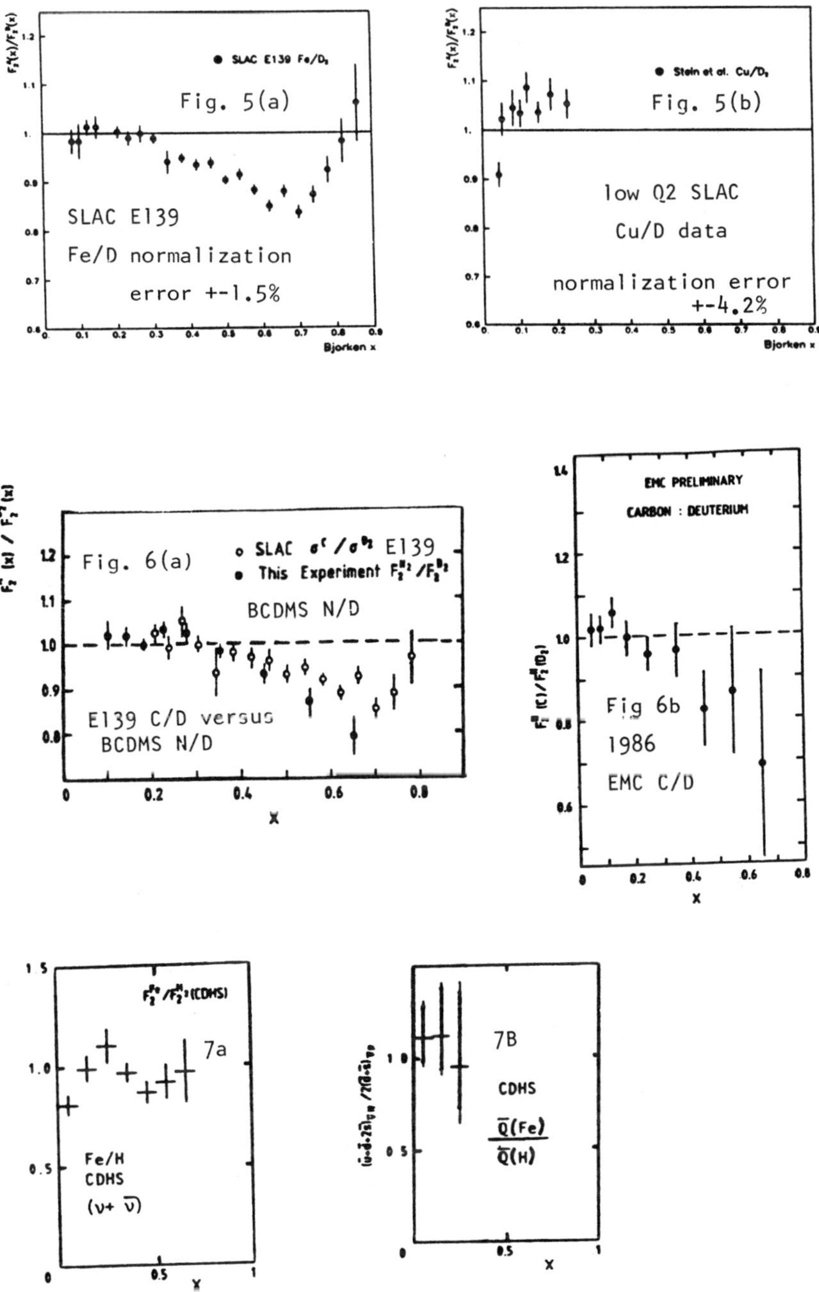
F₂ᴬ(x)/F₂ᴰ(x)
SLAC E139 Fe/D₂
Fig. 5(a)
SLAC E139
Fe/D normalization
error +-1.5%
Bjorken x

F₂ᴬ(x)/F₂ᴰ(x)
Stein et al. Cu/D₂
Fig. 5(b)
low Q2 SLAC
Cu/D data
normalization error
+-4.2%
Bjorken x

F₂ᴬ(x) / F₂ᴰ(x)
Fig. 6(a)
SLAC σᶜ/σᴰ² E139
This Experiment F₂ᴺ²/F₂ᴰ²
BCDMS N/D
E139 C/D versus
BCDMS N/D
X

EMC PRELIMINARY
CARBON : DEUTERIUM
F₂ᴬ(C)/F₂ᴰ(D)
Fig 6b
1986
EMC C/D
X

F₂ᶠᵉ/F₂ᴴ²(CDHS)
7a
Fe/H
CDHS
(ν+ ν̄)
X

7B
CDHS
Q̄(Fe)/Q̄(H)
X

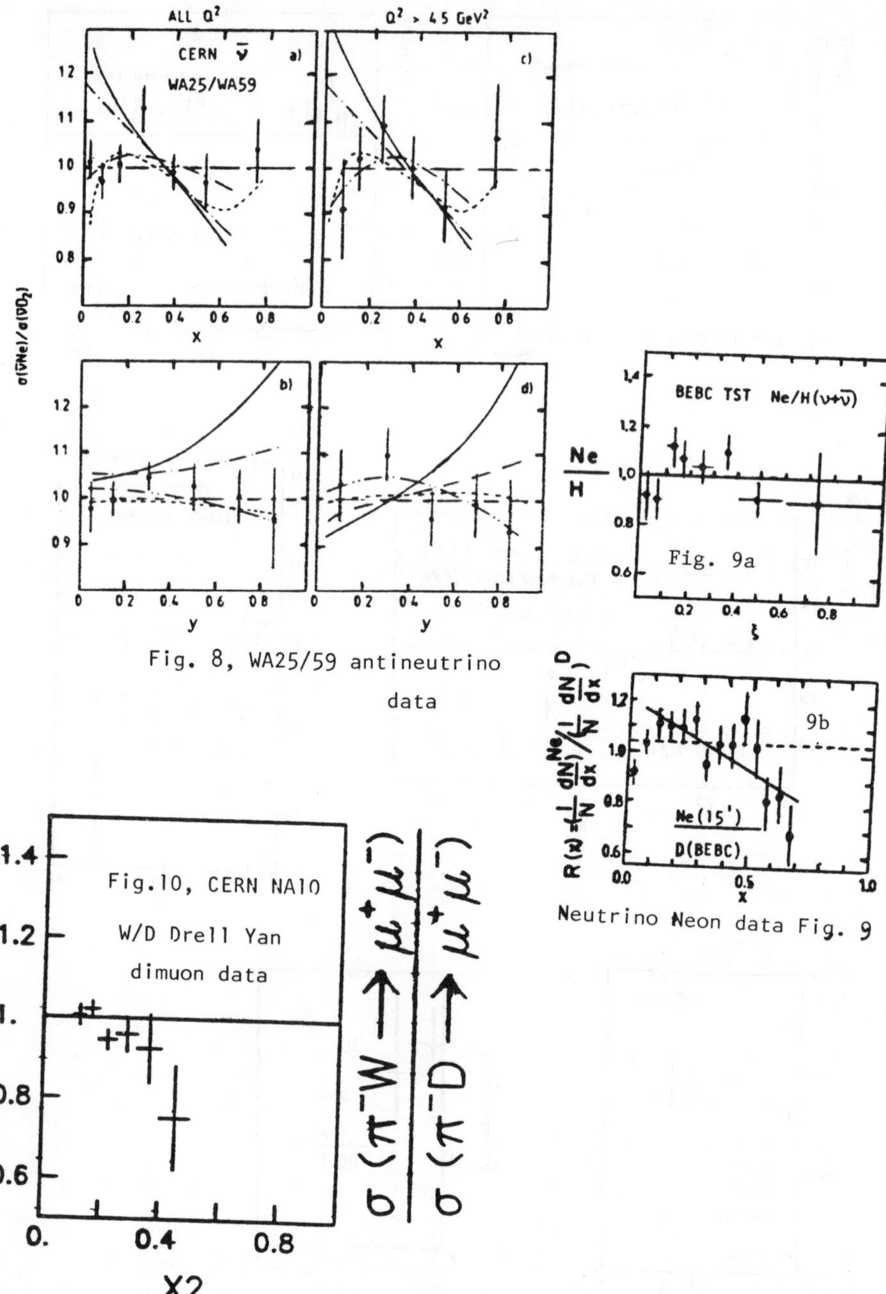

Fig. 8, WA25/59 antineutrino data

Neutrino Neon data Fig. 9

NUCLEAR EFFECT IN NEUTRINO INTERACTIONS ;

COMPARISON OF ν-Fe AND ν-D INTERACTIONS

E545 Collaboration

T. Kitagaki, S. Tanaka, H. Yuta, K. Abe, K. Hasegawa
A. Yamaguchi, K. Tamai, T. Onai and T. Yanokura
Tohoku University, Sendai 980, Japan

R. A. Burnstein, J. Hanlon and H. A. Rubin
Illinois Institute of Technology, Chicago,
Illinois 60616, USA

C. Y. Chang, S. Kunori, G. A. Snow, D. Son,
P. H. Steinberg and D. Zieminska
University of Maryland, College Park, Maryland 20740, USA

T. Kafka, W. A. Mann, A. Napier and J. Schneps
Tufts University, Medford, Massachusetts 02155, USA

(Presented by T. Kitagaki)

Abstract

We have analyzed neutrino-iron events occuring in the bubble chamber wall in the film from Fermilab neutrino-deuterium experiment E545. The X distributions of charged current events were obtained for both iron and deuterium. The ratio, $R_x = (d\sigma/dx)^{\nu Fe}/(d\sigma/dx)^{\nu D}$, shows the first example of a clear nuclear effect in neutrino interactions. The effect is also seen in the region $Q^2 < 10$ $(GeV/c)^2$, and $Q^2 > 10$ $(GeV/c)^2$ corresponding to different X regions. The nuclear effect is described by an apparent softening of valence quarks in the quark picture since it is seen in all region of X.

Introduction

The European Muon Collaboration has measured μ-Fe and μ-D scattering and found a difference in the structure function F_2 for iron and deuterium[1]. This nuclear effect on the structure function, the so called EMC effect, was also examined in electron scattering[2] and neutrino/ antineutrino interactions[3,4,5,6]. Intensive theoretical work followed.[7] However, the problem is still controversial. In experimental data, μ and e scattering data[1,2,8] showed a consistent linear drop in the ratio of F_2 or $d\sigma/dx$ as a function of X in the region of $0.3 < X < 0.6$. However, for the small X region, $X < 0.3$, the original EMC data (F_2) showed a rise of the ratio towards X=0 but the SLAC data ($d\sigma/dx$) did not. Neutrino interactions, CDHS ν, $\bar{\nu}$ experiment[3], BEBC ν, $\bar{\nu}$ experiments[4,5] and Fermilab νNe/νD experiments[6,9] have not reported clear evidence for the EMC effect.

In this paper we present a new result on the comparison of X distributions in ν-Fe and ν-D interactions obtained from a 15 foot bubble chamber experiment. The neutrino flux and analysis method are exactly the same for both iron and deuterium events. Therefore, any systematic error is minimized in this experiment.

Experimental procedure

The Fermilab 15 foot deuterium bubble chamber was exposed to a wide band neutrino beam produced by 350 GeV protons. A total of 330K pictures were taken and analyzed for ν-D interactions[10]. In addition we analyzed events inside the upstream bubble chamber wall generally refered to as wall on events. (Fig. 1 and 2) The chamber wall is a spherical shell of stainless steel with inner radius 190cm and 2.5cm thick. We can not see the vertices ; however we can measure the tracks which emerge in the bubble chamber liquid. Tracing back the measured tracks,

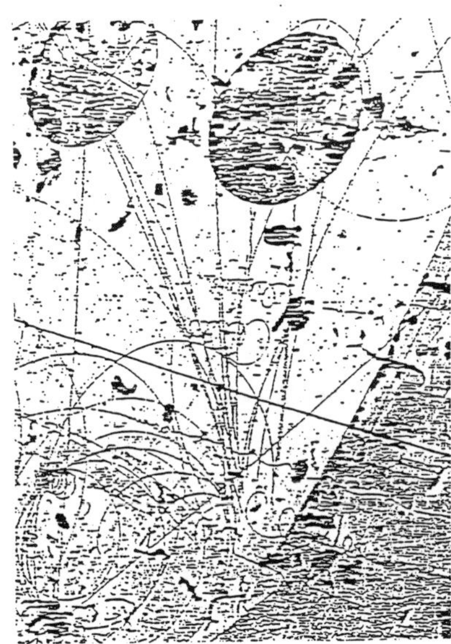

Fig. 1 ν-Fe events, wall on event.

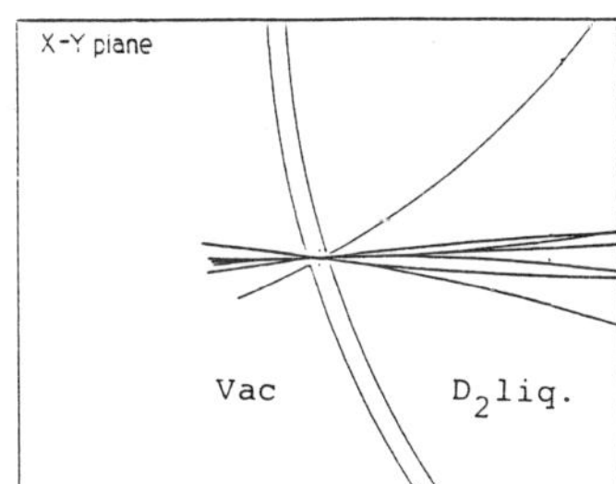

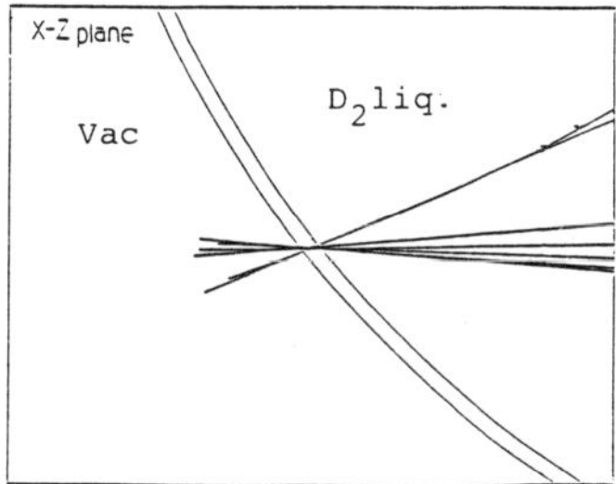

Fig.2 Display rejection of background.

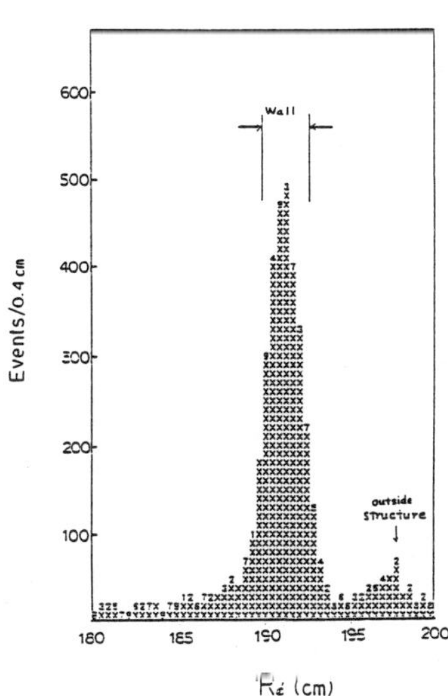

Fig.3 Vertex distribution of wall on events.

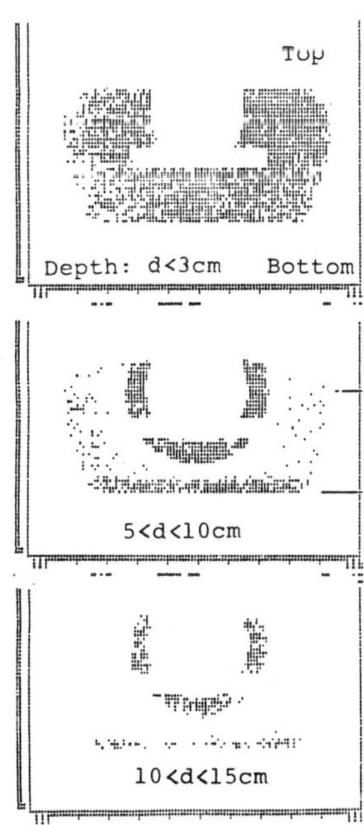

Fig.4 Neutrino tomography of 15 foot bubble chamber: Nose cone area.

we locate the position of vertices with a depth uncertainty of less than 5mm. Fig. 3 shows the radial distribution of iron event vertices from the center of chamber. Fig. 4 shows the neutrino-tomography of the chamber in three depth regions along the beam directions.

We set the fiducial volume of iron in the $\pm 45^{\circ}$ cone seen from the chamber center with additional cuts for height and the nose cone area. The iron thickness less than 7.5cm was used. It corresponds to approximately 1.26 tons of iron with an average thickness of 3.1cm. Each iron event was checked on a computer display after measurement and geometrical reconstruction of tracks and tracks which appeared to miss the reconstructed vertex by more than 5mm were rejected. (Fig. 2) The charged current events were defined by the same muon selection and cuts which were used in our neutrino-deuterium interaction analysis.

$$\Sigma P_L^{vis} > 5 \quad GeV/c$$

$$E_\nu \quad > 10 \ GeV$$

$$P_{TR} \quad > 1 \quad GeV$$

$$W \quad > 1.5 \ GeV/c^2$$

$$Q^2 \quad > 2 \ (GeV/c)^2.$$

After these cuts we obtained a sample of 4000 CC ν-Fe events in the fiducial volume compared to 12,000 CC ν-D events. We remark again that these ν-Fe and ν-D samples were obtained in a single experimental run, produced by the same neutrino flux and analyzed in the same way using the same cuts. Therefore, systematic biases are cancelled in the ratio, and the features of the ratio presented are very stable for a simultaneous change of various cuts in the two samples.

We used a momentum balance method to estimate the neutrino energy E_ν, which is a standard technique in the analysis of neutrino events in bubble chamber experiments.

$$E_\nu = P^\mu + P_L^{h,vis} + (P_L^{h,vis}/P_T^{h,vis})P_T^{mis}$$

After obtaining E_ν all other kinematical quantities such as Q^2, ν, X and etc. are derived.

$$Q^2 = -q^2 = -(k-k')^2$$

$$\nu = \frac{q\,p}{M} = E - E'$$

$$W^2 = (p + q)^2$$

$$X = \frac{Q^2}{2M\nu}, \qquad Y = \frac{\nu}{E}, \qquad V = X \cdot Y.$$

This method is statistically correct, and the result does not shift statistically even if we ignor some of the visible tracks. It is useful for the analysis of iron events.

We limit the depth of vertices within the iron to 7.5cm (Average thickness 3.1cm). Interactions in the iron could change the number of outcoming tracks and yield an apparent decrease of visible momentum $\sum \vec{P}^{h,vis}$ and an increase of missing momentum $\vec{P}^{mis}$. However, the momentum balance method is statistically correct, i.e. the average does not shift although the deviation increases. The deriation of E_ν and Q^2 increase with a decrease of visible momentum, but quantities such as X,Y,W are rather insensitive. Fig.5 shows the E_ν distribution obtained from deuterium events and from iron events. The two samples show a good agreement in all regions.

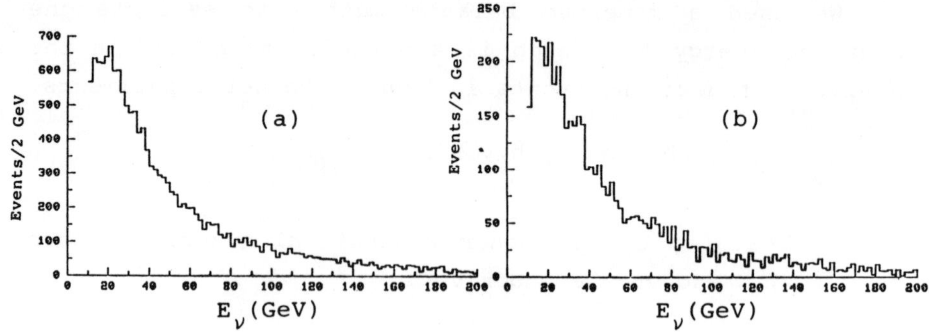

Fig. 5 E_ν distributions obtained from events
(a) ν-D, (b) ν-Fe.

Results

Fig. 6 (a) shows our result for the ratio

$$R_X = \left(\frac{d\sigma}{dX}\right)^{\nu'Fe'}_{RAW} \Big/ \left(\frac{d\sigma}{dX}\right)^{\nu D}_{RAW} , \quad \text{per nucleon,}$$

where 'Fe' shows isoscalar Fe corrected for the neutron excess by our deuteron data[10]. We compare the two X distributions under the assumption,

$$\sigma(\text{isoscalar}) = \sigma(D), \quad \text{per neucleon,} \tag{1}$$

for normalization. The assumption is reasonable in the first order and excludes the systematic errors due to the estimation of absolute cross sections. There may be a small additional coherent/quasi-coherent term in the iron cross section. However the deviation from (1) is negligibly small for the present discussion. This assumption implies that a small positive excess of R_X in the small X region must be cancelled by a large dip in the larger X region due to the shape of the X distribution.

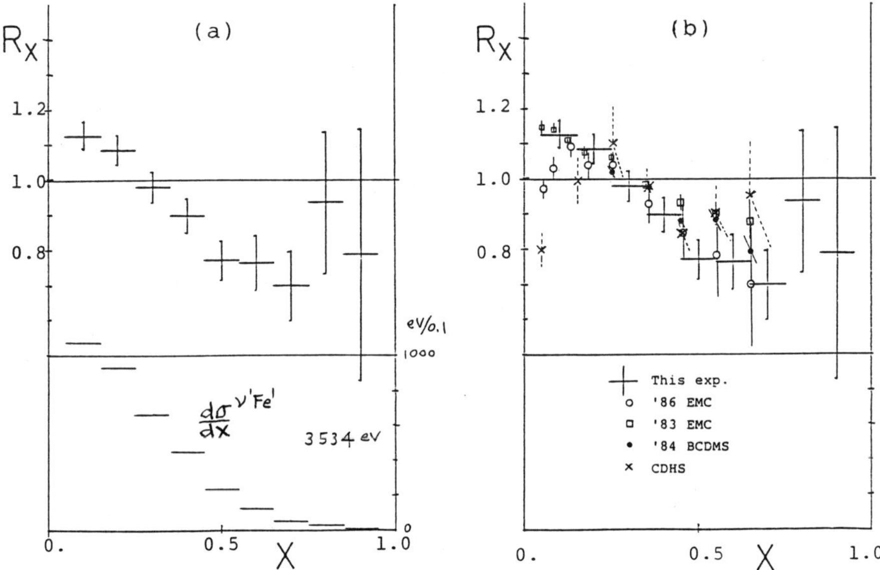

Fig. 6   (a) X distribution and $R_X = (d\sigma/dx)^{\nu'Fe'}/(d\sigma/dx)^{\nu D}$.
(b) Comparison with other data.

In Fig. 6, we omit the smallest X region, $X < 0.05$, since the region is quite sensitive to backgrounds such as muon induced events.

Fig. 6 (b) shows a comparison of our data with other iron data.[1,7,3,11] The latest data on the $d\sigma/dx$ ratio from EMC collaboration[11] (μ -Cu) shows an excellent agreement with our data. The '83 EMC and '84 BCDMS data were given for $F_2(x)$. However, the difference between $F_2(x)$ and $(d\sigma/dx)^{\nu N}$ is small except in the $X < 0.1$ region.

$$F_2(x) = 2x \left[Q(x) + \bar{Q}(x) \right] \frac{5}{18} \qquad \propto \text{Valance+Sea,}$$

$$\text{EMC '83, BCDMS '84}$$

$$(d\sigma/dx)^{\nu N} = \frac{2G^2 ME}{\pi} \times \left[Q(x) + \frac{1}{3}\bar{Q}(x) \right] \propto \text{Valance+}\frac{2}{3}\text{Sea,}$$

$$\text{This experiment}$$

In principle, our result includes all differences between iron and deuterium due to both experimental effect and to 'real physics'. The iron events could be different from deuterium events because of the following.

A. Experimental effects ;

  A-1 Differences in correction efficiencies

  A-2 Secondary interactions inside the iron layer

B. Nuclear effects

  B-1 Secondary interactions inside the nucleus

  B-2 Fermi motion

  B-3 Nuclear physics effects such as coherent or quasi-coherent interactions.

  B-4 Quark picture of nuclear effects such as deformation of quark distribution.

A-1. The data for R_x in Fig. 6 was directly derived from raw data for iron and deuterium. It is equivalent to applying the same correction efficiencies for both samples. This is valid for the measuring-pass efficiency and Monte Carlo correction for kinematical cuts, because we have analyzed the events in the same chamber liquid in the same way. Thus, the systematic bias due to the estimation of correction efficiencies for both samples is minimized. In this work absolute scanning efficiencies are not essential because of assumption (1). The only effect might be on the prong dependence of the efficiency and its subsequent effect on R_x. However, Fig. 7 shows that the cut on prong number does not affect the R_x result noticeablly. Therefore, we have used raw data for the derivation of R_x.

The effects of A-2 and B-1 are corrected by the momentum balance method used to estimate the missing momentum from the visible tracks. We note that the R_x distribution is not affected noticeablly by various experimental cuts such as upon depth in the iron layer, the small momentum cut as $P_h > 1$ GeV/c, large angle cuts, cuts on the E_ν region, and etc.

Therefore, we conclude that the observed effect in R_x is in fact due to the nuclear physics B-2, B-3 or B-4.

We have examined the Q^2 dependence of the effect. Fig. 8 is a comparison of Q^2-X regions in the various experiments. Fig. 9 shows the R_x distribution for $Q^2 < 10$ $(GeV/c)^2$ and $Q^2 > 10$ $(GeV/c)^2$, corresponding to different X regions. It is seen that both Q^2 regions show the effect stronqly. The local slopes are sharper than that of Fig. 6 but the sum of distributions makes Fig. 6. If we interpret these results by a deformation of quark behaviour, (B-4), then the quark distribution shifts towards smaller X everywhere, an apparent softening of the valence quark distribution.

Conclusion

We have compared the X distributions of charged current events in ν-Fe and ν-D interactions. The iron events were obtained from wall on events in Fermilab neutrino-deuterium experiment, E545. The neutrino flux, analysis and cuts are same for both iron and deuterium samples and the systematic error is small. The ratio $R_x = (d\sigma/dx)^{\nu Fe}/(d\sigma/dx)^{\nu D}$ clearly demonstrates the nuclear EMC effect in neutrino interactions. The effect is seen more strongly in $Q^2 < 10$ $(GeV/c)^2$ and $Q^2 > 10$ $(GeV/c)^2$ regions

corresponding to different X regions. The effect is due
to some physics inside the nucleus and an interpretation
in the quark picture is an apparent softening of valance
quarks.

References

1)   J. J. Aubert et al., Phys. Lett 123B, 275 (1983)

2)   A. Bodek et al., Phys. Rev. Lett. 50, 1431 (1983)
     R. G. Arnold et al., Phys. Rev. Lett. 52, 727 (1984)

3)   H. Abramowicz et al., Z. Phys. C25, 29 (1984)

4)   M. A. Parker et al., Nucl. Phys. B232, 1 (1984)

5)   A. M. Cooper et al., Phys. Lett. 141B, 133 (1984)
     A. M. Cooper, Neutrino 84, 394 (1984)

6)   V. V. Ammsov et al., JETP Lett. 39, 393 (1984)

7)   For example, O. Nachtman, Proceedings of the
     Neutrino '84, 405 (1984)

8)   R. Voss, Neutrino 84, 381 (1984)
     G. Bari et al., Phys. Lett. 163B, 282 (1985)

9)   J. Hanlon et al., Phys. Rev. D32, 2441 (1985)

10) For examples,
     J. Hanlon et al., Phys. Rev. Lett. 45, 1817 (1980)
     T. Kitagaki et al., Phys. Rev. Lett. 49, 98 (1982)

11) EMC Collaboration, Private communication, June (1986)

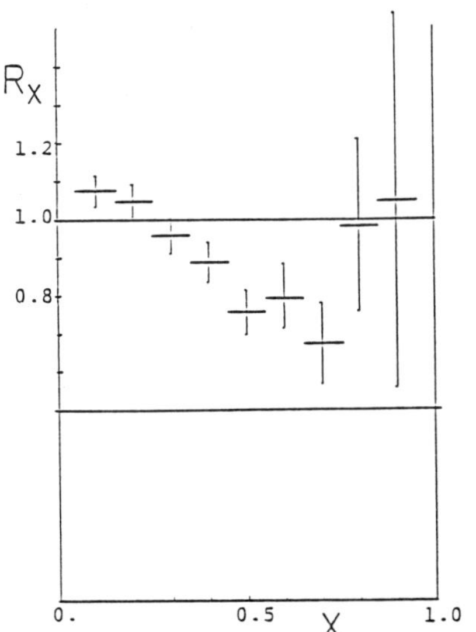

Fig. 7 R_X distribution with an additional cut, prong $\geqslant 3$.

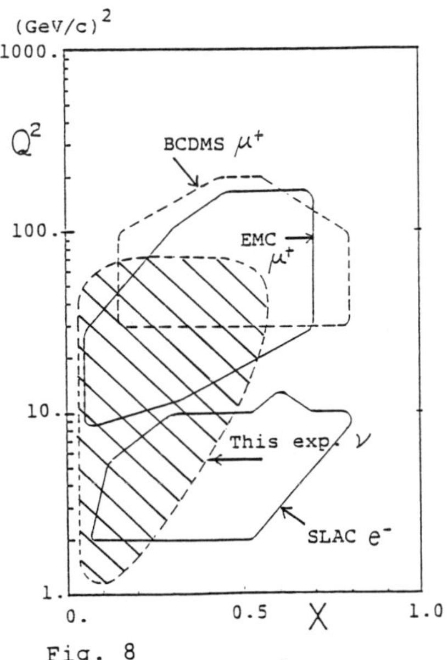

Fig. 8
Comparison of Q^2-X regions.

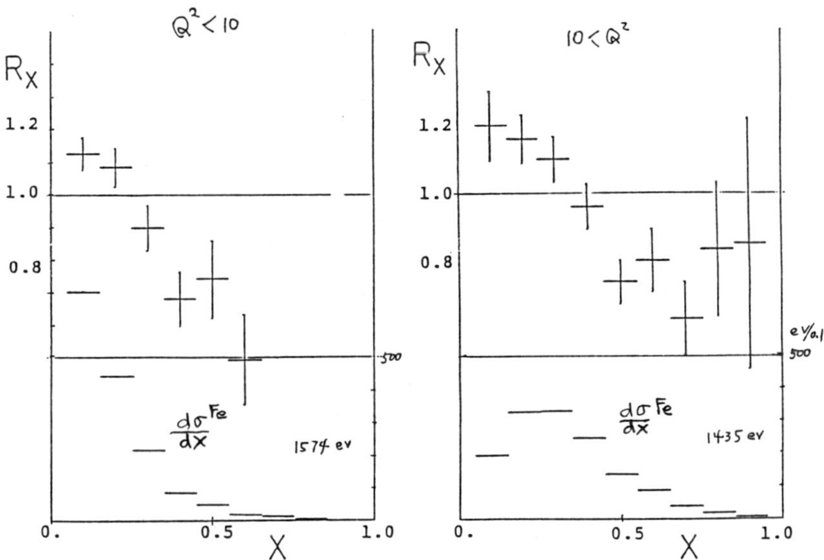

Fig. 9 R_X in different Q^2 regions (a) $Q^2 < 10$, (b) $Q^2 > 10$.

A STUDY OF $\nu_e\ e^-$ ELASTIC SCATTERING * [†]

R.C. Allen, V. Bharadwaj,[a] G.A. Brooks,[b] H.H. Chen, P.J. Doe,
R. Hausammann,[c] W.P. Lee, H.J. Mahler,[d] M.E. Potter,
A.M. Rushton,[e] and K.C. Wang[f]
University of California, Irvine, California 92717

T.J. Bowles, R.L. Burman, R.D. Carlini, D.R.F. Cochran, J.S. Frank,
E. Piasetzky[g] and V.D. Sandberg
Los Alamos National Laboratory, Los Alamos, New Mexico 87545

D.A. Krakauer, and R.L. Talaga
University of Maryland, College Park, Maryland 20742

Presented by H.H. Chen

Department of Physics, University of California
Irvine, California 92717

ABSTRACT

Recent results from the neutrino-electron elastic scattering
experiment at the LAMPF beam stop are presented. Based on the
data sample collected from September 1983 to December 1985, we
observed $121 \pm 25\ \nu_x\ e^-$ events of which 99 ± 25 are assigned to
$\nu_e\ e^-$ scattering. The resulting cross section agrees with
standard electroweak theory, rules out constructive interference
between weak charged-current and neutral-current interactions,
and favors the existence of interference between these two
interactions.

---

* Research supported in part by the U.S. National Science Foundation
  under grant No. PHY-8501559, and by the U.S. Department of Energy.

[†] Presented at NEUTRINO '86, June 3-8, Sendai, Japan.

INTRODUCTION

With the absence of observed structure for leptons, the study of purely leptonic reactions such as neutrino-electron elastic scattering can provide stringent tests for the validity of the electroweak theory of Weinberg,[1] Salam[2] and Glashow[3] (WSG). In contrast to $\nu_\mu e^-$, and $\bar{\nu}_\mu e^-$ scattering,[4,5] which have been studied extensively at high energy accelerators, and $\bar{\nu}_e e^-$,[6] which has been observed at a reactor, the first observation of $\nu_e e^-$ scattering was reported only very recently.[7] Unlike $\overset{(-)}{\nu}_\mu e^-$ scattering, which proceeds only via the weak neutral-current (NC) interaction, $\overset{(-)}{\nu}_e e^-$ scattering proceeds via both NC and the weak charged-current (CC) interactions. An accurate measurement of the cross section for $\overset{(-)}{\nu}_e e^-$ scattering can therefore provide fundamental information on the interference between these two interactions.[8]

The beam dump at the Clinton P. Anderson Meson Physics Facility (LAMPF) of the Los Alamos National Laboratory (LANL) is an intense source of ν_e's. In the beam dump, the decay of stopped π^+ followed by stopped μ^+ leads to the production of equal numbers of ν_μ, $\bar{\nu}_\mu$, and ν_e with well defined spectra. Production of $\bar{\nu}_e$ is highly suppressed ($<10^{-3}$) because of the absorption of π^- and μ^- by nuclei in the beam dump.

In this energy range (neutrino energy less than 53 MeV), the recoil electron from neutrino-electron scattering is confined to a forward kinematic cone of 10° if the detection threshold for the recoil electron is 20 MeV. To study $\nu_e e^-$ scattering, $\nu_\mu e^-$ and $\bar{\nu}_\mu e^-$ scattering events are subtracted from the total $\nu_x e^-$ scattering sample using the present world-averaged measured cross sections.[4,5] Backgrounds to $\nu_x e^-$ scattering produced by cosmic rays are expected to be roughly isotropic. An irreducible beam-associated background comes from $\nu_e + {}^{12}C \rightarrow e^- + {}^{12}N$, which is also close to isotropic given the ambiguity in electron track direction. With a LAMPF proton beam energy of 765 MeV and a beam current of 0.8 mA on the beam dump, the neutrino flux of each type is 4×10^7 /cm^2-sec at a mean detector distance of 9 m.

THE DETECTOR

The central neutrino detector, with a sensitive mass of 15 metric tons and dimensions 3 (w) x 3 (h) x 3.5 (l) m^3, is a fine-grained sandwich system arranged in 40 identical close-packed layers. It is located in a "neutrino cave" well shielded from cosmic rays and the beam stop. Each sandwich layer consists of a plane of plastic scintillator (2.6 gm/cm^2) and a polypropylene flash-chamber module, FCM, (1.3 gm/cm^2) for recording energy and position information, respectively. Each scintillation layer consists of four counters with a single photomultiplier for each counter.[9] This arrangement gives about 14 photoelectrons per MeV deposited in the scintillator. Each FCM contains 10 panels, alternating vertically and horizontally.[10] Each panel contains 520 flash tubes operating at an average efficiency of 55%. The resulting angular resolution of this fine-grained detector system, including multiple scattering, is $\pm 7^\circ$ for the low energy electrons to be detected. Additional details of this detector system and its performance will be provided elsewhere.[11]

The cosmic-ray shields surrounding the central neutrino detector consists of three components. From the outside inwards, a 900 gm/cm^2 steel and concrete layer attenuates the soft component (hadrons and gammas) of cosmic rays. This is followed by four layers of multiwire proportional chambers (MWPC's) to veto and tag incoming charged particles (mostly muons) with an inefficiency of a few times 10^{-5}. Then, another inert shield of steel and lead attenuates and converts the residual gammas which are produced in the outer inert shield.

The beam-stop shield consists of a steel wall 6.3 m thick which attenuates neutrons by a factor of about 10^{15}. A section of this steel was replaced by Uranium for the latter half of the data set reported here in order to test the neutrino origin of the observed signal, and to further reduce any residual non-neutrino beam-associated background.

The detector is triggered by a coincidence between at least 3 adjacent scintillation planes, with energy deposition between 1 and 16 MeV per plane, and with no veto from the MWPC system. If the MWPC veto is not used, cosmic-ray muons traversing the detector generate a trigger rate of more than 1 kHz. A 2 μs veto from the MWPC system removes these muons, but electrons from cosmic-ray stopped-muon decays produce a trigger rate of 20 Hz. Increasing the MWPC veto duration to 20 μs removes these events, and leaves a residual trigger rate of less than 0.1 per second. These are generated primarily by neutrals entering the detector system.

The trigger initiates the discharge of high voltage (HV) pulsers for the FCM's, and acquisition and transfer of data via CAMAC from the various detector components spanning a time period from 32 μs before to 64 ms after the trigger. The pre-trigger information is used in the off-line analysis for identifying stopped-muon decays; the post-trigger information, for ν_e capture by ^{12}C.

Data is taken during the LAMPF beam spill (Beam-On) and also between spills (Beam-Off) to allow a cosmic-ray background subtraction. In addition, triggers were included to take cosmic-ray through-muons and stopped-muon decays every ten minutes for calibration and monitoring purposes.

DATA REDUCTION

The data sample reported here is based on a total beam exposure of 3.49 A-hrs of protons (7.85×10^{22}) on the beam stop, with a Beam-Off/On live-time ratio of 3.88. Data reduction is performed in a series of stages. At the first stage, events with scintillator activity before the trigger or activity in the MWPC's are removed, along with events without a reconstructable track in either view of the FCM's. At the second stage, energy and dE/dx constraints were imposed to select a sample of electron events. At the third stage, a minimum track length requirement was made leaving 2,213 Beam-On and 6,747 Beam-Off events, and a net 479 beam-associated events. Finally, the forward 16 degree angle cut is applied. The results of this analysis is summarized in Table 1.

In Fig. 1a, the $\cos\theta_e$ distribution of all events after the third stage of data reduction is shown for both Beam-On and Beam-Off, where θ_e is the reconstructed angle of the electron relative to the direction of the incident neutrino. The forward angular cone, defined as $\cos\theta_e > 0.960$, contains 164.9 ± 20.0 beam-associated events as

Table 1

The number of events at each stage of the data reduction process.
The most significant cuts are indicated as typical examples.

Cut Level	Cut Description		Beam On	Beam Off	Beam Assoc.
0	TRIGGER:	Remove calib. events	163,152	446,671	48,326
1	MWPC:	No 2 of 4 layers hit	19,838	47,537	7,618
	PRETRIGGER:	No adjacent scint. activity			
	TRACKING:	One track in each view			
	FIDUCIAL:	Layers 2–39, 2" from edge			
2	MWPC:	No 2 adjacent MWPC	4,946	12,911	1,627
	ENERGY:	Total obs. E > 7 MeV			
		Total obs. E < 60 MeV			
		ΔE (mid) > 2.5 MeV			
		ΔE (end) > 2 MeV			
		dE/dx < 10 MeV/inch			
	# OF PLANES:	No more than 7 scint. planes			
3	TRACKING:	At least 3 FCM	2,213	6,747	479
4	ANGLE:	Forward 16 degrees	352	728	165

shown in Fig. 1b. A Monte Carlo simulation, described in the following section, gives the $\cos\theta_e$ distribution shown in Fig. 1c for
$\nu_x e^-$ scattering. Beam-associated background is estimated by averaging the number of beam-associated events in the $\cos\theta_e$ interval between 0.840 and 0.960, taking into account the fraction of $\nu_x e^-$
events outside the forward cone as determined from the Monte Carlo
simulation. The resulting beam-associated background in the forward
cone is 44.2 ± 14.2.[12] We attribute the remaining 120.7 ± 24.5
events in the forward cone to $\nu_x e^-$ elastic scattering.

MONTE CARLO SIMULATION

In order to understand the detector response to the various
physical processes of interest and to evaluate the detection efficiency for these processes, a detailed Monta Carlo simulation was
carried out. This simulation is based on the EGS4 (Electron-Gamma-
Shower) code released recently.[13] The simulation took into account

light attenuation in the scintillators, PMT photo-electron statistics, on-line trigger requirements, and detailed tracking in the
FCM's. Then the simulated events were translated into the on-line
data format and passed through the off-line analysis programs.

Cosmic-ray stopped-muon decay events were also simulated. In
Fig. 2, ΔE and angular distributions of the observed and the
simulated decays are shown. This agreement provides confidence in our
understanding of the detector performance.

Some aspects of the detector performance were simulated in the
Monte Carlo by using the cosmic-ray calibration events. Among these
are: pretrigger activity in the scintillators; and performance of the
MWPC veto system and the FCM system. In all, a total systematic
error of 6% was assigned to the absolute detection efficiency for
$\nu_x e^-$ scattering.

RESULTS

In order to determine an absolute cross section, the flux of
neutrinos from the beam stop must be known. This was measured in a
separate experiment using an instrumented beam stop at the Lawrence
Berkeley Laboratory (LBL) 184" synchro-cyclotron with 720 MeV protons.[14] A separate computer simulation is used to account for differences in the beam stops and to scale the LBL result to the LAMPF
beam energy. These adjustments resulted in a determination of the
neutrino flux to an uncertainty of 12%. Combining this with the 6%
uncertainty in the absolute detection efficiency results in an
overall systematic error of 13.4%.

To remove the contributions from $\nu_\mu e^-$ and $\bar{\nu}_\mu e^-$ scattering from
the total $\nu_x e^-$ scattering sample, we determined the expected number
of $\nu_\mu e^-$ and $\bar{\nu}_\mu e^-$ events from the measured cross sections[4,5] using
our Monte Carlo simulation. A total of 7.1 ± 1.8 and 14.2 ± 3.1
events are assigned to $\nu_\mu e^-$ and $\bar{\nu}_\mu e^-$ scattering, respectively. The
remaining 99.4 ± 24.8 events are assigned to $\nu_e e^-$ scattering. This
number is expected in the WSG electroweak theory with

$$\sin^2\theta_W = 0.24 \pm {}^{0.09}_{0.10} \text{ (stat)} \pm {}^{0.05}_{0.06} \text{ (syst)}$$

and a $\nu_e\,e^-$ cross section,

$$\sigma(\nu_e\,e^-)/E_\nu = [9.8 \pm {}^{2.7}_{2.6}\ (\text{stat}) \pm {}^{1.5}_{1.6}\ (\text{syst})] \times 10^{-45}\ \text{cm}^2/\text{MeV}.$$

The WSG electroweak theory predicts destructive interference between the NC and CC interactions. In Table 2, we compare our observed number of $\nu_e\,e^-$ events with the expected number, assuming "destructive interference" (WSG), "no interference", and "constructive interference", taking $\sin^2\theta_W = 0.22$. Our result rules out "constructive interference" by almost four sigma, disagrees with "no interference" by about 2.5 sigma, and is consistent with the WSG prediction.

We have also directly compared the $120.7 \pm 24.5\ \nu_x\,e^-$ events with the WSG prediction to determine

$$\sin^2\theta_W = 0.24 \pm {}^{0.06}_{0.09}\ (\text{stat}) \pm {}^{0.04}_{0.06}\ (\text{syst}).$$

The region on the C_A-C_V plane allowed for the NC coupling constants using our total event sample is shown in Fig. 3.

Table 2

The number of observed $\nu_e\,e^-$ scattering events compared with the expected number of events involving different NC and CC interference possibilities.

This measurement	99 ± 25
Expected number of events:	
(i) Destructive interference (WSG)	94 ± 13
(ii) No interference	187 ± 25
(iii) Constructive interference	293 ± 39

PLANS AND EXPECTATIONS

We plan to continue taking data in 1986 to increase our beam exposure by about 50%. Combined with analysis improvements currently under investigation, we anticipate reducing the overall statistical error to the 16% level. The present overall systematic error is dominated by the neutrino flux uncertainty of 12%. A new experiment is in progress at LAMPF which will reduce this uncertainty to about 6%.[15] We anticipate that with these efforts, the $\nu_e\,e^-$ cross section will be determined to better than 18% within a year.

REFERENCES AND FOOTNOTES

(a) Now at Fermilab, Batavia, Illinois 60510.
(b) Now at Baylor College of Medicine, Houston, Texas 77004.
(c) Now at University of Geneva, Geneva, Switzerland.
(d) Now at Cerberus A.G., Mannedorf, Switzerland.
(e) Now at Argonne National Laboratory, Argonne, Illinois 60439.
(f) Now at Rockwell International, Thousand Oaks, California 91360.
(g) Also at Tel-Aviv University, Ramat Aviv, Israel 69978.

1. S. Weinberg, Phys. Rev. Lett. 19, 1264 (1967).
2. A. Salam, in Elementary Particle Theory: Relativistic Groups and Analyticity (Nobel Symposium No. 8), edited by N. Svartholm (Almqvist and Wiksell, Stockholm, 1968), p. 367; A. Salam and J.C. Ward, Phys. Lett. 13, 168 (1964).
3. S.L. Glashow, Nucl. Phys. 22, 579 (1961).
4. L.A. Ahrens et al., Phys. Rev. Lett. 54, 18 (1985).
5. F. Bergsma et al., Phys. Lett. 117B, 272 (1982).
6. F. Reines et al., Phys. Rev. Lett. 37, 315 (1976).
7. R.C. Allen et al., Phys. Rev. Lett. 55, 2401 (1985).
8. B. Kayser, E. Fischbach, S.P. Rosen, and H. Spivack, Phys. Rev. D20, 87 (1979).
9. K.C. Wang and H.H. Chen, IEEE Trans. Nucl. Sci. 28, 405 (1981).
10. R.C. Allen, G.A. Brooks, and H.H. Chen, IEEE Trans. Nucl. Sci. 28, 487 (1981).
11. P.J. Doe et al., Nucl. Instr. & Meth. (to be published).
12. The quoted error includes additional systematic uncertainties associated with the shape and angular interval for estimating the beam-associated background.
13. W.R. Nelson, H. Hirayama, and D.W.O. Rogers, SLAC Report-265 (1985).
14. H.H. Chen, J.F. Lathrop, R. Newman, and J.C. Evans, Nucl. Instr. & Meth. 160, 393 (1979).
15. Experiment 866 at LAMPF.

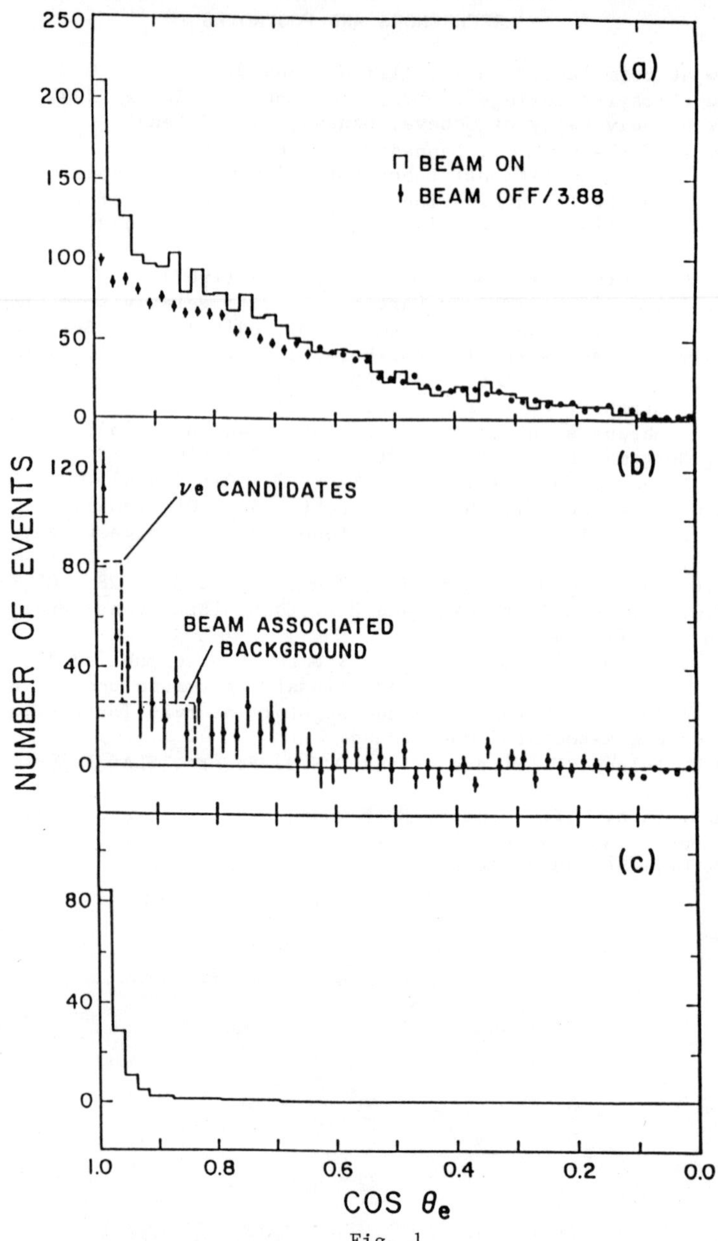

Fig. 1

Cosθ_e distributions in intervals of 0.02.
(a) For Beam-On and normalized Beam-Off events.
(b) For beam-associated events.
(c) For simulated ν_x e events.

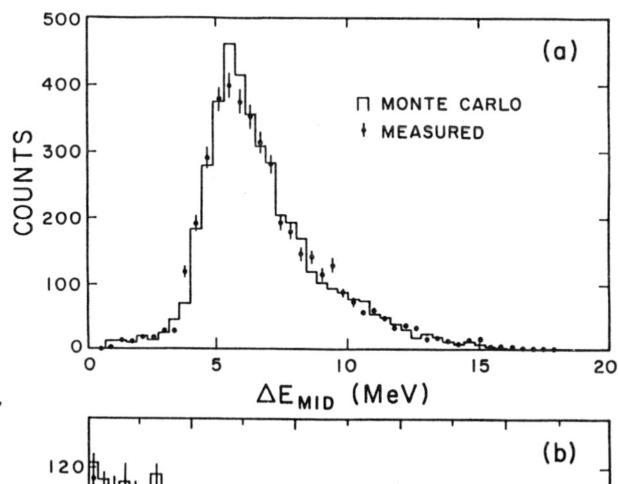

Fig. 2

Comparison of observed and simulated events from stopped-muon decays.

(a) ΔE distribution.

(b) $\cos\theta_e$ distribution.

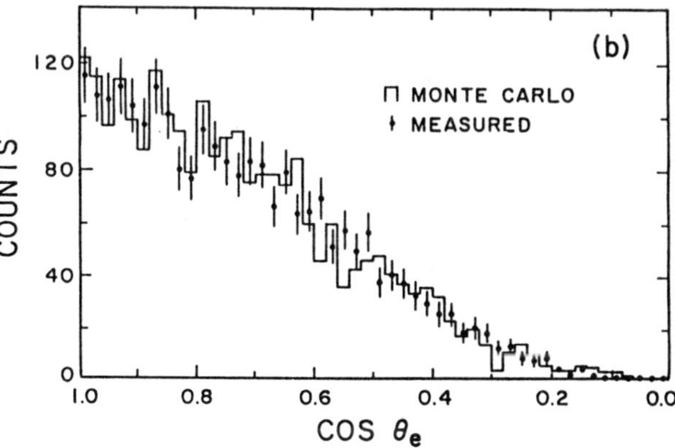

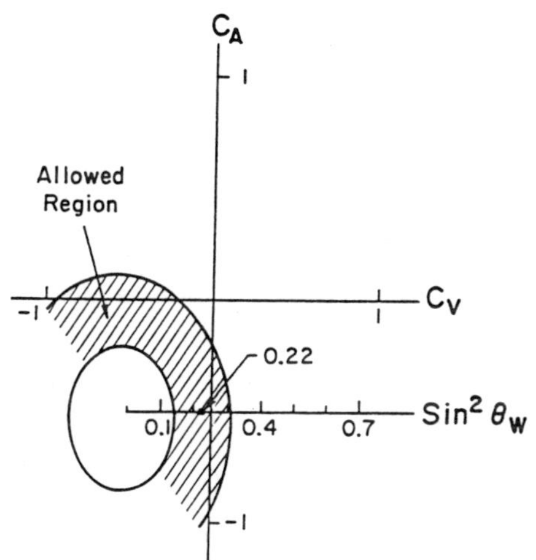

Fig. 3

Using our total event sample, the allowed region on the C_A-C_V plane is shown.   Here C_A and C_V refer to the NC coupling constants.

RESULTS ON THE UNIVERSALITY OF ν_e AND ν_μ COUPLING TO THE WEAK NEUTRAL CURRENT, ON THE RATIO OF PROMPT ν_e and ν_μ FLUXES AND A SEARCH FOR AN EXCESS OF MUONLESS NEUTRINO EVENTS

CHARM Collaboration

J.Dorenbosch

NIKHEF, Amsterdam, The Netherlands

J.V.Allaby, U.Amaldi, G.Barbielini, F.Bergsma, A.Capone, W.Flegel, L.Lanceri, M.Metcalf, C.Nieuwenhuis, J.Panman and K.Winter

CERN, Geneva, Switzerland

I.Abt, J.Aspiazu, F.W.Büsser, H.Daumann, P.D.Gall, T.Hebbeker, F.Niebergall, P.Schütt, and P.Stähelin

II. Institut fur Experimentalphysik, Universität Hamburg, Germany

P.Gorbunov, E.Grigoriev, V.Khovansky, S.Krushinin, A.Maslennikov and A.Rosanov

Institute for Theoretical and Experimental Physics, Moscow, USSR

A.Baroncelli, L.Barone, B.Borgia, C.Bosio, M.Diemoz, U.Dore, F.Ferroni, E.Longo, L.Luminari, P.Monacelli, F. de Notaristefani, R.Santacesaria, C.Santoni, L.Tortora and V.Valente

Instituto Nazionale di Fisica Nucleare, Rome, Italy

(Presented by F.Bergsma)

ABSTRACT

A beam dump experiment has been performed at the CERN SPS using the CHARM neutrino detector. The instrumentation and statistics have been significantly improved with respect to earlier experiments. The neutrinobeam, produced by dumping 400 GeV protons into a copper block, has been used to study electron neutrino interactions and prompt neutrino interactions. We have determined the ratio of electron neutrino induced neutral and charged current cross sections and deduced the ratio of neutral current coupling constants $g_{\nu_e \bar{\nu}_e}/g_{\nu_\mu \bar{\nu}_\mu} = 1.09 +0.15-0.18$. We compared the prompt muon neutrino and electron neutrino event rates, and searched for an excess of muonless events at low energy; we give an estimate for the charm-production cross section in proton collisions with Cu nuclei.

INTRODUCTION

In a beam dump experiment carried out at the CERN-Super-Proton-Synchrotron (SPS) in 1979, earlier evidence for prompt electron- neutrinos was confirmed and the existence of a flux of prompt ν_μ and $\bar{\nu}_\mu$ in roughly equal proportions was established [1-3]. Here we give results from a new beam dump experiment performed in 1982 at the CERN SPS, using a new beam installation. The target station was installed closer to the detector, thus increasing the solid angle of acceptance by a factor three. The beam monitoring was extended, especially the measurement of proton losses upstream of the target. The beam was guided through vacuum from the accelerator up to the target without any material (such as windows or monitors) in the beamline. We have made use of the relatively large fraction of electron neutrinos and antineutrinos($1/6$) in this beam to measure the ratio of neutral-current induced and charged-current induced ν_e and $\bar{\nu}_e$ events and to compare it to the same quantity for muon neutrinos to determine the ratio $g_{\nu_e \bar{\nu}_e} / g_{\nu_\mu \bar{\nu}_\mu}$. Next we have determined the flux of prompt neutrinos coming from sources with a short lifetime ($\leq 10^{-11}$ sec) . Contributions due to neutrinos from sources with a longer life- time (π, K etc.) are strongly suppressed by the high density of the dump target. We refer to this flux component as conventional. We separated the prompt flux from the conventional flux by extrapolating the rates to infinite target density, using the data from two targets with relative average density of 1 and $1/3$. For the prompt neutrino rates thus determined we investigated the flux ratio of ν_e and ν_μ, a possible excess of muonless events at low energy and determined the cross section for charm-production.

THE BEAM

The 400 GeV proton beam was extracted from the SPS in a single turn (23 μsec) and dumped into thick copper targets in the direction of the neutrino detector. The targets were installed at the downstream end of the decay tunnel for normal neutrino beam operation. The distance to the centre of the fiducial volume of the detector was 487.3 m. Targets with full and one-third density were irradiated by $13.26 \cdot 10^{17}$ and $6.35 \cdot 10^{17}$ protons, respectively. Prompt muon-neutrino's from decays of short-lived parent particles (mainly charmed mesons) and conventional neutrinos from long-lived parents (pions and kaons) were found to contribute in a ratio of about 1 to 2.7 (density 1) and 1 to 8 (density $^1/_3$) to the beam; for electron neutrinos these ratios are about 1 to 0.2 (density 1) and 1 to 0.6 (density $^1/_3$). The flux of prompt neutrinos is composed of electron and muon neutrinos with approximateley equal contributions of neutrinos and antineutrinos. The flux of neutrinos of conventional origins is composed of electron neutrinos again with equal contributions of neutrinos and antineutrinos, mainly from neutral kaon decays, and of muon-neutrinos with neutrinos and antineutrinos in the ratio of 1.6 to 1, mainly from pion decays. The mean energy of electron-neutrino induced events is 54.2 GeV and that of muon-neutrinos is 25.2 GeV .

THE DETECTOR

The CHARM neutrino detector was exposed to this dump neutrino beam in 1982. The detector [4] was composed of a fine-grain target calorimeter and of a muon spectrometer. The target calorimeter consisted of 72 marble plates of 8 cm thickness (about one radiation length) and of 3m x 3m cross-section which were interlaced between layers of scintillation counters, (3 cm thick, 15 cm wide and 300 cm long), proportional drift-tubes, (3cm x 3 cm x400 cm), and digital wire chambers working in the limited streamer mode [5] with wires oriented at 90° with respect to those of the proportional drift tubes. The proportional drift-tubes extended into the 45 cm wide magnetized iron frame that surrounded the calorimeter. The orientation of all detector elements alternated between horizontal and vertical in consecutive gaps. The total mass of the fiducial volume (59 planes, 2.4 x 2.4 m²) was 90.6 tons. The other parts of the calorimeter volume had only detector functions; the first two planes were used to veto incident muons, the lateral iron frames and the outer 30 cm of the calorimeter planes were used to veto cosmic rays and to perform shower measurements. The last 14 planes and the end calorimeter were used only for shower measurements and event classification. The response and resolution of the calorimeter to hadronic and electromagnetic showers has been measured in pion and electron beams in the energy range from 1 to 140 GeV [6]. Interactions in the CHARM detector were classified by an automatic pattern recognition program on an event-by-event basis. Events with an outgoing muon with more than 1 GeV energy were classified as charged-current muon-neutrino interactions (1μ), all others were called muonless (0μ). Muonless events are composed of neutral-current events initiated by muon- or electron-neutrinos and by charged-current electron-neutrino interactions (1e). The latter class of events was recognized directly by using the characteristic difference in lateral profile of electron and hadron induced showers [6].

UNIVERSALITY OF v_e and v_μ COUPLING TO THE NEUTRAL CURRENT

The notion of universality of the coupling of the electron, the muon and the tau family to the charged and the neutral weak currents is mainly based on experimental data [7].

The most precise comparison at low values of Q^2 of the relative coupling strength of the electron and muon families to the charged weak current has been derived from precise measurements of the ratio of the corresponding pion decay rates [8]

$$\Gamma(\pi \to e v_e)/\Gamma(\pi \to \mu v_v) = (1.218 \pm 0.014) \cdot 10^{-4}$$

with the result

$$(g_{ev_e}/g_{\mu v_\mu}) = .9939 \pm 0.0057 \qquad (1)$$

From measurements of muon-neutrino electron scattering [9] and of the forward-backward asymmetry in the reaction $e^+ e^- \to \mu^+ \mu^-$ [10] the relative axial-vector coupling strengths of the electron and the muon to the neutral-current have been derived, with the result

$$g_{ee}/g_{\mu\mu} = 0.96 \pm 0.11$$

No measurements of comparable accuracy have been reported on the ratio

$$g_{\nu_e\bar{\nu}_e}/g_{\nu_\mu\bar{\nu}_\mu}$$

of the coupling strengths of electron- and muon-neutrinos to the neutral weak current. It can be derived from measurements of the ratios of total cross section for semileptonic neutrino scattering

$$R^e = \sigma(\nu_e N \rightarrow \nu_e X)/(\sigma(\nu_e N \rightarrow eX) \qquad (2)$$

and

$$R^\mu = \sigma(\nu_\mu N \rightarrow \nu_\mu X)/(\sigma(\nu_\mu N \rightarrow \mu X)$$

in high energy neutrino beams using eq.(1). Values of R^μ for high energy ν_μ and $\bar{\nu}_\mu$ beams have been reported elsewhere [11] ; in a beam composed of equal fluxes of ν_μ and $\bar{\nu}_\mu$ we expect

$$R_\mu = 0.340 \pm 0.005 \quad , \qquad (3)$$

for E_h, the energy of the hadronic shower, larger than 2 GeV. Precise values for R^e have not yet been reported, because of the difficulty of producing high energy electron-neutrino beams. For the following analysis we summed up events from both densities. We determined the quantity R^e eq.(2) from the observed rates of these events after proper corrections for background losses and event class mixing, averaging over ν_e and $\bar{\nu}_e$ fluxes which are nearly equal [12]:

$$R^e = \frac{N(0\mu, E_{sh} > 2\text{GeV}) - \bar{R}^\mu N(1\mu, E_{sh} > 2\text{GeV}) - N(1e, E_{sh} > 2\text{GeV})}{N'(1e, E_h > 2\text{GeV})} \qquad (4)$$

$\bar{R}^{\mu}$ is obtained by taking into account the ratio of observed ν_{μ} and $\bar{\nu}_{\mu}$ induced CC events [11] observed in the dump neutrino beam. We unfolded the effects of different electron and hadron energy response and resolution, of the muon energy resolution and of the event selection criteria from the observed distributions using an iterative Monte Carlo method. We selected 0μ and 1μ events with $E_{shower} > 2$ GeV to evaluate the quantity R^e of eq. (2), and applied the corresponding selection of $E_{sh} = E_e + E_h > 2$ GeV for $1e$ events; in the denominator of eq. (4) we use, appropriately, $N'(1e, E_h > 2$ GeV$) = \alpha \times N(1e, E_{vis} > 2$ GeV$)$ with $\alpha = 0.865 \pm 0.025$ [13].

The corrected event numbers are summarized in table 1 Taking into account correlations of the errors, we deduced from eq. (4)

$$R^e = 0.406\ ^{+0.145}_{-0.135}\ ,$$

and combining with the value of R^{μ} from eq. (3) we derive, assuming $(g_{ev_e}/g_{\mu v_{\mu}}) = 1$ in agreement with eq. (1),

$$(g_{v_e \bar{v}_e}/g_{v_\mu \bar{v}_\mu}) = (R^e/R^{\mu})^{1/2} = 1.09\ ^{+0.18}_{-0.20}$$

Combining this value with a result from an exposure in a 160 GeV narrow band neutrino beam [14] we obtained

$$(g_{v_e \bar{v}_e}/g_{v_\mu \bar{v}_\mu}) = 1.05\ ^{+0.15}_{-0.18}$$

in good agreement with unity as expected from universality [15].

Table 1:

Number of events selected according to the definitions, used to evaluate the ratio $R^e = NC/CC(v_e + \bar{v}_e)$ from eq. (3). A value of $\bar{R}^\mu = 0.3334 \pm 0.0042$ is used [12].

EVENT DEFINITION	NUMBER OF SELECTED EVENTS
$N(1\mu, E_h > 2 \text{ GeV})$	$5005 \pm 69(\text{stat.}) \pm 47(\text{syst.})$
$N(0\mu, E_{sh} > 2 \text{ GeV})$	$2848 \pm 56 \qquad \pm 43$
$N(1e, E_{sh} > 2 \text{ GeV})$	$872 \pm 60 \qquad \pm 66$
$N'(1e, E_h > 2 \text{ GeV})$	$755 \pm 52 \qquad \pm 57$

COMPARISON OF PROMPT ELECTRON- AND MUON-NEUTRINO EVENT RATES

If e-μ universality holds, one expects equal fluxes of prompt electron and muon neutrinos to within a few percent [16]. We report here on a measurement of the prompt neutrino fluxes with $E_v > 20$ GeV, for which the systematic error is the smallest. We compared the fluxes by determining the charged-current events induced by electron neutrinos and muon neutrinos. The prompt and conventional events were separated by extrapolation to infinite target density. The electron neutrino induced charged-current events were identified by a pattern recognition program, which made use of the regular shape and high energy density of electromagnetic showers. [17]. Expressed as flux asymmetry with symmetric errors we find

$$\frac{N(1\mu)-N(1e)}{N(1\mu)-N(1e)} = 0.19 \pm 0.12(\text{stat.}) \pm 0.04(\text{syst.}) \qquad (5)$$

The event rates were corrected for the following effects: Insufficient length (ca. 6 proton interaction lengths) of the density $^1/_3$-target and material upstream of the targets; correction of the 0μ-event rates because the muon had less than 1 GeV energy or it was hidden in the hadronic shower or leaving the detector at the side;

misclassification of 1μ-events because a decaying pion or kaon faked a primary muon. The rates were corrected for acceptance losses using a Monte Carlo simulation. Cosmic-rays faking a neutrino event were determined outside the spill and subtracted; the deadtime of the detector was determined separately for both densities. Combining statistical and systematic errors in quadrature we find the asymmetry (5) is compatible with zero within 1.5 standard deviations.

We also evaluated the ratio of prompt electron (anti)neutrino and muon (anti)neutrino events to compare with other experiments and found:

$$\frac{N(1e)}{R(1\mu)} = \frac{\text{prompt}(\nu_e + \bar{\nu}_e)\text{flux}}{\text{prompt}(\nu_\mu + \bar{\nu}_\mu)\text{flux}} = 0.68^{+0.22}_{-0.13} \text{ (stat.) } \pm 0.06(\text{syst.})$$

In table 2 we compare our result with those of other beam dump experiments. There is, again, no compelling evidence for a deviation from e-μ universality.

Table 2:

Comparison of prompt ν_e/ν_μ-flux ratio
from different experiments

Experiment	flux ratio ν_e/ν_μ
THIS EXPERIMENT	$0.68^{+0.22}_{-0.13}$ (stat.) $\pm$ 0.06(syst.)
BEBC [18]	1.14 +0.19 -0.16
CDHS [19]	0.86 ± 0.13
FERMILAB [20]	0.95 ± 0.14

F. Bergsma

SEARCH FOR AN EXCESS OF MUONLESS EVENTS

In an earlier beam dump exposure we have searched for an excess of muonless events due to other neutrino species (v_x) which could produce final states without charged leptons. We subtracted the contribution of $v_\mu(\bar{v}_\mu)$ induced neutral-current events on the basis of the experimentally observed $v_\mu(\bar{v}_\mu)$ induced charged-current events and the known ratio of $R^\mu = \sigma(NC)/\sigma(CC) = 0.340 \pm 0.005$ for a beam composed of equal fluxes of v_μ and $\bar{v}_\mu$. Events induced by electron-neutrino NC and CC interactions were estimated by a model calculation assuming they are produced in $D\bar{D}$ production [16] and semileptonic decay. Normalizing the model to the events with $E_{sh} > 20$ GeV we found a possible excess of 54 ± 19 muonless events, with $2 < E_{sh} < 20$ GeV.

In the conditions of the new experiment we are reporting here we should have observed 170 ± 60 events/ton/10^{20} proton if the excess was genuine. However, we found only 103 ± 55 evts/ton/10^{20} proton, indicating a statistical fluctuation in the earlier experiment.

The present data analysis allows us to evaluate an excess in a model independent way by subtracting the $v_e(\bar{v}_e)$ induced CC events which are directly determined by a pattern recognition program [17]. The NC events induced by $v_e(\bar{v}_e)$ are determined by the ratio R^e, which we took equal to R^μ.

We applied a correction α for the condition $E_h > 2$ GeV used for NC events whereas the v_e induced CC events were selected by the condition $E_{sh} = E_e + E_h > 2$ GeV. The correction factor was calculated by Monte Carlo simulation, with the result $\alpha = 0.865 \pm 0.025$. The model independent excess for $E_h > 2$ GeV is found from the following expression (6) using the data from table 1

$$N(v_x) = N(0\mu) - \bar{R}^\mu N(1\mu) - (1 + \alpha \cdot \bar{R}^e) N(1e) = 32 \pm 118 \text{ evts/ton/} 10^{20} \text{proton} \quad (6)$$

We conclude that there is no evidence for an excess due to v_x interactions, at a level of 5.3% (95% c.l.) of all $v_\mu(\bar{v}_\mu)$ and $v_e(\bar{v}_e)$ induced CC events.

CHARM-PRODUCTION CROSS SECTION

If we assume $D\bar{D}$ production as the dominant mechanism of central($x\approx0$) charm production in proton collisions with Cu nuclei and parametrize the cross-section according to

$$E\partial^3\sigma/\partial x\partial P_t^2 \propto (1-x)^5 e^{-2P_t}$$

and semi-leptonic decay according to the spectator model of Altarelli et al. [21] we find, using the prompt ν_μ induced CC events with $E_\nu >$ 20 GeV:

$$\sigma(pCu \rightarrow D\bar{D}X)_{400GeV} \times BR_{semilept.} = 3.09\pm0.53\pm0.24 \text{ } \mu b/nucleon$$

We have assumed a linear A-dependence for central $c\bar{c}$-production and 8% contribution of $D\bar{D}$ production from the later generations of the hadron cascade induced by 400 GeV protons in copper.

For an average branching ratio of 10% for semileptonic decay we find $\sigma(pCu \rightarrow D\bar{D}X)=30.9 \pm 5.8$ $\mu b/nucleon$, in good agreement with the result of $\sigma=22\mu b$ obtained by the LEBC experiment in 400 GeV pp collisions [22].

References

[1]   P.Fritze et al.,ABCLOS Coll.,Phys.Let.96B(1980) 427

[2]   M.Jonker et al.,CHARM Coll.,phys.Let.96b(1980)435

[3]   H.Abramowicz et al.,CDHS Coll.,Z.Phys.13C(1982) 179

[4]   A.N.Diddens et al.,CHARM Coll, Nucl. Instr. Methods 178 (1980) 27

[5]   M.Jonker et al., Nucl. Instr. Methods 215 (1983) 361

[6]   M.Jonker et al., Nucl. Instr. Methods 215 (1983) 361
       F.Bergsma et al., CHARM Coll. (to be published)

[7]   J.D.Bjorken and S.Weinberg, Phys. Rev. Letters 38 (1977) 622

[8]   E. DI Capua et al., P.R. 133B (1964) 1333, see also
       D.A.Bryman et al., P.R. D11 (1975) 1337, and D.A.Bryman et al., Phys. Rev. Letters 50 (1983) 7.

[9]   F.Bergsma et al., CHARM COLL., Phys.Letters 147B (1984) 481
       L.A.Ahrens et al., Phys. Rev. Letters 55 (1985) 1814.

[10]   See e.g.   B.Naroska,   Proc.   Int.   Symp.   Lepton   Photon
       Interactions at High Energies, Ithaca 1983, eds.
       D.G.Cassel and L.Kreinick (Cornell Univ., Ithaca, N.Y. 1983).

[11]   A value of $\bar{R}^\mu$= 0.3334 ± 0.0042 is used in eq.(4). taking into
       account the observed flux ratio $\Phi(v_\mu)/\Phi(\bar{v}_\mu)$.
       F.Bergsma et al., CHARM Collaboration, Phys. Letters 99B
       (1981) 265 and Phys. Letters B (to be published) have mea-
       sured for E_h > 2 GeV
       R^v = 0.3132 ± 0.0031
       $R^{\bar{v}}$ = 0.395 ± 0.014

[12]   A full account of these corrections will be published in
       Z.Phys.C. The near equality of v_e and $\bar{v}_e$ fluxes follows from a
       Monte Carlo simulation of charged and neutral kaons, Λ and Σ
       for conventional origins and from the assumption of $D\bar{D}$ domi-
       nance for prompt origins.

[13]   The ratio α = N'(1e, E_h > 2 GeV)/N(1e, E_{sh} > 2 GeV) has been
       determined to 0.865 ± 0.025 by Monte Carlo simulation.

[14]   J.V.Allaby et al., CHARM Collaboration,test of the univerality
       of the electron-neutrino and muon-neutrino coupling to the
       charged weak current, Physics Letters B(in print).

[15]   J.V.Allaby et al., CHARM COLL., Experimental verification of
       the universality of v_e and v_μ coupling to the neutral weak cur-
       rent, Physics Letters B(in print).

[16]   E.L.Berger,L.Clavelli,and N.R.Wright,Phys. Rev. D 27 (1983)
       1080

[17]   K.Nieuwenhuis Phd thesis, "A study of the interactions of high
       energy electron-neutrinos" (1986) Univ. of Amsterdam
       (NIKHEF-H)

[18]   H.Grässler et al.,Nucl. Phys. B273 (1986) 253

[19]   result given by P.Debu(SACLAY) in a seminar at CERN on 11
       march 1986

[20]   M.E.Duffy et al., Phys. Rev. Letters 52 (1984) 1865

[21]   G.Altarelli,N.Cabibbo,G.Corbo,L.Maiani and G.Martinelli, Nucl.
       Phys. B208 (1982) 365

[22]   Proc. of 1985 Internat. Symp on Lepton and Photon Interactions
       at High Energies, Kyoto, p.488.

TEST OF THE UNIVERSALITY OF ELECTRON-NEUTRINO AND MUON-NEUTRINO COUPLING TO THE CHARGED WEAK CURRENT

The CHARM Collaboration

Presented by C. SANTONI

INFN Sezione Sanità and Istituto Superiore di Sanità, Rome, Italy.

Abstract

A measurement of the ratio of electron-neutrino and muon-neutrino charged-current (CC) cross-sections is reported. Using data collected in an exposure of the CHARM detector to a 160 GeV narrow-band neutrino beam at CERN we obtained $\sigma_{\nu_e}(CC)/\sigma_{\nu_\mu}(CC) = 1.20 \pm 0.11$. A ratio of the coupling constants $g_{e\nu_e}/g_{\mu\nu_\mu} = 1.10 \pm 0.05$ at $<Q^2> \simeq 16$ GeV^2 was derived in agreement with universal coupling strength of the electron and muon families to the charged weak current.

We report on a comparison between the strengths of the couplings of the electron and muon families to the charged weak current derived from the measurement of the ν_e and ν_μ charged-current (CC) cross-sections. The measurement was performed in an exposure of the CHARM detector [1] in a 160 GeV narrow-band beam at CERN [2] for events with a mean Q^2 of $\simeq 16$ GeV^2.

Electron-neutrino nucleon interactions have not yet been extensively studied at high energy [3] because the ν_e is a small fraction of high energy neutrino beams. In the narrow-band beam the fraction ($\simeq 5\%$) of ν_e coming from K_{e3} decays can be determined accurately by measuring the relative kaon and pion fluxes. Such a measurement has been performed using a helium filled differential Cerenkov counter [4], giving $K^+/\pi^+ = 0.1282 \pm 0.0028$. The energy spectrum of electron-neutrinos can be accurately calculated from the well-known K_{e3} decay kinematics, as well as that of muon-neutrinos originating mainly from two-body decays of pions and kaons and a smaller component from $K_{\mu 3}$ decays.

The CHARM neutrino detector comprises a fine-grain calorimeter surrounded by a magnetized iron frame and followed by a muon spectrometer. It has been described in detail elsewhere [1]. For this experiment the shower energy and vertex measurements are relevant. The energy resolution of the calorimeter can be parametrized according to [5] $\sigma(E)/E = 0.487/\sqrt{E(GeV)} + 0.0127$ for pions and $\sigma(E)/E = 0.18/\sqrt{E(GeV)}$ for electrons. The lateral vertex position of hadron showers was measured with a resolution [5] of $\simeq 5$ cm for pions of energy around 10 GeV. Above 50 GeV the vertex resolution was better than 3 cm.

The interactions in the CHARM detector were classified by an automatic pattern recognition program on an event-by-event basis [6]. Neutrino interactions are defined generically as events without entering charged tracks. A neutrino event is classified as a ν_μ-induced CC interaction if a muon of more than 1 GeV is originating from the event vertex. All other neutrino event candidates were classified as muonless. The classification errors on the two event classes was found to be $\pm 0.1\%$.

The event numbers of these two classes, thus obtained, have to be corrected for background and for losses. The background due to cosmic-rays was experimentally determined by using data taken outside the beam spill. Events due to the so-called wide-band background, caused by neutrinos from hadrons of undetermined charge and momentum, decaying upstream of the decay tunnel, were measured when the beam was blocked by a 1.5 m long iron absorber and subtracted. Events induced by CC ν_μ interactions in which the primary muon cannot be identified are classified as muonless. Some of these lost events have muons with an energy less than 1 GeV, others have muons leaving the sides of the detector or obscured by the hadronic shower. This is the largest correction, amounting to 3.5% of all CC ν_μ events, and the correction must therefore be made in a reliable way. In a previous publication [6] we have shown by two independent methods that the precision achieved is within 2% of the correction.

A small fraction of the muonless and therefore also of the ν_e-induced events contain a track which fulfils the requirements of a primary muon of a CC ν_μ event; these events are classified as CC ν_μ events. This background is caused by decays of pions or kaons which are part of the shower or by non-interacting, so-called punch-through, hadrons. The correction was evaluated from the data sample which has two tracks which satisfy the muon recognition criteria. We obtained at the end 106121 ± 408 muonic events and 35131 ± 274 muonless events. The details are described in a previous publication [6].

Figure 1a shows the shower energy distribution of the muonless event sample. We have chosen the variable $\sqrt{E}$ to ensure approximate independence from resolution effects. The contribution of events due to ν_e interactions produces a shoulder. The main contribution due to ν_μ neutral-current (NC) interactions is shown separately, as predicted by a Monte Carlo simulation. Figure 1b shows the radial distribution of ν_μ- and ν_e-induced muonless events.

The physics analysis, aiming at separating ν_e- and ν_μ-induced interactions on a statistical basis, exploited the differences which are apparent in Fig. 1. The muonless event sample was sorted into cells of the parameter space $[\sqrt{E}, r]$ with the dimensions of $0.5 \; (\mathrm{GeV})^{1/2} \times 20$ cm. Each generic cell has contributions from CC ν_e, NC ν_e and NC ν_μ, the respective numbers of events to be determined. The fractional contributions of CC ν_e and NC ν_e events to each cell have been determined by simulation using Monte Carlo methods. The fractional contribution of NC ν_μ events to each cell was derived from the CC ν_μ events. The Monte Carlo simulation was used in this case to correct for the different y distributions of CC ν_μ and NC ν_μ events, thus reducing the systematic errors due to the uncertainties in the beam spectra, in the detector response.

The simulation of elastic and deep-inelastic neutrino interactions was based on a quark parton model description of the nucleon and included radiative effects [6] and the different response of the calorimeter to electromagnetic and hadronic energy, in the ratio of 1.17 [5].

The y distribution of deep-inelastic neutrino scattering by NC or CC interactions was parametrized according to [7]

$$\mathrm{d}\sigma/\mathrm{d}y = A[(1 - \alpha) + \alpha \, (1 - y)^2] \, . \tag{3}$$

The coefficient α^{CC} was derived from measured y distributions of ν_μ- and $\bar{\nu}_\mu$-induced CC interactions [8], from the cross-sections and x distributions of ν_μ and $\bar{\nu}_\mu$ dimuon events [9] and from the measurements of semileptonic decays of charmed particles [10]. The coefficient α^{NC} was derived from $\alpha_{\nu_\mu}^{CC}$ and $\alpha_{\bar{\nu}_\mu}^{CC}$ and the value of $\sin^2\theta_w$ determined from leptonic ν_μ and $\bar{\nu}_\mu$ scattering [11]. We thus avoided making any assumptions on the ratio of deep inelastic ν_μ and ν_e cross-sections and found the following values

$$\alpha^{CC} = 0.11 \pm 0.04 \tag{4a}$$

$$\alpha^{NC} = 0.19 \pm 0.04 . \tag{4b}$$

Since in the expression of the NC cross-section the term proportional to the left-handed coupling constants of quarks dominates, the values of α^{CC} and α^{NC} are correlated.

The quality of the simulation has been verified by comparison with ν_μ-induced CC interactions. Forming 168 cells in the parameter space $[\sqrt{E}, r]$ the best fit gave $\chi^2/DF = 1.07$. The measured and simulated distributions projected onto $\sqrt{E}$ and r are shown in Figs. 2a and 2b. The agreement is excellent and confirms the validity of the simulation.

From the best fit of 168 cells of muonless events in the parameter space $[\sqrt{E}, r]$ we found [12]

$$N_{\nu_e}^{CC} = 2251 \pm 184 \text{ (stat.)} \pm 63 \text{ (syst.)} \qquad (E_\nu \geq 4 \text{ GeV}) \tag{5a}$$

$$N_{\nu_e}^{NC} = 575 \pm 386 \text{ (stat.)} \pm 114 \text{ (syst.)} \qquad (E_h \geq 4 \text{ GeV}). \tag{5b}$$

and a χ^2/DF of 0.9. The statistical errors were derived from the contour corresponding to one unit variation of χ^2 in varying both $N_{\nu_e}^{CC}$ and $N_{\nu_e}^{NC}$. Figures 3a and 3b show the measured and calculated best fit distributions for the projections $\sqrt{E}$ and r. We verified that the results are stable with respect to changes of the region of $[\sqrt{E}, r]$ and to variations of the size of the cells. The systematic errors in eq. (5) are due to uncertainties in the following quantities: (1) the narrow-band beam momentum, (2) the values of α^{NC} and α^{CC}, (3) the y distribution of

ν_e-induced NC events, (4) the energy calibration of the response of the calorimeter and (5) the radiative corrections.

Normalizing to the number of ν_μ-induced CC events we expected 1874 ± 76 CC ν_e events if the two cross-sections are equal. The error includes the uncertainties in the relative amounts of kaons and pions in the parent beam (3%), in the event acceptance, and in the normalization. We then obtained the ratio

$$\sigma_{\nu_e}(\text{CC})/\sigma_{\nu_\mu}(\text{CC}) = 1.20 \pm 0.11 , \tag{6}$$

where statistical and systematic errors have been combined in quadrature.

From the cross-section ratio eq. (6) we deduced a ratio of the coupling strengths to the charged weak current at $Q^2 \simeq 16$ GeV^2 of

$$g_{e\nu_e}/g_{\mu\nu_\mu} = [\sigma_{\nu_e}(\text{CC})/\sigma_{\nu_\mu}(\text{CC})]^{1/2} = 1.10 \pm 0.05 , \tag{7}$$

in agreement with the notion of universality and with previous measurements [3, 13, 14].

From eq. (5), we obtained the ratio of the NC and CC ν_e cross sections for deep inelastic scattering, with hadronic energy larger than 4 GeV:

$$R_{\nu_e} = \sigma_{\nu_e}(\text{NC})/\sigma_{\nu_e}(\text{CC}) = 0.27 \pm 0.19 . \tag{8}$$

We applied a correction to the number of CC ν_e events to account for the difference in the neutrino energy E_ν and the hadronic energy E_h. The error also takes into account the correlations between the measured values of $N_{\nu_e}^{\text{CC}}$ and $N_{\nu_e}^{\text{NC}}$. The ratio agrees, within the large errors, with $R_{\nu_\mu} = 0.3098 \pm 0.0031$ [6] as determined by this collaboration from the same set of data.

REFERENCES AND NOTES

[1]   A. N. Diddens et al. (CHARM Collab.), Nucl. Instrum. Methods 178 (1980) 27.

M. Jonker et al., Nucl. Instrum. Methods 215 (1983) 361.

[2]   A. Grant and J. M. Maugain, CERN/EF/BEAM 83-2 (1983).

[3]   J. Blietschau et al. (Gargamelle Collab.), Nucl. Phys. B133 (1978) 205.

A. M. Cnops et al., Phys. Rev. Lett. 40 (1978) 144.

H. Deden et al., Phys. Lett. 98B (1981) 310.

O. Erriquez et al., Rutherford Lab. report 81-023 (1981).

[4]   C. Bovet et al., report CERN 82-13 (1982).

[5]   M. Jonker et al. (CHARM Collab.), Nucl. Instrum. Methods 200 (1982) 183.

J. Dorenbosch et al. (CHARM Collab.), To be published in Nuclear Instruments and Methods.

[6]   J. Allaby et al. (CHARM Collab.), High precision measurement of the electroweak mixing angle from semi-leptonic neutrino scattering, to be submitted to Physics Letters.

[7]   M. Jonker et al. (CHARM Collab.), Phys. Lett. 102B (1981) 67.

[8]   J. G. H. de Groot et al., Z. Phys. C 143 (1979) 143.

M. Jonker et al. (CHARM Collab.), Phys. Lett. 109B (1982) 133.

[9]   H. Abramowicz et al., Z. Phys. C 15 (1982) 19.

[10]   J. Lee-Franzini, Proc. Europhysics Conf. on Flavour Mixing in Weak Interactions, Erice, 1984, ed. L. L. Chau (Plenum Press, New York and London, 1984), p. 217.

[11]   F. Bergsma et al. (CHARM Collab.), Phys. Lett. 147B (1984) 481.

L. A. Ahrens et al., Phys. Rev. Lett. 55 (1985) 1814.

[12]   By fixing the ratio of the NC and CC ν_e cross-sections to the value measured in muon-neutrino interactions for $E \geq 4$ GeV, $R_{\nu_\mu} = 0.3098 \pm 0.0031$ [10], the result of the fit was $N_{\nu_e}^{CC} = 2237 \pm 165$ (stat.).

[13]   E. Di Capua et al., Phys. Rev. 133B (1964) 1333.

         D. A. Bryman et al., Phys. Rev. D11 (1975) 1337.

         D. A. Bryman et al., Phys. Rev. Lett. 50 (1983) 7.

[14]   L. Di Lella, Proc. Int. Symp. on Lepton and Photon Interactions at High Energies,
         Kyoto, 1985, eds. M. Konuma and K. Takahashi (Kyoto Univ., Kyoto, 1986), p. 280.

Figure captions

Fig. 1   Experimental distributions of the muonless events as a function of $\sqrt{E}$, the square
          root of the shower energy (a), and of r, the radial vertex position of the neutrino
          interactions in the calorimeter with respect to the beam axis (b). The solid lines and
          the dashed regions represent the expected contributions of ν_μ and ν_e events. The ν_e
          events produce a shoulder in the $\sqrt{E}$ distribution.

Fig. 2   Experimental distributions of the CC ν_μ events as a function of $\sqrt{E}$ (a) and r (b) as
          defined in Fig. 1. The solid lines represent the results of a Monte Carlo simulation
          normalized to the data.

Fig. 3   Experimental distribution of the muonless events as a function of $\sqrt{E}$ (a) and r (b) as
          defined in Fig. 1. NC ν_μ events as determined by the fit have been subtracted. The
          solid lines show the expected distributions of the sum of ν_e-induced NC and CC
          events obtained from the fit. The dashed line in (a) represents the distribution of NC
          ν_e events.

a)
DATA
MC NC ν_μ
MC (CC+NC)ν_e
250
200
150
100
50
EVENTS (×10) / 0.5 (GeV)
2
4
6
8
10
12
$\sqrt{E}$ (GeV)$^{1/2}$

b)
DATA
MC NC ν_μ
MC (CC+NC)ν_e
80
60
40
20
EVENTS (×100) / 20 cm
40
80
120
160
r (cm)

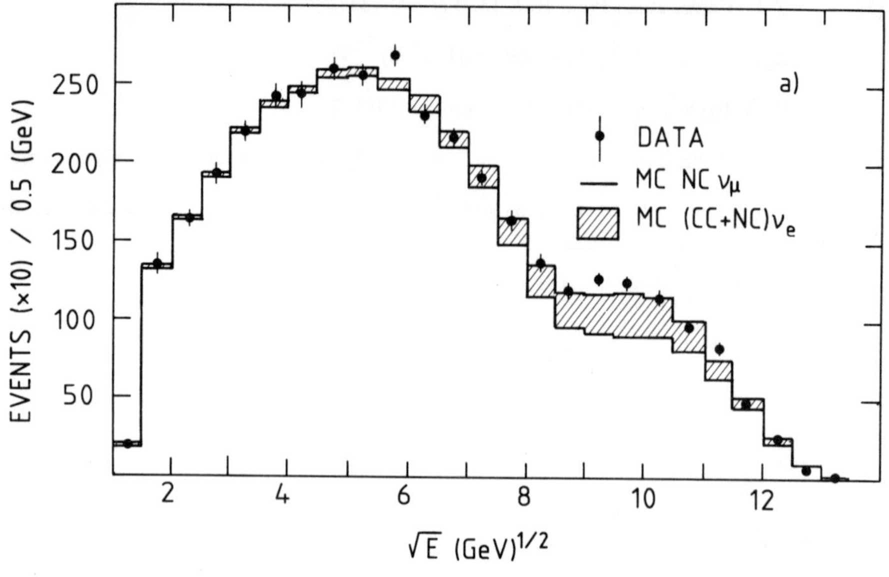

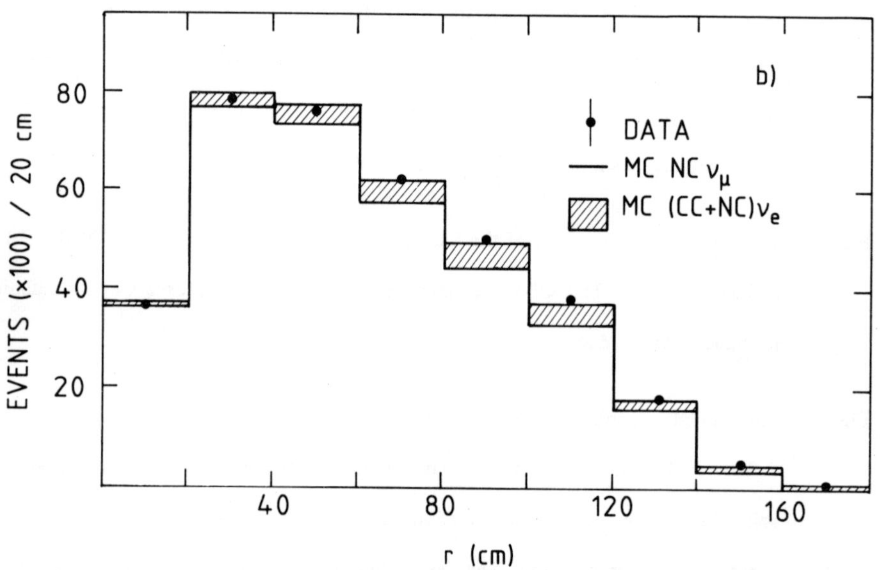

Fig. 1

C. Santoni
90
EVENTS (×100) / 0.5 (GeV)^1/2
a)
DATA
MC CCν_μ
60
30
0
2 4 6 8 10 12 14
√E (GeV)^1/2
30
EVENTS (×1000) / 20 cm
b)
DATA
MC CCν_μ
25
20
15
10
5
0 40 80 120 160
r (cm)

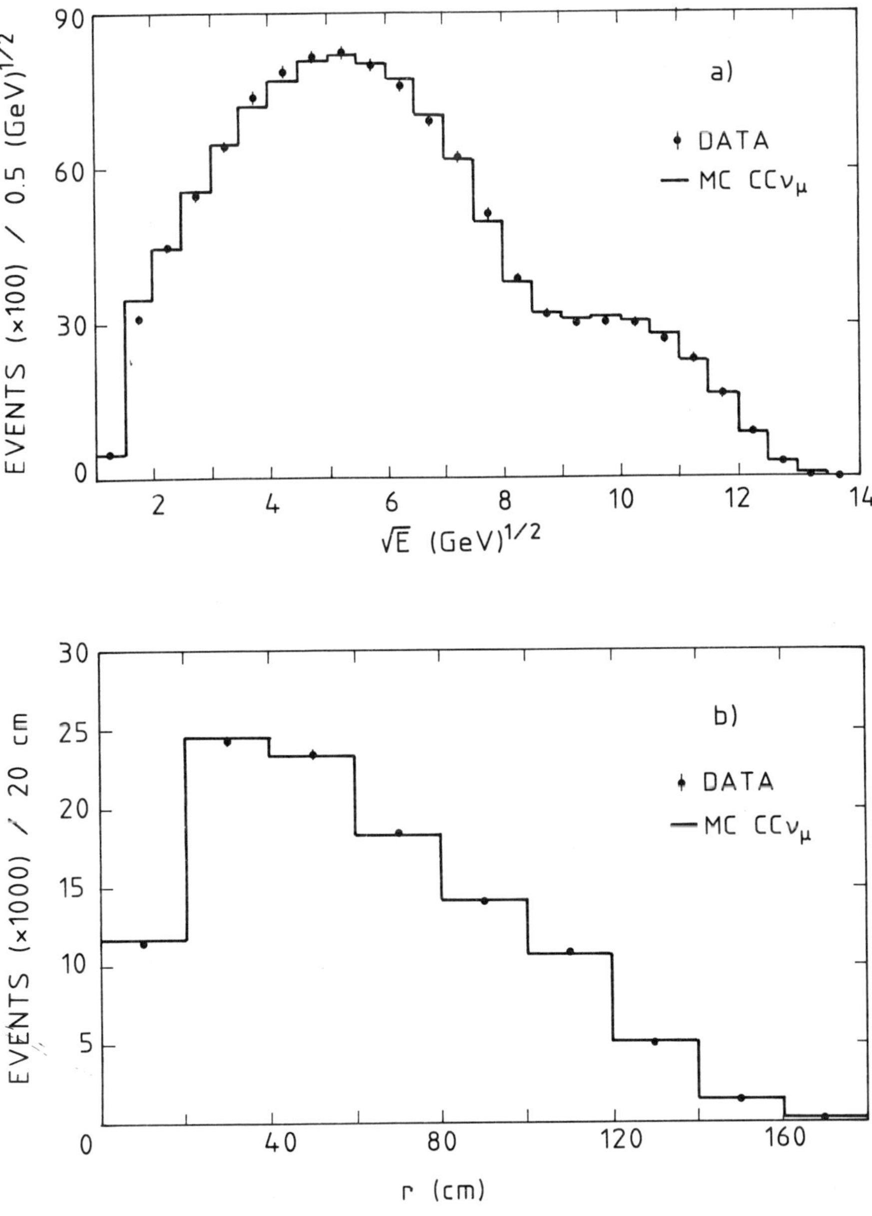

Fig. 2

500
400
300
200
100
0
EVENTS / 0.5 (GeV)^1/2
DATA
MC NC ν_e
MC (CC+NC)ν_e
a)
2 4 6 8 10 12 14
√E (GeV)^1/2

800
600
400
200
EVENTS / 20 cm
DATA
MC (CC+NC)ν_e
b)
0 40 80 120 160
r (cm)

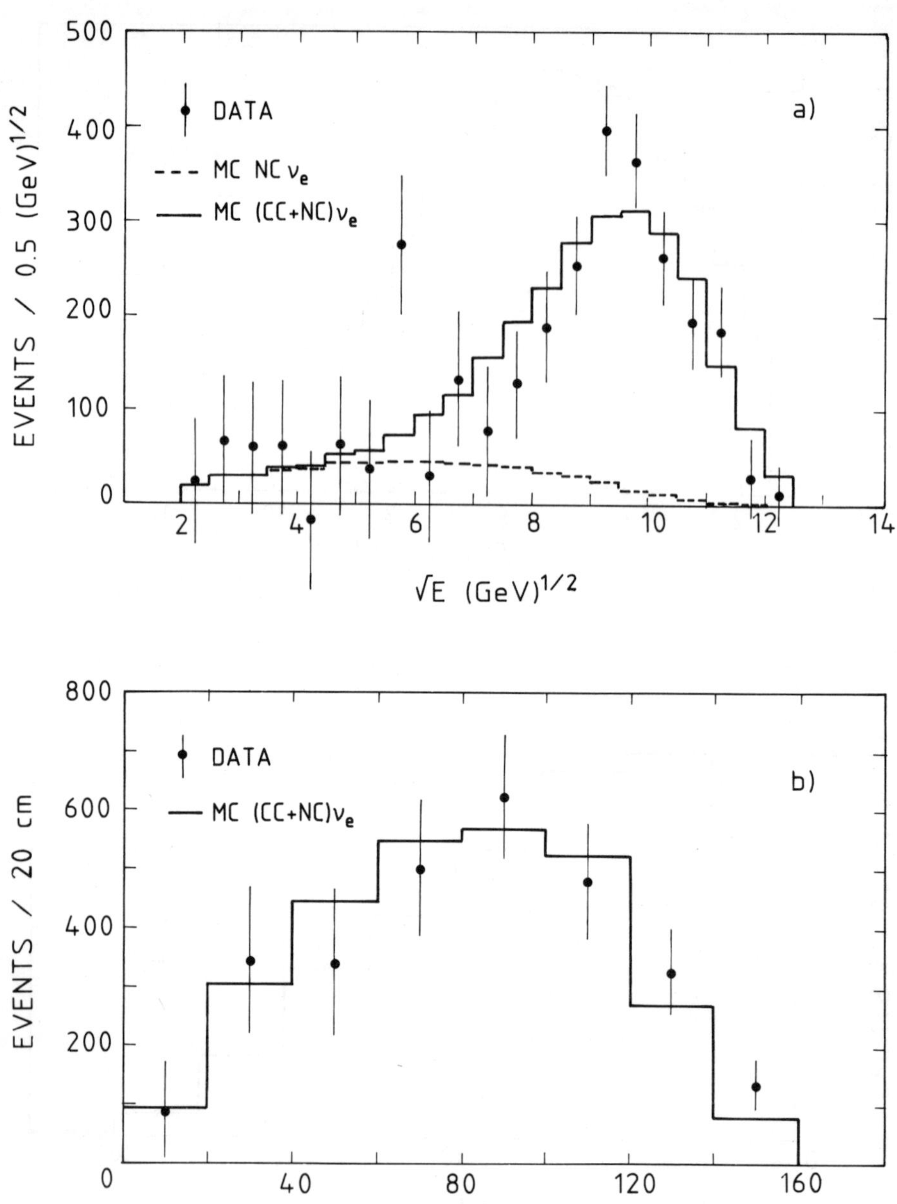

Fig. 3

$$\nu_\mu e \to \nu_\mu e \text{ and } \nu_\mu p \to \nu_\mu p$$

Exp. 734 Collaboration[1]

Presented by Y. Suzuki (Osaka University)

Abstract: Recent results on the neutrino electron and the
neutrino proton elastic scattering from Exp. 734 at BNL are
presented. The values of the $\sin^2\theta_w$ from the two measurements are in good agreement.

Since the discovery of the neutral current at CERN in
1973, tremendous experiments on the neutral current have
been carried out. All of then confirmed the standard
SU(2) × U(1) electroweak theory[2] and with the discovery
of W and Z bosons in 1983[3] the basic validity of the theory
is firmly established. It remains of interest, however, to
seek possible extensions of that model both by searching
for new phenomena and by making more precise measurements
of known phenomena. In the latter category are measurements
of the fundamental weak neutral current parameter, $\sin^2\theta_w$,
in all processes in which that parameter can be determined
accurately. The $\nu_\mu e \to \nu_\mu e$ scattering is one of the best
interactions to study those questions. In this talk we will
present recent results from an experiment on the measure-
ments of the cross sections of $\overset{(-)}{\nu}_\mu e \to \overset{(-)}{\nu}_\mu e$ and $\overset{(-)}{\nu}_\mu p \to$
$\overset{(-)}{\nu}_\mu p$ [4]. The experiments were performed at Brookhaven
National Laboratory A.G.S. with an average neutrino energy
of 1.4 GeV. The detector and the beam are described elsewhere in details[5].

$\nu_\mu e \to \nu_\mu e$ interactions are very simple theoretically.
The cross section is given directry by $\sin^2\theta_w$ and ρ,

$$\left(\frac{}{\sigma}\right) = \frac{G_F^2 m_e E_\nu}{2\pi} \left[(g_V \pm g_A)^2 + \frac{1}{3}(g_V \mp g_A)^2 \right]$$

$$g_V = \rho\left(-\frac{1}{2} + 2\sin^2\theta_W\right)$$

$$g_A = \rho\left(-\frac{1}{2}\right)$$

where g_V and g_A are the vector and the axial-vector coupling constants. $\rho = 1$ for the Glashow-Weinberg-Salam model. The radiative correction for $\sin^2\theta_W$ is very small, less than 0.001. If you take a ratio of the cross sections $\sigma(\bar{\nu}_\mu e)/\sigma(\nu_\mu e)$ which is independent of ρ, then the error on $\sin^2\theta_W$ is 1/8 of the error in the ratio around $\sin^2\theta_W = 0.22$. Therefore 10% measurement of the ratio gives the error on $\sin^2\theta_W$ of ± 0.012. However the experiment is very difficult because it has about four orders of magnitudes smaller cross section than the usual neutrino interactions on nucleons. We observe $1 \sim 2 \; \nu_\mu e$ events/day in the E734 detector. In figure 1 is shown a graph of integraged number of events observed all over the world starting from the year of 1973 when the first neutral current was observed. There are a few pioneering works around late 1970's, but the real phase transition has happened at 1982 when the two major counter experiments, CHARM at CERN and E734 at BNL became on the air. Most of the events have been collected by the two groups. The data summary of the E734 is shown in table 1. The experiment have finished on May, 1986. We have collected the data of 3.8×10^{19} Protons on Target(POT) for $\nu_\mu e$ interactions and 2.4×10^{19} POT for anti-neutrino interactions. The final analysis is in progress, we are expected to have about 190 neutrino electron events and 100 anti-neutrino events. In this report we present a result from 40% of the neutrino data and 60% of the anti-neutrino data[4]. We will also mention quickly about a possibility of measuring $d\sigma/dy$ with the higher statistic ν_μ data.

The signature of the neutrino electron scattering is very simple, which is an isolated very forward electromagnetic shower with

$$E_e \theta^2_e = 2m_e(1 - E_e/E_\nu) < 2m_e \sim 1 \text{ MeV.}$$

The emission angle is about 40 mrad for 500 MeV electron (BNL) and 14 mrad for 5 GeV electron (CERN). Backgrounds for the process are 1) neutral current π^o production (resonance/coherent), $\nu_\mu N \rightarrow \nu_\mu N'\pi^o$, where N' is missed the detection and one of the gammas from π^o decay converted in the very forward direction, 2) ν_e induced quasi-elastic interactions $\nu_e n \rightarrow e^- p$, $\bar{\nu}_e p \rightarrow e^+ n$. The distribution of γ's from the π^o decay through the reasonance production is flat in θ^2-distribution. And also the distribution through the coherent π^o production is almost flat at BNL energy. This is an advantage of the low energy experiment comparing to the high energy experiment for which the coherent π^o production is the most severe background. Since we have measured the ν_e spectrum[6] a calculation of the $e^- p$ and $e^+ n$ backgrounds are very reliable and they also distribute flat in θ^2-distribution.

The data reduction was very straight forward. We selected single forward electromagnetic showers, separated electrons and gammas by using the dE/dx measurements and made θ^2-distributions. Then the signal appears as a sharp forward peak and the background distributes in flat. In figure 2 is shown θ^2-distributions for both neutrino and anti-neutrino data. We see clean forward peaks in the distributions. From the distributions we extracted 51 $\pm$ 9 and 53 $\pm$ 10 neutrino and anti-neutrino electron scattering, respectively. The cross sections are

$$\sigma(\nu_\mu e) = 1.60 \pm 0.29(\text{stat}) \pm 0.26(\text{syst})$$

$$\sigma(\bar{\nu}_\mu e) = 1.16 \pm 0.26(\text{stat}) \pm 0.14(\text{syst}).$$

From the ratio, $\sigma(\nu_\mu e)/\sigma(\bar{\nu}_\mu e) = 1.38 \begin{smallmatrix} + 0.40 \\ - 0.31 \end{smallmatrix} \pm 0.17$, we obtained

$$\sin^2\theta_W = 0.209 \pm 0.029(\text{stat}) \pm 0.013(\text{syst})$$

and for the couplings,

$$g_V = -0.079 \pm 0.060$$

$$g_A = -0.483 \pm 0.042,$$

which is shown in figure 3. The results may be compared with the similar measurement by CHARM group at CERN[7]. They obtained $\sin^2\theta_W = 0.216 \pm 0.032(\text{stat}) \pm 0.010(\text{syst})$ from the ratio. If you perform a two parameter fit, treat ρ and $\sin^2\theta_W$ as free parameters, from the cross section measurements you get

$$\sin^2\theta_W = 0.209 \pm 0.030$$

$$\rho = 0.967 \pm 0.082.$$

The results of 68% C.L. of the two parameter space is shown in figure 4.

If we can measure the $d\sigma/dy$ of $\nu_\mu e \to \nu_\mu e$, then it will provide an additional test for the standard model. Also deviations from the standard model may be seen in $d\sigma/dy$. For example the effect of the magnetic moment of neutrinos have the y dependence of $(1-y)/y$[8], and if you see $(1-y)$ term in $d\sigma/dy$ distribution you might say if neutrinos are Dirac or Majorana[9]. We made a quick test to see a feasi-

bility of measuring $d\sigma/dy$ of $\nu_\mu e$ elastic scattering. We have
used 119 $\nu_\mu e$ data sample from 2.4×10^{19} POT. There is no way
to calculate $y = E_e/E_\nu$ directly, therefore we used a Monte
Carlo calculation to produce distributions for flat term and
$(1-y)^2$ term. We used θ^2 and E_e distributions for that purpose. The flat term and $(1-y)^2$ term were then fitted simultaneously in the θ^2 and E_e distributions and determined the
coefficiencies of the flat and $(1-y)^2$ terms. Both pure flat
term and pure $(1-y)^2$ term are exculded at 68% C.L. to reproduce the real distributions. The mixture of the two terms
was consistent with the standard model with 68% C.L.. To
obtain g_V and g_A from the total cross section, we need to
use both neutrino and anti-neutrino measurements, which is
shown figure 3, however the $d\sigma/dy$ measurement of neutrino
electron scattering alone gave you directly four separated
regions in g_V and g_A space. Unfortunately the analysis is
very preliminary, therefore there is not much more to say
about it, but the $d\sigma/dy$ measurement is very promising and
give a nice result in near future.

Now we turn to the subject of neutrino proton elastic
scattering. The cross section of the neutrino proton scattering is described by $\sin^2\theta_w$, vector form factors which is well
known and an axial vector form factor. If you assume M_A as
an input then you can get $\sin^2\theta_w$, or you can get M_A and
$\sin^2\theta_w$ simultaneously. For this analysis we also relax the
condition to look for an additional contribution to the axial
vector form factor $G_A(Q^2)$, we set

$$G_A(Q^2) = \frac{1}{2} \frac{g_A(0)}{(1+Q^2/M_A^2)^2} (1 + \eta).$$

If non-zero η was found, then there are, 1) heavy quark
contributions to the standard weak axial current, or 2) "non-
standard" primitive axial isoscalar current.

Neutral-current elastic proton scattering candidates
were single tracks fully contained within the detector and
not accompanied by any additional track. The minimum length

of an accepted track yielded up to nine measurements of dE/dx and corresponded to $Q^2 = 2MT > 0.35$ (GeV/c)^2 for elastic proton scatters. After proton candidates were selected using the dE/dx information, further criteria were applied to suppress background events exhibiting relatively large in-time extra energy depositions while maintaining high detection efficiency for the signal. The resulting sample contain backgrounds from neutrino interactions other than $\nu_\mu p \rightarrow \nu_\mu p$ and $\bar{\nu}_\mu p \rightarrow \bar{\nu}_\mu p$. The neutrino backgrounds were (a) $\nu_\mu n - \nu_\mu n$, (b) neutral-current single pion production and (c) a small contribution from the quasielastic reaction in which the muon was undetected. Also, in each sample corrections were neccessary for elastic proton scatters produced by the measured contamination of opposite helicity neutrinos in the incident beam. The backgrounds in which a μ or π^+ was observed to decay near the event vertex were directly subtracted. A Monte Carlo calculation was used to estimate the remaining backgrounds. After all subtractions, final samples of 951 $\nu_\mu p \rightarrow \nu_\mu p$ and 776 $\bar{\nu}_\mu p \rightarrow \bar{\nu}_\mu p$ events remain.

From the resultant Q^2 distribution shown in figure 5, one finds the ratio for $0.5 < Q^2 < 1.0$ (GeV/c)^2:

$$R_\nu = \frac{\sigma(\nu_\mu p \rightarrow \nu_\mu p)}{\sigma(\nu_\mu n \rightarrow \mu^- p)} = 0.153 \pm 0.007 \pm 0.017$$

$$R_{\bar{\nu}} = \frac{\sigma(\bar{\nu}_\mu p \rightarrow \bar{\nu}_\mu p)}{\sigma(\bar{\nu}_\mu p \rightarrow \mu^+ n)} = 0.218 \pm 0.008 \pm 0.023.$$

The 11% systematic uncertainty in R_ν and $R_{\bar{\nu}}$ in the equation above is the absolute scale uncertainty of each of the individual differential cross sections in figure 5. The differential cross section for $\nu_\mu p$ depends primarily on $\sin^2\theta_w$, while the differential cross section for $\bar{\nu}_\mu p$ is particulary sensitive to the axial-vector form factor, $G_A(Q^2)$. If it is assumed that $G_A(Q^2)$ is given by the axial-vector isovector dipole form factor with no corrections for heavy quark currents and no axial-vector isoscalar term, then extraction of $\sin^2\theta_W$ and M_A from $d\sigma(\nu_\mu p)/dQ^2$ and $d\sigma(\bar{\nu}_\mu p)/dQ^2$ yields $\sin^2\theta_W = 0.218 \begin{smallmatrix} + 0.039 \\ - 0.047 \end{smallmatrix}$ and $M_A = 1.06 \pm 0.05$ Gev, where the errors represent a 68% rectangular confidence area. This value of M_A is in good agreement with the present world-average value, $M_A = 1.032 \pm 0.036$ GeV.

Alternatively, if $\sin^2\theta_W$ is fixed at the value 0.22 and M_A constrained to the world-average value[11], a search may be made for additional term in $G_A(Q^2)$ due to neutral-current contributions from heavy quark currents or a "nonstandard" isoscalar axial-vector current. Writing $G_A(Q^2) = G_A^3(Q^2)(1+\eta)$, one finds $\eta = 0.12 \pm 0.07$, or equivalently, $0.00 < \eta < 0.25$ at 90%C.L.. This results for η is independent of the numerical vlaue assumed for $\sin^2\theta_W$.

To extract the most precise value of $\sin^2\theta_W$, we constrain M_A at the world average value and used the differential cross section data. The most precise value from this experiment is

$$\sin^2\theta_W = 0.220 \pm 0.016(\text{stat.}) \begin{smallmatrix} + 0.023 \\ - 0.031 \end{smallmatrix}(\text{syst}).$$

There is good agreement between the value of $\sin^2\theta_W$ from the pure leptonic interactions and from the semi-leptonic interactions. Furthermore, these values are in good agreement with the values of $\sin^2\theta_W$ determined from the masses of the W and Z bosons[12], from inelastic electron-deuteron scattering[13], and from deep-inelastic neutral-current neutrino reactions[14]. Hence, over a wide range of Q^2 and with significantly different assumptions and coorrections in the various experiments, the weak-neutral-current parameter $\sin^2\theta_W$ is within present experimental errors of about 10%, a universal constant. And after we will analyse the all of the data, we will have 190 $\nu_\mu e$ and 100 $\bar{\nu}_\mu e$ events that will give a result with $\Delta\sin^2\theta_w = \pm 0.015$ for pure leptonic neutral current interactions.

This work was supported in part by the U.S. Department of Energy, the Japanese Ministry of Education, Science and Culture through the Japan-U.S.A. Cooperative Research Project on High Energy Physics and the U.S. National Science Foundation.

References

1) E734 collaborators are K. Abe, L.A. Ahrens, K. Amako,
    S.H. Aronson, E.W. Beier, J.L. Callas, P.L. Connolly,
    D. Cutts, D.C. Doughty, L.S. Durkin, B.G. Gibbard,
    S.M. Heagy, D. Hedin, J.S. Hoftun, M. Hurley, S. Kabe,
    Y. Kurihara, R.E. Lanou, Y. Maeda, A.K. Mann, M.D. Marx,
    T. Miyachi, M.J. Murtagh, S. Murtagh, Y. Nagashima,
    F.M. Newcomer, T. Shinkawa, E. Stern, Y. Suzuki,
    S. Tatsumi, S. Terada, D.H. White, H.H. Williams,
    Y. Yamaguchi, and T. York from BNL, Brown Univ.,
    Hiroshima Univ., KEK, Osaka Univ., Pennsylvania Univ.
    and Stony Brook.
2) S. Glashow, Nucl. Phys., 22 (1961) 579; S. Weinberg,
    Phys. Rev. Lett., 19 (1967) 1264; A. Salam, in Elementa-
    ry Particle Theory, edited by N. Svartholn (Almquist &
    Wiksells, Stockholm, 1968), p.367.
3) G. Arnison et al., Phys. lett., 129B (1983) 273;
    P. Bagnaia et al., Phys. Lett., 129B (1983) 130.
4) L.A. Ahrens et al., Phys. Rev. Lett., 51 (1983) 1514;
    Phys. Rev. Lett., 54 (1985) 18; Phys. Rev. Lett., 56
    (1986) 1107.
5) L.A. Ahrens et al., Phys. Rev. D34 (1986) 75; sub-
    mitted to Nucl. Instl. and Meth.
6) L.A. Ahrens et al., Phys. Rev. D31 (1985) 2732.
7) F. Bersma et al., Nucl. Phys. B147 (1984) 481.
8) See for examples J.E. Kim, Phys. Rev. D14 (1976) 3000;

M.A.B. Beg et al., Phys. Rev. D17 (1978) 1395.

9) S.P. Rosen, PURD-TH-81-8.

10) J. Collins et al., Phys. Rev. D18 (1981) 242; R. Mohapatra et al., Phys. Rev. D19 (1979) 2165; S. Oneda et al., Phys. Lett. 88B (1979) 343; L. Wolfenstein, COO-3066-119 (1979).

11) T. Kitagaki et al., Phys. Rev. D28 (1983) 436; K.L. Miller et al., Phys. Rev. D26 (1982) 537; N.J. Baker et al., Phys. Rev. D23 (1983) 2499.

12) A. Clark, talk given in this conference.

13) C.Y. Prescott et al., Phys. Lett. 77B (1978) 347; 84B (1979) 524.

14) Talks given in the session on $\sin^2\theta_W$ at this conference.

E734   Data   Summary

	Neutrino		Anti-Neutrino	
	POT	No of $\nu_\mu e$	POT	No of $\bar{\nu}_\mu e$
1982	0.8×10^{19}	51 ± 9	---	---
1984	1.6×10^{19}	} 119 ± 11	1.8×10^{19}	53 ± 10
1986	1.4×10^{19}	70(Expected)	1.6×10^{19}	48(Expected)
Total		190(Expected)		100(Expected)

TABLE 1.

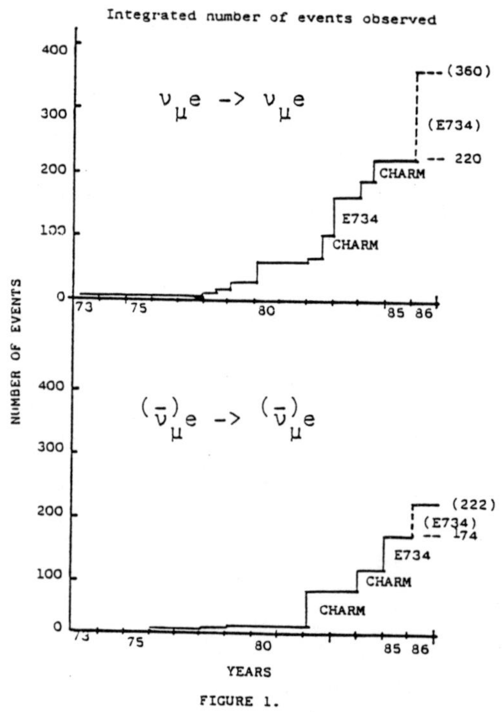

FIGURE 1.

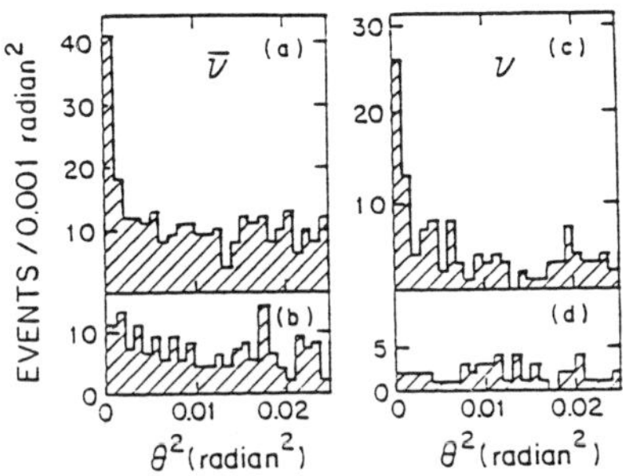

FIGURE 2. Distributions in θ_e^2 for (a) the primarily singly ionizing events from the $\bar{\nu}_\mu$ data and (b) the predominantly photon sample from the $\bar{\nu}_\mu$ data. (c),(d) The corresponding plots for the ν_μ-induced data.

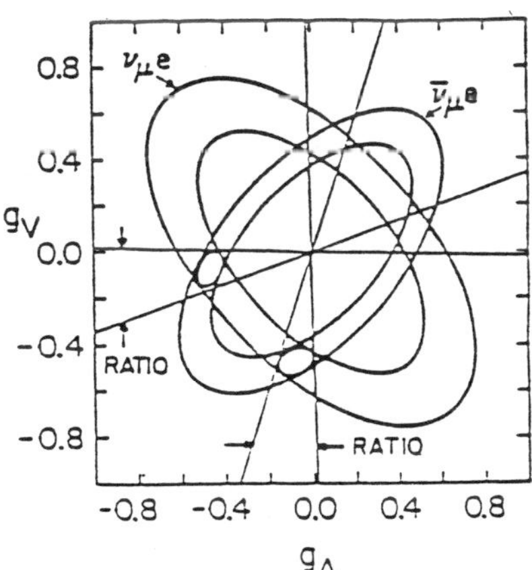

FIGURE 3. g_A versus g_V determined by $\sigma(\bar{\nu}_\mu e)$ and $\sigma(\nu_\mu e)$

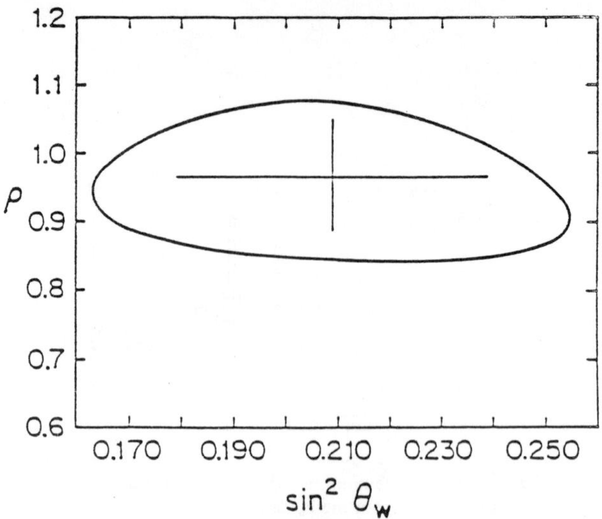

FIGURE 4 . 67% confidence level for the fit to ρ and $\sin^2 \theta_W$ from $\sigma(\bar{\nu}_\mu e)$ and $\sigma(\nu_\mu e)$

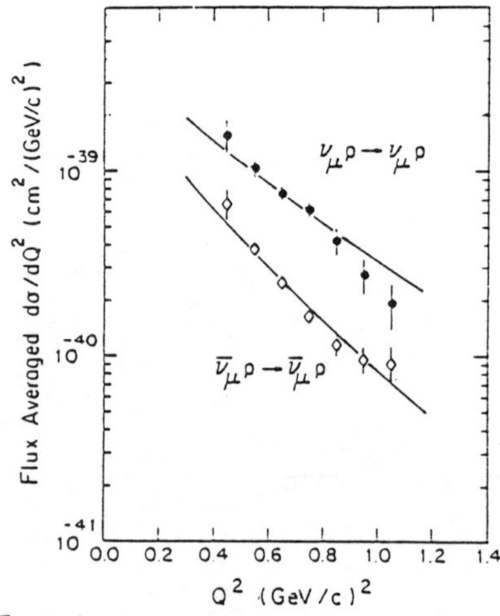

Figure 5 . The data points are the measured flux-averaged differential cross sections for $\nu_\mu p \rightarrow \nu_\mu p$ and $\bar{\nu}_\mu p \rightarrow \bar{\nu}_\mu p$ from this experiment. The solid curves are best fits to the combined data with the values $M_A = 1.06$ GeV and $\sin^2\theta_W = 0.220$. This fitting procedure imposes adjustment of the solid curves by scale factors of 1.05 for $\nu_\mu p$ and 1.09 for $\bar{\nu}_\mu p$, consistent with the absolute scale uncertainty of approximately 11% in each of the individual cross sections which was included in the fitting procedure

A PRECISE MEASUREMENT OF $\sin^2\theta_W$ USING THE FERMILAB NARROW BAND NEUTRINO BEAM

P. G. Reutens, D. Garfinkle, F. S. Merritt, M. J. Oreglia
Enrico Fermi Institute and Physics Department
University of Chicago, Chicago, Illinois 60637

P. Auchincloss, R. E. Blair, C. Haber, S. R. Mishra,
E. Oltman, M. Ruiz, F. J. Sciulli, M. H. Shaevitz, W. H. Smith, R. Zhu
Columbia University, New York, New York 10027

R. Coleman, H.E. Fisk, B. Jin, T. Kondo, D. Levinthal,
W. Marsh, P.A. Rapidis, H.B. White, D. Yovanovitch
Fermi National Accelerator Laboratory, Batavia, Illinois 60510

A. Bodek, F. Borcherding,
N. Giokaris, K. Lang, I. E. Stockdale
University of Rochester, Rochester, Rochester, New York 14627

Talk presented by F. S. Merritt

ABSTRACT

We report a measurement of $\sin^2\theta_W$ based on the ratios of NC to CC events measured in a narrow-band beam at energies of 30–240 GeV. The data are fully corrected for radiative effects, heavy-quark production, and other effects. The best value obtained, $\sin^2\theta_W = .239 \pm .010$, is consistent with the most recent values from W and Z production.

INTRODUCTION

The CCFR group has completed a measurement of $\sin^2\theta_W$ using the ratio of neutral to charged-current deep inelastic neutrino interactions. The data were taken in two runs in the narrow band neutrino beam at Fermilab. Results have already been published[1] for the first of these runs. The analysis presented here combines both runs and more than doubles the previous event sample.

$\text{Sin}^2\theta_W$ is determined from the ratios of NC to CC events using the equations[2]

$$R^\nu = \sigma^\nu_{nc}/\sigma^\nu_{cc} = \rho^2 \left(\frac{1}{2} - x + \frac{5}{9}x^2(1+r) \right) \qquad (1-a)$$

$$R^{\bar\nu} = \sigma^{\bar\nu}_{nc}/\sigma^{\bar\nu}_{cc} = \rho^2 \left(\frac{1}{2} - x + \frac{5}{9}x^2(1+\frac{1}{r}) \right) \qquad (1-b)$$

where $\bar{x} = \sin^2\theta_W$ and $r = \sigma^{\bar\nu}_{cc}/\sigma^\nu_{cc}$. These equations are valid even after cuts are applied in $y = E_{had}/E_\nu$, provided that R^ν, $R^{\bar\nu}$, and r are all evaluated within the same cuts. However the equations are exact only for a somewhat idealized target, and must be corrected for the non-zero charmed-quark mass, radiative effects, and the non-isoscalar target.

DETECTOR AND BEAM

The narrow band neutrino beam was produced from a high-energy momentum-selected π and K beam, with $\mu\nu$ decays occurring in a 350 m decay pipe. This produced a neutrino spectrum dominated by two bands, a high-energy band from K-decay ($E^{max}_\nu = .95E_{beam}$) and a lower-energy band from π-decay ($E^{max}_\nu = .43E_{beam}$) (see Figure 1). Due to two-body kinematics, the neutrino energy within each band decreased as a function of increasing decay angle and therefore of increasing radius r measured from the beam center in the detector (see Figure 2). This two-band structure and the ability to determine $E_\nu(r)$ from the radius of the event vertex were fully exploited in the neutral current analysis.

Interactions were observed in a 305 cm x 305 cm steel target-calorimeter[3], instrumented with 305 cm x 305 cm scintillation counters and spark chambers, with a counter after every 10 cm of steel and a chamber every 20 cm. The calorimeter was followed by a magnetized toroidal spectrometer used to measure E_μ for charged-current studies. The detector produced measured resolutions of $\delta E_{had}/E_{had} = .89/\sqrt{E_{had}}$ and $\delta E_\mu/E_\mu = .11$.

EVENT SELECTION AND ANALYSIS

From equations (1), $R^{\bar\nu}$ is insensitive to $\sin^2\theta_W$ (for $\rho = 1$ and $\sin^2\theta_W \approx .24$), so the measurement of $R^{\bar\nu}$ has little effect on the determination of $\sin^2\theta_W$. The important ratio is R^ν; an error of $\pm 1\%$ in R^ν produces an error of $\mp .005$

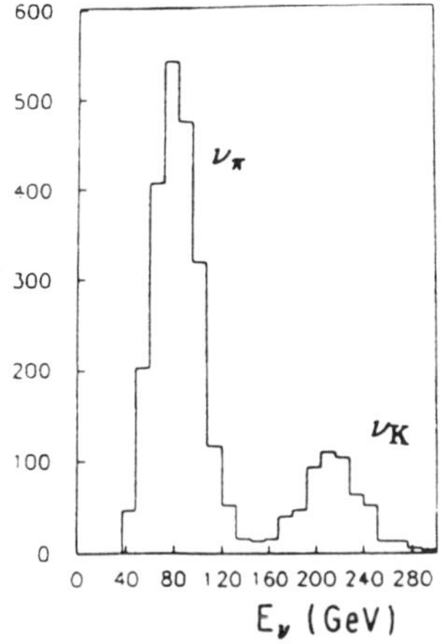

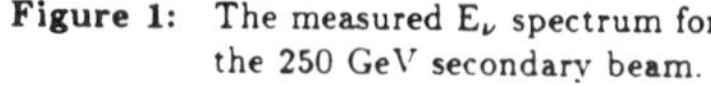

Figure 1: The measured E_ν spectrum for the 250 GeV secondary beam.

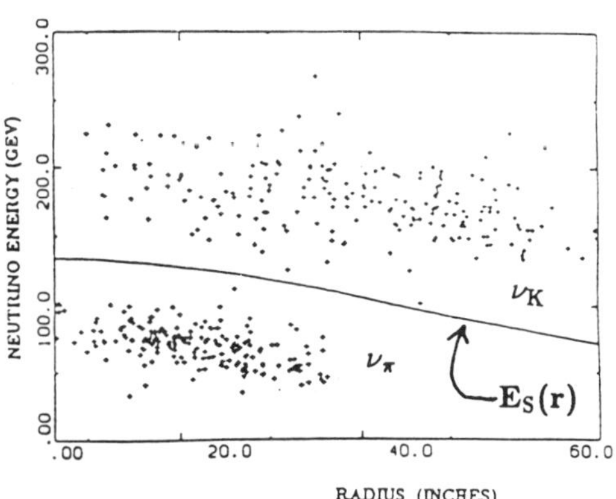

Figure 2: The measured E_ν spectrum versus radius of event vertex at 250 GeV (only part of the data sample is shown).

in $\sin^2\theta_W$. A precise measurement of R^ν is required, and so it is imperative to minimize any possible bias between NC and CC events, both in event selection and event analysis.

We have therefore used only events which trigger on total hadron energy (no final-state muon required), and have applied a software cut of 20 Gev to all events to ensure 100% trigger efficiency. The only quantities used in the analysis are those which we believe are measured equally well for neutral and charged-current events:

1) Longitudinal vertex position: the position of the most upstream counter in the hadron shower.

2) Transverse vertex position: the centroid of sparks in the 6 chambers immediately downstream of the vertex. Muon tracking was not used at all.

3) E_{had}: measured by summing the 20 counters immediately downstream of the vertex. CC events are corrected for the energy deposition of the muon.

4) Length: the number of counters traversed by the most penetrating charged particle (showing $\geq$ .5 minimum ionizing signal). This is the variable used to

distinguish NC events from CC events. For NC events, it is the length of the hadron shower; for CC events it is normally much larger due to the penetrating final-state muon.

In order to utilize the band structure of the neutrino beam, the measured E_ν spectrum of fully reconstructed CC events was used to determine, for each beam setting, a curve $E_s(r)$ lying in the valley between the π and K bands in the E_ν vs. r plane (see Figure 2). Since only E_{had} (not E_ν) could be used in the neutral current analysis, all interactions were classified as "π-band" if $E_{had} < E_s(r)$, and as "K-band" if $E_{had} > E_s(r)$. This left a background of ν_K events with low E_{had} in the π band, but since the ν_π events were far more numerous, the background was small and calculable. This procedure made it possible to measure $R^{\nu,\bar{\nu}}$ for each band and each beam setting.

DETERMINATION OF R^ν

NC and CC events were distinguished in each band on the basis of length. If all events at a given beam setting are histogrammed as a function of length, the neutral currents appear as a peak in the short-length region ($L \leq 21$ counters), while the CC events form a plateau extending up to very long length (see Figure 3-a). But several corrections must be made to these two regions in order to extract the true NC and CC signals.

Three of these corrections were measured (see Table I(a)). Cosmic ray events, which contributed mainly to the short-length region, were measured in a 10-msec gate (off beam spill) for each beam cycle throughout the run. Wide band background events, produced by ν's from π and K decays before momentum selection, were measured by running with the momentum slit closed at each beam setting. The probability that a hadron shower extended beyond the 21-counter cut was measured in a hadronic test beam at energies of 50–200 GeV. The probability of this hadronic "punch-through" was measured to be proportional to E_{had}; integrated over the measured E_{had} spectrum of our short-length events, it amounted to 0.7% of the total short-length event sample. Corrections for each of these measured backgrounds were made separately to the short- and long-length regions in each energy band and for each beam setting.

A more serious background came from CC events with very low-energy muons which either ranged out or left the sides of the calorimeter with L ≤ 21. These had to be extrapolated from the long-length CC distribution using a Monte Carlo, and both the short- and long-length regions had to be corrected for this effect. If no further cuts had been applied to the data, this would be the largest background and the largest source of experimental error in the analysis.

But both classes of short-length CC events — wide-angle muons and range-outs — necessarily had low-energy muons and therefore high-energy hadron showers. Since the maximum muon production angle increases with increasing E_{had}, almost all of this background could be eliminated by rejecting all events, both NC and CC, which had $E_{had} > E_{had}^{cut}(r)$. $E_{had}^{cut}(r)$ was defined so that an event with the expected energy $E_\nu(r)$ at radius r and with $E_{had} > E_{had}^{cut}(r)$ would necessarily have length > 21 counters.

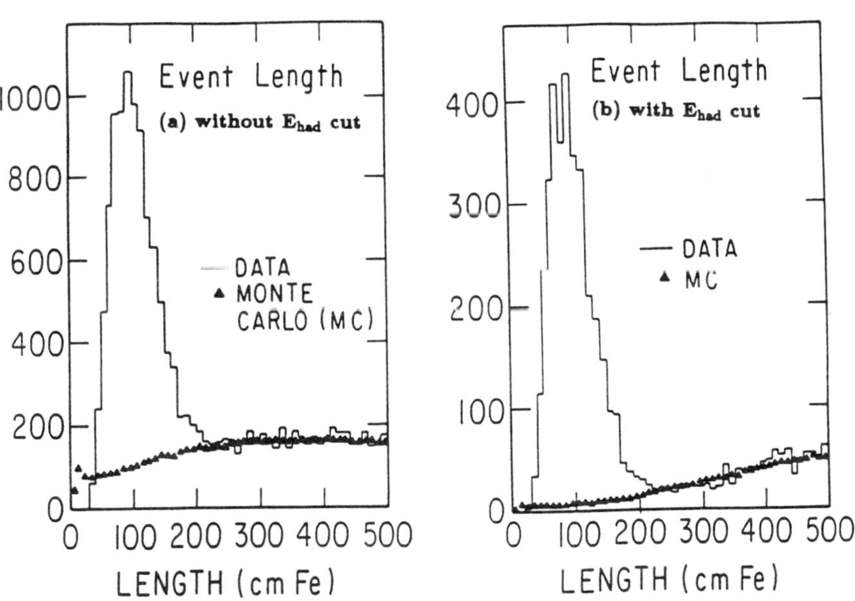

Figure 3: Event length for all events; (a) without the cut $E_{had} < E_{had}^{cut}(r)$, and (b) with the cut.

Table I: Corrections made to the short-length region ($L \leq 21$ counters). The hadron energy cut $E_{had} \leq E_{had}^{cut}(r)$ has been applied.

(a) Measured Corrections		(b) Calculated Backgrounds	
Cosmic rays: (from 10-msec CR gate)	-2.8%	ν_μ from $K_{\mu 3}$ decays: ν_e from K_{e3} decays:	-0.6% -6.6%
Wide band background: (measured with slit closed)	-1.8%	ν_π in K-band: ν_K in π-band:	-0.7% -8.9%
Hadronic punch-thru: (from test beam data)	$+0.7\%$	CC interactions with $L \leq 21$	-5.0%

The cut in E_{had} corresponded roughly to a cut in $y(= E_{had}/E_\nu(r)) \geq .7$, but the exact effect of the cut depended on beam setting and radius r of the event vertex. The data sample was reduced by about a factor of 2 by this cut, but the background was reduced by a factor of 10 (see Figure 3-b). The remaining CC background in the short-length region was due to the finite momentum spread and dispersion of the secondary beam, and to the finite resolution in measured E_{had} and vertex position, and this was calculated using the Monte Carlo. The subtraction was normalized to the events which had $L > 21$ and which ranged out or left through the sides of the calorimeter, since these events were kinematically most similar to the background events.

We estimate the error in the charged-current subtraction to be 3–5% of the subtraction at all energies. In estimating this, we have compared different normalization methods and have also tried varying the structure function parameters, beam profiles, and beam energies. The error includes an uncertainty in the length measurement due to back-scattered hadronic fragments.

The Monte Carlo program was also used to calculate backgrounds of neutrinos from $K_{\mu 3}$ and K_{e3}. The latter were important since ν_e interactions (both CC and NC) could not be experimentally distinguished from neutral current interactions of ν_μ interactions. For both of these, the subtraction was normalized to the charged-current ν_K rates; flux uncertainties do not affect any of these subtractions. Corrections were also made for the small spill-over of ν_π into the K-band, and the larger contamination of ν_K events with low E_{had} in the π-band. The corrections are summarized in Table I(b).

All of these corrections were made to both energy bands at each beam setting. A weighted average of all the R_ν values gave:

$$R^\nu = .297 \pm .005 \pm .004 \qquad \text{with } E_{had} \leq E_{had}^{cut}(r) \qquad (2\text{-a})$$

and

$$R^\nu = .292 \pm .004 \pm .008 \qquad \text{with no high-}E_{had} \text{ cut} \qquad (2\text{-b})$$

where the first error is statistical and the second is systematic.

The experimental systematic errors are summarized in Table II. Note that if no high-E_{had} cut is made, the systematic error in the CC subtraction is dominant. Otherwise, the systematic errors are in all cases less than half of the statistical error.

EXTRACTION OF $\text{SIN}^2\Theta_W$

Since equation (1) refers to an idealized target and does not include radiative effects, several corrections to equation (1) were required before $\sin^2\theta_W$ could be extracted. We introduced these corrections as multiplicative factors for each of the R^ν and $R^{\bar\nu}$ values in the equations, but for simplicity we will only quote the effects on the final value of $\sin^2\theta_W$.

For example, if we were to start with the ideal target of equation (1) and first

Table II: Experimental systematic errors in $\sin^2\theta_W$ (with and without the cut $E_{had} < E_{had}^{cut}(r)$)

Source	with cut	w/o cut
Short-length CC background:	±.003	±.010
Mean π, K momentum:	±.002	±.002
Momentum spread $\Delta p/p$:	±.002	±.001
Beam angular dispersion:	±.003	±.001
E_{had} calibration:	±.002	±.003
Miscellaneous:	±.003	±.002
Total	±.006	±.011

turn on the strange sea, then R^ν will be decreased by 2%. To correct for this reduction, we must increase our measured value of R^ν by 2%, and this decreases $\sin^2\theta_W$ by .010. When we raise the mass of the charmed quark m_c from 0.0 to 1.5 GeV, the charged-current production of c-quarks is suppressed and R^ν is increased; to correct for this we must decrease our measured R^ν and consequently we increase $\sin^2\theta_W$.

The effect of each of these corrections is given in table III. The corrections should be very similar for all deep-inelastic scattering experiments, although there will be differences due to different target materials, beam spectra, and cuts in E_{had}.

In correcting for the strange sea, we have assumed[4] that strange quarks have the same x distribution as $\bar{d}$-quarks, but that $s/\bar{d} = .5$, i.e, the sea is half SU3 symmetric. In simulating charmed quark production, we have assumed a fixed quark mass of 1.5 GeV and have used the slow-rescaling formalism[5]. For the non-isoscalar correction, we have used a calculated neutron excess of $(n-p)/(n+p) = .071$ for our target (mainly Fe).

The electromagnetic radiative correction, which takes into account the effect of photon production at the lepton vertex, was made using the analysis and formulas of ref [6]; this same correction was taken into account in extracting the structure functions used in this analysis. For the electroweak correction required by the definition $\sin^2\theta_W = 1 - M_W^2/M_Z^2$, we have used the analysis of Sirlin and

Table III: Theoretical corrections to $\sin^2\theta_W$ with and w/o the cut $E_{had} < E_{had}^{cut}$

Source	with cut	w/o cut
Strange sea ($s/\bar{d} = .5$):	−.010	−.011
Charmed quark mass: ($m_c = 1.5$ GeV)	+.015	+.014
Non-isoscalar correction:	−.009	−.009
Electromagnetic radiative correction (ref. [6]):	−.002	−.004
Electroweak radiative correction (ref. [7]):	−.009	−.009
Total	−.015	−.019

Table IV: Theoretical sources of error in $\sin^2\theta_W$

Source	$\delta\sin^2\theta_W$
q, $\bar{q}$ sea: (s and c content)	±.002
K-M mixing angles:	±.002
Radiative corrections: (E&M and electroweak)	±.003
m_c and slow rescaling: ($m_c = 1.5 \pm .4 GeV/c^2$)	±.005
Total	±.006

Marciano[7]. The dominant effect is due to the Wγ box diagram, which changes the relative scales of the CC and NC interactions by a factor of .983 — i.e., ρ^2 in equation (1) is replaced by $.983\rho^2$. The combined effect of the electromagnetic and electroweak corrections is to lower $\sin^2\theta_W$ by .013. After discussions[8] with several theorists about this method, we estimate the total uncertainty in $\sin^2\theta_W$ from the combined radiative corrections to be less than .003.

The largest and most uncertain of the theoretical corrections, given in Table IV, is the one for the charmed quark mass. Some authors[9] have used an effective mass as low as 1.1 GeV, or one that depends on Q^2, and indeed the whole slow-rescaling picture may not be correct in detail. We have assumed a fixed mass of 1.5 GeV with a large error of $\pm$ .4 GeV to include all of these uncertainties. This is the dominant theoretical error in this experiment, as well as in all other deep-inelastic neutrino measurements of $\sin^2\theta_W$.

RESULTS

Within the context of the standard model, $\rho^2 = 1$ and the neutral current interaction depends on the single parameter $\sin^2\theta_W$. Our best determination of $\sin^2\theta_W$ was obtained from fitting $\sin^2\theta_W$ to equation (1), using all beam settings and energy bands, and rejecting events with large E_{had}. This gave

$$\boxed{\sin^2\theta_W = .239 \pm .008 \pm .006 \ (\pm.006) \qquad \text{with } E_{had} \leq E_{had}^{cut}(r)} \qquad (3)$$

The first error is due to experimental statistics, the second is experimental systematics, and the third is the theoretical error. The same analysis but without the cut at large E_{had} gives almost the same result, but with a larger systematic error:

$$\sin^2\theta_W = .238 \pm .006 \pm .011 \ (\pm.006) \qquad \text{with no cut at large } E_{had} \qquad (4)$$

These values are in good agreement with the results from UA2 presented at this conference, as well as with the other neutrino experiments.

As a test of the standard model, we have simultaneously fitted both ρ and

$\sin^2\theta_W$ using equations (1), obtaining

$$\sin^2\theta_W = .248 \pm .025 \qquad \text{and} \qquad \rho = 1.005 \pm .023 \qquad (5)$$

The 1, 2, and 3 sigma contours of this fit are shown in Figure 4. This agrees with the standard model, which predicts $\rho = 1$.

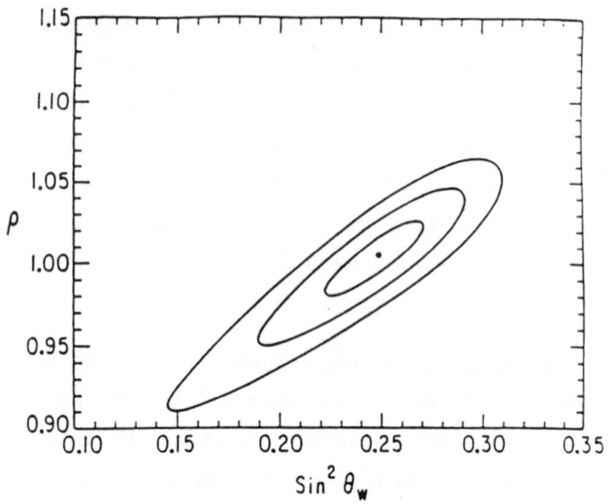

Figure 4: The 1, 2, and 3 sigma contours from a two-parameter fit of ρ and $\sin^2\theta_W$ to the corrected R_ν and R_ρ ratios, requiring $E_{had} < E_{had}^{cut}$

REFERENCES

1. P. G. Reutens et al., Physics Letters 152B, 404 (1985).
2. C. H. Llewellyn Smith, Nucl Phys B228, 205 (1983).
3. P. G. Reutens, Ph.D. Thesis, University of Chicago (1986).
4. K. Lang et al., Nevis Preprint #R1350 (1986); submitted to Zeit. fur Physik C.
5. R. M. Barnett, Phys. Rev. Lett. 36, 1163 (1976).
6. A. de Rújula, R. Petronzio and A. Savoy-Navarro, Nuclear Physics B154, 394 (1979).
7. A. Sirlin and W. J. Marciano, Nucl. Phys. B189, 442 (1981).
8. Private communications: A. DeRujula, A. Sirlin, E. Paschos.
9. H. Abramowicz et al., Zeit. fur Physik C28, 51 (1985).

PRECISION MEASUREMENT OF $\sin^2\theta_W$ FROM SEMILEPTONIC NEUTRINO SCATTERING

C. GUYOT

DPhPE, CEN-Saclay, 91191 Gif-sur-Yvette Cedex

for the CERN-Dortmund-Heidelberg-Saclay-Warsaw Collaboration

There is a considerable interest in improving the measurement of $\sin^2\theta_W$ with "low" energy neutrino experiments : a comparison with the measured heavy boson masses at the $\bar{p}p$ collider may allow for a test of electroweak radiative corrections [1]. In addition its value may be predicted by models of Grand Unification [2].

The present best value comes from the measurement of the ratio of the numbers of neutral current events to charged current events :

$$R_\nu = \left[\frac{NC}{CC}\right]_\nu = \frac{\nu + N \to \nu + X}{\nu + N \to \mu + X} \text{ for } E_{Had} > E_{H \text{ cut}}$$

by the CDHS Collaboration [3] which gives :

$$\sin^2\theta_W = 0.226 \pm 0.012 \text{ (experimental error)}.$$

In a dedicated run taken in 1984, the same collaboration has performed a new measurement with the following improvements :

i) The statistics have been increased by a factor 4 using a high flux 160 GeV Narrow Band beam.

ii) The background from neutrinos produced in the decay of pions or kaons upstream of the decay tunnel (the so called WBB background) was under better control. This was the source of the largest systematic uncertainty in the previous measurement.

iii) The CDHS detector has been upgraded in order to have a better localisation of muonless events.

The layout of the 160 GeV narrow band beam is shown on Fig.1. The contribution from the WBB background neutrinos is now fully measured by taking data with the 1.5 m iron dump moved in the hadron beam. A total of 6.10^{18} protons of 450 GeV/c momentum has been delivered on target by the CERN SPS. Eighty per cent of the data were taken with neutrinos and twenty per cent with antineutrinos.

The CDHS detector consists of 21 magnetized iron toroids of 1.875m radius instrumented with scintillators and drift chambers [14] (Fig. 2). For the present experiment the first ten modules were replaced by new ones, with 15 cm wide scintillator strips alternating in the horizontal and vertical direction (Fig. 3). This configuration allowed the determination of the transverse position of hadron showers to a precision of $\pm$ 5 cm. The detector was triggered on a local energy deposition (shower trigger) or, independently, on a particle penetrating at least 3 modules (muon trigger). To avoid any bias between NC and CC events, the event samples were defined from calorimetric and topological informations only, with no reference to the muon reconstruction. For each event, the event length L, defined as the longitudinal distance between the vertex and the last scintillator hit in the event, and the shower energy E_{sho}, from the pulse height recorded in the first 1.5 m of iron after the vertex, were measured.

For this analysis, only events from a shower trigger were retained. In order to ensure a full trigger and software efficiency for detecting showers, a cut E_{sho} > 10 GeV was performed.[1] A radius cut of 1.30m reduces the amount of events with a muon leaving at the side of the detector, with the help of the toroidal magnetic field which focuses muons from neutrino interactions. Thus, CC events with a penetrating

---

[1] The efficiency of the shower trigger for E_{sho} > 10 GeV was measured to be better than 99.9 %, using events taken with a muon trigger.

muon appear most of the time as long events and NC events appear as short showers as shown by the event length distribution (Fig. 4).

NC candidates are defined as events shorter than L_{cut} = 75. + 38. Ln E_{sho}(GeV) cm, choosen as a compromise which minimizes the amount of short CC events (L < L_{cut}) while reducing the number of long showers (L> L_{cut}) to less than 0.5 %.

For NC events, the measured shower energy E_{sho} is identical to the hadron energy E_H. For CC events, E_{sho} also contains the pulse height deposited by the muon in the first 1.5 m of iron. This pulse height, equivalent to typically 3 GeV of hadronic energy, is measured from isolated muon tracks with a precision of ± 5 %. The resulting uncertainty in the E_H cut for CC events induces a systematic error of ±0.3 % on R_ν.

After the small subtraction of cosmic ray events, using out-of-spill data, and WBB background events, several corrections have to be applied to the raw numbers of NC and CC candidates (see table 1).

The small correction for true NC events longer the L_{cut} comes from long showers and NC candidates events with a muon from the decay of a hadron in the shower. The latter correction, mostly from charm decay in short CC events, can be constrained by dimuon data.

The largest correction is the subtraction of the short CC events from the NC candidates and their addition to the CC events. As the length of CC candidates is given by the muon, this correction can be calculated by a Monte Carlo simulation of the muon length distribution of all CC events. To be almost independent of the understanding of the y distribution (y = 1 - E_μ/E_ν), the Monte Carlo is normalised to a monitor region (1.4 < L/L_{cut} < 2.4). Owing to the magnetic field and the radial cut, only 7 % of the short CC and the monitor events leaves at the side which makes possible this extrapolation from a region with < y> = 0.85 to the short CC region where < y > = 0.94.

In addition the Monte Carlo properly reproduces the data in the monitor region as shown on Fig. 5. The table 2 gives the systematic uncertainties to the correction when the beam parameters, the detector response and the physics inputs are varied within their known limits. The main uncertainty originates from an additive error on the measurement of the event length estimated to be ± 1.5 cm.

The events numbers must finally be corrected for CC events produced by electron-neutrinos from K_{e3} decay. These events are included in the NC sample since the final-state electron is hidden in the hadron shower. The absolute rate of CC ν_e events is determined from a Monte Carlo simulation, normalized to the number of fully reconstructed CC ν_μ events from $K_{\mu2}$ decay. The CC ν_e events are subtracted from the NC sample, and those with $E_H > 10$ GeV are added to the CC sample.

After all these corrections the NC to CC cross-section ratio for neutrino interactions in iron, with $E_H > 10$ GeV, is :

$$R_\nu = 0.3072 \pm 0.0025 \text{ (stat.)} \pm 0.0020 \text{ (syst.)}.$$

The determination of $\sin^2\theta_w$ in the standard model makes use of the following equation which relates CC and NC cross sections on an isoscalar target in a world with only $\overset{(-)}{u}$ and $\overset{(-)}{d}$ quarks [1] :

$$\frac{d^2\sigma_\nu^{NC}}{dxdy} = \left[\frac{1}{2} - \sin^2\theta_w + \frac{5}{9}\sin^4\theta_w\right]\frac{d^2\sigma_\nu^{CC}}{dxdy} + \frac{5}{9}\sin^4\theta_w \frac{d^2\sigma_{\bar\nu}^{CC}}{dxdy}$$

which gives, after integration over x and $E_H > 10$ GeV :

$$R_\nu = \frac{1}{2} - \sin^2\theta_w + \frac{5}{9}\sin^4\theta_w \ (1 + r)$$

where $r = \dfrac{CC^{\bar\nu}}{CC^{\nu}}$ is the ratio of the numbers of CC events which one

would get in this ideal world with identical neutrino and antineutrino fluxes.

To be applied to the real world, this relation needs corrections which have to be calculated in the Quark-Parton Model :

i) The non-isoscalarity of the iron target.

ii) The existence of a non zero strange sea together with the Kobayashi-Maskawa mixing and the threshold for charm production.

iii) The difference between the measured $r = 0.30 \pm 0.01$ and its value 0.39 in the ideal world described above.

The parameters of the Quark-Parton Model are given in table 3. The KM matrix is assumed to be unitary. Then the main uncertainty comes the charm threshold which is described by a "slow rescaling" model [7] with the charm mass m_c as a free parameter. We have choosen to give $\sin^2\theta_w$ in the on-shell scheme of Sirlin [8] where it is defined by :

$$\sin^2\theta_w = 1 - \frac{M_W^2}{M_Z^2}$$

The radiative corrections were calculated according to ref. [9]. As it is difficult to assign an error on m_c, $\sin^2\theta_w$ is given as a function of m_c with $m_c = 1.5$ GeV/c^2 as the central value :

$$\sin^2\theta_w = 0.225 \pm 0.005 \text{ (exp.)} \pm 0.003 \text{ (theo.)} + 0.013 \ (m_c - 1.5).$$

The experimental error results from the statistical and systematic errors on R_ν and r added in quadrature. The theoretical error excludes the uncertainty on m_c. With $m_c = (1.5 \pm 0.3)$ GeV/c^2 it increases to ± 0.005.

The value of $\sin^2\theta_w$ obtained in this experiment is compatible with earlier results obtained by this group [3]. It is also compatible with recent measurements from other semileptonic neutrino experiments [10, 11, 12], as well as with recent determinations of $\sin^2\theta_w$ from the W mass [13, 14] (see Fig. 6). The good agreement between the values of $\sin^2\theta_w$

obtained from neutrino scattering and from the W mass holds only if electroweak radiative corrections are applied. Without these corrections the values $\sin^2\theta_W = 0.211 \pm 0.008$ from the W mass and $\sin^2\theta_W = 0.236 \pm 0.007$ from this experiment would differ by more than two standard deviations.

References :

[1] C.H. Llewellyn Smith, Nucl. Phys. B228 (1983) 205.

[2] M.J. Marciano and A. Sirlin, Phys. Rev. D29 (1984) 945.

[3] H. Abramowicz et al., Z. Phys. C28 (1985) 51.

[4] A. Blondel, Proc. Int. Europhysics Conf. on High-Energy Physics,
     Bari, 1985 (European Physical Society, Geneva, 1985) p. 225.

[5] M. Holder et al., Nucl. Instrum. Methods 148 (1978) 235.

[6] H. Abramowicz et al., to be published in Phys. Rev. Lett.

[7] R.M. Barnett, Phys. Rev. D14 (1976) 70 ;
     H. Georgi and H.D. Politzer, Phys. Rev. D14 (1976) 1829.

[8] A. Sirlin, Phys. Rev. D22 (1980) 97 ;
     R.G. Stuart, preprint CERN-TH 4342/85 (1985).

[9] J.F. Wheater and C.H. Llewellyn Smith, Nucl. Phys. B208 (1982)
     27, erratum : Nucl. Phys. B226 (1983) 547.
     Reasonable agreement was also found with the following calculations:
     D. Yu. Bardin, P.Ch. Christova, O.M. Fedorenko,
     Nucl. Phys. B197 (1982) 1 ;
     D. Bardin, private communication ;
     H. Abramovicz, CDHS, internal note, unpublished ;
     A. Sirlin and W.J. Marciano, Nucl. Phys. B189 (1981) 442.

[10] P.G. Reutens et al., Phys. Lett. 152B (1985) 404 ;
     F. Merritt, talk at this Conference ;

[11] D. Bogert et al., Phys. Rev. Lett. 55 (1985) 1969 ;
     R. Brock, talk at this Conference.

[12] J. Panman, talk at this Conference.

[13] G. Arnison et al., Phys. Lett. 166B (1986) 484.

[14] J.A. Appel et al., preprint CERN-EP/85-166 (1985) ;
     A. Clark, talk at this Conference.

Table 1

Events numbers and correction of NC and CC events, for $E_h > 10$ GeV

	NC	CC	Change of NC/CC and systematic error (%)
Candidates	60986	137853	± 0.3
Cosmic rays	− 1120	− 9	− 1.8 ± 0.1
WBB background	− 2920	− 5187	− 1.2 ± 0.1
Long Shower	+ 159	− 158	+ 0.5 ± 0.2
Short CC	− 9642	+ 9526	− 22.5 ± 0.35
Ke correction 3	− 3016	+ 2488	− 8.0 ± 0.2
Corrected event numbers	44397	144513	± 0.65

Table 2

Systematic Error on the short CC Subtraction

	%
NBB beam energy ± 2 GeV	.02
K/π ratio ± 5 %	.08
Position of the beam center ± 1 cm	.08
Divergence ± .02 mrad	.10
Energy calibration	.10
Energy resolution	.10
Vertex resolution	.06
Momentum−range relation	.40
Longitudinal structure function	.60
Fraction of anti−quarks	.00
Total cross−section VS. energy	.16
Q^2 evolution	.05
Radiative corrections	.1
Statistics in the Monte Carlo	.75
Additive error on length	1.1
TOTAL ERROR	1.6 %

Table 3

Effect of the parameters of the model on $\sin^2\theta_W$. The change of $\sin^2\theta_W$ is given when one parameter is varied leaving all the others at their nominal value. All the corrections being correlated, they were applied at once.

Parameter	$\Delta \sin^2\theta_W$
Quark generation mixing $\left\|U_{ud}\right\|^2 = \left\|U_{cs}\right\|^2 = 0.947 \pm 0.006$	$+\ 0.0031 \pm 0.0003$
Longitudinal structure function $\sigma_L/\sigma_T = R_{QCD} \pm R_{QCD}$	$+\ 0.0006 \pm 0.0006$
Non–strange sea $(\bar{U} + \bar{D})/(U + D) = 0.13 \pm 0.02$	$+\ 0.0022 \pm 0.0003$
Strange sea $\bar{S}/\bar{D} = 0.45 \pm 0.10$ at $E_h = 30$ GeV	$+\ 0.0043 \pm 0.0010$
Charm sea $C/S = 0.15 \pm 0.15$	$+\ 0.0003 \pm 0.0003$
Non–isoscalar target (Fe) $D_V/U_V = 0.39 \pm 0.04$	$-\ 0.0090 \pm 0.0009$
Radiative Corrections $m_{top} = 45$ GeV/c^2 $m_{Higgs} = 100$ GeV/c^2	$-\ 0.011\ \ \pm 0.002$
Uncertainties for a fixed m_c	$\pm\ 0.004$
Charm–quark mass[1] $m_c = (1.5 \pm 0.3)$ GeV/c^2	$+\ 0.010\ \ \pm 0.004$
Total theoretical uncertainty	$\pm\ 0.005$

[1] also includes the influence of m_c on the determination of structure functions.

452

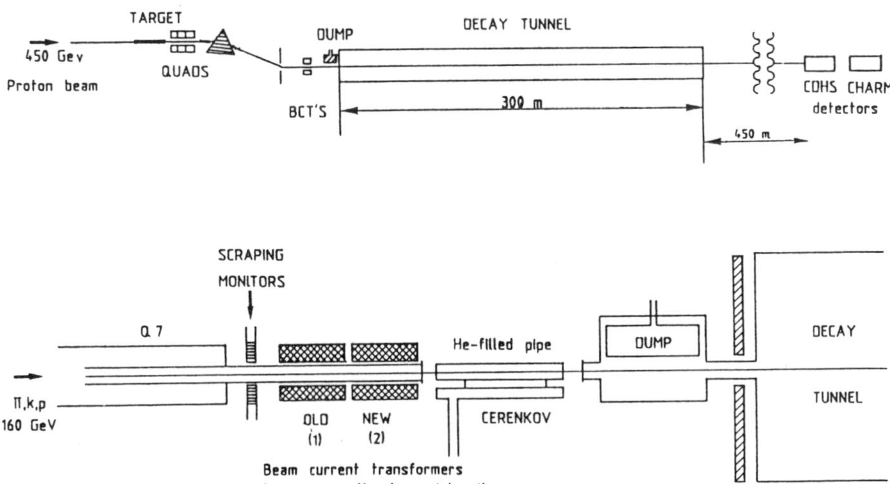

Fig.1.a) Overall view of the 1984 160 GeV Narrow band beam
     b) Details of the section before the decay tunnel showing the
        monitors in front of the dump.

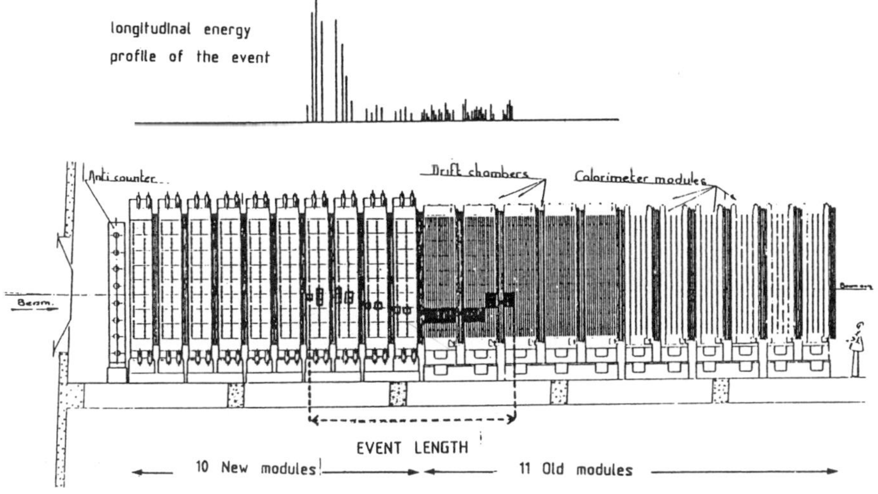

Fig. 2   Layout of the CDHS detector. The picture shows a charged current
         event with its μ^- focalised in the toroidal magnetic field.
         The event length is defined as the length of iron between the
         vertex and the stopping point of the most penetrating particule.

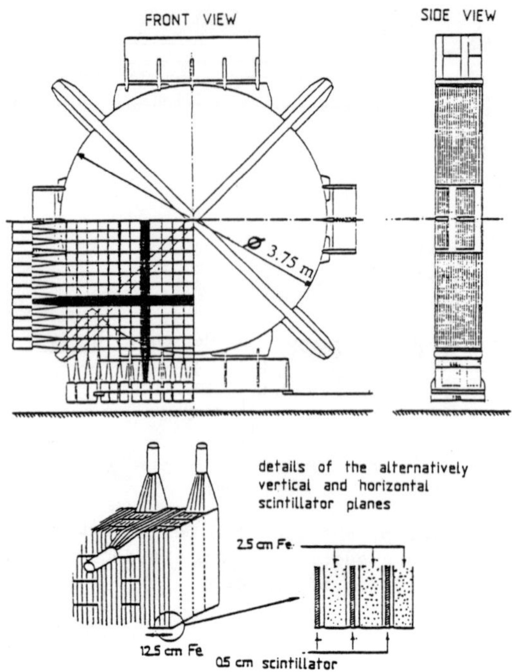

Fig. 3   Structure of the new type of CDHS modules.

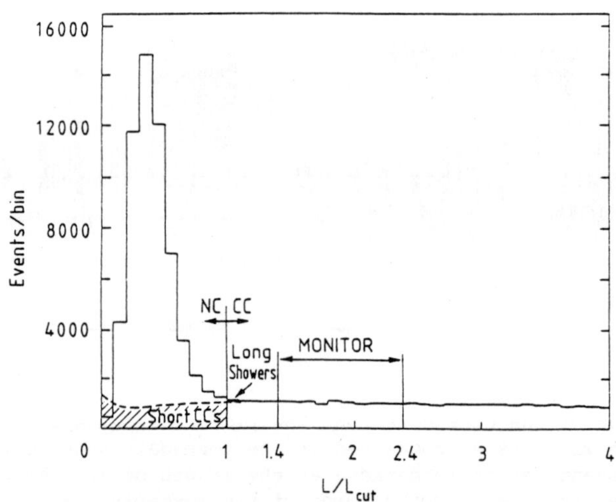

Fig. 4   Distribution of the event length L, in units of the cut-off
L_{cut} (see text), for neutrino events with $E_h > 10$ GeV.
The background from cosmic rays and WBB has been subtracted.

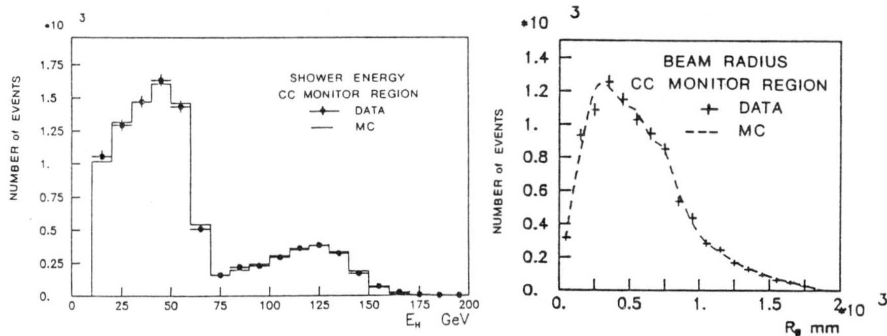

Fig. 5   Comparison between the data and the Monte Carlo simulation
for CC events in the monitor region.

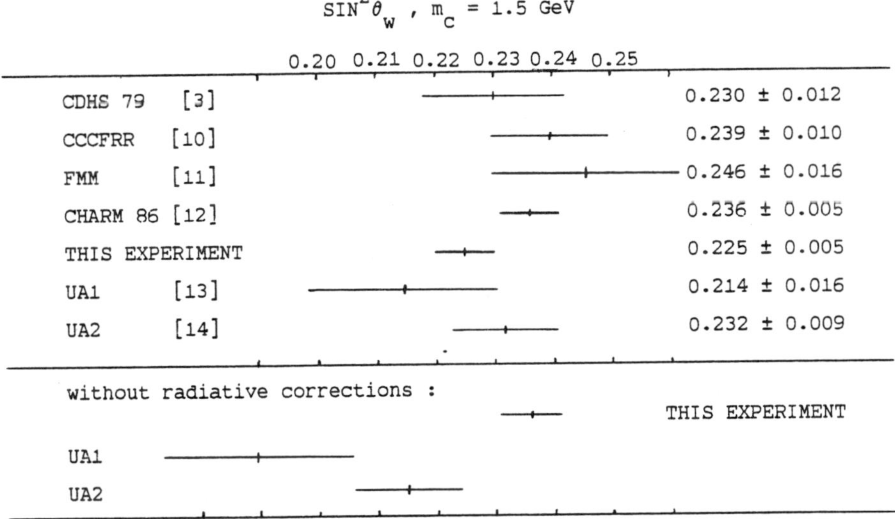

Fig.6   Comparison between the values of $\sin^2\theta_W$ as measured from the
most precise neutrino experiments and the values obtained from
the measurements of the W and Z masses. Only experimental errors
are shown (m_c = 1.5 GeV/c² for all neutrino experiments).

$\sin^2\theta_W$ From E594 at Fermilab

D. Bogert, R. Burnstein, R. Fisk, S. Fuess, J. Morfin,
T. Ohska, L. Stutte, and,
J. Bofill, W. Busza, T. Eldridge, J.I. Friedman, M.C. Goodman,
H.W. Kendall, I.G. Kostoulas, T. Lyons, R. Magahiz, T. Mattison,
A. Mukherjee, L. Osborne, R. Pitt, L. Rosenson, A. Sandacz,
F.E. Taylor, R. Verdier, S. Whitaker, G.P. Yeh, and
M. Abolins, R. Brock, A. Cohen, J. Ernwein, D. Owen,
J. Slate, M. Tartaglia, H. Weerts, and
J.K. Walker.

Fermi National Accelerator Laboratory
Massachusetts Institute of Technology
Michigan State University
University of Florida

Presented by Raymond Brock
Michigan State University
June 1986

ABSTRACT

A determination of the electroweak mixing parameter, $\sin^2\theta_W$, has been made by comparing deep-inelastic charged current events to neutral current events. The experiment was performed using a fine-grained neutrino detector at Fermilab in an exposure to a narrow band beam.

DETECTOR

The neutrino detector located in Laboratory C at Fermilab is a joint project of physicists, research associates, and graduate students from Fermilab, Michigan State University, Massachusetts Institute of Technology, and University of Florida (FMMF). It is a large calorimeter of 330 total tons which has been optimized by design to be capable of measuring the direction of hadron or electron energy flow. Measurement of the angle of hadron energy flow and distinguishing deep-inelastic charged current from neutral current events on an event-by-event basis requires that pattern recognition capability be high which can be achieved with very fine sampling. The physical characteristics of the FMMF detector are listed in Table I and the FMMF detector is shown schematically in Figure 1.

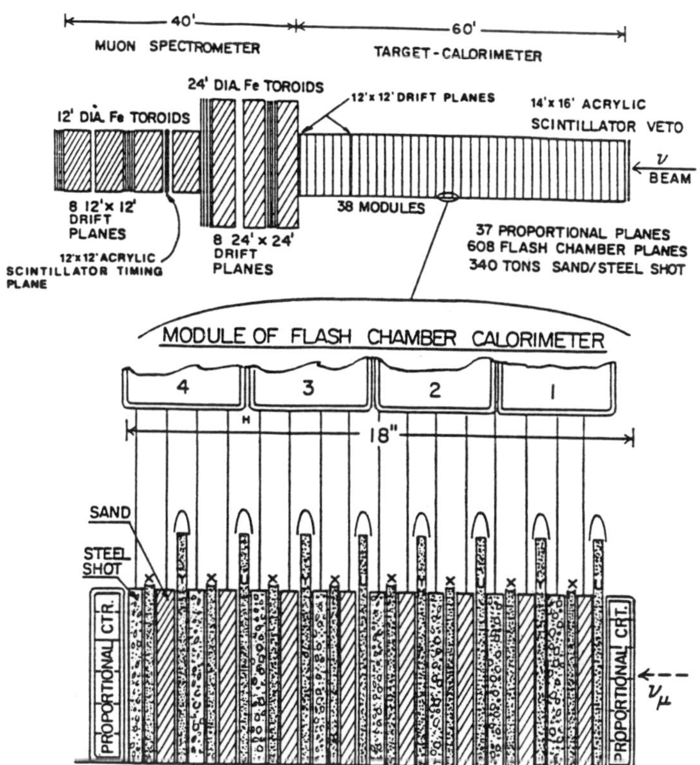

Figure 1

TABLE I

```
Target:
        330 tons (55 tons fiducial)
        X₀ = 12cm
        λ  = 90cm
Sampling:
        flash chambers 22% X₀ 0.3% λ
        Proportional tubes 3.5 X₀ 0.5λ
Muon momentum determination:
        4  stations of 12' x 12' and nearly 24' x 24' proportional
           tube planes
        40' magnetized iron toroids
Resolutions:
        Hadron energy
```

Target:

$X_0 = 12\text{cm}$

$\lambda = 90\text{cm}$

Sampling:

flash chambers 22% X_0 0.3% λ

Proportional tubes 3.5 X_0 0.5λ

Muon momentum determination:

4 stations of 12' x 12' and nearly 24' x 24' proportional tube planes

40' magnetized iron toroids

Resolutions:

Hadron energy	
flash chambers	$\sigma/E = 5.5 + 46E^{-1/2}\%$
proportional tube planes	$\sigma/E = 2.9 + 112E^{-1/2}\%$
Hadron angle	
flash chambers	$\sigma = 7 + 1008E^{-1}$ mrad
Muon momentum	
toroid spectrometer	$\Delta p/p = 10\%$

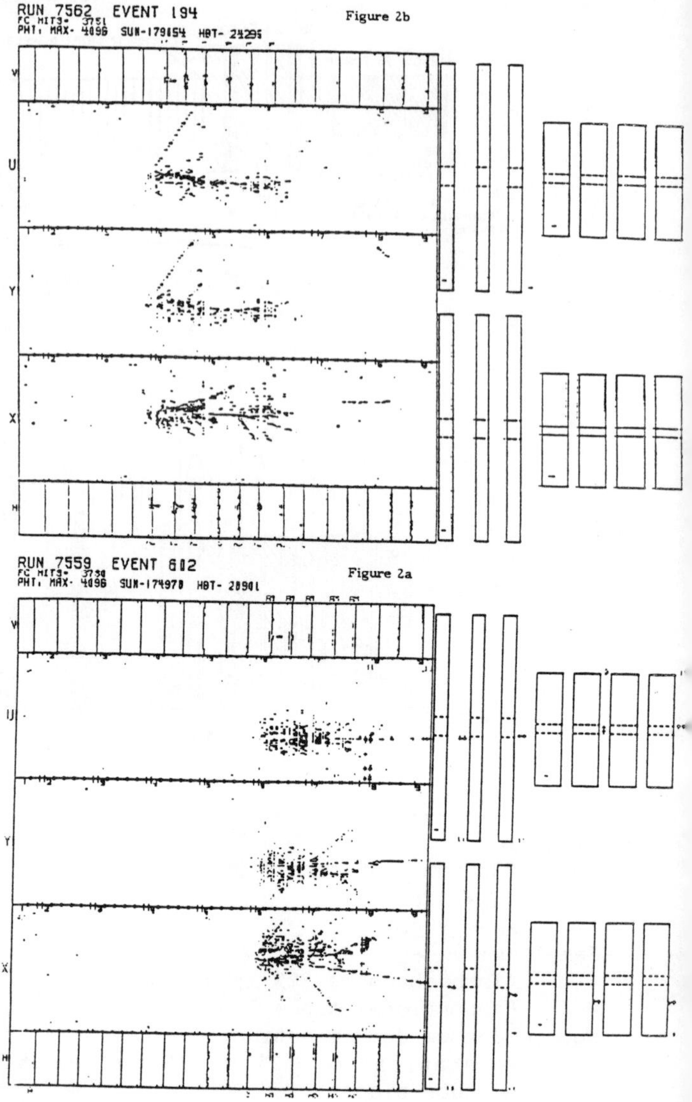

The fine granularity is achieved by using polypropylene flashchambers which provide digital (4mm x 5mm cell) information on the passage of ionizing particles.[1] The target-calorimeter is a modular "sandwich" construction of 38 units of approximately 9 metric tons each of material. The target calorimeter is approximately 20m long and presents a 3.7m x 3.7m square surface to the incoming neutrino beam for a total mass of approximately 330 metric tons. Each module consists of 16 3.7m x 3.7m x 1.6cm target-conversion planes filled in alternating units of sand and steel-shot pellets (average packing-fraction of 1/2). The average neutron excess is only 2%, meaning that the detector is iso-topically scalar to good approximation. A detail of one such module is also shown in Figure 1 and examples of computer-reconstructed charged current and neutral current events are shown in Figure 2. Clearly apparent is the outgoing single muon track indicating its presence both by the single trail of ionization in the flash chambers as well as the hits in the proportional tube chambers in 4 stations in the large iron spectrometer downstream of the target.

Energy is measured in the flash chambers by counting the hit cells. Raw hit cells are corrected for efficiency and multiplicity in segments of the individual chambers by an analysis of cosmic ray muons which are taken throughout each running period. In addition to the efficiency and multiplicity corrections, a statistical correction is applied to account for the possibility that more than one track could have traversed a single cell. The energy resolution for the flash chambers alone is well-fit below 200 GeV and is shown in Table I. The unique capability of a second generation detector is the ability to measure the direction of energy flow. Table I also shows the angular resolution for the projected angle for hadrons. These resolutions persist for recently measured energies from 20 to 440GeV (including the preliminary 1984-1985 Tevatron results).

EXPERIMENTAL DETAILS

The exposure in the narrow band running period of 1982 (January through June) constituted the data-taking for experiment E594. This running was done at three positive secondary momentum settings (165, 200, and 250 GeV) for neutrinos and one negative setting (165 GeV) for antineutrinos. Results from this experiment have been published.[2]

The general philosophy in this experiment was to minimize systematic uncertainties due to misclassification of either CC events as NC or and to minimize contamination from events which were induced by neutrinos from the 3-body decays of kaons. Because of the kinematical features of the CC cross section and the special characteristics of the dichromatic beam, these requirements were satisfied by accepting events in which the primary vertex was within 1m of the beam center, and by forming the quantity

$$y(R) = E_{hadron}/\langle E_\nu(R)\rangle$$

where the denominator is the mean neutrino energy at the event radius as determined from complete beam simulation Monte Carlo. This quantity functions essentially like the standard inelasticity, y. By eliminating events for which $y(R)>0.7$, events which have a ranging muon and, hence, simulate NC interactions are essentially eliminated. In addition, a minimum hadronic energy of 10GeV was demanded in order to be assured of 100% trigger efficiency.

To determine R (=NC/CC for neutrinos) and $\bar{R}$ (=NC/CC for antineutrinos), deep inelastic neutrino events were classified as NC or CC according to pattern recognition codes formulated by studying data as well as Monte Carlo events. The ratio of the identified NC-CC events was then formed and corrections for ambiguities were applied. In order to determine a correction for the contamination NC->CC, hadron showers in a calibration test beam of varying energies were modified in software by eliminating the incoming hadron track (thereby simulating a NC event) and applying the classification algorithms. In addition, identified CC events were reprocessed through the classification programs ignoring the original muon. These tests suggest that with the cuts, the probability of misidentifying NC->CC was 4%. A correction for CC->NC is minimal because of the y(R) cut and was determined by analyzing neutrino events generated by a full-shower Monte Carlo program. Figure 3 shows a CC event generated by this program showing that the general features of these events are very similar to those of real interactions. Transition curve comparisons between events from a simulation of the hadron test beam and the actual hadron exposure show that the sizes of hadron showers are well represented by this program. This analysis suggests that 1% of the time an identified NC event was really a CC event with a hidden muon. Table II shows the NC/CC ratio for Raw and Corrected data at each of the beam settings as well as for the combined positive running. These corrections include the contributions from all sources of neutrinos after the cuts: $K_{\mu 2}$

neutrinos are about 11% of the $\pi_{\mu 2}$ sample, K_{e3} and wide band background each contribute roughly 1% to the NC and NC + CC samples respectively. Also shown, is the fractional change in R ($\bar{R}$) due to these corrections, demonstrating that they were minor.

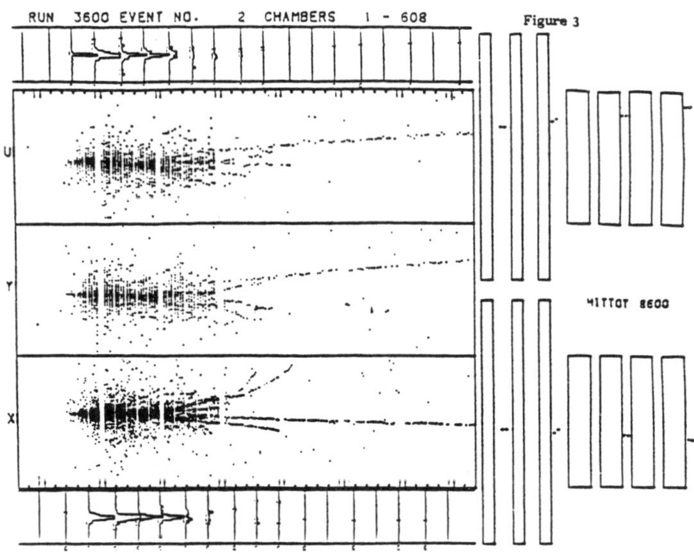

TABLE II

BEAM SETTING	R,raw	R,corrected	$\Delta R/R$
-165	723/1940	740/1928	3%
165	950/3235	966/3219	2%
200	638/2184	647/2175	2%
250	656/2093	677/2072	4.5%
all +	2244/7512	2290/7466	3%

The actual determination of $\sin^2\Theta_W$ is in general straightforward and can be estimated from the relation

$$R = [1/2 - \sin^2\Theta_W + 5/9 \sin^4\Theta_W(1+r)]\rho^2$$
$$\bar{R} = [1/2 - \sin^2\Theta_W + 5/9 \sin^4\Theta_W(1+r^{-1})]\rho^2$$

where ρ is an overall modification of the strength of the NC interaction beyond that expected from the Standard Model and r is the ratio of CC interations for $\bar{\nu}$ and ν. In order to precisely determine $\sin^2\theta_W$ these relations must be supplemented with additional Parton Model ingredients and corrections. These ingredients include the addition of the strange and charmed quark-sea (which are not know particularly well), higher ordered radiative corrections (which are calculable), and modifications to the Parton Model to account for the non-scaling s->c and d->c transitions. These latter contributions to the CC cross section are non-scaling because the experiment was necessarily performed at low energy and because there are substantial thresholds encountered in attempting to produce a final state charmed quark from a nearly massless quark. Indeed, at 40GeV, the strange sea is barely 40% excited in this particular transition, which because of its Cabbibo-favored status, constitutes a significant portion of the total strange quark contribution to the CC cross section.

Uncertainties in these latter issues result in a "theoretical" systematic error in the final result for $\sin^2\theta_W$. In all experiments of this type these theoretical uncertainties actually compete in importance with the experimental systematic uncertainties from the confusion of NC<->CC and muon removal. It's important to note that the importance of the theoretical uncertainties (especially the charmed quark production) as well as the uncertainties due to misidentification are in large part the price of performing these high precision experiments in (predominantly) the pion band at 400GeV accelerators.

The determination of $\sin^2\theta_W$ was done by treating each of the 4 data sets independently and minimizing χ^2 by comparing R($\bar{R}$) with a detailed Monte Carlo simulation of the experiment incorporating experimental details such as particulars of the beams and backgrounds, experimental cuts, and measured resolutions. The physics simulation included parton Q^2 evolution as parameterized by Duke and Owens[3], the addition of the strange sea (s/$\bar{u}$ = 0.5), quark mixing as described by Kobayashi and Maskawa[4], and radiative corrections employing the scheme of Sirlin and Marciano[5]. In particular, this latter correction leads to an overall reduction of R to 98.3% of the tree-level value due to the QED box diagram correction to the CC amplitude which we include as a non-unity value for ρ of 0.9915. To account for heavy quark production within the Parton Model, we have employed the so-called "slow rescaling" idea of Barnett[6] with $m_c = 1.5 \text{GeV}/c^2$. The cumulative effect of all of these corrections is shown in Figure 4. Included in Figure 4 is the correction of $\sin^2\theta_W$ due to the non-isoscalar nature of the target, which amounts to a reduction by 0.002. This correction is not applied in Reference 1.

The result for the one-parameter fit is then (without the non-isoscalar correction),

$$\sin^2\theta_W = 0.246 \pm 0.012(\underline{\text{stat}}) \pm 0.011(\underline{\text{syst}}) \pm 0.005(\underline{\text{theory}})$$

for $\chi^2/\text{dof} = 3.9/3$.

The summary of the contributions to the systematic errors is included in Table III.

TABLE III

	$\delta\sin^2\theta_W$	%
EXPERIMENTAL SOURCES		
NC<->CC confusion	±0.011	±4.2
muon elimination	±0.005	±2.0
THEORETICAL SOURCES		
$xs(x) = 0.5 \pm 0.2\, x\bar{u}(x)$	±0.001	±0.4
$m_c = 1.5 \pm 0.4$	±0.005	±2.0
$\Lambda = 200\,^{+200}_{-100}$ MeV/c	±0.0002	±0.1%

Including the weak radiative corrections leads to only minor changes for the two popular subtraction schemes:

$$1\text{-}M^2_W/M_Z \text{ scheme:} \quad \sin^2\theta_W = 0.247 \pm 0.012 \pm 0.013$$
$$\overline{\text{MS}} \text{ scheme:} \quad \sin^2\theta_W = 0.245 \pm 0.012 \pm 0.013.$$

If $m_c = 0$, then

$$\sin^2\theta_W = 0.228 \pm 0.012$$

with a substantially poorer fit ($\chi^2/\text{dof} = 8.0/3$). Parameterizing the m_c dependence as quadratic as was done in the 1985 CDHS publication[7] gives

$$\sin^2\theta_W = 0.236 + 0.002 m_c^2.$$

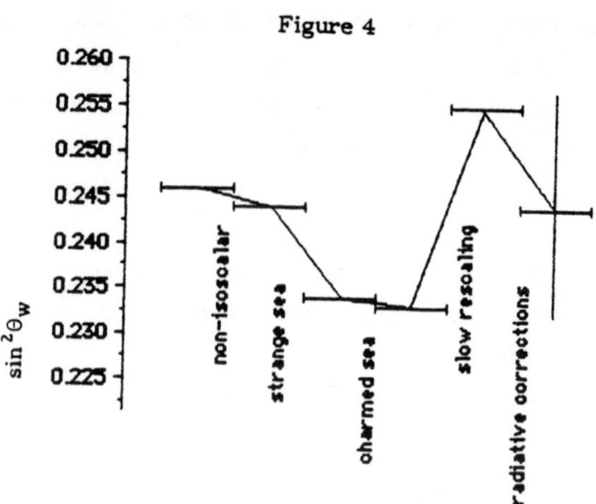

Comparison with other high precision experiments shows good agreement, both in the final value for $\sin^2\theta_W$, but also the relative importance of the very difficult corrections that each experiment applies to their data. Experiments have now reached the level of precision necessary for detailed comparison with the Collider results. However, the problems associated with the theoretical uncertainties will complicate this important comparison.

----

[1] F.E. Taylor et al., "The Construction and Performance of Large Flash Chambers", IEEE Transactions on Nucl. Sci., NS-27 (1980) 30. D. Bogert et al., "The Operation of a Large Flash Chamber Neutrino Detector at Fermilab", IEEE Transaction on Nucl. Sci., NS-29 (1982) 363.

[2] D. Bogert et al., Phys. Rev. Lett. 55, 574 (1985). D. Bogert et al., Phys. Rev. Lett. 55, 1969 (1985).

[3] D.W. Duke and J.F. Owens, Phys. Rev. D30, 39 (1984).

[4] M. Kobayashi and K. Maskawa, Prog. Theor. Phys. 49, 652 (1973). The values used for this anlaysis are from F.J. Gilman, Rev. Mod. Phys. 56 S296 (1984).

[5] A.Sirlin and W.J. Marciano, Nucl. Phys. B189, 442 (1981).

[6] R.M. Barnett, Phys. Rev. D14, 70 (1976); J. Kaplan and F. Martin, Nucl. Phys. 115, 333 (1976); R. Brock, Phys. Rev. Lett. 44, 1027 (1980).

[7] H. Abramowicz et al., Z. Phys. C28, 51 (1985).

A PRECISE DETERMINATION OF THE ELECTROWEAK MIXING ANGLE
FROM SEMI-LEPTONIC NEUTRINO SCATTERING

The CHARM Collaboration

J.V. Allaby, U. Amaldi, G. Barbiellini, M. Baubillier , F. Bergsma,
A. Capone, W. Flegel, F. Grancagnolo , L. Lanceri , M. Metcalf,
C. Nieuwenhuis, R. Pain , J. Panman, R. Plunkett and K. Winter
CERN, Geneva, Switzerland

I. Abt, J. Aspiazu, A. Bungener, F.W. Büsser, P.D. Gall
T. Hebbeker, F. Niebergall, P. Schütt and P. Stähelin
II. Institut für Experimentalphysik , Universität Hamburg, Hamburg, FRG

P. Gorbunov, E. Grigoriev, V. Khovansky and A. Rosanov
Institute for Theoretical and Experimental Physics, Moscow, USSR

A. Baroncelli , L. Barone , B. Borgia , C. Bosio , M. Diemoz ,
C. Dionisi , U. Dore , F. Ferroni , E. Longo , P. Loverre ,
L. Luminari , P. Monacelli , S. Morganti , F. de Notaristefani ,
C. Santoni , L. Tortora and V. Valente
Istituto Nazionale di Fisica Nucleare, Rome, Italy

(presented by J. Panman)

Abstract

The cross-section ratio of neutral-current and charged-current
semi-leptonic interactions of muon-neutrinos on isoscalar nuclei has been
measured with the result: $R^{\nu} = 0.3093 \pm 0.0031$ for hadronic energy larger
than 4 GeV. From this ratio we determined the electroweak mixing angle
$\sin^2\theta_W = 0.236 + 0.012(m_c - 1.5) \pm 0.005$ (exp.) ± 0.003 (theor.), where m_c
is the charm-quark mass in GeV/c^2.

Following the discovery of the neutral weak-current [1] great efforts have been spent to test the prediction of the Glashow-Salam-Weinberg model that the coupling in all neutral-current phenomena depends on one single parameter, $\sin^2\theta_W$. This prediction is based on the Born approximation of the theory [2]. However, the theory being founded on gauge symmetries, radiative corrections are calculable to all orders and can be tested by experiments. The corrections to the Born terms are different for individual processes. Therefore, measurements of $\sin^2\theta_W$ made with sufficient precision and extracted from the data using the simple zero-order approximations should show characteristic differences for different reactions. It is this feature which provides a test of the gauge nature of the theory. A precise measurement of $\sin^2\theta_W$ in semileptonic neutrino scattering, combined with a direct measurement of the Z and W boson masses provides such a test [3].

We report on a high precision measurement of R^ν, the ratio of cross-sections of deep inelastic neutral-current (NC) and charged-current (CC) semileptonic interactions of neutrinos:

$$R^\nu = \sigma_{NC}^\nu / \sigma_{CC}^\nu, \tag{1}$$

which may be used to determine $\sin^2\theta_W$. For isoscalar targets it follows from isospin invariance alone that the contributions of u and d quarks to the NC and CC cross-sections are related to $\sin^2\theta_W$ by the equation [4]:

$$R^\nu = \sigma_{NC}^\nu / \sigma_{CC}^\nu = 1/2 - \sin^2\theta_W + 5/9\sin^4\theta_W(1+r) \tag{2}$$

A measurement of the ratio of CC cross-sections for antineutrino and neutrino scattering, r, is needed to determine the small correction term, $5/9\sin^4\theta_W \cdot r$

The results reported here were obtained in a high statistics exposure of the CHARM neutrino detector in a 160 GeV/c narrow band beam (NBB), developed by Grant and Maugain [5] to provide a higher flux, resulting in an event sample that is a factor of 17 larger than that obtained during a previous exposure [6]. Neutrinos were produced in the decay of charge- and momentum-selected pions and kaons at the 450 GeV CERN Super Proton Synchrotron. The 120 m long beam line was tuned to a central momentum of 160 GeV/c and accepted particles with a momentum spread of $\pm$ 6 % (r.m.s.). About 8 m behind the last quadrupole, hadrons entered a 300 m long decay tunnel, which was evacuated to a pressure of 1 torr. At the end of the decay tunnel all surviving hadrons were absorbed in a 185 m long iron shield. Muons were ranged out by a combination of the iron shield and a 220 m long earth shielding following the iron.

The ejected proton intensity was measured using a beam current transformer and the beam position on the 3 cm diameter carbon target was carefully monitored. The detector was exposed to neutrino and antineutrino fluxes produced by 3.27×10^{18} and 0.90×10^{18} protons on target, respectively. Two beam current transformers placed just in front of the entrance of the decay tunnel, after the last quadrupole lens, measured the intensity of the secondary beam. The relative amounts of pions and kaons in the beam were determined experimentally using a helium filled differential Cerenkov counter. The measurements yielded the ratios K^{+}/π^{+} = 0.1282 $\pm$ 0.0028 and K^{-}/π^{-} = 0.0630 $\pm$ 0.0033.

The CHARM neutrino detector is a fine-grain calorimeter followed by an iron spectrometer with a toroidal magnetic field, and surrounded by a magnetized iron frame. The calorimeter has a sampling step corresponding to one radiation length or 0.22 absorption length, with scintillators, proportional drift tubes and streamer tubes as detecting elements. It is described in detail elsewhere [7]. The response of the calorimeter was recently calibrated in electron and pion beams in the range from 0.5 to 140 GeV/c [8]. The energy of hadron showers is measured by the scintillators with a resolution of $\sigma(E)/E = 0.487/\sqrt{(E/\text{GeV})} + 0.0127$. The lateral vertex position of hadron showers is determined by combining the measurements of all detection elements; it reaches a precision of better than 3 cm above 50 GeV. For this experiment the fiducial volume extended over 55 target plates leaving five plates at the beginning of the detector to reject incoming tracks and 18 plates at the end to ensure good shower containment. This resulted in a fiducial mass of 87 tons.

A special feature of the CHARM detector is the low detection threshold, which makes it possible to measure showers down to an energy of 2 GeV with a very high trigger efficiency. Care was taken to avoid problems caused by two interactions within the conversion time needed by the electronics of the detector elements. After each trigger the time between the trigger and the following interaction in the detector was recorded. Events followed by another interaction within 1.2 μsec, indicating the presence of a second interaction within the sensitive time of the proportional drift tubes, were rejected. This procedure induced an additional effective deadtime of ~ 2% which does, however, not affect the measurement of R^{ν}. The average total dead time was 17.1% in the neutrino exposure and 12.8% in the antineutrino one.

Interactions in the CHARM detector were classified by an automatic pattern-recognition program on an event-by-event basis. Selection criteria were applied which attempt to identify optimally the physical processes and to minimize the corrections needed to relate the visible cross-sections to the physical cross-sections.

Neutrino interactions are defined as events which have no entering charged tracks. A neutrino event is called a charged-current interaction if a muon is originating from the event vertex. A muon is assumed to originate from the event vertex if its extrapolated intercept with the vertex plane of the event lies within 30 cm of the measured shower vertex, or if sufficient energy is measured in a box around the extrapolated point. The muon has to be visible over a range corresponding to an energy-loss of at least 0.67 GeV. Its total range has to exceed 1 GeV energy-loss, corresponding to a penetration of 20 calorimeter planes. The visible range of a muon track is defined as the length, measured in equivalent range, between the first clearly visible point outside the shower and the last point on the track, as seen by the proportional tube system. Tube hits are deemed to be clearly visible points on a track if they are separated laterally by at least 5 quiet tubes from their neighbouring hits (15 cm isolation criterion). In addition they are required to be isolated from other hits, which do not belong to the muon track, in the longitudinal direction. These isolation criteria are much more strict than the ones used in a previous publication on the same subject [6] and tend to reject more tracks on the basis of the requirement of 0.67 GeV visible range.

All other neutrino event candidates were classified as neutral-current interactions. Only those events which have their vertex inside the fiducial volume and which have a shower energy of at least 2 GeV were analysed. Representations of a typical NC and CC event in the CHARM detector are shown in figure 1.

In order to obtain the final CC and NC event numbers from the automatic classification, a number of corrections have to be applied which will now be discussed. In table 1, these corrections to the data are summarized, for hadron energies larger than 4 GeV. We have chosen this value to eliminate events wich are not due to deep inelastic scattering, reducing contributions to R^ν by pseudo-elastic and coherent pion production in NC and CC interactions to the level of $\Delta R^\nu/R^\nu \leq 10^{-4}$.

Table 1: Neutrino event numbers ($E_h \geq 4$ GeV).

	Neutral current	Charged current
Uncorrected data sample	39239 ± 198	108472 ± 329
Trigger + filter efficiency	7 ± 4	0 ± 0
Scan correction	40 ± 40	60 ± 44
Cosmic and WB correction	-2310 ± 87	-4311 ± 119
Difference in energy cut	-	0 ± 129
Lost muons correction	-3737 ± 105	3735 ± 105
π and K decay	1892 ± 50	-1835 ± 50
Ke3 charged-current	-1768 ± 68	-106 ± 6
Ke3 neutral-current	-532 ± 20	-33 ± 2
Corrected event numbers	32831 ± 283	105982 ± 408

The trigger rate induced by cosmic-ray events is $\simeq$ 4 kHz. A fast filter program reduced the cosmic ray induced background to the level of a few Hz, while retaining nearly all NC events. The combined trigger and filter efficiency of all NC events with a hadron energy above 4 GeV was found to be 99.98%.

The program recognized and flagged event topologies for which the automatic procedure could fail. These flagged events (~ 4000) were visually inspected in a double scan and corrected on a event-by-event basis. In addition to these flagged events a randomly selected sample totalling 5% of all triggers surviving the cosmic ray filter was scanned in order to determine the efficiency of the automatic pattern recognition program and to correct for it on a statistical basis. Out of $\approx$ 20000 events, 7 events were found to be misclassified, of which 3 CC and 2 NC events were wrongly assigned to be outside the fiducial volume. The results of this scanning allowed us to correct the errors introduced by the classification code to $\Delta R^{\nu}/R^{\nu} \leq 0.1\%$.

The corrections for wide-band (WB) and cosmic ray background are combined in table 1. The cosmic ray background contribution in the live time of the experiment was found to be $(1.42 \pm 0.02)\%$ for NC events and negligible for CC events. Events due to the so-called wide-band background, caused by neutrinos from hadrons of undetermined charge and momentum, decaying upstream of the decay tunnel were measured when the beam was blocked by a 1.5 m long Fe absorber. The procedure consists of defining the neutrino flux generated before the dump to be background, and the flux from decays downstream of the dump as "beam". A small correction for non-containment in the dump and for contributions of secondary particles created in the dump and in the residual gas of the decay tunnel was applied. The relative normalization of the periods when the dump blocked the entrance of the decay tunnel with respect to the standard running conditions was performed using the signal of the secondary hadron beam current transformer.

The uncertainty introduced by the subtraction of the cosmic and wide-band backgrounds amounts to $\Delta R^{\nu}/R^{\nu} \simeq \pm\ 0.3\%$.

The energy deposition of the muon in the hadron shower region was subtracted from the hadron energy. The additionnal uncertainty in the hadron energy measurement introduced by this subtraction was taken into account as a source of systematic uncertainty for CC events.

CC events in which the primary muon cannot be identified are automatically classified as NC. Some of these lost CC events have muons with an energy less than 1 GeV. Another contribution to the loss of CC events is caused by muons of more than 1 GeV either leaving the sides of the detector before their energy loss is sufficiently large, or being obscured by the hadronic shower. A correction is required for these losses. The precision which can be reached in measuring R^{ν} depends in an essential way on the reliability of this correction. We therefore determined the correction by two independent methods. In one method the efficiency of recognizing primary muons is obtained by overlaying by computer simulated muons on real showers obtained by removing the original muon from CC events. In the other method the effective shower length distribution is calibrated by measuring the length of the hidden part of primary muon tracks in CC events, and using this distribution in a Monte-Carlo program to determine the muon recognition efficiency. The total correction induced by all sources of unidentified muons is 3.5% of the CC event rate. The two methods give consistent results to within 2% of the correction. In order to minimize systematic errors in the calculation of the correction a "reference-region" is defined relative to which the unseen muons are calculated. Charged-current events belong to the reference-region if the primary muon has a computed total energy in the range between 3 and 5 GeV. The ratio of all CC events with an unseen muon and identified CC events in the reference region is calculated using the Monte-Carlo program described above. The absolute number of CC events where the muon was not recognized is obtained by multiplying this ratio with the actual number of CC events found in the reference region. Backgrounds in the reference region were taken into account. This procedure eliminates all errors introduced by uncertainties in the beam energy, the energy-loss of muons, the density of the detector and the slope of the y-distribution. This cancellation is exact for all such scale errors because the simulated events in the reference region scale with the number of events with lost muons. The systematic error is dominated by the determination of the shower length ($\pm$ 0.5 calorimeter samplings) and is estimated to be $\pm$ 2.8% of the correction, contributing an error of $\Delta R^{\nu}/R^{\nu} \simeq \pm 0.4\%$.

A small fraction of NC events contains a track which fulfills the requirements of a primary muon of a CC event; these events are classified as CC events. This background is caused by decays of pions or kaons in the shower or by non-interacting hadrons, the so-called punch-through tracks. The correction is evaluated by analysing CC events in which the primary muon has been removed. The fraction of these events where another track is found which satisfies the muon recognition criteria yields the correction. This procedure is done in bins of the hadron energy in order to correct for the difference of CC and NC hadron energy distributions. A small contribution, present in CC but not in NC events is due to prompt muons from charm decays and can be subtracted with sufficient precision. The size of this correction, integrated for hadron energy above 4 GeV, is $\simeq$ 4.8% of the NC rate, determined with a fractional error of $\pm$ 2.6%, contributing an error of $\Delta R^{\nu}/R^{\nu} \simeq \pm 0.2\%$.

Both NC and CC interactions of electron-neutrinos originating from K_{e3}-decays in the beam are classified as NC events. The contribution of these events amounts to ~ 7% of the muon-neutrino induced NC events. A small correction was applied for events with hadron showers generating a decay muon which fulfills the requirements of a primary muon, causing these events to be classified as CC events. Measurement of the K/π ratio with an accuracy of ~ ± 3%, determines the Ke3 correction with a fractional error of ± 4%, corresponding to $\Delta R^{\nu}/R^{\nu} \simeq \pm 0.3\%$.

Taking these corrections into account we obtained for events with hadronic energy greater than 4 GeV: $R^{\nu} = 0.3098 \pm 0.0031$ for the calorimeter material and, correcting for the small deviation from isoscalarity:

$$R^{\nu} = 0.3093 \pm 0.0031 \tag{3}$$

For the ratio of the total antineutrino and neutrino charged-current cross-sections we obtained: $r = 0.439 \pm 0.011$. Correcting for the different spectra in neutrino and antineutrino exposures gives:

$$r = 0.456 \pm 0.012 \tag{4}$$

The uncertainty on r is dominated by the event-statistics (1.4%), WB background subtraction (0.8%) and the relative normalization error of the $\bar{\nu}$ to ν flux (2.0%).

Using equation (2) and neglecting all corrections to this equation due to the presence of other than only u and d quarks, a raw value of $\sin^2\theta_W$, assuming $\rho = 1$, is obtained:

$$\sin^2\theta_W\big|_{raw} = 0.2356 \pm 0.0050 \tag{5}$$

The result (5), deduced in the approximations of equation (2), has to be corrected for various effects in a quark-parton model (QPM) framework. The different components of the corrections applied to obtain $\sin^2\theta_W$ from R^ν and r measured with a 4 GeV hadron energy cut are given in table 2 and will now be discussed.

Table 2:   Corrections to $\sin^2\theta_W$ ($E_h \geq 4$ GeV).

Source	$\Delta\sin^2\theta_W$	theor. uncertainty
muon mass	+ 0.0011	± 0.0001
W^2 thresholds, F_L	+ 0.0005	± 0.0005
K.M. mixing matrix		± 0.0010
Strange sea at $m_c=0$	− 0.0074	± 0.0010
Charm sea at $m_c=0$	+ 0.0015	± 0.0010
Radiative corrections	− 0.0092	± 0.0020
Total uncertainty (fixed m_c)		± 0.0030
Charm mass ($m_c = 1.5$ GeV/c^2)	+ 0.0140	
Total ($m_c = 1.5$ GeV/c^2)	+ 0.0005	± 0.0030

The muon mass causes a small phase space suppression in CC events and the final state invariant mass threshold depends on the produced quark flavour. A small correction for the longitudinal structure function (F_L) has been applied; the F_L distribution was taken from a measurement done by this collaboration [9]. The strange quark sea and the charm quark sea in the nucleons have been taken into account. In addition, charged current interactions changing a light quark into a charm-quark are kinematically suppressed because of the mass of the charm-quark, m_c. A fixed mass, m_c = 1.5 GeV/c², was chosen, and the threshold effects were computed using the slow rescaling procedure [10]. The distributions were determined from dimuon production data in deep inelastic neutrino and muon scattering, respectively [11] [12]. In these calculations, we used Kobayashi-Maskawa (K.M.) mixing matrix elements obtained [13] assuming three families of quarks and requiring unitarity of the matrix.

The preceding corrections change the electro-weak mixing angle by $\Delta \sin^2 \theta_W$ = + 0.0097.

The major theoretical uncertainties of these corrections are due to the limited knowledge of the momentum-weighted contents of s and c quarks in the nucleon, the K.M. mixing elements and threshold-effects in charm-production [4]. These uncertainties are summarized[1] in table 2. The largest uncertainty is in the choice of the charm-quark mass m_c which affects the threshold suppression of charm-production. We, therefore, present the result on $\sin^2 \theta_W$ as a function of m_c with m_c = 1.5 GeV/c² as the central value. This linear approximation is valid in a range: 1 GeV/c² $\leq m_c \leq$ 2 GeV/c².

---

[1] The uncertainty on the strange sea is estimated taking into account m_c threshold suppression.

The effect of QED and electroweak radiative corrections on $\sin^2\theta_W$ was estimated following Bardin et al. [14], in the on-shell renormalization scheme [15]. The calculation was performed using a set of computer programs made avalaible to us by D.Yu. Bardin. Setting the top and Higgs masses to 45 GeV and 100 GeV, respectively, we found: $\Delta\sin^2\theta_W = - 0.0092$, and estimated a precision of $\pm$ 0.002 for this computation.

The total theoretical uncertainty in the determination of $\sin^2\theta_W$, excluding the uncertainty on m_c, is estimated to be of the order of $\pm$ 0.003. Higher twist effects are supposed to be small and were neglected. For simplicity, we assumed the slow rescaling procedure to be valid. We do not quote additionnal systematic errors for these assumptions.

Applying both quark-parton model and radiative corrections, we obtain the final result:

$$\sin^2\theta_W = 0.236 + 0.012(m_c - 1.5) \pm 0.005 \ (\text{exp.}) \pm 0.003 \ (\text{theor.}) \qquad (6)$$

where m_c is the charm-quark mass in GeV/c^2.

This result is compared with those obtained in other recent semi-leptonic neutrino scattering experiment and reported in this conference in Table 3 for m_c = 1.5 GeV. The agreement is good and significant because of the different experimental procedure used.

Table 3

Comparison of values of $\sin^2\theta$ derived from semi-leptonic neutrino scattering experiments. The common theoretical error is ±0.005.

Group	Method	$\sin^2\theta$
FMMF	event-by-event	0.246 ± 0.016 (exp)
CCFR	event length	0.239 ± 0.010 (exp)
CDHS	event length	0.225 ± 0.005 (exp)
CHARM	event-by-event	0.236 ± 0.005 (exp)

The mean value derived from these experiments

$$\sin^2\theta\big|_{average} = 0.233 \pm 0.004(exp) \pm 0.005(theo)$$

can be combined with the value of m_W and m_Z recently published by the UA2 Collaboration [17] to determine the radiative shift Δr of the W mass. We find

$$\Delta r = 1 - (m^2_W \sqrt{2}\, G_\mu \sin^2\theta)/\pi\alpha = 0.077 \pm 0.025\ (exp) \pm 0.038\ (syst)$$

in good agreement with calculated electroweak radiative corrections [15]. The systematic error is the quadratic sum of the theoretical uncertainties in semi-leptonic neutrino scattering and of the energy scale error entering into the m_W mass determination. Future experiments at the upgrated CERN $p\bar{p}$ collider and the CHARM II experiment on $\bar{\nu}_\mu e \rightarrow \bar{\nu}_\mu e$ scattering will help to reduce both the systematic and the experimental error and allow a more sensitive test of the radiative shift.

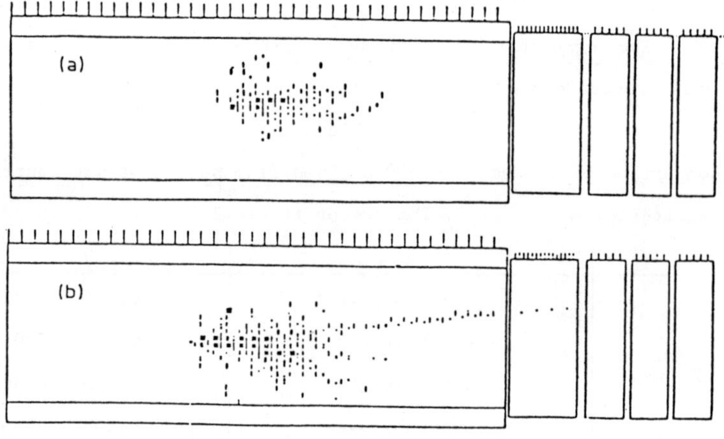

Figure 1:   Neutral-current (a) and Charged-current (b) events recorded in the CHARM detector.

References

[1]   F.J. Hasert et al., Phys. Lett. 46B (1973) 138.

[2]   S.L. Glashow, Nucl. Phys. 22 (1961) 579;

        A. Salam and J. Ward, Phys. Lett. 13 (1964) 168;

        S. Weinberg, 1964. Phys. Rev. Lett. D5 (1964) 1264.

[3]   W.J. Marciano and A. Sirlin, Phys. Rev. D29 (1984) 945, erratum

        Phys. Rev. D31 (1985) 213.

[4]   C.H. Llewellyn-Smith, Nucl. Phys. B228 (1983) 205.

[5]   A. Grant and J.M. Maugain, CERN/EF/BEAM 83-2.

[6]   CHARM Collab., M. Jonker et al., Phys. Lett. 99B (1981) 265.

[7]   CHARM Collab., A.N. Diddens et al., Nucl. Instr. & Meth. 178 (1980)

        27.

        C. Bosio et al., Nucl. Instr. & Meth. 157 (1978) 35.

        CHARM Collab., M. Jonker et al., Nucl. Instr. & Meth. 200 (1982)

        183.

[8]   CHARM Collab., F. Bergsma et al., (to be published in Nucl. Instr. &

        Meth.)

[9]   CHARM Collab., F. Bergsma et al., Phys. Lett. 141B (1984) 129.

[10]   R.M. Barnett, Phys. Rev. D14 (1976) 70;

        H. Georgi and H.D. Politzer, Phys. Rev. D14 (1976) 1829.

[11]   CDHS Collab., J. Knobloch et al., Private communication.

[12]   EMC Collab., J.J. Aubert et al., Nucl. Phys. B213 (1983) 31.

[13]   K. Kleinknecht and B. Renk, Phys. Lett. 130B (1983) 459.

[14]   D.Yu. Bardin and O.M. Dokuchaeva, Preprint JINR-E2-86-260 (1986);

        D.Yu. Bardin and O.M. Federenko, Sov. Yad. Phys. 30 (1979) 811 [Sov.

        J. Nucl. Phys. 30 (1979) 418] and D.Yu. Bardin and O.M. Dokuchaeva,

        Sov. Yad. Phys. 36 (1982) 482 [Sov. J. Nucl. Phys. 36 (1982) 282].

[15]   A. Sirlin, Phys. Rev. D22 (1980) 971.

[16]   CDHS Collab., H. Abramowicz et al., submitted to Phys. Rev. Lett.

[17]   UA1 Collab., G. Arnison et al., Phys. Lett. 166B (1986) 484.

        UA2 Collab., XXIth Rencontre de Moriond, Les Arcs, France, March

        1986.

Preliminary results on μe-pair production
and internal μ-bremsstrahlung in ν-inter-
actions in the energy range 3 - 30 GeV

SKAT-Collaboration

(IHEP Berlin-Zeuthen - IHEP Serpukhov)

presented by R. Nahnhauer

1. Introduction

The data we present in the following come from an
exposure of the heavy freon filled bubble chamber SKAT to
the wide band neutrino beam of the Serpukhov 70 GeV
proton accelerator. The average neutrino energy in our
experiment is about 7 GeV. The properties of the chamber
liquid (hadronic interaction length $\lambda_H \approx 60$ cm, radiation
length $x_0 = 11$ cm, photon conversion length $l_\gamma = 14$ cm)
provide a good identification of muons, electrons and
photons in our experiment. It is therefore well suited to
search for the dilepton production final states

$$\nu_\mu N \longrightarrow \mu^- e^+ X \tag{1}$$

and

$$\nu_\mu N \longrightarrow \mu^- e^- X \tag{2}$$

and the muon inner bremsstrahlungs process

$$\nu_\mu N \longrightarrow \mu^- (\gamma) X \tag{3}$$

The statistics we use for the investigation of reactions
(1) - (3) are about 10000 neutrino induced charged current
events with energies $E_\nu \geqslant 3$ GeV.

As shown by a lot of other experiments /1/ opposite
sign dilepton production like in reaction (1) can be
explained by the semileptonic decay of charmed particles.
We want to calculate from our data the total charm particle
production ratio and the ratio of Λ_c to D production in
our special kinematical region near to the charm production threshold.

For like sign dilepton production the experimental
situation is quite contradictory /2-10/ and a unique
theoretical explanation does not exist. New experimental
data are therefore welcome. Our kinematical region at low
energies in the hadronic rest frame excludes possible
effects of associated charm production or production of
beauty particles.

The detection of photons coming from muon internal
bremsstrahlung gives the possibility to investigate
exclusively one part of electromagnetic radiative corrections in neutrino induced reactions.

2. /ue-event selection

The criteria of charged current event selection the
neutrino energy determination and the background reduction
procedure have been described elsewhere /11/. For a
further increase in the efficiency of muon selection for
reactions (1) and (2) only events with $p_\mu \geqslant 1$ GeV have
been used. Electrons have been in general identified by
the following conditions: characteristic spiralization or
strong multiple scattering at the final part of the track,
observation of high energetic δ-electrons at the track,
tridents at the track, observation of one or more

Brems γ's pointing to the track, annihilation (for e^+)

Finally ue-events have been selected due to the following procedure:

- for an electron track there have to be satisfied at least two of the above criteria

- electron or positron momentum $p_e \geqslant 400$ MeV/c

- the electron track should be clearly visible at a distance of 1.5 cm in space from the primary vertex

- no additional e^+e^- pair at the primary vertex.

For further background reduction in the $\mu^- e^-$ sample we applied two additional cuts

- $p_\mu \to p_{e^-}$

- $1/2 \, |\lambda_\mu - \lambda_e| > \Delta\lambda_e$ with $\langle \Delta\lambda_e \rangle = 0.1$ rad

  $1/2 \, |\varphi_\mu - \varphi_e| > \Delta\varphi_e$ with $\langle \Delta\varphi_e \rangle = 0.05$ rad

  where λ and φ are the dip and azimuthal angles of the corresponding tracks.

Finally among our 10060 charged current neutrino events we have found 18 $\mu^- e^+$ events and 7 $\mu^- e^-$ events which fulfil the above criteria. The loss of ue -events due to our cuts comes out to be 15.1 % of the true $\mu^- e^+$ signal and 8.6 % of the true $\mu^- e^-$ signal.

The background for reactions (1) and (2) is mainly due to

- asymmetric γ-quants converting near the primary vertex or asymmetric Dalitz pairs

- $\bar{\nu}_e N$ ($\nu_e N$) charged current interaction with chamber leaving π^- mesons simulating a μ^-

- compton electrons produced near the primary vertex

- δ-electrons (from μ or π) produced near the primary vertex.

The results of our calculation for these background
processes are given in the following table

back-ground to	asym. γ's Dalitz pairs	$\bar{\gamma}_e N(\nu_e N)$ cc-int.	compt. electr.	δ-electr.	total back-ground
$\mu^- e^+$-ev.	0.8 ± 0.3	0.1 ± 0.1	–	–	0.9 ± 0.3
$\mu^- e^-$-ev.	0.8 ± 0.3	0.8 ± 0.4	1.6 ± 0.5	0.2 ± 0.1	3.4 ± 0.7

3. Results for $\nu N \longrightarrow \mu^- e^+ X$

Taking into account background and losses we get the
following corrected number of $\mu^- e^+$-events:

$$N(\mu^- e^+) = 20.1 \pm 5.2$$

The ratio of $\mu^- e^+$-events above the charm threshold, i.e.
for the hadronic mass $W > m_{\Lambda_c^+} \approx 2.3$ GeV and $p_\mu > 1$ GeV,
$p_{e^+} > 0.4$ GeV/c comes out to be

$$R_{\mu^- e^+} = \frac{N(\nu_\mu N \longrightarrow \mu^- e^+ X)}{N(\nu_\mu N \longrightarrow \mu^- X)} = (4.6 \pm 1.2) \cdot 10^{-3}$$

Among the 18 $\mu^- e^+$ events found there are 3 events with
a neutral strange particle. Correcting for the corresponding registration efficiency we estimate an average
number of neutral strange particles per $\mu^- e^+$-event of
about 0.7.

To interpret our data as due to charm particle production we compare them with Monte Carlo results for quasielastic Λ_c^+ production /12/

$$\nu_\mu N \longrightarrow \mu^- \Lambda_c^+ \tag{3}$$

and inclusive D-production /13/

$$\nu_\mu N \longrightarrow \mu^- D X \tag{4}$$

This has been done in figs. 1 and 2 for the positron
momentum distribution and the four momentum squared Q^2.
As one can see neither reaction (3) (dashed lines) nor
reaction (4) (dashed dotted lines) can describe the data.
Only the sum of both with a ratio Λ_c : D = 1 : 1.5 gives
a good description of the distributions (solid lines).

Taking into account a ratio of D^+ : D^0 of 1 : 3 and
corresponding branching ratios we observe a total charm
particle ratio corrected for the cut in the positron
momentum of

$$R^c = \frac{N(\nu_\mu N \longrightarrow \mu^- c X)}{N(\nu_\mu N \longrightarrow \mu^- X)} = (10.3 \pm 4.0) \cdot 10^{-2}$$

in good agreement with our previous result /14/.

4. Results for $\nu_\mu N \longrightarrow \mu^- e^- X$

After correction for background and losses we get from
our 7 $\mu^- e^-$-events on the 1 σ level

$$N (\mu^- e^-) = 3.9 \begin{smallmatrix} + 4.2 \\ - 2.1 \end{smallmatrix}$$

This number leads to a ratio to all charged current events above the charm threshold W > 2.3 GeV of

$$R/^{u^-e^-} = \frac{N(\bar{\nu}_\mu N \longrightarrow /^{u^-e^-}X)}{N(\bar{\nu}_\mu N \longrightarrow /^{u^-}X)} = (8.9 \; {}^{+ \; 9.6}_{- \; 4.8}) \cdot 10^{-4}$$

The number of $/^{u^-e^-}$-events is about five times smaller than that of $/^{u^-e^+}$ events in our experiment resulting in a ratio

$$R/^{u^-e^-}//^{u^-e^+} = \frac{N(\bar{\nu}_\mu N \longrightarrow /^{u^-e^-}X)}{N(\bar{\nu}_\mu N \longrightarrow /^{u^-e^+}X)} = 0.19 \; {}^{+ \; 0.22}_{- \; 0.12}$$

In fig. 3 our result is compared to other neutrino experiments. As one can see our data are in reasonable agreement with that of the Gargamelle group /4/ at slightly higher average energy whereas two other bubble chamber experiments /2,6/ and counter experiments observe lower rates at small E_ν . One should keep in mind however the different cuts in all experiments and the up to now rather large experimental errors.

Among our 7 $/^{u^-e^-}$-events there is one with an additional neutral strange particle. The probability for this event to be due to backbround processes is estimated to be smaller than 15 %.

5. Results for internal muon bremsstrahlung

The process of internal muon bremsstrahlung can be illustrated by the diagram given in fig. 4. One has to detect a photon pointing to the primary vertex of the

event and having a small angle to the muon in the final
state.

For the event selection we used our general definitions for charged current events /11/. Photons are
identified by their e^+e^--conversion pairs and linked by
a kinematical fit procedure to the primary vertex /15/.
For background reduction minimal and maximal conversion
length cuts of 1 cm and 28 cm have been applied. Together
with geometrical and scanning efficiencies this lead to a
total photon registration efficiency of $\varepsilon_{tot} = 0.57$.
The minimal photon energy considered was $E_\gamma = 50$ MeV.

In fig. 5 we show the $\cos \Theta_{\mu\gamma}$ distribution where $\Theta_{\mu\gamma}$
is the angle between muon and photon direction in the
laboratory frame. The resolution in this angle comes out
to be $\Delta\Theta_{\mu\gamma} = 18$ mrad for $\cos \Theta_{\mu\gamma} > .99$. In fig. 5 a clear
peak is seen in this region which can be attributed to
muon assiciated photons. The result of a Monte Carlo calculation with a Phase Space Model (dashed histogram in
fig. 5) behaves flat for $\cos \Theta_{\mu\gamma} > .99$. Within this model
which describes our data otherwise rather well /11,15/,
photons originate only from π^0-decays. Background
sources for the observed signal like outer muon bremsstrahlung or bremsstrahlung from δ-electrons and
tridents can be neglected for our experimental conditions
/16/.

From fig. 5 we estimate a net signal of 140 ± 12 muon
assóciated photons leading to a ratio to all charged
current events of

$$R_{\mu\gamma} = \frac{N(\nu_\mu N \rightarrow \mu^-(\gamma) X)}{N(\nu_\mu N \rightarrow \mu^- X)} = (2.0 \pm 0.3) \cdot 10^{-2}$$

To compare our data with theoretical predictions we
have calculated the integral of a semiclassical formula

/17/ of the ratio $R^{\mu\gamma}$ for $E_\gamma > 50$ MeV, $p_\mu > .5$ GeV/c and our muon momentum spectrum.

$$R^{\mu\gamma} = \frac{\alpha}{2\pi} \int \frac{dE_\gamma}{E_\gamma} \cdot \int \frac{\beta_\mu \cdot \sin\Theta_{\mu\gamma}\, d(\cos\Theta_{\mu\gamma})}{(1 - \beta_\mu \cdot \cos\Theta_{\mu\gamma})^2}$$

The result is $R^{\mu\gamma}_{theo} = 2.2$ % for $\cos\Theta_{\mu\gamma} > .99$ and $R = 5.5$ % for the whole region of $\cos\Theta_{\mu\gamma}$ taking into account a lepton phase space factor of $E_\mu / (E_\mu + E_\gamma)$. The predicted distribution in $\cos\Theta_{\mu\gamma}$ is given in fig. 5 as solid line. It describes the data well.

As one can see from fig. 6 the ratio R increases with increasing neutrino energy. This was found also at higher energies for other kinematical cuts /18/. Applying these cuts both data sets agree well with each other assuming a roughly linear dependence on E_ν (see fig. 7).

Finally in figs. 8 and 9 the distributions in the scaling variables x_B and y_B are given. One observes that the effect is concentrated at small x-values but behaves nearly flat in y.

6. Summary

- Studying neutrino induced dilepton production in the energy range $3 \leq E_\nu \leq 30$ GeV we have observed for $p_\mu \geq 1$ GeV/c and $p_e > .4$ GeV/c 18 $/\mu^- e^+$ and 7 $/\mu^- e^-$ events within 10060 charged current events.

- After correction for background and losses this leads to corrected numbers of 20.1 ± 5.2 $/\mu^- e^+$ and $3.9 ^{+4.2}_{-2.1}$ $/\mu^- e^-$ events.

- The corresponding ratios to all charged current events above the charm threshold $W > 2.3$ GeV are $\mu^- e^+ / \mu^- = (4.6 \pm 1.2) \cdot 10^{-3}$ and $\mu^- e^- / \mu^- = (8.9 \,{}^{+\,9.6}_{-\,4.8}) \cdot 10^{-4}$. For $\mu^- e^- / \mu^- e^+$ we found therefore a value of $0.19 \,{}^{+\,0.22}_{-\,0.12}$.

- The properties of the observed $\mu^- e^+$ events could be described assuming that they come from semileptonic decays of quasi-elastic Λ_c^+ production and inclusive D-meson production in a ratio $1 : 1.5$.

- For the total charm particle production ratio to charged current events we found $R^C = (10.3 \pm 4.0) \cdot 10^{-2}$.

- We observed also a signal of muon associated photons which has been interpreted as due to muon internal bremsstrahlung and could be well described by theoretical expectations.

References

/1/   N.J. Baker et al., Phys. Rev. D32 (1985) 531 and
        references therein
/2/   H.C. Ballagh et al., Phys. Rev. D24 (1981) 7
/3/   V.V. Ammosov et al., Phys. Lett. 106B (1981) 151
/4/   A. Haatuft et al., Nucl. Phys. B222 (1983) 365
/5/   P. Marage et al., Z.f.Phys. C21 (1984) 307
/6/   A.C. Shaffer Nevis rep. R-1326 (1985)
/7/   T. Trinko et al., Phys. Rev. D23 (1981) 1889
/8/   M. Jonker et al., Phys. Lett. 107B (1981) 241
/9/   K. Nishikawa et al., Phys. Rev. Lett., 54 (1985) 1336
/10/   H. Burkhardt et al., CERN-EP/85-191 (1985)
/11/   D.S. Baranov et al., Z.f.Phys. C21 (1984) 189 and 197
/12/   G. Köpp et al., Nucl. Phys. B123 (1977) 61
/13/   A.B. Krebs et al., Sov. Nucl. Phys. 31 (1980) 713
/14/   V.V. Ammosov et al., Z.f.Phys. C30 (1986) 183
/15/   D.S. Baranov et al., Sov. Nucl. Phys. 41 (1985) 1520
/16/   W. Lohmann et al., CERN 85-03 (1985)
/17/   C.S.W. Chang and D. Falkoff, Phys. Rev. 76 (1949) 365
/18/   H.C. Ballagh et al., Phys. Rev. Lett. 50 (1983) 196

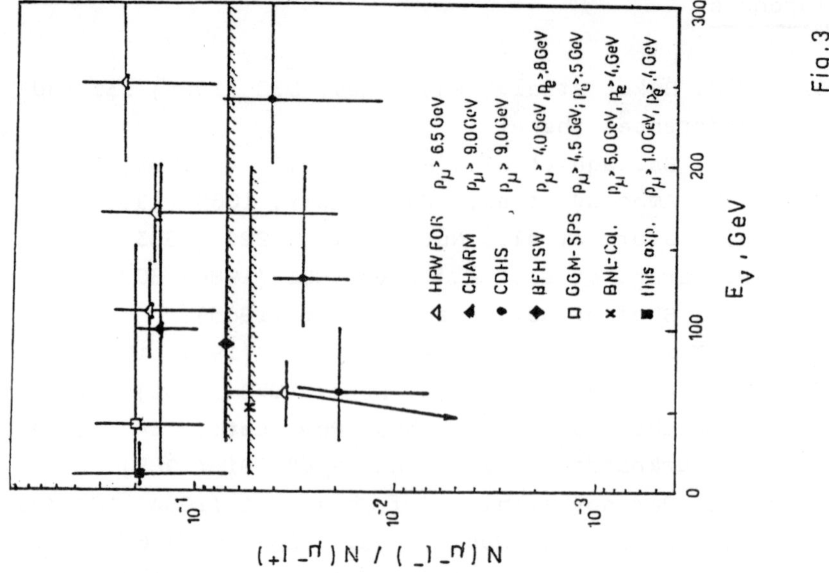
HPW FOR $p_\mu > 6.5$ GeV
CHARM $p_\mu > 9.0$ GeV
CDHS $p_\mu > 9.0$ GeV
BFHSW $p_\mu > 4.0$ GeV, $p_e > 8$ GeV
GGM-SPS $p_\mu > 4.5$ GeV; $p_e > 5$ GeV
BNL-Col. $p_\mu > 5.0$ GeV, $p_e > 4$ GeV
this exp. $p_\mu > 1.0$ GeV, $p_e > 4$ GeV
$N(\mu^- l^-)/N(\mu^- l^+)$
E_ν , GeV
Fig. 3

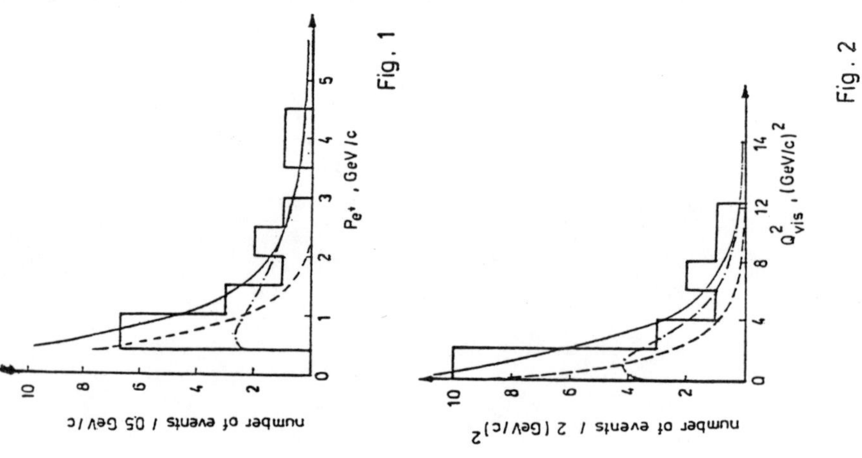
number of events / 0.5 GeV/c
P_{e^+} , GeV/c
Fig. 1
number of events / 2 [GeV/c]^2
Q^2_{vis} , (GeV/c)^2
Fig. 2

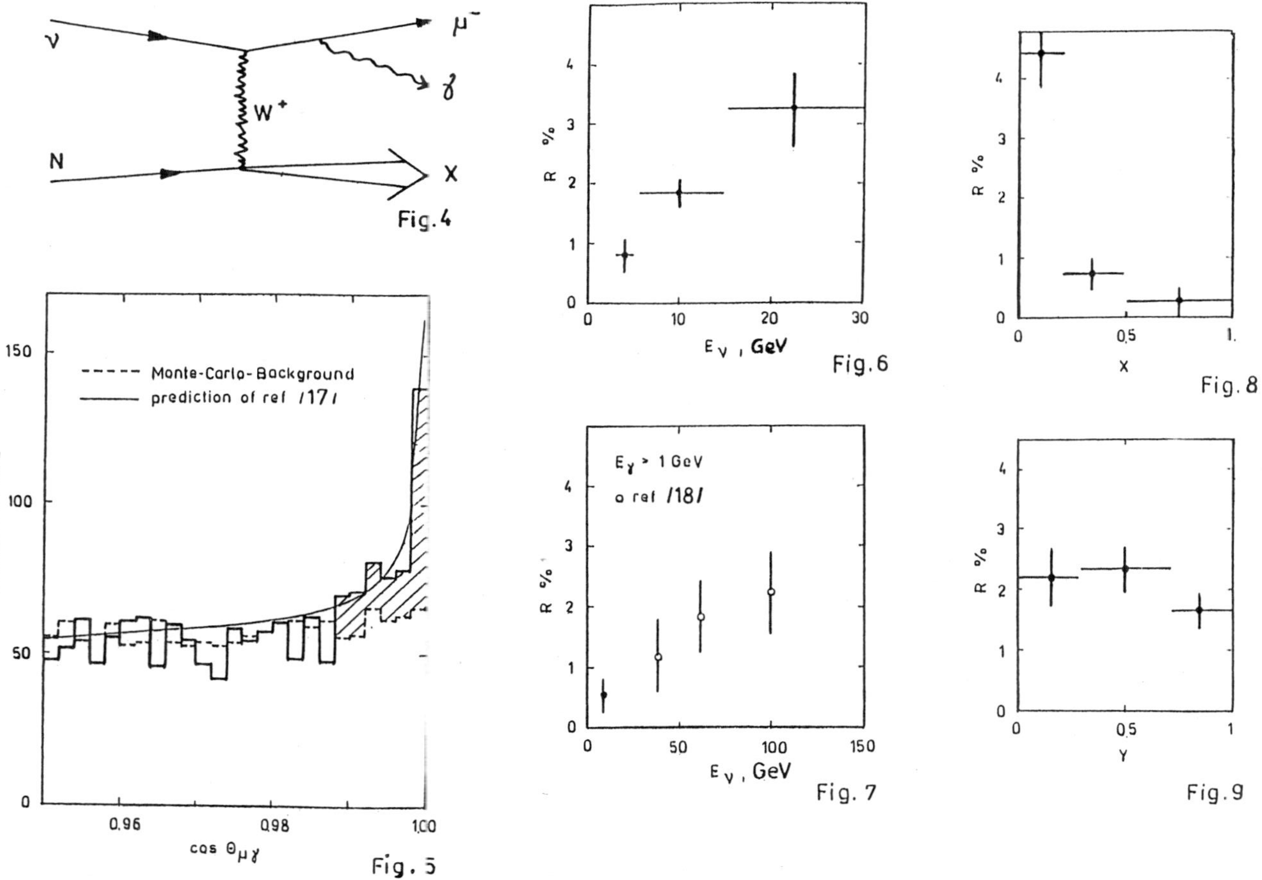
ν
μ⁻
γ
W⁺
N
X
Fig. 4
Gammas / 0.002
Monte-Carlo-Background
prediction of ref /17/
cos Θ μ γ
Fig. 5
R %
E ν , GeV
Fig. 6
R %
E γ > 1 GeV
o ref /18/
E ν , GeV
Fig. 7
R %
X
Fig. 8
R %
Y
Fig. 9

A MEASUREMENT OF THE PRODUCTION CROSS-SECTION OF PROMPT LIKE-SIGN

DIMUON EVENTS IN NEUTRINO AND ANTINEUTRINO BEAMS

CERN-Dortmund-Heidelberg-Saclay-Warsaw Collaboration

presented by

K. Kleinknecht
Institut für Physik
Universität Mainz
Fed. Rep. Germany

Abstract

From an exposure of the CDHS detector in wide-band and narrow-band neutrino beams at the CERN SPS, 367 $\mu^-\mu^-$ events induced by neutrinos and 73 $\mu^+\mu^-$ events produced by antineutrinos were observed. After subtraction of non-prompt backgrounds, a prompt signal of like-sign dimuon events remains. For neutrino beams in the energy range $100 < E_{vis} < 300$ GeV, 40 ± 14.4 prompt $\mu^-\mu^-$ events with muon momenta $p_\mu > 9$ GeV/c remain, for antineutrino beams there are 10.4 ± 4.3 prompt events. The ratios of prompt like-sign to opposite-sign events are $(3.2 \pm 1.2)\%$ for neutrinos and 5.1 ± 2.3 % for antineutrinos.

K. Kleinknecht

1. INTRODUCTION

Prompt like-sign dimuon production is one of the remaining mysteries
in neutrino physics, not easily explained in the standard model. Theoretically, the only possible source for like-sign dimuons invoking
only known particles is the creation of a charm-anticharm pair at the
hadronic vertex of a standard charged-current neutrino interaction.
But perturbation QCD calculations of this process [1] yield rates which
are lower that the experimentally observed ones [2-9] by one or two
orders of magnitude. This new experiment has a luminosity about 5 times
larger than previous experiments.

2. BEAMS

Three neutrino beams at the CERN 450 GeV Proton-Synchrotron were used:
i)   a 300 GeV narrow-band neutrino beam (NBB) with a total exposure
     of 2.2×10^{18} protons
ii) a 400 GeV wide-band (WBB) neutrino exposure, where positive par-
     ticles were focussed by a Van der Meer horn, with 1.3×10^{18}
     protons on target
iii) a 400 GeV wide-band antineutrino beam focussing negative particles,
     with 2.6×10^{18} protons on target.

3. DETECTOR

The detector used in the neutrino NBB exposure is the CDHS detector
described before [10]. It consists of 19 magnetized iron toroids used
as hadron calorimeters separated by three-plane drift chambers for
momentum measurement. For the WBB exposures, eight of the toroids with
5 cm iron plates were replaced by new modules with 2.5 cm iron sampling
thickness and finer scintillator granularity.

The advantages of this detector for like-sign dimuons are threefold:
the geometrical acceptance for both muons is near unity since they are

both focussed towards the detector axis (acceptance for like-sign dimuons is > 0.98 for muon momenta p_μ > 9 GeV/c); the detector is dense,
the mean free path for hadrons is $\sim$ 29 cm in the iron-scintillator
sandwich (this reduces the most important background of pions decaying
to muons); the detector is massive (650 tons fiducial volume).

4. EVENT SELECTION

Dimuon events were retained if
- the vertex is inside a cylinder around the beam axis with radius 1.6 m
- both muon tracks traverse at least six drift chambers
- the distance between the two muon tracks projected back to a plane
  perpendicular to the beam axis at the vertex is less than 20 cm.

Events were selected and reconstructed by programme. Measured quantities include both muon momenta, $p_\mu^{(1)}$ and $p_\mu^{(2)}$, and the energy is then
$E_{vis} = E_1 + E_2 + E_H$. For p_μ > 6 GeV/c and E_{vis} > 30 GeV, there are
350 $\mu^-\mu^-$ and 2736 $\mu^-\mu^+$ events in the neutrino WBB, correponding to
1.65×10^6 μ^- charged-current events, 17 $\mu^-\mu^-$ and 128 $\mu^-\mu^+$ events in
the 300 GeV neutrino NBB for 43684 μ^- CC events, and 73 $\mu^+\mu^+$ and 772
$\mu^+\mu^-$ events in the antineutrino WBB, with 0.667×10^6 μ^+ CC events.

These raw event numbers were corrected for geometrical acceptance and
for reconstruction efficiency. The latter was obtained by double scanning, and was found to be, for p_μ > 9 GeV, 0.83 ± 0.07 for like-sign
dimuons, 0.90 ± 0.04 for opposite-sign dimuons, and 0.99 ± 0.01 for
single muon events.

5. BACKGROUND SUBTRACTION

The main background source is the decay of a pion or kaon in the hadronic shower of a single muon charged-current neutrino interaction. This
process was simulated in a Monte Carlo calculation in which charged-
current hadron shower energies are generated according to the observed
distributions, and the individual hadrons in each shower are generated
according to both experimental and theoretical information available.

The uncertainty in this calculation is mainly due to the uncertainty
in the hadron fragmentation functions f(z); the other relevant parameters (detector density and pion and kaon lifetimes) are accurately
known. The data used come from bubble chamber exposures, viz. the 15 ft
FNAL chamber with neon filling, and BEBC at CERN with neon and deuterium fillings [11] . The shape of the pion z distributions was obtained
by a fit to all bubble chamber data available, of the form A exp (-Bz)
for $z > z_o$ and C exp (-Dz) for $z < z_o$. The rate of kaons was obtained
using K^o production data, and assuming $N_{K+} + N_{K-} = N_{KO} + N_{\bar{K}O}$. The z
dependences of the charged kaons were obtained by fitting exponentials
to the Lund model calculations. Fig.1 shows the contribution of this
π/K decay background to the total event sample. The uncertainty in this
calculations is estimated to be ± 15 %.

A second, much smaller background comes from neutrino induced trimuon
events. For incident neutrinos, these events of the type $\nu + N \rightarrow \mu^-\mu^+\mu^- X$
are understood as electromagnetic production of a muon pair either by
the leading μ^- in a charged-current reaction or by a particle in the
hadronic shower. If the range of the μ^+ is so short that the muon es-
capes detection, the event can simulate a like-sign dimuon. This background has been estimated by using the calculation of J.Smith and G.
Valenzuela [12] based on the most recent experiment [13]. The amount of
this background is about 8% of the raw like-sign dimuon sample.

The third background source is due to a coincidence in time and space
of two single muon charged-current events. In order to understand this
back-ground, two successive real, single-muon events were artificially
combined. In this way, a large sample of "overlay" events was generated.
The distribution in the radial distance of the two muons at the vertex,
Δr, is shown in fig.2 for artificial overlay events and for real like-
sign two-muon data. It is evident that the data for Δr > 40 cm are well
reproduced by the overlay events, while at Δr < 20 cm there is a clear
signal for true dimuon data. This background is usually smaller than
the trimuon background, however is becomes important at high transverse

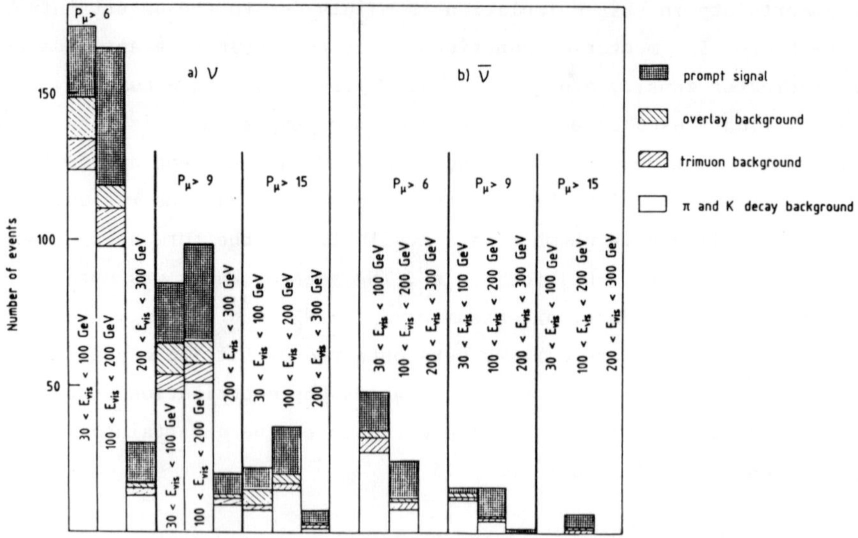

Fig. 1 Histogram to show the different background subtractions for neutrino and antineutrino like-sign dimuon events in the three bins of E_{vis} and for the three minimum muon momentum cuts.

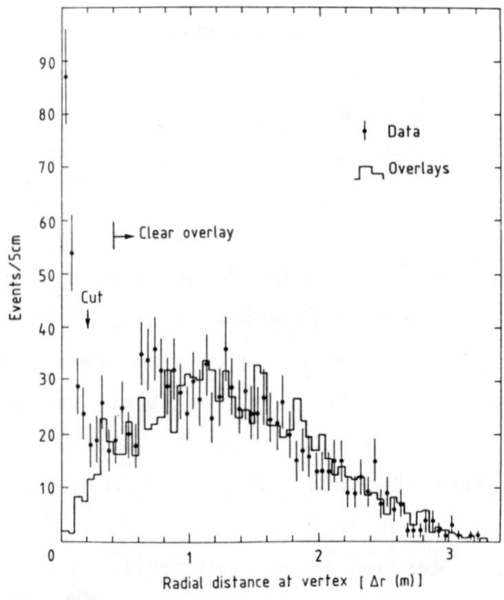

Fig. 2 Distribution in Δr, the separation of the two reconstructed muons in the vertex plane, for raw like-sign dimuon events and "overlay" events. The two curves are normalized in the region $\Delta r > 40$cm. The event acceptance criterion is $\Delta r < 20$cm.

momenta of the second muon relative to the beam direction. Such events could be of great interest, but this background impedes the sensitivity of the experiment in this domain.

The relative importance of the background can be seen in fig.1. The remaining prompt signal is most significant for muon momenta above 9 GeV/c and visible energies above 100 GeV. We obtain for these conditions the following ratios of the prompt like-sign rates relative to the charged-current rate and relation to the opposite-sign rate in neutrino beams:

$$\frac{\sigma(\mu^-\mu^-\ \text{prompt})}{\sigma(\mu^-)} = (1.16 \pm 0.42) \times 10^{-4}$$

and

$$\frac{\sigma(\mu^-\mu^-\ \text{prompt})}{\sigma(\mu^-\mu^+\ \text{prompt})} = (3.2 \pm 1.2)\ 10^{-2}$$

This signal is 2.8 standard deviations above zero.

For antineutrinos, the corresponding numbers are

$$\frac{\sigma(\mu^+\mu^+\ \text{prompt})}{\sigma(\mu^+)} = (1.7 \pm 0.7)\ 10^{-4}$$

$$\frac{\sigma(\mu^+\mu^+\ \text{ptompt})}{\sigma(\mu^+\mu^-\ \text{prompt})} = (5.1 \pm 2.3)\ 10^{-2}$$

with a significance of 2.4 standard deviations. The complete results for different cuts in muon momentum and in visible energy are given in Table 1. The energy dependence of the ratio of like-sign dimuon to charged-current rates is shown in fig.4 together with earlier results. The agreement is marginal, due to different and less reliable methods of subtracting the π/K decay background.

The nature of the like-sign dimuon events is illustrated in fig.6 and 6. Fig.5 shows the transverse momentum p_T of the second muon relative

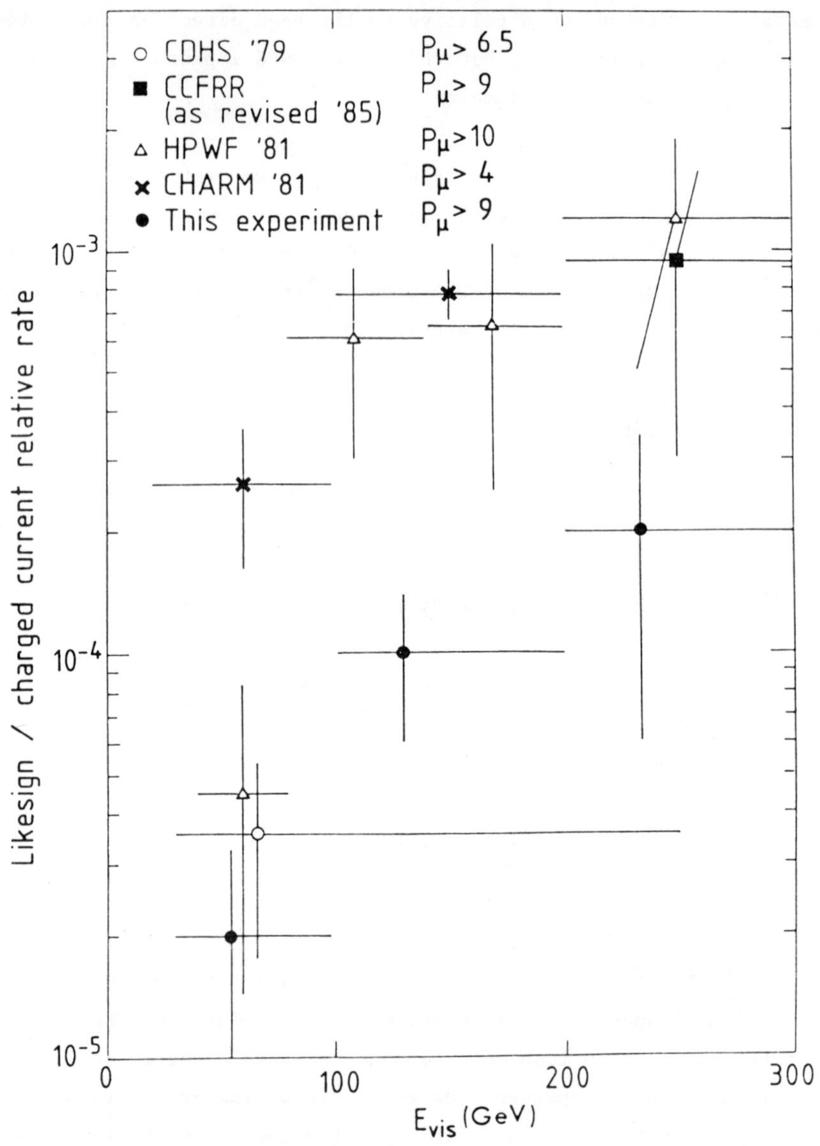

Fig. 3 Ratio of like-sign to charged-current cross-sections in neutrino interactions as a function of visible energy.

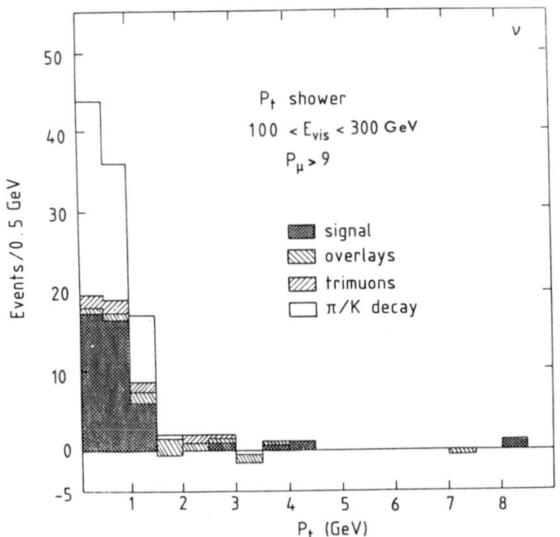

Fig. 4 Distribution of transverse momentum of the second muon in $\nu N \to \mu^- \mu^- X$.

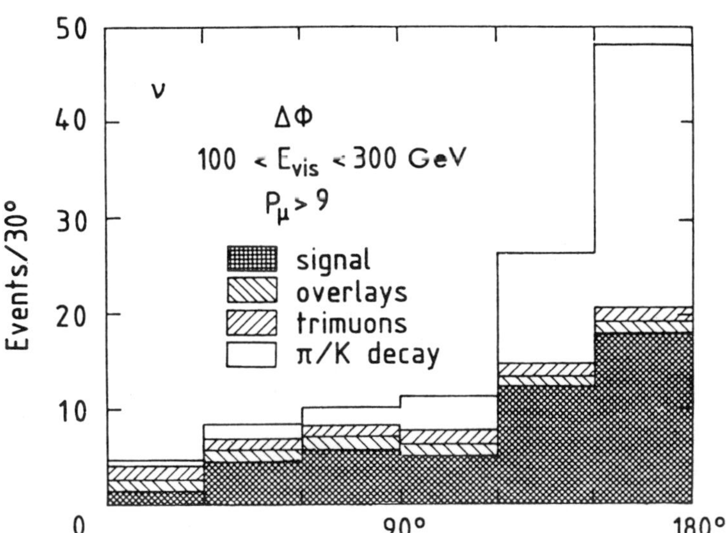

Fig. 5 Distribution in $\Delta\Phi$, the angle between the two muons in the plane normal to the neutrino direction, for $\nu N \to \mu^- \mu^- X$.

to the hadron shower axis, calculated with the definition that the first muon is the one which has the larger p_T relative to this axis. Fig.6 gives the distribution of events in the azimuthal angle ϕ between the two muons in the plane perpendicular to the beam axis. Both distributions show that the second muon is associated with the hadron shower, while the first one originates at the lepton vertex of the neutrino interaction.

We are therefore left with the conclusion that prompt like-sign dimuons are most likely due to charm pair production, but that the observed rate is still one order of magnitude above the most recent result of perturbative QCD calculations [1]. Non-perturbative contributions to the process are estimated [14] to be compatible with experiment. But this possibility would imply the existence of diffractive charm production in hadron collisions with cross-sections larger than the upper limits from several experiments, and is therefore not a likely explanation [15].

REFERENCES

[1] K.Hagiwara, Nucl.Phys. B173(1980)487

[2] A.Benvenuti et al., Phys.Rev.Lett. 35(1975)1202

[3] M.Holder et al., Phys.Lett. 70B(1977)396

[4] A.Benvenuti et al., Phys.Rev.Lett. 41(1978)725

[5] J.G.H.deGroot et al., Phys.Lett. 86B(1979)103

[6] T.Trinko et al., Phys.Rev. D23(1981)1889

[7] K.Nishikawa et al., Phys.Rev.Lett. 46(1981)1555

[8] M.Jonker et al., Phys.Lett. 107B(1981)241

[9] K.Nishikawa et al., Phys.Rev.Lett. 54(1985)1336

[10] M.Holder et al., Nucl.Instr. 148(1978)235

[11] Priv. communications of C.Baltay, A.M.Cooper, B.Nellen, and R.Wigmans

[12] J.Smith and G.Valenzuela, Phys.Rev. D28(1983)1071

[13] T.Hansl et al., Phys.Lett. 77B(1978)114, Nucl.Phys. B142(1978)381

[14] R.M.Godbole and D.P.Roy, Phys.Rev.Lett. 48(1982)1711, Z.Phys.C22 (1984)39;
F.A.Choban, Z.Phys.C25(1984)269

[15] J.R.Cudell, F.Halzen and K.Hikasa, Preprint MAD/PH/276 (May 1986), Univ. of Wisconsin, Madison

Table 1

Background corrected dimuon signals on ratios of prompt like-sign to charged-current and opposite-sign dimuon events.

a)   (NBB and WBB combined)

	E_{vis} GeV	Prompt event numbers		$\dfrac{\mu^-\mu^-}{\mu^-}\times 10^{-4}$	$\dfrac{\mu^-\mu^-}{\mu^-\mu^+}\times 10^{-2}$
		$\mu^-\mu^-$	$\mu^-\mu^+$		
$P_\mu > 6$	30 − 100	24. ± 26	1238 ± 41	.26± .28	1.3 ± 1.4
	100 − 200	46.5 ± 22	1131 ± 37	1.6 ± .74	3.5 ± 1.6
	200 − 300	13 ± 6.1	166 ± 14	3.8 ± 2.0	7.3 ± 3.4
$P_\mu > 9$	30 − 100	20.5 ± 13.5	898 ± 34	.20 ± .13	1.9 ± 1.2
	100 − 200	33 ± 13.5	1020 ± 14	1.05 ± .43	2.9 ± 1.2
	200 − 300	7 ± 4.9	158 ± 13	2.1 ± 1.5	4.2 ± 3.0
$P_\mu >15$	30 − 100	7 ± 5.2	317 ± 20	.08 ± .06	2.1 ± 1.6
	100 − 200	16 ± 6.7	703 ± 28	.52 ± .22	2.3 ± 1.0
	200 − 300	4.5 ± 3	131 ± 12	1.3 ± .9	3.6 ± 2.4

b) $\bar{\nu}$

	E_{vis} GeV	Prompt event numbers		$\dfrac{+\,+}{+}\times 10^{-4}$	$\dfrac{+\,+}{+\,-}\times 10^{-2}$
		$\mu^+\mu^+$	$\mu^+\mu^-$		
$P_\mu > 6$	30 − 100	13 ± 9	483 ± 24	.32 ± .22	1.9 ± 1.35
	100 − 200	11.7 ± 5.3	218 ± 15	2.2 ± 1.0	4.5 ± 2.0
	200 − 300	.6 ± 1	9.5 ± 3.2	4.5 ± 7.5	5.7 ± 6
$P_\mu > 9$	30 − 100	3 ± 4.4	336 ± 20	.06 ± .09	.7 ± 1.1
	100 − 200	9.6 ± 3.9	195 ± 15	1.7 ± .7	4.4 ± 1.8
	200 − 300	.8 ± 1	8 ± 3	5.5 ± 7	10 ± 15
$P_\mu >15$	30 − 100	−1.1 ± 1.8	108 ± 11	−.02 ± .04	1.1 ± 1.8
	100 − 200	4.5 ± 2.5	110 ± 11	.8 ± .45	4.1 ± 2.3
	200 − 300	−	8 ± 3	−	−

NEUTRINO PRODUCTION OF DIMUONS

K. Lang, A. Bodek, F. Borcherding,
N. Giokaris, I.E. Stockdale
University of Rochester, Rochester, NY 14627

P. Auchincloss, R.E. Blair, C. Haber, W.C. Leung,
S.R. Mishra, E. Oltman, M. Ruiz,
F.J. Sciulli, M.H. Shaevitz, W.H. Smith, R. Zhu
Columbia University, NY, NY 10027

Y.K. Chu, D.B. MacFarlane, R.L. Messner,
D.B. Novikoff, M.V. Purohit
California Institute of Technology, Pasadena, CA 91125

D. Garfinkle, F.S. Merritt, M. Oreglia, P. Reutens
University of Chicago, Chicago, IL 60637

R. Coleman, H.E. Fisk, Y. Fukushima, Q. Kerns, B. Jin,
D. Levinthal, T. Kondo, W. Marsh, P.A. Rapidis, S. Segler,
R. Stefanski, D. Theriot, H.B. White, D. Yovanovitch
Fermi National Accelerator Laboratory, Batavia, IL 60510

O. Fackler, K. Jenkins
Rockefeller University, NY, NY 10021

Talk presented by Wesley H. Smith

Abstract

Neutrino interactions with two muons in the final state have been studied using the Fermilab narrow band beam. A total of 437 ν_μ and 31 $\overline{\nu}_\mu$ opposite sign dimuon events with $P_\mu > 4.3$ GeV/c are used to extract the strange quark content of the nucleon. A sample of 18 ν_μ like sign dimuon events with $P_\mu > 9$ GeV/c yields 6.6±4.8 events after background subtraction. The kinematics of these events is compared with those of non-prompt sources. Limits are obtained on the coupling of neutral heavy muons between 1 to 3×10^{-4} of Fermi strength in the mass range 0.25-14 GeV/c^2.

Experimental Setup

Neutrino data were taken in the Fermilab dichromatic neutrino beam[1] during two runs of the CCFRR detector[2] with an integrated flux of 8.8×10^{18} 400 GeV protons on target. The neutrino beam was produced by the decays of sign and momentum selected ($\Delta P/P = \pm11\%$) hadrons in a 352m long decay pipe. Usable neutrinos and antineutrinos with energies between 30 and 230 GeV arrived at the detector located

1292m downstream of the beginning of the decay pipe.

The neutrino detector consists of an iron target calorimeter instrumented with scintillation counters and spark chambers, followed by an iron toroidal muon spectrometer. The 690-ton calorimeter is constructed of 168 3m x 3m x 5cm steel plates, 84 3m x 3m liquid scintillation counters (located every 10 cm of steel) and 36 3m x 3m spark chambers. The average density of the calorimeter is 4.25 g/cm^3. The hadron energy resolution is .89 $\sqrt{E \text{ (GeV)}}$.

The muon spectrometer consists of three 3m long solid iron toroidal magnets (containing 1.6m of steel each) with 1.8m outer radius and 12.7cm inner radius for the coils. The toroids were instrumented with spark chambers in the toroid gaps. The total transverse momentum kick of the toroids is 2.4 GeV/c and the rms fractional momentum resolution is 11%.

Analysis

Candidate dimuon events were chosen[3] from the entire sample of charged current triggers. Computer drawn pictures of the dimuon candidate sample were scanned by hand and 3208 dimuon events were selected. The efficiency for a dimuon to reach this sample was greater than 99%. The computer reconstruction of these events was examined by physicists and, if necessary, spark chamber hits were reassigned to the correct tracks and their momenta were recalculated.

Dimuon and single muon events passed the same analysis cuts. Events were required to have a transverse vertex within a 2.5m x 2.5m square and a longitudinal vertex at least 3.8m upstream of the downstream end of the target. At least one track must have had an angle of less than 250 mrad with respect to the beam axis. The straight line extrapolation of this track from the target must have passed through the first toroid magnet and intersected the trigger counter placed after it.

Opposite Sign Dimuons

Neutrino interactions with two muons of the opposite sign, $\nu_\mu + N \rightarrow \mu^- + \mu^+ + X$, are believed to come primarily from the production of a single charmed particle (e.g. $\nu_\mu + N \rightarrow \mu^- + D + X'$) with the subsequent semileptonic decay of the charmed hadron.

There is an additional source of opposite sign dimuons from decays to muons of primary π's or K's at the hadron vertex in charged current events and the production of prompt or non-prompt muons from the secondary interactions of the primary hadrons. This background was calculated using primary hadron spectra and multiplicities from the Lund Monte Carlo[4] and checked against BEBC ν-Ne[5] and EMC μ-p data[6]. The contribution of the subsequent interactions of these hadrons was calculated using the measured prompt and non-prompt muon production by hadrons in the Fermilab experiment E379 variable density target[7].

After analysis cuts there are 437 opposite sign di-muons with both muon momenta greater than 4.3 Gev/c from neutrinos and 31 from antineutrinos. The calculated π and K decay background is 106±21 for neutrinos and 7±1 for anti-neutrinos. Figure 1a presents a comparison of the energy dependence of the neutrino-induced $2\mu/1\mu$ rate for our exp-eriment (CCFRR) and CDHS wide band data[8]. Both samples are background subtracted and corrected for acceptance. The fully corrected CCFRR prompt $2\mu/1\mu$ rate rises to (9.0 ± 1.1) x10^{-3} at $E_\nu = 200$ Gev. Figure 1b shows the same rates cor-rected in addition for the effect of the charm mass thres-hold of 1.5 GeV/c^2 (slow rescaling).

It should be noted that the correction for slow re-scaling should remove most of the energy dependence from the $2\mu/1\mu$ ratio. Figure 2 shows the energy dependence of this ratio from our experiment after correction for three different assumptions for the mass parameter, m_c. While the data are not statistically strong and other interpretations are possible, it is clear that within the context of slow rescaling, the data prefer values of the mass parameter higher than m_c = 1.5 Gev/c^2.

Charm production of dimuons arises from the quark sub-processes,

$$\nu_\mu + \binom{d}{s} \rightarrow \mu^- + c + X \qquad \bar{\nu}_\mu + \bar{s} \rightarrow \mu^+ + \bar{c} + X$$

with the d-quark reaction suppressed through the Cabbibo mechanism by $\sin^2\theta_c$. The $2\mu/1\mu$ ratio then provides inform-ation about the momentum in the nucleon carried by strange quarks relative to that carried by lighter quarks: κ = $2s/(\bar{u} + \bar{d})$. The calculation of κ requires: (1) the fragmen-tation function for c $\rightarrow$ D; (2) the D-decay kinematics; (3) the branching ratio B = $\Gamma(D\rightarrow\mu\nu X)/\Gamma(D\rightarrow all)$ for the appro-priate mixture of D$^+$ and D^o; and (4) the parameterization of the threshold behavior of charm production (slow rescal-ing). The charm fragmentation of D^{*+} mesons in e$^+$e$^-$ reac-tions from the ARGUS group[9] is used. Most recent data on charmed particle lifetimes from the MARK III e$^+$e$^-$ experi-ment[10] applied to the charm particle composition from neut-rino interactions[11] give an average branching ratio for our case, B = .109±.014.

To extract the value of κ, the x-distributions of the background-subtracted dimuon data are compared with a Monte Carlo calculation based on the input discussed above. We use data with E_ν > 100 GeV to reduce the charm threshold uncertainties. The charm branching ratio is allowed to vary with the input constraint of 10.9±1.4% mentioned above. The χ^2 surface as a function of κ and the branching ratio is shown in Fig. 3 with the 1, 2, and 3 standard deviation contours. The CDHS value from wide-band dimuon data is also shown. We find κ = 0.52 $^{+0.17}_{-0.15}$ for E_ν > 100 GeV. The CDHS group finds κ = .52±.09 for E_ν > 80 GeV independent of a derived branching ratio of .071±.013.

Like Sign Dimuons

There are 18 neutrino-induced like sign dimuon events with both muon momenta greater than 9 GeV/c. The hadron background for these events is calculated to be 10.7 ± 2.1 events. The background from misclassified trimuon events for which the third muon is hidden in the shower is 0.6 ± 0.2 events. The background from spatially and temporally coincident charged current neutrino events is 0.1 ± 0.1 events. The total background is 11.4 ± 2.2 events. This leaves an observed like sign dimuon excess above background of 6.6 ± 4.8 events.

If we interpret this excess as a prompt signal and use as a model for the prompt dimuon events that of the non-prompt background, the acceptance-corrected ratio of prompt like sign dimuon production to single muon production is $(1.0\pm0.7)\times10^{-4}$. The energy dependence of the rate of prompt like sign dimuons to charged current events for our experiment and several others[3] is shown in Fig. 4. The solid line in Fig. 4 is the prompt signal calculated from a charm-anticharm gluon bremsstrahlung model[12] using $\alpha_s = 0.2$ and $m_c = 1.5$ GeV/c^2. The calculated rate scales with α_s^2 and predicts 0.2 events for our experiment.

Beyond the indications of a slight excess from the like-sign production rates, the kinematics of the observed events may be compared with the kinematics expected from the non-prompt background. Distributions of several kinematic variables for the like sign dimuons and the hadron background are shown in Fig. 5. The hadron background histogram, shown as a dashed line, is normalized to 10.7 events. The distribution of the angle between the two muon tracks projected on a plane perpendicular to the incident neutrino is shown in Fig. 5a. The peaking of this distribution at 180° indicates that the second muon is associated with the hadron vertex, as is expected for π and K decay.

For further comparison with the hypothesis of π and K

decay, the second muon is chosen to be the muon which has the smaller momentum in the direction perpendicular to the axis of the hadron shower (P_{t2min}). The hadron shower direction is determined from the incoming ν beam properties and the measured vector momentum of the chosen first muon. Figure 5b shows the $(P_{t2min})^2$ distribution. There is one event with a $(P_{t2min})^2$ of 6.0 $(GeV/c)^2$, which is unlikely to be from π and K decay but could originate from the trimuon or coincident background of 0.7 events. Figure 5c displays the momentum of the chosen second muon. Although the data are somewhat different in shape from that of the hadron background, the means are not statistically different.

Figure 5d shows the distribution of missing energy for all like sign dimuon events with visible energy greater than 100 GeV. The missing energy is the difference between the neutrino energy determined by the transverse vertex radius and the measured energy. The mean missing energy for the data is 2.2 standard deviations greater than that expected for the hadron background. With the possible exception of the missing energy and P_t distributions, none of the kinematics suggests an inconsistency with the hypothesis of π and K decay as the source.

Heavy Neutral Leptons

We have searched in the neutrino-induced data for neutral heavy leptons[13] with two distinct signatures[14]: an opposite sign dimuon, or two separated interaction vertices. The identification of first and second muon in dimuon events is done by charge assignment since the antineutrino contamination of the neutrino data is 10^{-3}. The conventional sources of opposite sign dimuons are charm production and hadronic shower production. For these mechanisms, the second muon is produced at or near the hadron vertex. These sources were eliminated by three requirements: (1) the azimuthal angle between first and second muon be less

than 120° in the plane perpendicular to the beam direction,
(2) the transverse momentum of the second muon with respect
to the beam axis be less than 1.2 GeV/c, and (3) the dimuon
momentum asymmetry, P_{μ^+}/P_{μ^-} be greater than 0.1.

A double vertex signature might be expected of a low
mass (<2.5GeV/c^2) neutral heavy lepton with sufficiently
small coupling to muons. There were no events with a second
vertex of energy greater than ten times minimum ionizing
located beyond 1m downstream of the first vertex and with
an average pulse height of less than half times minimum
ionizing between the vertices.

Figure 6 shows the 90%-confidence-level upper limit on
heavy lepton production as a function of mass and coupling
with respect to Fermi strength, U^2. The double vertex
search, also shown, yields a limit of 3×10^{-4} in U^2 near a
mass of 2 GeV/c^2. The dimuon search mode probes the mass
range between 2 and 14 GeV/c^2 with U^2 down to 5×10^{-3}.

References

1. R. Blair et al., Nucl. Inst and Meth. <u>226</u>,281(1984);
    Phys. Rev. Lett. <u>51</u>,343(1983).
2. D. Macfarlane et al. Z. Phys. <u>C26</u>,1(1984).
3. K. Lang et al., Nevis Preprint #R1350 (1986);
    submitted to Z. Phys. C.
4. T. Sjostrand, Computer Phys. Comm. <u>27</u>,243(1982).
5. P.C. Bosetti et al.,Nucl. Phys. <u>B149</u>,13(1979;
    <u>B209</u>,29(1982).
6. J.J. Aubert et al., Phys. Lett. <u>114B</u>, 373 (1982).
7. J. Ritchie, Ph.D. Thesis, UR861, U. of Rochester(1983);
    K.W.B. Merritt, Ph.D. Thesis, Caltech(1981).
8. H. Abramowicz et al., Z. Phys. <u>C15</u>,19(1982).
9. H. Albrecht et al., Phys. Lett. <u>150B</u>,235(1985).
10. R.M. Baltrusaitis et al., Phys. Rev. Lett.
    <u>54</u>,1976(1985).
11. N. Ushida et al., Phys. Lett. <u>121B</u>,292(1983).
12. B. Young, T.F. Walsh and T.C. Yang, Phys. Lett. <u>74B</u>,111
    (1978); V. Barger, W.Y. Keung, R.J.N. Phillips, Phys.
    Rev. <u>D25</u>,1803(1982).
13. M. Gronau, C.N. Leung and J.L. Rosner, Phys. Rev.
    <u>D29</u>,2359(1984); C.N. Leung and J.L. Rosner, Phys. Rev.
    <u>D28</u>,2205(1983).
14. S.R. Mishra et al., Nevis Preprint #R1358 (1986), to be
    submitted to Z. Phys. C.

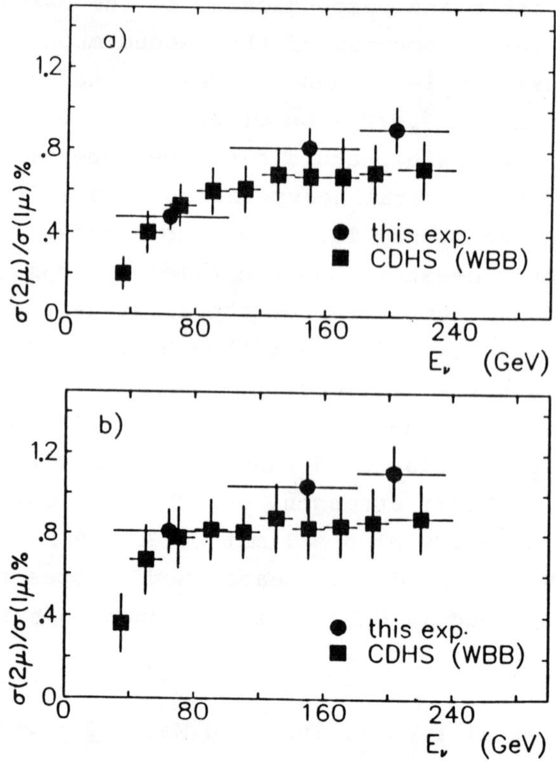

Fig. 1. Energy dependence of prompt opposite sign dimuons:
a) after correction for acceptance.
b) after additional correction for charm threshold.

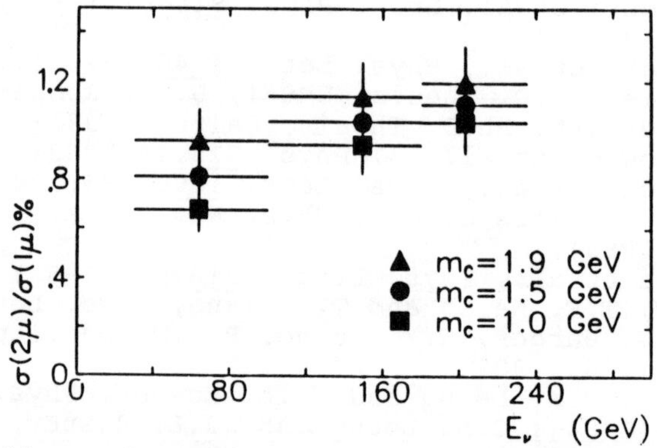

Fig. 2. Effect of slow rescaling with different m_c.

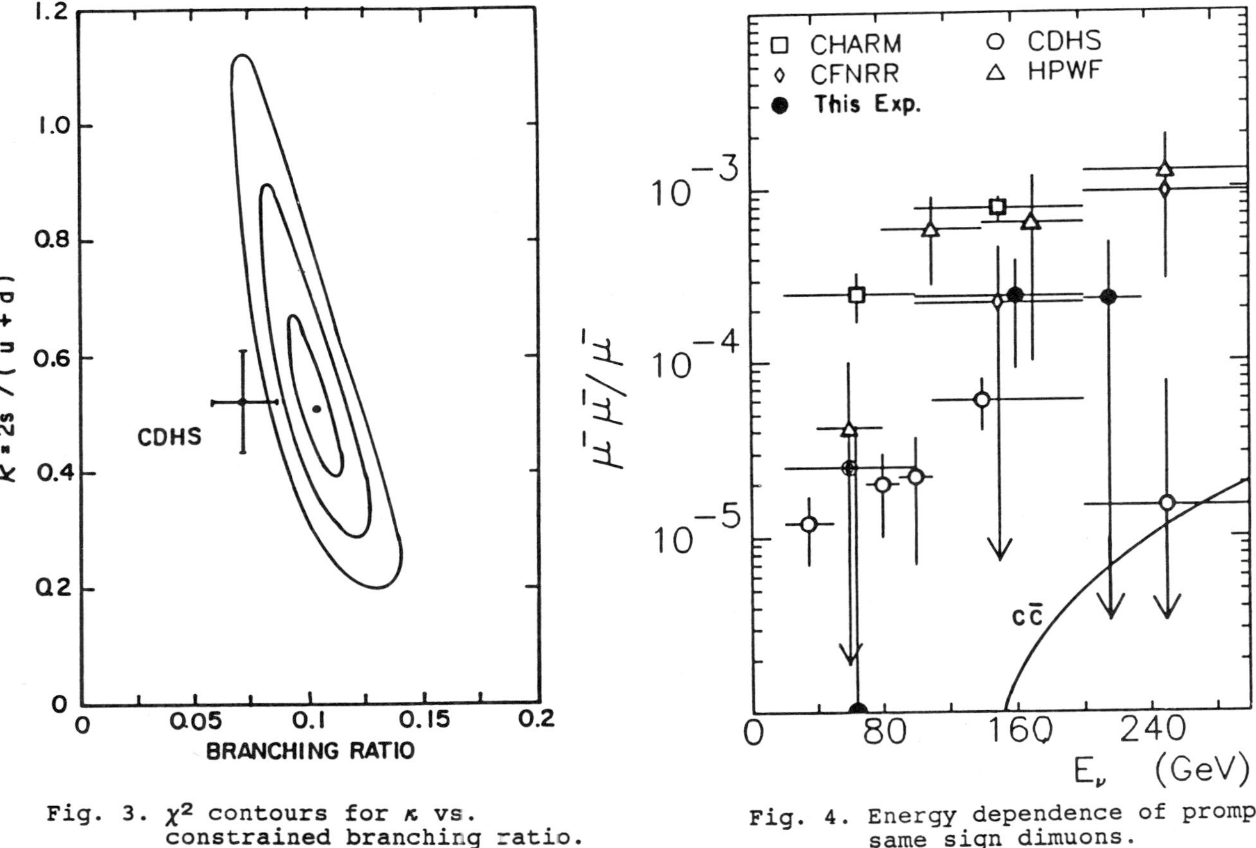

Fig. 3. χ^2 contours for κ vs. constrained branching ratio.

Fig. 4. Energy dependence of prompt same sign dimuons.

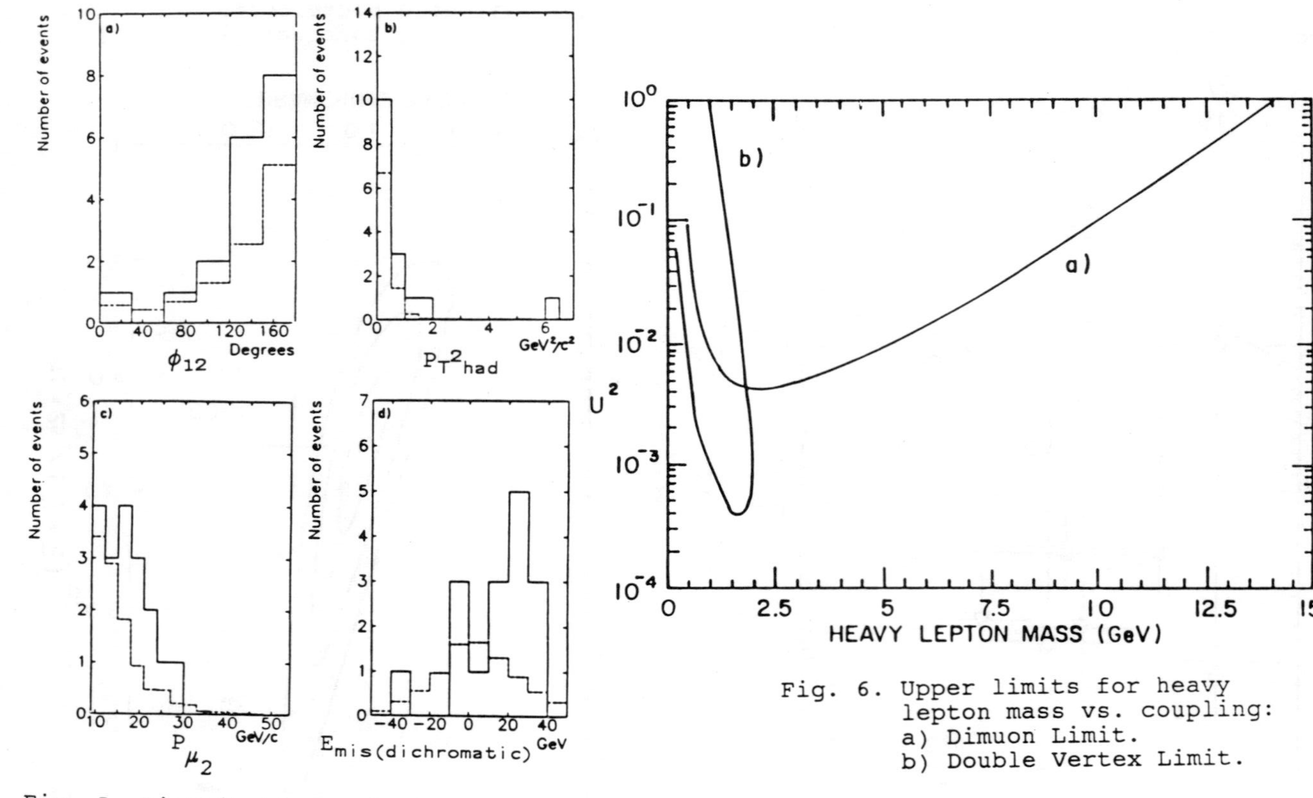

Fig. 6. Upper limits for heavy lepton mass vs. coupling: a) Dimuon Limit. b) Double Vertex Limit.

Fig. 5. Distributions of like sign dimuons (solid line) and background (dashed).

STRANGE PARTICLE PRODUCTION IN NEUTRINO-NEON
CHARGED CURRENT INTERACTIONS

Presented by R. Plano, Rutgers University at The 12th International
Conference on Neutrino Physics and Astrophysics, Sendai,Japan, June 86

N. J. Baker, P. L. Connolly, S. A. Kahn, M. J. Murtagh, R. B. Palmer,
N. P. Samios, and M. Tanaka, Brookhaven National Laboratory

C. Baltay, M. Bregman, L. Chen, M. Hibbs, R. Hylton, J. Okamitsu, and
A. C. Schaffer, Columbia University

M. Kalelkar, Rutgers University

(Expanded version to be published in Physical Review D, 1 Sep, 1986)

I. INTRODUCTION.

Neutral strange particle production in charged-current ν_μ
interactions have been studied in the Fermilab 15-foot neon bubble
chamber. The reactions studied included:

$$\nu_\mu + Ne \rightarrow \mu^- K_s^o X, \qquad K_s^o \rightarrow \pi^+ \pi^- \qquad 2279 \text{ decays}$$
$$\nu_\mu + Ne \rightarrow \mu^- \Lambda X, \qquad \Lambda^o \rightarrow \pi^- p \qquad 1843 \text{ decays}$$
$$\nu_\mu + Ne \rightarrow \mu^- \bar{\Lambda} X, \qquad \bar{\Lambda}^o \rightarrow \pi^+ \bar{p} \qquad 93 \text{ decays}$$

Associated production is expected to be the major source of
strange particles in charged-current neutrino interactions since the
d-quark is the only valence quark the neutrino interacts with, and
makes a strange quark only via a Cabibbo disfavored process; strange
quarks must be ejected from the sea. Other processes, such as charm
production and interactions with sea quarks are expected to be small.

The sample described here, from 61,800 charged-current events, is
used to provide insight into these ideas. Other topics include:

the first clear observation of Σ^0 production (94 examples) and Ξ^- production (4 examples) by neutrinos.

the dependence on various leptonic and hadronic variables is investigated.

a fit to single and associated production of s, s/$\bar{s}$, and c quarks is described based on the numbers of single and double strange particle production events.

II. EXPERIMENTAL PROCEDURE

The beam used in this experiment was the Fermilab double horn-focussed wide band ν_μ beam. The primary proton beam was extracted at an energy of 400 GeV in a fast spill ($\sim$20 μsec) with an average intensity of 10^{13} protons per pulse incident on the production target. The resulting neutrino spectrum peaked at 23 GeV, with about 10% of the events above 100 GeV; the average neutrino energy was 46 GeV.

The detector was the Fermilab 15-foot bubble chamber filled with a "heavy" neon-hydrogen mixture (61% atomic neon). The chamber liquid had a radiation length of 40 cm, so that electrons were easily identified providing a clear distinction between γ pair production and neutral strange particle decay (V^0). All γ's were measured to estimate the π^0 energy in each event. The interaction length was 125 cm providing a clean identification of muons; the fastest negative leaving track in each event was taken to be the muon. The "punch-through" of pions which thereby simulated muons was 9.1%.

The film was scanned for all charged current events having at least one V^0. To be accepted, an event had to meet the following criteria: μ^- momentum greater than 2 GeV/c, V^0 length greater than 1 cm, V^0 potential length greater than 10 cm, and visible energy in excess of 10 GeV.

All such events were measured and processed through our version of the geometrical reconstruction program TVGP and the kinematic fitting program SQUAW. For each V^0 the following hypotheses were tried as constrained fits to the production vertex: $K_s \to \pi^+\pi^-$, $\Lambda \to p\pi^-$, $\bar{\Lambda} \to \pi^+\bar{p}$, and $\gamma p \to (p)e^+e^-$. Fits were required to have two or three degrees of freedom, and a χ^2 less than five times the number of constraints.

This sample of events, consisting of 2279 K_s, 1843 Λ, and 93 $\bar{\Lambda}$, will be referred to as the strange charged-current (SCC) sample. It will be compared with a random charged-current (RCC) sample, selected randomly, without regard to the presence or absence of strangeness. After all cuts, the RCC sample consisted of about 2500 events.

The measurement of events in both samples included all the charged tracks at the primary vertex, and all the visible associated neutrals (γ's, V^o's and neutron interactions). We then corrected for the estimated 20% of the hadronic energy missing[1].

III. PRODUCTION RATES

To calculate V^o production rates, it was necessary to correct the observed number of V^o's for various efficiencies and backgrounds. The ten correction factors included corrections for geometric detection efficiency, interaction before decay, low-lifetime loss, slow decay prong, punch-through background, fake fits, random scan efficiency, reconstruction efficiency, branching ratio, and ambiguity resolution. The major contribution comes from the branching ratio to undetected (all neutral) decay modes which is well known; it includes a factor of 2 for K_L, so that our "K^o" rates include K^o and $\bar{K}^o$. The average total weights are K^o: 4.56 ± 0.23, Λ: 2.18 ± 0.12, and $\bar{\Lambda}$: 3.08 ± 0.41.

Inclusive V^o production rates as a fraction of all charged-current events are shown in Table I. They are based on a total charged-current sample of 61800 ± 2800 events with $E_\nu > 10$ GeV and $P_\mu > 2$ GeV/c.

In Table II we give the observed number of events and corrected rates for 11 specific single-, double-, and triple-V^o production channels. The rates are the solutions of 11 simultaneous equations;

for example, the number of observed single-K^o events is written as the sum over the true rates for each of the 11 channels multiplied by the probability for that channel to be observed as a single K^o and no other neutral strange particle.

TABLE I

Total Inclusive Strange Particle Production Rates. The Λ rate includes Λ's from Σ^o decay. The K^o rate includes ($K^o + \bar{K}^o$).

Particle	Observed	Corrected	Rate per charged-current event
K^o	2279	10392	0.168 ± 0.012
Λ	1843	4018	0.065 ± 0.005
$\bar{\Lambda}$	93	286	0.0046 ± 0.0008
Σ^o	94	713	0.011 ± 0.003
Ξ^-	4	36	$(6 \pm 4) \times 10^{-4}$

TABLE II

Single, Double and Triple V^o Production Rates. Note that X does not contain a visible V^o. K^o includes $K^o + \bar{K}^o$.

Reaction	Observed events	Rate per charged-current event
	Single-V^o	
$\nu_\mu Ne \to \mu^- K^o X$	1834	$(8.0 \pm 0.8) \times 10^{-2}$
$\mu^- \Lambda X$	1569	$(3.2 \pm 0.3) \times 10^{-2}$
$\mu^- \bar{\Lambda} X$	79	$(2.4 \pm 0.6) \times 10^{-3}$
	Double-V^o	
$\mu^- K^o K^o X$	100	$(1.6 \pm 0.8) \times 10^{-2}$
$\mu^- K^o \Lambda X$	205	$(2.3 \pm 0.4) \times 10^{-2}$
$\mu^- K^o \bar{\Lambda} X$	6	$(1.4 \pm 0.5) \times 10^{-3}$
$\mu^- \Lambda\Lambda X$	20	$(1.6 \pm 6.9) \times 10^{-4}$
$\mu^- \Lambda\bar{\Lambda} X$	8	$(9 \pm 3) \times 10^{-4}$
	Triple-V^o	
$\mu^- K^o K^o K^o X$	4	$(6 \pm 3) \times 10^{-3}$
$\mu^- K^o K^o \Lambda X$	8	$(6 \pm 2) \times 10^{-3}$
$\mu^- K^o \Lambda\Lambda X$	5	$(2 \pm 1) \times 10^{-3}$

Strange particle production has been studied previously both in neutrino[3-10,13] and antineutrino[8-12] experiments. In Table III we compare our results with some of these earlier measurements. The statistical richness of this experiment should be emphasized; we have more events than all the other experiments combined.

TABLE III

Comparison of Results on K^o and Λ production by ν and $\bar{\nu}$

Reference		$\langle E_\nu \rangle$ (GeV)	N_K	R_K	N_Λ	R_Λ	R_Λ/R_K
This exp.	νNe	46	2279	0.168±0.012	1843	0.065±0.005	0.39±0.04
All others combined:			1980		1257		
9	νNe	103	203	0.230±0.017	98	0.057±0.007	0.25±0.03
10	νn	62	234	0.208±0.016	157	0.071±0.007	0.34±0.05
10	νp	62	154	0.177±0.016	77	0.043±0.006	0.24±0.04
8	νp	~45	23	0.15 ±0.04	4		
7	νp	43	359	0.175±0.009	180	0.045±0.004	0.26±0.03
5	νp	~45	89		58		
4	νp	~45	19		20		
6	νd	~ 3	13	0.024±0.009	26	0.028±0.010	1.17±0.61
3	νFr	~ 3	14		40		
13	νFr	8	82	0.071±0.008	76	0.031±0.004	0.43±0.07
12	$\bar{\nu}$Ne	81	64	0.219±0.028	37	0.065±0.012	0.30±0.05
13	$\bar{\nu}$n	45	95	0.219±0.025	68	0.082±0.012	0.37±0.07
13	$\bar{\nu}$p	45	193	0.222±0.018	113	0.070±0.008	0.32±0.05
15	$\bar{\nu}$Ne	~45	350	0.164±0.009	257	0.063±0.004	0.38±0.03
11	$\bar{\nu}$p	~45	88	0.151±0.032	46	0.045±0.009	0.30±0.09

R_K and R_Λ are inclusive rates as a fraction of all charged-current events. N_K and N_Λ are the raw, observed numbers of V^o's.

IV. Σ^O and Ξ^- Production.

The $\Lambda\gamma$ effective mass distribution, shown in Figure 1, exhibits a very clear signal at the Σ^O mass. A fit to the distribution, using a Gaussian shape for the Σ^O and a third-order polynomial background, which had a χ^2 of 23.5 for 24 degrees of freedom, is shown as the solid line, while the dashed line indicates the fitted background. The fit yielded a total of 94 $\pm$ 25 Σ^O's with a mass of M - 1195 $\pm$ 2 MeV and a width of σ - 8 $\pm$ 2 MeV. The mass is consistent with the world average of 1192.5 MeV, and the width coincides with our mass resolution in the $\Lambda\gamma$ system at the Σ^O mass.

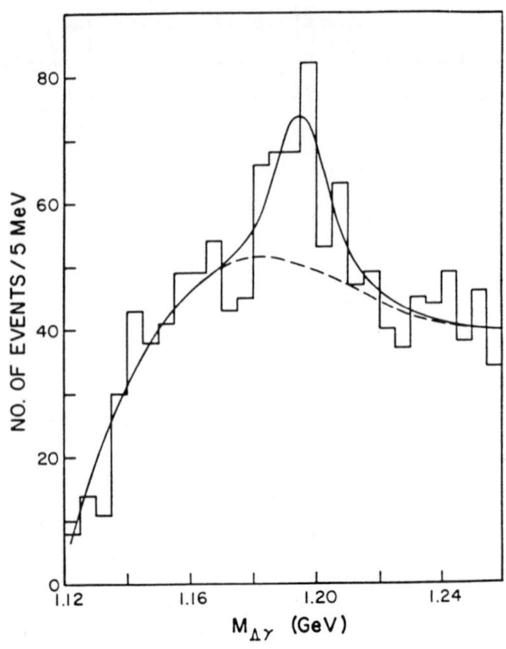

Figure 1:   Effective mass of the $\Lambda\gamma$ system, using fitted Λ and γ.

To obtain the Σ^O rate, it is necessary to weight the Σ^O for both the Λ and γ detection efficiencies. The sources of γ losses include (average weights given in parentheses): geometric detection efficiency (1.33), low-energy $e^{\pm}$ prongs (1.16), Short projected $e^{\pm}$ tracks on film (1.37), poor $e^{\pm}$ track visibility (1.03), and random γ scan efficiency (1.69). The overall average weight for γ's was 3.18 $\pm$ 0.31. The total inclusive Σ^O production rate is then (1.1 $\pm$ 0.3)% of all charged current events. The fraction of all Λ's coming from Σ^O decay is (16 $\pm$ 5)%. To our knowledge, this is the first measurement of Σ^O production in neutrino interactions.

Ξ^- production in neutrino interactions has not been previously reported. The cross section is expected to be small since double s-quark production is needed to produce a strangeness -2 particle.

Approximately one-third of the film was scanned for events in which a V^o pointed to a kink on a negative track. Candidates were measured and a kinematic fit made to the hypothesis $\Xi^- \to \Lambda\pi^-$, $\Lambda \to p\pi^-$. In addition to the selection criteria imposed on the entire SCC sample, two additional requirements were imposed: a) the Ξ^- track length must exceed 1 cm and b) the muon candidate was required to be the fastest leaving negative track as well as the track with the largest transverse momentum with respect to the vector sum of the other track momenta. This was intended to reduce the background due to K_L interactions.

Five events satisfied all these criteria of which one was induced by $\bar{\nu}$. The four ν events correspond to a rate of $(6 \pm 4) \times 10^{-4}$ of all charged-current events including a correction of 0.5 ± 0.5 events for the only significant background -- that of K_L interaction. This rate is comparable with the rate for $\Lambda\Lambda$ production $(1.6 \pm 6.9) \times 10^{-4}$, which also requires two s-quarks. None of the events is compatible with Ξ_c production giving a 90%-confidence-level upper limit on the probability for producing Ξ_c times the branching ratio for the decay $\Xi_c \to \Xi^-$ plus charged pions of 4×10^{-4}, expressed as a fraction of the cross section for all charged-current events.

V. DIFFERENTIAL DISTRIBUTIONS.

To investigate strange particle production mechanisms we consider next the differential distributions for K^o and Λ production. To obtain the rates for these variables, we obtained distributions of each variable for the SCC and RCC samples separately, and then divided them after appropriate normalization. $W > 2$ GeV was required in both samples to avoid the strangeness threshold.

There are striking differences between the K^o and Λ as illustrated by Figures 2 (E_ν), 3 (W^2), and 4b (y_B). The production rate of Λ is essentially independent of E_ν, Q^2, W^2 and $y_B = (E_\nu - E_\mu)/E_\nu$, while the K^0 rate increases sharply with all of these variables. A simple interpretation is that the Λ is produced in the target fragmentation region, and is generally independent of the energy of the W^+ coming from the leptonic vertex. By contrast, the K^0 is in the current jet where the s (or $\bar{s}$) quark is picked up from the sea, and this process is increasingly favored as the W^+ energy rises. This is in agreement with the independence of both the K^0 and Λ rates of x_B (Figure 4). x_B is the fraction of the nucleon momentum carried by the struck quark and the production of K^0 is determined primarily not by the struck quark, but rather by the W^+.

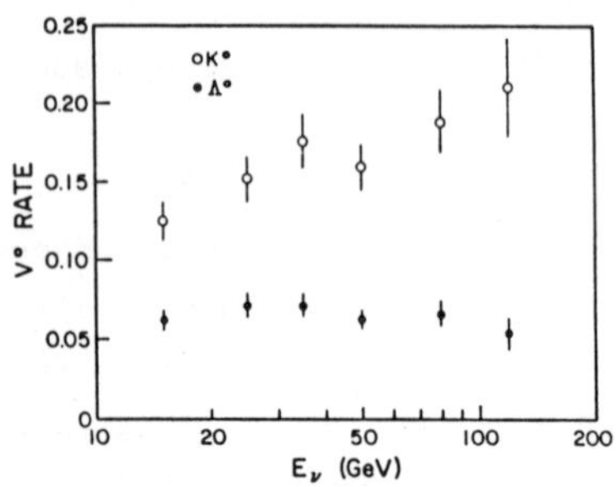

Figure 2: Inclusive V^0 production rates as a function of incident neutrino energy.

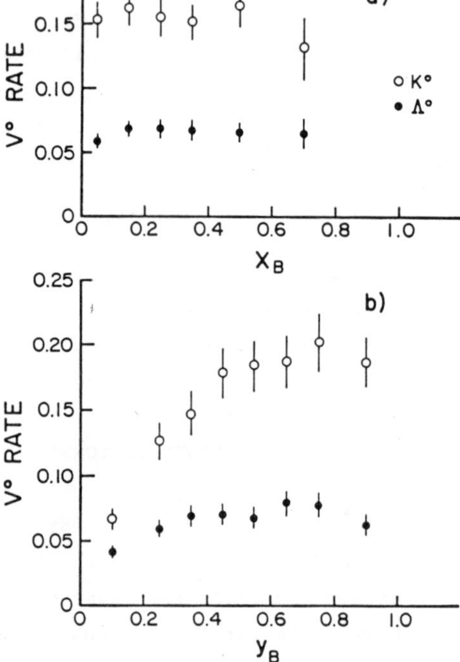

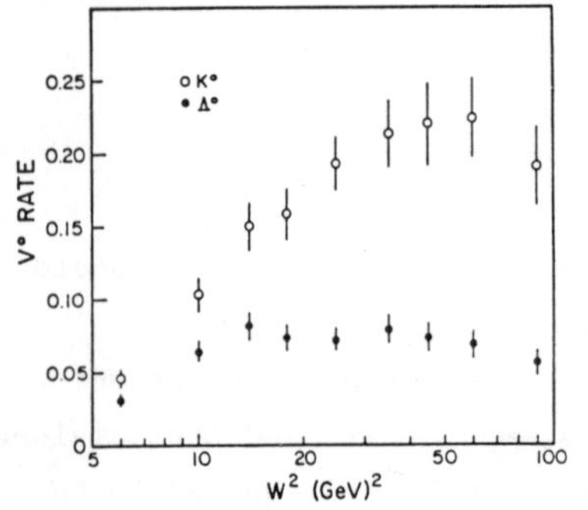

Figure 3: Inclusive V^0 production rates as a function of W^2, the invariant hadronic mass squared.

Figure 4: Inclusive V^0 production as a function of the Bjorken scaling variables x_B and y_B.

The x_F distributions (Figure 5) show the Λ to be backward, characteristic of a target fragmentation process, and the K^o central but somewhat forward, as expected from its production in the current jet. The asymmetry parameters

$$A = (N_F - N_B)/(N_F + N_B)$$

are $A_{K^o} = 0.16 \pm 0.02$ and $A_\Lambda = -0.71 \pm 0.02$. These are comparable to or significantly less than other experiments. The pions are produced in a narrow region around $x_F = 0$; their asymmetry parameters are $A_+ = +0.004 \pm 0.012$ and $A_- = -0.082 \pm 0.017$.

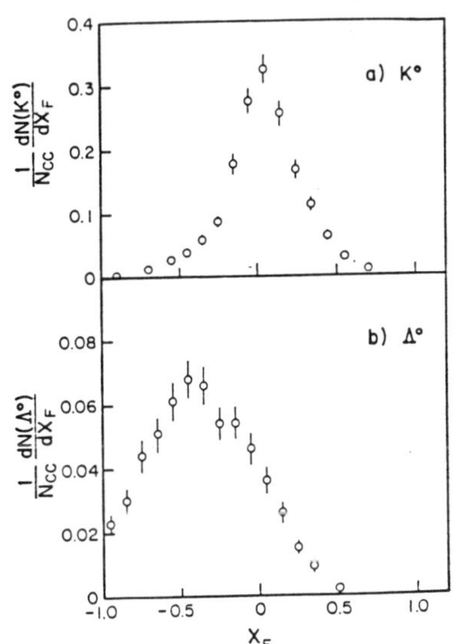

Fig. 5: Feynman x distributions for a) K^o, and b) Λ^o.

Fig. 6: Normalized distributions in the fragmentation variable z for K^o, Λ, π^+, and π^-.

Figure 6 shows the normalized z distribution for K^o, Λ, π^+, and π^-, where the fragmentation variable z is the fraction of the corrected hadronic energy carried by the particle. Note the statistically good information over more than four decades. The strange particle distributions turn over at low z, due in part to mass effects.

In the region $0 < P_T^2 < 0.5$ (GeV/c)^2, the P_T^2 distributions for K^o and Λ can be fit by a simple exponential of the form A exp ($-BP_T^2$). QCD predicts that $<P_T^2>$ should increase with Q^2 and decrease with x_B. We see no such effect, as illustrated in Figure 7, whether P_T is measured relative to the W^+ direction or to the experimentally more precise "lepton plane", the plane containing the neutrino and the muon, $<P_T^2>_{OUT}$. There is a clear increase with W^2 at least for the K^o.

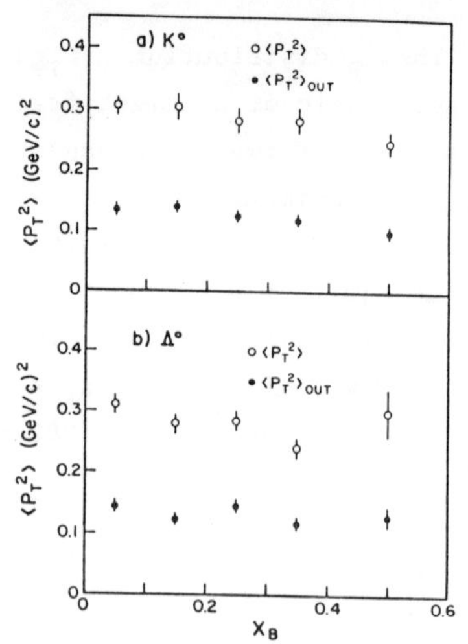

Fig. 7: Average transverse momentum squared as a function of x_B.

VI. RATES FOR SINGLE AND ASSOCIATED S-QUARK PRODUCTION.

As described in the Introduction, the primary mechanism for strange-particle production is expected to be associated production of an $s\bar{s}$ quark pair. It is possible to produce a single s-quark via charm production, although single $\bar{s}$-quark production has no simple production mechanism.

The aim of the fit described here is to use our measured rates for single- and double-V^o production to derive:

R_{AP}, the rate for associated production, and

R_S, the rate for single s-quark production (via charm).

We define the following probabilities:

P_{SM} — the probability that an s-quark appears as a meson (not baryon)

P_{ASM} — the probability that an $\bar{s}$-quark appears as a meson (not baryon)

P_{MV} — the probability that a strange meson will appear electrically neutral (not as a K^+ or K^-)

P_{BV} — the probability that a strange baryon will appear electrically neutral (not as a Σ^+ or Σ^-)

We have fit the seven appropriate expressions involving these parameters to the seven single and double V^o production rates shown in Table II, excluding $\Lambda\Lambda$. The fit was reasonable ($\chi^2 = 1.4$ for 2 degrees of freedom), and yielded the following results:

R_{AP} = 0.195 ± 0.014   Associated Production is dominant.

R_S = 0.050 ± 0.015   Charm Production. Consistent with the charged-current $\mu^- e^+$ rate of 0.5% and a roughly correct 10% branching rate for charm into e^+.

P_{SM} = 0.445 ± 0.040

P_{ASM} = 0.945 ± 0.010   Almost one; for an $\bar{s}$-quark to make an antibaryon, it must pick up two antiquarks from the sea whereas to make a meson, it needs only one u- or d-quark.

P_{BV} = 0.414 ± 0.029

P_{MV} = 0.500 ± 0.040   Expected, since neon is isoscalar.

Finally, we note that our fitted results, assuming associated production and charm production both take place, but double associated production does not, predict 2.4×10^{-4} for the $\Lambda\Lambda$ channel (which was not used in the fit), whereas the measured rate is $(1.6 \pm 6.9) \times 10^{-4}$.

VII.   CONCLUSIONS.

We have studied strange-particle production in ν_μ-Ne charged-current interactions, using a sample of 4215 fitted V^o's corresponding to 61,800 charged-current events. This sample is larger than all previous samples combined. Inclusive V^o production rates as a fraction of all charged-current events are measured and are shown in Tables I and II. The Λ/K ratio is 0.39 ± 0.04 and the fraction of Λ coming from Σ^o is (16 ± 5) %.

The single- and double-V^o production was used to determine that associated $s\bar{s}$ production occurs at a (19.5 ± 1.4)% rate while single s-quark production, presumably via charm production, occurs at a (5.0 ± 1.5)% rate.

The Λ comes from target fragmentation and its rate is independent of E_ν, Q^2, W^2, x_B, and y_B. The K^o is produced in the current jet and increases with E_ν, Q^2, W^2, and y_B, but is independent of x_B.

The P_T^2 distribution for both the K^o and Λ can be described by a simple exponential. Neither $<P_T^2>$ or $<P_T^2>_{OUT}$ shows significant variation with Q^2 or x_B, as predicted by QCD, but an increase with W^2 is observed.

ACKNOWLEDGMENTS

We thank the scanning and measuring staffs at our institutions, and the operating crews of the 15-ft chamber and the neutrino beam at Fermilab. This research was supported by the Department of Energy (Contract No. DE-AC02-76CH00016) and by the National Science Foundation.

REFERENCES

1. A. Grant, Nucl. Inst. and Methods <u>127</u>, 355 (1975).

2. Particle Data Group, Rev. Mod. Phys. <u>56</u>, S1 (1984).

3. H. Deden et al., Phys. Lett. <u>58B</u>, 361 (1975).

4. J. P. Berge et al., Phys. Rev. Lett. <u>36</u>, 127 (1976).

5. J. P. Berge et al., Phys. Rev. <u>D18</u>, 1359 (1978).

6. N. J. Baker et al., Phys. Rev. <u>D24</u>, 2779 (1981).

7. H. Grassler et al., Nucl. Phys. <u>B194</u>, 1 (1982).

8. R. Brock et al., Phys. Rev. <u>D25</u>, 1753 (1982).

9. P. Bosetti et al., Nucl. Phys. <u>B209</u>, 29 (1982).

10. D. Allasia et al., Nucl. Phys. <u>B224</u>, 1 (1983).

11. O. Erriquez et al., Nucl. Phys. <u>B140</u>, 123 (1978).

12. V. V. Ammosov et al., Nucl. Phys. <u>B177</u>, 365 (1981).

13. V. V. Ammosov et al., Z. Phys. C. <u>30</u>, 183 (1986).

COMPARISON OF QUASIELASTIC SCATTERING $\nu_\mu N \longrightarrow \mu^- P$ AND
Δ^{++} PRODUCTION REACTION $\nu_\mu p \longrightarrow \mu^- \Delta^{++}$ IN THE
BNL 7-FOOT DEUTERIUM BUBBLE CHAMBER

T. Kitagaki, H. Yuta, S. Tanaka, A. Yamaguchi, K. Abe,
K. Hasegawa, K. Tamai, S. Kunori*, Y. Otani**,
H. Hayano***, H. Sagawa***, K. Akatsuka**** and K. Furuno
Tohoku University, Sendai, Japan

N. J. Baker[+], A. M. Cnops[++], P. L. Connolly[+++],
S. A. Kahn, H. G. Kirk, M. J. Murtagh, R. B. Palmer,
N. P. Samios and M. Tanaka
Brookhaven National Laboratory, Upton, NY 11973

M. Higuchi and M. Sato
Tohoku Gakuin University, Sendai, Japan

(Presented by H. Yuta)

I. INTRODUCTION

We have studied the neutrino quasielastic and Δ^{++} (1232) production reactions in $\nu_\mu d$ interactions

$$\nu_\mu d \longrightarrow \mu^- p p_s \qquad (1)$$

$$\longrightarrow \mu^- \Delta^{++} n_s \qquad (2)$$

by using high statistics data obtained from a bubble chamber experiment at Brookhaven National Laboratory (BNL). These reactions are the most basic and simple exclusive ν-nucleon reactions and provide information on the nucleon structure of the hadronic weak axial vector current.

In this report, results on a detailed study of the axial form factor in these two reactions are presented. In particular, assuming the dipole axial form factor the behaviour of the axial-vector mass, M_A is examined from the Q^2 distributions for these reactions. We also compare the reaction mechanisms and show the similarity between these reactions.

II. EXPERIMENTAL PROCEDURE

The data used here came from an exposure of the BNL 7-foot bubble chamber filled with deuterium to the AGS wide band neutrino beam. The neutrino beam flux in the chamber peaked at 1.0 GeV and extended up to $\sim$15 GeV. A total of 1,800,000 photographs were taken. The film was scanned for neutral-induced events with more than one track visible in the chamber. Approximately, 32% of the film was rescanned yielding the scanning efficiences of 0.90 ± 0.01 and 0.95 ± 0.01 for the two and three prong events, respectively. All events were measured and processed through both the geometry program, TVGP, and the kinematic fitting program, SQUAW. Events which failed in the first measurement were remeasured to minimize the experimental bias.

The μ^-p and $\mu^-p\pi^+$ events were selected by using 3-constraint kinematical fits (3C) and particle identification. If the spectator nucleon was not measured, an initial value of (0 ± 45) MeV for P_x, P_y and P_z of the spectator was assigned in the fit. From this selection, 2684 μ^-pp_s events and 1610 $\mu^-p\pi^+n_s$ events were obtained in the fiducial volume of $4m^3$. The background in the μ^-pp_s events comes primarily from the reaction $\nu d \longrightarrow \mu^-p\pi^o p_s$. The background was estimated to be less than 6%. The overall correction factor was found to be 1.07 ± 0.05. For the $\mu^-p\pi^+$ channel, the corrections were made for contaminations from other reaction channels, H_2 contamination in the deuterium liquid, event loss due to fast neutron spectators in the kinematical fit and the scan-measuring inefficiency. The overall correction factor was estimated to be 1.13 ± 0.07.

The final sample for 2617 quasielastic events was obtained by selecting events in the neutrino energy range $0.3 \leq E_\nu \leq 6.0$ GeV. The final sample of 1385 Δ^{++} events was selected by requiring the hadronic mass W to be less than 1.4 GeV and the neutrino energy to be $0.5 \leq E_\nu \leq 6.0$ GeV. Table I shows the data used for the present analysis together with those from other recent experiments.[1-7]

III. FORM FACTOR ANALYSIS

To extract the nucleon weak form factors from reactions (1) and (2), we fit the experimental data to the theoretical predictions by using a maximum likelihood method. The predictions of the cross sections, $d\sigma/dQ^2$ and $d\sigma/dQ^2 dW$, are formulated based on the standard V-A theory with the following assumptions ; (i) time-reversal invariance and charge symmetry, (ii) PCAC and (iii) CVC. Under these assumptions, the axial vector form factors become only unknown form factors.

In the quasielastic scattering, the axial vector form factor, $F_A(Q^2)$, is conventionally parametrized by a dipole form;

$$F_A(Q^2) = -1.254/(1+Q^2/M_A^2)^2, \qquad (3)$$

where M_A is the axial vector mass. The experimental studies[1-7] have so far indicated that the dipole form of $F_A(Q^2)$ provides the best description of the data. The maximum likelihood function L^Q used in this analysis is given by

$$L^Q(M_A) = \prod_{i=1}^{N} \left[\frac{R(Q_i^2)\, d\sigma/dQ^2}{\int_{Q_{min}^2}^{Q_{max}^2} d\sigma/dQ^2} \right]^{W(Q_i^2)} \qquad (4)$$

where N is the total number of events in the Q^2 range from Q^2_{min} to Q^2_{max}, $R(Q_i)$ is a correction factor[8] for the free neutron cross section due to the effects of Pauli exclusion principle and deuteron binding and $w(Q^2)$ is the weight based on the scanning efficiency.

For Δ^{++} production reactions, several theoretical studies[9-12] have been performed assuming the above hypotheses. Detailed comparisions[7,13] of these predictions to the data have shown that the Adler model[10] describes the data most adequately. Thus, we use the Adler model developed by Schreiner et al.[13] The axial vector form factors are parameterized as

$$F^A_i(Q^2) = \frac{C^A_i(0)\left[1+a_i Q^2/(b_i+Q^2)\right]}{(1+Q^2/M_A^2)^2} \qquad (i = 3,4,5) \qquad (5)$$

where $C_i(0)$, a_i and b_i are the model-dependent parameters and given as

$$c_3(0) = 0, \; c_4(0) = -0.3, \; c_5(0) = 1.2$$

$$a_3 = b_3 = 0, \; a_4 = a_5 = -1.21, \; b_4 = b_5 = 2.0,$$

for the Adler model. The likelihood function, L^D, is defined as

$$L^D(M_A) = \prod_{i=1}^{N} \left[\frac{d^2\sigma/dQ_i^2 dW_i^2}{\int_{Q^2_{min}}^{Q^2_{max}} \int_{W^2_{min}}^{W^2_{max}} d^2\sigma/dQ_i^2 dW_i^2}\right]^{w_i(Q_i^2)} \qquad (6)$$

where W^2_{min} and W^2_{max} are taken to be 1.08 and 1.4 GeV. No Fermi motion effects in deuteron was included. To compare the M_A values for reactions (1) and (2), we use events with Q^2_{max} less than 3 GeV^2.

Figs. 1 (a) and (b) show the Q^2 distributions for reactions (1) and (2). The maximum likelihood fits to the data with the dipole axial vector form factors give

$$M_A = 1.07 \pm 0.05 \text{ GeV} \tag{7}$$

for reaction (1) and

$$M_A = 1.28 \pm 0.12 \text{ GeV} \tag{8}$$

for reaction (2) for the Q^2 range from 0.1 to 3 GeV^2. These M_A values deviate more than one standard deviation from each other. The ANL experiment[7] measured the M_A for both reactions in the same exposure and found the consistent values of M_A (see Table I). If the M_A corresponds to the mass of the existing axial vector meson, it is, in principle, expected to have the same M_A value for both reactions.

Figs. 2 (a) and (b) show the variation of M_A as a function of Q^2_{min} for reactions (1) and (2). An apparent decrease of M_A is observed as Q^2_{min} increases for reaction (1). A similar behaviour is also observed for reaction (2). The values of M_A given in (7) and (8) were obtained for $Q^2_{min} = 0.1$ GeV^2 where the deuteron effects are expected to be small according to the calculation of Singh[8] for reaction (1).

The curves in Figs. 1 (a) and (b) are the predictions with $M_A = 1.07$ and 1.28 GeV for reactions (1) and (2), normalized at $Q^2 = 0.25$ GeV^2. Good agreement is observed between the data and the predictions for $Q^2 > 0.25$ GeV^2, while disagreement is apparent for $Q^2 < 0.15$ GeV^2. The deviations, for instance, at $Q^2 = 0.1 - 0.15$ GeV^2 are more than one standard deviation and amount to $\sim 15\%$ of the data points for both reactions. The similar discrepancy is observed in the other deuterium bubble chamber experiments[1-3,6,7], but it is less obvious in the hydrogen bubble chamber experiments[4,5].

We have first examined the possible experimental bias for $Q^2 \lesssim 0.15$ GeV^2. Since quasielastic events with $Q^2 \simeq 0$ appear as 1 prong events, such events are lost in the scanning. We evaluated the event loss using the Monte Carlo generated events and the proton detection efficiency obtained from the observed spectator proton momentum distributions in reaction (1) which are plotted in Fig. 3. The circle dots in Fig. 1 (a) include the correction, but the agreement is not satisfactory. The correction for $Q^2 > 0.1$ GeV^2 is less than 1% which is expected because the proton track length in deuterium is longer than 15cm for $Q^2 > 0.1$ GeV^2. This suggests that the deviation in the small Q^2 region may comes from other sources such as deuteron effects or/and the form of $F_A(Q^2)$ used in this analysis.

To examine the deuteron effects, we display in Figs. 4 (a) and (b) the M_A variations for events with the spectator nucleon momentum, $P_s < 50$ MeV. The events with $P_s < 50$ MeV is expected to be less affected by Fermi motion and to be close to free nucleon scattering except for the effects of the Pauli exclusion principle. The similar decreasing behaviours of the M_A are observed for both samples of the events. However, the M_A values for the Δ^{++} are shifted down compared with those without p_s cut. This may suggests that the Fermi motion effect is present in the Δ^{++} channel where the deuteron effect is not included in the fit. The M_A value for $Q^2 = 0.1 - 3$ GeV^2 is now

$$M_A = 1.14 \pm \begin{smallmatrix} 0.16 \\ 0.14 \end{smallmatrix} \text{ GeV}$$

for $p_s < 50$ MeV, consistent with the M_A value of

$$M_A = 1.09 \pm 0.08 \text{ GeV}$$

for reaction (1). We have also used for reaction (1) the new calculations of $R(Q^2)$ on the deuteron effects by Singh and Arenhovel[14] and found similar behaviour of the M_A variation.

Another possible source for these discrepancies comes from the deviation of the axial vector form factor, $F_A(Q^2)$ from the dipole form. Fig. 5 (a) shows the ratio of $(dN/dQ^2)_{Data}/(dN/dQ^2)_{Dipole}$ with $M_A = 1.07$ GeV for reaction (1), normalizing to the events with $Q^2 = 0.2 - 1.05$ GeV^2. Also shown in Fig. 5 (b) is the similar distribution for reaction (2). As seen in Fig. 5, the dipole $F_A(Q^2)$ describes the data well for $Q^2 \gtrsim 0.2$ GeV^2, whereas similar deviations are observed for both reactions for $Q^2 \lesssim 0.2$ GeV^2. If this deviation comes from the deformation of the dipole $F_A(Q^2)$, we can, then, modify the dipole factor as

$$F_A(Q^2) = F_A^{Dipole} \cdot f(Q^2).$$

The function, $f(Q^2) = 1 + a(1-e^{-bQ^2})$ with $a = 0.3$ and $b = 8$, for instance, discribes the data satisfactory for small Q^2 region.

IV. COMPARISION OF $\mu^- p$ AND $\mu^- \Delta^{++}$ CHANNELS

To compare the two reactions, we show in Figs. 6 (a) and (b) the E_ν distribution of $\sigma(\mu^- \Delta^{++})/\sigma(\mu^- p)$ and the Q^2 distribution of $(d\sigma(\mu^- \Delta^{++})/dQ^2)/(d\sigma(\mu^- p)/dQ^2))$. The curves in Figs. 6 (a) and (b) are the ratios of the corresponding predictions with $M_A = 1.28$ GeV for the Δ^{++} and $M_A = 1.07$ GeV for the quasielastic reactions. The prediction ratio of $\sigma(\mu^- \Delta^{++})/\sigma(\mu^- p)$ describes the data

H. Yuta

well except the points for $E_\nu > 2.3$ GeV where the deviations are slightly more than one standard deviation.

In contrast to Fig. 6 (a), the ratio of $(d\sigma(\mu^-\Delta^{++})/dQ^2)/(d\sigma(\mu^-p)/dQ^2)$ stays constant for $Q^2 > 0.05$ GeV^2. The reduced χ^2 is 0.7 for the constant ratio and 2.6 for the prediction ratio, favouring the constant ratio for $Q^2 > 0.05$ GeV^2.

In a simple quark model, the processes $W + d \rightarrow u$ and $u + (du) \rightarrow p$ can be considered to be similar to the processes $W + d \rightarrow u$ and $u + (uu) \rightarrow \Delta^{++}$. The only difference between the quasielastic and Δ^{++} production reactions is the recombination of the recoiled u-quark to different diquark states ; (uu) for Δ^{++} (I = 3/2) and (ud) for p(I = 1/2). The observed constant ratio in Fig. 6 implys that the recombination of the u-quark to the different diquark does not affect the momentum transfer to the d-quark and that the Q^2 dependence for the Δ^{++} production reaction is the same as the neutrino quasi-elastic reaction.

ACKNOWLEDGEMENTS

We would like to thank the AGS staff, the crews of the 7-foot bubble chamber, and the scanning and measuring personnel both at BNL and at Tohoku University for their dedicated efforts. This research was supported by the US-japan Cooperative Research Project on High Energy Physics under the Ministry of Education, Science, and Culture of Japan and the U.S. Department of Energy under Contract No. DE-AC02-76-CH00016.

Table I

Axial-vector mass M_A in the dipole form factor from $\nu + H_2/D_2$ experiments.

Reactions	E_ν(GeV)	Raw Events	M_A(GeV)	Ref.
$\nu + n \longrightarrow \mu^- p$				
	0.3 -6.0	1138	1.070 ± 0.057	BNL(1981)[1]
	0.15-3.0	1737	1.00 ± 0.05	ANL(1982)[2]
	5.0 -200	362	$1.05 \pm^{0.12}_{0.16}$	FNAL(1983)[3]
	0.3 -6.0	2617	1.07 ± 0.05	This exp.
$\nu + p \longrightarrow \mu^- \Delta^{++}$				
	5.0 -100	138	$1.25 \pm^{0.15}_{0.13}$	FNAL(1978)[4]
	5.0 -200	551	0.85 ± 0.10	BEBC(1980)[5]
	0.5 -6.0	871	$0.98 \pm^{0.06}_{0.03}$	ANL(1982)[6,7]
	0.5 -6.0	1385	1.28 ± 0.11	This exp.

References

1. N. J. Baker et al., Phys. Rev. D23, 2499 (1981)
2. K. L. Miller et al., Phys. Rev. D26, 537 (1982)
3. T. Kitagaki et al., Phys. Rev. D28, 436 (1983)
4. J. Bell et al., Phys. Rev. Letters 41, 1008 (1978)
5. P. Allen et al., Nucl. Phys. B176, 269 (1980)
6. G. M. Radecky et al., Phys. Rev. D25, 1161 (1982)
7. S. J. Barish et al., Phys. Rev. D19, 2521 (1975)
8. S. K. Singh, Nucl. Phys. B36, 419 (1971)
9. S. L. Adler, Ann. of Phys. (N.Y.) 50, 189 (1968); S. L. Adler, Phys. Rev. D12, 2644 (1975)
10. P. Zucker, Phys. Rev. D4, 3350 (1971)
11. P. Salin, Nuovo Cimento 48A, 506 (1967)
12. J. Bijtebier, Nucl. Phys. B21, 158 (1970)
13. P. A. Schreiner and F. Von Hippel, Nucl. Phys. B58, 333 (1973) ; P. A. Schreiner and F. Von Hippel, Phys. Rev. Letters, 30, 339 (1973)
14. S. K. Singh and H. Arenhorel, Johannes Gutenberg University Report, MKPH-T-86-1 (1986)

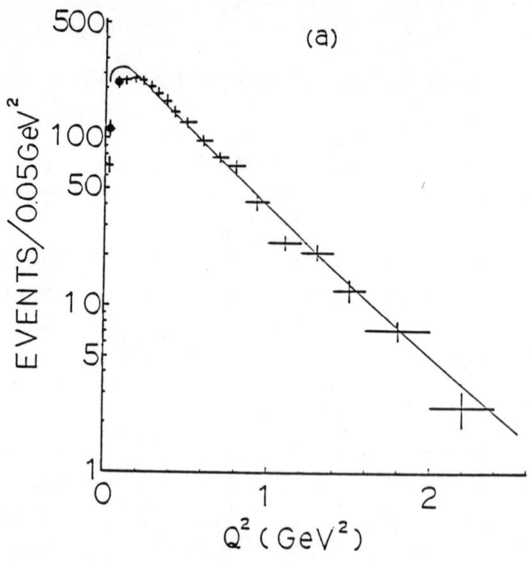

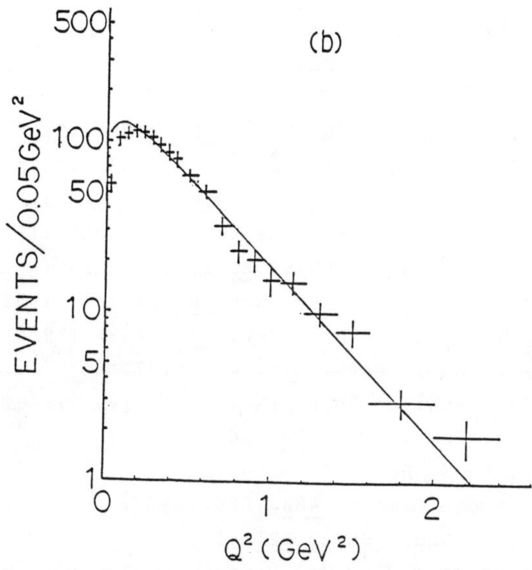

Fig. 1   The Q^2 - distribution for (a) the quasielastic
and (b) the $\triangle^{++}$ production reactions. The
circle dots are the points corrected for the
1 prong event loss. The curves are the theoretical predictions with the fitted M_A values.

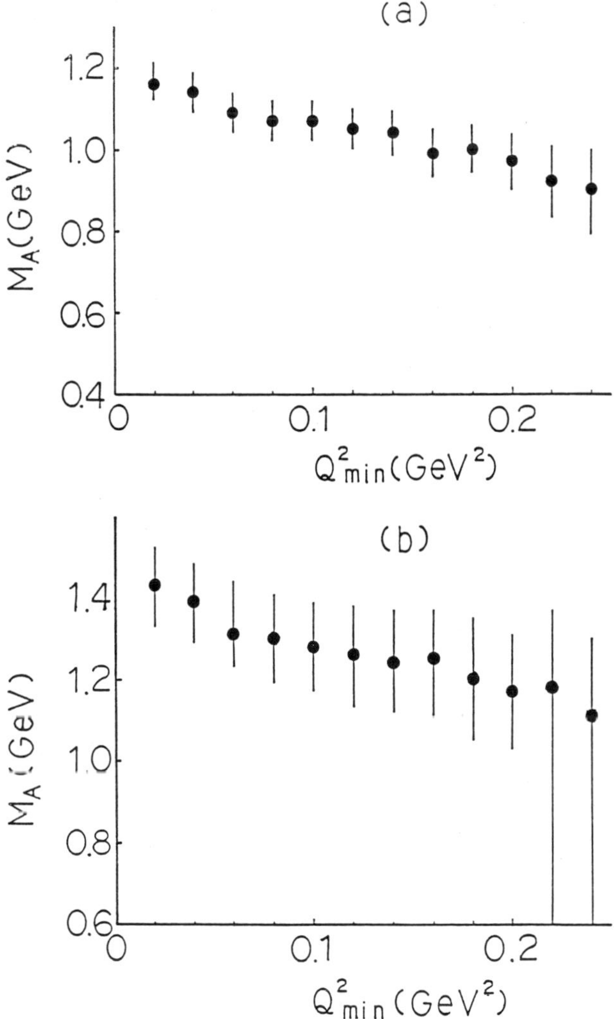

Fig. 2   The M_A distributions as a function of Q^2_{min} for (a) the quasielastic and (b) the Δ^{++} production reactions.

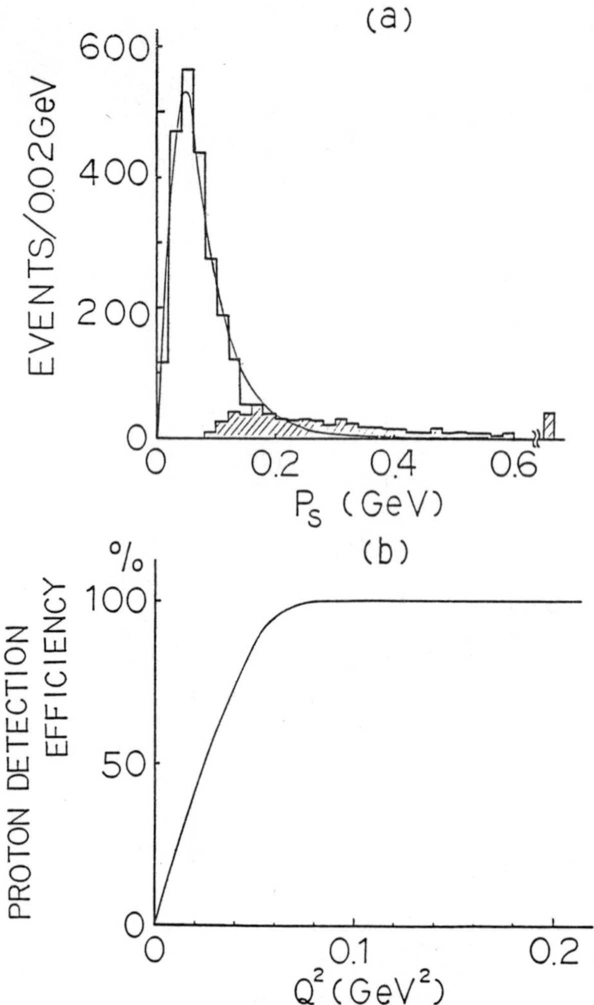

Fig. 3   (a) The spectator proton momentum distribution with the prediction from the Hulthen wave function. The shaded and the unshade areas correspond to the measured and the fitted spectator momenta, (b) The proton detection efficiency as a function of Q^2.

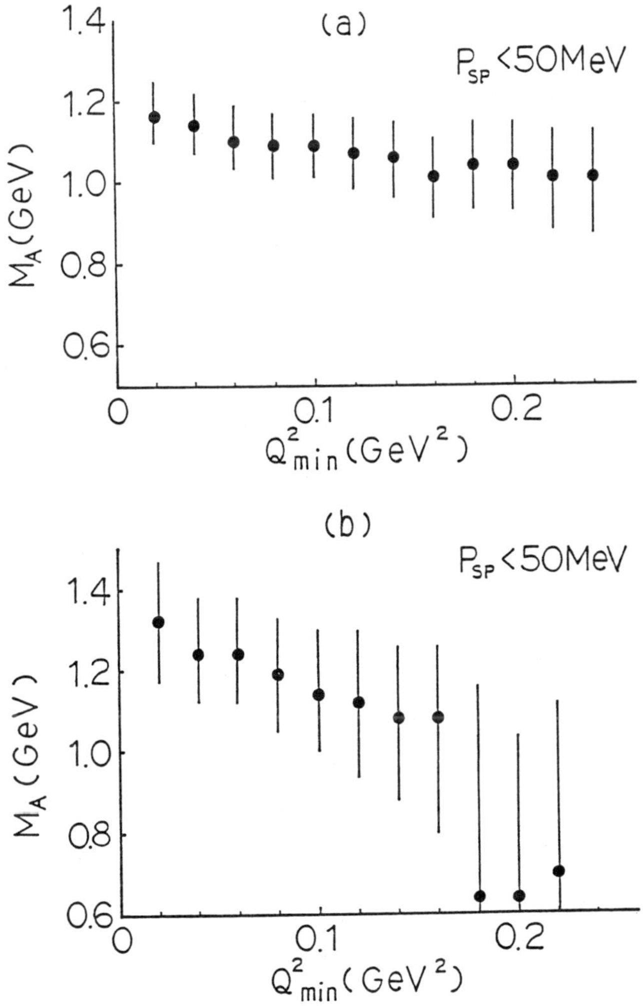

Fig. 4   The M_A distributions as a function of Q^2_{min} for events with p_s 50 MeV ; (a) for quasielastic and (b) the Δ^{++} production reactions.

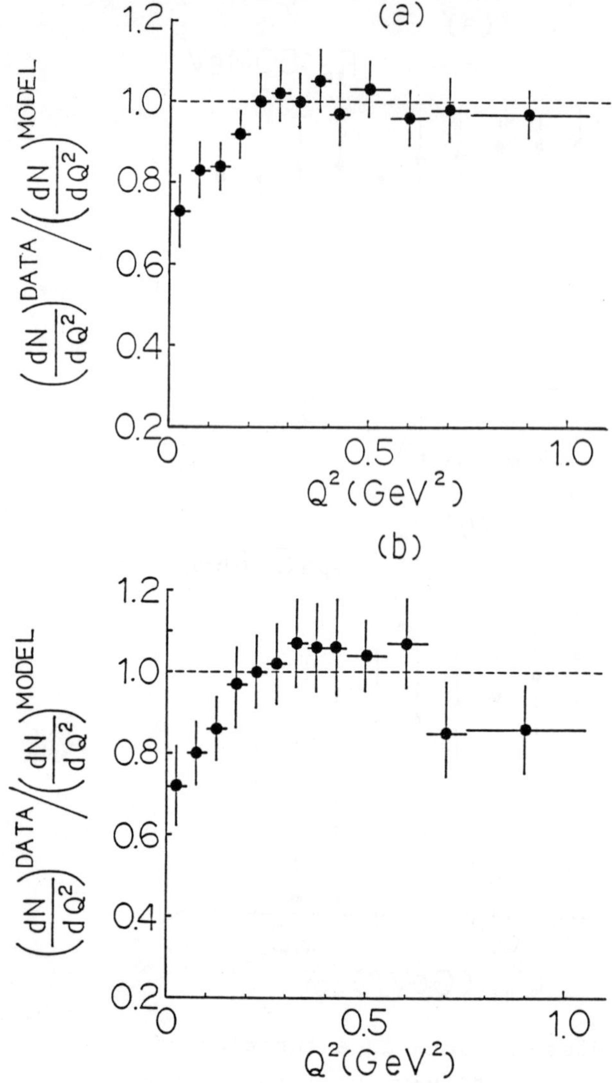

Fig. 5   The ratios of $(dN/dQ^2)^{Data}/(dN/dQ^2)^{Model}$ for
(a) the quasielastic and (b) the Δ^{++} production
reactions.

(a)

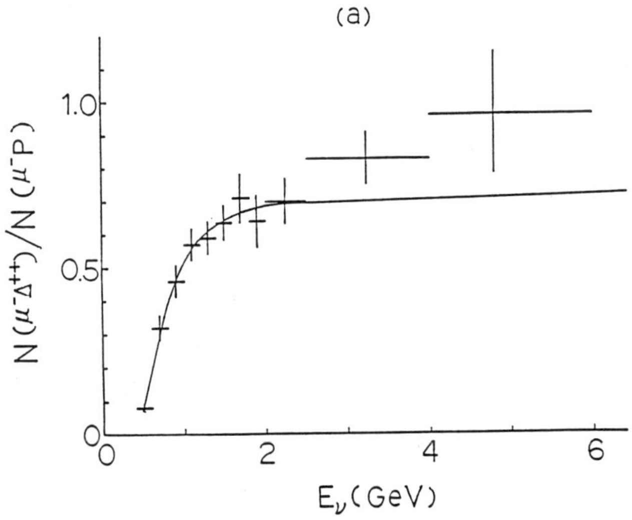

(b)

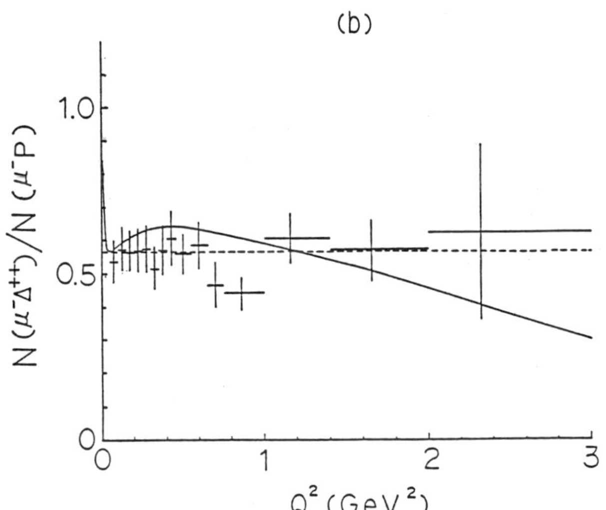

Fig. 6   The ratios of $N(\mu^-\Delta^{++})/N(\mu^- p)$ for (a) E_ν and (b) Q^2 distributions.

HADRONIC COMPONENT IN NEUTRINO INTERACTIONS

P. Marage

Inter-University Institute for High Energies (ULB-VUB)
ULB - CP 230 - Boulevard du Triomphe
B-1050 Bruxelles - Belgium

ABSTRACT

A review is presented of the manifestations of the hadronic component
in neutrino interactions. Detailed studies give support to the ideas
of meson dominance in the case of the weak current, and in particular
provide tests of the PCAC hypothesis at high energy.

1. INTRODUCTION

In neutral or charged current
interactions of (anti)neutrinos
on nucleons or nuclei, the intermediate vector boson (Z° or
$W^\pm$) may couple to a $q\bar{q}$ pair,
which interacts strongly with
the target. Similar processes
are described by the vector
meson dominance (VMD) models
in the case of (real or vitual)
photon interactions [1].

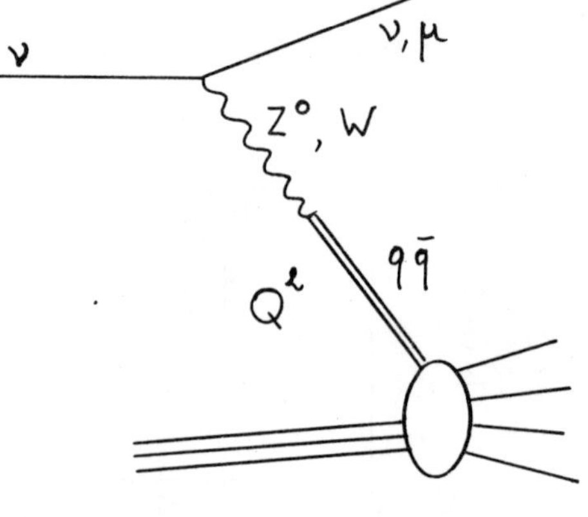

Kinematically, these processes are restricted to the low-Q^2,
low-x region [1a], which corresponds to interactions with large impact parameters, and where the concepts of deep inelastic scattering
do no apply.

Table I. Hadronic states coupling to the weak current

	Neutral current	Charged current
Vector current	ρ° (ω, ϕ, ψ, ...)	$\rho^{+,-}$
Axial-vector current	A_1°	$A_1^{+,-}$
	π°	$\pi^{+,-}$

As presented in Table I, the vector currents couple to the $J^{PC} = 1^{--}$ mesons (dominated by the ρ mesons). The axial-vector currents, which do not exist in the case of electromagnetic interactions, couple not only to axial-vector mesons (dominated by the A_1), but also to the pion field; this follows from the non-conservation of the axial-vector current. Indeed, the Partial Conservation of the Axial-vector Current (PCAC) hypothesis states that the divergence of the weak axial current is proportionnel to the pion field. On grounds of the CVC (Conservation of the Vector Current) and PCAC hypotheses, Adler has shown [2] that, at $Q^2 = 0$, the axial-vector weak current behaves like a pion.

2. DIFFRACTIVE PRODUCTION OF (AXIAL-)VECTOR MESONS

The diffractive production of (axial-)vector mesons by high energy neutrinos has been reported by two experiments using bubble chambers filled with hydrogen, and by a counter experiment (see Table II).

Table II. Observation of diffractive production
of (axial-)vector mesons

Detector (ref)	Channel	Signal
15' (FNAL) [3] (1978)	$\nu p \to \mu^- p \rho^+$	signal/background = 16/3
	$\mu^- p A_1^+$	13/8
	$\nu\ p\ \rho^\circ$	8/6
BEBC (SPS) [4] (1978)	$\nu p \to \mu^- p \rho^+$	seen
CDHS (SPS) [5] (1980)	$\nu N \to \nu\ \psi\ X$	signal = (45 + 13) events

3. COHERENT INTERACTIONS OF (ANTI)NEUTRINOS ON NUCLEI

The bulk of the data concerning the hadronic component in $\overset{(-)}{\nu}$ interactions comes from the observation of coherent interactions on nuclei. Several bubble chamber experiments have reported the study of inclusive and exclusive channels in neutral and in charged current interactions, and two non-bubble chamber experiments have reported the observation of neutral current production of a single π° (see Table III).

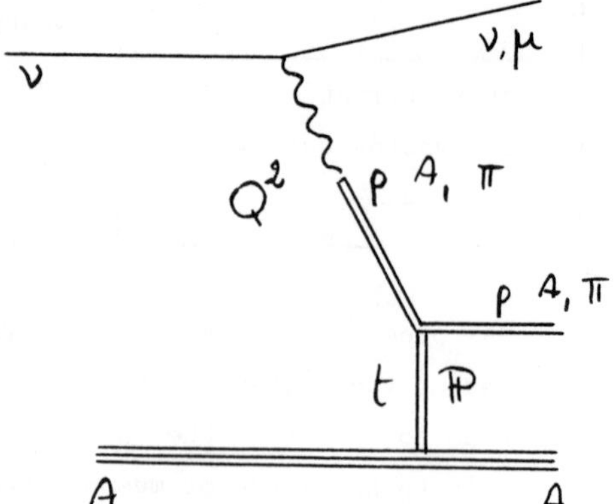

Table III. Observation of coherent interactions of $\overset{(-)}{\nu}$ on nuclei

Detector	(ref)	Beam	Channel
Gargamelle+Freon	[6] 1979	ν - PS	ρ°
Gargamelle+Prop./Freon	[7] 1980	ν - SPS	$A_1^{+,\circ}$, $\rho^{+,\circ}$
Aachen-Padova Spark ch.	[8] 1983	ν - PS	π°
Gargamelle+Freon	[9] 1983	ν - PS	$\pi^\circ \to 1\,\gamma$
BEBC+Neon	[10a] 1984	$\bar{\nu}$ - SPS	Inclusive CC
	[10b] 1985		π^-
	[10c] 1986		ρ^-
CHARM	[11] 1984	$\nu, \bar{\nu}$ - SPS	π°
SKAT+Freon	[12] 1985	$\nu, \bar{\nu}$ - Serp.	π^+, π^-, π°
15'+Neon	[13] 1986	ν - FNAL QTB	ρ^+
15'+Neon	[14] 1986	$\nu, \bar{\nu}$-Tev. QTB	Inclusive CC

3.1. Inclusive Signal

In coherent interactions [15], the nucleus behaves as a whole, without breakup. In consequence, no nucleons or stubs are emitted at the interaction vertex (stubs are defined as nuclear fragments or evaporation protons of momentum less than 300 MeV). In addition, the nucleus remains undetected even in bubble chamber experiments, the number of prongs being thus even and the final state charge zero.

Kinematically, coherent interactions are characterized by an exponentially falling $|t|$- distribution, where t is the square of the 4-momentum transfer to the nucleus :

$$\frac{d\sigma}{dt} \sim e^{-b|t|} \tag{1}$$

$b^{1/2}$ being of the order of the transverse dimensions of the nucleus.

In neutral current interactions, coherent events are characterized by a small emission angle of the hadronic system [8,11] (see Fig. 1).

In charged current interactions, the $|t|$-distributions for 2-prong events with and without stubs have very different shapes, as can be seen on Fig. 2a, in the case of $\bar{\nu}$ interactions in BEBC [10a]. The coherent signal is the peak at $|t| < 0.1$ GeV^2 for the events without stub, the shape of the incoherent background under the peak being obtained from the $|t|$-distribution of the events with stubs. The signal extracted from Fig. 2a is (234 ± 26) events, on a total of 17000 charged current interactions, or (1.4 ± 0.1) % of the $\bar{\nu}$ cross section at the SPS energy. A small signal is also reported for the observation of coherent 4-prong events.

The E632 experiment at the Tevatron Quadrupole Triplet Beam has reported to this Conference [14] preliminary results for the observation of 2-prong coherent events at $|t| < 0.1$ GeV^2, both in ν and $\bar{\nu}$ interactions (see Fig. 2b).

3.2. Single Pion Coherent Production

The study of the reactions

$$\overset{(-)}{\nu} A \rightarrow \overset{(-)}{\nu} \pi^\circ A; \quad \overset{(-)}{\nu} A \rightarrow \overset{(+)}{\mu}{}^- \overset{(-)}{\pi}{}^+ A \tag{2}$$

allows a test of the meson dominance model, and in particular of the PCAC hypothesis.

The dominant contribution to single pion coherent production is due to the scattering of the longitudinal component of the A_1 [16]. At $Q^2 = 0$, according to Adler's theorem, the A_1 behaves like a pion, and the strength of its coupling to the W is given by the pion decay constant f_π; the Q^2 dependence of this process is given by the A_1 propagator.

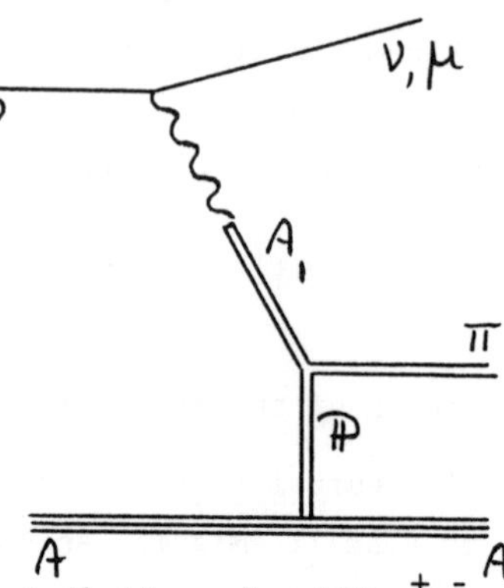

Fig. 3 and 4 show the differential distributions for 130 $\mu^+\pi^-$ events without stub at $|t| < 0.1$ GeV^2 from the WA59 experiment [10b]: (90 ± 4) % of these events are coherent; the incoherent background is obtained from the 6 events with stubs and is subtracted from the distributions (shown hatched on the figures). The experimental distributions are compared with the predictions of a Monte Carlo simulation taking into account the measurement errors.

The $|t|$-distribution (Fig. 3a) shows a dip at very small $|t|$-values, due to the lower kinematic limit of $|t|$: $t_{min} \simeq [(Q^2 + m_\pi^2)/2\nu]^2$; this threshold effect is eliminated by the use of the variable $t' = |t| - t_{min}$ (Fig. 3b). The widths of the $|t|$ and t' distributions reflect the transverse dimensions of the nucleus, broadened by the measurement errors; they are in good agreement with the model predictions.

In Fig. 4, differential distributions are displayed for the coherent events, and compared to the normal charged current events.

Fig. 4c shows that the Q^2-values for the coherent events are lower than those for other charged current events, but do extend up to 2 GeV^2; $<Q^2>$ was used to extract the value of the mass m_A which describes the data best, giving $m_A = (1.35 \pm 0.18)$ GeV, close to the A_1 mass, for the axial-vector mass.

The cross section for reaction (7) is related, through the PCAC hypothesis, to the elastic pion-nucleus cross section; the absolute value of the cross section for single pion coherent production provides thus a test of the basic assumption of the model. Fig. 5 compares, for several energy bins, the measured cross section to the model prediction, including a factor F_{abs} , which accounts for the nuclear reinteractions of the pion ($<F_{abs}> = 0.43$). Fig. 5 also presents the cross sections measured in other experiments [8-9, 11-12], scaled to correspond to charged current interactions on neon. A nice agreement with the model predictions is observed for all experiments, providing a consistent test of PCAC in the neutrino energy range 2 - 150 GeV.

The SKAT experiment has the unique feature to provide data in three different channels in the same experiment, indicating that the absolute cross sections for charged current coherent pion production by ν and $\bar{\nu}$ are equal within errors, and consistent with twice the neutral current cross section; this is in agreement with the predictions of the standard model [12].

3.3 ρ Meson Coherent Production

Two experiments have reported to this Conference the observation of the processes

$$\overset{(-)}{\nu} A \to \overset{(+)}{\mu}{}^{-} \overset{(-)}{\rho}{}^{+} A; \overset{(-)}{\rho}{}^{+} \to \pi^{+} \pi^{\circ}. \tag{3}$$

At $|t| < 0.1$ GeV^2, the BEBC experiment [10c] observes 37 $\mu^{+}\rho^{-}$ events without stub and 5 events with stubs, giving a coherent signal of (28 ± 7) events; the 15' experiment [13] observes 11 events without stub and 1 event with stubs. Fig. 6 shows the coherent ρ signal [10c].

The absolute value of the coherent ρ^- cross section in two energy bins for the WA59 experiment is shown on Fig. 7, and compared with Monte Carlo predictions based on VMD [16a,17]; two different values for R, -the ratio of the longitudinal to transverse ρ^- cross sections-, have been tried [18].

Differential distributions for the coherent signal in WA59 are shown on Fig. 8 and 9. A good agreement is observed between the data and the predictions of the VMD model (R = 0), for all variables presented. In particular, the Q^2 distribution supports the prediction of a vanishing cross section at $Q^2 = 0$, as required by CVC; this is in strong contrast with the sharp peaking at $Q^2 = 0$ of the cross section for the single pion production (Fig. 4c). On the other hand, the x-distribution of the ρ^- events is similar to that for the π^- events, being restricted to small x-values. This is kinematically linked to the small $|t|$-values selected for the coherent events; more generally, it is a necessary consequence of the fact that long distance hadronic fluctuations have to develop from the intermediate W boson in order to interact coherently with the nucleus.

Fig. 10 presents the $\cos\theta$ distributions for the ρ coherent signal in the two experiments, θ being the angle between the directions of the pion and of the (ρ, nucleus) system in the ρ rest frame. The shape of these distributions indicates a substantial longitudinal ρ production.

4. TOTAL CROSS SECTIONS

4.1. PCAC Test

The study of the total $\overset{(-)}{\nu}$ cross section at Q2 -> 0 allows to test Adler's theorem, which predicts the $\overset{(-)}{\nu}$ cross section to be proportional to the pion cross section; on the other hand, if the axial-vector current was conserved, -as the vector current is-, the cross section should vanish at Q2 -> 0. As appears on Fig. 11, data from charged current ν and $\bar{\nu}$ interactions on hydrogen (BEBC WA21 Collab.) and on neon (BEBC WA59 Collab.) are in agreement with the PCAC predictions [19].

4.2. Shadowing

The hadronic component in $\overset{(-)}{\nu}$ interactions, if it exists, must be partially absorbed on the outer part of the nucleus, leading to the observation of shadowing, with a cross section of the form :

$$\sigma(A) = A^{\gamma} \sigma(N), \text{ with } \quad 2/3 < \gamma < 1 \tag{15}$$

Two experiments performed at the CERN PS in the late '60 did not observed shadowing [20]; however, they suffered of theoretical uncertainties on the subtraction of the quasi-elastic events, and did not subtract the coherent interactions.

A comparison of ν and $\bar{\nu}$ interactions on deuterium (BEBC WA25 Collab.) and on neon (WA59 Collab.), performed in view of the study of the EMC effect, now shows clear evidence for shadowing at $x < 0.15$ and $Q^2 < 1$ GeV^2 [21] (see fig. 12).

5. CONCLUSIONS

1.a Coherent interactions of ν and $\bar{\nu}$ on nuclei have been observed in several experiments. The observed rate for $\bar{\nu}$ charged current interactions is (1.4 ± 0.1) % of the total cross section at the CERN SPS energy; correcting for the reinteractions inside the nucleus of the coherently produced particles, they account for about 3 % of the total $\bar{\nu}$ cross section, and as much as about 10 % for $x <$ 0.1;

1.b Detailed studies have been performed of the coherent production of single pions and of ρ mesons, providing tests of the meson dominance models and of the PCAC hypothesis;

1.c Results concerning the A_1 coherent production could be obtained in a near future; at the Tevatron, coherent production of F and F* mesons -which have Cabbibo - favoured couplings to the W boson- is also expected.

2.a The study of the total $\overset{(-)}{\nu}$ cross sections at Q^2 -> 0 has also provided new tests of the PCAC hypothesis;

2.b Shadowing has been observed in the comparison of $\overset{(-)}{\nu}$ cross sections on deuterium and neon.

3. These results give support to comments formulated at this Conference about high precision neutrino experiments, especially measurements of $\sin^2 \theta_W$ [22] : given the actual level of precision, PCAC effects and nuclear effects -in addition to the celebrated EMC effect- should be taken into account.

REFERENCES

[1] a. T.H. Bauer et al., Rev. Mod. Phys. $\underline{50}$ (1978) 261
    b. A. Donnachie and G. Shaw, ed., Electromagnetic Interactions
       of Hadrons (Plenum, New York, 1978).
[2] S.L. Adler, Phys. Rev. $\underline{135B}$ (1964) 963.
[3] J.Bell et al., Phys. Rev. Lett. $\underline{40}$ (1978) 1226.
[4] D.R.O. Morrison, Proc. 1978 Intern. Meeting on Frontier of Phy-
    sics (Singapore, 1978) 205.
[5] H. Abramowicz et al., Phys. Lett. $\underline{109B}$ (1982) 115.
[6] D. Picard, Production Cohérente des mesons vecteurs dans l'expé-
    rience wide-band du PS, Orsay, LAL-79/38 (1979).
[7] A. Bouchakour, Thesis (Strasbourg, 1980).
[8] H. Faissner et al., Phys. Lett. $\underline{125B}$ (1983) 230.
[9] E. Isiksal, D. Rein and J.G. Morfin, Phys. Rev. Lett. $\underline{52}$ (1984)
    1096.
[10] a. P. Marage et al., Phys. Lett. $\underline{140B}$ (1984) 137
     b. P. Marage et al., Zeit. Phys. $\underline{C31}$ (1986) 191
     c. WA59 Collab., Study of Coherent Production of ρ Mesons by
        Charged Current Antineutrino Interactions in BEBC, paper sub-
        mitted to this Conference.
[11] F. Bergsma et al., Phys. Lett. $\underline{157B}$ (1985) 469.
[12] H.J. Grabosch et al., Zeit. Phys. $\underline{C31}$ (1986).
[13] H.C. Ballagh et al., Observation of Coherent ρ Production in
     Neutrino-Neon Interactions, paper submitted to this Conference.
[14] E632 Collab., Observation of Coherent Charged Current Neutrino
     Interactions in the 15' Bubble Chamber at the FNAL Tevatron,
     paper submitted to this Conference.
[15] L. Stodolsky, Coherence in High Energy Reactions, in : Methods
     in Subnuclear Physics, Proc. Intern. School of Elem. Part. Phys.
     (Herceg-Novi, 1968), ed. M. Nikolic, vol. IV, part 1 (Gordon
     and Breach, New York, 1970) 259.
[16] a. C.A. Piketty and L. Stodolsky, Nucl. Phys. $\underline{B15}$ (1970) 571
     b. D. Rein and L.M. Sehgal, Nucl. Phys. $\underline{B223}$ (1983) 29.
[17] a. M.K. Gaillard, S.A. Jackson and D.V. Nanopoulos, Nucl. Phys.
        $\underline{B102}$ (1976) 326; $\underline{B112}$ (1976) 545
     b. M.S. Chen, F.S. Henyey and G.L. Kane, Nucl. Phys. $\underline{B118}$ (1977)
        345
     c. M.K. Gaillard and C.A. Piketty, Phys. Lett. $\underline{68B}$ (1977) 267
     d. A. Bartl, H. Fraas and W. Majerotto, Phys. Rev. $\underline{D16}$ (1977) 2124.
[18] J.J. Aubert et al., Phys. Lett. $\underline{161B}$ (1985) 203, and ref. therein.
[19] P. Allport, WA21 and WA59 BEBC Collab. priv. comm. (preliminary).
[20] a. K. Borer et al., Phys. Lett. $\underline{30B}$ (1969) 572
     b. M. Holder, Lett. Nuovo Cim. III, $\underline{14}$ (1970) 445.
[21] WA25 and WA59 BEBC Collab., priv. comm. (preliminary).
[22] L.M. Sehgal, this Conference.

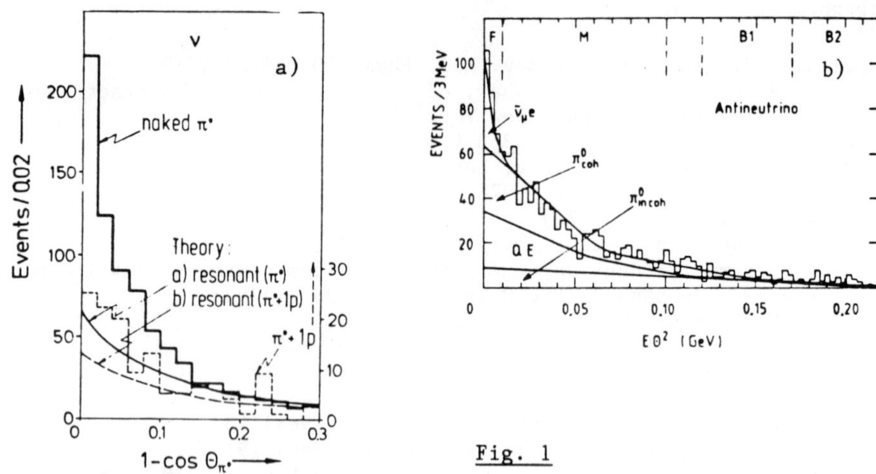

<u>Fig. 1</u>

Coherent production of a single π° meson a) ν in Aachen-Padova experiment [8] ; b) ν̄ in CHARM experiment [11]

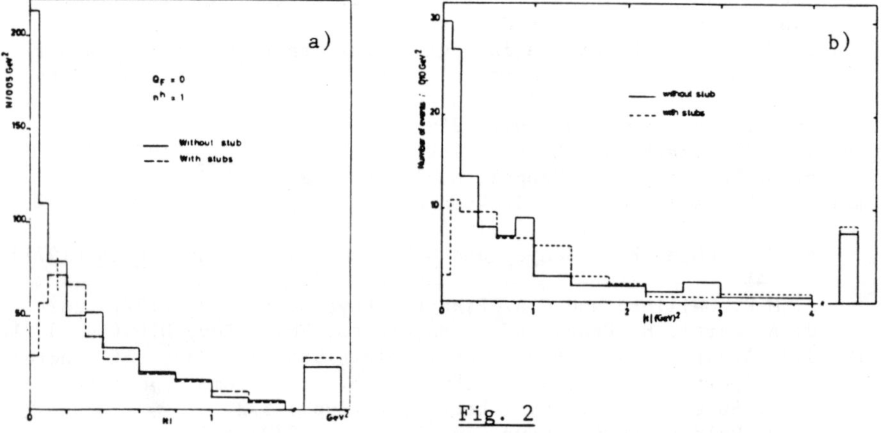

<u>Fig. 2</u>

|t|-distributions for 2-prong events with and without stubs a) BEBC
WA59 experiment [10a]; b) 15' Tevatron E632 experiment [14]

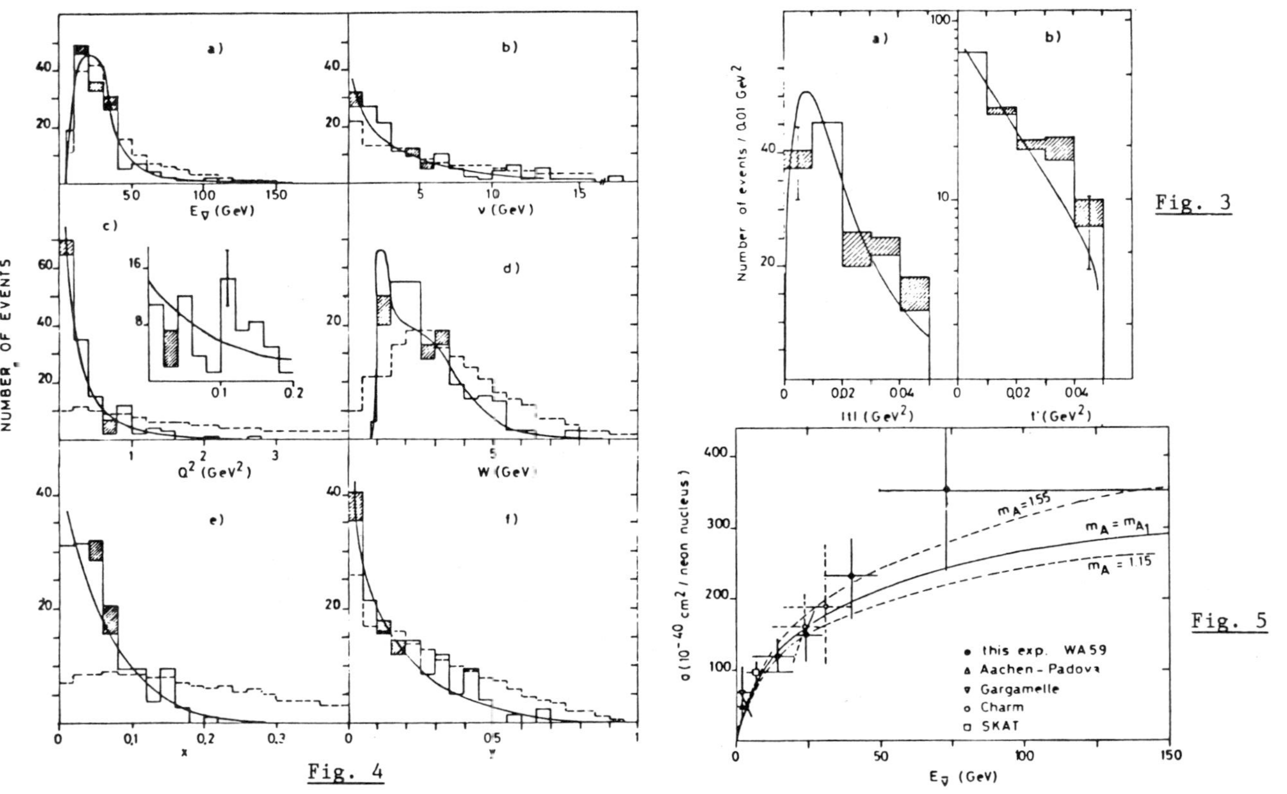

Single pion coherent production by antineutrinos in BEBC [10b] : differential cross sections for coherent events at $|t| < 0.05$ GeV^2 (dashed = 1 % of charged current) (Fig. 3 and 4), and total cross section (Fig. 5)

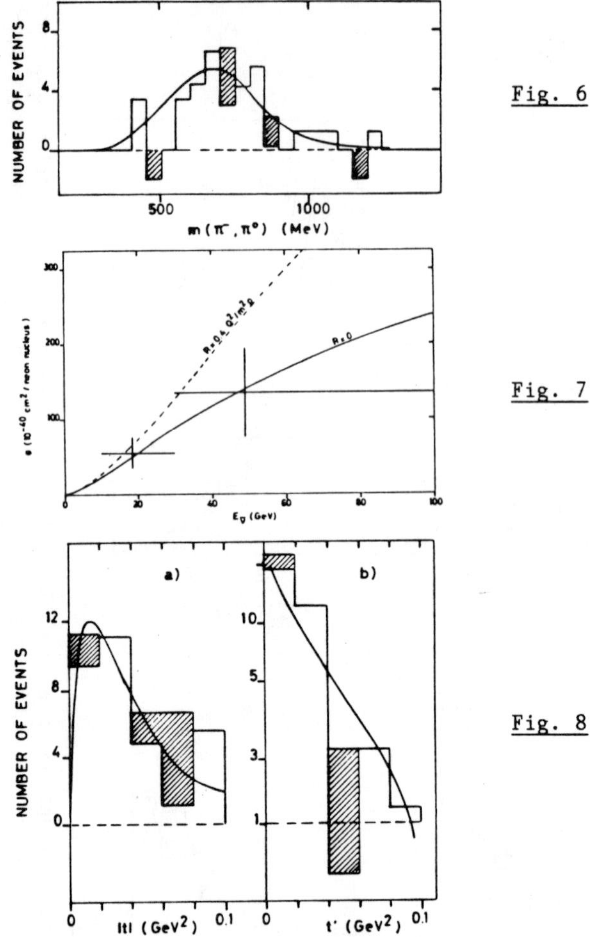

Fig. 6

Fig. 7

Fig. 8

ρ meson coherent production by antineutrinos in BEBC [10c] : (π^-, π^0) mass for coherent events at $|t| < 0.10$ GeV^2 (Fig. 6); total ρ coherent cross section (Fig. 7); $|t|$ and t' distributions at $|t| < 0.10$ GeV^2 (Fig. 8)

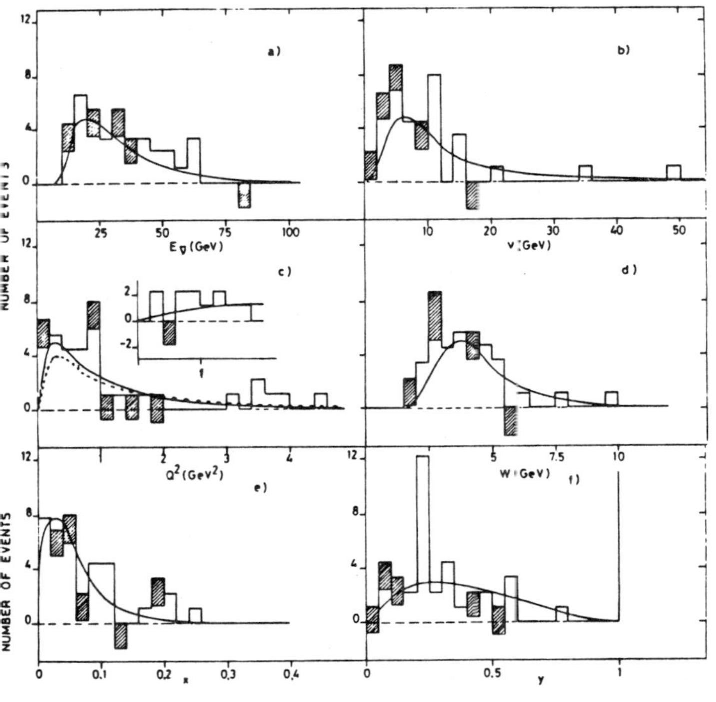

Fig. 9 : Differential cross sections for coherent ρ production at $|t| < 0.1$ GeV^2 by antineutrinos in BEBC [10c]

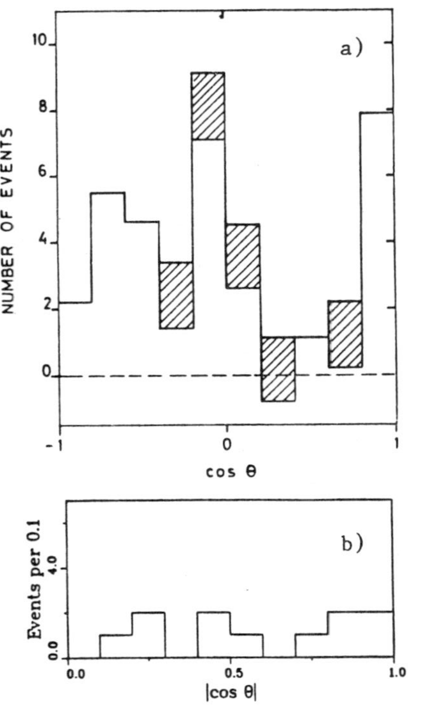

Fig. 10 : θ distributions for coherent ρ events a) in BEBC [10c] b) in the 15' b.c. [11]

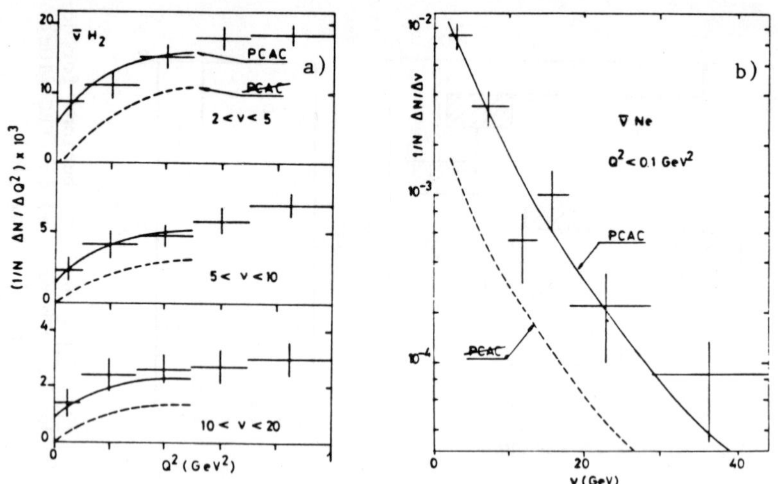

Fig. 11 : Total $\bar{\nu}$ cross sections at $Q^2 \to 0$ a) $\bar{\nu}$ H$_2$; b) $\bar{\nu}$ Ne [19] (preliminary)

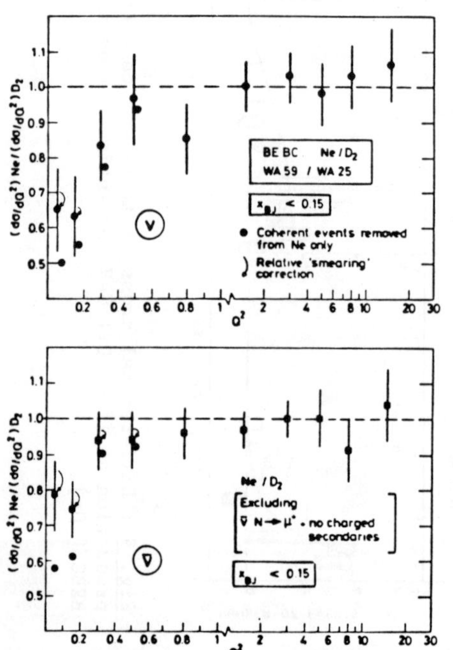

Fig. 12 : Ratio of ν and $\bar{\nu}$ cross sections on Ne and D$_2$ (preliminary)

Observation of Coherent ρ^+ Production in
Neutrino-Neon Interactions

H. C. Ballagh, H. H. Bingham, T. J. Lawry, J. Lys,
G. R. Lynch, and M. L. Stevenson
Department of Physics and Lawrence Berkeley Laboratory
University of California, Berkeley, California 94720

F. R. Huson,[a] E. Schmidt, W. Smart, and E. Treadwell
Fermilab, Batavia, Illinois 60510

R. J. Cence, F. A. Harris, M. D. Jones, A. Koide,
M. W. Peters, and V. Z. Peterson
Department of Physics, University of Hawaii at Manoa
Honolulu, Hawaii 96822

H. J. Lubatti, K. Moriyasu,[b] and E. Wolin[c]
Visual Techniques Laboratory, Department of Physics
University of Washington
Seattle, Washington 98195

U. Camerini, W. Fry, D. Gee,[d] M. Gee,[d] R. J. Loveless and D. D. Reeder
Department of Physics, University of Wisconsin
Madison, Wisconsin 53706

__Abstract__. Evidence is presented for coherent ρ^+ production from neon in the reaction $\nu_\mu + Ne \rightarrow \mu^- + \rho^+ + Ne$. A preliminary estimate of the rate is $(0.3 \pm 0.1)\%$ of all charged current interactions of average energy 90 GeV. The data come from Fermilab E546, an exposure of the 15-foot Ne-H$_2$ bubble chamber system to a quad-triplet-focussed broad band ν beam produced by 400 GeV protons.

__Introduction.__ We report here observation of coherent $\rho+$ production
$$\nu_\mu + Ne \rightarrow \mu^- + \rho^+ + Ne$$
$$\hookrightarrow \pi^+\pi^0$$
$$\hookrightarrow 2\gamma \qquad\qquad (1)$$

Coherent π^+ and π^- production from Neon by neutrinos and antineutrinos has been reported recently[1] as has coherent π^0 production in neutral current ν and $\bar{\nu}$ interactions on nuclei.[2] Some time ago M. K. Gaillard, et.al. and M. S Chen, et.al.,[3] particularly, stressed the interest of these reactions as probes of the space time structure of charged and neutral currents and their quantum numbers, and as potential sources of new higher vector, axial vector, etc., mesons. Their papers explain also the analogy with coherent electro- and muo-production of vector mesons from nuclei on which there is much literature[4].

__This experiment__ (Fermilab E546) used the 15-foot bubble chamber, filled with a neon-hydrogen mix (47% atomic neon, radiation length 53 cm, pion absorption length 193 cm), and equipped with a two-plane external muon identifier (EMI) and internal picket fence (IPF). The chamber was exposed to a quadrupole-triplet-focussed neutrino beam produced by 400 GeV protons. Using the EMI, 8,485 charged current events (with $E_\mu > 4$ GeV) were identified. Each was fully measured, including $\gamma \rightarrow e^+e^-$ conversions in the neon and V^0 decays. The average event energy was ~90 GeV.

    <u>Theory</u>. We have made some calculations for reaction (1) starting with the cross section formulae given by Chen, et al.[3] for diffractive vector meson production on a nucleon target. We keep the transverse virtual gamma cross section for which we use the naive vector dominance form (case iv of Chen, et al.), and set the longitudinal cross section to zero. We take an exp(bt) dependence with b = 79 GeV^{-2}, as in ref. (1). We assume that the ρ-meson-nucleus cross section is proportional to atomic number and we neglect any absorption. We compare our results below with these calculations.

    <u>Coherent ρ^+ production candidates</u> were selected by requiring that the event have: i) one and only one positive hadron that is not an identified proton (at this stage, events were not yet rejected if there were evaporation protons as well), ii) no negative hadrons, iii) no primary vees, iv) exactly two primary gammas. To avoid large errors, we also require fractional momentum errors of <0.30 for the μ^- and <0.60 for the positive hadron. Note that protons are identified only if they stop in the bubble chamber, possibly after a scatter.

    These cuts leave us with 76 events. A histogram of the gamma-gamma mass shows a clear peak at the π^0 mass (Fig. 1).

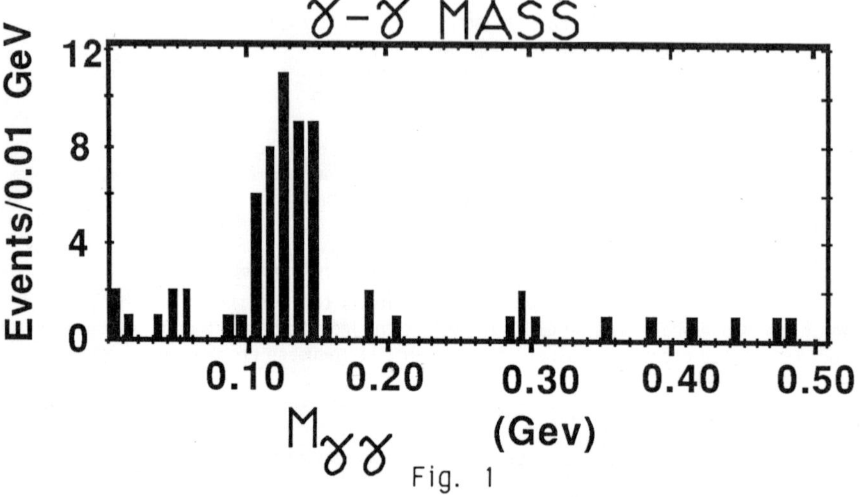

Fig. 1

We now select as π^0 events those with a gamma-gamma mass in the range 100-155 MeV/c^2. The $\pi^+\gamma\gamma$ effective mass for the remaining 44 events (we assume the positive hadron is a pion) shows a clear ρ^+ peak and very little background (Fig. 2). Removing 3 events with $\pi^+\gamma\gamma$ mass > 1200 MeV/c^2 leaves 41 events which we take to be $\mu^-\rho^+$ events. Of these events, 23 have no identified primary proton and so are candidates for reaction (1). Generally, stopping protons were measured if their range in the bubble chamber was > 1 cm, corresponding to momenta > 200 MeV/c.

    To look for coherent events, we want to look at t, the four-momentum transfer from the current to the nucleus. If we assume that an event is coherent, with recoil kinetic energy T_{Ne} of the Ne nucleus, then we have:

$$-t = p_T^2 + (E_\mu + E_\rho - p_{\mu L} - p_{\rho L})^2 + 2\,T_{Ne}(E_\mu + E_\rho - p_{\mu L} - p_{\rho L})$$

where p_T is the magnitude of the observed transverse (to the neutrino direction) momentum imbalance, $p_{\mu L}$ and $p_{\rho L}$ are longitudinal momenta of the μ^- and the ρ^+, respectively. Since we do not observe the Ne recoil, and can only estimate the neutrino energy from the final state particles' energies, we calculate an approximate value for $-t$ by neglecting the last term. (This approximation is good to better than 5% if $-t < 1$ GeV^2.)

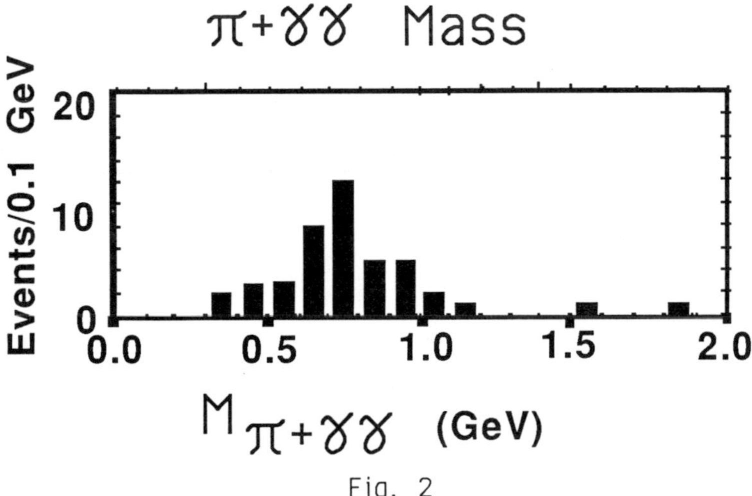

Fig. 2

Cross Section. A histogram of $-t$ for the 23 coherent event candidates shows a peak at very small values (Fig. 3a); 11 events have $-t < 0.1$ GeV^2. In contrast, for the 18 events with identified protons, if we calculate $-t$ as above with neglect of the protons, we find just 1 event at < 0.1 GeV^2 (Fig. 3b).

Following Marage et al.,[1] we assume that the events with protons give a good estimate of the non-coherent background at $-t < 0.1$ GeV^2 (after normalizing at $-t > 0.1$ GeV^2). Therefore we conclude that we have a net signal of 10 events at $-t < 0.1$ GeV^2, which we interpret as from reaction (1).

To estimate a rate for reaction (1), we need to make some corrections: (a) the scanning efficiency for two-prong events with neutrals was 0.86, (b) the measuring, reconstructing and pointing efficiency per gamma was 0.90, (c) the probability that both gammas convert in the liquid in the fiducial volume was 0.70, (d) the probability that the π^+ not interact before travelling far enough to pass the $\Delta p/p$ cut was 0.91, (e) the probability that true coherent events pass the $-t < 0.1$ GeV^2 cut (in spite of measurement errors) was 0.93. Then we find a rate for reaction (1), relative to all charged current events, of (0.3 ± 0.1)%, in accord with predictions,[3] within their uncertainty of at least a factor two. The incident neutrino energy in the experiment ranged from ~ 6 to 320 GeV, with a median value of 80 GeV. Of the 11 events with $-t < 0.1$ GeV^2, 10 have energies < 80 GeV. Therefore it is likely that the rate, relative to all charged current interactions, decreases with increasing energy, contrary to our estimate of approximately equal numbers of coherent events above and below 80 GeV, based on ref. (3). A decrease in the relative rate with energy is observed for coherent π^- production by $\bar{\nu}$[1], also.

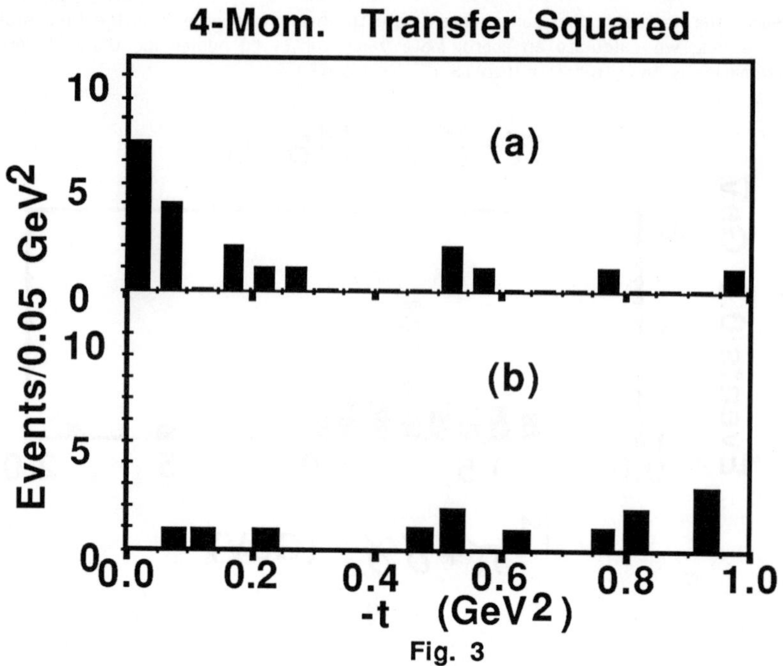

Fig. 3

<u>Characteristics of Coherent Events</u>. Distributions of the usual scaling variables $x = Q^2/2M_p\nu$ and $y = \nu/E_\nu$, ($Q^2 = (p_\nu-p_\mu)^2$, $\nu = E_\nu-E_\mu$, M_p = proton mass; the neon mass may be more appropriate for coherent events) are shown in Figs. 4 and 5. The coherent candidates ($|t| < 0.1$ GeV^2 and no extra protons) are shown separately from the incoherent events in these and the following figures. As predicted by ref. (3), x for the coherent candidates is sharply peaked below 0.1 while x for the incoherent candidates is much broader. For the coherent candidates y has a mean value of 0.35 ± 0.07. The mean value and the distribution agree with our calculation using the formulae of ref. (3), which predict $<y> = 0.36$. This is in contrast to the approximately flat distribution between 0. and 1. for the normal charged-current events and the much softer distribution for the incoherent $\mu^-\pi^+\pi^0$ events, a subsample of the total with relatively low hadron energy, thus relatively low y. These characteristics of the coherent events are expected. The nuclear form factor tends to restrict $|t|$ and thus also x; it also favors high energy for the virtual vector meson, i.e., high y, relative to incoherent events of the same topology. As predicted, the Q^2 distribution (Fig. 6) for the coherent events peaks near M_ρ^2 while Q^2 for the incoherent events extents to much higher values.

X DISTRIBUTION
(a)
(b)
(c)
Events/0.05
4
0
4
0
4
0.0 0.2 0.4 0.6 0.8 1.0
X

Fig. 4

Y DISTRIBUTION
(a)
(b)
(c)
Events/0.10
4
0
4
0
5
0
0.0 0.2 0.4 0.6 0.8 1.0
Y

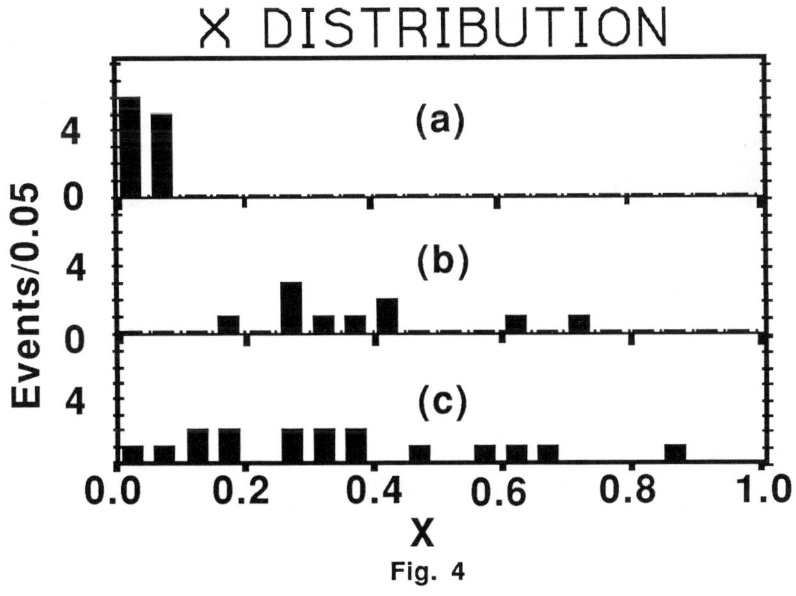

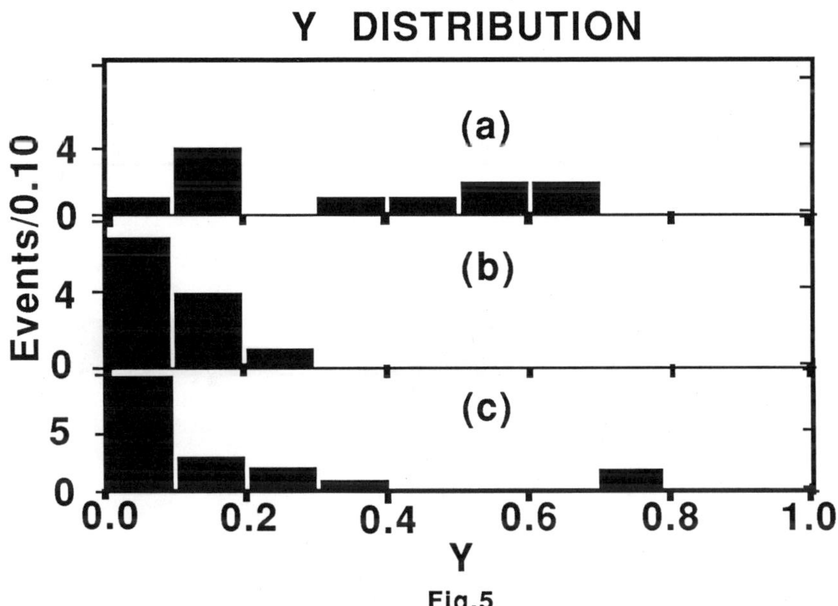

Fig.5

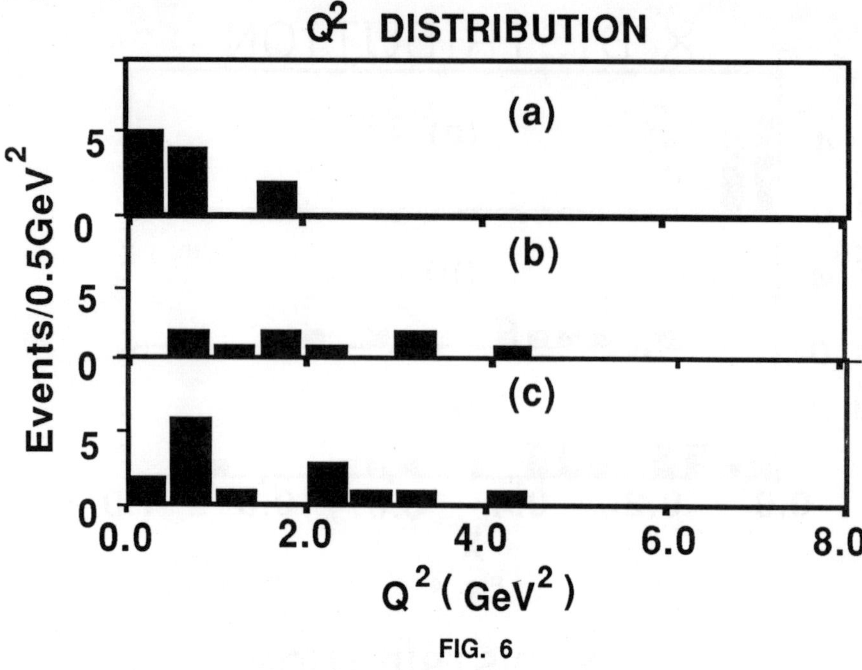

FIG. 6

The helicity polar and azimuthal angular distributions for the ρ^+ decay are shown in Figs. 7 and 8. In principle, these distributions provide information on the relative magnitude of the longitudinal and transverse pieces of the coherent cross section. Our statistics are too limited for a meaningful estimate, but the flatness of the $\cos\theta$ distribution suggests comparable contributions of longitudinal and transverse terms. The fraction of events with

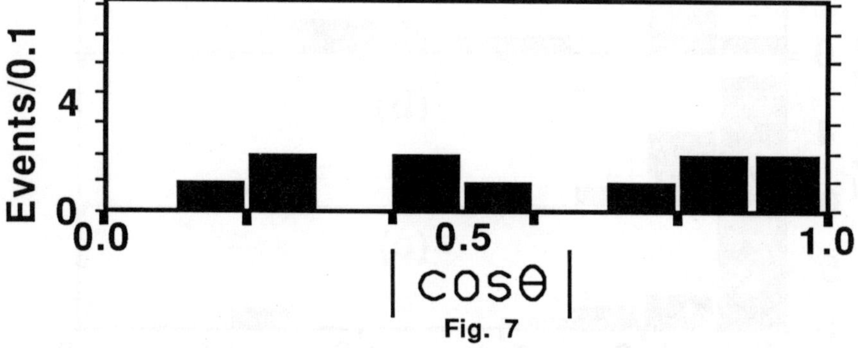

Fig. 7

$|\cos\theta| > 0.5$ is 0.55 ± 0.15 compared with 0.125 expected for pure $\sin^2\theta$. The transverse term should show a $\sin^2\theta \cos^2\psi$ dependence, analogous to S-channel helicity conservation in vector meson photo- and electro-production. ψ is the difference of two azimuthal angles: i) the angle between the lepton (ν-μ) plane and the "production" plane (Q-ρ) measured in the hadron

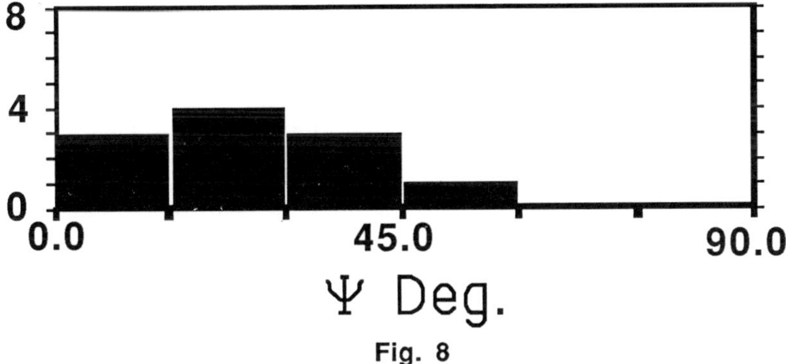

Fig. 8

rest system, and ii) the angle between the latter plane and the ρ decay plane measured in the ρ rest system. Fig. 8 is preliminary; the lepton and production planes are poorly defined experimentally, for coherent events.

In <u>conclusion</u>, we have presented evidence for·coherent ρ$^+$ production from neon by high energy neutrinos. Within our limited statistics the characteristics of the coherent candidates are in accord with the predictions of ref. (3).

We are grateful to the Fermilab Accelerator and 15-foot bubble chamber staffs for this successful and productive run, and to our scanners and measurers for their careful work on this difficult film. We thank M. K. Gaillard for illuminating discussions.

(a) Now at Texas Accelerator Center, Woodlands, TX. 77380
(b) Now at Arete Associates, P.O. Box 350, Encino, CA. 91316
(c) Now at Yale University, New Haven, CT. 06511
(d) Now at University of California, Riverside, CA. 92521

<u>References</u>

1) P. Maraqe, et al., this Conference;
    P. Marage, et al., Phys. Lett. <u>140B</u> (1984) 137.
2) E. Isiksal, et al., Phys. Lett. <u>52</u> (1984) 1096;
    H. Faissner, et al., Phys. Lett. <u>125B</u> (1983) 230.
3) M. K. Gaillard, et al, Phys. Lett. <u>68B</u> (1977) 267;
    M. S. Chen, et al., Nucl. Phys. <u>B118</u> (1977) 345.
4) T. H. Bauer, et al., Rev. Mod. Phys. <u>50</u> (1978) 261.

<u>Figure Captions</u>

1) γ-γ mass for $\mu^-\pi^+\gamma\gamma$ events.
2) $\pi^+\gamma\gamma$ mass for the events with 100 < $M_{\gamma\gamma}$ < 155 MeV.
3) 4-momentum transfer squared, −t, distribution: for a) coherent candidates with no primary protons; and b) incoherent events with one or more primary protons in addition to the $\mu^-\pi^+\gamma\gamma$ system.
4) x distribution: for a) coherent candidates; and for b) incoherent events, with −t > 0.1 GeV^2, or c) with extra protons.
5) y distribution: for a) coherent candidates with prediction of ref. (3) superposed; and for b) incoherent events, with −t > 0.1 GeV^2, or c) with extra protons.
6) Q^2 distribution: for a) coherent candidates; and for b) incoherent events, with −t > 0.1 Gev^2, or c) with extra protons.
7) Helicity polar angular distribution for the ρ$^+$ decay
8) Helicity azimuthal angular distribution for the ρ$^+$ decay, relative to the ν-μ plane.

Coherent Neutrino Scattering. Some Applications

L. M. Sehgal

III. Physikalisches Institut, Technische Hochschule,
Aachen.

The coherent scattering of neutrinos from nuclei
came into the limelight several years ago with the
suggestion that it might play a role in supernova
development [1]. More recently, there has been a re-
surgence of interest in this reaction as a result of
suggestions that it might form a basis for new types
of detectors suitable for low energy neutrinos or other
types of weakly interacting neutral particles [2,3].
A neutrino of energy E_ν scatters from a nucleus
of mass M_A $(M_A \gg E_\nu)$ with an angular distribution

$$\frac{d\sigma}{d\cos\theta} = \frac{G_F^2 \, E_\nu^2}{2\pi} \, [C_V^2 \, (1 + \cos\theta) + C_A^2 \, (3 - \cos\theta)] \qquad (1)$$

The matrix elements C_V and C_A are coherent sums of the
vector and axial vector charges of the constituent
nucleons:

$$C_V = [Z \, (\tfrac{1}{2} - 2 \sin^2 \Theta_w) - \tfrac{1}{2} N] \, F_{nuc} \, (Q^2)$$

$$C_A = \tfrac{1}{2} \, g_A \, [(Z_+ - Z_-) - (N_+ - N_-)] \, F_{nuc} \, (Q^2) \qquad (2)$$

where $g_A = 1.25$, $Q^2 = 2E_\nu^2 \, (1 - \cos\theta)$, and $Z_\pm$ $(N_\pm)$ are
the proton (neutron) numbers with spins parallel (+) or
antiparallel (-) to the nuclear spin. The nuclear form
factor $F_{nuc} \, (Q^2)$ ensures that coherence is limited to
$Q^2 R^2 \lesssim 1$, R being the nuclear radius. The axial vector
coupling C_A remains small because of the cancellation of
nucleon pairs with opposite spins, and vanishes for

spin-zero nuclei. The vector coupling, on the other hand, is significantly amplified by the coherence. For neutrino energies $E_\nu << R^{-1}$ (~ 50 MeV for $A \sim 30$) the nuclear form factor may be ignored, and the cross section takes the approximate form ($N=Z$, $C_A << C_V$)

$$\frac{d\sigma}{d\cos\theta} \approx \frac{G_F^2\, E_\nu^2}{2\pi}\, A^2\, (-\sin^2\theta_W)^2\, (1 + \cos\theta) \qquad (3)$$

The solitary signature of a coherent reaction is the small nuclear recoil energy $E_{rec} = Q^2/2M_A$. Averaged over scattering angles, this is

$$<E_{rec}> = \frac{2}{3}\, E_\nu^2/M_A \simeq \frac{2}{3A}\, (\frac{E_\nu}{MeV})^2\ keV \qquad (4)$$

Typically, for a ^{27}Al nucleus, the recoil energy is $<E_{rec}> = 100$ eV (2.5 keV) for $E_\nu = 2$ MeV (10MeV). Ideas have been put forward for low-temperature devices in which this tiny recoil could be converted into a measurable signal. One proposal [2] envisages a target consisting of superconducting grains (each a few microns in size) held in a metastable state in a magnetic field. A coherent reaction occurring in a single grain can cause a temperature rise, enough to flip the grain to a normal conducting state. The accompanying collapse of the magnetic field could be picked up by a coil surrounding the grain. Another proposal is to use a macroscopic target of crystalline silicon cooled to a few mK [3]. A recoiling nucleus in such a target would convert its energy into a pulse of phonons, which, because of the very low specific heat of Si, could result in a detectable thermal signal. Such devices would have obvious application to the study of low-energy neutrino sources such as reactors or the sun.

At the low momentum transfers relevant to coherent

scattering, there are some interesting refinements to the cross section given by Eq. (3). One of these [4] is a radiative correction associated with the "neutrino charge- radius" diagram shown in Fig. 1. The effect of this correction is essentially to replace the parameter $\sin^2 \theta_W$ by

$$\sin^2 \Theta_{eff} = \sin^2 \Theta_W \left[1 - \Delta (Q^2) \right]$$

where
$$\Delta(Q^2) = \frac{\alpha}{2\pi \sin^2 \Theta_W} \left[\frac{1}{3} \ln \frac{M_W^2}{m_\ell^2} - 2 I_\ell (Q^2) \right],$$

$$I_\ell (Q^2) = \int_0^1 dx \; x \; (1-x) \; \ln \left[1 + \frac{Q^2}{m_\ell^2} x (1-x) \right] \tag{5}$$

Here m_ℓ denotes the mass of the charged lepton associated with the neutrino ν_ℓ. This correction induces a flavour-dependence into the coherent scattering, and an apparent violation of universality between ν_e, ν_μ and ν_τ (Fig. 2). For scattering close to the forward direction, the cross sections are in the ratio

$$\sigma(\nu_i)/\sigma(\nu_j)\Big|_{Q^2 \approx 0} = 1 + \frac{\alpha}{3\pi \sin^2 \Theta_W} \ln \frac{m_i^2}{m_j^2}, \tag{6}$$

$$\lim_{Q^2 \to 0} \; \sigma(\nu_e) : \sigma(\nu_\mu) : \sigma(\nu_\tau) = 1 : 1.04 : 1.06.$$

As one practical consequence, a neutrino detector based on coherent scattering would not be insensitive to neutrino oscillations : such oscillations would reflect themselves in changes in the fraction of low energy recoils.

For very low momentum transfers such that the impact parameter is of the order of atomic dimensions (i.e. $Q^2 R^2_{atom} \lesssim 1$), neutrino scattering will occur coherently on the whole atom [5]. One should then expect interference beween scattering on the nucleus and on the surrounding electron shell. The vector matrix element is modified to

$$C_V^{atom} = C_V + Z \left[\pm \tfrac{1}{2} - 2 \sin^2 \Theta_W \right] F_{el}(Q^2), \quad (7)$$

$F_{el}(Q^2)$ being the form factor associated with the electron cloud and C_V being given by Eq. (2). The + (-) sign in Eq. (7) applies to ν_e (ν_μ, ν_τ) scattering. In the case of ν_e, the nucleus and the electrons interfere destructively, producing a deep minimum in $d\sigma/dQ^2$. The minimum occurs at a value of Q^2 such that

$$F_{el}(Q^2) = \frac{-\left(\tfrac{1}{2} - \sin^2 \Theta_W\right) + \tfrac{1}{2} N/Z}{\tfrac{1}{2} + 2 \sin^2 \Theta_W} \approx \tfrac{1}{2} \quad (8)$$

By contrast, the interference effect for ν_μ and ν_τ is constructive and small. These remarks are illustrated in Fig. 3 for scattering from two typical atoms, ^{28}Si and ^{56}Fe.

The effects mentioned above could have implications in astrophysical contexts. The different charge radii of ν_μ and ν_τ, for example, can lead to slightly different indices of refraction in a medium [4]. Because of the electrical neutrality of matter, an effect can occur only at the level $G_F \alpha^2$; an order of magnitude estimate would be

$$k (n_\mu - n_\tau) \sim G_F N_e \left(\tfrac{\alpha}{\pi}\right)^2 \ln \frac{m_\tau^2}{m_\mu^2} \quad (9)$$

(k being the neutrino momentum and N_e the density of electrons). This is considerably smaller than the difference in refractive index of ν_e and ν_μ associated with the charged current coupling of ν_e to electrons [6]

$$k (n_e - n_\mu) = - \sqrt{2} \, G_F N_e \quad (10)$$

Nevertheless, an effect of the type (9) could have an influence at the high densities prevailing in collapsing stars. Similarly, the different behaviour of $d\sigma/dQ^2$ for ν_e compared to ν_μ or ν_τ , especially when $Q^2 \approx 1/R^2_{atom}$ (Fig. 3), implies differences in the radiation pressure exerted by different types of neutrinos in an atomic medium.

Finally, detectors based on coherent scattering can also register quanta of "dark matter" that may be present in the terrestrial neighbourhood. A discussion of the expected rates, for various forms of dark matter, is given in Ref. 7.

References

1. D.Z. Freedman, Phys. Rev. D9 (1974) 1389
2. A. Drukier and L. Stodolsky, Phys. Rev. D30 (1984) 2295.
3. B. Cabrera, L.M. Krauss and F. Wilczek, Phys. Rev. Lett. 55 (1985) 25.
4. L.M. Sehgal, Phys. Lett. 162 B (1985) 370
5. L.M. Sehgal and M. Wanninger, Phys. Lett. 171 B (1986) 107.
6. L. Wolfenstein, Phys. Rev. D17 (1978) 2369, D20 (1979) 2634.
7. M.W. Goodman and E. Witten, Phys. Rev. D 31 (1985) 3059.

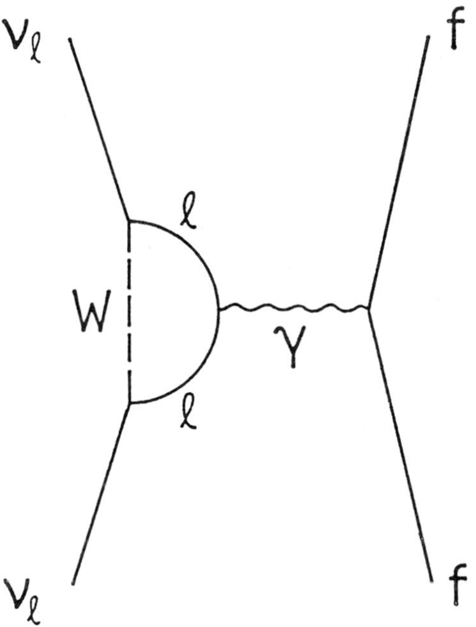

Fig. 1. The neutrino charge radius diagram

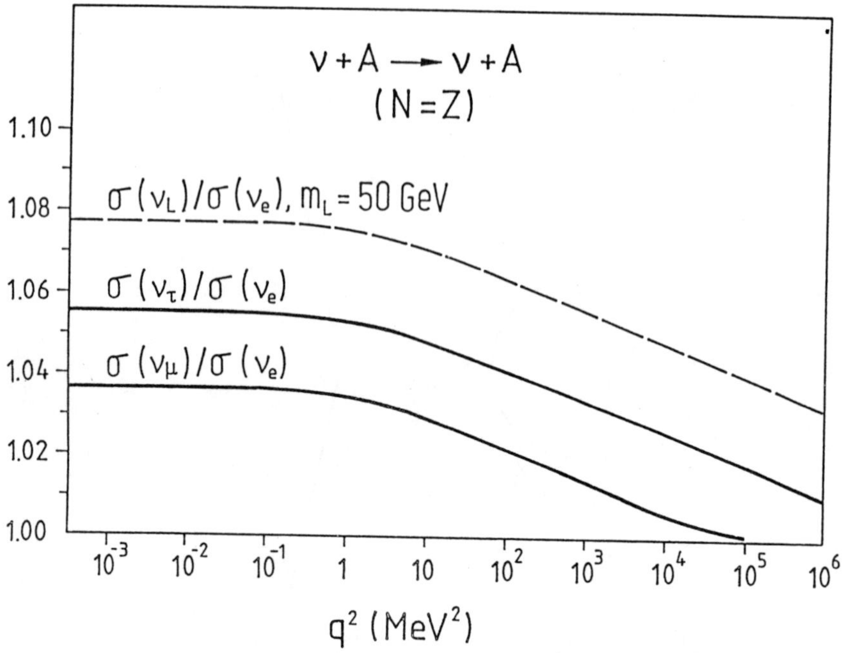

Fig. 2. Differences in the coherent cross section $\nu + A \to \nu + A$ for different neutrino flavours (N=Z).

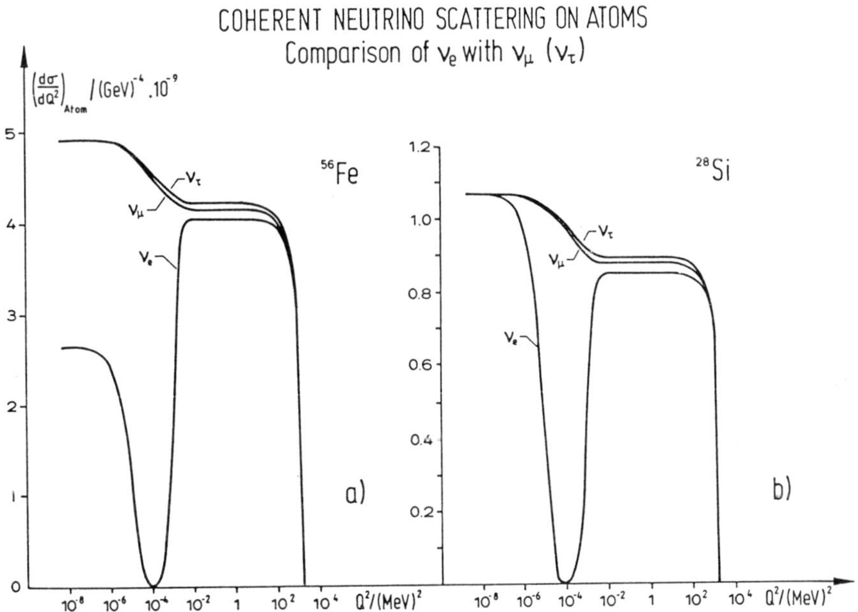

Fig. 3. Coherent neutrino scattering from ^{56}Fe and ^{28}Si atoms (E_ν = 20 MeV).

5 August 1986

RESULTS FROM UA1 AND UA2 RELEVANT TO
THE STANDARD ELECTROWEAK MODEL

Allan G. Clark

CERN, 1211 Geneva 23, Switzerland

Invited talk at the 12th. Intl. Conference
on Neutrino Physics and Astrophysics,
Sendai, Japan, 3 - 8 June 1986.

ABSTRACT

A summary is presented of those results from the UA1 and UA2 Collaborations that are relevant to the verification of the Standard Model of electroweak interactions. On the basis of presently available data, there is no evidence for deviations from the expectations of the Standard Model. The experimental uncertainties relevant to existing and future measurements of the Standard Model parameters at the $\bar{p}p$ Collider are discussed.

1. INTRODUCTION

Since the start of CERN Collider operation in 1981, the UA1 and UA2 experiments have collected data samples on the leptonic (electron and muon) decays of the $W^{\pm}$ and Z^0 gauge bosons, of respectively 599 $W^{\pm}$ and 88 Z^0 events.

At this stage, both experiments are being modified prior to operation at the improved CERN Collider in 1987, and the full data samples are being analysed.

In this talk, we concentrate on measurements of $W^{\pm}$ and Z^0 production and decay that are relevant to verification of the Standard Electroweak Model [1]). In doing so, we attempt to emphasise the (complementary) strengths and weaknesses of measurements from UA1 and UA2, and we indicate the expected accuracy of measurements related to the Standard Model beyond 1987.

In Section 2 we discuss the identification of leptons in the UA1 [2] and UA2 [3] detectors, and we tabulate the event samples. Details of the analysis procedures used by each experiment have already been extensively described for $W^{\pm}$ and Z^0 decay [4-5], and we concentrate in that section on aspects of the analysis essential to the data quality.

Using data available by early July 1986 (and therefore updated by reports since this Conference at the Aachen $\bar{p}p$ Workshop [6−8]), we summarize in Section 3 measurements which are relevant to the verification of the Standard Model:

i. the $W^{\pm}$ and Z^0 production cross − section,

ii. the charge asymmetry for $W^{\pm}$ production and decay,

iii. tests of lepton universality at the $\bar{p}p$ Collider,

iv. measurements of the $W^{\pm}$ and Z^0 mass and width, and consequent limits on the number of additional neutrinos,

v. the status of the Standard Model parameters, including radiative corrections, and

vi. the decays $Z^0 \rightarrow e^+e^-\gamma$ and $Z^0 \rightarrow \mu^+\mu^-\gamma$, and possible deviations from the Standard Model.

Finally, in Section 4 we briefly describe (using the upgraded UA2 experiment as an example), the expected measurement accuracy of the Standard Model parameters in future CERN $\bar{p}p$ Collider operation [9].

2. LEPTON IDENTIFICATION IN THE UA1 AND UA2 DETECTORS.

The UA1 detector [2] is a large general-purpose detector equipped to identify electrons and muons from a background of QCD jets. With similar identification quality, and a superior measurement accuracy, the more specialised UA2 detector [3] is equipped to identify electrons but has no muon detection capability. Subject to analysis restrictions specific to each detector, both experiments can infer the existence of non-interacting particles (characteristic of the neutrino for $W \rightarrow \ell\nu_\ell$ decay), and from arguments of transverse momentum balance to estimate the missing transverse momentum, $\vec{p}_T^{\,m}$, due to these particles. Given the expected signature for the decay (W $\to \tau\nu_\tau$; $\tau \to \nu_\tau$ + hadrons), the UA1 Collaboration has recently exploited the missing$-p_T$ capability of their detector to isolate a sample of events compatible with this decay [8,10].

2.1 Electron Identification

The analysis criteria of each experiment require for electron identification the deposition of energy in a calorimeter, that is consistent with that expected for electrons, and the spatial matching of this energy deposition with a reconstructed charged track expected from the decay electron(s) [4-7]. In detail, however, the analysis criteria differ significantly :

i. The UA1 calorimeter provides four longitudinal samplings of electromagnetic showers, and a compatibility comparison is made with the profile expected for electrons. Because of the limited lateral granularity of their calorimeter, an energy isolation requirement is made (typically 3 GeV transverse energy and charged track momentum in a cone of 0.7 radians about the electron track). The UA2 calorimeter, on the other hand, has only one longitudinal sampling but good lateral granularity; this allows a comparison with the expected lateral energy profile, and a less stringent isolation requirement.

ii. The existence of a dipole magnetic field over the full UA1 detector acceptance allows a measurement of the particle charge and a comparison of the reconstructed track momentum with the energy deposition. This is only possible in the UA2 detector for the region $20° < |\theta| < 37.5°$ with respect to the beam line. A superior measurement of the charge asymmetry for W$^\pm$ production is therefore available from UA1.

iii. The dominant background to the electronic W- and Z^0- decay samples is due to QCD jets which fake the electron signature. These jets include high-p_T converted π^0's, and low multiplicity jets in which a high-p_T π^0 and a charged track overlap. The latter background is minimised in the case of UA2 by the existence of a position-sensitive preshower detector preceding the calorimeter.

iv. The UA1 detector is equipped with calorimetry to within angles $\theta = 0.2°$ of the beam line, while the UA2 detector no particle detection capability at angles $\theta < 20°$ to the beam. Since the dominant QCD background is characterised by two jets back-to-back in the direction transverse to the beam, one of the two jets is less likely to escape detection in the UA1 detector. This provides a topology selection to preferentially reduce the hadronic background in processes such as W $\to e\nu$ decay (see subsection 2.3).

It will be shown (subsection 2.5) that essentially background-free samples of decays $W \rightarrow e\nu_e$ and $Z^0 \rightarrow e^+e^-$ are available from each detector. Of relevance then is the accuracy of the measured energy. Uncertainties of the measured electron energy result from the intrinsic resolution, smeared by cell-to-cell calibration differences and (in the case of UA1 as a result of limited calorimeter granularity) pile-up due to the deposition of energy by additional particles near the electron. Contributions to the resolution are listed in Table 1. Most important, however, is the uncertainty of the overall energy scale of the calorimeter, both after initial calibration and after the evolution of the calorimeter response (including changes of the scintillator attenuation length) with time. The smaller uncertainty quoted by the UA2 experiment is mainly due to the initial calibration in a test beam of each calorimeter cell; in the case of UA1, test-beam calibrations were made using a prototype module. The energy-scale uncertainty reflects directly on measurements of the $W^{\pm}$ and Z^0 masses, and consequently on the evaluation of Standard Model parameters. It is this uncertainty which for each experiment limits the accuracy of existing measurements (see subsections 3.4, 3.5), and will continue to limit the accuracy beyond 1987 (see section 4).

Table 1: Uncertainties of calorimetric energy measurement in UA1 and UA2.

	UA1 [4]	UA2 [5]
Intrinsic calorimeter resolution K, $\sigma_E = K \sqrt{E}$ (GeV)	0.16	0.14
Cell-to-cell calibration uncertainties (%)	3.0	2.5
Pile-up effects from calorimeter granularity (GeV)	0.5	—
Calibrated energy-scale uncertainties (%)	3.0	1.5
Time-dependent uncertainty of energy scale (%)	1.0	0.3

2.2 Muon Identification

Only the UA1 detector is equipped to detect muons. The signature for muon detection is the existence of a charged track aligned spatially with signals from muon chambers that follow the electromagnetic and hadronic calorimeters. To reduce the background from QCD processes, a strong isolation of the muon from other hadronic (generally jet) activity is required. For the isolation of $W \to \mu\nu_\mu$ and $Z^0 \to \mu^+\mu^-$ signals, less than 3.0 GeV transverse energy deposition (and 1 GeV transverse track momentum) is typically required in a cone of 0.4 radians about the muon candidate, and less than 10 GeV summed calorimeter energy and track momentum in a cone of 0.7 radians about the muon.

Depending on the cleanliness of the sample required, increasingly strong track-alignment selections are applied to reduce remaining punch-through and decay contaminations. Details of the analysis are described elsewhere [11].

The muon momentum, p^μ, is measured with a typical accuracy $\delta p^\mu / p^\mu \sim 0.005\, p^\mu$, where p^μ is in units of GeV/c. As a result, high-p_T muons resulting from $W^\pm$ or Z^0 decay are not well measured. Measurements of the partial widths for the decays $W \to \mu\nu_\mu$ and $Z^0 \to \mu^+\mu^-$ are nevertheless essential for tests of lepton universality (see subsection 3.3).

2.3 Neutrino Identification

For an ideal detector, the existence of non-interacting particles can be inferred from a measurement of the momentum of all detected particles. However, in typical $\bar{p}p$ interactions a large fraction of the total collision energy is carried by particles that do not leave the vacuum pipe and are therefore undetected. Consequently, only that component, $\vec{p}_T^{\,m}$, of missing momentum transverse to the beam axis can be reliably measured. From transverse momentum conservation,

$$\Sigma\vec{p}_{T\text{detected}} + \vec{p}_T^{\,m} = 0. \tag{1}$$

Depending on the interpreted physics process, $\vec{p}_T^{\,m}$ may be ascribed to one or more neutrinos, or to some other non-interacting particle. In particular, decays $W \to \ell\nu_\ell$ are characterised by an identified charged lepton with significant undetected energy (that is $|\vec{p}_T^{\,m}|$ or $E_T^{\,m}$) directed approximately back-to-back to the charged lepton.

In the UA1 experiment p_T^ν is defined for the purposes of $W^\pm$-decay by

$$\vec{p}_T^{\,\nu} = \vec{p}_T^{\,m} = -\vec{p}_T^{\,\ell} - \Sigma\, \vec{p}_T^{\,i}, \tag{2}$$

where $\vec{p}_T^{\,i}$ is a vector with magnitude equal to the transverse energy $E_T^{\,i}$ deposited in the ith calorimeter cell (excluding the charged lepton), and direction defined by the calorimeter impact position. However no detector is perfect and in particular the UA1 central region (pseudo$-$rapidity $|\eta| < 1.5$) has only partial jet detection within $\pm\, 4°$ of the vertical direction (Fig. 1). By rejecting events for which $\vec{p}_T^{\,\nu}$ points within $\pm\, 15°$ of the vertical direction or for which $\vec{p}_T^{\,\nu}$ points in the direction of other high-p_T activity, the $|\vec{p}_T^{\,\nu}|$ distribution has a

Gaussian resolution to at least the 4σ level. The measurement accuracy of each $\vec{p}_T^{\,\nu}$ component is $0.4\sqrt{\Sigma E_T^i}$ (folded in quadrature with the error of the charged lepton energy measurement), where E_T^i is measured in units of GeV.

As noted above, the UA2 detector is not equipped to identify muons. Furthermore, there is no particle detection at angles $\theta < 20°$ to the beam and only partial detection in the region $20° < |\theta| < 40°$ (see Fig. 1). As a result the $|\vec{p}_T^{\,\nu}|$ resolution is slightly deteriorated and more importantly has non-Gaussian tails. In the UA2 experiment $\vec{p}_T^{\,\nu}$ is defined as

$$\vec{p}_T^{\,\nu} = -\vec{p}_T^{\,\ell} - \Sigma\,\vec{p}_T^{\,j} - \lambda\vec{P}_T,\tag{3}$$

where j extends over all reconstructed calorimeter energy depositions (jets) of $p_T^j > 3$ GeV/c excluding the charged lepton and $\vec{P}_T$ is the contribution of all other particles. The correction factor $\lambda = 1.5 \pm 0.6$ takes account of the non-linearity of the calorimeter response to low energy particles, and is estimated from a sample of well-measured Z^0-decays.

Figure 2 shows for each detector the p_T^e vs. p_T^ν distributions for electron samples satisfying $p_T^e > 15$ GeV/c. At large values of p_T^e, signals from $W \to e\nu_e$ ($p_T^\nu \approx p_T^e$) and $Z^0 \to e^+e^-$ ($p_T^\nu \approx 0$) decay are clearly visible above a background of mis-identified low-p_T hadrons. The background rejection provided by the p_T^ν- cut is evidently superior in the UA1 detector.

2.4 Tau Identification

The isolation of a sample of events consistent with the decay $W \to \tau\nu_\tau$ was initially reported in 1985 [10], and new data have been reported since this Conference [8]. The UA1 Collaboration exploits the excellent $|\vec{p}_T^{\,m}|$ resolution of their detector to isolate a sample of events consistent with the decay

$$W \to \tau\nu_\tau,\ \tau \to \nu_\tau + \text{hadrons}.\tag{4}$$

Topologically, the events are characterised by a single jet approximately back-to-back with significant missing energy from the ν_τ's. The jet has a low charged-particle track multiplicity and significant electromagnetic energy deposition, because of the dominant decays

$$\tau \to \nu_\tau + \pi^\pm + n\pi^0$$
$$\tau \to \nu_\tau + 3\pi^\pm + n\pi^0\tag{5}$$

Therefore, the UA1 Collaboration initially select events satisfying

i. $|\vec{p}_T^{\,m}| > 15$ GeV/c and $> 4\sigma$ where σ is the resolution in E_T^m,

ii. one and only one isolated and collimated jet satisfying $E_T > 12$ Gev, and of low track multiplicity but with at least one charged track within a cone of 0.4 radians about the jet axis, and

iii. no non-associated high-p_T (> 5 GeV/c) track or energy deposition within $\pm 30°$ in azimuth of $\vec{p}_T{}^m$.

After rejecting some events by scanning, a likelihood function is constructed which takes account of the expected toplogy and the resulting detector response; this has been demonstrated to efficiently isolate the τ-decay sample from other mono-jet signatures. It is stressed, however, that the ability to isolate this event sample is due primarily to the good rejection afforded by the $\vec{p}_T{}^m$ measurement.

2.5 The $W^\pm$ and Z^0 Event Samples.

The UA1 and UA2 Collaborations have collected data samples at centre-of-mass energies $\sqrt{s} = 540$ Gev and $\sqrt{s} = 630$ Gev for the integrated luminosities shown in Table 2. The uncertainty on the measured luminosity, which dominates the error on cross-section measurements (section 3.1), is respectively $\pm 15\%$ and $\pm 8\%$ for the UA1 and UA2 data, and arises from uncertainties of the total non-diffractive $\bar{p}p$ cross-section measurement and uncertainties in the monitoring of the luminosity. The UA2 detector benefits from precise measurements in the same intersection region by the UA4 Collaboration [12].

Table 2: Integrated luminosity for the UA1 and UA2 experiments (nb^{-1}).

Process	UA1 $\sqrt{s} = 546$ GeV	UA2 $\sqrt{s} = 546$ GeV	UA1 $\sqrt{s} = 630$ GeV	UA2 $\sqrt{s} = 630$ GeV
$W \rightarrow e\nu_e$ [6,7]	136	142	593 (263) [1]	738
$\mu\nu_\mu$ [6,7]	108	—	584 (254) [1]	—
$\tau\nu_\tau$ [8]	118	—	597	—
$Z^0 \rightarrow e^+e^-$ [6,7]	136	141	593 (263) [1]	738
$\mu^+\mu^-$ [6,7]	108	—	584(254) [1]	—

[1] Excludes data from the 1985 run

2.5.1 The W sample.

The final data samples for each experiment are listed in Table 3, together with associated background estimates. Because of the large event samples now available, each experiment makes an event selection to optimise for a given Physics measurement either the background or the measurement accuracy. The numbers quoted are therefore only indicative. The sample quoted for UA2 is final (and updates that presented at this Conference), while the UA1 event sample remains preliminary for 1985 electron data.

i. $W \to e\nu_e$. Following a selection $p_T^e > 15$ GeV/c and $p_T^\nu > 15$ GeV/c, the UA1 Collaboration present a total sample of 262 events. The dominantly low-p_T background from misidentified jets is estimated from the $|\vec{p}_T^\nu|$ resolution and does not significantly influence the measurement of m_W. Because of the poorer $|p_T^\nu|$ resolution, the UA2 Collaboration presents data (255 events) satisfying $p_T^e > 20$ GeV/c and transverse mass $m_T^{e\nu} > 50$ GeV where

$$m_T^{e\nu} = 2p_T^e p_T^\nu (1 - \cos\Delta\Phi) \tag{6}$$

and $\Delta\Phi$ is the angle between $\vec{p}_T^e$ and $\vec{p}_T^\nu$. The hadronic background contribution to the UA2 data is estimated from a sample of high-p_T jets consistent with π^0's, normalised in the low-$m_T^{e\nu}$ region. As shown in Table 3, the background from τ-decay is significant in UA1 (because of the low-p_T threshold), and the background from Z^0-decay is significant in UA2 (because of the lack of particle detection at angles $|\theta| < 20°$ to the beam). Figures 3a and 3c show respectively the $m_T^{e\nu}$ and p_T^e distributions of UA1 and UA2 data. In these figures, the expected background contributions are superimposed.

ii. $W \to \mu\nu_\mu$. The final event sample for UA1 is 82 events. To minimise the background, a subsample of 68 events satisfying stringent track-matching criteria is used for Physics analysis. The $m_T^{\mu\nu}$ distribution of this sample is shown in Fig. 3b. The shape of this distribution does not show the distinctive Jacobian peak characteristic of $W^\pm$ decay, because of the poor muon measurement accuracy.

iii. $W \to \tau\nu_\tau$. The UA1 Collaboration report a sample of 32 events, corresponding to an integrated luminosity of 715 nb$^-$1. They estimate a total background of 2.3 ± 0.6 events in this sample.

2.5.2 The Z^0 sample.

Final data samples are listed for each experiment in Table 4. Because of the increased rejection against hadronic background resulting from a requirement of 2 high-p_T leptons in the event, the detailed selection criteria can be relaxed to maintain a good detection efficiency.

Table 3: Event samples from $W^{\pm}$ decay for the UA1 and UA2 experiments.

	$\sqrt{s} = 546$	$\sqrt{s} = 630$
UA1 : W → eν		
Event sample	59	207
: QCD Backgrounds		6.1 ± 1.4
: W → τν(τ → e)	(scaled)	13.1 ± 0.9
: W → τν(τ → hadron)		1.3 ± 0.2
: $Z^0 \to e^+e^-$		—
Signal	53 ± 8	186.5 ± 14.5
UA2 : W → eν		
Event sample	41	214
: QCD background	1.9 ± 0.4	9.7 ± 2.6
: W → τν(τ → e)	0.8 ± 0.2	4.4 ± 0.4
: W → τν(τ → hadrons)	—	—
: $Z^0 \to e^+e^-$	1.5 ± 0.4	8.1 ± 1.6
Signal	36.7 ± 6.3	191.8 ± 14.5
UA1 : W		
Event sample	11	57
: backgrounds W → τν(τ → μ)	0.6 ± 0.1	3 ± 2
: signal	10.4 ± 3.3	54 ± 8
UA1 : W → τν		
: events	32	
: backgrounds	2.3 ± 0.3	
: signals	29.7 ± 5.7	

i. $Z^0 \to e^+e^-$. By requiring (Fig. 4a) at least two electromagnetic calorimeter energy depositions satisfying $E_T^e > 10$ Gev and consistent with expectations for an electron, the UA2 experiment records 48 events satisfying $m^{ee} > 76$ Gev, with an estimated background of 12 events (Fig. 4c). If additional loose tracking criteria are applied to at least one cluster, they obtain 37 events with an estimated background of one event (Fig. 4b). In a similar analysis the UA1 experiment obtain a sample of 32 events, with a background of less than one event (Fig. 4c).

ii. $Z^0 \to \mu^+\mu^-$. A total of 19 decays are reported by UA1, with a small background.

iii. $Z^0 \to \tau^+\tau^-$. This decay mode has not been demonstrated at the Collider. However, the UA1 Collaboration has noted one event [13] not incompatible with the decay $Z^0 \to \tau^+\tau^-$; $\tau^- \to e^-\bar{\nu}_e\nu_\tau$; $\tau^+ \to \mu^+\nu_\mu\bar{\nu}_\tau$.

Table 4: Event samples for Z^0 decay from the UA1 and UA2 experiments.

	$\sqrt{s} = 546$	$\sqrt{s} = 630$
UA1 : $Z^0 \to e^+e^-$		
: events	4	28
: background	<0.1	<0.7
UA2 : $Z^0 \to e^+e^-$		
: events	9	28
: background	0.2 ± 0.1	0.8 ± 0.2
: signal	8.8 ± 3.0	27.2 ± 5.4
UA1 : $Z^0 \to \mu^+\mu^-$		
: events	4	15
: background	<0.1	<1

3. PHYSICS RESULTS RELATED TO THE STANDARD MODEL.

3.1 Production Cross-section.

A measurement of the $W^\pm$ and Z^0 production cross-section at collision energies $\sqrt{s} = 540$ and 630 GeV provides a weak and indirect comparison of the quark coupling to predictions of the Standard Model, since at these collision energies $\approx 65\%$ of the cross-section is expected to result from quark fusion. Theoretical estimates are compared with the data after higher-order QCD corrections are made. Uncertainties of the theoretical estimates mainly result from QCD effects (for example the evolution of structure functions measured at low energy, the QCD momentum scale used, the non-inclusion of some higher order diagrams, etc.). The value of any validity test is limited by these uncertainties as well as by the error on the luminosity measurement. As shown in Tables 5 and 6, available data are in good agreement with recent calculations of Altarelli et al. [14], though systematically larger. Although the individual σ_Z^ℓ and σ_W^ℓ measurements have limited accuracy, most theoretical and experimental uncertainties cancel in the ratio $R = \sigma_Z^\ell/\sigma_W^\ell$ (see subsections 3.4 and 3.5).

Table 5: *Cross-section σ_W^ℓ for $\bar{p}p \to W + X$, $W \to \ell\nu_\ell$.*

Cross-section		σ_W^ℓ (nanobarns)		$R = \dfrac{\sigma_W^\ell\ (\sqrt{s} = 630\ \text{GeV})}{\sigma_W^\ell\ (\sqrt{s} = 540\ \text{GeV})}$
		$\sqrt{s} = 540$ GeV	$\sqrt{s} = 630$ GeV	
σ_W^e	UA1	$0.55 \pm 0.08 \pm 0.09^*$	$0.60 \pm 0.05 \pm 0.09^*$	1.09 ± 0.18
	UA2	$0.57 \pm 0.1\ \pm 0.07$	$0.61 \pm 0.05 \pm 0.07$	1.07 ± 0.20
σ_W^μ	UA1	$0.56 \pm 0.18 \pm 0.12$	$0.67 \pm 0.08 \pm 0.14$	1.18 ± 0.44
σ_W^τ	UA1		$0.61 \pm 0.12 \pm 0.11$	
Theory		$0.36\ ^{+0.11}_{-0.05}$	$0.45\ ^{+0.14}_{-0.08}$	1.26 ± 0.02

* Published result [4] excluding 1985 data.

Table 6: *Cross-section σ_Z^ℓ for $\bar{p}p \to Z + X$, $Z \to \ell^+\ell^-$.*

Cross-section		σ_Z^ℓ (picobarn)	
		$\sqrt{s} = 540$ GeV [1]	$\sqrt{s} = 630$ GeV [1]
σ_Z^e	UA1	$42\ ^{+25}_{-18} \pm 6$	$73 \pm 14 \pm 11$
	UA2	$112 \pm 37 \pm 9$	$69 \pm 13 \pm 6$
σ_Z^μ	UA1	$100 \pm 50 \pm 15$	$73 \pm 19 \pm 13$
Theory		$42\ ^{+13}_{-\ 6}$	$51\ ^{+16}_{-10}$

[1] Published result [4] excluding 1985 data.

3.2 The charge asymmetry of $W^\pm$ production and decay.

As noted above, the $W^\pm$ production cross-section at Collider energies is dominated by quark fusion involving at least one valence quark. Therefore, if as expected from the Standard Model the exchanged W couples to left-handed doublets ($V - A$ coupling), the charged decay lepton should be aligned with respect to the incident proton in the $W^\pm$ rest frame with angle θ^* given by

$$dN/d\cos\theta^* \propto (1 - q\cos\theta^*)^2 + 2\alpha\cos\theta^*, \qquad (7)$$

where q is the lepton charge (-1 for electrons) and for V-A coupling $\alpha = 0$. The value of $\cos\theta^*$ is not definded for $p_T^W \neq 0$, and in this case both experiments use the Collins-Soper convention [15], with the constraint $m^{e\nu} = m_W$. If two solutions of $\cos\theta^*$ are allowed, the solution of smallest longitudinal momentum $|p_L^W|$ is chosen.

The most accurate data are from the UA1 experiment (shown excluding 1985 data in Fig. 5) and the superimposed V-A prediction is in good agreement with the data. Similar data from the UA2 experiment exist in the range $20° < |\theta| < 37.5°$ with respect to the beam, and these data are also in good agreement with $V-A$ predictions. The UA2 experiment quotes a value $\alpha < 0.37$(68% confidence level). The comparison of both UA1 and UA2 data with Standard Model predictions is presently limited by statistics.

Since two weak interaction vertices are involved in $W^\pm$ production and decay, it is not possible to distinguish $(V-A)$ and $(V+A)$ coupling. The inclusion of a third weak interaction vertex (as for example τ-decay of the W) would enable these couplings to be distinguished.

For a particle of arbitrary spin J, one expects [16]

$$\langle \cos\theta^* \rangle = \langle\lambda\rangle.\langle\mu\rangle/J(J+1) \tag{8}$$

where $\langle\mu\rangle$ and $\langle\lambda\rangle$ are respectively the global helicities of the production (u$\bar{\text{d}}$) system and the decay (e$\bar{\nu}$) system. The published UA1 result, excluding 1985 data, is $\langle\cos\theta^*\rangle = 0.43 \pm 0.07$, consistent with the value $\langle\cos\theta^*\rangle = 0.5$ expected for $J = 1$.

Because the Z^0 coupling is not left-handed, the expected asymmetry is small, and the limited available statistics precludes an accurate comparison with the Standard Model. The UA1 Collaboration quotes an asymmetry of $A_Z = 0.30 \pm. 0.15$; this precludes at present a meaningful measurement of the couplings.

3.3 Studies of lepton universality.

From the Standard Model, the cross-sections σ_W^ℓ and σ_Z^ℓ are independent of the nature of the charged final-state lepton ℓ (ℓ = e, μ, or τ). As shown in Table 7, this is well verified by data from the UA1 Collaboration. The present accuracy is limited by statistics, and should be significantly improved in Collider operation beyond 1987. The systematic error on the cross-section ratio σ_W^μ/σ_W^e results from uncertainties of the muon cut efficiencies, and with increased statistics is likely to be reduced. It is noted that in a different kinematic region, lepton universality is already well tested [17] using the reactions $e^+e^- \to \mu^+\mu^-$ and $e^+e^- \to \tau^+\tau^-$.

We have already noted (subsection 3.1) that at the $\bar{p}p$ Collider, tests of the $W^\pm$ and Z^0 couplings to the quark sector are poor. Since this Conference, however, the UA2 Collaboration [18] has reported evidence for an enhancement in the jet-jet invariant mass

Table 7: Tests of lepton universality from UA1.

Measured Quantity	Value
$\sigma_W^\tau / \sigma_W^e$	$1.02 \pm 0.35(\text{stat.}) \pm 0.15(\text{syst.})$ ($\sqrt{s}$ = 540 GeV)
	$1.12 \pm 0.15(\text{stat.}) \pm 0.15(\text{syst.})$ ($\sqrt{s}$ = 630 GeV)
$\sigma_W^\tau / \sigma_W^e$	$1.02 \pm 0.22(\text{stat.})$ ($\sqrt{s}$ = 630 GeV)
$\sigma_Z^\mu / \sigma_Z^e$	$2.38 \pm 1.85(\text{stat.})$ ($\sqrt{s}$ = 540 GeV)
	$1.0 \pm 0.32(\text{stat.})$ ($\sqrt{s}$ = 630 GeV)

distribution, m^{JJ}, consistent with the decays W $\rightarrow$ jet + jet and $Z^0 \rightarrow$ jet + jet. These data are qualitative only. As last reported in rapporteur talks by Dobrzynski [19] and diLella [20], the existence of the top quark is not yet firmly established.

3.4 Measurements of the $W^\pm$ and Z^0 mass and width.

Both the UA1 and UA2 experiments use (see for example Fig. 3a) the distribution of transverse mass, $m_T^{e\nu}$, to evaluate the $W^\pm$ mass. In each case a Monte Carlo simulation which takes account of the expected $W^\pm$ longitudinal and transverse motion, and expected detector characteristics, is used to generate $m_T^{e\nu}$ distributions that are compared with the data. Details of the both the fit and the theoretical input differ slightly and the reader is referred to the most recent publications [4,5]. To improve the measurement accuracy, the UA1 Collaboration selects a well measured event sample satisfying $p_T^e > 30$ GeV/c and $p_T^\nu > 30$ GeV/c. The UA2 Collaboration uses a fit to the full data sample. The result of these fits is shown in Table 8. Using a fit to the $m_T^{e\nu}$ distributions that includes the $W^\pm$ width, Γ_W, as a free parameter, limits on that quantity are also quoted (Table 8) by each Collaboration.

In each experiment, systematic uncertainties limit the accuracy of the data. These include theoretical uncertainties of the generated distributions and the dominant uncertainty of the calorimeter energy scale. Since in measurements of m_W/m_Z the scale uncertainty cancels, this error is quoted separately by the UA2 Collaboration in Table 8.

A fit to the W $\rightarrow \mu\nu_\mu$ sample from UA1 has also been presented, but as shown in Table 8, the mass uncertainty is large.

Each of the experiments has presented background-free samples of the decay $Z^0 \rightarrow e^+e^-$, and UA1 has presented a sample of decays $Z^0 \rightarrow \mu^+\mu^-$. The di-lepton mass enhancements are fitted to a relativistic Breit-Wigner shape, distorted by the experimental mass resolution. Results of the fits are shown in Table 8.

A first estimate of the Z^0 width, Γ_Z, can be obtained by a direct measurement of the width of the mass peak. However, given the limited statistics and the poor measurement accuracy, the determination of Γ_Z depends critically on a precise knowledge of the measurement accuracy. Resultant limits on Γ_Z are given in Table 8.

Table 8: Measurements of the W and Z^0 mass and width from UA1 and UA2.

	UA1	UA2
a) Electron decay		
m_W (GeV)	$83.5\pm1.0(\text{stat})\pm2.7(\text{syst})^{*}$	$80.1\pm0.6(\text{stat})\pm0.5(\text{syst})^{1}$ $\pm1.3(\text{syst})^{2}$GeV
Γ_W (GeV) (90%CL)	< 6.5 *	< 7.0
m_Z (GeV)	$93.0\pm1.4(\text{stat})\pm3.0(\text{syst})^{*}$	$92.1\pm1.1(\text{stat})\pm1.5(\text{syst})$
Γ_Z (GeV) (90%CL)	< 8.3 *	< 7.1
b) Muon decay		
m_W (GeV)	$80.7\ ^{+4.5}_{-4.1}(\text{stat})\ ^{+9.0}_{-8.1}(\text{syst})$	
m_Z (GeV)	$96.8\ ^{+3.3}_{-3.1}(\text{stat})\ ^{+4.3}_{-4.0}(\text{syst})$	

$(\text{syst})^{1}$: Systematic error resulting from uncertainty in Γ_W ($\pm$ 0.2 GeV), uncertainty in the measurement of P_T^{ν} ($\pm$ 0.3 GeV) and cell-to-cell calibration uncertainties ($\pm$ 0.3 GeV).

$(\text{syst})^{2}$: Systematic error resulting from energy scale uncertainty.

$*$: published result [4] excluding 1985 data.

Within the context of the Standard Model, the value of Γ_Z is related to the number of addition fermion doublets for which the decay $Z^0 \rightarrow f\bar{f}$ is kinematically allowed. If any additional $W^{\pm}$ and Z^0 decay products result from new fermion doublets in which only the neutrino is significantly less massive than $m_Z/2$,

$$\Gamma_Z(\text{meas}) = \Gamma_Z(3 \text{ fermion families}) + 0.177\Delta n_{\nu}, \tag{9}$$

where Γ_Z is in units of Gev and Δn_{ν} is the number of additional neutrinos. Since Γ_W is independent of Δn_{ν} if the associated charged lepton mass exceeds m_W, a model-dependent [21] estimate is available from the ratio

$$R = \sigma_Z^e / \sigma_W^e$$

A. Clark

$$= (\sigma_Z/\sigma_W)(\Gamma_Z^{ee}/\Gamma_W^{e\nu})(\Gamma_Z/\Gamma_W^{e\nu}) \tag{10}$$

Subject to limitations noted above, the measurement error of R is at present statistics limited. The ratio σ_Z/σ_W has recently been estimated [14] to be 0.3 ± 0.02, and $\Gamma_Z^{ee}/\Gamma_W^{e\nu}$ is given directly from the Standard Model (results quoted below assume measured values for m_W and m_Z, and $m_{top} \approx 40$ Gev). The UA1 and UA2 Collaborations respectively measure

R > 0.090 (90% confidence level), and
R > 0.098 (90% confidence level) $\tag{11}$

to be compared with the theoretical estimate $R = 0.11 \pm 0.02$ [14]. Corresponding to the quoted limits on R, the UA1 and UA2 Collaborations measure respectively

$\Delta n_\nu < 6.6 \qquad$ (90% confidence level), and
$\Delta n_\nu < 2 \pm 2$ (90% confidence level). $\tag{12}$

In the UA2 measurement, the quoted errors reflect theoretical uncertainties, and the stringent limit on Δn_ν results in part from the large value of σ_Z^e measured at $\sqrt{s} = 540$ Gev (see Table 6).

The evaluated Δn_ν-value depends sensitively on the assumed value of m_{top} because of the effect of kinematic factors. We repeat that the quoted limits depend on very specific conditions as noted above.

3.5 Status of the Standard Model parameters.

Ignoring the fermion and Higgs scalar masses, and the elements of the Kobayashi-Maskawa matrix [22], the minimal Standard Model is characterised by three parameters, taken here to be α_{EM} (the fine structure constant), and the masses m_W, m_Z. To compare the UA1 and UA2 measurements with predictions of the Standard Model, suitably renormalised and radiatively corrected theoretical quantities must be used [23]. The definition for which

$$\sin^2\theta_W = 1 - (m_W/m_Z)^2 \tag{13}$$

is used [24], leading to the following predictions:

$$m_W^2 = A^2/[(1-\Delta r)\sin^2\theta_W] \tag{14}$$

$$m_Z^2 = 4A^2/[(1-\Delta r)\sin^2 2\theta_W] \tag{14'}$$

where $A = (\pi\alpha/\sqrt{2}G_F)^{1/2} = (37.2810 \pm 0.0003)$ GeV/c² using the measured values of α_{EM} and G_F [25]. In the above equations, the value Δr reflects the effect of one-loop radiative corrections on the $W^{\pm}$ and Z^0 masses and has been computed to be [24]

$$\Delta r = 0.0696 \pm 0.0020 \tag{15}$$

for $m_t = 36$ GeV/c² and assuming that the mass of the Higgs boson, M_H, is equal to M_Z. Although the quoted theoretical error in Eq. (13) is quite small, Δr can be significantly changed [24,26] in the case of a very heavy t-quark ($\Delta r \approx 0$ for $m_t = 240$ GeV/c²), or if a new fermion family exists with a large mass splitting between the two members of an SU(2) doublet.

From equations (14) and (14') a model-dependent estimate of $\sin^2\theta_w$ can be evaluated. The results are shown in Table 9 and are in agreement with the mean of results from the CDHS [27], CHARM [28], CCFRR [29] and FMMF [30] Collaborations presented at this Conference.

In the above discussion, the ρ parameter [31]

$$\rho = m_W^2/(m_Z^2 \cos^2\theta_w) \tag{16}$$

is defined to be $\rho = 1$. By combining equations (13) and (16) an estimate of ρ is obtainable and the results listed in Table 9 are compatible with the expectaion $\rho = 1$ of the minimal Standard Model with scalar Higgs particles.

A measurement of the radiative correction Δr is also possible from these data. Eliminating $\sin^2\theta_w$ from equations (14) and (14'), the results (called Δr^c in Table 9) are obtained. Alternatively, by using low-energy measurements of $\sin^2\theta_w$ and the measured values of m_W and m_Z, the results (Δr^d) of Table 9 are obtained. Within the present statistical and systematic limits, the presence of radiative corrections cannot be demonstrated. This conclusion is very evident from Fig. 6 where the 68% c.l. contours of $(m_Z - m_W)$ vs. m_Z for UA2 data are compared with the Standard Model predictions and the ranges allowed by the low-energy data presented at this Conference. Moreover, the contribution of the interesting weak radiative contribution to Δr (for example from fermion loops or Higgs corrections) is small (<1%) for values of $m_{top} \leq 100$ GeV., and is beyond the capability of planned Collider measurements.

3.6 Deviations from the Standard Model.

From the data collected in 1983, corresponding to an integrated luminosity of ≈ 140 nb⁻¹, 3 of the 17 Z^0 events observed by the UA1 and UA2 Collaborations could most easily be explained by the decay $Z^0 \rightarrow \ell^+\ell^-\gamma$. In data collected in 1984 ($L \approx 310$ nb⁻¹) no additional event of this type was recorded. The UA1 Collaboration [32] has reported preliminary evidence for a second $Z^0 \rightarrow \mu^+\mu^-\gamma$ decay in data collected during 1985. However, because of the poor

Table 9: Measurements of the Standard Model parameters in UA1 and UA2.

Quantity	UA1[†]	UA2
$\sin^2\theta_w$ [a]	0.194 ± 0.03	$0.242 \pm 0.023 \pm 0.009$
$\sin^2\theta_w$ [b]	$0.214\ ^{+0.005}_{-0.006} \pm 0.015$	$0.232 \pm 0.004 \pm 0.008$
ρ	$1.026 \pm 0.037 \pm 0.019$	$0.988 \pm 0.027 \pm 0.006$
Δr [c]		$0.105 \pm 0.077 \pm 0.029$
Δr [d]		$0.06 \pm 0.03 \pm 0.03$

a. evaluation from the ratio (m_w/m_z) assuming $\rho = 1$

b. evaluation using as input measured values of α_{EM} and G_F, and 1-loop radiative correction calculations

c. using Collider data alone

d. including as input low-energy measurements of $\sin^2\theta_w$, using $\sin^2\theta_w = 0.228 \pm 0.006$

†   results quoted exclude data collected in 1985.

mass resolution of the muon channel, the conventional decay $Z^0 \to \mu^+\mu^-$ cannot be excluded for either candidate. For the electron decay channel, there are now 69 Z^0 candidates, including two decays of the type $Z^0 \to e^+e^-\gamma$. The probability of observing at least two such decays in the measured (or less likely) configurations is expected for the internal bremsstrahlung hypothesis to be $\approx 13\%$, and therefore no unconventional mechanism is required to explain these events.

The analysis of data collected during 1985 is continuing, and cross-section limits on the production of new particles (for example additional high-mass $W^\pm$ or Z^0 particles) or unconventional decays of the $W^\pm$ and Z^0 (for example decays into heavy leptons or super-symmetric particles) are preliminary. In particular UA1 [7] quote preliminary limits for the mass of additional bosons coupling as for the Z^0 and $W^\pm$ of respectively 166 GeV and 220 GeV with 90% confidence level. So far, there is no evidence of any deviation from the predictions of the minimal Standard Model; existing limits are published by the UA1 [4] and UA2 [5] Collaborations.

4. CONCLUSIONS AND FUTURE COLLIDER MEASUREMENTS.

Results collected from the $\bar{p}p$ Collider during the past four years have exceeded all expectations. In this period each of the UA1 and UA2 Collaborations have identified limitations of the existing detector designs, and are now revamping their detectors [33,34] to take advantage of the improved CERN Collider [9]. Furthermore, the CDF detector [35] will collect similar data at the Fermilab Collider, and the D0 experiment [36] at the Fermilab Collider is under construction.

From comparisons described in the previous sections, we can conclude that for measurements related to the Standard Model:

i. The existing energy-scale uncertainty of calorimeter calibrations is a major limitation. In the upgraded UA2 detector, an energy-scale calibration uncertainty of $\pm 1\%$ is anticipated. Any improvement on that is difficult but if achieved would be valuable (for example the planned UA1 and D0 calorimeters).

ii. The p_T^{ν}- resolution of the existing UA2 detector is inadequate, and is being improved by adding full calorimetric coverage to within $\approx 5°$ of the beam line. The accuracy of Standard Model measurements will be largely unaffected by this change, but the analysis will be simplified, background contributions will be reduced, and possible deviations from the Standard Model will be more accessible.

iii. In both the UA1 and UA2 detectors, the rejection power of the electron identification criteria to hadronic background is marginal, especially for low cross-section processes such as

$$\bar{p}p \rightarrow W + X, \quad W \rightarrow t + b, \quad t \rightarrow \ell \nu_\ell b. \qquad (17)$$

The upgraded UA1 detector will benefit in this respect from calorimetry with much improved lateral granularity. The inclusion of a new central tracking detector, with a transition-radiation detector to reject overlaps of π^0's with charged particles and an array of silicon counters to reject π^0 conversions, will improve the rejection of background in the UA2 detector by at least one order of magnitude.

Given these improvements, the expected accuracies of the Standard Model parameters are listed, using the upgraded UA2 detector as an example, in Table 10. In this Table, an integrated luminosity of 10 pb^{-1} is assumed.

The following comments may be appropriate.

i. The present (σ_Z/σ_W) measurement is limited by statistics; an improved measurement will place a better constraint on the number of additional neutrinos.

Table 10: Accuracy of Standard Model parameters in the improved UA2 detector.

Quantity	Value	Existing accuracy (fraction)		Expected accuracy (fraction)	
		statistical	systematic	statistical	systematic
σ_w^e	530 pb	.075	.085	$\sim$.028	$\sim$.085
σ_z^e	69 pb	.19	.085	$\sim$.071	$\sim$.085
σ_z^e/σ_w^w	0.13	.30	small	$\sim$.11	small
m_w	80.1	.007	$\sim$.005(analysis) .016(energy s	$\sim$.002	$\sim$.003 $\sim$.01
m_z	92.1	.012	.016	$\sim$.004	$\sim$.01
m_w/m_z	0.87	.013	$\sim$.005	$\sim$.004	$\lesssim$.002

Derived Quantities	Value	Existing accuracy (absolute)		Approx. expected accuracy (absolute)	
		stat.	syst.	stat.	syst.
$\sin^2\theta_w$ [a]	.242	.023	.009	$\sim$.006	$\sim$.004
$\sin^2\theta_w$ [b]	.232	.004	.008	$\sim$.001	$\sim$.006
Δr [c]	.105	.08	.03	$\sim$0.024	$\sim$0.02
Δr [d]	.06	.03	.03	$\sim$0.017	$\sim$0.02
ρ	.988	.027	.006	$\sim$0.01	$\sim$0.003

a – d : see Table 9

ii. Given the accurate data on the decay $Z^0 \to \ell^+\ell^-$ expected from future e^+e^- machines, an accurate measurement of the ratio (m_W/m_Z) is of importance.

iii. Using Collider measurements alone, expected deviations $\Delta r \neq 0$ will be accessible at approximately the 3 s.d. level, but a measurement of the weak radiative contribution will remain inaccessible.

iv. Measurements of $\sin^2\theta_W$ will be much improved, but will remain limited by analysis systematics (in the case of $\sin^2\theta_W$) or by energy-scale uncertainties (in the case of $\sin^2\theta_W$ as evaluated from Eq. 14).

v. Measurements of lepton universality by the UA1 Collaboration will be much improved, since the accuracy of present measurements is limited by statistics.

vi. An accurate measurement of the top-quark mass is an important priority for each experiment.

So far, there is no evidence for any deviation from the minimal Standard Model. The thrust of future measurements will be to identify deviations if they exist. Within the context of the Standard Model, the isolation of weak gauge coupling remains possible, particularly at the Fermilab Collider.

Acknowledgements.

Thanks are due to Prof. T. Kitagaki and his colleagues for organising a very enjoyable and productive Conference in Sendai.

I wish to thank my colleagues in the the UA1 and UA2 Collaborations for their friendly co-operation and support since the start of $\bar{p}p$ Collider operation. Special thanks are due to S. Wimpenny for communicating UA1 data prior to this Conference.

The transformation of this script into legibility is due to Mme. M. Prost, to whom I am grateful.

References.

1. S.L. Glashow, Nucl. Phys. 22 (1961) 579 ;
   S. Weinberg, Phys. Rev. Lett. 19 (1967) 1264 ;
   A. Salam, Proc. 8th Nobel Symposium, Aspenäsgårchen, 1968, p.367.

2. J. Timmer, Proc. 3rd Moriond Workshop on $\bar{p}p$ Physics, La Plagne, 1983, p.593
   (Editions Frontières, Paris. Ed. J. Tran Thanh Van) ;
   E. Locci, ibid, p.31
   K. Eggert et al., Nucl. Inst. and Methods, 176 (1980) 217.

3. B. Mansoulié, Proc. 3rd Moriond Workshop on $\bar{p}p$ Physics, La Plagne, 1983, p.609.
   (Editions Frontières, Paris. Ed. J. Tran Thanh Van) ;
   A. Beer et al., Nucl. Instrum. Methods 224 (1984) 360.

4. G. Arnison et al., Phys. Lett. 122B (1983) 103 ;
   G. Arnison et al., Phys. Lett. 126B (1983) 398 ;
   G. Arnison et al., Phys. Lett. 129B (1983) 273 ;
   G. Arnison et al., Nuovo Cimento Lett. 44 (1985) 1 ;
   G. Arnison et al., Phys. Rev. Lett. 166B (1985) 484.

5. M. Banner et al., Phys. Lett. 122B (1983) 476 ;
   P. Bagnaia et al., Phys. Lett. 129B (1983) 130 ;
   P. Bagnaia et al., Z. Phys. C - Particles and Fields 24 (1984) 1 ;
   P. Bagnaia et al., Z. Phys. C - Particles and Fields 30 (1986) 30.

6. UA2 Collaboration, presented by S. Loucatos. Proc. of 6th Workshop on $\bar{p}p$ Collider
   Physics, Aachen, July 1985. To be published.

7. UA1 Collaboration, presented by D. Denegri. Proc. of 6th. Workshop on $\bar{p}p$ Collider
   Physics, Aachen, July 1985. To be published.

8. UA1 Collaboration, presented by R. Batley. Proc. of 6th. Workshop on $\bar{p}p$ Collider
   Physics, Aachen, July 1985. To be published.

9. Staff of the CERN proton-antiproton project, Phys. Lett. 107B (1981) 306 ;
   B. de Raad, Proc. of 4th Topical Workshop on $\bar{p}p$ Collider Physics, Berne, 1984.
   Cern Yellow Report 84-09 (1984) p.344 ;
   R. Billinge, ibid. p.357.

10. A. Savoy − Navarro, Proc. of 5th. Workshop on $\bar{p}p$ Collider Physics, St Vincent, 1985,
    p.196. (World Scientific Publishing Co., Singapore, Ed. M. Greco) ;

C. Rubbia, Proc. of the 1985 Intl. Symposium on Lepton and Photon Interactions, Kyoto, 1985. (Yukawa Hall, Kyoto 606. Eds. M. Kanuma and K. Takahashi).

11. G. Arnison et al., Phys. Lett. 134B (1984) 469 ;
    G. Arnison et al., Phys. Lett. 147B (1984) 241.

12. UA4 Collaboration, M. Bozzo et al., Phys. Lett. 147B (1984) 392 ;
    UA4 Collaboration, G. Sanguinetti, Proc. 3rd. Moriond Workshop on $\bar{p}p$ Physics (1984) p.25. (Editions Frontières, Paris. Ed. J. Tran Thanh Van).

13. UA1 Collaboration, presented by R. Leuchs. Proc. Intl. Symposium on Physics of $p\bar{p}$ Collisions, Tsukuba, 1985, p.178. KEK Report 85-5.

14. G. Altarelli et al., Nucl. Phys. B246 (1984) 12 ;
    G. Alteralli et al., Z. Phys. C - Particles and Fields 27 (1984) 617.

15. J.C. Collins and D.E. Soper, Nucl. Phys. B193 (1983) 381 ;
    J.C. Collins and D.E. Soper, Nucl. Phys. B194 (1984) 4445
    J.C. Collins and D.E. Soper, Nucl. Phys. B197 (1982) 446.

16. M. Jacob, Nuovo Cimento 9 (1958) 826.

17. P. Langacker, Proc. XXII Intl. Conf. on High-Energy, Leipzig, 1985, p.215. (Ed. A. Meyer and E. Wisczorek) ;
    L. Mariani, Proc. Intl. Euro Physics Conf. on High-Energy Physics, Bari, 1985, p.639. (Laterzau, Bari 1985. Eds. L. Natti and G. Preparata).

18. UA2 Collaboration, presented by P. Bagnaia. Proc. of the 6th. Workshop on on $\bar{p}p$ Collider Physics, Aachen, July 1985. To be published.

19. L. Dobrzynski, Proc. of 5th. Intl. Conference on Physcis in Collision, Autun, France, 1985, p.25. (Editions Frontières, Paris, 1985. Ed. B. Aubert and L. Montanet).

20. L. DiLella, Proc. of the Int. EuroPhysics Conf. on High−Energy Physics, Bari, 1985, p.761. (Laterza, Bari 1985. Ed. L. Natti and G. Preparata).

21. F. Halzen and K. Marsula, Phys. Rev. Lett. 51 (1983) 857.

22. M. Kobayushi and K. Maskawa, Prog. Theor. Phys. 49 (1973) 652.

23. A. Sirlin, Phys. Rev. D22 (1980) 971 ;
    W. Marciano, Phys. Rev. D20 (1979) 274
    M. Veltman, Phys. Lett. 91B (1980) 95
    F. Antonelli et al., Phys. Lett. 91B (1980) 90.

A. Clark

24. W. Marciano and A. Sirlin, Phys. Rev. D29 (1984) 945.

25. Particle Data Group, Rev. Mod. Phys. 56 (1984) 56.

26. M. Veltman, Nucl. Phys. B123 (1977) 89 ;
    Z. Hioki, Nucl. Phys. B229 (1983) 284.

27. CDHSW Collaboration, presented by C. Guyot. See these proceedings.

28. CHARM Collaboration, presented by J. Panman. See these proceedings.

29. CCFR Collaboration, presented by F. Merrit. See these proceedings.

30. FMMF Collaboration, presented by R. Brock. See these proceedings.

31. A. Ross and M. Veltman, Nucl. Phys. B95 (1975) 135 ;
    P. Hung and J. Sakurai, Nucl. Phys. B143 (1978) 81.

32. S. Wimpenny, Proc. of the XXIth. Rencontres de Moriond, Les Arcs, France, 186. To
    be published.

33. UA1 Collaboration, presented by J. Dowell, Proc. of the 6th. Workshop on $\bar{p}p$ Collider
    Physics, Aachen, July 1985. To be published. See also references therein.

34. The UA2 Collaboration, CERN/SPSC 84-30 (1984), CERN/SPSC 84-95 (1984), UA2
    Collaboration, presented by C. Booth, Proc. of the 6th. Workshop on $\bar{p}p$ Collider
    Physics, Aachen, July 1985. To be published.

35. The CDF Collaboration, presented by M. Eaton, Proc. of the 6th. Workshop on $\bar{p}p$
    Collider Physics, Aachen, July 1985. To be published. See also references therein.

36. The D0 Collaboration, presented by M. Marx. Proc. of the 6th. Workshop on $\bar{p}p$
    Collider Physics, Aachen, July 1985. To be published. See also references therein.

Figures Captions

1. Regions of partial particle identification or nonexistent particle detection in the UA1 and UA2 detectors (see text).

    a. A region in azimuth ($\pm 4°$ to the vertical) covered by light-guides in the UA1 calorimeters; a cut of $\pm 15°$ is made at the analysis level.

    b. A region $|\theta| < 20°$ to the beam line of no particle identification in the UA2 detector, and a region $20° < |\theta| < 37.5°$ of poor jet energy measurement.

2. The distribution p_T^e is p_T^ν of all events for which there exists at least one candidate electron of $p_T^e > 15$ GeV/c. In events for which two or more electron candidates exist, that of highest p_T^e is chosen.

    a. UA1 (1984 data only).

    b. UA2 (full data sample).

3. Final data samples for the process $W^\pm \to \ell^\pm \nu_\ell$.

    a. The distribution $dN/dm_T^{e\nu}$ for events satisfing $p_T^e > 15$ GeV/c and $p_T^\nu > 15$ GeV/c (UA1, excluding 1985 data [4]).

    b. The distribution $dN/dm_T^{\mu\nu}$ for events satisfying $p_T^\mu > 15$ GeV/c and $p_T^\nu > 15$ GeV/c (UA1 [7]).

    c. The distribution dN/dp_T^e for events satisfying $p_T^e > 15$ GeV/c, $m_T^{e\nu} > 50$ GeV (UA2 [6]).

The superimposed curves show expectations for W-decay. The background contributions from QCD processes (   ) and the processes $W \to \tau\nu_\tau$ or $Z^0 \to e^+e^-$ (///) are also shown (see text and Table 3).

4. The distribution m^{ee} (m^{ee} > 20 GeV) for candidate electron pairs.

    a. Data from UA2 [6] after calorimetric selections only.

    b. Data from UA2 [6] after final event selection criteria are applied.

    c. Data from UA1 [7] after final event selection criteria are applied.

    In each case the expected background contribution from QCD processes is superimposed (---).

5. The distribution $dN/d\cos\theta^{*}_{e}$ from the UA1 experiment for the decay $W \rightarrow e\nu_{e}$.

6. The 68% contour lines (both statistical (i) and with systematic uncertainties added in quadrature (ii) for the measured values of m_{z} and (m_{z}-m_{w}) from the UA2 experiment. Superimposed ((a) and (b)) is the band of allowed values ($\sin^{2}\theta_{w}$ = 0.233 ± 0.006) obtained from an average of recent low-energy data [27-30]. Also shown the values of m_{z} and m_{w} expected from the minimal Standard Model with (c) and without (d) radiative corrections.

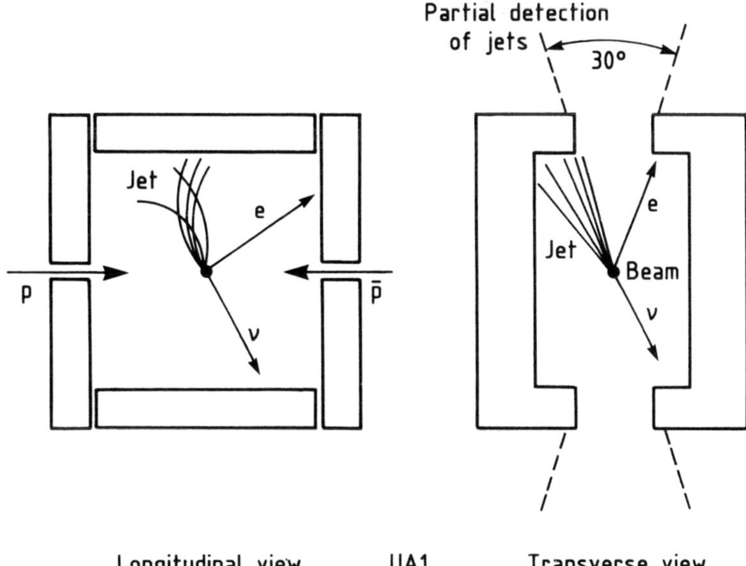

Longitudinal view UA1 Transverse view

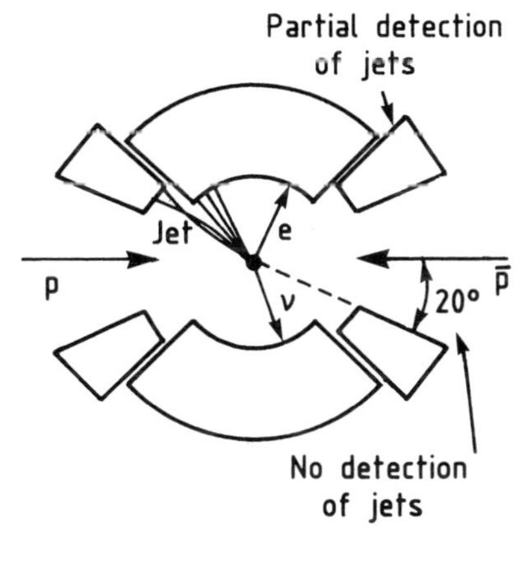

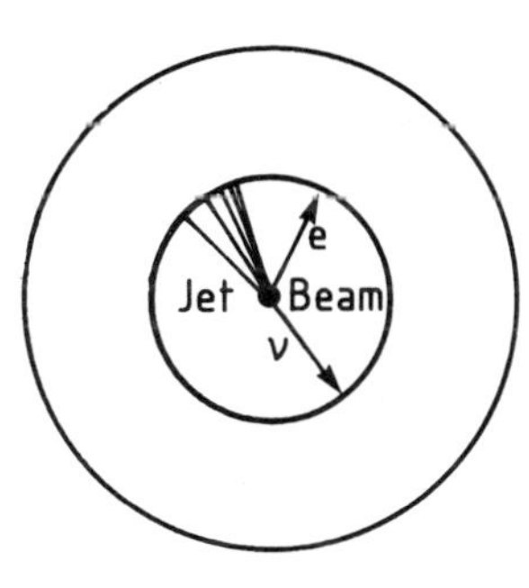

Longitudinal view UA2 Transverse view

Fig. 1

a)
UA1
774 events   ∫Ldt=263nb⁻¹
p_T^ν GeV/c
p_T^e GeV/c
b)
UA2
1809 entries
∫L dt= 880 nb⁻¹
p_T^ν (GeV/c)
p_T^e GeV/c)

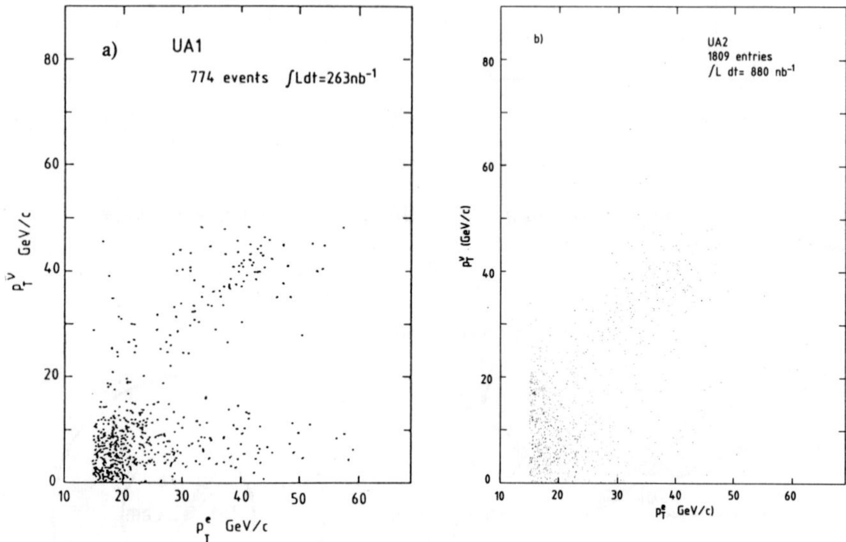

Fig. 2

a)
W → ev
Transverse mass
148 events
UA1
Events per 4 GeV/c²
16
12
8
4
0
20   40   60   80   100
m_T^{ev} (GeV)

b)
UA1
W^± → μ^± v
57 events
√s= 630 GeV
Events per 8 GeV/c²
10
8
6
4
2
0   50   100   150   200
m_T^{μv} (GeV)

c)
UA2
880 nb^{-1}
255 events
Events per 2 (GeV/c)
32
24
16
8
20   30   40   50   60   70   80
p_T^e (GeV/c)

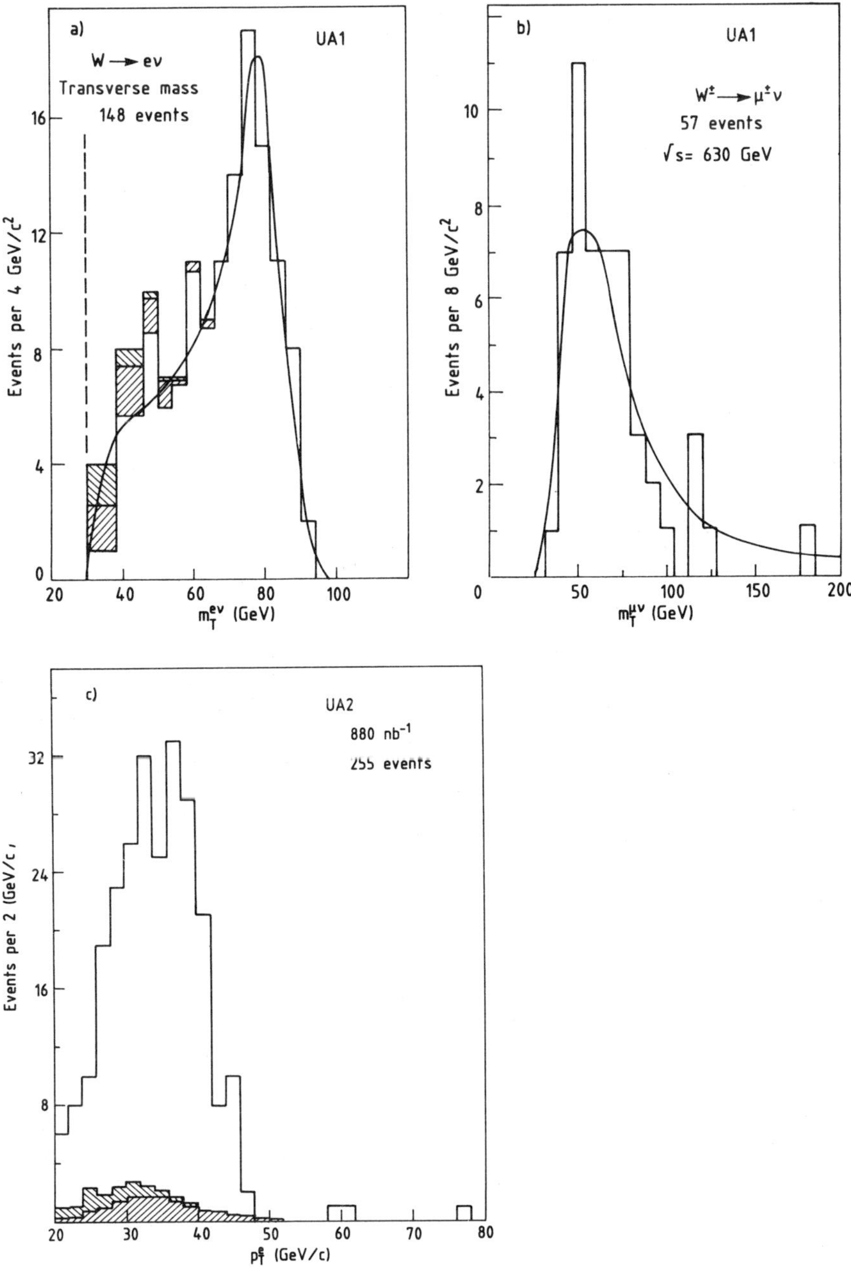

Fig. 3

UA2: 880 nb⁻¹
: calorimeter selections
a)
48 events
(12 events background)
Events per 2 GeV
60
40
20
UA2: 880 nb⁻¹
: final selections
b)
37 events
(1 event background)
Events per 2 GeV
12
8
4
UA1: 770 nb⁻¹
: final selections
c)
32 events
(≤0.8 events background)
Events per 4 GeV
12
8
4
30
50
70
90
110
mᵉᵉ (GeV)

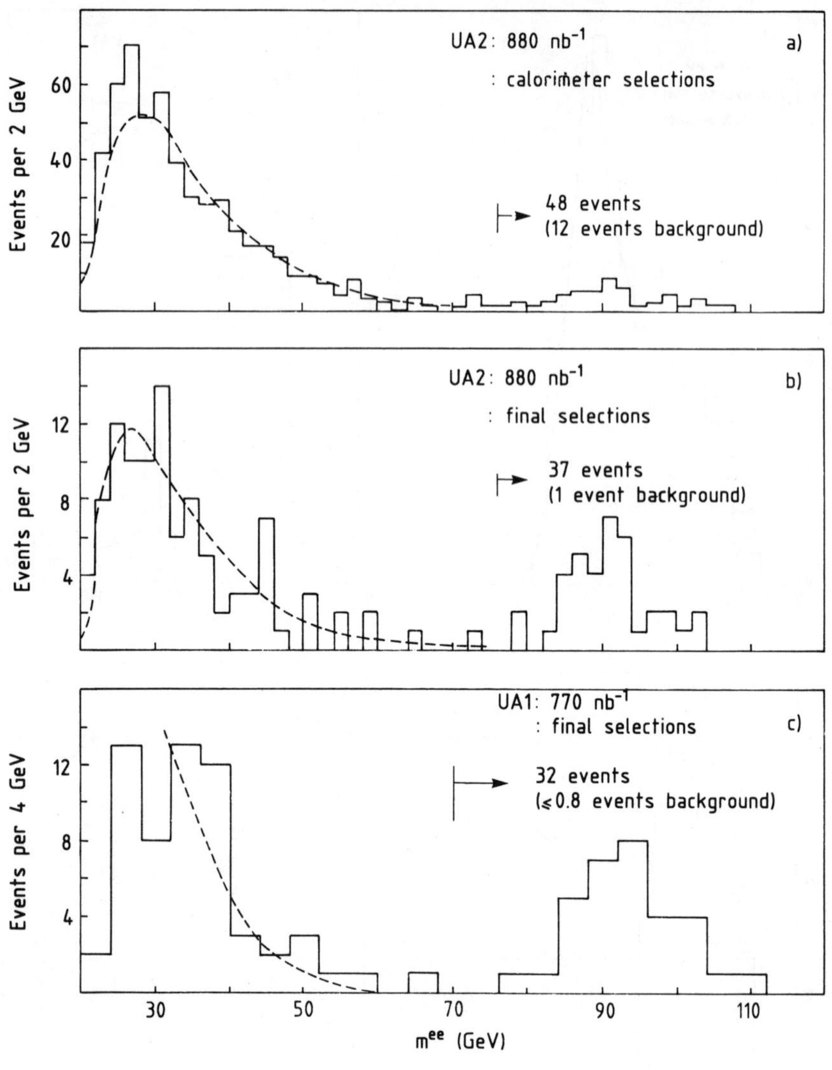

Fig. 4

UA1
75 events
Background subtracted
and
acceptance corrected
16
12
0.8
0.4
(1+cos θ*_e)²
1/N dN/d cos θ*_e
10 06 -02 02 06 10
cos θ*_e

Fig. 5

UA2
14
12
8
(ii)
(i)
a)
b)
c)
d)
(m_Z - m_W) [GeV]
88 90 92 94 96 98 100
m_Z (GeV)

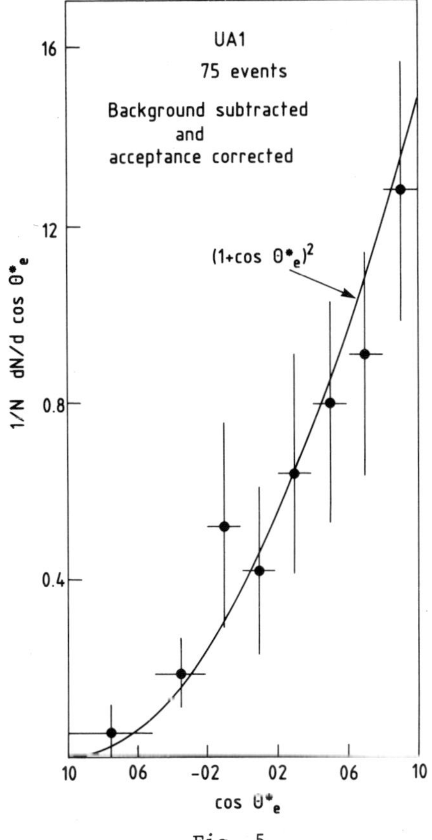

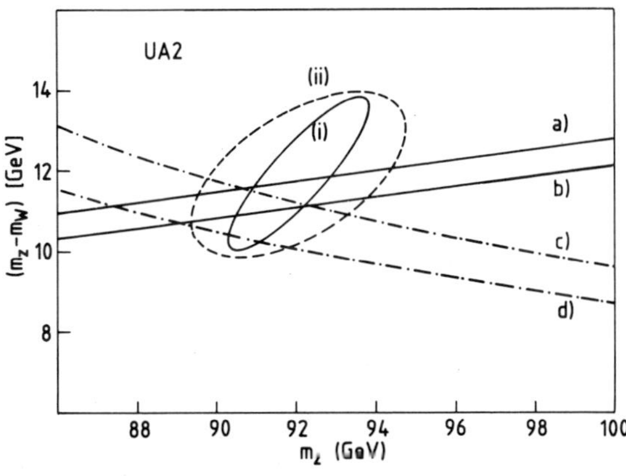

Fig. 6

Complete Determination of the
Weak Interaction in Muon Decay
and
Comparison with the Standard Model

W. Fetscher

Institut für Mittelenergiephysik der ETHZ,
CH-5234 Villigen, Switzerland

Muon decay, as a reaction involving four different leptons, is especially well suited to investigate the structure of the charged leptonic weak interaction. Most of the experimental results are in good <u>agreement</u> with the V-A interaction. In a recent paper by H.-J. Gerber, K.F. Johnson and myself [1], however, we have shown that the V-A interaction <u>follows</u> (within defined experimental errors) from present experiments under very general assumptions. This talk explains our reasoning.

The experimental data for normal muon decay have been obtained by detecting the charged leptons. Although these data generally agree well with V-A, they are not sufficient to determine the interaction, as was shown by C. Jarlskog in 1966 [2], who also pointed out the need to include coincidence measurements between the electron and one neutrino from the decay. In fact, the data from normal muon decay do not rule out a scalar type of interaction <u>instead</u> of the generally assumed V-A [3].

The hamiltonian of the interaction has already been formulated by L. Michel in 1950 [4]. Assuming a local, derivative-free and lepton-number conserving interaction, and allowing violation of parity, charge conjugation and time reversal invariance, one arrives at the respectable number of ten complex coupling constants, which, due to the freedom to choose an arbitrary phase, amount to 19 real parameters to be determined by experiment.

The hamiltonian was generally written in the <u>charge retention</u> form:

$$H = \sum_i \{C_i (\bar{e}\Gamma_i \mu)(\bar{\nu}_\mu \Gamma^i \nu_e) + C_i' (\bar{e}\Gamma_i \mu)(\bar{\nu}_\mu \Gamma^i \gamma_5 \nu_e)\} + \text{h.c.}$$

This form is closer to experiments in which only the charged leptons are measured. Parity violation is treated as in nuclear β decay by using the parity representation with parity-even and -odd terms (T.D. Lee and C.N. Yang, 1956 [5]). The Fierz transformations [6], however, allow various equivalent forms for the hamiltonian. The use of a different form is suggested by the V-A interaction, as known from nuclear beta decay and predicted by the standard model also for muon decay. It is characterized as a <u>charge-changing</u> interaction with left-handed fermions.

We have therefore used a charge-changing hamiltonian in the "helicity projection form" [7], which uses fields of definite handedness, to perform an analysis of normal and inverse muon decay.

We denote our matrix element by

$$M \sim \sum_{\substack{\gamma=S,V,T \\ \epsilon,\mu=L,R \\ (n,m)}} g^\gamma_{\epsilon\mu} \langle \bar{e}_\epsilon | \Gamma^\gamma | (\nu_e)_n \rangle \langle (\bar{\nu}_\mu)_m | \Gamma_\gamma | \mu_\mu \rangle$$

γ labels the type of interaction: Γ^S, Γ^V, Γ^T (scalar, vector, tensor). The indices ϵ and μ indicate the chiral projections (left-handed, right-handed) of the spinors of the experimentally observed particles, $\epsilon \hateq$ electron, $\mu \hateq$ muon. n indicates the helicity of the electron antineutrino, m that of the muon neutrino. n and m are uniquely determined for given γ, ϵ, μ.

The decay rate determines the strength of the interaction. In the following we are only interested in the relative contributions of the different couplings. The $g^\gamma_{\epsilon\mu}$ are therefore normalized to

$$A \equiv 4(|g^S_{RR}|^2 + |g^S_{LR}|^2 + |g^S_{RL}|^2 + |g^S_{LL}|^2)$$
$$+ 16(|g^V_{RR}|^2 + |g^V_{LR}|^2 + |g^V_{RL}|^2 + |g^V_{LL}|^2)$$
$$+ 48(|g^T_{LR}|^2 + |g^T_{RL}|^2): \quad = 16$$

"V–A" corresponds to $g^V_{LL} = 1$ and all other $g^\gamma_{\epsilon\mu} = 0$.

We now define the four quantities $Q_{\epsilon\mu}$ corresponding to the four possible combinations of handedness ϵ of the electron and of handedness μ of the muon, disregarding the helicities of $\bar{\nu}_e$ and ν_μ, since the neutrini have not been detected in the experiments of normal muon decay:

$$Q_{RR} \equiv \tfrac{1}{4}|g^S_{RR}|^2 + |g^V_{RR}|^2 \qquad = 2(b+b')/A$$

$$Q_{LR} \equiv \tfrac{1}{4}|g^S_{LR}|^2 + |g^V_{LR}|^2 + |g^T_{LR}|^2 = [(a-a') + 6(c-c')]/2A$$

$$Q_{RL} \equiv \tfrac{1}{4}|g^S_{RL}|^2 + |g^V_{RL}|^2 + |g^T_{RL}|^2 = [(a+a') + 6(c+c')]/2A$$

$$Q_{LL} \equiv \tfrac{1}{4}|g^S_{LL}|^2 + |g^V_{LL}|^2 \qquad = 2(b-b')/A$$

We note: $0 \leq Q_{\epsilon\mu} \leq 1$ and $\sum_{\epsilon,\mu} Q_{\epsilon\mu} = 1$. A $Q_{\epsilon\mu}$ is thus the probability for a transition from a muon of handedness μ to an electron of handedness ϵ.

The parameters $\{a/A, \ldots, c'/A\}$ have been introduced [8] to express all possible results of the measurements on the electron (positron) in the decay of polarized (and unpolarized) muons. Their values have been derived for positive muon decay [1,9] from the following, complete set of measurements:

<u>Shape of positron energy spectrum</u>

(1) $\rho = 0.752 \pm 0.0027$ [10]

<u>Decay asymmetry between μ spin and e momentum</u>

At spectrum end point:

(2) $\xi\delta/\rho = 0.9989 \pm 0.0023$ [11]

Differential:

(3) $\delta = 0.7502 \pm 0.0043$ [12]

The polarization vector (P_L, P_{T_1}, P_{T_2}) of the positron yields the remaining six parameters:

<u>Longitudinal polarization:</u>

(4) $<P_L> \equiv \xi' = 0.998 \pm 0.045$ [13]

<u>Angular dependence of P_L:</u>

(5) $\xi'' = 0.65 \pm 0.36$ [13]

<u>Energy dependence of P_{T_1}:</u>

(6) $\alpha/A = 0.015 \pm 0.052$ [9]
(7) $\beta/A = 0.002 \pm 0.018$ [9]

<u>Energy dependence of P_{T_2}:</u>

(8) $\alpha'/A = -0.047 \pm 0.052$ [9]
(9) $\beta'/A = 0.017 \pm 0.018$ [9]

We find upper limits for Q_{RR}, Q_{LR} and Q_{RL}, which in turn yield upper limits for the absolute values of eight complex coupling constants $g^{\gamma}_{\epsilon\mu}$ (see Table 1). Since Q_{LL} is bounded by a lower limit, it is not possible to deduce an upper limit for $|g^{S}_{LL}|$ from normal muon decay without detecting the neutrini. In fact, with the data from normal muon decay we can not tell if

$$g^{S}_{LL} = 0, \quad g^{V}_{LL} = 1 \quad (V\text{–}A),$$

$$\text{or} \quad g^{S}_{LL} = 2, \quad g^{V}_{LL} = 0 \; !$$

This kind of ambiguity has been noted by C. Jarlskog [2] in the context of a different form of the hamiltonian. She proposed to measure electron–neutrino correlations to resolve it, experiments which have not been performed to date. We use instead the data from inverse muon decay:

$$\nu_{\mu} + e^{-} \to \mu^{-} + \nu_{e} \; .$$

The total rate S, normalized to the rate predicted by V–A, is found to be $S = 0.98 \pm 0.12$ [14]. S has been calculated in terms of the charge changing hamiltonian in the parity representation [15]. In terms of our $g^{\gamma}_{\epsilon\mu}$ we get

$$S = (1/2)(1-h)\{|g^{V}_{LL}|^2 + (3/8)|g^{V}_{RL}|^2$$

$$+ (3/32)|g^{S}_{LR}-(10/3)g^{T}_{LR}|^2 + (3/32)|g^{S}_{RR}|^2 + (4/3|g^{T}_{LR}|^2\}$$

$$+ (1/2)(1+h)\{|g^{V}_{RR}|^2 + (3/8)|g^{V}_{LR}|^2$$

$$+ (3/32)|g^{S}_{RL}-(10/3)g^{T}_{RL}|^2 + (3/32)|g^{S}_{LL}|^2 + (4/3)|g^{T}_{RL}|^2\} \; ,$$

where h is the helicity of the ν_{μ} from pion decay. The deviation of $|h|$ from 1 is known very precisely: $1 - |h| < 4.1 \times 10^{-3}$ [16]; the sign of h has been determined by <u>electromagnetic</u> interactions for ν_{μ} and $\bar{\nu}_{\mu}$ [17,18]. Thus S gives information about the first 5 coupling constants g^{V}_{LL}, g^{V}_{RL}, g^{S}_{LR}, g^{T}_{LR} and g^{S}_{RR}, all of which couple to left–

handed ν_μ. The influence of four of them on S is found to be negligible with the upper limits derived from <u>normal</u> muon decay. We obtain

$$S = |g_{LL}^V|^2 \quad ,$$

which yields a <u>lower</u> limit for $|g_{LL}^V|$, and through the normalization requirement we get an upper limit for the remaining $|g_{LL}^S|$:

$$|g_{LL}^S|^2 < 4(1-S)$$

(see Table 1).

We have thus completely determined the weak interaction between the electron and the muon and their neutrini in normal and inverse muon decay using only leptonic data. Our results are also shown in <u>Fig. 1</u>, where each of the ten coupling constants is given within one of the squares defined uniquely by the handednesses of electron and muon and by the type of interaction. The outer circles display the mathematical limits for the $g_{\epsilon\mu}^\gamma$ in the complex plane, the inner circles for nine of the $g_{\epsilon\mu}^\gamma$ show the areas still allowed by experiment (90 % c.l.). For g_{LL}^V, which we have chosen to be real, we get the small line close to $g_{LL}^V = 1$ in agreement with V-A.

From the $g_{\epsilon\mu}^\gamma$ we can deduce nine independent limits for the violation of T-invariance in muon decay. They are also given in Table 1.

We furthermore find the minimum amount of measurements necessary to determine the interaction by calculating the probabilities P_R^e and P_R^μ for right-handed electrons and muons, respectively:

$$P_R^e \equiv Q_{RR} + Q_{RL} = \frac{1}{2}\,\Delta\xi'$$

(electron longitudinal polarization)

and $\quad P_R^{\mu} \equiv Q_{RR} + Q_{LR} = \frac{1}{2} \Delta\xi + \frac{2}{3} \Delta\delta \ (1+\Delta\xi)$

(electron decay asymmetry relative to muon spin), where $\xi' = 1 - \Delta\xi'$, $\xi = 1 - \Delta\xi$ and $\delta = (3/4)(1-\Delta\delta)$.

The μ decay interaction can therefore be determined from <u>five</u> different measurements, one of which gives information about the absolute magnitude, while the other four give information about the handedness of electron, muon and of one neutrino:

(1) τ_{μ} Fermi coupling constant

(2) δ

(3) ξ 16 decay parameters

(4) ξ'

(5) $S(\nu_{\mu}+e^- \rightarrow \mu^- +\nu_e)$ 2 decay parameters: <u>V-A</u>

Total   19 decay parameters

In summary, we have completely determined the weak interaction in muon decay from existing experiments. Present experimental errors, however, still allow substantial contributions from interactions other than V-A. These limits can in the future be reduced most efficiently by precise measurements of the muon decay asymmetry, the electron longitudinal polarization and the reaction cross section for inverse muon decay.

<u>Acknowledgements:</u>

I wish to thank H.-J. Gerber, K.F. Johnson and F. Scheck for many enjoyable discussions.

References

[1] W.Fetscher, H.-J.Gerber and K.F.Johnson, Phys. Lett. 173B (1986) 102.

[2] C.Jarlskog, Nucl. Phys. 75 (1966) 659.

[3] K.Mursula and F.Scheck, Nucl. Phys. B253 (1985) 189.

[4] L.Michel, Proc. Phys. Soc. A63 (1950) 514.

[5] T.D.Lee and C.N.Yang, Phys. Rev. 104 (1956) 254.

[6] M.Fierz, Z. Phys. 101 (1937) 553.

[7] F.Scheck, Leptons, hadrons and nuclei (North-Holland, Amsterdam, 1983).

[8] T.Kinoshita and A.Sirlin, Phys. Rev. 108 (1957) 844.

[9] H.Burkard, F.Corriveau, J.Egger, W.Fetscher, H.-J.Gerber, K.F.Johnson, H.Kaspar, H.J.Mahler, M.Salzmann and F.Scheck, Phys. Lett. B160 (1985) 343.

[10] S.E.Derenzo, Phys. Rev. 181 (1969) 1854.

[11] J.Carr, G.Gidal, B.Gobbi, A.Jodidio, C.J.Oram, K.A.Shinsky, H.M.Steiner, D.P.Stoker, M.Strovink and R.D.Tripp, Phys. Rev. Lett. 51 (1983) 627;
H.M.Steiner, private communication.

[12] B.Balke et al., LBL-18320 (1984), unpublished.

[13] H.Burkard, F.Corriveau, J.Egger, W.Fetscher, H.-J.Gerber, K.F.Johnson, H.Kaspar, H.J.Mahler, M.Salzmann and F.Scheck, Phys. Lett. 150B (1985) 242.

[14] CHARM Collab., F.Bergsma et al., Phys. Lett. B122 (1983) 465.

[15] K.Mursula, M.Roos and F.Scheck, Nucl. Phys. B219 (1983) 321.

[16] W.Fetscher, Phys. Lett. 140B (1984) 117.

[17] L.Ph.Roesch et al., Helv. Phys. Acta 55 (1982) 74.

[18] A.I.Alikhanov et al., JETP 11 (1960) 1380;
G.Backenstoss et al., Phys. Rev. Lett. 6 (1961) 415;
M.Bardon et al., Phys. Rev. Lett. 7 (1961) 23;
A.Possoz et al., Phys. Lett. B70 (1977) 265;
R. Abela et al., Nucl. Phys. A39 (1983) 413.

Table 1

The complete set of coupling constants $g^\gamma_{\epsilon\mu}$ of the muon-decay interaction determined from experiments without model assumptions. The "standard model" sets $g^V_{LL} = 1$, all others to zero. An additional V+A interaction is represented by g^V_{RR}. The $Q_{\epsilon\mu}$ are the relative rates. The $\tau^\gamma_{\epsilon\mu}$ are normalized T-violating amplitudes. 68% (90%) CL is given.

Handednesses of electron (ϵ) and muon (μ)	Relative abundance	Type of interaction	Handednesses of ν_μ(m) and $\bar{\nu}_e$(n)	Coupling constant	Main input	Violation of T-invariance ($\lvert\tau^\gamma_{\epsilon\mu}\rvert \lesssim 1$)
$\epsilon\mu$	$10^3 \times Q_{\epsilon\mu}$	γ	mn	$10^3\lvert g^\gamma_{\epsilon\mu}\rvert$		$10^3 \times \lvert\tau^\gamma_{\epsilon\mu}\rvert$
RR	<1.4(2.0)	S	LR	<74(91)	$\xi\delta/\rho$	<74(91)
		V	RL	<37(45)		<74(91)
LR	<2.7(3.9)	S	LL	<116(137)		<116(137)
		V	RR	<52(62)	$\xi\delta/\rho,\rho,\delta$	<104(125)
		T	LL	<34(40)		<116(137)
RL	<36(45)	S	RR	<378(448)		<371(436)
		V	LL	<97(114)	$P_L(\xi')$	<192(227)
		T	RR	<95(112)		<325(383)
LL	>960(949)	S	RL	<743(961)	$S, h_{\nu_\mu},$	<690(843)
		V	LR	>928(877)	Q_{RR}, Q_{LR}, Q_{RL}	--

$$\mu^- \rightarrow e^- + \bar{\nu}_e + \nu_\mu$$

90% c.l. for the $g^\gamma_{\varepsilon\mu}$. Antifermions are of opposite handedness.

$g^\gamma_{\varepsilon\mu}$	S	V	T
	$2i$ ... -2 ... 2 ... $-2i$; $e\ \mu\ \nu_e$: $R\ R\ L$; ν_μ : L	i (V+A) ... -1 ... 1 ... $-i$; $e\ \nu_e$: R ; ν_μ : R	
	$2i$ ... -2 ... 2 ... $-2i$; $e\ \mu\ \nu_e$: $L\ R\ R$; ν_μ : L	i ... -1 ... 1 ... $-i$; $e\ \nu_e$: L ; ν_μ : R	$i/\sqrt{3}$ ... $\frac{1}{\sqrt{3}}$ ... $\frac{1}{\sqrt{3}}$ ... $-i/\sqrt{3}$; $e\ \nu_e$: R ; ν_μ : L
	$2i$ ... -2 ... 2 ... $-2i$; $e\ \mu\ \nu_e$: $R\ L\ L$; ν_μ : R	i ... -1 ... 1 ... $-i$; $e\ \nu_e$: R ; ν_μ : L	$i/\sqrt{3}$ ... $\frac{1}{\sqrt{3}}$ ... $\frac{1}{\sqrt{3}}$ ... $-i/\sqrt{3}$; $e\ \nu_e$: L ; ν_μ : R
	$2i$ ... -2 ... 2 ... $-2i$; $e\ \mu\ \nu_e$: $L\ L\ R$; ν_μ : R	i (V−A) ... -1 ... 1 ... $-i$; $e\ \nu_e$: $R\ L$; ν_μ : L	

<u>Figure 1.</u>

90% c.l. limits for the coupling constants $g^\gamma_{\varepsilon\mu}$. Each coupling is uniquely determined by the handednesses ε and μ of the electron and the muon, respectively, and the type of interaction γ = S, V or T.

EXPERIMENTAL STATUS OF QUARK MIXING MATRIX

K. Kleinknecht

Institut für Physik, Universität Mainz

Mainz, Fed. Rep. Germany

Abstract

In this review, available data on weak decays and weak production of heavy and light quarks are summarized. They are used
to obtain allowed ranges of the elements of the quark mixing
matrix and of the weak mixing angles, both for the case of
three and four generations of quarks.

1. INTRODUCTION

In the last few years, much work has been done on extracting the weak
mixing angles between quarks [1] from experimental data [2-8]. This
subject is of interest because of its connection to the family problem
of quarks and their masses [9]. Efforts were directed towards measurements of the weak quark couplings and the extraction of the mixing
angles from these data, and towards models deriving relations between
the mixing angles and the quark masses and thus explaining the observed
pattern of mixing angles. We present here an updated version [26] of
our former analysis of experimental data [8], taking into account recent improvements. In addition, we extend the analysis to <u>four</u> <u>genera-
tions</u> of quarks, thus obtaining experimental information on the weak
couplings of a possible fourth pair of seqential quarks. The weak
mixing of six quarks is described by a unitary 3 x 3 matrix V_{ik}. This
matrix can be parametrized in terms of three angles and one phase,
connected to CP violation. Instead of the original parametrization of

Kobayashi and Maskawa [1], we use the one of Maiani [10] (table 1),
Chau [11]and Fritzsch [12]. These are identical apart from the definition of the phase. Since we deal here only with absolute values of
matrix elements, results for the mixing angles apply to these three
parametrizations in the same way.

Table 1

Maiani parametrization [10]

$$V = \begin{pmatrix} c_\beta c_\theta & c_\beta s_\theta & s_\beta \\ -s_\gamma c_\theta s_\beta e^{i\delta'} - s_\theta c_\gamma & c_\gamma c_\theta - s_\gamma s_\beta s_\theta e^{i\delta'} & s_\gamma c_\beta e^{i\delta'} \\ -s_\beta c_\gamma c_\theta + s_\gamma s_\theta e^{-i\delta'} & -c_\gamma s_\beta s_\theta - s_\gamma c_\theta e^{-i\delta'} & c_\gamma c_\beta \end{pmatrix}$$

As discussed before [8], information on the weak quark couplings comes
from measurements of weak decays of light and heavy quarks and from
neutrino production of charm quarks. We first go through these experimental constraints, and then proceed to derive bounds on the mixing
angles.

2.   CONSTRAINTS ON MATRIX ELEMENTS
2.1   Light quark couplings
2.1.1   Coupling V_{ud}

This parameter is obtained by comparing decay rates of nuclear beta
decays and muon decays. A new analysis of radiative corrections in the
nuclear beta decays has been presented by Marciano and Sirlin [13].
They conclude that the error on V_{ud} can be decreased, and their result
is

$$V_{ud} = 0.9729 \pm 0.0012 \tag{1}$$

2.1.2   Coupling V_{us}

Two kinds of experimental information on the coupling of strange and

up-quarks exist: one is the self-consistent analysis of weak semileptonic hyperon decays, as done by the WA2 collaboration [14], with the
result V_{us} = 0.231 ± 0.003. The other comes from an analysis [15] of
K_{e3} decays, $K_L \to \pi e \nu$ and $K^+ \to \pi e \nu$. This result V_{us} = 0.221 ± 0.002
differs from the one obtained in hyperon decays; Leutwyler and Roos [15]
argue that SU(3) breaking effects are larger in hyperon decays than in
K meson decays, and that therefore the results from K_{e3} are more reliable
from a theoretical point of view. We therefore use their value

$$V_{us} = 0.221 \pm 0.002 \tag{2}$$

2.2 Charm couplings
2.2.1 Coupling V_{cd}

This coupling has been determined from measurements of single charm
production in neutrino and antineutrino reactions. The coupling parameter is obtained from the measured ratios of dimuon to single muon
production cross-sections in neutrino reactions ($R^\nu = \sigma^\nu_{\mu\mu^+}/\sigma^\nu_\mu$) and
antineutrino reactions ($R^{\bar{\nu}} = \sigma^{\bar{\nu}}_{\mu^+\mu^-}/\sigma^{\bar{\nu}}_\mu$):

$$\beta V^2_{cd} = \frac{R^\nu - RR^\nu}{1 - R} \cdot \frac{2}{3}$$

where β is the semileptonic branching ratio of the mixture of charmed
particles produced in the neutrino reactions, and $R = \sigma^{\bar{\nu}}/\sigma^\nu$ is the
ratio of total neutrino cross-sections.

The result of the CDHS collaboration, [16]

$$\beta V^2_{cd} = (0.41 \pm 0.07)10^{-2} \tag{3}$$

has been used to extract V_{cd}. Since the Mark III collaboration has obtained new measurements of the semileptonic branching ratio, we re-
analyze this number. We start from the observed semileptonic branching
ratio B_μ on the ψ'' resonance; the average of four experiments [17] is
B_μ = 9.9 ± 0.9 %. The fraction of D^o's produced on the ψ'' is 57 ± 6 %.
On the other hand, the fraction of D^+ and D^o mesons amongst all charmed

particles produced in neutrino reactions are evaluated from the emulsion experiment [18]. For visible energies above 30 GeV, they are $f(D^+) = 36.2 \pm 8.9$ % and $f(D^o) = 50.7 \pm 9.0$ %, and the remaining fraction is due to charmed baryons. Using, in addition, the ratio of observed D^+ and D^o lifetimes [19], we obtain for the branching ratios $B_\mu(D^+) = (6.6 \begin{smallmatrix} + & 0.9 \\ - & 0.8 \end{smallmatrix})$% and $B_\mu(D^+) = (14.7 \begin{smallmatrix} + & 1.8 \\ - & 1.5 \end{smallmatrix})$%, and for the semileptonic branching ratio of the mixture of charmed particles produced in neutrino reactions

$$\beta = (9.3 \pm 1.0) \text{ \%} \tag{4}$$

Together with eq.(3) we obtain the new value

$$|V_{cd}| = 0.21 \pm 0.03 \tag{5}$$

<u>2.2.2   Coupling V_{cs}</u>

In the past, this coupling was obtained from neutrino and antineutrino dimuon production data. Since here the charm production from strange sea quarks is considered, the quantity measured is the product $|V_{cs}|^2 \cdot 2S$, where $S = \int xs(x)dx$ is the integral of the strange sea structure function. In the absence of an independent determination of S, one obtains as an upper limit $2S \leq \bar{U} + \bar{D}$, where $\bar{U}$ and $\bar{D}$ are the momentum fractions of non-strange sea antiquarks. We call $\alpha = 2S/(\bar{U} + \bar{D})$ the ratio of these momentum fractions, and $\alpha^* = 2S(|V_{us}|^2 + |V_{cs}|^2/r_s)/(\bar{U} + \bar{D})$ the same ratio modified by the threshold suppression factor r_s for the charm quark mass. Then from the x-distribution of neutrino dimuons [16] one obtains

$$\frac{|V_{cs}|^2}{|V_{cd}|^2} = (6.26 \pm 0.73) \frac{1 + \alpha^*}{\alpha} \tag{6}$$

and from the cross-section ratios R^ν and $R^{\bar\nu}$ of dimuon production [5]:

$$\frac{|V_{cs}|^2}{|V_{cd}|^2} = (5.9 \pm 1.5 + (8.5 \pm 1.7)\alpha^*)/\alpha \tag{7}$$

An alternative new method of obtaining V_{cs} is based on the measurement and calculation of the decay rate for $D^+ \rightarrow \bar{K}^o e^+ \nu_e$ [20]. This rate can be expressed as

$$\Gamma(D^+ \to \bar{K}e^+\nu_e) = |f_+^D(0)|^2|V_{cs}|^2 \ (1.54 \cdot 10^{11} \ s^{-1})$$

where $|f_+(0)|$ is the form factor for $D_{\ell 3}$ decay for zero momentum trans-
fer. Combining data on the branching ratios of the decays $D^+ \to K^{-o}e^+\nu$
and $D^o \to K^-e^+\nu$ and with the world average values of D^+ and D^o lifetimes
[21], the experimental decay width is $\Gamma(D \to \bar{K}e^+\nu) = (0.79 \pm 0.11)$ x
$10^{11}s^{-1}$. This leads to

$$|f_+^D(0)|^2|V_{cs}|^2 = 0.51 \pm 0.07.$$

In a recent theoretical calculation of the transition form factor using
QCD sum rules the value

$$|f_+^D(0)| = 0.6 \pm 0.1$$

was obtained [22]. This, together with the experimental rate, gives

$$|V_{cs}|^2 = 1.42 \pm 0.31 \tag{8}$$

If, on the other hand, this calculation is not relied upon, at least
the limit

$$|f_+^D(0)| < 1$$

can be used [23] to obtain a lower limit on $|V_{cs}|$.

2.3 Bottom-quark couplings

2.3.1 Ratio $|V_{ub}|/|V_{cb}|$

The inclusive electron spectrum from B decays is the most sensitive
method for obtaining a limit on this ratio. The energy of electrons
from semileptonic decays of the kind $B \to e\nu X_u$ extends beyond the maximum energy possible in decays with a charmed hadronic state, $B \to e\nu X_c$.
The limit on $|V_{ub}|/|V_{cb}|$ deduced from experimental electron spectra
depends on the calculated shapes for these two decay methods. The original calculation of Altarelli et al., [24], used in earlier analyses,
seems not to be completely adequate for the present data [21]. In order

to avoid this difficulty, it is possible to use the momentum region above the endpoint for $B \to e\nu X_c$, i.e. from 2.4 to 2.6 GeV/c. The limit

$$|V_{ub}|/|V_{cb}| < 0.19 \qquad \text{with 90 \% C.L.} \tag{9}$$

has been derived [21].

2.3.2 B lifetime

From the measured B lifetime $\tau_B = 1.26 \pm 0.16$ ps [21] and the average semileptonic B meson branching ratio of 12.1 $\pm$ 0.8 %, the partial semileptonic decay width of B mesons is obtained: $\Gamma(b \to c\ell\nu) = (9.6 \pm 1.4) \, 10^{10}\text{s}^{-1}$. This width is related to the b coupling parameters by

$$\Gamma(b \to c\ell\nu) = \frac{m_b^5}{m_\mu^5} \frac{1}{\tau_\mu} (|V_{ub}|^2 + 0.45 \, |V_{cb}|^2)$$

and therefore

$$(|V_{ub}|^2 + 0.45 \, |V_{cb}|^2) = (8.9 \pm 1.3) \, 10^{-4} \tag{10}$$

2.4 Determination of allowed ranges for coupling constants and angles

Three generations

Using the constraints of eqs.(1), (2), (5) - (10) we first do a fit [26] using the six-quark parametrization of table 1, along the lines of ref. 8 . We obtain, for a χ^2 = 6.28, the values for the angles $\sin\theta$ = 0.221 $\pm$ 0.002, $\sin\gamma$ = 0.044 $\pm$ 0.003 and $\sin\beta$ < 0.0088, and $2S/(\bar{U} + \bar{D})$ = 0.41 $\pm$ 0.06. The corresponding 90 % C.L. allowed ranges for the mixing matrix elements are given in table 1. The contours for the allowed values of $\sin\gamma$ and $\sin\beta$ are shown in fig. 1.

Table 2

Elements of 3 x 3 quark mixing matrix V_{ik} from fit of experimental
constraints (90 % C.L. allowed ranges)

	d	s	b
u	0.9743 – 0.9757	0.219 – 0.225	0.000 – 0.009
c	0.219 – 0.225	0.9733 – 0.9748	0.039 – 0.050
t	0.002 – 0.017	0.037 – 0.048	0.9987 – 0.9993

Four generations

We use the parametrization of Bose and Paschos [25]. The allowed range
of parameters was examined by a numerical scan of the parameter space.
The free parameters of this scan were the six angles θ_1 to θ_6, the
three phases δ, ψ and ϕ, and the values of α and $f^+(0)$. The parameters
were varied in a region around the solution with the minimal $\chi^2 = \chi_o^2$.
As experimental constraints, eqs.(1),(2),(5) to (7), (9),(10) were used
with the theoretical value for $f_+^D(0)$ as an optional constraint. All
combinations of parameters leading to a χ^2 value below $\chi_o^2 + 1.6^2$ were
accepted as allowed values. Table 3 gives the resulting bounds for the
elements of the 4 x 4 unitary matrix without the constraint on $f_+^D(0)$,
and table 4 the bounds including this constraint. Also, in fig.2 contour plots for allowed regions of each pair of angles θ_1 to θ_6 are
given [26].

Table 3

Elements of 4 x 4 quark mixing matrix V_{ik} from fit of experimental
constraints (90 % C.L allowed ranges). Constraint $f_+^D(0) < 1$ used.

	d	s	b	b'
u	0.9708–0.9750	0.218–0.224	0. –0.009	0.028–0.092
c	0.169 –0.237	0.66 –0.98	0.039–0.050	0. –0.78
t	0. –0.17	0. –0.77	0.030–0.999	0. –0.997
t'	0. –0.15	0. –0.35	0. –0.999	0.05 –0.9993

allowed ranges for angles:

$0.218 < \sin\theta_1 < 0.245$ $0 \leq \sin\theta_4 \leq 1$

$0.039 < \sin\theta_2 < 0.07$ $0 \leq \sin\theta_5 < 0.125$

$0. \quad \leq \sin\theta_3 < 0.40$ $0 \leq \sin\theta_6 < 0.55$

other parameters:

$0.25 < 2D/(\bar{U} + \bar{D}) < 0.84$ $0.65 < f_+^D(0) < 1$

Table 4

Elements of 4 x 4 quark mixing matrix V_{ik} from fit of experimental constraints (90 % C.L allowed ranges). Constraint $|f_+^D(0)| = 0.6\pm0.1$ used.

	d	s	b	b'
u	0.9708–0.9750	0.218–0.224	0. −0.009	0.024–0.092
c	0.180 −0.236	0.86 −0.98	0.039–0.050	0. −0.46
t	0. −0.17	0. −0.46	0.040–0.999	0. −0.998
t'	0. −0.1	0. −0.33	0. −0.999	0.05 −0.9997

allowed ranges for angles:

$0.218 < \sin\theta_1 < 0.240$ $0 \leq \sin\theta_4 \leq 1$

$0.039 < \sin\theta_2 < 0.60$ $0 \leq \sin\theta_5 < 0.1$

$0. \quad \leq \sin\theta_3 < 0.25$ $0 \leq \sin\theta_6 < 0.44$

other parameters:

$0.28 \leq 2S/(\bar{U} + \bar{D}) < 0.61$ $0.63 \leq |f_+^D(0)| \leq 0.80$

If we use instead of this parametrization the one based on a generalization of Maiani's form, as proposed by Gronau and Schechter [25] and by Türke et al. [25], the sines of the six angles, s_{ik}, correspond approximately to the coupling strengths between pairs of generations i and k, for i < k. We then obtain for the couplings between the first three generations:

$$0.218 \leq s_{12} \leq 0.224$$
$$s_{13} < 0.009$$
$$0.035 < S_{23} < 0.059$$

In addition, limits on the angles related to the fourth generation are obtained. They are:

$$s_{14} < 0.09$$
$$s_{24} < 0.46$$
$$s_{34} < 1.0$$

Discussion

In the six-quark case, the influence of the changes in the input parameters on the matrix elements and angles compared to ref.8 is visible, but not dramatic. For the matrix elements relevant to the fourth generation in the eight-quark case, the result shows that, with present epxerimental data, only the elements $V_{t'd}$ and $V_{ub'}$ and perhaps $V_{t's}$, are significantly constrained. Marciano and Sirlin have remarked [13] that $V_{ub'}$ is consistent with zero only at the 2σ level, and that there is room for a fourth generation. However, the knowledge on the other elements, V_{ts}, $V_{cb'}$, V_{tb}, $V_{t'b}$ and $V_{t'b'}$ is very limited, and in particular there is no reason to believe that the angles θ_2, θ_4 and θ_6 are small. It is even not known whether the bottom right-hand 2 x 2 submatrix of the mixing matrix V is diagonal or antidiagonal. In the case of three generations the pattern of mixing angles is rather clear: the sequence of decreasing angles is $s_{12} > s_{23} > s_{13}$. However, it is not known yet if such a pattern would also hold in the case of four generations. The upper limits obtained here follow the sequence $s_{34}^{max} > s_{24}^{max} > s_{14}^{max}$. It is therefore possible that s_{34} and s_{24} are larger than s_{12}, contrary to the expectation of theoretical models relating these angles to the ratios of quark masses [9]. The observation of the top quark and the measurement of its weak couplings in W decays at LEP 200 will improve this situation significantly.

References

[1] M.Kobayashi and K.Maskawa, Prog. Theor. Phys. 49 (1973) 652

[2] K.Kleinknecht, Proc.10th Int.Neutrino Conf,, Balatonfüred, 1982
(Central Res.Inst.Physics, Budapest, 1982), Vol.1, p.115

[3] E.A.Paschos and U.Türke, Phys. Lett. 116B (1982) 360

[4] S.Pakvasa, Proc.12st Int.Conf.on High Energy Physics, Paris 1982,
J.Phys.43, Suppl.12 (1982) C3-234

[5] K.Kleinknecht and B.Renk, Z.Phys. C16 (1982) 7; Z.Phys. C20 (1983) 67

[6] L.L.Chau et al., Phys.Rev.D27 (1983) 2145; L.L.Chau, Phys.Rep. 95
(1983) 3

[7] A.Buras, W.Slominski and H.Steger, Nucl.Phys. B238 (1984) 529

[8] K.Kleinknecht and B.Renk, Phys.Lett. 130B (1983) 459; Comm.Nucl.
Part.Phys. 13 (1984) 219

[9] H.Fritzsch, Nucl.Phys. B155 (1979) 189
B.Stech, Phys.Lett. 130B (1983) 189

[10] L.Maiani, Int.Symp.on Lepton and Photon Interactions at High Energies, Hamburg 1977 (DESY, Hamburg 1977), p.877

[11] L.L.Chau and K.Keung, Phys.Rev.Lett.53 (1984) 1908

[12] H.Fritzsch, Phys.Rev.D32 (1985) 3058

[13] W.J.Marciano and A.Sirlin, Phys.Rev.Lett. 56 (1986) 22

[14] M.Bourquin et al., Z.Phys. C21 (1983) 27

[15] H.Leutwyler and M.Roos, Z.Phys. C25 (1984) 91

[16] H.Abramowicz et al., Z.Phys. C15 (1982) 19

[17] J.M.Feller et al., Phys.Rev.Lett. 40 (1978) 274
W.Bacino et al., Phys.Rev.Lett. 42 (1979) 1073
R.H.Schindler et al., Phys.Rev. D24 (1981) 78
R.M.Baltrusaitis et al., Phys.Rev.Lett. 54 (1985) 1976

[18] N.Ushida et al., Phys.Lett. 121B (1983) 292

[19] Average obtained from R.M.Baltrusaitis et al., Phys.Rev.Lett. 54
(1985) 1976 and references quoted in K.R.Schubert, Proc. 11th Int.
Conf. on Neutrino Physics and Astrophysics, Nordkirchen 1984, ed.
by K.Kleinknecht and E.A.Paschos, World Scientific, Singapore 1984,
p.760

[20] We thank F.J.Gilman for calling our attention to this point.

[21] E.H.Thorndike, talk at the Int.Symp. on Lepton and Photon Interactions at High Energies, Kyoto 1985, ed. by M.Konuma and K.Takahashi,
Kyoto 1986, p.406

[22] T.M.Aliev et al., Sov.J.Nucl.Phys.40 (1984) 527 (Yad.Fiz.40(1984)
823); M.Wirbel,B.Stech and M.Bauer obtain $f_+^D(0) = 0.76$, Z.Phys.
C29 (1985) 637

[23] F.J.Gilman and K.Kleinknecht, in: Review of Particle Properties, Appendix III, 1986 edition

[24] G.Altarelli et al., Nuc.Phys. B208 (1982) 365

[25] S.K.Bose and E.A.Paschos, Nucl.Phys. B169 (1980) 384; see also M.Gronau and J.Schechter, Phys.Rev. D31 (1985) 1668 and U.Türke et al., Nucl.Phys. B285 (1985) 313 for another 4 x 4 parametrization following Maiani (ref.10).

[26] K.Kleinknecht and B.Renk, preprint University of Mainz, MZ-ETAP/ 86-3 (May 1986)

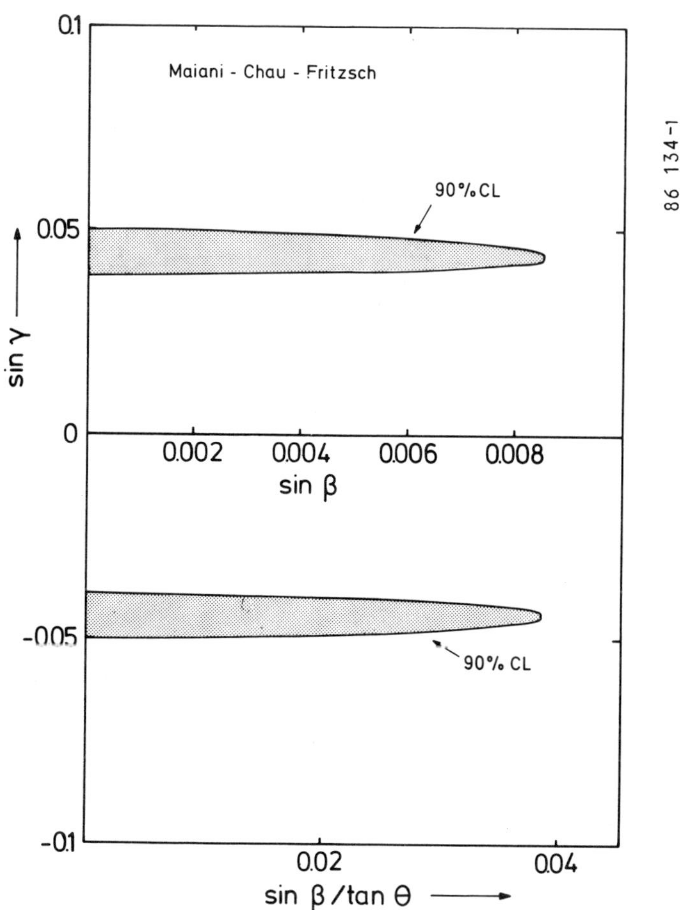

Fig. 1: Allowed range (shaded) in the plane of mixing parameters
sinγ and sinβ in the Maiani parametrization. Contour
corresponds to 90 % C.L.

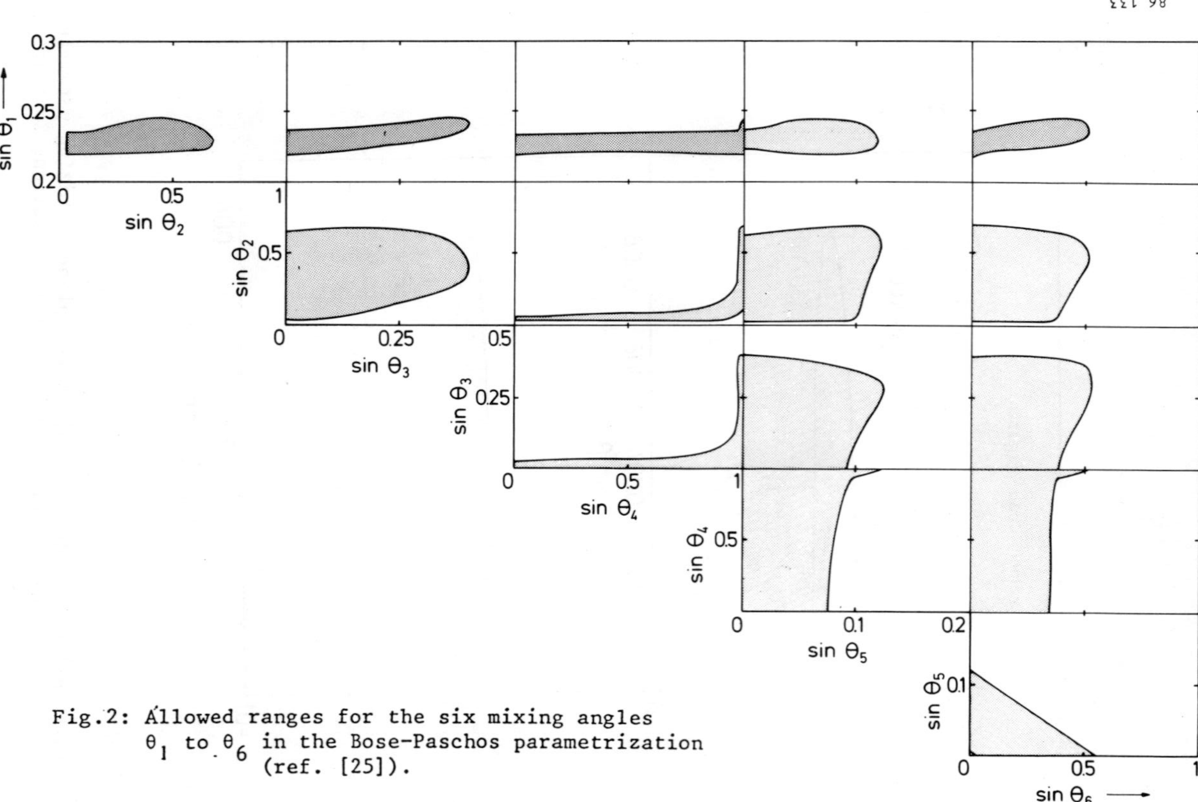

Fig. 2: Állowed ranges for the six mixing angles θ_1 to θ_6 in the Bose–Paschos parametrization (ref. [25]).

CP VIOLATION AND MIXING

A.I. Sanda[*]
The Rockefeller University, New York, N.Y. 10021

Abstract

A review of recent progress in (i) measurements of CP violating effects; (ii) theoretical predictions of these effects; (iii) meson-antimeson mixing effects is presented.

I Introduction

At present, essentially all experimental observations can be explained within the context of the standard model, $SU(3)_C \times SU(2)_L \times U(1)$ gauge theory of strong and electro-weak interactions with the KM quark mixing matrix. The observed CP noninvariance in the K meson system is one exception. The KM matrix presents a nice parametrization, but not a basic understanding of the phenomenon. Several alternative models have been proposed. These predict intercoting physics at high energies some of which may be observable at up-coming accelerators, and others may have profound consequence toward our understanding of the early universe. In this sense, CP noninvariance offers a valuable clue toward going beyond the standard model.

In this talk I shall briefly discuss the essence of various models:

   1) The Kobayashi-Maskawa model,

   2) Weinberg's model of spontaneous CP violation with natural flavor conservation,

3) Left-right symmetric model of CP violation,

4) Superweak model,

then go on to talk about their predictions for:

a) ϵ'/ϵ,

b) neutron electric dipole moment (edm),

c) μ meson polarization in $K \to \pi\mu\nu$ decay.

I will also discuss the present status of $B^0-\bar{B}^0$ and $D^0-\bar{D}^0$ mixing. Finally I shall end with the future prospects of the field.

II. Preliminaries

If the Hamiltonian

$$H = \sum_{i,j} \left(g_{ij} O_{ji} + g^{*}_{ij} O^{*}_{ji} \right)$$

describes the interaction of quarks with flavor indices i and j, it is easily seen that[1]

$$(CP)^{-1} O_{ji} CP = O^{*}_{ji}$$

so that H commutes with the CP operator if

$$g_{ij} = g^{*}_{ij} \; .$$

The fact that K_L and K_S are not CP eigenstates implies that there is at least one observable complex phase in some interaction of quarks. There are several ways in which the observable phase can be introduced:

1) Kobayash-Maskawa model[2]:

With three families of quarks, the charged current interaction can be written as

$$H = \frac{g}{2\sqrt{2}} K_{ij} \, \bar{U}_i \, \gamma_\mu (1-\gamma_5) D_j W^\mu + h.c. \qquad (1)$$

where $U^T=(u,c,t)_L$; $D^T=(d,s,b)_L$; W stands for the charged gauge boson; K is a unitary matrix. It should be noted that for any 3x3 general unitary matrix K, all but one complex phase can be absorbed in the phase definition of D_i and U_i. I find Wolfenstein's parametrization[3] of K particularly easy to remember:

$$K = \begin{bmatrix} 1-\lambda^2/2 & \lambda & A\lambda^3(\rho-i\eta) \\ -\lambda & 1-\lambda^2/2 & A\lambda^2 \\ A\lambda^3(1-\rho+i\eta) & -A\lambda^2 & 1 \end{bmatrix}. \tag{2}$$

The values of the parameters $\lambda=\sin\theta_c$, $A\sim1$, and $\rho^2+\eta^2<.8$ follow from[4] B lifetime ~1psec, and $\Gamma(b\to u)/\Gamma(b\to c)<.08$, respectively.

2) <u>The Weinberg's Model[5]</u>:

The most novel feature of this model is that the CP symmetry is broken spontaneously. The necessary complex phases come from the vacuum expectation values of the Higg's field. The simplest potential[6] which exhibit this behavior can be written in terms of three $SU(2)_L$ doublets of Higgs fields ϕ_1, ϕ_2, and ϕ_3:

$$V(\phi) = \sum_{i=1}^{3} \left(-\mu^2\phi_i\phi_i^* + \lambda(\phi_i\phi_i^*)^2\right)$$

$$+ C\left(\phi_1^*\phi_2^*\phi_3^2 + \phi_1^*\phi_3^*\phi_2^2 + \phi_2^*\phi_3^*\phi_1^2\right) + c.c.$$

$$\tag{3}$$

and set

$$\phi_1^T = (0, ve^{i\alpha_1}), \quad \phi_2^T = (0, ve^{i\alpha_2}), \quad \phi_3^T = (0, ve^{i\alpha_3})$$

Now write

$$V = -3\,v^2 + 3\lambda v^4 + Cv^4\left[\cos(2\alpha_3-\alpha_2-\alpha_1)\right.$$

$$\left.+ \cos(2\alpha_2-\alpha_1-\alpha_3) + \cos(2\alpha_1-\alpha_2-\alpha_3)\right]. \tag{4}$$

With some algebra, one can show that the minimum of the potential is at

$$\cos(2\alpha_3 - \alpha_2 - \alpha_1) = \cos(2\alpha_2 - \alpha_1 - \alpha_3) = \cos(2\alpha_1 - \alpha_2 - \alpha_3) = -1/2 \tag{5}$$

so that indeed the ground state is given by nonvanishing $\arg\langle\phi_i\rangle$ and there will be some observable phases.

3) <u>Left-Right symmetric models</u>[7]:

One can write a P and CP ivariant hamiltonian based on the $SU(2)_L \times SU(2)_R \times U(1)$ gauge group. Parity is broken spontenaously and the mass of W_R is arranged to be much larger than that of W_L. The Higg's potential can be arranged so that CP is also broken spontaneously. Then the KM matrix for the left handed fermion fields, U_L, and that for the right handed fermion fields, U_R, satisfy

$$U_L = U_R^{+} \tag{6}$$

In simple cases, the KM matrix can be chosen orthogonal and a phase can be introduced in the mass eigenstates of the gauge bosons

$$W_1 = \cos\xi\, W_L + e^{i\eta}\sin\xi\, W_R$$
$$W_2 = -e^{i\eta}\sin\xi\, W_L + \cos\xi\, W_L . \tag{7}$$

From an analysis of $\Delta S=1$ decays and from deep inelastic anti-neutrino scattering one can conclude

$$\tan\xi \lesssim .09, \qquad M_2 \gtrsim 300\ \text{GeV}, \tag{8}$$

where W_1 is identified with the $W^{\pm}$ boson of the standard model.

In all existing models with pontaneous CP violation, there is a neutral flavor changing scalar particle which may contribute to the K_L-K_S mass matrix.

4. <u>Superweak models[9]</u>:

A class of models in which the observable phase occurs only in the $\Delta S = 2$ transition amplitude is defined to be superweak models. One such example is a model with more than one Higgs doublets some of which have flavor changing interactions. In order to be consistent with the strength of $\Delta S = 2$ transition implied by the size of the $K_L - K_S$ mass difference, the masses the of neutral Higg's particles with $\Delta S \neq 0$ interactions have a lower bound of 5~30TeV. A complex phase can be introduced in the flavor changing interactions though, for example, the spontaneous symmetry breaking[10] with a potential similar to Eq. 3. The only effect predicted by the model at low energy is CP violation in mixing; therefore, it predicts that $\varepsilon'/\varepsilon = 0$ and neutron edm=0.

III Models and Experiments

In this section I compare experimental results with theoretical predictions.

(a) $\underline{\varepsilon'/\varepsilon}$:

Recently, two measurements have been reported:

$$\frac{\varepsilon'}{\varepsilon} = \begin{array}{ll} (-4.6 \pm 5.3 \pm 2.4) \times 10^{-3} & \text{Chicago-Saclay[11]} \\ (1.7 \pm 8.4) \times 10^{-3} & \text{BNL-Yale[12]} \end{array} \qquad (9)$$

If $K \to 2\pi$ decays and $K^0 - \bar{K}^0$ was mixing amplitudes are denoted by

$$a_I e^{i\delta_I} = \langle (\pi\pi)_I | H | K^0 \rangle$$
$$M_{12} - \tfrac{i}{2}\Gamma_{12} = \langle K^0 | H | \bar{K}^0 \rangle \qquad (10)$$

where I denotes the isospin quantum number of the 2π state and δ_I denotes the $\pi\pi$ phase shift. Then ε and ε' can be defined as

$$\varepsilon' = i e^{i(\delta_2 - \delta_0)}/20\sqrt{2} \cdot (\arg a_2 - \arg a_0)$$

$$\varepsilon = e^{i\pi/4}/2\sqrt{2} \left(-\arg \langle \bar{K}^0 | H | K^0 \rangle + 2 \arg a_0 \right). \qquad (11)$$

Note that the phase convention has not been specified at this stage.

1) <u>The KM model prediction for ε'/ε</u>[13]:

In this model, a dominant contribution to the argument of a_I comes from the diagram shown in Fig. 1 plus its QCD corrections. As the gluon is flavor SU(3) singlet, an effective interaction generated from this diagram is strictly $\Delta I = 1/2$ and

$$\mathscr{I}m \, a_2 = 0$$

$$\mathscr{I}m \, a_0 = \sqrt{2} \cdot 5.3 \times 10^{-4} \, \eta A^2 G_F \chi_5 \langle \pi\pi \mid O_5 \mid K^0 \rangle \qquad (12)$$

where

$$\chi_5 = (.016 \pm .003) \, \ln \, m_t^2/m_c^2 \, ,$$

$\langle \pi\pi \mid O_5 \mid K^0 \rangle$ is a matrix element of four quark operator and κ_5 contains the QCD corrections.

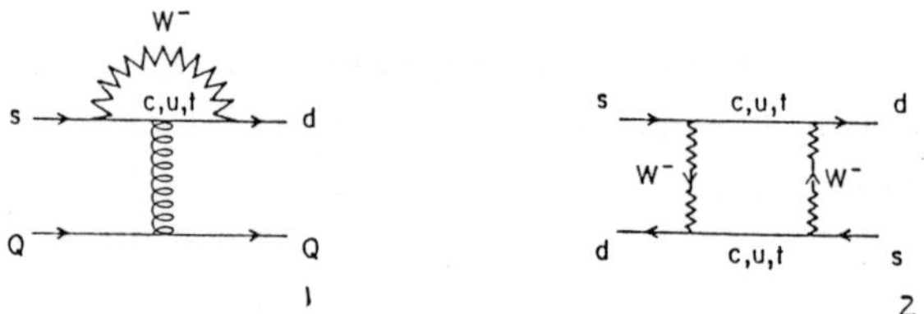

Fig. 1. Examples of diagrams cntributing to ε'/ε in the KM model.

Fig. 2. One of the dagrams which contributes to $\langle K^0 \mid H \mid \overline{K}^0 \rangle$.

The argument of $\langle \overline{K}^0 \mid H \mid K^0 \rangle$ gets its major contribution from the diagram shown in Fig. 2 and the result is

$$\arg \langle \overline{K}^0 \mid H \mid K^0 \rangle = \frac{G_F^2}{8\pi^2} \frac{m_c^2}{m_K \Delta m_K} (1.22 \times 10^{-4}) \, \eta A^2 \chi \langle K^0 \mid O \mid \overline{K}^0 \rangle$$

$$\chi = -\eta_1 + \eta_3 f_1 + 2.46 \times 10^{-3} (1-\rho) A^2 \eta_2 f_2 \qquad (13)$$

where η_1, η_2, and η_3 are the numerical coefficients obtained from QCD corrections and f_1, and f_2 are known functions

of quark masses, and $\langle K^o|0|\bar{K}^o\rangle$ is the matrix element
of the four quark operator.

Finally, we obtain

$$\varepsilon'/\varepsilon = .11(2\sim9)\mathcal{K}_5/(B\mathcal{K}).\qquad(14)$$

The factor (2-9) represents our ignorance of $\langle\pi\pi|0_5|K^o\rangle$
where the value 9 corresponds to the vacuum saturation
approximation and the value 2 corresponds to a "reasonable"
lower limit to the value of the matrix element. B is the
bag constant which represents our ignorance about the matrix
element $\langle K^o|H|\bar{K}^o\rangle$. Fig. 3 shows the lower limit

$$B\varepsilon'/\varepsilon > .11(2)\mathcal{K}_5/\mathcal{K}\qquad(15)$$

as a function of m_t.

On Fig. 3, I have shown the experimental values, Eq.9
as well as the mass range in which experiments at TRISTAN
can produce the top quark. It is clear that if $m_t < 50\,GeV$,
new generation of experiments will measure nonvanishing
ε'/ε.

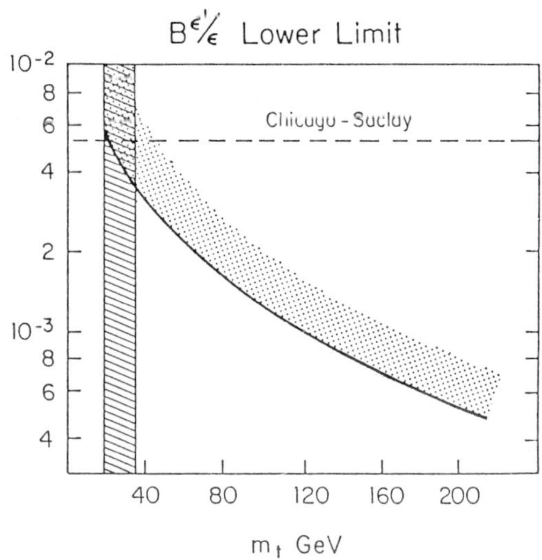

Fig. 3. The KM model lower limit to ϵ'/ϵ as a function of the top quark mass. The value "2" in Eq. 14 is used. The horizonthal line is the upper limit given by Chicago-Saclay group. The vertical shaded area specifies the top quark mass region in which the top quark can be produced at TRISTAN. The experimental limit will soon come down and information in the top quark mass should be available in the near future.

2) <u>The Weinberg model prediction for ϵ'/ϵ:</u>

In this model, the dominant contributions to arg a_0 and $\arg\langle K^0|H|\widetilde{K^0}\rangle$ come from Figs. 4a and 4b, respectively[14]. If the vacuum saturation approximation is used to evaluate the matrix elements, it has been shown that

$$\arg a_0 \gg \arg \langle K^0|H|\bar{K}^0\rangle \tag{16}$$

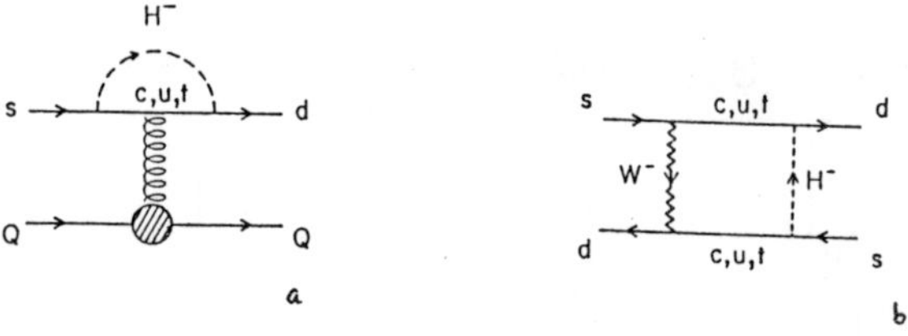

Fig. 4a. A dominant diagram contributing to arg a_0 in the Weinbergs model.

Fig. 4b. A dominant diagram contributing to $\langle K^0|H|K^0\rangle$ in the same model.

In such a case, as it can be easily deduced from Eq. (11),

$$\epsilon'/\epsilon \sim 1/20 \,. \tag{17}$$

which contradicts the experimental upper limit. There is, however, and indication that the vacuum saturation approximation used in evaluating the matrix elements must be reexamined.[15] The operator resulting from the diagram shown in Fig. 4a has the form $\bar{s}\sigma_{\mu\nu}dG^{\mu\nu}$, where $G^{\mu\nu}$ is the gluon field strength, and it transforms like a $(3,\bar{3})$ under chiral SU(3) x SU(3). The desired matrix element, therefore, vanishes in the chiral limit. This is not, however reflected on the vacuum saturation approximation. The chiral symmetry is broken and it might be that the vacuum saturation approximation gives an accurate answer. For now we follow the estimate of reference[15] and obtain[16]

$$\epsilon'/\epsilon \sim .016 \,. \tag{18}$$

The result of Donoghue and Holstein[15] is

$$\epsilon'/\epsilon \sim .006 \,. \tag{19}$$

While this calculation ignores $\eta-\eta'$ mixing the discrepancy between Eqs. (18) and (19) reflects the uncertainty in this type of computation.

It is clear, however, that new set of measurements will have definite verdict on this model.

3) <u>The L-R symmetric model prediction for ϵ'/ϵ</u>[17]:

The $\arg\langle K^0|H|\bar{K}^0\rangle$ gets contributions from diagrams shown in Fig. 5a,b and arg a_0 and arg a_2 get contributions from diagrams shown in Fig. 5c. Note that the $\Delta S=2$ neutral scalar exchange contribution can dominate the arg $\langle K^0|H|K^0\rangle$ and therefore ϵ. In such a case ϵ' vanishes as in the superweak model. In all versions of L-R models in which there is no restriction on the contribution from diagram shown in Fig. 5b, one can only give an upper limit to ϵ'/ϵ which is saturated when the mass of

the flavor changing neutral scalar is taken to be infinity.
The result is

$$(20)$$

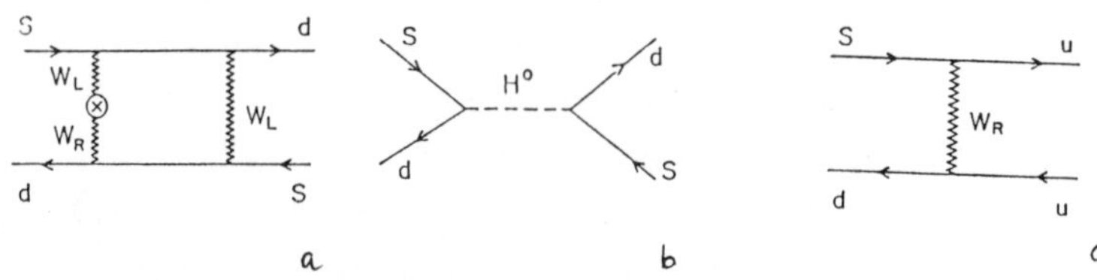

a   b   c

Fig. 5a. A box diagram contributing to $\arg\langle K^0 |H| K^0\rangle$ in the L-R symmetric model.

Fig. 5b. In the class of models in which CP is broken spontaneously, there is necessarily at least one neutral flavor changing sclar particle. This may dominate over other contributionssuch as the one shown in Fig. 5a. In this case the model reduces to a super weak model.

Fig. 5c. An Example of diagrams contributing to $\arg a_0$ and $\arg a_2$.

(b) <u>Neutron electric dipole monent[18]</u>:

Many measurements of the neutron edm have been made and the Particle Data Group gives a combined result:

$$d_N = (2.3 \pm 2.3) \times 10^{-25} \, e\,cm.$$

There are two recent measurements which are not included in the above result

$$d_N = (-2.0 \pm 1.0) \times 10^{-25} \, e\,cm \quad \text{Leningrad}[20]$$

$$(21)$$

$$d_N = (-1.8 \pm 2.9) \times 10^{-25} \, e\,cm \quad \text{Inst. Laue Langevin}[21]$$

It is likely that the latest measurements, using ultra cold neurons, will improve by at least an order of magnitude.

1) <u>The KM Model prediction for d_N:</u>

Examples of diagrams which contribute to the neutron edm in the KM model are shown in Fig. 6. Note that at least two W boson propagator is present and their contribution is very small compared to the expeimental upper limit.[22]

$$d_N < 10^{-32} - 10^{-30} \; ecm.$$ (22)

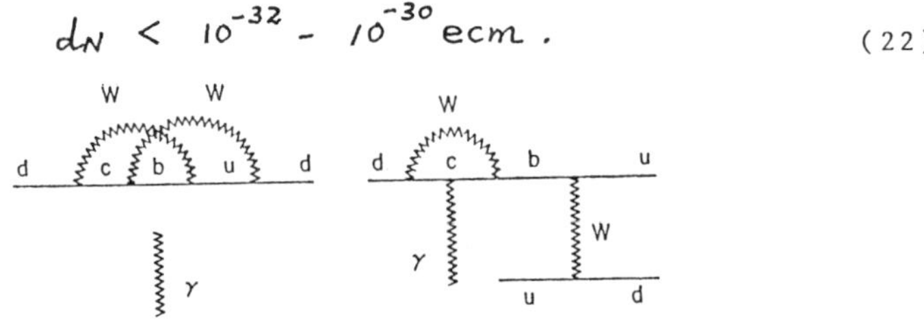

Fig. 6. An example of diagrams which contribute to the neutron edm. Note that multibody nature of the neutron influences the prediction.

2) <u>The Weinberg Model prediction for d_N:</u>

Fig. 7 shows the diagram which contributes to the d quark edm. Note that there are two independent couplings for H^-

$$\alpha \; \bar{J}_L u_R H^-, \qquad \beta \; \bar{d}_R u_L H^-.$$ (23)

Fig. 7. The diagram which dominates the neutron edm in the Weinberg model.

For this reason, diagram with a single H^- propagator can contribute and it gives much larger prediction compared to the KM model. Arg $\alpha\beta$ is fixed by computing the prediction for ε. The result is[23,24]

$$d_N \sim 10^{-25} \, ecm. \tag{24}$$

The actual value depends on the top quark mass and QCD corrections, for details see Ref. 24.

3) <u>The L–R symmetric model prediction for d_N</u>:
The diagram contributing to the d quark edm moment is shown in Fig. 8. Its contribution is[25]

$$d_N < 2 \times 10^{-26} \, ecm, \tag{25}$$

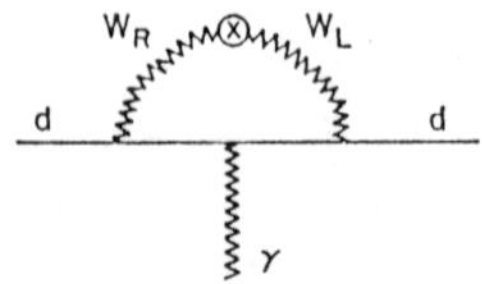

Fig. 8. The diagram which contributes to the neutron edm in the L–R symmetric model.

As its prediction on ε'/ε, the upper limit is saturated only if the mass of scalar flavor changing neutral particle is infinite.

(c) <u>The μ polarization in $k \to \mu\nu\pi$ decay.</u>

A correlation which is odd under T can be measured in $K \to \mu\nu\pi$ decays, namely the muon polarization normal to the muon and pion plane: $P = \langle \sigma_\mu \cdot (P_\pi \times P_\mu) \rangle$. Experimentally one finds[19]

$$K_L \to \pi^- \mu^+ \nu_\mu \; : \; P = (1.7 \pm 5.6) \times 10^{-3},$$
$$K^+ \to \pi^0 \mu^+ \nu_\mu \; : \; P = (-3.7 \pm 4.7) \times 10^{-3}, \tag{26}$$

i.e., numbers consistent with zero.

An important sublety has to be kept in mind: although the correlation P is odd under T a nonvanishing value for it does not automatically establish T violation. Final state interactions can also generate such a correlation even if T is conserved. In the case of $K_L \to \mu^+ \nu_\mu \pi^-$ electromagnetic final state interactions can generate a nonzero P. It can be calculated as a function of the kinematic variables and is found to range between $(2-3) \times 10^{-3}$. A nonvanishing P in $K^+ \to \mu^+ \nu_\mu \pi^0$ on the other hand immediately establishes T violation since no final state interaction can occur. Theoretical estimates of P in the KM and LR symmetric models are very small compared to the experimental values given in Eq. (26). In the Weinberg model, however,

$$P \sim 10^{-4}$$

is obtained.[26]

IV Meson-Antimeson Mixing

There have been some observations which are consistent with existence of substantial $D^0 - \bar{D}^0$ mixing. Since this is unexpected in the standard model, I would like to discuss the background of this subject and its implications in the remainder of my talk. Mixing can be searched for by two different methods: (i) like sign dipletons from semi-leptonic $D\bar{D}$ decays; (ii) search for decays with wrong sign K mesons.

Crucial quantities which govern mixing effects can be seen from[27]

$$\text{Prob}(P \to \bar{P}) = \frac{BR(P \to \bar{\ell} + X)}{BR(P \to \ell + X)} = \frac{1}{2}(x^2 + y^2), \qquad (27)$$

where ℓ is a lepton, $x = \Delta m / \Gamma$, $\Delta M = |m_2 - m_1|$, $y = \Delta\Gamma / 2\Gamma$, $\Delta\Gamma = \Gamma_2 - \Gamma_1$. The decay of P or $\bar{P}$ is probed by observing the leptonic decay. For K mesons $\Delta\Gamma \sim \Delta M$; however; this is due to a peculiar accident of nature that $m_K \sim 3 m_\pi$ and the width of K_L is highly suppressed due to the phase space. For heavy mesons such as D and B, y is expected to be small. The mixing thus depends on x which is interpreted as

$$x = \Delta m / \Gamma = \text{life time} / \text{mixing time}.$$

(28)

It is clear that this ratio must be large in order for the mixing effects to be observable.

In the standard model, a major contribution to meson (P) and antimeson ($\bar{P}$) mixing (ΔM_P) is given by a box diagram shown in Fig. 9. The box graph contribution to ΔM is roughly proportional to the mass square of the exchanged quark. Using a generic language of the weak SU(2) doublets

$$\begin{pmatrix} U \\ D \end{pmatrix} = \begin{pmatrix} c \\ s \end{pmatrix}, \begin{pmatrix} t \\ b \end{pmatrix}, \cdots$$

(29)

one finds q=(d,s)

$$\left(\Delta M\right)_{D\bar{q}} \sim m_U^2 \quad ; \quad \left(\Delta M\right)_{U\bar{u}} \sim m_D^2.$$

(30)

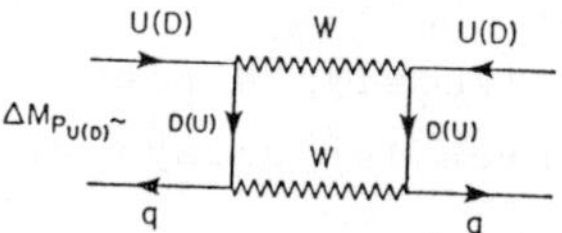

Fig. 9. A box diagram which contributes to ΔM_P Note that due to the GIM cancellation, the contribution is proportional to the mass^2 of the internal quark.

The only assumption made here is that the quark mixing angles are small compared to the ratio of quark masses in succeeding doublets. Thus

$$(\Delta M)_{D\bar{q}} \gg (\Delta M)_{U\bar{u}}$$

Second $\Gamma_{U\bar{u}} \gg \Gamma_{D\bar{q}}$ since $(U\bar{u})$ states being much heavier have much more phase space to decay than $(D\bar{q})$ states. The latter are even more suppressed by small KM angles. Thus

$$\left(\Delta M / \Gamma\right)_{D\bar{q}} \gg \left(\Delta M / \Gamma\right)_{U\bar{u}} \tag{31}$$

by orders of magnitude.

(a) $\underline{D^0 - \bar{D}^0 \text{ mixing}}$:

The MARK III collaboration[28] has recently presented data which can be interpreted as suggesting $D^0 - \bar{D}^0$ mixing at the 1% level. Events of the type

$$e^+ e^- \rightarrow \psi(3770) \rightarrow X_1 + X_2$$

where

$$X_1, X_2 = K^{\pm} \pi^{\mp}, \ K^{\pm} \pi^{\mp} \pi^0, \ K^{\pm} \pi^{\mp} \pi^{\pm} \pi^{\mp}$$

$$\tag{32}$$

with $m_{X_1} \approx m_{X_2} \approx m_D$ have been studied. In particular there were

$$162 \ \text{events with} \ K^+ K^- + \pi\text{'s} \ (\Delta S = 0)$$

and

$$3 \quad " \quad K^{\pm} K^{\pm} + \pi\text{'s} \ (\Delta S = 2).$$

As only $.4 \pm .2$ events out of 3 can be attributed to mis-identification of particles, the rest must come from either (a) $D^0 - \bar{D}^0$ mixing or from (b) doubly Cabibbo suppressed decays. If doubly Cabibbo suppressed decays are down by $\sim \tan^4 \theta_c = 3 \times 10^{-3}$, one such event out of 3 can be attributed to this mechanism. It has been pointed out[29] that the quantum statistics effect which tend to suppress the background due to doubly suppressed Cabibbo decays should be present. Thus above estimate of the background is likely to be an over estimate.

With the general argument presented at the beginning of this section, D^o-$\bar{D}^o$ mixing is not expected to be observable within the context of the standard model. It is unlikely that D^o-$\bar{D}^o$ mixing would be significantly enhanced by the presence of the forth family, the right-handed currents or supersymmetry states. A best candidate model to explain mixing at 1% level could be based on a nonminimal Higgs sector. Such a model could give

$$x_D \sim .1 .$$

The CDHS collaboration[30] has recently published data on prompt like-sign dimuons in neutrino scattering:

$$N(\mu^{\pm}\mu^{\pm}) / N(\mu^+\mu^-) \sim .02 \sim .04 .$$

$$(33)$$

It should be kept in mind that the signal barely reaches the 3σ level; nevertheless it is worthwhile to consider the possibility that the signal is real. The "wrong-sign" muons are said to be quite consistent with coming from D decays as far as their kinematical distributions are concerned. Invoking $c\bar{c}$ production from gluons would reproduce this pattern; however perturbative calculations yield a rate roughly 30 times smaller than the observed signal.

It should also be mentioned that Chicago, Princeton, Iowa group[31] has obtained a 90%CL limit

$$\frac{\Gamma(\bar{D}^0 \to \mu^+ + X)}{\Gamma(\bar{D}^0 \to \mu^- + X)} < 5.6 \times 10^{-3} \tag{34}$$

which implies

$$x_D < .1 .$$

Clearly much experimental work is needed in order to establish D^0-$\bar{D}^0$ mixing. Reward however is worth the hard work needed for the pursuit.

    b. $\underline{B^0\text{-}\bar{B}^0 \text{ mixing}}$:

According to the general argument, B_d and B_s, the bound state of $b\bar{d}$ and $b\bar{s}$, respectively have much better chance of being observed. A precise predictions for x_{Bd} and x_{Bs} cannot be given at this time since the KM angles are not fixed. An intelligent guess is that[32]

$$\tag{35}$$

$$x_{Bd} \sim .1 \quad , \quad x_{Bs} \sim .5 .$$

We expect B meson mixing effects to be observed in the future.

For now the limit for CLEO on B_d-$\bar{B}_d$ mixing measured at $T(4S)$ is[33]

$$N(\ell^\pm \ell^\pm)/N(\ell^+\ell^-) \lesssim .2 \quad \text{or} \quad x_{Bd} < .6 . \tag{36}$$

For observing B_s-$\bar{B}_s$ mixing at CESR, it is necessary to know the position of resonances which decay to B_s-$\bar{B}_s$. The ratio of B_s to B_s^* also influences the production rate of neutral B_s meson. These were computed in the potential model[34]. In Fig. 10, the potential model prediction which is expressed as the ratio

$$\frac{\sigma^{++} + \sigma^{--}}{\sigma^{+-} \, BR(B_s \to \ell + X)^2} \tag{37}$$

is shown. Here $\sigma^{\pm\pm}(\sigma^{\pm})$ is the production cross section for the same sign (opposite sign) lepton events, and $BR(B_s \to \ell + X)$ is the leptonic branching ratio.

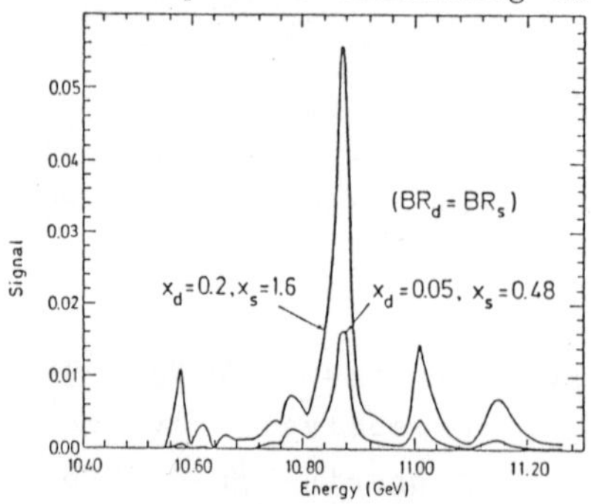

Fig. 10. A Prediction of the potential model for the ratio shown in Eq. (37) as a function of the e^+e^- beam energy.

The absence of $B_s - \bar{B}_s$ mixing at the level significantly below $X_{B_s} \sim .5$ will be a blow to the standard model.

V Future Prospects

Propsects for future experimental progress is bright. The measurement of ϵ'/ϵ by Chicago-Saclay group is expected to reach a sensitivity of 10^{-3}. Experiments along the similar direction is in progress also at CERN. The Rutgers group has taken 3.2 million events in search for non-vanishing η_{+-0} of $K_s \to 3\pi$ decay. At present, their result based on 10^5 events is [35]

$$|\eta_{+-0}| = .023 \pm .029. \tag{38}$$

Measurements of neutron electric dipole moment using ultra-cold neutrons will soon reach the sensitivity of $d_N \sim 10^{-26}$ ecm level and it may eventually lead to 10^{-27} ecm sensitivity.

Another direction in which future progress is expected is the measurement of the electric dipole moment of an atom. Recent measurement gives[36]

$$d(^{129}Xe) = -(.3 \pm 1.1) \times 10^{-26} \; ecm. \tag{39}$$

There is a suppression factor of 10^{-3} in relating an atomic and neutron electric dipole moments[37]

$$d(^{129}Xe) \sim 10^{-3} \, d_N \tag{40}$$

and the above measurement is not sensitive to the neutron electric dipole moment. It is however sensitive to CP violating interaction of the type

$$i \frac{G_F}{\sqrt{2}} \eta_0 \, \bar{N} \gamma_5 N \, \bar{N} N. \tag{41}$$

The standard model predicts[38]

$$\eta_0 = .6 \times 10^{-8}, \tag{42}$$

while above measurements implies

$$\eta_0)_{exp} = .07 \pm .24. \tag{43}$$

Fortson[37] points out that the upper limits on the ratios

$$\frac{\eta_0)_{exp}}{\eta_0)_{KM}} \quad \& \quad \frac{d_N)_{exp}}{d_N)_{KM}} \tag{44}$$

are both around 10^7. It should be stressted that this type of measurements have possibility of improving the sensitivity by three orders of magnitude.[37] It is important, therefore, to examine theoretical predictions for η_0. Also the experiment is sensitive to the electron edm. This type of experiment has a potential of attacking the problem from a different direction.

VI Summary

Much progress has been made toward search for CP violating effects outside of $K_L \to 2\pi$ decay – yet none is found. Much progress in theoretical descriptions of CP violating effects has been made – yet there is no understanding of the effect.

It is clear that CP violating phenomena offer a major clue to go beyond the standard model. Theoretical understanding of the effect should lead to a deeper understanding of the universe.

Stakes are getting high enough to warrant a major push in this direction both experimentally and theoretically.

Acknowledgements

I acknowledge enjoyable and fruitful collaboration with I.I. Bigi over the years. Much of the material covered here comes from our collaboration.

*

Work supported in part by the Department of Energy under Contract Grant Number DE-AC02-81ER40033B.000

References

1. See, for example, J.J. Sakurai, "Invariance Principles and Elementary Particles", Princeton Univ. Press (1964).

2. M. Kobayashi and K. Maskawa, Prog. Theor. Phys. $\underline{49}$, 652 (1973).

3. L. Wolfenstein, Phys. Rev. Lett. 51, 1945 (1983).

4. E.H. Thorndike, Proceedings of the 1985 Int. Symp. on Lepton and Photon Interactions at High Energy", Kyoto, Edited by M. Konuma and K. Takahashi.

5. S. Weinberg, Phys. Rev. Lett. $\underline{31}$, 657 (1976).

6. G.C. Branco, Phys. Rev. Lett. $\underline{44}$, 504 (1980).

7. For a long list of references see M.A.B. Beg and S. Sirlin, Phys. Rep. $\underline{88}$, 1 (1982).

8. I.I. Bigi and J.M. Frere, Phys. Lett. B110, 255 (1982); J.F. Donoghue and B.R. Holstein, Phys. Lett. B113, 382 (1982); H. Abramowicz et al., Z. Phys. C12, 225.

9. L. Wolfenstein, Phys. Rev. Lett. $\underline{13}$, 180 (1964).

10. C. Branco and A.I. Sanda, Phys. Rev. $\underline{D26}$, 3176 (1982).

11. R. Bernstein et al., Phys. Rev. Lett. $\underline{54}$, 1631 (1985).

12. R.C. Larsen et al., Phys. Rev. Lett. $\underline{54}$, 1628 (1985).

13. F.J. Gilman and M.B. Wise, Phys. Lett. $\underline{838}$ (1979) 83; Phys. Rev. $\underline{D20}$ (1979) 2392; B. Guberina and R. Reccei Nucl. Phys. $\underline{B163}$ (1980) 289; F.J. Gilman and J.S. Hagelin, Phys. Lett. $\underline{126B}$ (1983) 111. We shall use numerical values obtained by A.J. Buas, W. Slomminski and H. Steger, Nucl. Phys. $\underline{B238}$ (1984) 529.

14. A.I. Sanda, Phys. Rev. $\underline{D23}$, 2647 (1981); N. Despande, Phys. Rev. $\underline{D23}$, 2653 (1981).

15. Y. Dupont and T.N. Pham, Phys. Rev. $\underline{D28}$, 2169 (1983).

16. See lectures by A.I. Sanda and also by T.N. Pham, Proceedings of the Fifth Moriond Workshop, La Playue-Savoie, France, 1985. ed. J. Tran Thanh Van.

17. C. Cheng, Nucl. Phys. B214, 435 (1983); G. Ecker, W. Grimus and H. Neufeld, Nucl Phys. B239, 421 (1983); G.C. Branco, J.-M. Frere, and J.-M. Gerad, Nucl. Phys. B221, 317 (1983); J. Basecq et al., Carnegie Mellon preprint CMU-HEP85-14; G. Ecker and W. Grimus Wein preprint UWTHPH-1985-7.

18. For the definition of the edm and experimental background, see N. Ramsey, Rep. Prog. Phys. 45, 95 (1982).

19. Review of Particle Properties, Rev. of Mod. Phys. 56, S1 (1984).

20. V.M. Lobashev adn A.P. Serebrov, J. de Physique, C3, suppl. 45, 11 (1984).

21. J.M. Pendlebury et al., Phys. Lett. 136B, 327 (1984).

22. J.O. Eeg, Proceedings of the Fifth Moriond Workshop, La Plague-Savoie, France, (1985), ed. J. Tran Thanh Van.

23. H.-Y. Cheng, Brandeis Preprint BRX-TH-180 (1986).

24. I.I. Bigi and A.I. sanda, in preparation.

25. J. Liu, Carnegie Mellon preprint CMU-HEP85-15.

26. L. Wolfenstein, private communication.

27. A. Pais and S.B. Treiman, Phys. Rev. D12, 2744 (1975); L.B. Okun, V.I. Zakharov, and B.M. Pontecorvo, Lett. Nuovo Cimento 13, 218 (1975).

28. R.H. Schindler, SLAC Summer School on Physics of the High Energy Accelerators, Stanford, Calif. (1985).

29. I.I. Bigi and A.I. Sanda, Phys. Lett. B171, 320 (1986).

30. H. Burkhardt et al., CERN preprint CERN-EP/85-191.

31 W.C. Louis et al., Phys. Rev. Lett. 56, 1027 (1986).

32. I.i. Bigi and A.I. Sanda, Phys. Rev. D29, 1393 (1984).

33. A. Bean, et al., CLNS 86/741, (1986).

34. S. Ono, A.I. Sanda, N.A. Tornqvist and J. Lee Franzini, Phys. Rev. Lett. 55, 2938 (1985); S. Ono, A.I. Sanda, N.A. Tornqvist, Phys. Rev. D34, 186 (1986).

35. R. Thompson, private communication.

36. T.G. Vold et al., Phys. Rev. Lett. 52, 2229 (1984).

37. E.N. Fortson private communication.

38. I.B. Khriplovich, Zh. Eksp. Teor. Fiz. 71, 51 (1976) [Sov. Phys. JETP 44, 25 (1976); I.S. Alterev et al., Phys. Lett. 102B, 13 (1981).

OBSERVATION OF CORRELATED NARROW PEAK STRUCTURES IN POSITRON AND ELECTRON SPECTRA FROM SUPERHEAVY COLLISION SYSTEMS

John, Schweppe*

Yale University
A.W. Wright Nuclear Structure Laboratory
P.O. Box 6666
New Haven, CT 06511, USA

We have observed that the positrons associated with a narrow peak in the positron spectrum from U+Th collisions are correlated with the simultaneous emission of electrons whose energy spectrum also contains a narrow peak. The mean energies and widths of the two peaks are equal within measurement errors. Neither the coincidence-peak intensity nor the energy distributions of the positrons and electrons can be accounted for by known nuclear internal conversion processes. Similar observations have also been made in the Th+Th and Th+Cm collision systems. New measurements confirm the emission of correlated positrons and electrons and establish the existence of more than one set of correlated energies.

1   Introduction

This report describes the latest developments in a series of experiments performed at GSI Darmstadt by the EPOS collaboration[1-3] to study the production of positrons in collisions of very heavy ions and atoms. The original motivation for this effort was an experimental search for spontaneous positron production[4-6], a new QED process predicted to occur in the large electromagnetic fields produced transiently in these collsions. It is not necessary to describe spontaneous positron production in any greater detail here because that was apparently not what was found.

---

*The experiments described here were carried out by the EPOS collaboration at the Gesellschaft für Schwerionenforschung, Darmstadt, and reflect the combined efforts of its members: Hartmut Backe, Klaus Bethge, Helmut Bokemeyer, Tom Cowan, Helmut Folger, Jack S. Greenberg, Kiyo Sakaguchi, Dirk Schwalm, Kurt E. Stiebing, and Paul Vincent.

What was found was the following. Shown in Figure 1 is the energy distribution of positrons emitted in six very-heavy-ion collision systems[2]. The six systems cover a range of combined nuclear charge, $Z_u \equiv Z_{proj} + Z_{targ}$, the important parameter for spontaneous positron production, from $Z_u = 188$ down to $Z_u = 163$. For the purposes of spontaneous positron production, the heavier systems in parts (a) through (e) are supercritical, while the Th+Ta collision system shown in part (f) is well below the critical point. The bombarding energies of the collisions are indicated and can be summarized as being collisions near the Coulomb barrier, corresponding in each case to a partial overlap of the nuclear density distributions. In addition to the smooth background of dynamically produced positrons, a narrow peak is clearly visible in each spectra at a energy between 300 and 400 keV (see also References 7 and 8). The spectra correspond to kinematic conditions chosen empirically to enhance the peak with respect to the background by taking advantage of an apparent difference in the heavy-ion scattering-angle correlations for the events associated with the positron peak with respect to the dominant Rutherford-scattering events[1,2]. The width of the peak in each case is about 70 keV. This is consistent with the Doppler broadening expected for monoenergetic emission of positrons from a system moving with the velocity of the center of mass of the heavy-ion collision system. It also corresponds by the Uncertainty Relationship to emission from a system surviving more than 10^{-20} seconds, or at least an order of magnitude longer than the Rutherford scattering time.

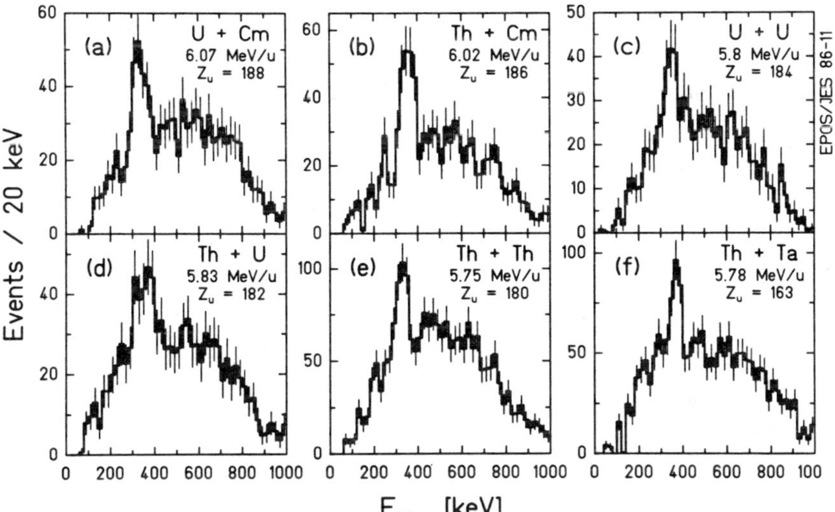

FIGURE 1. Positron energy distributions for the six collision systems and projectile energies indicated. Kinematic constraints have been chosen as discussed in the text.

Perhaps the most striking feature of the data is the near constancy of the energy of the peak, despite the range of collision systems studied. This is summarized in Figure 2, where the peak position is plotted as a function of the combined nuclear charge, Z_u. This constancy is in direct contradiction to the predictions[9] of the simplest model of spontaneous positron production in which it is assumed that the nuclear charge configuration of the two colliding nuclei does not change appreciably during the collision. This prediction[9] is shown by the dashed line (a) in Figure 2. The curves (b) and (c) show the effect of assuming that the two nuclei coalesce into a spherical form and that additionally all electrons are removed during the collision, respectively. It is obviously difficult to explain the data within a model based on spontaneous positron production unless one assumes that the nuclear charge configuration and perhaps also the electronic charge state track with the combined charge of the collision system in just such a way as to maintain a nearly constant binding energy for the $1s\sigma$ electrons of the quasimolecule. Furthermore, the coalesced system must somehow reform a binary system with kinematics resembling that of Rutherford scattering in order to match the experimental detection trigger.

The difficulties with an explanation based on spontaneous positron production led us to consider other possible sources for the similar, approximately monoenergetic, positron emission found in all these collsion systems. An earlier analysis of the positron line shape had already shown us the difficulty of explaining the narrow positron structures on the basis of the relatively broad energy distribution of positrons from the internal pair conversion of excited nuclear states populated during the collsion. Indeed, three-body decay processes in general, in which the positron shares kinetic energy statstically with other particles, produce an energy distributions of positrons which is too wide to match the observed peaks. As

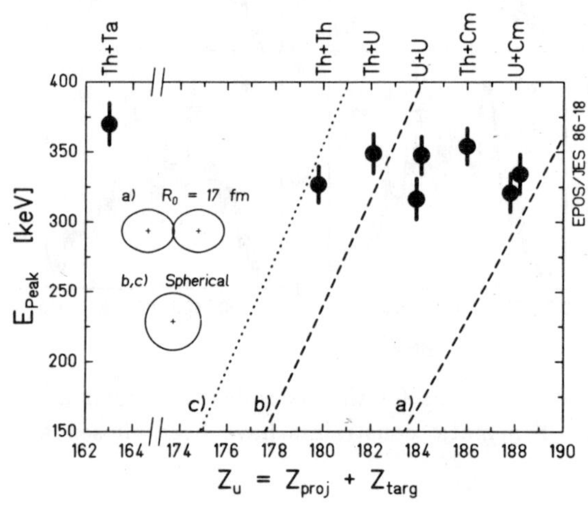

FIGURE 2. The mean energies of the positron peaks in Figure 1 are plotted as a function of Z_u. The calculated[9] lines are described in the text.

a result, we began to consider the possibility that the experimentally-observed positron peaks are narrow because the positron is the product of the two-body decay of a neutral state into a positron-electron pair[2]. In this case , of course, the symmetric positron and electron would be emitted back to back in the rest frame of the decaying state and share equally the available energy. Obviously the signature of such a decay process would be the emission of monoenergetic electrons of approximately the same energy as the observed positron peaks in coincidence with these events.

2   Apparatus

In order to look for coincident positron-electron emission in heavy-ion collisions, we modified our positron spectrometer, EPOS[10], to include the detection of electrons. The spectrometer is shown in Figure 3 below in three views: from the side in part (a), in three cross-sectional views through the new electron counter, through the target chamber, and through the positron detector in part (b), and as a perspective view of the main elements in part (c). The original positron spectrometer, built for the detection of positrons emitted in heavy-ion collisions, is based on a solenoidal-magnetic-field transport system which combines a large positron detection efficiency with good separation of the positrons from the overwhelming backgrounds of gamma rays and delta electrons copiously produced in heavy-ion collisions. The positrons are transported at right angles to the beam direction by the $\sim$ 2 kilogauss magnetic field to an axially positioned, 10 mm $\times$ 100 mm, cylindrical Si(Li) detector some 83 cm away from the target. The intrinsic resolution of the liquid-nitrogen-cooled Si(Li) counter is $\sim$ 12 keV. The axial-focussing property of the solenoidal field for charged particles produces the separation of positrons from gamma rays, and a baffle system suppresses electrons. Backgrounds are further suppressed by the detection of the characteristic positron annihilation radiation in a cylindrical arrangement of NaI(Tl) crystals positioned around the positron detector.

The kinematic parameters for each positron-producing collision are determined by measuring the scattering angles of both heavy nuclei in a pair of position-sensitive, parallel-plate, avalanche counters located in the target chamber. They cover the laboratory polar-angular region from 25° to 65° with respect to the beam axis and a total azimuthal-angular region of 120°.

A major background of positrons, that due to the internal pair conversion of excited nuclear states, was determined by folding the gamma-ray yield monitored in two 3"×3" NaI(Tl) crystals, positioned just outside the target chamber, with theoretical internal-pair-conversion coefficients according to an empirical method based on measurements of low-Z collision systems where this source of positrons dominates.

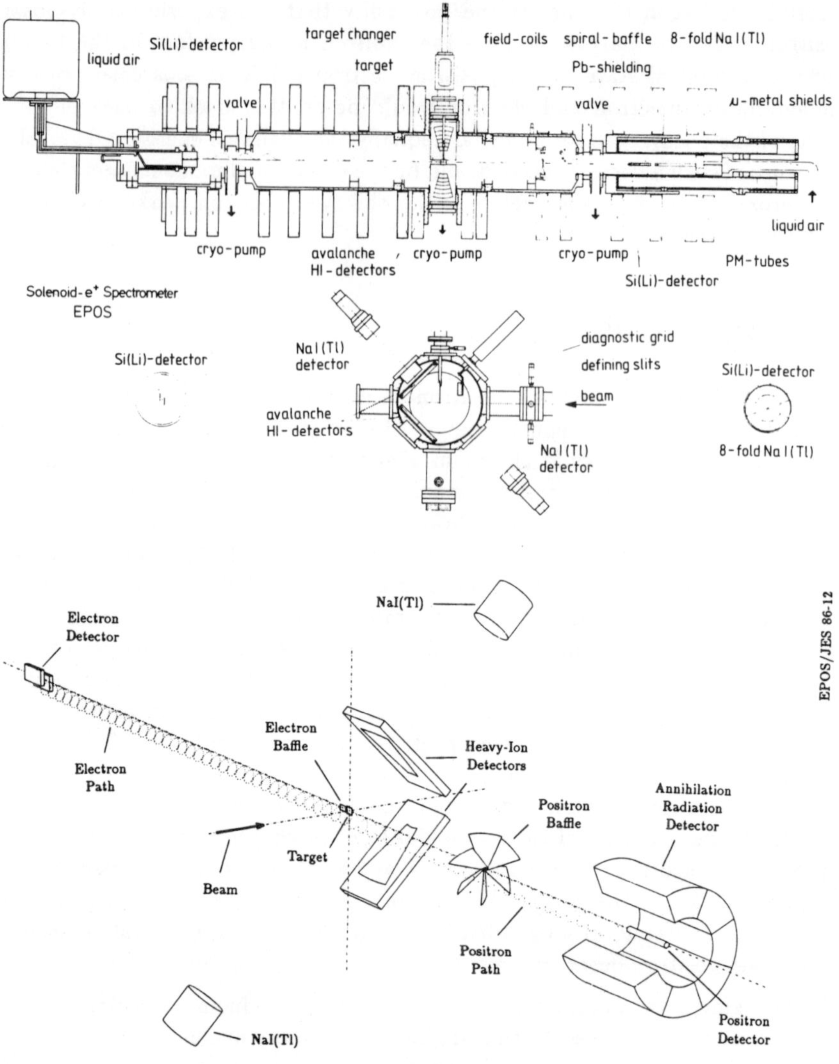

FIGURE 3. Schematic view of the EPOS spectrometer (upper and middle panels) and a perspective drawing of the main components (lower panel). (Two of the eight segments of the annihilation-radiation detector are not drawn in the lower panel in order to allow a better view of the positron detector.)

.The new electron detector is installed in the opposite arm of the solenoid from the positron detector, about 112 cm from the target. The electron detector consists of two $65 \times 32 \times 3$ mm^3 planar Si(Li) crystals. Their arrangement[11] is shown in Figure 3 (b). As opposed to the radially symmetric and axially positioned positron-detection system, the electron detectors are positioned symmetrically off the solenoid axis to allow the large flux of low-energy delta electrons to spiral past between the two detectors. They are oriented with their sensitive faces parallel to the solenoid axis and facing inward to suppress the detection of oppositely spiralling positrons. The intrinsic resolution of the alcohol-cooled Si(Li) counters was ~ 35 keV.

3 Monte Carlo Simulations

In order to gain an understanding of the response of our apparatus to different electron-positron production processes, Monte Carlo simulations were performed[12]. Of these, two important processes will be discussed here. The simplest process in many ways is the previously mentioned, two-body decay of a neutral state into an electron-positron pair. The results of a simulation[12] of this process are shown in Figure 4 for the case that a neutral state existing at rest in the center of mass of the colliding heavy nuclei decays into an electron and a positron, which are emitted back to back in this reference frame, each claiming half of the available total energy of 1.8 MeV (chosen to reproduce the energy of the peaks in the positron spectra of Figure 1). The input into the Monte Carlo calculation is not only the calculated kinematics of such a decay from the center of mass of a heavy-ion collision as observed in the laboratory reference frame, but also the measured transport efficiencies and detector response matrices for the detection of the positron and the electron. Figure 4 (a) shows the calculated distribution of positron energies measured in singles. Part (b) shows the two-dimensional distribution of coincidence events as a function of the kinetic energy of the emitted positron (horizontal scale) and of the electron (vertical scale). As expected, in the laboratory reference frame both the positron and electron show the full Doppler broadening of ~ 70 keV. This is seen even more clearly in Figure 4 (c) and (d) for positrons and electrons, respectively. In anticipation of the comparison with the measured data below, the peak has been isolated in these projections of Figure 4 (b) by gating in each case on the peak in the other lepton channel in the energy interval $340 < E < 420$ keV. These are the two gates marked A and B, respectively, in Figure 4 (b).

As can be seen in Figure 4 (b), because of the particularly simple form of back-to-back emission in the center-of-mass rest frame, the Doppler broadening has a simple, diagonal form in the laboratory rest frame. As a result, a more

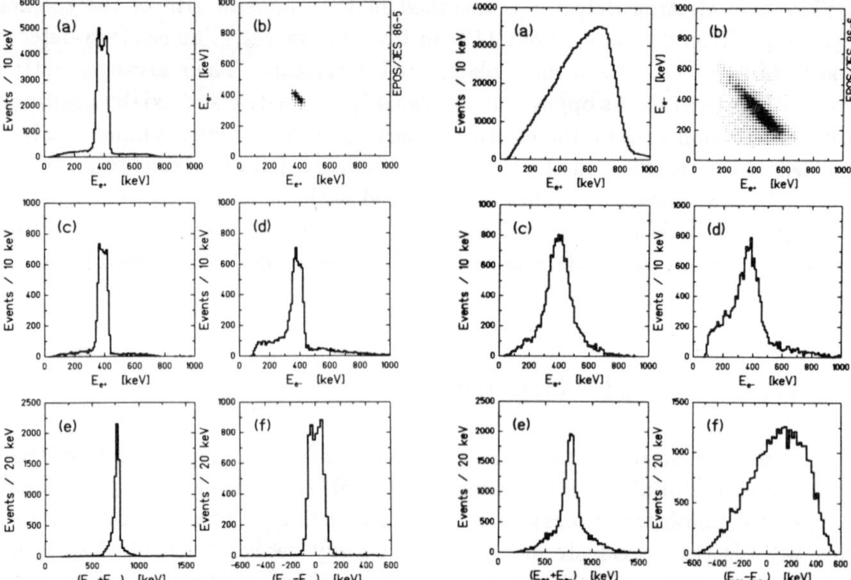

FIGURE 4. Monte Carlo simulation of the two-body decay of a neutral state of mass 1.8 MeV into an electron-positron pair in the EPOS spectrometer.

FIGURE 5. Monte Carlo simulation of the internal pair conversion of an E1 nuclear transition for $Z = 92$ in the EPOS spectrometer.

natural coordinate system is the sum and the difference of the positron and electron energies. This gives new coordinate axes at 45° to the E_{e^+} and E_{e^-} axes. The projections onto these variables are shown in parts (e) and (f), respectively. As above, the peak area has been picked out by the gates marked C and D in Figure 4 (b). Gate C is wedge shaped to compensate approximately for the increase in Doppler broadening for larger positron and electron energies. The naturalness of these projections can be seen most clearly in Figure 4 (c) where, due to the event-by-event cancellation of the Doppler shifts of the positron and the electron, the width of the peak is determined essentially by the resolution of the detection systems and is significantly less than the Doppler-broadened widths measured in the individual positron and electron spectra of parts (c) and (d).

For comparison, Figure 5 shows the same projections of a Monte Carlo simulation[12] of the most obvious source of electron-positron pairs in heavy-ion collisions, the internal pair conversion of excited nuclear states. An E1 transition for $Z = 92$ is assumed and the energy of the transition is again 1.8 MeV. This three-body decay process is representative in general of normal positron-producing processes in heavy-ion collisions. The differences between it and the two-body decay of Figure 4 are clear. The singles positron spectrum in part (a) is much broader and the distribution of coincidence events in the plane of positron and electron energies of part (b) is also much larger. This last point is important not only for the form of the various projections, but also for the relative detection efficiency of such events within the narrow gates A through D. Although one can recover relatively narrow peaks by placing narrow gates on the electron or positron energies, as shown in parts (c) and (d), respectively, the projections onto the energy sum and difference axes, shown in parts (e) and (f), respectively, once again bring out the large difference to the two-body decay considered above. The energy-sum spectrum of Figure 5 (e) is about three times as wide as the corresponding projection for two-body decay, shown in Figure 4 (e) above and the difference distribution in part (f) is about four times as wide as that in Figure 4 (f). Clearly these two types of production processes can be distinguished.

Of course, the peak events are only a small fraction of the electrons and positrons produced in heavy-ion collsions, so these events have to be added to the smooth backgrounds of electrons and positrons from dynamic quasimolecular processes and from nuclear processes. This sum is shown below in Figure 7, parts (e) through (l), for the two processes and the four projections discussed above in Figures 4 and 5. The ratio of peak events to dynamic and nuclear background is chosen for comparison to the measured data.

4   Data

These two Monte Carlo simulations form a basis for examining the measured data. The first measurements of coincident electron-positron emission in heavy-ion collisions with the new apparatus were performed in July, 1985[3]. The U+Th collision system was measured at a bombarding energy of 5.83 MeV/u. The $\sim$ 1830 detected coincident events are shown in Figure 6 (a) as a function of the measured kinetic energy of the positron and the electron. The total projections onto the positron and electron energy axes are shown in parts (b) and (c), respectively. Also indicated on the two-dimensional plot of part (a) are the same four gates A through D discussed above in Figure 4. The projections of the measured coincident events onto the positron-energy, electron-energy, energy-sum, and energy-difference axes using gates A through D are shown in Figure 7, parts (a) through (d), respectively. In each measured spectrum, a peak is clearly visable above the smooth background. The peak in the energy-sum spectrum of part (c) contains 35.3 ± 9.4 counts above the continuous background. A statistical fluctuation of this size is excluded at the $\sim 6\sigma$ level.

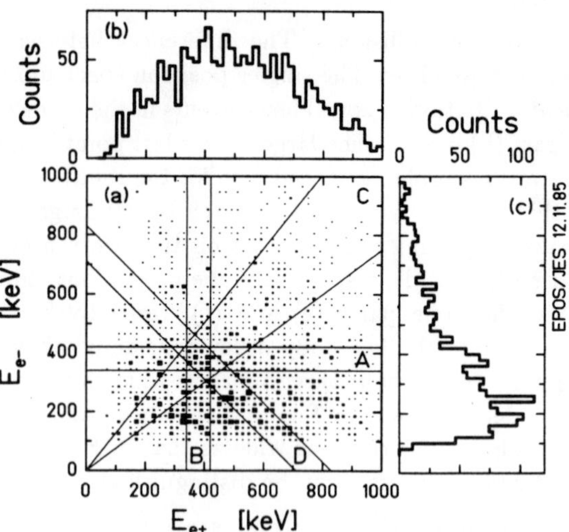

FIGURE 6. Measured coincident electron positron data for the U+Th collision system at a bombarding energy of 5.83 MeV/u. Part (a) is the distribution of coincident events as a function of the positron and the electron kinetic energies. Parts (b) and (c) are the total projections of this data onto the positron-energy and the electron-energy axes, respectively.

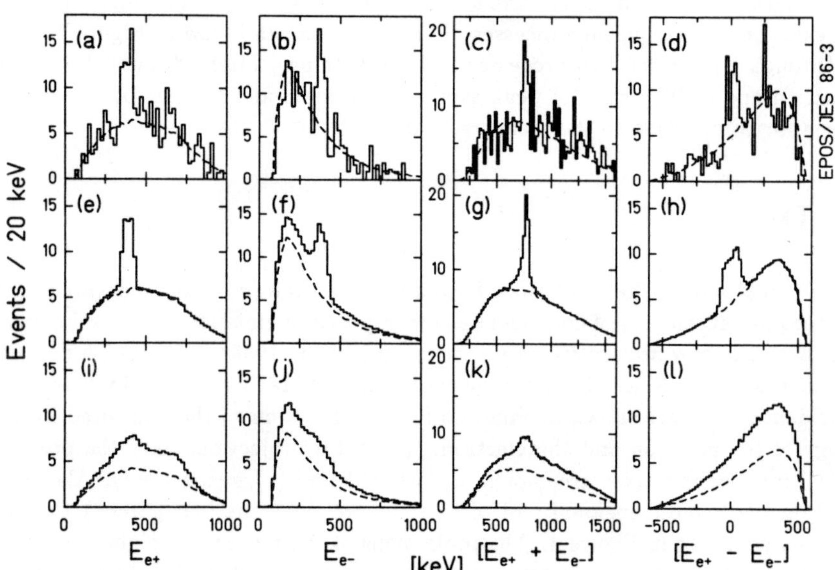

FIGURE 7. Comparison of projections (described in the text) of a measurement of the U+Th collision system at a bombarding energy of 5.83 MeV/u in parts (a) through (d) with the Monte Carlo simulation of the decay of a neutral state in parts (e) through (h) and of the internal pair conversion of a nuclear transition in parts (i) through (l).

The results of the two Monte Carlo simulations described above are shown below the data in parts (e) through (l) of Figure 6. The peak events have been added to a Monte Carlo calculation[12] of the dynamic and nuclear background of positrons and electrons with a single, overall normalization to the total number of detected events. This part of the calculation is indicated by the dashed line in all of the panels of Figure 6. Parts (e) through (h) show the two-body decay of a neutral state, whereby it has been assumed that 3% of the positrons emitted in the heavy-ion collisions are due to this decay process in order to reproduce the observed peak intensity in the measured coincident spectra. Parts (i) through (l) show the internal pair conversion of an E1 nuclear transition. Here it is necessary to assume that fully 30% of the positron yield is due to the internal pair conversion of this single nuclear transition in order to produce the peak intensity shown. This is due to the reduced detection efficiency mentioned above for detecting these events within the narrow gates used to produce these plots. (To reproduce the 35 counts measured in the peak would actually require more like 60% of *all* measured positrons to come from *one* nuclear transition, in contradiction with the observation that *all* nuclear internal pair conversion accounts for only 20–30% of the positron yield.) Clearly the agreement between the measured spectra and the simulation of a two-body decay process is striking.

A study has been made to check that the observed peaks in the four projections shown in Figure 7 (a) through (d) correspond to events truly concentrated in one spot in the two-dimensional distribution of Figure 6 (a), and not manufactured by the narrow gates A through D. As an example, Figure 8 shows three projections onto the energy-sum axis. The three gates used are indicated in the two-dimensional plot of part (c). Part (b) repeats Figure 7 (c) above. Parts (a) and (d) are projections of the two neighboring regions. No significant peak is obvious in either distribution. As in Figure 6 above, the dashed lines are the same projections of a Monte Carlo simulation. Figure 9 shows projections onto neighboring regions for the other three coordinate axes: positron energy, electron energy, and energy-difference. In each case, the two neighboring projections, indicated in Figure 9 (a), are added together. Again, no prominent structure is evident in these neighboring regions.

Measurements were also made on two other collision systems, Th+Th and Th+Cm, at similar bombarding energies[3]. Similar correlated structures are seen in both systems, although at a lower level of statistical certainty. There is one additional feature worth noting which appears in these data: In the Th+Th system, peaks were observed in the positron spectra both at $\sim$ 310 and at $\sim$ 370 keV. There is evidence that, in both cases, an electron peak accompanies the positron peak with $E_{e^-} \simeq E_{e^+}$. This observation may be related to possible differences in the energies of the positron peaks shown in Figure 1 above, reflecting the possibilty of more than one structure.

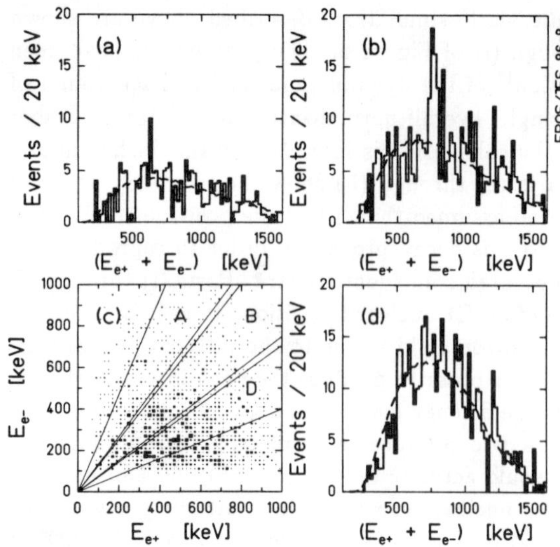

FIGURE 8. Three projections of a measurement of the U+Th collision system at a bombarding energy of 5.83 MeV/u onto the energy-sum axis. The projections shown in parts (a), (b), and (d) of the data in part (c) are gated on the regions marked A, B, and D in part (c), respectively.

FIGURE 9. Projections of a measurement of the U+Th collision system at a bombarding energy of 5.83 MeV/u onto the positron-energy, electron-energy, and energy-difference axes. The projections shown in parts (b), (c), and (d) of the data in part (a) are gated on the pairs of regions marked B, C, and D in part (a), respectively.

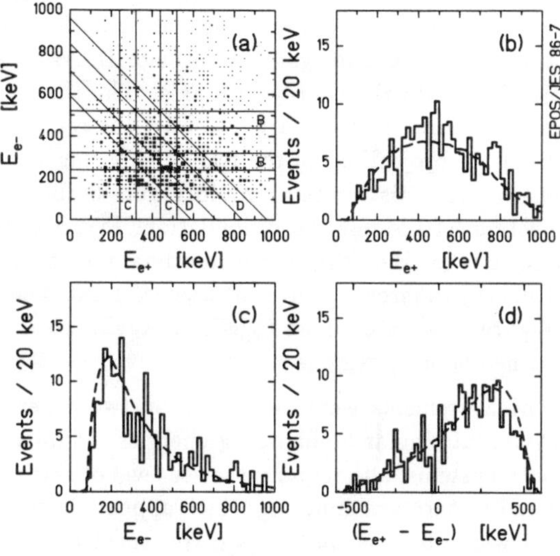

These measurements were repeated in the Spring and the Summer of 1986. For these runs, the resolution of the electron counters was improved to ~ 10 keV by cooling with liquid nitrogen. The collision system U+Th was measured again at bombarding energies around 5.87 MeV/u. The analysis of the data collected is still proceeding, but a preliminary analysis has reaffirmed the existence of the coincident electron-positron events, indicates that the energy-sum lines are significantly narrower than than the corresponding positron and electron lines, and confirms the evidence of the earlier measurement that there are more than one set of discrete energies. Some preliminary results are shown in Figure 10. The upper panels (a) and (b) and the lower panels (c) and (d) show energy-sum and -difference spectra for two subsets of the data gated on beam energy, heavy-ion scattering angle, and positron/electron time of flight chosen to enhance these two sum lines, respectively.

5   Conclusions

The experimental situation can be briefly summarized as follows. A narrow structure is observed in the distribution of electron energies correlated to the narrow peaks already reported in the energy distribution of positrons emitted in very-heavy-ion collisions near the Coulomb barrier. The energy of the electron structure is approximately that of the correlated positron structure. The distribution of the event-by-event sum of the positron and electron energies also contains a structure which is significantly narrower than that in either the positron or electron distributions. This seems to indicate that the Doppler shifts of the positron

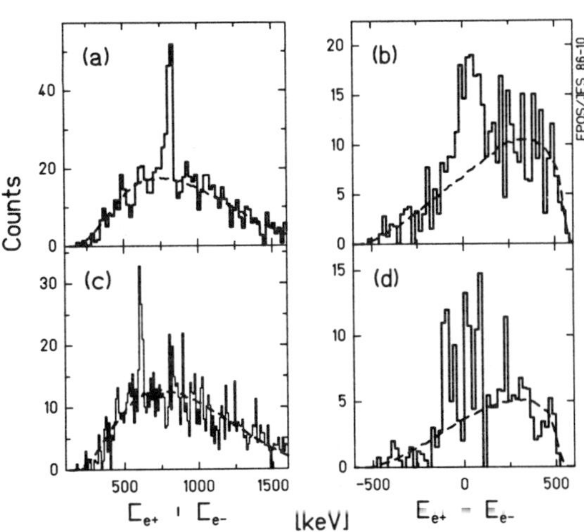

FIGURE 10. Results of a preliminary analysis of measurements in the U+Th collision system at bombarding energies around 5.87 MeV/u. The upper panels and lower panels show energy sum and energy difference distributions for two subsets of the data gated on beam energy, heavy-ion scattering angle, and positron/electron time of flight chosen to enhance these two sum lines, respectively.

and the electron partially cancel event by event, implying that the positron and the electron are emitted approximately back to back in the rest frame of their production. The measured widths of the peaks in the four projections of Figure 7 (a) through (d) are consistent with a velocity for the source of the correlated electrons and positrons comparable to the velocity of the center of mass of the colliding heavy ions. The coincidence intensity and the shape of the spectra cannot be reproduced by known nuclear internal-conversion processes. There are at least two sets of correlated electron-positron energies, and the lines have been seen in three different collision systems: U+Th, Th+Th, and Th+Cm. There is also some evidence for a dependence of the correlated electron-positron production on the projectile energy, the scattering angles of the colliding nuclei, and the measured time of flight of the positron and the electron in the spectrometer. The last point may indicate the existence in the production process of either a time delay or a dependence on the emission angle of the positron and electron.

Theoretical efforts to understand this experimental effect have have included considerations of conventional atomic physics effects[13] as well as speculations on the breakdown of the QED vacuum[14], the production of a new, previously undetected, neutral particle[2,3,15] with a mass of $\sim 3m_{e-}$, in particular the axion[16], the formation of polyelectronic complexes[17], or new bound states of electron-positron pairs[18], or solitons in the QED vacuum[19]. So far, no completely consistent description has been found which can account for all of the observed properties — in particular the multiplicity of lines, the observed intensity, or the apparent low velocity of the decaying state. It appears that an extensive experimental program studying the various characteristics listed above will be necessary in order to shed light on this unexplained experimental effect.

Acknowledgements

As stated at the begining of this report, the experiments described here were carried out by the members of the EPOS collaboration at the Gesellschaft für Schwerionenforschung, Darmstadt. We are very grateful to the operating and engineering staff at the GSI UNILAC for their dedication in operating the accelerator and assistence in performing these measurements, and to the Transplutonium Program of the U.S Department of Energy for the loan of the curium material. Particular appreciation is due to Werner Kreuzer for his dedication and ingenuity in the design and construction of our spectrometer. This work was supported in part by the Bundesministerium für Forschung und Technologie of the Federal Republic of Germany, by the U.S. Department of Energy through Contracts No. DE–AC02–76ER03074 and No. DE–AC02–76CH00016, and by the U.S. National Science Foundation.

References

(1) J. Schweppe *et al.*, Phys. Rev. Lett. **51**, 2261 (1983).

(2) T. Cowan *et al.*, Phys. Rev. Lett. **54**, 1761 (1985).

(3) T. Cowan *et al.*, Phys. Rev. Lett. **56**, 444 (1986).

(4) V.V. Voronkov and N.N. Kolesnikov, ZETF **39**, 189 (1960)
     (Sov. Phys. JETP **12**, 136 (1961)).

(5) W. Pieper and W. Greiner, Z. Phys. **218**, 327 (1969)

(6) S.S. Gershtein and Ya.B. Zel'dovich, ZETF **57**, 654 (1969)
     (Sov. Phys. JETP **30**, 358 (1970));
     Lett. Nuovo Cimento **1**, 835 (1969)).

(7) M. Clemente *et al.*, Phys. Lett. **B137**, 41 (1984).

(8) H. Tsertos *et al.*, Phys. Lett. **B162**, 372 (1985).

(9) J. Reinhardt *et al.*, Z. Phys. **A303**, 173 (1981).

(10) H. Bokemeyer *et al.*, in *Quantum Electrodynamics of Strong Fields*,
      edited by W. Greiner (Plenum, New York, 1983), pp. 273–292;
      J.S. Greenberg and P. Vincent, in *Treatise on Heavy-Ion Science*,
      edited by D.A. Bromeley (Plenum, New York, 1985), pp. 139–421.

(11) H. Backe *et al.*, Phys. Rev. Lett. **50**, 1838 (1983).

(12) T. Cowan, to be published in *Physics of Strong Fields*,
      edited by W. Greiner (Plenum, New York).

(13) W. Lichten and A. Robatino, Phys. Rev. Lett. **54**, 781 (1985);
      P. Schlüter *et al.*, Z. Phys. **A323**, 139 (1986);
      P. Schlüter *et al.*, Phys. Rev. **C33**, 1816 (1986).

(14) G. Soff *et al.*, Nuc. Ins. Meth. Phys. Res. **B9**, 747 (1985);
      B10 214 (1985)

(15) A.B. Balentekin *et al.*, Phys. Rev. Lett. **55**, 461 (1985);
      A. Schäfer *et al.*, J. Phys. **G11**, L69 (1985);
      J. Reinhardt *et al.*, Phys. Rev. **C33**, 194 (1986);
      A. Chodos and L.C.R. Wijewardhana, Phys. Rev. Lett. **56**, 302 (1986);
      K. Lane, Phys. Lett. **B169**, 97 (1986);
      B. Müller and J. Reinhardt, Phys. Rev. Lett. **56**, 2108 (1986);
      E. Ma, Phys. Rev. **D34**, 293 (1986);
      D. Carrier *et al.*, Phys. Rev. **D34**, 1332 (1986);
      G. Mageras *et al.*, Phys. Rev. Lett. **56**, 2672 (1986);
      T. Bowcock *et al.*, Phys. Rev. Lett. **56**, 2676 (1986);
      A. Zee, Phys. Lett. **B172**, 377 (1986);
      A. Schäfer *et al.*, Mod. Phys. Lett. **A1**, 1 (1986);
      A. Schäfer *et al.*, Z. Phys. **A324**, 243 (1986);
      U.E. Schröder, Mod. Phys. Lett. **A1**, 157 (1986).

(16) N.C. Mukhopadhyay and A. Zehnder, Phys. Rev. Lett. **56**, 206 (1986);
      R.D. Peccei *et al.*, Phys. Lett. **B172**, 435 (1986);

L.M. Kraus and F. Wilczek, Phys. Lett. **B173**, 189 (1986);
M. Suzuki, Phys. Lett. **B175**, 364 (1986);
L.M. Kraus and M.B. Wise, Phys. Lett. **B176**, 483 (1986).
(17) C.-Y. Wong, Phys. Rev. Lett. **56**, 1047 (1986);
M.-C. Chu and V. Pönisch, Phys. Rev. **C33**, 2222 (1986).
(18) B. Müller *et al.*, J. Phys. **G12**, L109 (1986).
(19) L.S. Celenza *et al.*, Phys. Rev. Lett. **57**, 55 (1986).

Search for Neutral Particles in Electron
Beam-Dump Experiment*

N. Sasao

Dept. of Physics, Kyoto University
Kyoto Japan, 606

Introduction

This report describes the results of an experiment
which has searched for neutral penetrating particles with
a 2.5-GeV electron beam. The experiment has been
performed at National Laboratory for High Energy Physics
(KEK) in Japan by a group from Kyoto University and KEK[1].
The results reported here come mainly from the data taken
in March, 1986.[2]

This experiment has been motivated by the recent
observations of the anomalous electron-positron peaks in
heavy-ion collisions at GSI[3]. One of the most
interesting interpretations of these peaks is production
of a new neutral "particle" with mass around 1.8 MeV.[4]
In particular there are extensive investigations as to
whether this "particle" can be interpreted as the axion
which was introduced to suppress P and CP violations in
gauge theories of the strong interaction.[5]

It is expected that the neutral "particle", if it
exists, is produced in a strong electromagnetic field

produced by two heavy-ions colliding each other. If that is the case, an electron beam must be able to produce it too via the Primakoff process and/or the bremsstrahlung process. In the Primakoff process, as shown in Fig. 1(a), the neutral "particle" is supposed to couple with two photons produced by a target nuclei and incoming electrons. Fig. 1(b) shows the bremsstrahlung process, in which the "particle" directly couple to electrons. Therefore, an electron beam experiment can provide a clean and crucial test for the particle interpretation.

Experimental Apparatus

Fig. 2 shows our experimental apparatus. It is located inside the beam dump tunnel at the end of the 2.5-GeV electron linear accelerator. The machine has been operated with an average peak current of 13 mA and ~1 μsec spill at the reptetion rate of 10 pps. A total of 0.027 Coulomb has been injected into a 3.5-cm-thick tungsten target. The dump consists of iron, lead and plastic blocks, which attenuates the neutrons, the major background particles, to a tolerable level. The neutral "particles" are supposed to penetrate through the dump and to decay into electron-positron pairs inside a 2.2-m-long decay volume. The distance between the target and an entrance of the decay volume is 2.4 m. The detector system for electron-positron pairs consists of scintillation counters, multi-wire proportional chambers (MWPC), a pair magnet and a lead glass counter. In order to veto charged particles, the dump is followed by two identical scintillation counters (VETO). There is an scintillation counter refered to as S3 in front of the magnet, and a scintillation hodoscope in front of the lead glass counter. The sensitive regions of the chamber are

64 mm in height and 112 mm in width for MWPC 1 ~ 7 and
256 mm in height and 512 mm in width for MWPC8. The lead
glass counter is composed of 18 identical modules, each
being 7.5cm × 7.5cm in cross section and 26 cm (14Xo) in
depth, and is placed in a form of 3×6 matrix. Roughly
speaking, an acceptance of the detector is determined by
the chambers and the lead glass counter.

A hardware trigger signal is generated whenever there
exits an energy deposite of 100 MeV or more in the lead
glass counter. Then dizitized information is read into a
computer and written onto magnetic tapes. This includes
information on pulse heights from individual lead glass
modules, pulse heights and timing information from all
counters, hit positions of all chambers etc. The magnetic
field has provided a horizontal momentum kick of 13.5 MeV
for ~70% of the total running time and 40.5 MeV for the
rest.

It is found that the two veto scintillators are the
weakest counter, and therefore the electron beam intensity
is adjusted so that the counting rates for those counters
are less than one per a single spill. It turns out that
the cosmic-ray background is completely negligible.

Analysis

The neutral "particles" are supposed to penetrate
through the dump and to decay into electron-position paris
in the decay volume. Therefore, candidate events are
demanded to pass through the following criteria : (i) no
signal in the veto counters, (ii) a hit in the S3 counter,
(iii) two charged tracks in the MWPC6 through 8, (iv)
hit(s) in the hodoscope, and (v) energy deposite(s) in the
lead glass which match the momentum determined by the MWPC
system. Moreover the reconstructed momentum vector for

the neutral "particle" should point back to the target. In the actual analysis, however, no event is left after the conditions from (i) to (iv) are imposed. Fig. 3 shows an example of a charged particle event coming from the target. It demonstrates that all the detector elements are working properly.

Now it is necessary to estimate expected event rates in order to interpret the result. First of all, it is assumed that the effective couplings of the neutral "particle" to photons and electrons are given, respectively, by

$$\frac{g_\gamma}{M_e} \, F_{\mu\nu} \, \tilde{F}^{\mu\nu} \, \phi$$

and

$$g_e \, \bar{e} \, \gamma_5 \, e \, \phi \, .$$

Here M_e is the electron mass, and g_e and g_γ denote the dimensionless coupling constants. Once the interaction Lagrangians are given, then the production cross section for the Primakoff process[6] and the bremsstrahlung process[7] can be calculated. The decay width for the two-photon and electron-position mode may also be calculated from the Lagrangians above, and are given, respectively, by

$$\Gamma(\phi \to \gamma\gamma) \; = \; (g_\gamma^2/4\pi)\,(M_a^3/M_e^2)$$

and

$$\Gamma(\phi \to e^+ e^-) \; = \; (g_e^2/4\pi)\,\{(M_a/2)^2 - M_e^2\}^{1/2}$$

where M_a denotes the mass of the nuetral "particle". Next it is necessary to know the electron and photon energy spectra in the tungsten target. To this end, the first generation electron and photon spectra due to Tsai et al.[8] are used. Then, detailed Monte Carlo simulations are performed to estimate the expected event rates.

Results and Discussions

It is covenient to discuss the results according to the relative size of coupling parameters g_e and g_γ; (i) $g_e > g_\gamma$, (ii) $g_e < g_\gamma$, and (iii) $g_e \sim g_\gamma$. If the coupling between the neutral "particle" to electron is fundamental, then the coupling to photon is induced by the Feynman diagram shown in Fig. 4(a). Then the coupling constant g_γ is roughly given by $g_\gamma \sim \alpha g_e$ ($\alpha = 1/137$), and it corresponds to the case (i). This is true in most of the axion models. On the other hand, if the coupling between the neutral "particle" to photon is fundamental, then the coupling to electron is induced by the Feynman diagram shown in Fig. 4(b). In this case the coupling constant g_e is roughly given by $g_e \sim \alpha g_\gamma$. This situation corresponds to the case (ii). The third possibility is to assume that both coupling constants g_e and g_γ are sizable and free. In this report, only the cases (i) and (iii) are considered.

For the case (i), it is appropriate to consider the bremsstrahlung production process and the subsequent decay into electron-positron pairs. Assuming $g_\gamma = 0$, not only the production cross section but also the life time of the neutral "particle" are determined by the constant g_e alone. The excluded parameter range for $\alpha_e = g_e^2/4\pi$ at the 90% confidence level (C. L.) is shown in Fig. 5 by the hatched area as a function of the mass M_a. The dash-dotted line in the figure indicates the prediction by the standard axion model with 3 generations as a function of the parameter X, the ratio of the vacuum expectation values for two Higgs doublets. For the axion mass of 1.8 MeV, suggested by the heavy-ion experiment, the parameter X is estimated to be either ~24 or ~1/24. The large value of X, implying a long axion life time, is ruled out by many experiments in the past. The small value of X, on the other hand, leads to a shorter life time. In this case, the measurement of the axion production in the radiative Υ decay is the only high-energy experiment to date that can exclude the existence of such an axion.[9] The present experiment has excluded the interval $0.022 < X < 0.074$, reinforcing the statement that the 1.8-MeV "particle" cannot be interpreted as the standard axion.

To avoid this difficulty, several variants of the standard model have been proposed.[10],[11] In these new theoretical models, axions are postulated to couple preferentially to electrons and light quarks, and to have short life time to accomodate the negative results of earlier search experiments. The point indicated by a cross in Fig. 5 is the predicted value for a 1.8-MeV axion in the model by Krauss and Wilczek.[11] The models by Peccei et al. also predict the simelar values.[10] It is clear that the 1.8-MeV "particle" is not one of the variants of the axion models postulated by these authors.

For the case (iii), it is appropriate to consider the Primakoff production process. Fig. 6 shows the region excluded by the present experiment in the two dimensional parameter space α_e and $\alpha_\gamma = g_\gamma^2/4\pi$. Here it is assumed that the mass of the neutral "particle" is 1.8 MeV. The dash-dotted lines in the figure show the life time. Roughly speaking, this experiment is sensitive to 10^{-13}~10^{-12} sec, which is essentially determined by the electron beam energy and the dump length. The horizontal dashed line indicates the upper limit on α_e allowed by the measurement of the electron anomalous magnetic moment.[11]

The vertical dashed line, on the other hand, represents the upper limits on α_γ allowed by the measurement of the Delbruck scattering.[12] Further more, it is naively expected that the ratio of the partial widths for the two modes should satisfy

$$(\frac{\alpha}{\pi})^{-2} < \frac{\Gamma(\phi \to \gamma\gamma)}{\Gamma(\phi \to ee)} < (\frac{\alpha}{\pi})^2$$

if there is no special conspiracy. These two limits are represented by the two declined solid lines in Fig. 6. Finally, it is safe to assume that the life time region longer than 10^{-8} sec is ruled out by the low energy electron beam experiment in the past[13]. In conclusion, most of the parameter region is excluded, and very tiny space is left available for the new neutral "particle" to exist.

Conclusions

(i)   The "particle" observed at GSI is not the standard axion.

(ii)   It is also unlikely that the "particle" is one of the variants of the standard model.[14]

(iii) In the two dimensional space of α_e and α_γ, there is a very tiny region left for experimental investigations, where the life time corresponds to about 10^{-13} sec.

N. Sasao

References

1) A. Konaka, K. Imai, H. Kobayashi, A. Masaike,
   K. Miyake, T. Nakamura, N. Nagamine, and N. Sasao,
   Department of Physics, Kyoto University
   Kyoto 606, Japan, and
   A. Enomoto, Y. Fukushima, E. Kikutani, H. Koiso,
   H. Matsumoto, K. Nakahara, S. Ohsawa, T. Taniguchi,
   I. Sato and J. Urakawa, National Laboratory for High
   Energy Physics, Ibaraki 305, Japan

2) A. Konaka et al., KEK Preprint 86-9, May (1986).

3) J. Schweppe, Proceedings of this Conference;
   J. Schweppe et al., Phys. Rev. Lett. $\underline{51}$, 2261 (1983);
   M. Clemente et al., Phys. Lett. $\underline{137B}$, 41 (1984);
   T. Cowan et al., Phys. Rev. Lett. $\underline{54}$, 1761 (1985);
   T. Cowan et al., Phys. Rev. Lett. $\underline{56}$, 444 (1986)

4) A. B. Balantekin et al., Phys. Rev. Lett. $\underline{55}$, 461
   (1985); A. Schafer et al., J. Phys. $\underline{G11}$, L69 (1985)

5) R. D. Peccei and H. R. Quinn, Phys. Rev. Lett. $\underline{38}$,
   1440 (1977); S. Weinberg, Phys. Rev. Lett. $\underline{40}$, 223
   (1978); F. Wilczek, Phys. Rev. Lett. $\underline{40}$, 279 (1978)

6) A. Halprine, C. M. Andersen and H. Primakoff, Phys.
   Rev. $\underline{152}$, 1295 (1966); M. Yoshimura, private
   communication

7) Y. S. Tsai, SLAC-PUB-3926, April (1986)

8) Y. S. Tsai and V. Whitis, Phys. Rev. $\underline{149}$, 1248 (1966)

9) T. Bowcock et al., Phys. Rev. Lett. $\underline{56}$, 2676 (1986);
   M. S. Alam et al., Phys. Rev. $\underline{D27}$, 1665 (1983);
   G. Mageras et al., Phys. Rev. Lett. $\underline{56}$, 2672 (1986);
   M. Sivertz et al., Phys. Rev. $\underline{D26}$, 717 (1982)

10) R. D. Peccei, T. T. Wu and T. Yanagida, Phys. Lett.
    $\underline{172B}$, 435 (1986)

11) L. M. Krauss and F. Wilczek, Phys. Lett. $\underline{173B}$, 189
    (1986)

12) Schafer et al., GSI-86-12

13) D. J. Bechis et al., Phys. Rev. Lett. $\underline{42}$, 1511 (1979)
14) R. D. Peccei, Proceedings of this Conference

Figure Captions

Fig. 1. Two processes contributing to the production of neutral "bosons" : (a) the Primakoff process, and (b) the bremsstrahlung process.

Fig. 2. Schematics of the experimental apparatus.

Fig. 3. An example of a charged particle event. The upper half shows the plan view, and the lower the elevation view. The numbers in the figure represent ADC or TDC channels, and X shows the hit positions on MWPC or scintillation counters.

Fig. 4. The Feynman diagrams representing the induced coupling of the neutral "particle" (a) to two photons, and (b) to electron-positron pairs.

Fig. 5. Constraints on α_e as a function of the "particle" mass. The hatched region is excluded (90%C.L.) by the present experiment assuming the axion production via the bremsstrahlung process. The dash-dotted line indicates the prediction of the standard axion model. The values of the parameter X are also shown by numbers at several points. The point indicated by a cross is the prediction by Ref. 11.

Fig. 6. Constraints on α_e and α_γ. The hatched region is excluded (90% C. L.) by the present experiment assuming the Primakoff process. The dash-dotted lines indicate the life time of the hypothetical particle of mass 1.8 MeV. The horizontal dashed line shows the upper bound allowed by the measurement of the electron anomalous magnetic moment while the vertical one by the measurement of the Delbruck scattering cross section. The two declined solid lines represent the boundaries of the region in which the two coupling constants α_e and α_γ are expected to exist.

Fig. 1

IRON
LEAD
PLASTIC
TARGET
VETO
DECAY VOLUME
MWPC 1~5
S3
MWPC 6,7
PAIR MAGNET
MWPC 8
HODOSCOPE
LEADGLASS
10 cm
100 cm

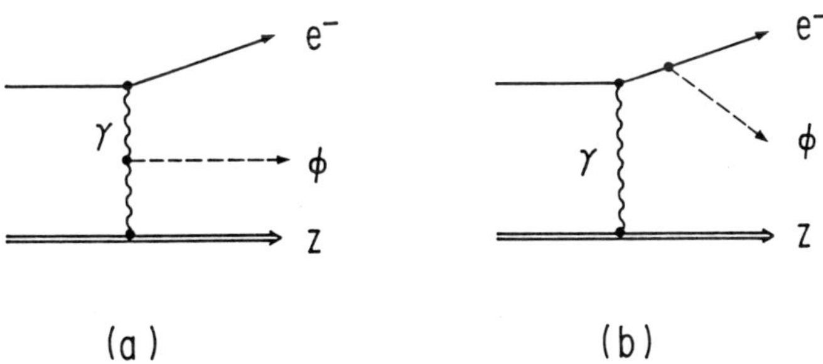

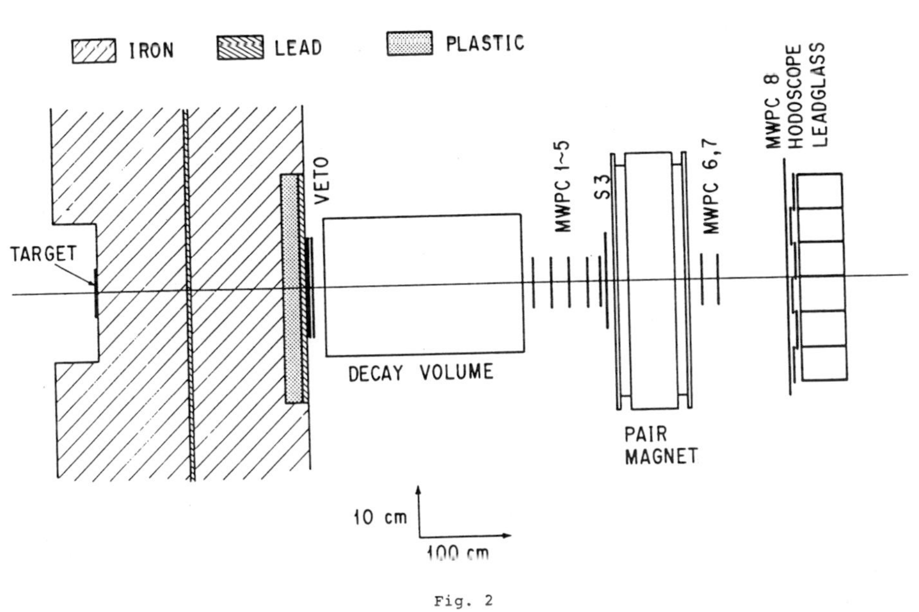

Fig. 2

Run#   45   CUR   727
Event#   95615   MØ1   543
Magnet   MØ2   557
                      MØ3   87
LEAD GLASS
8   2   3
2   -3   1
5   1   1
0   90   -4
2   3   3
1   -1   0
TARGET
DECAY VOLUME
MWPC
PAIR MAGNET
        ADC   TDC   ADC   TDC
V11   411   763 V12   311   772
V21   266   780 V22   200   790
S3   538   941
HO1   1 4110 HO2   -1 4130
HO3   -15 4130 HO4   -2 4110
HO5   289   925 HO6   254   938
HO7   -1 4130 HO8   -10 4130
HO9   6 2422 H10   0 2442
H11   -5 2442 H12   43 2442

Fig. 3

φ   γ
e⁻   γ
(a)
φ   γ   e⁻
(b)

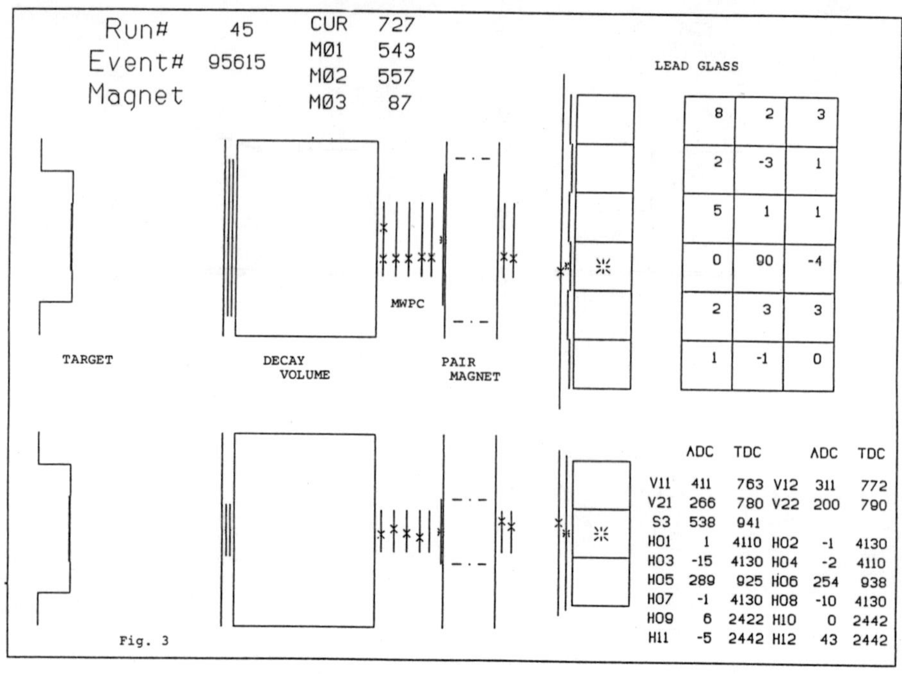

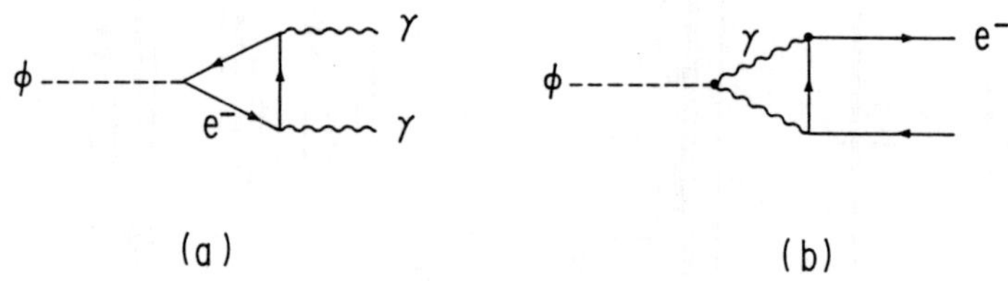

Fig. 4

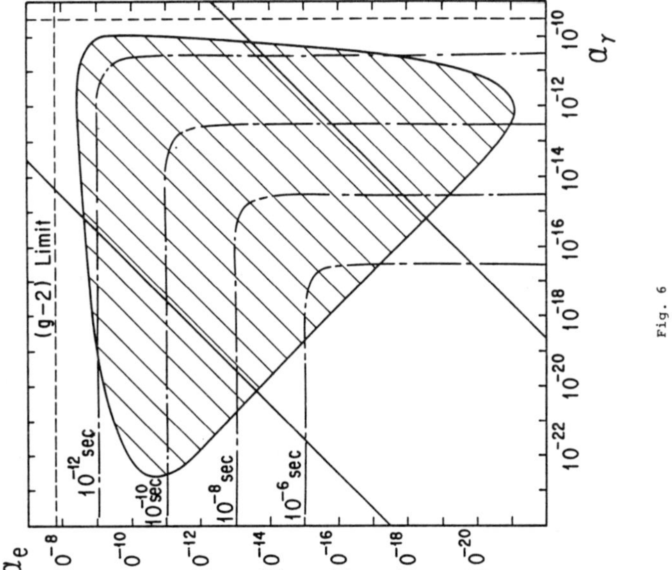

Fig. 6

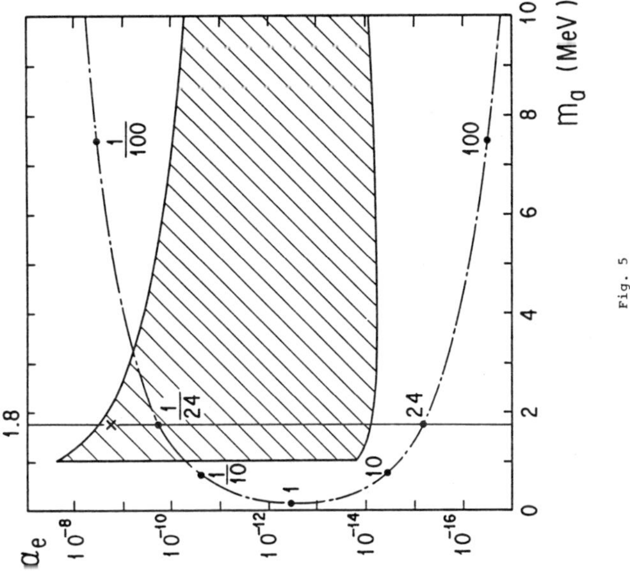

Fig. 5

DEEP UNDERGROUND EXPERIMENTS

ON NUCLEON DECAY

Hinrich Meyer

FB-Physik der Universität*

5600 Wuppertal-1

INTRODUCTION

The experiments I am going to discuss in this talk are located in
deep mines – KOLAR (Au), HOMESTAKE (Au), SILVERKING (Ag), KAMIOKA
(Pb, Zn), SOUDAN (Fe), IMB (NaCl) – roadtunnels – NUSEX (Mont Blanc
tunnel, FREJUS (Fréjustunnel), ICARUS, LVD (Gran Sasso tunnel) – or
specifically constructed underground laboratories – BAKSAN (Baksan
valley) as shown in figure 1. Most of them are close to the 45° latitude line (dashed), however somewhat more relevant is the position with
respect to the earth magnetic field (dotted line) that has some non-
negligible influence on the rate of ν-events in the detectors.

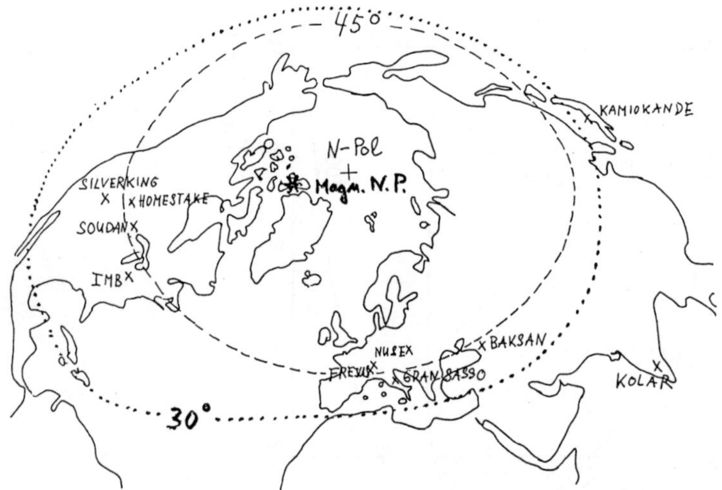

Figure 1: A map of the underground laboratories

The questions of physics being studied are rather fundamental in general, this explains to some extend why experimenters have chosen to
work in rather remote and sometimes really unpleasant places. The following (incomplete) list gives an overview:

1. Nucleon instability
2. $n\bar{n}$ - oscillations
3. Solar neutrinos
4. Cosmic ray sources and composition
5. Unusual matter, (Monopols, WIMPS, strange matter, Cygnets)
6. Neutrino - oscillations
7. Double β - decay

Most of these subjects are covered in other talks at this conference, I
will talk only on searches for nucleon decay. Most of the main conferences
in recent years had summary talks on nucleon decay, I will emphasize the
more recent developments and ask the reader to consult the respective conference proceedings for earlier results.

Nucleon decay was specifically searched for after the hypothesis of
"Grand Unification" of the fundamental forces led to definite predictions
[1], [2], [3] for lifetimes and decay modes of the nucleon which also
seemed to be rather accessible to simple experiments.

Simultanously it was realised [4],[5], that nucleon instability could
explain - at least in principle - the apparent matter/antimatter asymetry
in our universe.

The lifetime predicted by SU5 for example was about 10^{30} years with
the main decay mode $p \rightarrow e^+ \pi^o$ ($n \rightarrow e^+ \pi^-$), a 'gold rush' prediction, but now
12 years later the lower limit for this decay mode has gone up by a
factor of 100, but it also is clear that the original motivations to
find nucleon instability still hold up today. And in fact the notion of
an universal instability of all matter touches almost everybody able
to think. Table 1 gives an overview of current experiments and new projects in terms of the luminosity L measured in kiloton-years (kty) with
sensitivities reached up to now (L(6/86'' and a projection till the end
of 1988 (L(12/88)).

EXPERIMENT	STATE	START	FID.MASS M_f (to)	M_f/M_{tot}	L(6/86) (kty)	L(12/88) (kty)
KOLAR I	INDIA	Oct. 80	65	0.46	.325	
KOLAR II		Dec. 85	120			.600
NUSEX	ITALY	July 82	120	0.86	.384	
IMB I	USA	Aug. 82	3300	0.47	3.770	
IMB III		June 86				8.000
KAMIOKANDE I	JAPAN	July 83	880	0.29	1.470	
KAMIOKANDE II		Jan. 86	800			2.500
FREJUS	FRANCE	March84	750	0.83	.700	2.500
SOUDAN II	USA	1987	900	0.83		1.000
SUPER KAMIOKANDE	JAPAN	project	22000	0.46		
ICARUS	ITALY	project	4000	0.66		

TABLE 1

THE EXPERIMENTS

The running experiments have been described in detail earlier [6],[7], [8], [9], [10], [11], [12], [13] let me mention here only the recent improvements of KOLAR [14], IMB [15] KAMIOKANDE [16] and the new projects SOUDAN II [17], SUPERKAMIOKANDE [18], ICARUS [19], and LVD [20].

In the Kolarmine a new experiment has been completed - KOLAR II [14]- it has dimensions $(6\times6\times6,5)m^3$ and a mass of 260 to. The sampling length has been decreased from 10 mm to 6 mm Fe, the proportional tubes for tracking are unchanged in cross section, $(10\times10)cm^2$ and dE/dx is measured on each wire. This experiment is somewhat higher up in the mine at a level of 6045 hg/cm^2 as compared to 7000 hg/cm^2 of KOLAR I.

The IMB experiment (IMB III) [15] has changed all photomultipliers from 5" tubes to 8" tubes, and a wavelength-shifter plate on each tube gives a further increase in light collection efficiency together a factor of 4 will be achieved. The size of the detector is unchanged.

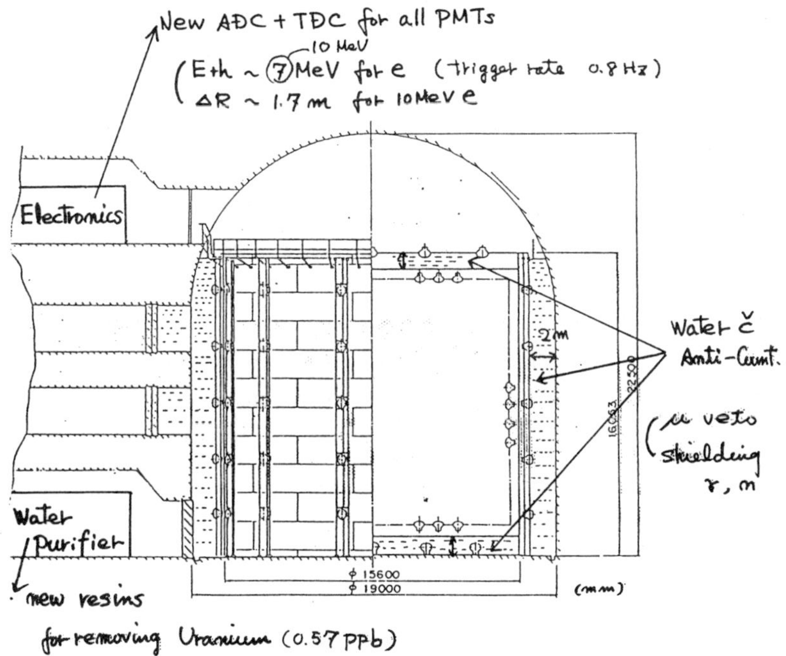

<u>FIG. 2</u>: A sketch of the recent changes in the
KAMIOKANDE experiment

The Kamiokande collaboration was enlarged (KAMIOKANDE II [16]) and all
20" tubes are equipped with TDC's to record the time evolution of the
Cerenkov ligth in great detail. This in particular improves the accuracy
in vertex reconstruction by about a factor of 4 to 6 depending on the
event configuration. The experiment is now surrounded by an active veto
shield consisting of about 2 meters of water with photomultipliers partly
pointing to the outside. The top- and bottom-parts of this shield are taken
away from the original detector which now has a somewhat reduced fiducal-
volume (see figure 2). The vetoshield is not to important for nucleon-
decay searches, however it helps reducing the background on solarneu-
trinos (^{8}B ν's)[21].

From the future projects SOUDAN II [17] is actually under construction
wlth parts of it expected to go into operation next year. It has 2mm Fe-

plates for sampling and drifttubes for particle tracking and - most important - dE/dx measurements. The detector is not very deep below ground and the total trigger rate may be not to easy to handle.

The low candidate rate for nucleon decay (see below) and the good performance of KAMIOKANDE leads to the suggestion to build a much larger detector 15 times the size of KAMIOKANDE II using about 11000 20" photomultipliers (SUPERKAMIOKANDE)[18]. The total volume foreseen is 45000 m^3 with about 20000 to fiducal volume. It presents no particular technical problems as a detector however it would be rather expensive. If placed at the same depth as KAMIOKANDE (2700 hg/m^2) a further problem may be the rate due to atmospheric muons, specifically as a background source for solar neutrino detection (induced radioactivity- see A. Suzuki's talk at this conference[21]).

For the new GRAN SASSO laboratory a very ambitious project as been proposed a Cryogenic Imaging Chamber filled with liquid Ar or liquid CH_4 placed in a magnetic field of several 100 mT. High spatial resolution and dE/dx measurements should make it a superior instrument, the size foreseen is about 6000 m^3 total volume, however is it really large enough in view of the low candidate rate for nucleon decay?

Another detector in the Gran Sasso laboratory [22], primarily designed to detect the ν-burst from Supernova explosions might be useful for some specific decay modes of nucleons [20]. It is planned to be very large, 8000 to of liquid scintillator and its potential as a detector for $p \to \nu K^+$ has been described recently in considerable detail [20].

The main parameter in the ongoing searches for nucleon decay - as it appears now - is the luminosity of a given experiment. Figure 3 shows the integrated luminosity (kty) for the running experiments as it has developed over the years. The Fe-calorimeter experiments (KOLAR I, NUSEX and FREJUS) are continuously running since their start up, NUSEX may however stop at the end of this year. The two water experiments, IMB I and KAMIOKANDE I have, after little more than one year of effective running time made major modifications and recently started again as IMB III [15] and KAMIOKANDE II [16]. The expected luminosity integrated by the end of 1988 in the different experiments can be taken from figure 3 as well (see also table 1). For background free decay modes a level of 20 kty could be available and the lifetime limit for p→ e$^+$π^o can be pushed beyond 10^{33} years, where the background from electron neutrino produced single π^o production would be reached [23].

A new project, to efficiently compete should have at least 20 kto fiducal mass it will however work in the neutrino background limited region unless it is also qualitatively improved over the present generation of experiments.

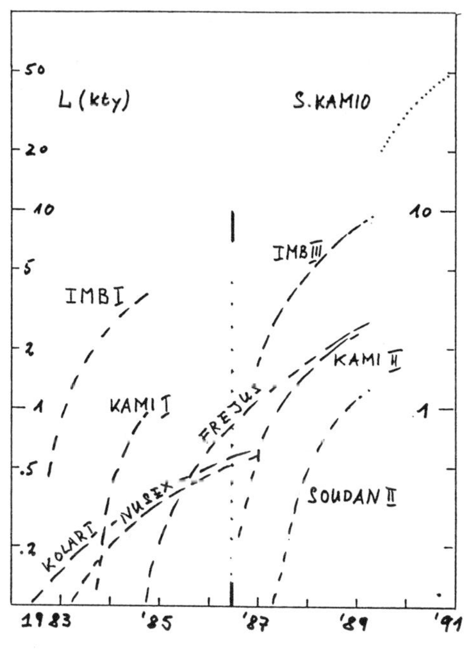

FIG. 3: The integrated luminosity (L) in kty versus time for the proton decay experiments. The dotted line indicates what might be achieved by a project like SUPERKAMIOKANDE

BACKGROUND

The only source of contained events – we believe – is interactions of neutrinos $(\nu_\mu, \bar\nu_\mu, \nu_e, \bar\nu_e)$ inside the detector. The neutrinos originate from interactions and decays of cosmic ray particles in the earth atmosphere. The flux of neutrinos has been repeatedly calculated [24] and is known with moderate (30%) accuracy. Figure 4 shows the results from T. Gaisser et.al. [25], the spectrum is very steep and, most important, there is a strong contribution from electron neutrinos at a level of about 35%. The

energy spectrum shows further dependences on the earth magnetic field
and on the solar spot cycle however mainly below 1 GeV neutrino energy.

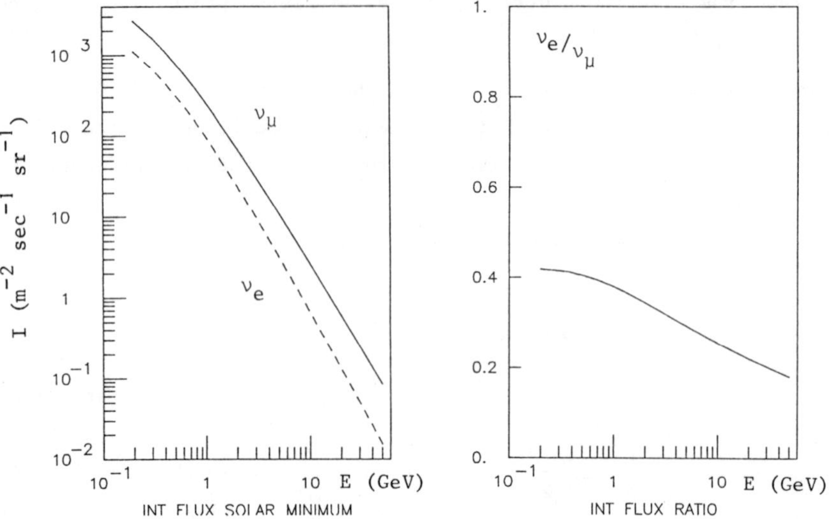

FIG. 4:   The integral energy spectrum of atmospheric neutrinos [24]

The final states are dominated by quasielastic scattering (e.g. $\nu_\mu + n \rightarrow \mu^- p$)
and single π-production (e.g. $\nu_e + n \rightarrow e^- \pi^0 p$), figure 5 shows the relation
contributions of the various final states. Table 2 gives a breakdown with
respect to charged (CC) and neutral (NC) current events as well as for

ν_μ	CC	75	15
$\bar\nu_\mu$	CC	25	5
$\nu_e + \bar\nu_e$	CC	50	9
$\nu_\mu + \bar\nu_\mu$ $\nu_e + \bar\nu_e$	NC	10	3

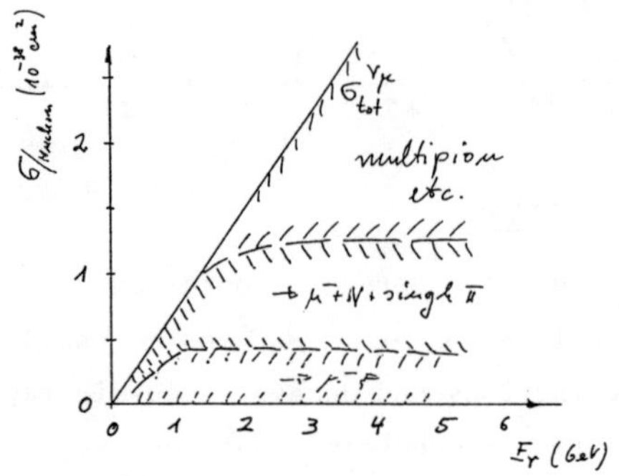

TABLE 2: ν-event rates for
L = 1,5 kty

FIG. 5:   The cross section for
different typs of neu-
trino events

neutrino typs. The numbers are for about 1,5 kty in the neutrino energy range 0.5(GeV)<E<5 (GeV), also shown is the number events with visible
energy in the range of the nucleon mass ($\pm$ 200 MeV) (last column).

The neutrino background has to be calculated for each experiment individually since the properties of the detector as well as the location
on earth (geomagnetic field cutoff) strongly influence the event pattern.
It would be most natural to choose an appropriate neutrino beam at an
accelerator and expose a model detector to study the neutrino background
directly. This has actually been done by the Nusex collaboration [26].
In the end however a very good Monte Carlo simulation is unavoidable
since the atmospheric neutrinos are coming from all direction with about
the same intensity and there is the high level (35%) of electron neutrinos
in contrast to accelerator beams (few % only). The background condition
at accelerators is certainly much worse (n, K_L^0) and finally a very high
statistics experiment (>10000 events) should be performed. The experiments therefore rely on Monte Carlo simulations based on an accurate
knowledge of the detector properties and on published data from neutrino
experiments [27],[28],[29]. The Fréjus collaboration is presently reana-
lysing the Aachen Padova experiment [30], since it has very large statistics
and a detector of similar structure. The exposure was taken in the CERN-
Gargamelle neutrino beam which has a suitable energy spectrum at rather
low background level at the position of the Aachen Padova experiment.

Figure 6,7,8 show some typical neutrino events as observed in the Fré-
jusdetector. The high resolution reflects the very large number of individual cells per view (470000) and the small sampling length (3mm Fe =
1/5 r.l.), see figure 9,10 [11].

The neutrino background is very well described by the current Monte
Carlo simulations, as an example the total energy observed in
IMB (Fig. 11) is compared to a high statistics M.C. simulation.

Finally I like to compare the observed multiplicities from KAMIOKANDE
(number of resolved Cerenkov rings) and FREJUS (number of tracks) in figure 12
where the FREJUS data have been scaled up by the ratio of luminosities,
the agreement is strikingly good except for 1 track events because KAMIOKANDE

H. Meyer

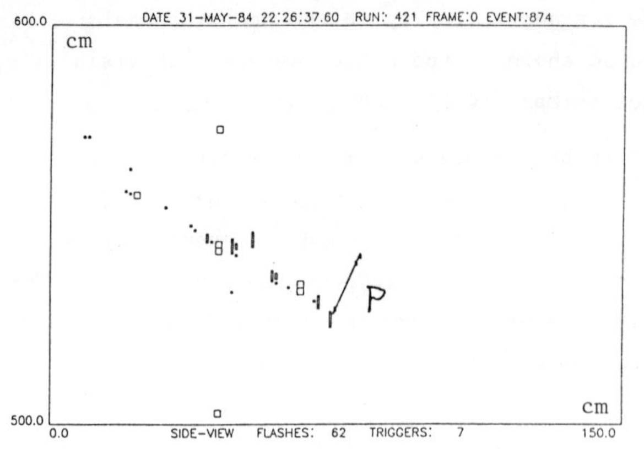

FIG. 6: An example of a quasi elastic scattering, $\nu_e n \to p e^-$

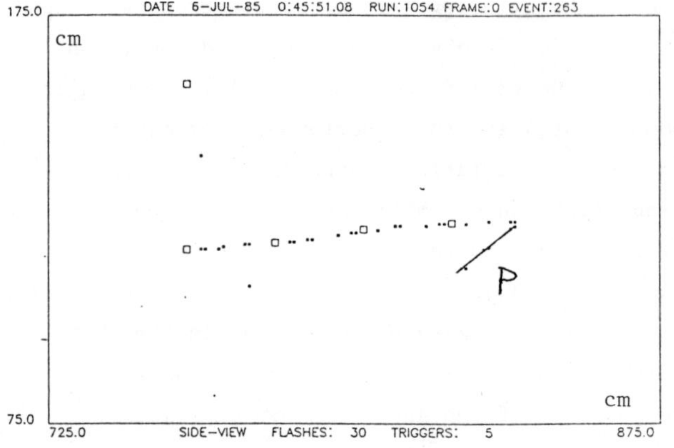

FIG. 7: An example of a quasi elastic scattering $\nu_\mu n \to p \mu$ The recoil proton is actually observed like in Fig. 7

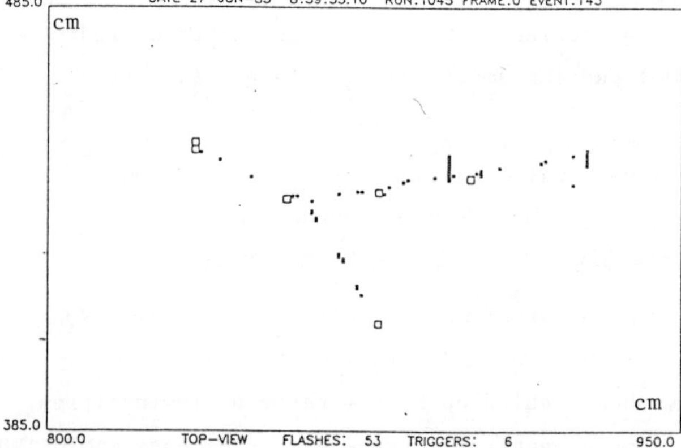

FIG. 8: A 3 prong neutrino event. The scales in the figures are in [cm], at a density of $2 g/cm^3$

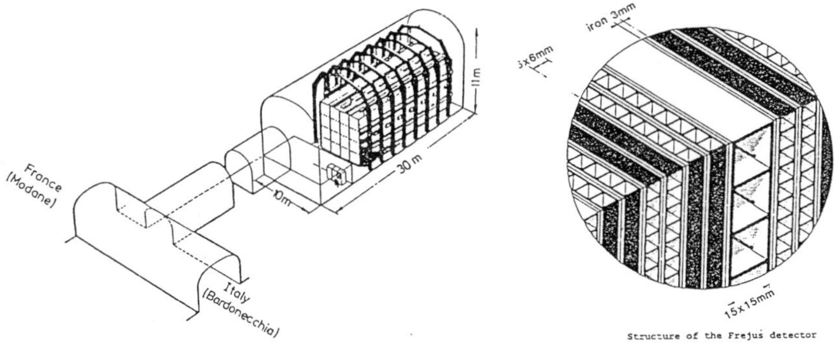

FIG. 9: The Fréjus detector in
the Frejusroadtunnel

FIG.10: Details of the sampling
structure of the Frejus-
detector

has a very much lower trigger threshold. IMB I has not explicitly given
a multiplicity distribution the observed fraction of 2 ring events is
however about a factor of 5 lower than in FREJUS and KAMIOKANDE indicating
a serious loss of particles per event.

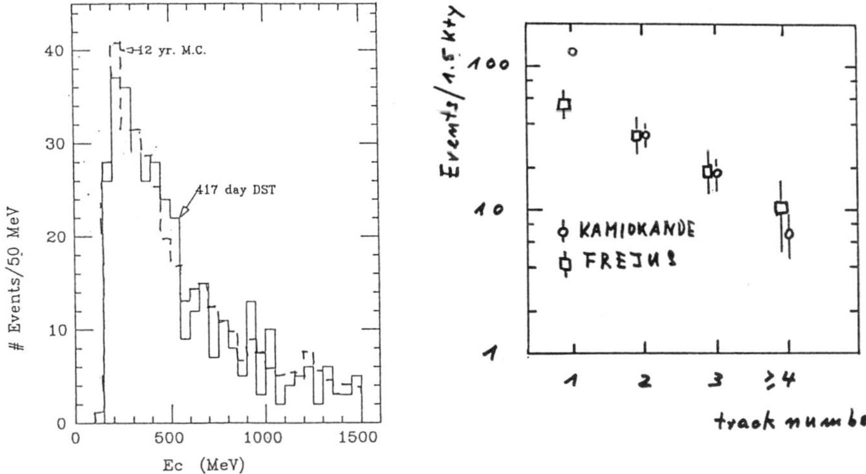

FIG. 11: The visible energy distribution of ν-events compared
to a M.C. simulation

FIG. 12: A comparison of observed
multiplicities for ν-events
(KAMIOKANDE[28] and FREJUS
(preliminary)

SEARCH FOR NUCLEON DECAY CANDIDATES

Candidate events for nucleon decay should be searched for by applying the decay kinematics of a particle at rest. This implies:

$$\sum_i P_i \simeq P_f$$

$$\sum_i E_i = E_N$$

$$E_N^2 - \left(\sum_i P_i\right)^2 = M_N^2$$

where P_i, E_i are the momenta and energies of the particles in an event, P_f is the Fermimomentum of nucleons in the respective nucleus (O,Fe) and M_N the nucleon mass. The basic experimental information is Cerenkov rings for H_2O-detectors and sample points on the particle tracks for Fe-detector. A $\hat{C}$-ring uniquely determines the direction of a track with further details from the time development of the $\hat{C}$-rings. The number of photoelectrons (pulseheight) determines the $\hat{C}$-energy loss of the particle. To convert this information into momentum and energy of the particle, it's nature has to be assumed. From figure 13 which shows the visible $\hat{C}$-energies for different particles and momenta one can see how strongly particle identification influences the kinematics, a particle with a total number of photoelectrons about 1000 could be an electron of 300 MeV/$_c$ or a muon of 500 MeV/$_c$ or a noninteracting pion of 600 MeV/$_c$. For the KAMIOKANDE experiment and for simple event patterns, e.g. 1 or 2 ring events, can a distinction between a showering and a non showering particle be made. Further information comes from time delayed signals they do indicate the decay of muons. Particles with $\beta \leq 0.72$ are not detected, e.g. protons in neutrino events in general, or pions and muons below ~200 MeV/$_c$ momentum. This causes a serious loss of particles in events more complicated than just the two body decay modes of nucleons.

In fine grain tracking detectors - like the Fréjus experiment with 3mm Fe sampling-tracks are very well defined and showering and non-showering tracks can be distinguished with high reliability. The detection threshold for particle tracks is well below 100 MeV/$_c$ in general, e.g. a μ-decay electron (35 MeV) produces 6 hits on average. The track direction can not always be determined, except for showering tracks and in favour-

able cases using the increase of multiple scattering towards the end of
a track. Muon decay is seen with moderate (~30%) efficiency detecting
time delayed hits (NUSEX, FREJUS). Containement of the events is defined
by requiring the particles to stop well inside the detector surface. For
H_2O detectors this requires ~2m cut and a considerable loss in detector
volume (~50%), in Fe-detectors one can go closer to the detector face
(~10 cm)   with correspondingly reduces losses   (10 %)   useful
detector mass.

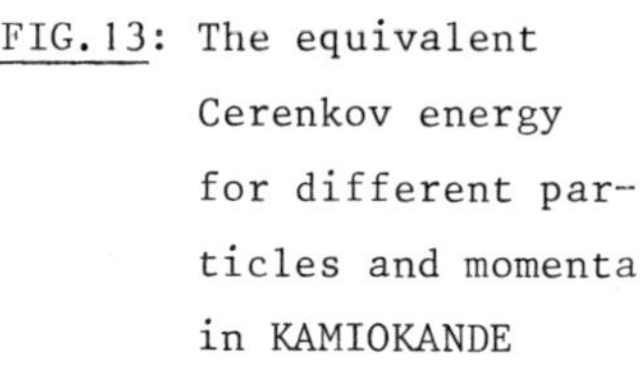
FIG.13: The equivalent
Cerenkov energy
for different par-
ticles and momenta
in KAMIOKANDE

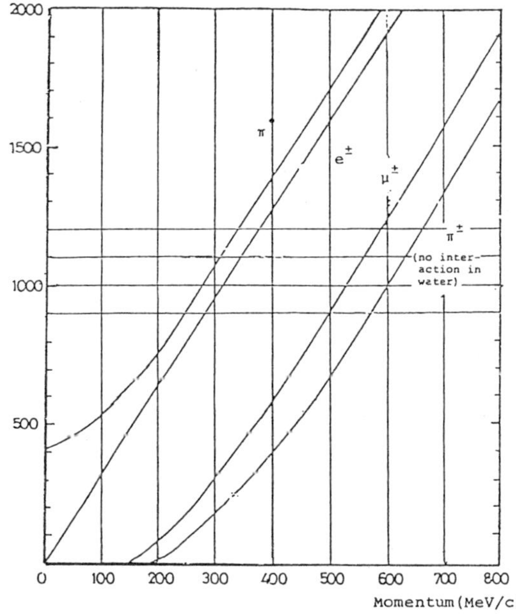

IMB I

Since particle identification on an event by event basis is not
available, the data are compared to M.C. predictions on the assumption
that the detected light originates from a showering particle [31].
For each event then the total visible Cerenkov-energy E_C is calculated.
The momentum balance is approximated by a quantity called anisotropy A_C
it measures the energy weighted vectorsum of the directions of PMT's
fired w.r.t. the vertex of the event. Single prong events have
~A_C ~0.75, momentum balanced events have $0 < A_C \lesssim 0.4$. Various nucleon de-

cay modes cover regions in E_c vs. A_c depending on the assumed final state. If all energy appears in showering particles like $p \to e^+ \pi^o$, $p \to e^+ \omega^o \to e^+ + \pi^o + \gamma$, the visible energy corresponds to the nucleon mass M_N, while for modes with nonshowering particles like $n \to \mu^+ \pi^-$ the visible energy is in general much lower. In the real data sample there are candidates events in the nucleon decay regions defined by M.C. calculation for almost all decay modes searched for.

Neutrino events simulated on the basis of accelerator data, account for essentially all of the candidate events and therefore only lower limits on nucleon lifetime have been derived. Figure 14 shows the data (a) the neutrino simulation (b) and the simulation of a particular nucleon decay mode $p \to \mu^+ \pi^o$ (c). In general IMB I has reached the background level for the majority of the interesting decay modes of nucleons [32].

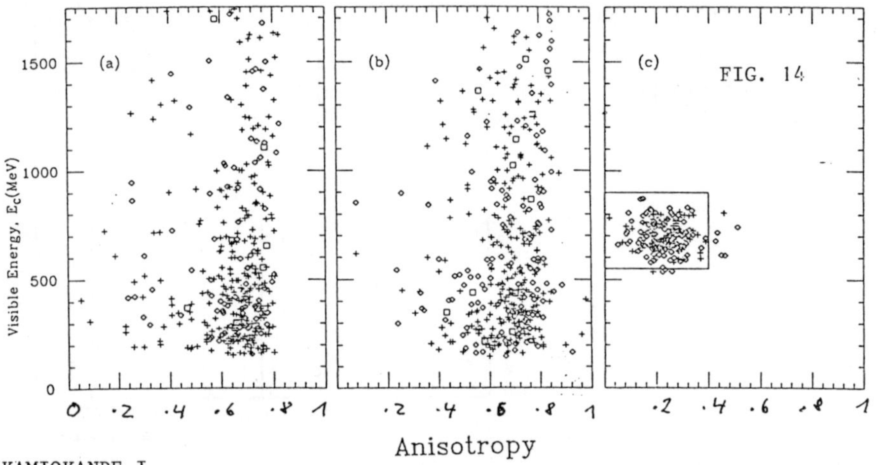

KAMIOKANDE I

KAMIOKANDE has much higher light collection efficiency for each event due to the very large diameter (20") of the phototubes. This gives good energy measurement and in particular allows distinction between showering and nonshowering particles in favourable cases (e.g. 1 ring event). Also individual particle tracks are identified by observing their Cerenkovring pattern. Therefore more specific kinematic tests can be applied. The neutrino background is again simulated on the basis of accelerator data by M.C.[28]. The comparison is shown in figure15(2 rings) and figure16(3 rings)[33].

In the two ring sample no candidate appears in the nucleon decay region, but also the M.C. calculation (with 7times more statistics) does not predict a background event. Clearly an improvement over the IMB data.

In the 3 ring data sample some events appear in the nucleon decay region, but also the M.C. simulation predicts ν-events as nucleon decay candidates. The background level therefore is reached also in KAMIOKANDE for more complicated events and only lower limits for nucleon life time can be quoted.

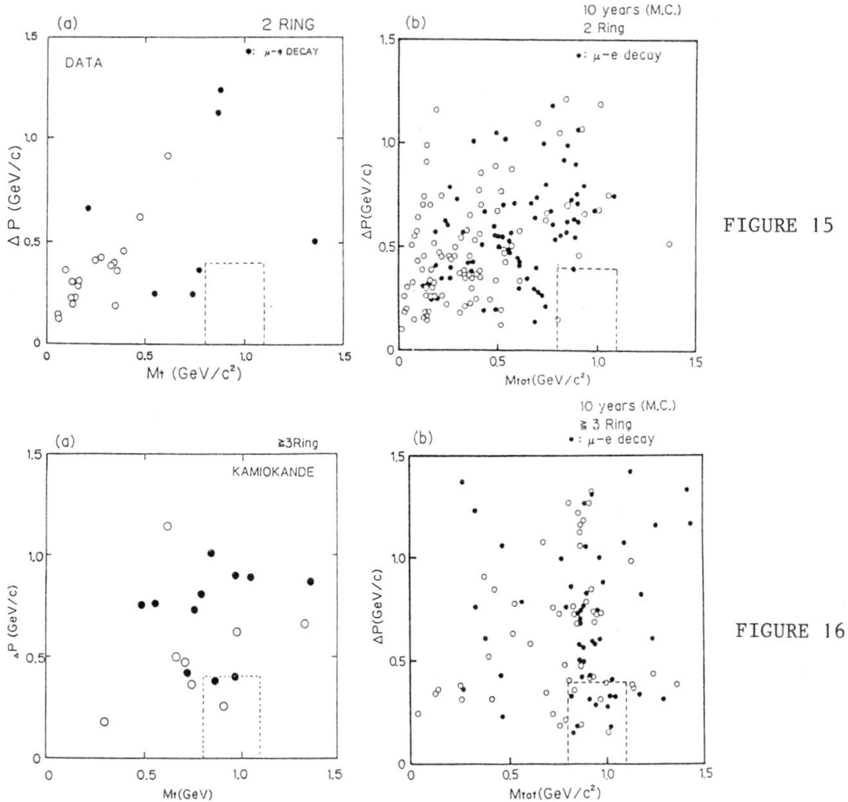

FIGURE 15

FIGURE 16

H. Meyer

KOLAR I

The KOLAR I experiment claimes several nucleon decay candidates [35].
A general kinematical analysis of all contained events has however not been
presented [36], rather individual events, considered compatible with
nucleon decay have been discussed in considerable detail [14]. If true
nucleon decays, they should have by now been confirmed by the larger
detectors, which is clearly not the case, in particular not in the Fré-
jusexperiment (see below).

NUSEX

The Nusex group has described several nucleon decay candidates[3],
however the neutrino background expectation is at a similar level.There-
fore only lower limits for nucleon lifetime have been quoted [36]. This
is also in disagreement with the KOLAR I experiment since the respective luminosities are similar.

FREJUS

The FREJUS experiment is both larger (6times NUSEX) and has finer
granularity compared to the other Fe-calorimeters, e.g. a 1 GeV showering
particle has ~18 hits (KOLAR I) and 35 hits (NUSEX) as compared to
145 hits (FREJUS). Showering from nonshowering particles can be distinguished with high reliability, event vertices can be determind with mm
precision. For the two, three and more prong contained events momentum
balance (P_{mis}) and visible energy E_{vis} are determined as shown in fig. 17
$E_{vis} < 1,7$ GeV.   Only one event can be considered as a nucleon
decay candidate. It has four prongs and two vertices as shown in figure 18.

Under the assumption that vertex B is the primary vertex (3prongs) a
nucleon decay hypothesis is unlikely and vertex A is a secondary scattering
e.g. $\pi p \to \pi p$. If vertex A is the primary vertex 4 particles emerge and vertex B is a secondary π-nucleus elastic scattering. The event then is rather
well balanced (<200 MeV/$_c$) and has a total energy – from range measurements – close to the nucleon mass. In general a nucleon decay with four
particles in the final state is strongly influenced by the nucleus (Fe)
like in a final state $e\pi\pi$, the probability to observe all three pions
without loss through π interactions in the nucleus is clearly very small,
however the FREJUS event nominally is balanced. A balanced nucleon decay
event with all energy visible is rather easily obtained in final states
with 3 leptons like $e\mu\mu\pi$, however thesis clearly a nonstandard decay

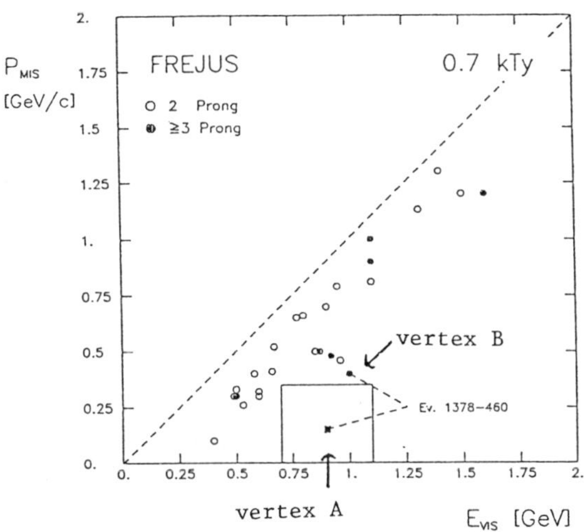

FIG. 17: The visible energy and the momentum balance for the Fréjus
detector. The event 1378-460 is shown in Fig. 18

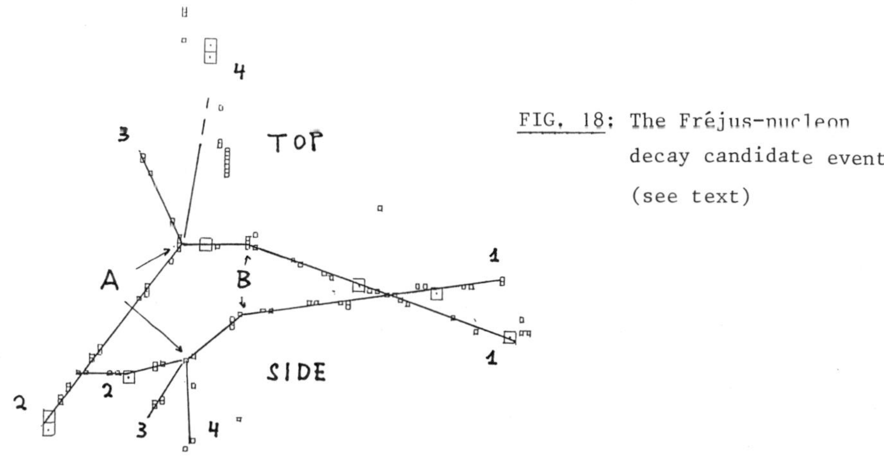

FIG. 18: The Fréjus-nucleon
decay candidate event
(see text)

mode of a neutron. More events of this kind need to be observed to consider a case for nucleon decay, one event is just not enough, it could in
principle be a (very unlikely though) ν-background event. On the other
hand this type of event is very difficult to observe in H_2O Cerenkov de-
dectors, since in almost all kinematical variations one or two particles
would be below Cerenkov-threshold and the event appears unbalanced.

LIFETIME LIMITS

A lifetime limit for a particular decay mode is calculated according
to
$$\tau_N \cdot \frac{1}{B} = \frac{X \cdot 10^{32} \cdot L}{N} \cdot \varepsilon$$

X is 2,9 if N is the proton and 3.2 if N is the neutron for Fe as the detector material, L is the luminosity in kiloton year, N the number of observed events (2.3 if no candidate is seen, 3.8 if one candidate is con-
sidered- at 90% confidence level), B the nucleon decay branching ratio
for the decay mode considered and the efficiency ε can vary considerably
among different experiments and for different decay modes, it is certain-
ly high for a decay mode like $p \to e^+ \pi^0$ and rather low for $p \to \nu\nu\nu\pi^+$. Also the
background level for a given decay mode can be rather high therefore re-
ducing the lifetime limit accordingly. A detailed comparison of limits
given by IMB and KAMIOKANDE (background subtracted on the basis of M.C.
simultaion of ν-events) is shown in fig. 19 [27],[33].They are based on
1.47 kty (KAMIOKANDE) and 3.77 kty (IMB) and are quite compatible. The
channels $p \to e^+ \pi^0, e^+ \eta^0$ are hardly background limited and the limits reflect
the luminosities. The difference for the channels $e^+ \pi^-$ and $e^+ \rho^-$ reflect
very different event efficiencies. The mode $n \to \nu\pi^0$ has very high background
for IMB and rather low background for KAMIOKANDE. The limits from NUSEX
are in general lower by the ratio of luminosities (factor 4). From the
FREJUS experiment no detailed information is available yet, since the
M.C. for ν-background as well as nucleon decay detection efficiency has
not been completed. The limits will however be $\sim 3 \cdot 10^{31}$ years at least.
From IMBI there are rather stringent limits for a number of more unusual
decay modes they are shown in table 3 . Although not very much discussed
at present they may at some later time provide very useful constraints
for new models beyond the standard model of particle physics. More limits of this kind should be given by the current experiments.

For two much discussed decay modes, $p \to e^+ \pi^0$ the expected dominant
mode in SU5-GUT and $p \to \nu K^+$ in SUSY-GUT the combined limits are

τ/B $(p \to e^+ \pi^0)$ $> 4 \cdot 10^{32}$ y $\qquad$ (IMB,FREJUS,KAMIOKANDE)

and

τ/B $(p \to \nu K^+)$ $\quad > 6 \cdot 10^{31}$ y $\qquad$ (IMB,KAMIOKANDE)

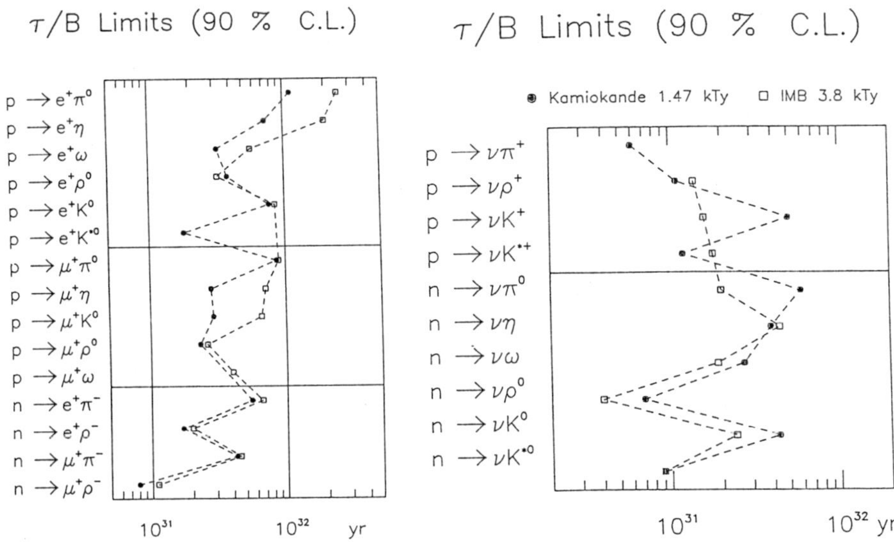

FIG. 19: Lifetime limits on nucleon decay modes [27][33]

Unusual decay modes   IMB   (Toyama)

Decay-Mode	Candidates	Background	$\tau/B \cdot 10^{31}$ y
p→ e⁺γ	0	.28	36
μ⁺γ	0	1.9	28
e⁺e⁻e⁺	0	.28	51
μ⁺μ⁻μ⁺	1	.10	19
n→ ν γ	73	60	3
e⁺e⁻ν	5	5	7.6
μ⁺μ⁻ν	14	7.3	2.4

TABLE 3

They will (probably) be improved by about a factor 4 by the combined impact of the third generation detectors (KAMIOKANDE II, IMB III, FREJUS, SOUDAN II, KOLAR II) by 1.1.1989 (see table 1). Limits are of considerable interest e.g. SU5 may already be ruled out with the high upper limit for p→e⁺π⁰ however it is much more interesting to ask for the chances to observe nucleon decay within the next 3 years with the new detectors. A look at table 4 provides a guideline, it shows the luminosities, ν-event rates and the candidate level and the corresponding background level of the running experiments. One observes:

EXPERIMENT	$\int Ldt$ (kty)	# of con-tained events*	CANDIDATES	BACKGROUND
KOLAR I	.325		5	.6
NUSEX	.380	37	3 (1)	1.6 (0.15)
FREJUS	.610	51	1	prob. small
KAMIOKANDE I	1.470	173	5 (1)	3.2 (0.03)
IMB I	3.770	401	20	17

* Not corrected for trigger threshold !

TABLE 4

1. The candidate level is only ~(2-5)% of the ν-event rate, that is
   <5 events/kty
2. The rate for <u>good</u> candidates is even smaller only <u>one each</u> from NUSEX,
   FREJUS and KAMIOKANDE, this means ~1 event/kty

This allows then by 1989 for $\gtrsim 10$ good candidates from all experiments taken together in my mind hardly enough (leaving aside some lucky circumstances) to convincingly prove nucleon decay.

FUTURE EXPERIMENTS

Extrapolating the present states of the running experiments till about 1989 (see figure 3 and table 1) it is obvious that the candidate level for nucleon decay events is rather low, about 1 event/kty. Neutrino background is probably not the limiting factor it is the rate. New experiments therefore should be very large, with fiducal mass larger than the integrated mass of the present experiments to provide a real gain. SUPERKAMI-OKANDE [18] is a proposal that just meets this requirement. It would extent the range of lifetimes by about an order of magnitude. If placed at the proper depth it could address some of the other physics issues from my list in the introduction as well. One does not need to go to the moon, as it has recently been suggested [37].

However we should keep in mind that nucleon decay might just shown up in the near future in present experiments that would completely change the picture and - because of the rate problem - require a much bigger detector anyway.

Acknowledgement:

I would like to thank, E. Fiorini, M. Koshiba, J. Lo Secco,
T. Suda, A. Suzuki, Y. Totsuka and in particular H.J. Daum
for very useful discussions.

* Supported by BMFT, Bonn

REFERENCES

1. J. PATI, A. SALAM : Phys. Rev. Lett. $\underline{31}$, 661 (1973)
2. H. GEORGI, S. GLASHOW: Phys. Rev. Lett. $\underline{32}$, 438 (1974)
3. H. GEORGI, H. QUINN, S. WEINBERG: Phys. Rev.Lett. $\underline{33}$, 451 (1974)
4. A.D. SAKHAROV: JETP Lett. $\underline{5}$, 24 (1976)
5. M. YOSHIMURA: Phys. Rev. Lett. $\underline{41}$, 281 (1978)
6. M.R. KRISHNASWAMY et.al.: Pramana $\underline{19}$, 525 (1982)
7. G. BATTISTONI et.al.: preprint, LNF-85/48 24.Oct.1985
8. R. BIONTA et.al.: Phys. Rev.Lett. $\underline{51}$, 27 (1983)
9. J. BARTELT et.al.: Phys. Rev.Lett. $\underline{50}$, 651 and 655 (1983)
10. KAMIOKANDE: J. of the Phys. Soc. of Japan $\underline{54}$, 3213 (1985)
11. R.BARLOUTAUD: Fréjus-Collaboration: Contribution to the 7th Work-
    shop on Grand Unification, ICOBAN '86, Toyama (Japan)
12. M.L. CHERRY et.al.: Phys. Rev. Lett. $\underline{47}$, 1507 (1981)
13. J.A. GAIDOS et.al.: Proceedings of the 1982 Summer Workshop on
    Proton Decay Experiments - ANL-HEP-PR-82-24
    and A.MILLS MORE: Thesis, Univ. of Wisconsin (1984)
14. M.R. KRISHNASWAMY et.al.: Contribution C 52 to this
    conference
15. J.M. LO SECCO: Contribution to this conference
16. A. SUZUKI: Contribution to this conference
17. D. AYRES et.al.: The SOUDAN II Nucleon Decay Experiment,
    ANL-HEP-PR-84-30 (1984)
18. Y. TOTSUKA: SUPERKAMOKANDE, Contribution to the 7th Workshop on
    Grand Unification, ICOBAN '86, Toyama, Japan
19. ICARUS Proposal, INFN/AE-85-7, September 1985

20. LVD-Collaboration, Underground Physics '85 St. Vincent (1985)

21. A. SUZUKI, KAMIOKANDE, Contribution to this conference

22. E. FIORINI: talk at this conference

23. D. REIN: Phys. Rev. $\underline{D28}$, 1800 (1983 and Phys. Rev. $\underline{D30}$ (E) 242(1984)
    T.K. GAISSER, M. NOWAKOWSKI and E.A. PASHOS: Phys. Rev. $\underline{D33}$,1233(1986)

24. T.K. GAISSER et al.: Phys. Rev.Lett. 51, 223 (1983)

25. T.K. GAISSER: private communication

26. G. BATTISTONI et.al.: Nucl. Inst. Meth. $\underline{219}$, 300 (1984)

27. T. HAINES, IMB: Contribution to this conference, Toyama, Japan

28. M. NAKAHATA et.al., KAMIOKANDE: Preprint UT-ICEPP-86-05

29. M. DERRICK et.al.: Phys. Rev. $\underline{D30(R)}$, 1605 (1984)

30. H. FAISSNER et.al.: Phys. Rev. Lett. $\underline{41}$, 213 (1979)

31. G. BLEWITT et.al.: Phys. Rev.Lett. $\underline{55}$, 2114 (1985)

32. G. BLEWITT: Thesis, Calt-68-1327

33. T.SUDA, KAMIOKANDE Collaboration: Contribution to the 7th Workshop
    on Grand Unification, ICOBAN'86, Toyama, Japan

34. M.R. KRISHNASWAMY et.al.: Phys.Lett. $\underline{106B}$, 339 (1981) $\underline{115B}$,349(1982)

35. G. BATTISTONI et.al.: Phys. Lett. $\underline{133B}$, 454 (1983)

36. G. BATTISTONI et.al.: Contribution to the XXII Int.Conf. on High
    Energy Physics, Vol.I, 246, Leipzig, July 1984

37. J. PATI, A. SALAM, B. SREEKANTAN: I. J. Mod. Phys. A1, 147 (1986)

Proton Decay Rate and Monte Carlo Simulation
in Lattice QCD

Yasuo Hara

Institute of Physics, University of Tsukuba,
Ibaraki 305 Japan

Proton decay rates in grand unified theories are calculated by using numerical results obtained by Monte Carlo simulation in lattice QCD.

For simplicity I discuss only $p \to \pi^0 e^+$ and $n \to \pi^0 \bar{\nu}_e$ in simple SU(5) model[1]. Application to other decay modes and to other grand unified theories[2] is straightforward.

In simple SU(5) model proton decay is mediated by superheavy leptoquark gauge bosons X and Y, and the effective Lagrangian which induces $p \to \pi^0 e^+$ and $n \to \pi^0 \bar{\nu}_e$ is given by

$$L_{eff} = (g^2/m_X^2) \, \varepsilon_{ijk} [2(\bar{u}_{kR}^c d_{iL})(\bar{e}_L^+ u_{jR})$$

$$+ (\bar{u}_{kL}^c d_{iR})(\bar{e}_R^+ u_{jL} + \bar{\nu}_{eR}^c d_{jL})] \, , \tag{1}$$

if we neglect the Higgs conribution and generation mixing.

Radiative corrections due to virtual hard gluons with momenta between μ and m_X renormalize the effective Lagrangian by a factor[3]

$$A_3 = \left[\frac{\alpha_3(\mu)}{\alpha_3(m_X)} \right]^{2/(11-2N/3)} \tag{2}$$

in one-loop order, where N is the number of quark flavor.

Hence the effective Lagrangian renormalized at $q^2 = \mu^2$ is expressed as

$$(L_{eff})_\mu = 4\pi(\alpha_{GUT}/m_X^2) A_3 \varepsilon_{ijk} [2A_{12}^L (\bar{u}_{kR}^{-c} d_{iL})(\bar{e}_L^{-+} u_{jR})$$

$$+A_{12}^R (\bar{u}_{kL}^{-c} d_{iR})(\bar{e}_R^{-+} u_{jL} + \bar{\nu}_{eR}^{-c} d_{jL})] \quad , \tag{3}$$

where A_{12}^L and A_{12}^R are renormalization factors due to the electroweak interaction ($A_{12}^L \approx 1.6$ and $A_{12}^R \approx 1.5$)[4], and $\alpha_{GUT} = \alpha_3(m_X) \approx 1/41$.

Now the strong interaction problem is to calculate $<\bar{\ell}, \text{meson}|(L_{eff})_\mu|N>$ with appropriate value of μ.

In this talk I assume the validity of current algebra and PCAC relation[5]. In this model all nucleon$\rightarrow$antilepton + pseudoscalar meson ($N \rightarrow \bar{\ell}+P$) decay amplitudes are expressed as sums of current commutator terms and nucleon pole terms (Fig. 1). For example,

$$<\pi^0\bar{\ell}|(L_{eff})_\mu|N> \approx (-i/f_\pi \sqrt{E_\pi})<\bar{\ell}|[Q_5^{(3)}, (L_{eff})_\mu]|N>$$

$$+ \text{ nucleon pole term} \quad , \tag{4}$$

where $f_\pi = 132$ MeV and $Q^{(3)} = \int d^3x q^\dagger(x)\gamma_5(\lambda_3/2)q(x)$.

Both current commutator terms and nucleon pole terms are related to three-quark annihilation matrix element (Fig. 2),

$$< 0|\varepsilon_{ijk}[u_k^T(0)C^{-1}(\gamma_5+\text{const.})d_i(0)]u_j(0)|p;\vec{q}_p=0>$$

$$=<0|\varepsilon_{ijk}[u_k^T(0)C^{-1}\gamma_5 d_i(0)]u_j(0)|p;\vec{q}_p=0>\equiv 2\beta u_p(0) \quad , \tag{5}$$

where $u_p(0)$ is the Dirac spinor of a proton with zero momentum, and space reflection invariance is used.

If we assume the validity of current algebra and PCAC relation, $N \rightarrow \bar{\ell}\pi^0$ decay amplitudes and widths are expressed as

$$(\alpha_{GUT}/m_X^2)\,A\beta\,[\bar{u}_\ell(b+d\gamma_5)u_N\,] \tag{6}$$

and

$$\Gamma \simeq (\alpha_{GUT}/m_X^2)^2 A^2\beta^2 m_N(b^2+d^2)/16\pi \;, \tag{7}$$

respectively, where $A \equiv A_{12}^R A_3 \simeq A_{12}^L A_3$,

$$b = -d/3 = \sqrt{2}\pi(1+g_A)/f_\pi \quad \text{for } p\to\pi^0 e^+ \tag{8}$$

and

$$b = d = \sqrt{2}\pi(1+g_A)/f_\pi \quad \text{for } n\to\pi^0\bar{\nu}_e \;. \tag{9}$$

In (8) and (9) terms proportional to $g_A=1.25$ are the nucleon pole terms and the rests are the current commutator terms.

While there is an objection against this model based on the fact that the extrapolation from the soft pion limit to the physical region is very long, i.e., $E_\pi/m_N \simeq 1/2$, there are arguments that the model may be at least semi-quantitative[6]. I consider that the model is reliable semi-quantitatively with a correction factor of $\sim 2^{\pm 1}$ since we find that the magnitude of the matrix element is significant and not suppressed at all at the end of the extrapolation (see, Fig. 5).

In the past bag models, nonrelativistic SU(6) model and other models with $\mu \simeq 1$ GeV as a typical hadronic scale were used to calculate the matrix element β. However, in these models it is difficult to calculate the matrix element β reliably. For example, there is no reliable connection to the renormalization factor A_3 in these models.

In the following I report present status of our information on the parameters, β, A_3 and m_x based on the numerical results of Monte Carlo simulation in lattice QCD obtained by Y. Iwasaki, T. Yoshié and S. Itoh[7),8)]

(1)β; The matrix element β, which is related to the wave function of the nucleon in the quark model $\psi_N[\vec{r}_a+\vec{r}_b-2\vec{r}_c,\vec{r}_a-\vec{r}_b]$ through $\psi_N(\vec{0},\vec{0})=(\sqrt{2}/3)\beta$ in the nonrelativistic limit, has been calculated by Iwasaki, Yoshié and Itoh in the quenched approximation to lattice QCD with a renormalization group improved lattice SU(3) gauge action and Wilson's quark action on a $16^3 \times 48$ lattice.

They have calculated propagators of hadrons[7)]

$$A(\tau) = \sum_{\vec{n}} <P(n)P^{\dagger}(0)> \qquad [n=(\vec{n},\tau),0=(\vec{0},0)] \tag{10}$$

in Euclidian lattice space, where $P(n)$ is an annihilation operator of a hadron at site n. For the proton the annihilation operator is

$$P(n) = \varepsilon_{ijk}[u_k^T(n)C^{-1}\gamma_5 d_i(n)]u_j(n) , \tag{11}$$

which corresponds to the operator $\varepsilon_{ijk}[u_{k\uparrow}(n)d_{i\downarrow}(n) -u_{k\downarrow}(n)d_{i\uparrow}(n)]u_j(n)$ in the nonrelativistic limit.

The proton propagators calculated for five values of the hopping parameter K=0.14, 0.145, 0.15, 0.1525 and 0.154 are shown in Fig. 3. The hopping parameter K is related to the bare quark mass m_0 through the relation,

$$2m_0a = \left(\frac{1}{K} - \frac{1}{K_c} \right) , \tag{12}$$

where a is the lattice spacing and K_c is the value of the hopping parameter for which the pion mass vanishes.

At K=0.1525 and 0.154 they have not been able to obtain values of the baryon propagators for large τ because of large fluctuation. Because of large fluctuation they

have not calculated the baryon propagators for the hopping
parameter between 0.154 and K_c.

They have fitted the hadron propagators for large τ
to $A_0 e^{-m\tau}$. The results for the masses are shown in Fig.4.
By assuming that the masses (mass squared for the pion)
are quadratic functions of $1/K$, they have determined

$$K_c = 0.1569(2) , \tag{13}$$

and

$$a^{-1} = 1810(60)\,\text{MeV} , \quad a = 0.109(4)\,\text{fm} \tag{14}$$

from the condition, $m_\rho(K=K_c)a=0.426(15)a=770$ MeV.

Thus, they have obtained theoretical values $m_p =$
1080(80) MeV and $m_\Delta=1370(120)$ MeV, which should be compared
with the experimental values, 940 and 1232 MeV, respectively.

In Fig. 5 we show the amplitudes of the proton propagators,

$$A_0 = (\text{volume}) |\langle 0| c_{ijk}[u_k^T(0)C^{-1}\gamma_5 d_i(0)]u_j(0)|\text{proton};$$

$$\vec{q}_p = 0\rangle|^2 = 4\beta^2 . \tag{15}$$

The expectation values and the statistical errors are
estimated by neglecting the statistical errors of the mass
at each K.

By **extrapolating** the results for A_0 to $K=K_c$ by assuming
that the amplitude A_0 is a quadratic function of $1/K$, we
obtain

$$\beta^2 = [0.029(\text{GeV})^3]^2 \times (1 \pm 0.44) . \tag{16}$$

(See, Fig.5). **As** is shown in Fig. 5, the amplitude
decreases as K increases. The behavior indicates that

attraction among three quarks at short distance decreases
as the bare quark mass decreases.

As a reference I show a compilation[6] of previous
calculations of β based on the bag model and other models
in Table 1. We find that our result (16) is much bigger
than the results in the bag model and quark model. These
models are not suitable for the estimation of three-quark
correlation in the nucleon, while they may be useful for
studying single-quark and two-quark properties of hadrons
such as magnetic moments and mass differences.

(2) A_3; In the past there was an ambiguity in the
choice of $\alpha_3(\mu)$ in A_3. This ambiguity is removed in
lattice QCD.

Since the decay amplitude is independent on the renormalization point μ, μ dependence of $(L_{eff})_\mu$, i.e.,
that of A_3, must be cancelled by μ dependence of the matrix
element β.

The matrix element β evaluated in lattice QCD,
$\beta_{lattice}$, is related to the matrix element β evaluated
in QCD with a renormalization point μ, β_μ, through a renormalization factor $[\alpha_3(lattice)/\alpha_3(\mu)]^\gamma$ $[\gamma=2/(11-2N/3)]$.
Thus, because of the relation,

$$A_3(\mu)\beta_\mu = \left[\frac{\alpha_3(\mu)}{\alpha_3(m_x)}\right]^\gamma \beta_\mu = \left[\frac{\alpha_3(lattice)}{\alpha_3(m_x)}\right]^\gamma \beta_{lattice}, \quad (17)$$

we have to use $\alpha_3(lattice)$ as $\alpha_3(\mu)$ in Eq. (2) for A_3 when
we use the lattice QCD to evaluate the matrix element β.

Since $\alpha_3(lattice)=0.20$ is used in the monte Carlo
simulation[7], the renormalization factor A_3

$$A_3 \approx [0.20/(1/41)]^{2/7} \approx 1.8 \qquad (18)$$

in our case.

(3) M_x; In simple SU(5) model m_x is predicted to be[9]

$$m_x = 1.3 \times 10^{14} \times (1.5)^{\pm 1} [\Lambda_{\overline{MS}}/100 \text{ MeV}] \text{GeV} \tag{19}$$

for three generations, a single Higgs doublet and $m_t = 50$ GeV.

The strong scale parameter in the minimum subtraction scheme $\Lambda_{\overline{MS}}$ has been estimated to be[8]

$$\Lambda_{\overline{MS}} = (99 \pm 1) \text{ MeV} \tag{20}$$

by studying the string tension by Monte Carlo simulation in lattice QCD in the quenched approximation (with $\sqrt{\sigma} = 420$ MeV).

The result (20) does not include systematic errors due to the quenched approximation, finite lattice size etc. By assigning ample systematic error to (20), I conclude

$$(100 \text{ MeV}) \times (2/3) \lesssim \Lambda_{\overline{MS}} \lesssim (100 \text{ MeV}) \times (3/2) . \tag{21}$$

The result (21) is consistent with recent experimental values[10] of $\Lambda_{\overline{MS}}$.

The matrix element β and the renormalization factor A_3 are independent on the grand unified theories, while m_x is dependent on the grand unified theories.

By making use of the estimated values of the parameters, β and A_3, (16) and (18), we find

$$\Gamma(p \to e^+ \pi^0) = (5\pi/4)(1+g_A)^2 A^2 \beta^2 (m_N/f_\pi^2)(\alpha_{GUT}/m_x^2)^2 \text{GeV}^5$$

$$= 4.1 \times 10^{-3} \times (4)^{\pm 1}/m_x^4 (\text{GeV}^5) , \tag{22}$$

$$\Gamma(n \to \bar{\nu}_e \pi^0) \approx (1/5)\Gamma(p \to e^+ \pi^0) , \tag{23}$$

where we have assumed that the errors associated with the decay amplitudes (8) and (9) in the current-algebra PCAC model are $\sim 2^{\pm 1}$ and that the sum of systematic and statistical errors of $\beta^2 = (0.029 \text{ GeV}^3)^2$ is also $\sim 2^{\pm 1}$.

From (19) and (22) we obtain

$$\tau(p \to e^+ \pi^0) = 1.45 \times 10^{27} \times (4)^{\pm 1} (m_x/1.3 \times 10^{14} \text{ GeV})^4 y$$

$$= 1.45 \times 10^{27} \times (20)^{\pm 1} (\Lambda_{\overline{MS}}/100 \text{ MeV})^4 y \; . \tag{24}$$

From (21) and (24) we obtain

$$\tau(p \to e^+ \pi^0) < 1.5 \times 10^{29} y \; . \tag{25}$$

By comparing this result with recent experimental result on the proton decay nate[11],

$$\tau(p \to e^+ \pi^0) > 3.3 \times 10^{32} y \; , \qquad (90 \text{ \% C.L.}) \tag{26}$$

we find that the simple SU(5) model with three generations, a single Higgs doublet and $m_t = 50$ GeV is excluded.

Even for $\Lambda_{\overline{MS}} < 400$ MeV we find

$$\tau(p \to e^+ \pi^0) < 7.4 \times 10^{30} y \; , \tag{27}$$

and (27) is still incompatible with the experimental result (26).

Even if we allow existence of various Higgs particles with mass m_x, they do not prolong the life of the proton by more than[9] $\times 150$ and the SU(5) model with $\Lambda_{\overline{MS}} < 150$ MeV is still incompatible with the experimental result (26).

I thank Y. Iwasaki, T. Yoshié, and S. Itoh for informing me of their numerical results on the proton propagators before publication.

Authors	$\beta\,(\text{GeV}^3)$
Donoghue and Golowich	0.003
Thomas and McKellar	0.02
Milosevic, Tadic, Trampetic	0.005
Ioffe	0.009
Krasniikov, Pivovarov, Tavkhelidze	0.006
Ioffe and Smilga	0.006
Tomozawa	0.009
Brodsky, Ellis et al.	0.03
This paper	0.03

Table 1. Compilation of calculations of β. The first three values are obtained by making use of the bag model. The results depend sensitively on the size of the bag assumed. Next three values are obtained using QCD and finite energy sum rules. Tomozawa's result is based on the relativistic quark model with harmonic oscillator potential ($\hbar w = 250$ MeV and $m_u = 330$ MeV). Brodsky et al's value is based on the knowledge about the short-distance structure of baryon wave functions gleaned from QCD form factor calculations and the $J/\psi \to p\bar{p}$ decay rate.

References

1) H. Georgi and S.L. Glashow, Phys. Rev. Lett. 32 (1974) 438.

2) As a review,
P. Langacker, Phys. Rep. 72 (1981) 185,
P. Langacker, Proc. 1985 Int. Symp. on Lepton and Photon Interactions at High Energies, eds. M. Konuma and K. Takahashi (Kyoto, 1985), p.186.

3) A.J. Buras, J. Ellis, M.K. Gaillard and D.V. Nanopoulos, Nucl. Phys. B135 (1978) 66.

4) J. Ellis, M.K. Gaillard and D.V. Nanopoulos, Phys. Lett. 88B (1979) 320,
F. Wilczek and A. Zee, Phys. Rev. Lett. 43 (1979) 1571.

5) Y. Tomazawa, Phys. Rev. Lett. 46 (1981) 463,
V.S. Berezinsky, B.L. Ioffe and Ya.I. Kogan, Phys. Lett. 105B (1981) 33,
M. Wise, R. Blankenbecler and L.F. Abbott, Phys. Rev. D23 (1981) 1591.

6) S.J. Brodsky, J. Ellis, J.S. Hagelin and C.T. Sachrajda, Nucl. Phys. B238 (1984) 561.

7) S. Itoh, Y. Iwasaki and T. Yoshié, Univ. of Tsukuba preprint, UTHEP-155.

8) S. Itoh, Y. Iwasaki and T. Yoshié, Univ. of Tsukuba preprint, UTHEP-154.

9) W.J. Marciano, Proc. Fourth Workshop on Grand Unification, ed .H.A. Weldon et al. (Birkhauser, Boston, 1983), p.13.

10) G.P. Lepage, Proc. 1983 Int. Symp. on Lepton and Photon Interactions at High Energies, eds. D.G. Cassel and D.L. Kreinick (Cornell, 1983), p.565,
F. Sciulli, Proc. 1985 Int. Symp. on Lepton and Photon Interactions at High Energies, eds. M. Konuma and K. Takahashi (Kyoto, 1985), p.8.

11)   Y. Totsuka, Proc. 1985 Int. Symp. on Lepton and Photon
      Interactions at High Energies, eds. M. Konuma and
      K. Takahashi (kyoto, 1985), p.120.

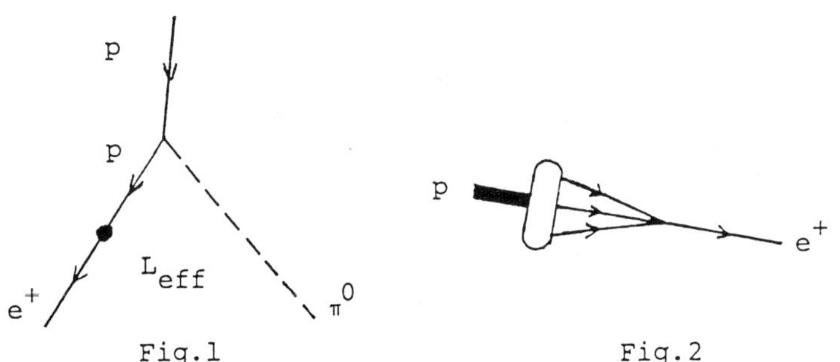

Fig.1

The proton pole term

Fig.2

The three-quark annihilation
matrix element

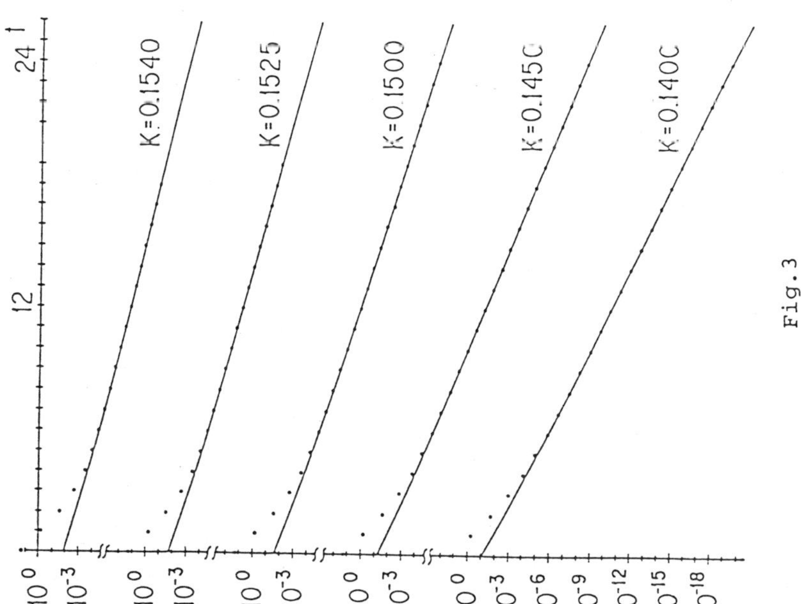

Fig.3

ma   (m_π a)²
1.6
1.4
1.2
1.0
0.8
0.6
0.4
0.2
m_Δ
m_P
m_ρ
m²_π
m_Δ
m_p
6.4
6.5
6.6
6.7
6.8
6.9
7.0
7.1
1/K_c
1/K

Fig. 4

β² × 10³ (GeV⁶)
10
5
1
6.5
7.0
1/K_c
1/K

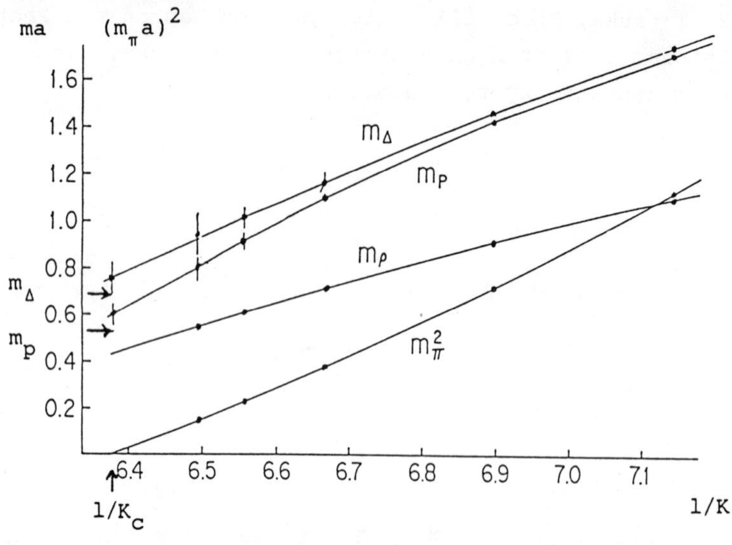

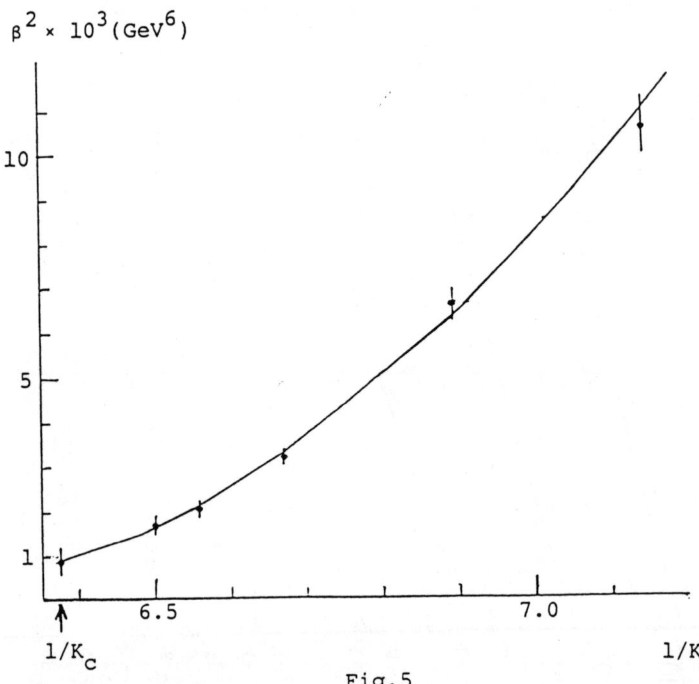

Fig.5

PROTON DECAY IN UNIFIED THEORIES
WITH SUPERSYMMETRY AND SUPERGRAVITY

Pran Nath[*] and R. Arnowitt
Northeastern University, Boston, Massachusetts 02115 U.S.A.

ABSTRACT

An analysis of proton decay in unified theories with supersymmetry and supergravity is presented.

1. INTRODUCTION

Proton decay experiments[1] have progressed to a point where it is reasonable to conclude that the minimal SU(5) GUT model[5] is safely ruled out. The characteristic signal of proton decay in the minimal SU(5) GUT model is the decay mode $p \rightarrow e^+ \pi^0$ which theroetically has the upper limit

$$\tau_{p \rightarrow e^+ \pi^0} \leq 2.3 \times 10^{31} \text{ yr (Minimal SU(5) GUT)} \qquad (1.1)$$

The latest experimental data for this decay mode is shown in Table I[3] which gives a combined lower limit of

$$\tau_{p \rightarrow e^+ \pi^0} \geq 3.6 \times 10^{32} \text{ yr (Experimental)} \qquad (1.2)$$

Table I. $p \rightarrow e^+ \pi^0$ Lifetime.

Experiment	Lower Limit on $p \rightarrow e^+ \pi^0$ Life-time
KGF	2×10^{31} yr
NUSEX	1.5×10^{31} yr
IMB	25×10^{31} yr
Kamiokande	8.4×10^{31} yr

From Eqs. (1.1) and (1.2) we find that the theoretical upper limit of the minimal SU(5) GUT for the $p \rightarrow e^+ \pi^0$ decay mode lies below the experimental lower limit. Thus the minimal SU(5) GUT is ruled out. Further, the experimental constraint of Eq. (1.2) is not easily overcome in other GUT models.[6]

[*] Conference speaker.

2. SUPERSYMMETRIC GUTS

Supersymmetry provides a resolution to the experimental proton decay constraint of Eq. (1.2). In supersymmetry the interesting proton decay[7,9] proceeds via dimension five operators mediated by the heavy Higgs fields.[10,11] We consider here an SU(5) SUSY GUT theory with a superpotential for Yukawa interactions of the form

$$g_Y = -\frac{1}{8} \varepsilon_{uvwxy} H^u M_i^{vw}(f_1^+)_{ij} M_j^{xy} + H_x' M_{yi}' \cdot (f_2^+)_{ij} M_j^{xy} \tag{2.1}$$

where H^x and H_x' are 5_L and $\bar{5}_L$ of Higgs fields, and M^{xy} and M_y' are 10_L and $\bar{5}_L$ of quark-lepton fields. The indices $i,j=1,2,3$ are generation indices while f_1, f_2 are Yukawa-coupling-constant matrices. After spontaneous breaking of SU(5) at the GUT mass into SU(3)×SU(2)×U(1) one eliminates the heavy Higgs triplet fields to obtain the baryon-number-violating dimension five operator

$$\mathcal{L}_5 = \mathcal{L}_5^L + \mathcal{L}_5^R \tag{2.2}$$

where $\mathcal{L}_5^L$ contains left-handed matter and SUSY-matter fields in the form LLLL while $\mathcal{L}_5^R$ contains the right-handed matter and SUSY-matter fields in the form RRRR. $\mathcal{L}_5^L$ is given by[12-14]

$$\mathcal{L}_5^L = \frac{\varepsilon_{abc}}{M} (Pf_1^u V)_{ij}(f_2^d)_{k\ell} \{\tilde{u}_{Lbi}\tilde{d}_{Lcj}[\bar{e}_{Lk}^c(Vu_L)_{a\ell} - \bar{\tilde{v}}_k^c d_{La\ell}]$$

$$+ \bar{u}_{Lbi}^c d_{Lcj}[\tilde{e}_{Lk}(V\tilde{u}_L)_{a\ell} - \tilde{v}_k \tilde{d}_{La\ell}] + \bar{u}_{Lbi}^c d_{Lcj}[e_{Lk}(V\tilde{u}_L)_{a\ell} - v_k \tilde{d}_{La\ell}]$$

$$+ \bar{u}_{Lbi}^c \tilde{d}_{Lcj}[\tilde{e}_{Lk}(Vu_L)_{a\ell} - \tilde{v}_k d_{La\ell}] + \tilde{u}_{Lbi}\bar{d}_{Lcj}^c[e_{Lk}(V\tilde{u}_L)_{a\ell} - v_k \tilde{d}_{La\ell}]$$

$$+ \tilde{u}_{Lbi}\bar{d}_{Lcj}^c[\tilde{e}_{Lk}(Vu_L)_{a\ell} - \tilde{v}_k d_{La\ell}]\} + H.c. \tag{2.3}$$

In Eq. (2.3) M is the Higgs triplet mass, V is the K-M matrix and P is a diagonal phase matrix while $f^{u,d}$ are diagonal Yukawa matrices which are related to the up- and down-quark masses m_i^u, m_i^d.

Similarly $\mathcal{L}_5^R$ is given by

$$\mathcal{L}_5^R = -\frac{\varepsilon_{abc}}{M}(V^\dagger f^u)_{ij}(PVf^d)_{k\ell}(\bar{e}_{Ri}^c u_{Raj}\tilde{u}_{Rck}\tilde{d}_{Rb\ell}+\tilde{e}_{Ri}\tilde{u}_{Raj}\bar{u}_{Rck}^c d_{Rb\ell}$$

$$+\tilde{e}_{Ri}\bar{u}_{Raj}^c\tilde{u}_{Rck}d_{Rb\ell}+\bar{e}_{Ri}^c\tilde{u}_{Raj}\tilde{u}_{Rck}d_{Rb\ell}+\tilde{e}_{Ri}\bar{u}_{Raj}^c u_{Rck}\tilde{d}_{Rb\ell}$$

$$+\bar{e}_{Ri}^c\tilde{u}_{Raj}u_{Rck}\tilde{d}_{Rb\ell})+\text{H.c.}\tag{2.4}$$

The dimension five operators of Eqs. (2.2), (2.3) and (2.4) are dressed by gaugino exchange to generate the dimension-six nucleon-decay Lagrangian. The computation of the dressing diagrams are carried out in the framework of the low energy supergravity models which are described briefly in Sec. 3.

3. N=1 SUPERGRAVITY MODELS

We assume that after elimination of the superheavy fields and the heavy-Higgs fields, one is left with an effective low energy theory which contains a pair of light Higgs doublets H_α', H^α and possibly also a Higgs singlet U so that the low energy effective superpotential has the form[12,13,15,16]

$$g(Z) = \mu H^\alpha H_\alpha' + \lambda' U H^\alpha H_\alpha' - \frac{1}{6}\lambda'' U^3 + g_Y \ , \tag{3.1}$$

where g_Y is the Yukawa superpotential. N=1 Supergravity models depend on the following (soft breaking) parameters in the low energy domain:[15] the gravitino mass $m_{3/2}$, the Higgs-mixing parameter μ, the Polony constant A and gaugino masses $\tilde{m}_\gamma$ (photino) and $\tilde{m}_g$ (gluino). After SU(2×U(1) breaking, the low energy theory would also contain the new parameter α_H defined by $\tan\alpha_H = \langle H_5'\rangle/\langle H^5\rangle$. α_H is determined by the mechanism of SU(2)×U(1) breaking. In models where SU(2)×U(1) breaking occurs at the tree level (T.B. models)[17] and in the dimensional transmutation (D.T.)[18] and "no-scale" (N.S.) models[19], one has $\alpha_H \simeq 45°$. In renormalization group (R.G.) models[20] where SU(2)×U(1) breaking is induced by the effects of a heavy top quark through renormalization group corrections from the GUT mass to the W mass, α_H is generally small, e.g., $\alpha_H \sim 10° - 25°$ Sometimes it is convenient to categorize supergravity models as models with small D-terms (such as the T.B., D.T. and N.S. models) or models with large D-terms (such as the R.G. model).

4. DRESSING OF DIMENSION-FIVE OPERATORS

In previous analyses[9,11] the dressing of dimension-five operators has been carried out using only Wino exchange and two generations of s-matter fields in the dressing loop integrals. In such calculations $\bar{\nu}K$ modes dominate[9,21] the nucleon decay. However, there exist a number of additional effects which can significantly modify this result. These are: (i) Wino dressing diagrams including the third generation effects in the dressing loop integrals, (ii) L-R mixing effects in the squark mass matrices arising from the soft breaking terms, (iii) contributions of the RRRR dimension five operators and (iv) gluino and Zino dressing contributions.

The effects of (i), (ii) and (iii) can be significant if the top quark mass obeys $m_t \gtrsim 30$ GeV. Regarding (iv), the gluino contributions cancel among each other if the up and the down squarks are degenerate in the first two generations but can contribute[22] when the squark mass degeneracy is broken since in general

$$\tilde{m}_{dL}^2 - \tilde{m}_{uL}^2 = \cos^2\theta_W \cos 2\alpha_H M_Z^2 \qquad (4.1)$$

Thus the gluino dressing contributions are enhanced in models with large D-terms and suppressed in models with small D-terms. The situation is quite the opposite for zino dressing contributions[12,13]. In models with large D-terms one has small zino dressing contributions while sizable zino dressing contributions can arise in models with small D-terms.

5. NUCLEON DECAY AMPLITUDES

We discuss now the full analysis including the Wino, Zino and gluino dressing exchange diagrams with LLLL and RRRR dimension five operators and third generation effects. The dimension six operator governing the process $N \to \bar{\nu}_i K (i=e,\mu,\tau)$ is given by[12,13]

$$\mathcal{L}_6(N \to \bar{\nu}_i K) = [(\alpha_2)^2 (2MM_W^2 \sin2\alpha_H)^{-1} P_2 m_c m_i^d V_{11}^\dagger V_{21} V_{22}][F(\tilde{c};\tilde{d}_i;\tilde{W})$$

$$+ F(\tilde{c};\tilde{e}_i;\tilde{W})] \times \{[1+y_i^{tK}+(y_{\tilde{g}}+y_{\tilde{Z}})\delta_{i2}+\Delta_i^K]\alpha_i^L$$

$$+ [1+y_i^{tK} - (y_{\tilde{g}}-y_{\tilde{Z}})\delta_{i2}]\beta_i^L + (y_1^{(R)}\alpha_3^R + y_2^{(R)}\beta_3^R)\delta_{i3}\}, \qquad (5.1)$$

where $\alpha_i^L \equiv \varepsilon_{abc}(d_{aL}\gamma^0 u_{bL})(s_{cL}\gamma^0 \nu_{iL})$. In Eq. (5.1) α_i^R is α_i^L with $(d_L, u_L \to d_R, u_R)$ and $\beta_i^{L,R}$ is $\alpha_i^{L,R}$ with $d \leftrightarrow s$. y_i^{tK} is the third-generation contribution with Wino dressings and is given by[25]

$$y_i^{tK} = \frac{P_3}{P_2} \left[\frac{m_t V_{31} V_{32}}{m_c V_{21} V_{22}} \right] \left[\frac{F(\tilde{t};\tilde{d}_i;\tilde{W})+F(\tilde{t};\tilde{e}_i;\tilde{W})}{F(\tilde{c};\tilde{d}_i;\tilde{W})+F(\tilde{c};\tilde{e}_i;\tilde{W})} \right] \qquad (5.2)$$

In Eqs. (5.1) and (5.2) F are the dressing loop integrals and $y_{\tilde{Z}}$ and $y_{\tilde{g}}$ are the dressing contributions arising from the Zino and the gluino exchange diagrams. y_1^R and y_2^R in Eq. (5.1) are the contributions from the RRRR dimension five operators. We note here that the RRRR dimension five operators contribute to Eq. (5.1) only through Wino dressings while the gluino and Zino dressing contributions from RRRR dimension five operators vanish identically due to their flavor diagonal couplings. The conventional result, i.e., the dominance of the $\bar{\nu}_i K$ modes in nucleon decay holds if one neglects all the correction terms y_i^t, $y_{\tilde{g}}$, etc. However, significant deviations from the conventional results can occur in certain domains of the parameter space. Consider specifically the third generation contribution y_i^{tK} in Eq. (5.1). From Eq. (5.2) one finds that it is the product of the top quark mass and the third generation KM matrix elements that enter the decay. Thus while the third generation KM matrix elements are small, their product including the top quark mass is not necessarily negligible. For example, for $m_t=50$ GeV one has $m_u V_{11} : m_0 V_{21} : m_t V_{31} \approx 1:60:250$. In fact there exist domains in the parameter space where $|y_i^{tK}| \sim 0(1)$. Further with $P_3/P_2 = +1$, the sign of y_i^{tK} is negative leading to a destructive inference between the second and the third generation effects. This has the effect of suppressing the $\bar{\nu}_i K$ modes. Simultaneously, in the decay channel $N \to \bar{\nu}_i \pi$, the third generation effects enter constructively enhancing the $\bar{\nu}_i \pi$ modes.

A qualitative statement of the conditions under which $\bar{\nu}_i K$ modes are suppressed and the $\bar{\nu}_i \pi$ modes are enhanced is given in Refs. (12) and (13) while a quantitative analysis has been performed in Ref. (26). Results of Ref. (26) are shown in Fig 1 where the contour maps of $\Gamma(p \to \bar{\nu}\pi^+)/\Gamma(p \to \bar{\nu}K^+)$ are given for the input parameters $\alpha_H = 15°$, $\mu = 20$ GeV, $m_{3/2} = 178$ GeV, $\tilde{m}_\gamma = 3$ GeV and $P_1/P_2 = -P_2/P_3 = -1$. From Fig. 1, we find that the $\bar{\nu}\pi$ modes are comparable to or dominate over the $\bar{\nu}K$ modes over a significant domain of the parameter space of the Polony constant A and the top quark mass m_t. The analysis of Ref. 26 also exhibits the phenomena of the enhancement of $\bar{\nu}\rho$ and $\bar{\nu}\omega$ modes and a relative suppression of the $\bar{\nu}K^*$ modes precisely in domains where the $\bar{\nu}\pi$ modes dominate the $\bar{\nu}K$ modes.

Nucleon decay branching ratios are exhibited in Table II for the input parameters $m_t = 50$ GeV, $\tilde{m}_\gamma = 3$ GeV, $m_{3/2} = 178$ GeV and $A = 1.2$.[27,29] We see from Table II that the non-strange modes $\bar{\nu}\pi$, $\bar{\nu}\rho$ and $\bar{\nu}\omega$ dominate over the strange modes $\bar{\nu}K$ and $\bar{\nu}K^*$. Further, while in supersymmetric theories the absolute lifetimes cannot be predicted theoretically (due to the arbitrariness of the Higgs triplet mass) one may use the experimental lower bound for one decay mode to predict the theoretical lower limits for all other modes. In Table II, the

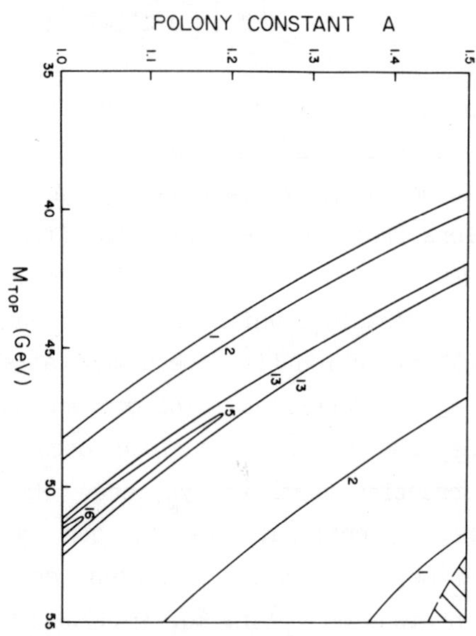

Fig. 1.   The ratio $\Gamma(b \to \bar{\nu}\pi^+)/\Gamma(b \to \bar{\nu}K^+)$.

Table II.

Decay Modes	Branching Ratios(%)	Theoretical Lifetimes (10^{31})yrs	Experimental Lower Bounds(10^{31})yrs
$P\to\bar{\nu}\pi^{+}$	66.2	0.4	0.4
$\to\bar{\nu}\rho^{+}$	18.8	1.6	0.9
$\to\bar{\nu}K^{+}$	14.4	2.0	1.5
$\to\bar{\nu}K^{*+}$	6.2×10^{-2}	4.8×10^{2}	–
$n\to\bar{\nu}\pi^{0}$	43.3	0.9	2.1
$\to\bar{\nu}K^{0}$	24.3	(1.6)	(1.6)
$\to\bar{\nu}\rho^{0}$	12.4	3.1	0.4
$\to\bar{\nu}\omega^{0}$	12.2	3.2	2.1
$\to\bar{\nu}K^{*0}$	0.6	64.8	–

experimental lower bound on $\eta\to\bar{\nu}K^{0}$ mode is used as an input. It is seen that the predicted theoretical lower bounds are close to and generally somewhat larger than the experimental lower bounds (Koshiba (Ref. 4)). The experimental lower bound on $p\to\bar{\nu}K^{+}$ can also be used to determine a lower bound on M/β. From the results of Table II one finds $M/\beta=2.0\times10^{18}$ GeV^{-2}. Theoretical evaluations of β lie in the range $\beta=0.003$ GeV^{3} to 0.03 GeV^{3} (see talk by Y. Hara at this Conference) which gives for M the bound

$$M \gtrsim (0.6-6.0)\times10^{16} \text{ GeV} \tag{5.3}$$

The evaluation of Eq. (5.3) is consistent with an upper bound on M found in Ref. 30 arising from an analysis of the heavy Higgs sector[7]

$$M \leq M_X \tag{5.4}$$

The possible existence of a fourth generation can further modify nucleon decay. This can happen in two ways. First a fourth generation would produce additional contributions through the fourth generation squark-slepton fields in the dressing loop diagrams. Second, there would be additional channels $\bar{\nu}_\tau$, π, $\bar{\nu}_\tau$, K, $\bar{\nu}_\tau$, ρ, etc., for nucleon decay for a massless fourth sequential neutrino. The relative strength of the fourth generation contributions depends of course on the fourth generation mass spectra and on the detailed nature of the fourth generation KM matrix elements.[31] However, the situation is not as arbitrary as might appear. For example, substantial nucleon decay contributions from the fourth generation require an enhancement of the lower limit of Eq. (5.3) (for nucleon stability) and conflict with Eq. (5.4). Thus the constraint of Eq. (5.4) restricts the acceptable fourth generation parameters such as the mass spectrum and the KM matrix elements.[32]

REFERENCES

1. For a review of proton decay experiments see Refs. 2-4.

2. H. Meyer, talk at this conference.

3. Y. Totsuka, in Proceedings of the 1985 International Symposium on Lepton and Photon Interactions at High Energy, edited by M. Konuma and K. Takahashi.

4. See talks by M. Koshiba (Kamiokande), J. LoSecco (Irvine-Michigan-Brookhaven), E. Fiorini [nucleon-stability experiment (NUSEX)], V.S. Narasimham (Kolar Gold Fields), D. Cline (Harvard-Pennsylvania-Wisconsin), and S. Julien (Frejus), in Proceedings of the XXII International Conference on High Energy PHysics, Leipzig, 1984, edited by A. Meyer and E. Wieczorek (Academie der Wissenschaften der DDR, Zeuthen, DDR, 1984).

5. H. Georgi and S.L. Glashow, Phys. Rev. Lett. <u>32</u>, 438 (1974).

6. See e.g., S. Dimopoulos and L. Hall, in Proc. of the XVI International Conf. on High Energy Physics, Leipzig, 1984.

7. Exchange of the heavy gauge bosons does not produce interesting proton decay in Supersymmetry. This arises from the fact that due to the contributions of the superpartners to the beta function the unification scale is enlarged so that $M_x \equiv 0(1) \times 10^{16}$ GeV. Thus the proton life time which is scaled by M_x^4 is now $\tau_p \sim 0(10^{37})$ yr and is outside the realm of observability.[8]

8. Neutrino interactions generated by cosmic rays introduce an upper limit on the experimentally accessible proton lifetime. This limit is estimated to be $\sim 0(1) \times 10^{33}$ yr. Realistic

simulations of the background may allow one to extend further the experimentally accessible upper limit on the proton lifetime. See in this context W.A. Mann, et al., ANL-HEP-PR-86-04/TUFT-HEP-8601 (1986).

9. For a review see W. Lucha, Fortschr. Phys. $\underline{33}$, No. 10 (1985) and UWThPh-1985-24; K. Enqvist and D.V. Nanopoulos, CERN-TH 4066/84.

10. S. Weinberg, Phys. Rev. $\underline{D26}$, 287 (1982); N. Sakai and T. Yanagida, Nucl. Phys. $\underline{B197}$, 533 (1982).

11. A detailed set of references can be found in Refs. 9 and 21.

12. R. Arnowitt, A.H. Chamseddine and P. Nath, Phys. Lett. $\underline{156B}$, 215 (1985).

13. P. Nath, A.H. Chamseddine and R. Arnowitt, Phys. Rev. $\underline{32D}$, 2348 (1985).

14. Eqs. (2.3) and (2.4) are in the representation where the quark-lepton mass matrices are diagonal.

15. For a review of supergravity models, see P. Nath, R. Arnotwitt and A.H. Chamseddine, "Applied N=1 Supergravity," (World Scientific Pub. Co., Singapore, 1984) and HUTP-83/A077-NUB#2488 (1983).

16. $H^{4,5}$ and $H'_{4,5}$ are the isodoublet components of the 5 and $\bar{5}$ of Higgs.

17. A.H. Chamseddine, R. Arnowitt and P. Nath, Phys. Rev. Lett. $\underline{49}$, 970 (1982); R. Barbieri, S. Ferrara and C. Savoy, Phys. Lett. $\underline{119B}$, 342 (1982).

18. C. Kounnas, A.B. Lahanas, D. Nanopoulos, and M. Quiros, Phys. Let. $\underline{127B}$, 82 (1983); Nucl. Phys. $\underline{B236}$, 438 (1984).

19. J. Ellis, A.B. Lahannas, D.V. Nanopoulos and K. Tamvakis, Phys. Lett. $\underline{134B}$, 429 (1984).

20. L. Ibanes, Phys. Lett. $\underline{118B}$, 73 (1982); J. Ellis, D.V. Nanopoulos, and K. Tamvakis, Phys. Lett. $\underline{121B}$, 123 (1983); L. Alvarez-Gaume, J. Polchinski and M.B. Wise, Nucl. Phys. $\underline{B221}$, 495 (1983).

21. P. Nath and R. Arnowitt, in Proc. of the Workshop on Physics Simulations at High Energy held at the University of Wisconsin-

Madison, May 5-16, 1986.

22. J. Milutinovic, P.B. Pal, and G. Senjanovic, Phys. Lett. <u>140B</u>, 215 (1983); see also, Refs. 23 and 24 and J. McDonald and C.E. Vayonakis, Phys. Lett. <u>144B</u>, 199 (1984).

23. R. Arnowitt, A.H. Chamseddine and P. Nath, in Proceedings of the Fifth Workshop on Grand Unification, Brown University, 1984, edited by K. Kang, H. Fried, and P. Frampton (World Scientific, Singapore, 1984).

24. S. Chadha and M. Daniel, Phys. Lett. <u>137B</u>, 374 (1984).

25. Δ_i^K are aditional first- and third- generation contributions. Δ_i^K are generally small.

26. T.C. Yuan, Phys. Rev. <u>D33</u>, 1894 (1986).

27. We use the Chiral-Lagrangian approach of Ref. 28 to convert the quark decay amplitudes to nucleon decay rates. Here the nucleon decay rates are determined in terms of two hadronic parameters α and β. For example β is the three quark matrix element of the nucleon wavefunction u_L^γ ($\gamma \equiv 1.2$) such that

$$\varepsilon_{abc} \langle 0|\varepsilon_{\alpha\beta}\, d_{aL}^\alpha\, u_{bL}^\beta\, u_{cL}^\gamma|p\rangle = \beta u_L^\gamma$$

28. M. Claudson, M.B. Wise and L. Hall, Nucl. Phys. <u>B195</u>, 43 (1982).

29. The analysis of nucleon decay branching ratios is given in Refs. 12,13 and 26. The results of Table II are from Ref. 26 which also includes the vector boson decay modes.

30. K. Enqvist, A. Masiero, and D.V. Nanopoulos, Phys. Lett. <u>156B</u>, 209 (1985).

31. KM matrix with four generations has been investigated by a number of authors. See, e.g., X. He and S. Pakvasa, Phys. Lett. <u>156B</u>, 236 (1985) and references quoted therein.

32. A full analysis of the effects of the fourth generation on nucleon decay shall be presented elsewhere.

LEPTON FLAVOR CONSERVATION TESTS WITH MUONS

M. Blecher[*]

Virginia Polytechnic Institute and State University
Blacksburg, Virginia 25061, USA

INTRODUCTION

It has often been remarked that it is a pity the standard model works so well. Although it is in agreement with experiment there is no satisfactory explanation for many of its features: the presence of adjustable constants, the three observed but possible additional flavors of leptons and quarks, the mass spectrum, and the assumption of zero mass neutrinos.

Massless neutrinos are assumed to account for the null branching ratios of lepton flavor nonconserving processes such as:

$$R_\gamma = \mu \to e\gamma/\mu \to \text{decay} \tag{1}$$

$$R_e = \mu \to eee/\mu \to \text{decay} \tag{2}$$

$$R_2 = \mu \to e\gamma\gamma/\mu \to \text{decay} \tag{3}$$

$$R_- = \mu^- Z \to e^- Z/\mu^- \to \text{capture} \tag{4}$$

$$R_+ = \mu^- Z \to e^+(Z-2)/\mu \to \text{capture} \tag{5}$$

$$R_L = K_L \to \mu e/K_L \to \text{decay} \tag{6}$$

$$R_K = K^+ \to \pi^+\mu e/K^+ \to \text{decay} \ . \tag{7}$$

Many other examples exist, but these branching ratios provide the most restrictive limits, and new results are or will soon be available. Small neutrino masses would allow these reactions to proceed, but at rates far below present detection levels. Extensions of the standard model, which predict lepton flavor nonconservation, require additional particles, some with mass far too high to be produced with present or proposed accelerators. Thus these low energy experiments, performed with ever greater sensitivity, are potential indicators of very high energy phenomena.

[*]On leave at TRIUMF, 4004 Wesbrook Mall, Vancouver, B.C., Canada V6T 2A3   Supported by USNSF· grant Phys-8503845.

Altarelli et al.[1] examined the effect of an additional lepton generation with a heavy neutrino, while keeping the same gauge algebra.
For muon reactions (1), (2), and (4), muon electron conversion, reaction (4), was the most sensitive process due to the coherence effect in
nuclei. This reaction could proceed as shown in Fig. 1a. Marciano and
Sanda[2] also examined the muon processes. They considered gauge theories
which allow their occurrence through one loop diagrams involving intermediate charged and neutral leptons. Possible muon-electron conversion
diagrams are shown in Fig. 1. Models with left- or right-handed currents as well as left-right mixing were considered. In all cases $R_- >$
R_γ and R_e. However, the ratio R_γ/R_e was model dependent. Thus,
to determine the correct model, and the necessary coupling constants
and masses, all the muon reactions need to be studied.

Models without additional leptons but with exotic particles are
also of interest. Shanker[3] and Ellis[4] studied the effects of an
expanded Higgs sector, while technicolor models in which the reactions
could proceed by exchange of leptoquarks have been studied by Shanker[3]
and Hall and Randall.[5] From Shanker possible diagrams are shown in
Fig. 2. As no quarks are involved in the muon decays these reactions
do not provide information on leptoquark models to leading order. Cahn
and Harari[6] grouped quarks and leptons in generation multiplets:
$(G=1;\nu_e,e,u,d)$, $G=2;\nu_\mu,\mu,c,s)$, and $(G=3;\nu_\tau,\tau,t,b)$. An overall gauge
theory $G \times H$ was assumed, where G contains the standard model algebra
and H is an $SU(2)$ algebra that connects different generations through
exchange of horizontal gauge bosons. Possible diagrams, grouped
according to whether or not the process is diagonal (the gauge boson
connects quarks to quarks or leptons to leptons) and the magnitude of
the generation change, δG, are shown in Fig. 3. In these models
reaction (4) is also highly favored and at present it sets most of the
mass limits for the exchange particle.

Muon positron conversion, reaction (5), is of interest because in
addition to flavor violation, lepton number is not conserved. This
process may be induced by a heavy Majorana neutrino.[7] If so, the
observation of a positive rate would imply nonzero rates for the other
muon processes while the reverse is not true as the others could

proceed whether or not the intermediate neutrino is Majorana. Zee[8] has shown that an enlarged Higgs sector may also bring about a positive rate for reaction (5). Muon positron conversion is not expected to be coherent and there are many possible final states. Due to the complicated nuclear physics no reliable model calculation of the positron momentum spectrum is currently available.

RECENT RESULTS

New results for the muon reactions (1)-(5) have recently become available. These results and their impact on the extensions of the standard model discussed above will be reviewed. No new results are available concerning the kaon decays. However, there are experiments proposed or running, the goals and impact of which will be briefly discussed.[9]

Throughout the years during which the muon processes have been studied remarkable progress has been made in the measurement of the upper limits for the branching ratios. For example, Fig. 4 shows that the upper limit for R_- has been dropping by about 2.5 orders of magnitude per decade. The new results have come from the meson factories: LAMPF in the United States, SIN in Switzerland, and TRIUMF in Canada.

The apparatus used by the SIN group[10] is shown in Fig. 5. A hollow double cone target is surrounded by 5 cylindrical wire chambers and a hodoscope of 64 scintillators. The experiment searched for $\mu \to eee$ with an efficiency of $\approx 14\%$ at a μ^+ stop rate of $\approx 10^7$/s by seeking triple coincidences with negative particle production. The three charged particle tracks were reconstructed and candidate events sought from among those for which:

$$\Sigma \ \overline{P} = 0 \tag{8}$$

$$\Sigma \ E = m_\mu c^2 \ . \tag{9}$$

The experiment has an unavoidable background,

$$\mu \to e\nu\nu ee \ , \tag{10}$$

the rate of which is below the sensitivity of reaction (2) in the region where Eqs. (8) and (9) are satisfied. In fact 7300 such background events were observed and shown to be consistent with predictions of the standard model. The experimental spectrum near the endpoint region is shown in Fig. 6a and a Monte Carlo spectrum of reaction (2) events, assuming a constant amplitude, in Fig. 6b. Also shown are the 68% and 90% contours. The result of the experiment was that no candidate event was found and $R < 2.4 \times 10^{-12}$ (90% C.L.). The experiment was not background limited and more data-taking leading to a final limit, $R < 10^{-12}$, is expected.

The LAMPF experimental setup[11] used to search for $\mu \to e\gamma$ is shown in Fig. 7. An 8-plane stereo drift chamber, a 36 scintillation counter hodoscope, and 396 NaI(Tl) crystals view a thin stopping target. A new upper limit for reaction (1) was produced at a μ^+ stop rate of $<0.5 \times 10^6$/s and efficiency of $\approx 30\%$. The trigger requirement was 1 track (hit) in the drift chamber (hodoscope) and 2 NaI hits in coincidence. Candidate events should satisfy Eqs. (8) and (9). However, events were accumulated if $E_e, E_\gamma > 44, 40$ MeV, $E_e + E_\gamma + |\overline{P}_e + \overline{P}_\gamma| < 135$ MeV, and $\theta_{e\gamma} > 160°$, where the photon direction was determined from the conversion point and the target position of the positron.

This experiment must also contend with an unavoidable background, radiative muon decay,

$$\mu \to e\nu\nu\gamma \ . \tag{11}$$

However, as LAMPF is a low duty cycle ($<10\%$) facility the main background comes from random coincidences between photons from reaction (11) and Michel positrons near their end point energy.

The timing spectrum shown in Fig. 8 contains a prompt peak in which candidate events are searched for sitting atop a broad random background. An upper limit for R_γ was obtained using the method of maximum likelihood together with the measured random spectra, the standard model prediction for the radiative decay background, and the detector response. The likelihood function is shown in Fig. 9. The 90% confidence level occurs for $N_{e\gamma} < 11$ events yielding $R_\gamma < 4.9 \times 10^{-11}$. The LAMPF group has also produced the current upper limit for reaction (3), $R_2 < 3.8 \times 10^{-10}$ (90% C.L.).[12]

The TRIUMF detection system for muon electron (positron) conversion centered around a time projection chamber[13] (TPC) viewing an extended shredded Ti target as shown in Fig. 10. Scintillation counters and wire chambers interior and exterior to the TPC served as event triggers and cosmic-ray vetos. Three-dimensional position information was obtained for up to 12 segments of a charged particle's track. Reactions (4) and (5) were simultaneously studied. Most of the data taking occurred at a stop rate of $\approx 10^6$/s and the total number of muon captures accumulated was $N \approx 9 \times 10^{12}$. For reaction (4), expected to be coherent, electrons with $P \approx 101$ MeV/c including energy loss were searched for. An efficiency of $\approx 6\%$ was achieved. For reaction (5) an excess of positrons with $P > 90.5$ MeV/c over those expected from the background served as the reaction signature. The momentum dependence of the acceptance and more selective cuts to remove background events reduced the efficiency to $\approx 0.4\%$ in this case.

Figure 11 shows the information obtained for each event. The xy coordinates of the track and the mean energy loss in the TPC are deduced from the pulse height distribution on the cathode pads of each struck wire, while the z coordinates are determined from the drift times to the wires. The charge is determined from the curvature of the track in the xy plane, and the momentum from a helical fit to the track coordinates. A χ-squared per degree of freedom is calculated for use as a goodness of fit indicator. The projection of the track to the target and angle of the track with respect to the beam are helpful in elimination of background events. The projected position of the track in the internal and external wire chambers can be compared with the measured positions, which also aids in background reduction. The scintillators and wire chambers in which energy is deposited are also indicated. Multiple counters or tracks in two TPC sectors are indications of cosmic rays. In this particular event there are no multiples and the cosmic ray veto counters shown as short solid and dashed lines outside the TPC have not fired. Pulse heights in the struck counters also provide energy loss information aiding in the identification of electrons.

In this experiment there were two sources of unavoidable background: (1) μ decay from bound atomic orbits (the probable source of the 94 MeV/c event shown in Fig. 11),

$$"\mu" \rightarrow e\nu\nu \, , \tag{12}$$

and (2) radiative μ capture with asymmetric gamma conversion (both internal and external),

$$\mu^- Z \rightarrow \nu(Z-1)\gamma \rightarrow e^+ e^- \, , \tag{13}$$

which was the only unavoidable background for reaction (5). In addition cosmic rays posed a problem because only single tracks were searched for. A serious background was due to pion contamination in the beam and radiative pion capture with asymmetric gamma conversion. Beam contamination was brought to the level $\pi/\mu \approx 10^{-4}$, $e/\mu \approx 10^{-2}$ with an RF separator and pion (prompt) events further rejected by vetoing events for which energy was deposited in the beam counters. The normalization of the experiment was obtained by stopping π^+, detecting the positrons from $\pi \rightarrow e\nu$, and using, $(\pi \rightarrow e\nu)/(\pi \rightarrow \mu\nu) = 1.23 \times 10^{-4}$, the observed branching ratio.

The observed e^- momentum spectrum for $87 < P(\text{MeV}/c) < 107$ is shown as the solid histogram in Fig. 12. The dashed and dotted histograms are Monte Carlo calculations of the decay in orbit background[14] and conversion process, respectively. Radiative capture, reaction (13), makes a negligible contribution in this momentum region. The conversion process was generated with $R_- = 7 \times 10^{-11}$, the previous upper limit.[15] No candidate events were found in the range $96.5 < P(\text{MeV}/c) < 106$, where 85% of the Monte Carlo conversion events were located. This results in the limit $R_- < 4.5 \times 10^{-12}$ (90% C.L. preliminary).

The e^+ experimental spectrum is shown as the solid histogram in Fig. 13. Due to the complicated nuclear physics of the final state both the main background, reaction (13), and the process itself are difficult to calculate. The background, dashed histogram, was calculated from muon radiative capture theory[16] assuming 10% of the high energy photon ($E > 70$ MeV) events left the final state nucleus with less than 10 MeV excitation energy,[17] and then normalized to the experimental rate.[18] Internal conversion estimated to be $\approx 15\%$ of the external conversion contribution is not included. For the process itself, dotted histogram, $R_+ = 9 \times 10^{-10}$ and a Lorentzian distribution of excitation energies with $\langle E_x \rangle = \Gamma = 20$ MeV were assumed in order to make a direct comparison with the previous experiment.[15] For P > 90.5 MeV/c no events are observed. The 90% confidence level upper limit obtained from Poisson statistics is $< 6 \times 10^{-11}$ (90% C.L. preliminary).

DISCUSSION

In extended models with additional heavy leptons or an enlarged Higgs sector reaction (4) could proceed via isoscalar or isovector coupling,

$$R_- = f(Z)\{g^0 + g^1(N-Z)/3A\}^2 \tag{14}$$

where $g^{0,1}$ are hypothetical isoscalar and isovector coupling constants, $f(Z)$ is a nuclear structure factor, and Z,N,A are the number of target protons, neutrons, and nucleons. Shanker[19] has calculated $f(Z)$ for various targets. The effect on $g^{0,1}$ of the present upper limit as compared with limits obtained previous to the TRIUMF experiment[15,20] is shown in Fig. 14.

These coupling constants can be expressed in terms of a heavy neutrino mass and mixing coefficients with the lighter neutrinos in the model of Altarelli et al.[1] The result[21] is shown in Fig. 15. As the branching ratios are roughly proportional to m^4, where m is the neutrino mass, the decay limits would have to improve by factors >100 to compete with the conversion limit. It should be noted that for a given coupling and branching ratio upper limit the neutrino mass is an upper limit. For Dirac neutrino masses $\lesssim 14$ GeV/c^2 other experiments[22] provide better mass limits. In the case of Majorana neutrinos the lifetime limits for neutrinoless double beta decay provide better mass limits.[23]

In terms of models in which Higgs, or leptoquark exchange, is responsible for reactions (1)-(7) the branching ratios are proportional to m^{-4}, where m is the mass of the exchange particle. Following the theory of Shanker[3] the results shown in Table 1 can be obtained. Aside from the K_L, K_S mass difference, which may be due to Higgs particle mass differences,[3] the conversion experiment sets the mass limit. The results are very model dependent. A different choice of coupling strengths as given by Ellis[4] would place the Higgs mass in the GeV range rather than the TeV range. In a recent preprint Hall and Randall[5] have provided a simple solution to the strong CP problem by introduction of scalar leptoquarks. Flavor violation is a natural consequence of such a model. In terms of the coupling constant $\beta < 1$, the mass limits of the exchanged leptoquark are given in Table 2. R_e is not given as it is expected to occur at a much lower level than R_γ, which proceeds via a box diagram. Here too the conversion process sets the mass limit. The mass limit from the allowed $K_L \to \mu^+\mu^-$ decay is calculated assuming the contribution from exotic particle exchange can be no larger than the observed rate. The much larger mass bound from the conversion process implies the latter contribution is small.

Finally from the model of Cahn and Harari[6] the mass limits for horizontal gauge bosons mediating flavor changing processes are given in terms of coupling constant ratio, g_H/g_W, and mixing angles in Table 3. The branching ratios depend on the mixing angle factor to the power $n = |\delta G|^2$ and on m^{-4}, where m is the boson mass. For processes which are not diagonal all boson masses m_+, m_-, m_3 are assumed equal. For diagonal processes the mass parameter δ is obtained from $\delta^{-2} = m_3^{-2} - m_\pm^{-2}$, so that these limits are meaningless for degenerate masses. If the angular parts of the mass limits are $\approx (\sin\theta_c)^{n/4}$, where θ_c is the Cabibbo angle, the conversion experiment sets the limit. The limit from the K_L, K_S mass difference is potentially important, but requires knowledge of both mixing angle and δ.

In the case of muon positron conversion a variety of specific model calculations have been made. See, for example, Ref. 24 and the references therein. In that reference Leontaris and Vergados considered the conversion to be mediated by massive Majorana neutrinos, light Higgs particles, and intermediate charged Higgs particles. Both left-handed and left-right symmetric models were employed and realistic nuclear wave functions used. The calculated branching ratios are all very small and beyond the range of experiment. For the coherent process in which the final state nucleus is in the ground state, R_+ (coherent) $< 10^{-27}$, and for the total rate, $R_+ < 10^{-20}$.

FUTURE PROSPECTS

New proposals[25] at LAMPF and SIN to extend the sensitivity of the limits on R_γ and R_- have been submitted, but these are in a very early stage of development. However, results from kaon experiments[9] proposed or running at BNL in the United States and KEK in Japan can be expected in the next few years. These are aiming to lower the current branching ratio upper limits by factors of 500 for $R_K(e^-)$ and $(6{-}60) \times 10^5$ for R_L. If successful reaction (6) will determine the mass limits in the models discussed above. The observation of nonzero branching ratios would require measurement of the other lepton flavor nonconserving reactions to unravel the correct theory.

ACKNOWLEDGMENTS

Thanks are due to G. Azuelos, D.A. Bryman, R.A. Burnham, M.D. Hasinoff, R.E. Marshak, J.A. MacDonald, J.N. Ng, T. Numao, J.M. Poutissou, and R. Poutissou for helpful discussions.

REFERENCES

*On leave at TRIUMF, 4004 Wesbrook Mall, Vancouver, B.C., Canada V6T 2A3. Supported by USNSF: grant Phys-8503845.

1. G. Altarelli et al., Nucl. Phys. B125, 285 (1977).
2. W.J. Marciano and A.I. Sanda, Phys. Rev. Lett. 38, 1512 (1977).
3. O. Shanker, Phys. Rev. D 20, 1608 (1979).
4. J. Ellis, AIP Conf. Proc. 102, 191 (1983).
5. L.J. Hall and L.J. Randall, Harvard University preprint HUTP-86/A002.
6. R.N. Cahn and H. Harari, Nucl. Phys. B176, 135 (1980).
7. R.E. Marshak, Riazuddin, and C.P. Ryan, Theory of Weak Interactions in Particle Physics, J. Wiley, 1969, p. 66.
8. A. Zee, Phys. Lett. 93B, 389 (1980).
9. i) A Search for the Rare Decay $K^+ \rightarrow \pi^+\mu^+e^-$, AGS experiment no. 777, Brookhaven, Washington, Yale collaboration; the goal is

10^{-11} for $R_K(e^-)$. ii) A Search for Flavor Changing Neutral Currents $K_L \to \mu e$ and $K_L \to e^+ e^-$, AGS experiment no. 780, Brookhaven, Yale collaboration; the goal is 10^{-10} for R_L. iii) Study of Very Rare K_L Decays, AGS proposal no. 791, Pennsylvania, Princeton, Stanford, Temple, William and Mary collaboration; the goal is 10^{-12} for R_L. iv) T. Inagaki et al., KEK proposal 85-1 (1985); the goal is 10^{-11} for R_L. v) A recent brief review of these experiments is given by D.A. Bryman, Proceedings, Winter Institute on Particle Physics, F. Khanna ed., Lake Louise, Canada, 1986.

10. H.K. Walter, Nucl. Phys. A434, 409c (1985).

11. R.D. Bolton et al., Phys. Rev. Lett. 56, 2461 (1986).

12. D. Grosnick et al., Proc. of the Santa Fe DPF meeting, T. Goldman and M. Nieto ed., Santa Fe, New Mexico, 1985.

13. C.K. Hargrove et al., Nucl. Instr. Meth. 219, 461 (1984).

14. F. Herzog and K. Alder, Phys. Acta. 53, 53 (1980).

15. A. Badertscher et al., Nucl. Phys. A377, 406 (1982).

16. H.P.C. Rood and H.A. Tolhoek, Nucl. Phys. 70, 658 (1965).

17. D. Beder, Can. J. Phys. 63, 154 (1985).

18. M. Doubeli et al., presented at 3rd International Conference on Meson in Nuclei, Bechyne, Czechoslovakia, June, 1985.

19. O. Shanker, Nucl. Phys. B206, 253 (1982).

20. D.A. Bryman, M. Blecher, K. Gotow, and R.J. Powers, Phys. Rev. Lett. 28, 1469 (1972).

21. T. Numao, TRIUMF, private communication.

22. See the contributions of F. Bergsma and T. Maruyama to this meeting.

23. See the contribution of D. Caldwell to this meeting.

24. G.K. Leontaris and J.D. Vergados, Nucl. Phys. B224, 137 (1983).

25. i) M.D. Cooper et al., LAMPF proposal 969 (1986); the goal is $R_\gamma < 10^{-13}$.
ii) A. Badertscher et al., SIN letter of intent R85-07.0; the goal is $R_- < 10^{-13}$.

Table 1. Mass [TeV/C^2] bounds from different processes.

Process	Higgs scalars	Pseudoscalar leptoquarks		Vector	Experimental limit	
		a)	b)			
R_γ	0.3	—	—	—	4.9×10^{-11}	c)
R_e	2	—	—	—	2.4×10^{-12}	c)
R_-	25	3	23	124	4.5×10^{-12}	c)
R_L	1	0.2	0.7	13	6×10^{-6}	d,e)
$\dfrac{K_L \to \mu^+\mu^-}{K_L \to \text{all}}$	5	1	4	62	9.1×10^{-9}	e,f)
$R_K(e^+)$	0.7	0.1	0.3	4	7×10^{-9}	e)
(e^-)	0.8	0.1	0.3	4	5×10^{-9}	e)
$\delta m(K_L - K_S)$	150	—	—	—	3.5×10^{-15} GeV/c^2	e)

a)Average fermion mass = 1 GeV/c^2

b)Average fermion mass $\approx \sqrt{(m_t m_\tau)}$ = 7.8 GeV/c^2 for uμ couplings and $\approx \sqrt{(m_b m_\tau)}$ = 3.1 GeV/c^2 for dμ couplings.

c)Text.

d)Upper limits, $R_L < 1.6 \times 10^{-9}$ and $(K_L \to \mu^+\mu^-/K_L \to \text{all}) < 1.8 \times 10^{-9}$ were reported by Clark et al., Phys. Rev. Lett. $\underline{26}$, 1667 (1971). Since the latter decay has been observed at a level of 9.1×10^{-9} and the reason for the discrepancy unknown, the previous limit has been used.

e)Particle Properties Data Booklet.

f)Observed.

Table 2. Mass [TeV/c^2]/β a) bounds from different processes.

Process	Scalar leptoquark b)
R_γ	53
R_-	304
R_L	8
$\dfrac{K_L \to \mu^+\mu^-}{K_L \to \text{all}}$	46
$R_K(e^+)$	16
$R_K(e^-)$	18

a)$\beta < 1$.

b)For branching ratio limits see Table 1.

Table 3. Mass $[TeV/c^2]/[g_H/g_W]$[a)] bounds from various processes.

Process	$|\delta G|$	Diagonal d)	Horizontal gauge boson b,c)	Interaction
R_L	0	no	5	V–A
$R_K(e^-)$	0	no	18	V
R_γ	1	yes	$15\,\sqrt{\left|\sin\beta_L\cos\beta_L\right|}$ e)	V
R_e	1	yes	$69\,\sqrt{\left|\sin\beta_L\cos\beta_L\right|}$	V–A
R_-	1	no	$210\,\sqrt{\left|\sin\beta_{LU} - \sin\beta_{LD}\right|}$	V
$\delta m(K_L-K_S)$	2	yes	$400\,\sin\beta_D$ f)	V–A

a) g_H is the horizontal gauge coupling strength, g_W is the usual weak coupling strength.

b) For branching ratio limits see Table 1.

c) β's are mixing angles, L for leptons, U and D for quarks.

d) For processes which are not diagonal all boson masses m_+, m_-, m_3 are assumed equal. For diagonal processes the mass parameter $\delta^{-2} = m_3^{-2} - m_+^{-2}$.

e) The calculation was made for a pure vector theory with a single gauge boson of mass δ to mimic the effect of symmetry breaking. The two contributions, see Fig. 3, are separately divergent, but a GIM mechanism makes the sum finite.

f) Using the theoretical Lagrangian of M.K. Gaillard and B.W. Lee, Phys. Rev. D $\underline{10}$, 897 (1974).

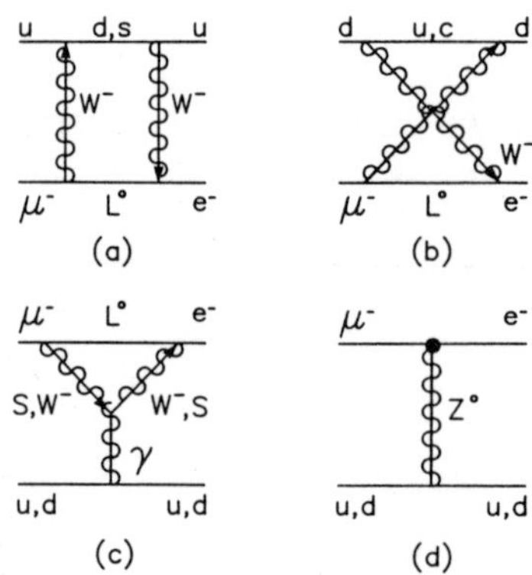

Fig. 1. Diagram contributing to muon electron conversion due to an additional heavy neutrino L in models of Refs. 1 and 2. From Ref. 2, S is an unphysical Higgs added to maintain gauge invariance.

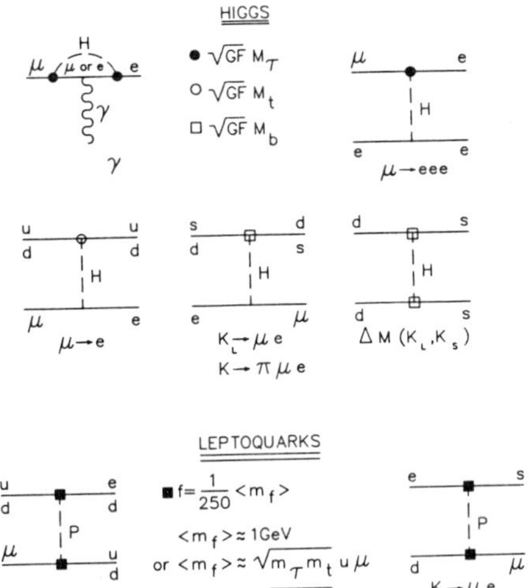

Fig. 2. Diagrams contributing to reactions (1)–(7) due to Higgs and leptoquark exchange in the model of Ref. 3.

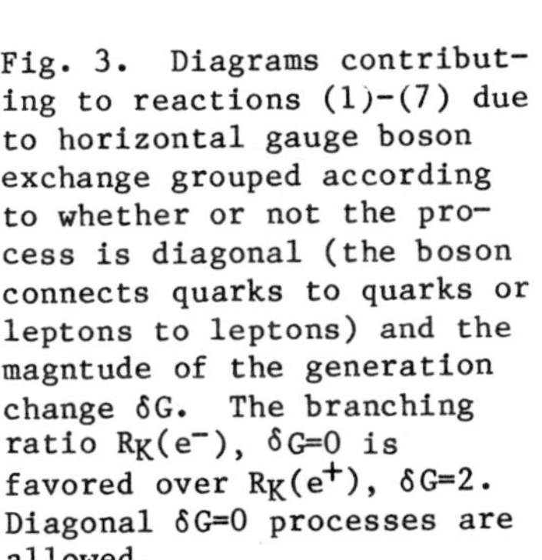

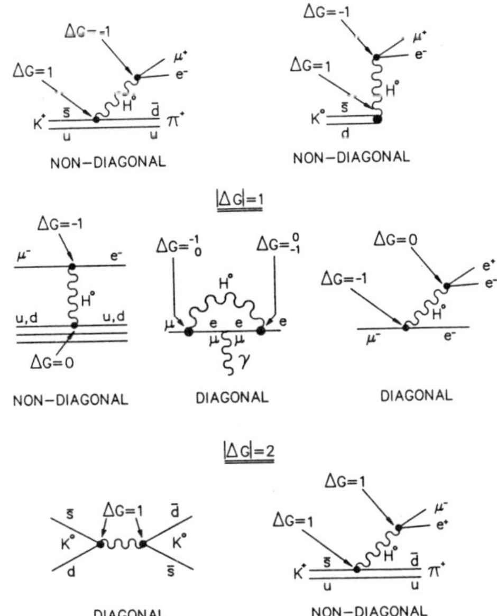

Fig. 3. Diagrams contributing to reactions (1)–(7) due to horizontal gauge boson exchange grouped according to whether or not the process is diagonal (the boson connects quarks to quarks or leptons to leptons) and the magntude of the generation change δG. The branching ratio $R_K(e^-)$, δG=0 is favored over $R_K(e^+)$, δG=2. Diagonal δG=0 processes are allowed.

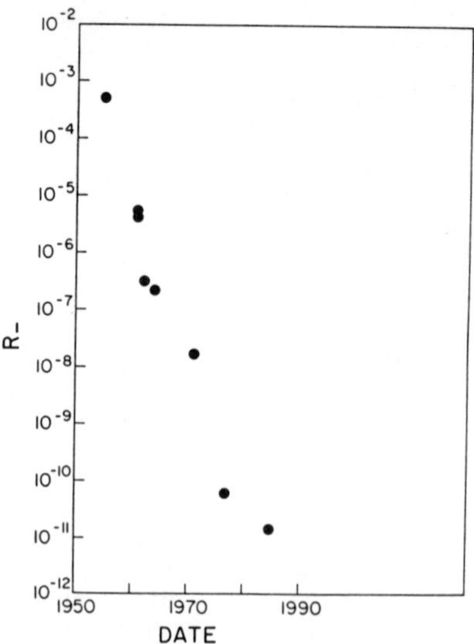

Fig. 4.   Historical survey of the measurement of the upper limit for R_-.

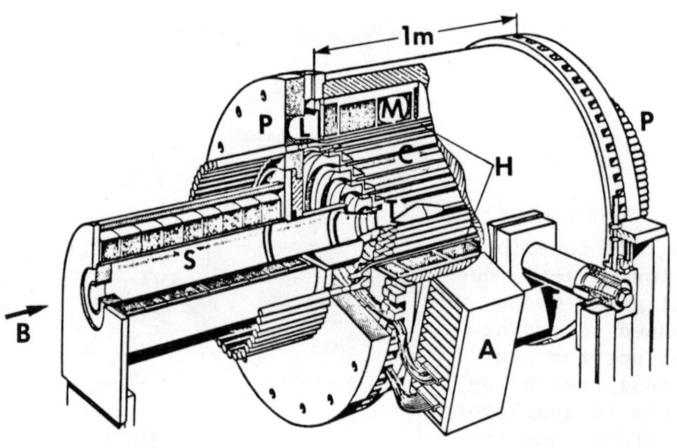

Fig. 5.   Apparatus (SINDRUM) for the SIN experiment to search for $\mu \rightarrow 3e$, reaction (2).   Beam (B) is focused on conical target (T) with solenoid magnet (S).   Charged particles from the target travel in helical paths in magnetic field (coils M) and are detected with hodoscope (H) of scintillation counters (light guides L, photomultipliers P) and wire chambers (C) with preamplifiers A.

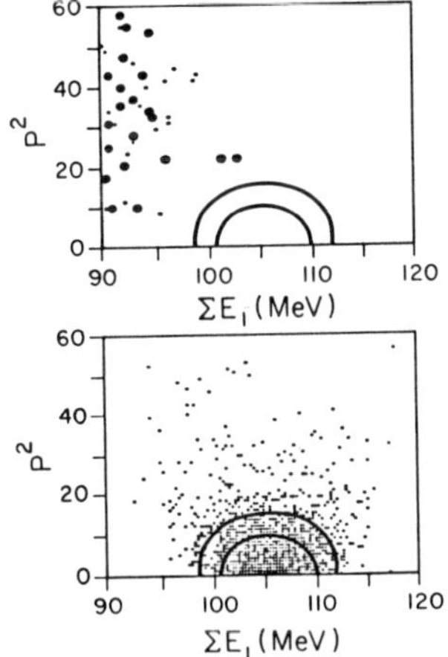

Fig. 6. Top: SIN spectrum of momentum vs energy for the three charged particles near where reaction (2) events are expected [Eqs. (8) and (9) satisfied]. Bottom: Monte Carlo calculation of reaction (2) events. Also shown are the 90% and 68% contours.

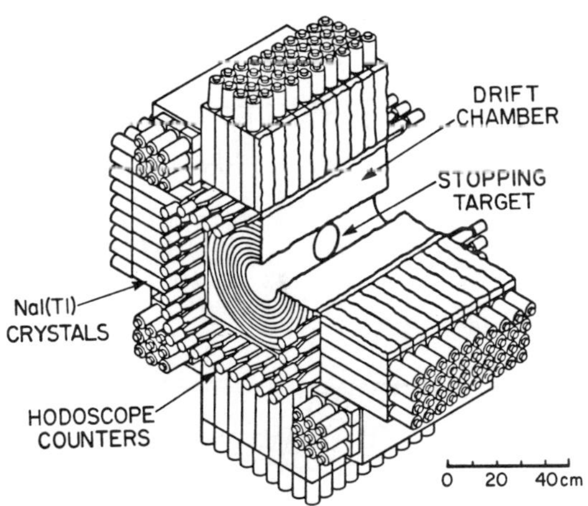

Fig. 7. Apparatus (crystal box) for the LAMPF experiment to search for flavor nonconserving muon decays, reactions (1-3).

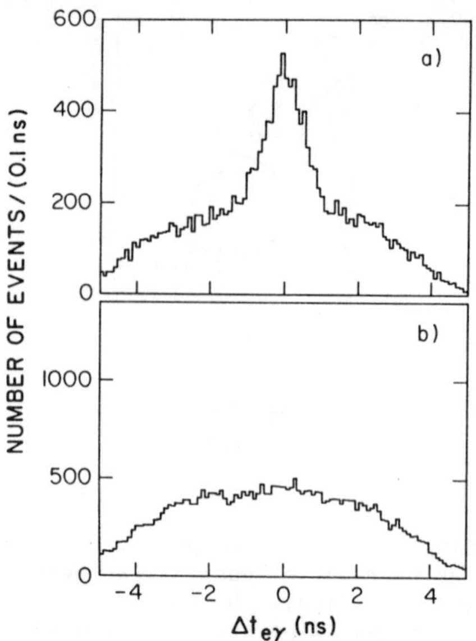

Fig. 8.   LAMPF timing spectrum for reaction (1) which must contend with a large random background of photons from radiative muon decay and positrons from normal decay.

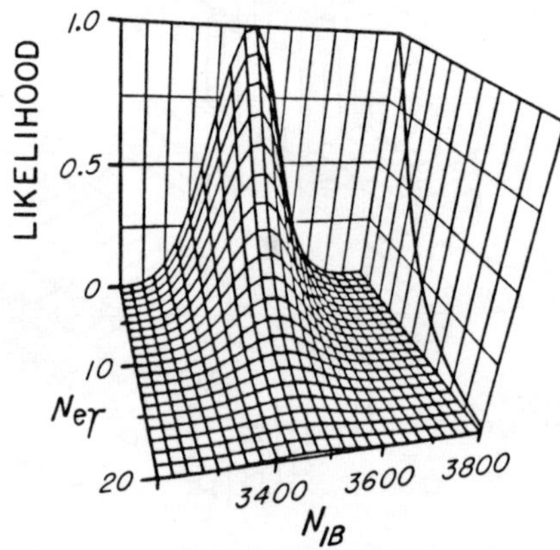

Fig. 9.   LAMPF likelihood function for radiative muon decay events N_{IB}, calculated from the standard model and detector response and reaction (1) events, $N_{e\gamma}$. The 90% confidence level is reached for $N_{e\gamma} < 11$.

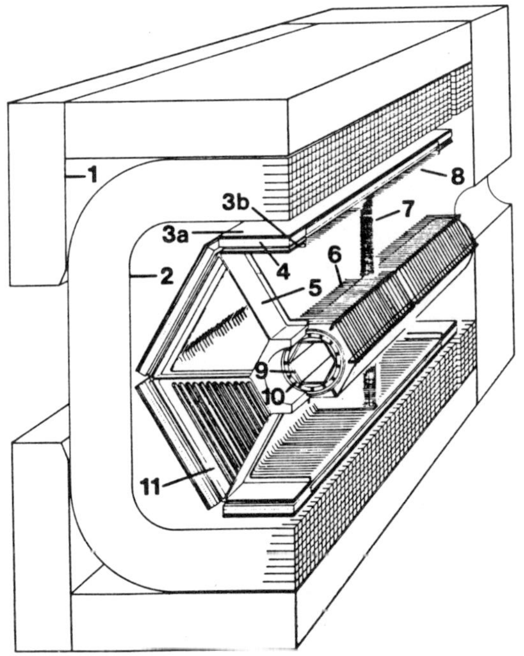

Fig. 10. Apparatus for the TRIUMF experiment to search for
muon conversion, reactions (4) and (5). A time projection
chamber (TPC) of hexagonal symmetry with 6 sectors of 12
proportional wires (11) surrounds an extended shredded
titanium target (not shown). Scintillation counters and wire
chambers internal 9,10 and external 3a,b,4 to the chamber and
also divided into 6 sectors provide an event trigger. Beam-
defining counters which count muon stops and veto prompt pion
events and cosmic-ray scintillators and drift chambers atop
magnet 1 (coils 2) are not shown. The TPC frame, central
high voltage plane, and field wires are shown as 5-8,
respectively.

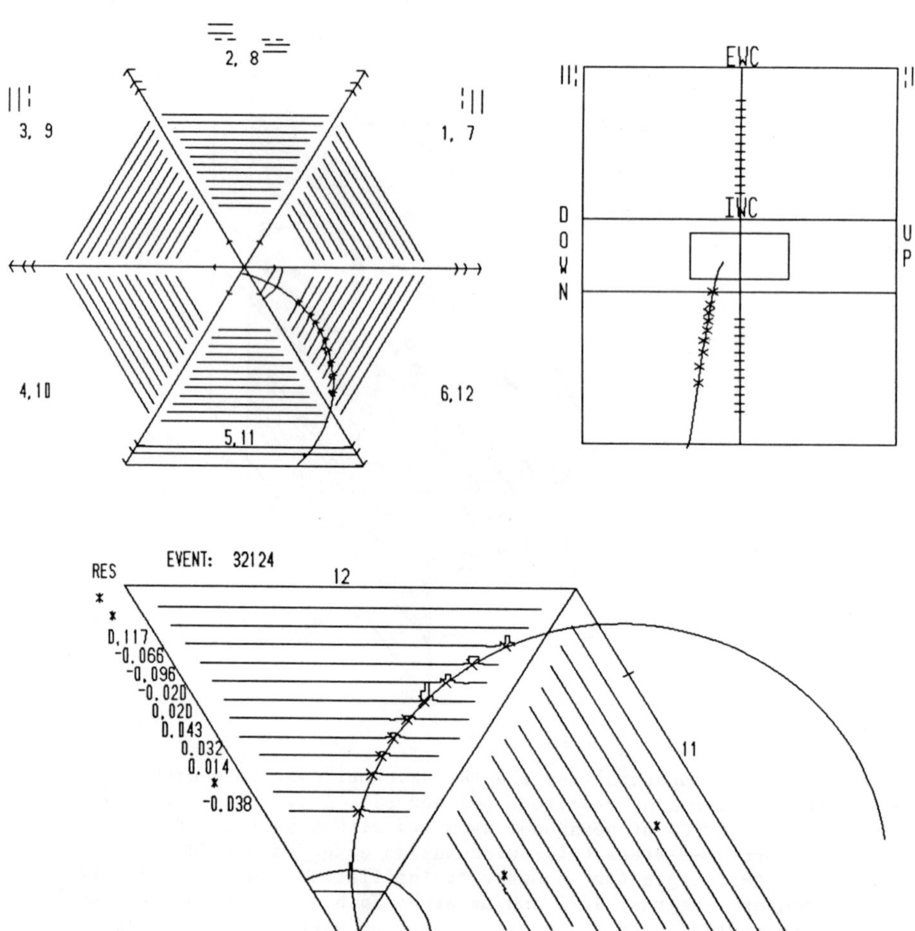

Fig. 11. A reconstructed event in the TPC $[P(e^-) = 94.2$ MeV/c] showing the typical information obtainable (see text for for details).

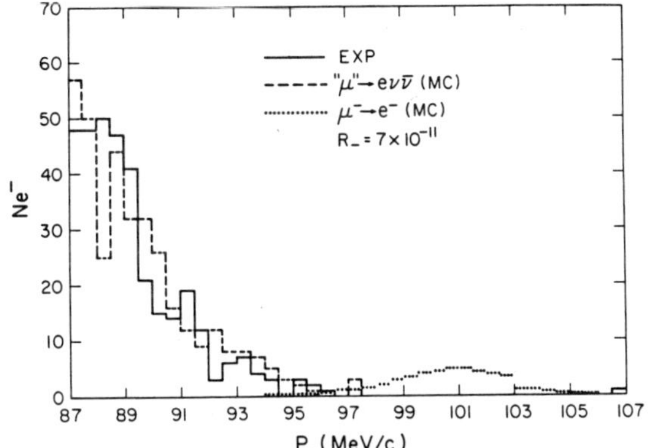

Fig. 12. TRIUMF experimental spectrum for reaction (4), solid histogram, compared with Monte Carlo calculations of the background from muon decay in orbit, dashed histogram. Radiative muon capture with asymmetric gamma conversion makes a negligible contribution in this momentum region. The dotted histogram is a Monte Carlo calculation of the muon electron conversion spectrum if the branching ratio R_- was as large as the previously measured upper limit,[15] $R_- < 7 \times 10^{-11}$.

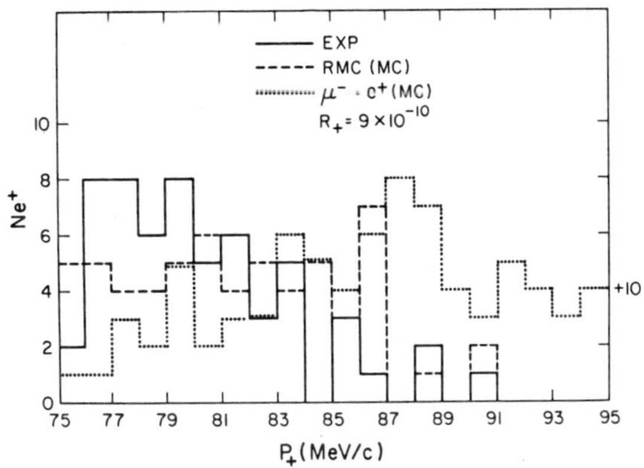

Fig. 13. TRIUMF experimental spectrum for reaction (5), solid histogram, compared with a Monte Carlo calculation of the background from radiative muon decay with asymmetric gamma conversion, dashed histogram. The dotted histogram is a Monte Carlo simulation of the expected positron conversion spectrum assuming Lorentzian excitation with $\langle E_x \rangle = \Gamma = 20$ MeV and branching ratio, $R_+ = 9 \times 10^{-10}$, the previously measured upper limit.[15]

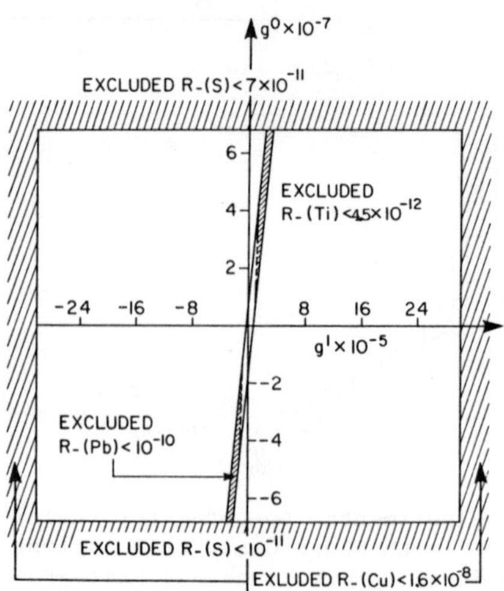

Fig. 14.   Phase space of hypothetical isoscalar and isovector coupling constants $g^{0,1}$ responsible for R_-. The regions excluded by the new TRIUMF result for R_- and previous results are shown. The shaded region indicates what would be excluded by a measurement of $R- < 10^{-10}$ in lead.   Such data taken at TRIUMF is under analysis.

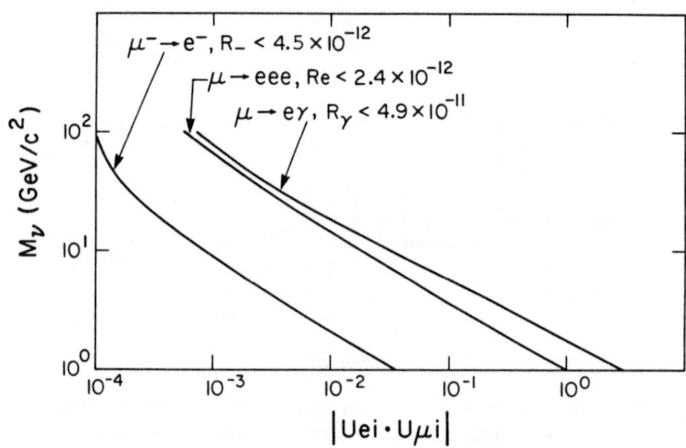

Fig. 15.   Calculation[21] of the product of the mixing coefficients of the light neutrinos with the heavy neutrino in terms of the heavy neutrino mass using the limits R_γ, R_e, and R_-.

PRESENT STATUS OF BAIKAL DEEP UNDERWATER
NEUTRINO EXPERIMENT

L.B.Bezrukov, B.A.Borisovets, E.V.Bugaev,
G.V.Domogatsky, A.A.Doroshenko, M.D.Gal'perin,
A.V.Golikov, Zh.-A.M.Jilckibaev, A.M.Klabukov,
S.I.Klimushin, A.V.Lopin, B.K.Lubsandorzhiev,
A.E.Ovchinnikov, A.I.Panfilov, I.A.Sokal'sky,
L.N.Stepanov, I.I.Trofimenko
Institute for Nuclear Research, Academy of Sciences
of the USSR, 60-th October Anniversary Prospect 7a,
Moscow 117312, USSR

V.I.Bogdanov, N.M.Budnev, L.I.Bumazhkin,
A.G.Chensky, V.I.Dobrynin, V.A.Durasov,
A.P.Koshechkin, G.A.Kushnarenko, V.A.Naumov,
M.I.Nemchenko, S.A.Nickiforov, Yu.V.Parfenov,
P.A.Pokalev, V.A.Poleschuk, V.V.Popov,
A.A.Schestakov, V.A.Taraschansky, Yu.M.Vetrov,
V.L.Zurbanov
Irkutsk State University, Irkutsk

V.B.Kabikov, L.A.Kuzmichöv, E.A.Osipova
Moscow State University, Moscow

A.I.Averin, V.A.Fialkov, E.B.Karabanov,
P.P.Scherstyankin
Limnology Institute, Siberian Division of the
USSR Academy of Sciences, Irkutsk

G.N.Dudkin, M.N.Gushtan, A.S.Lukanin, V.N.Padalko
Tomsk Polytechnical Institute, Tomsk

D.D.Kiss, P.Kakas , S.Feher
Central Research Institute of fundamental Physics,
Hungarian Academy of Sciences, Budapest

G. Domogatsky

Presented by G.V.Domogatsky

ABSTRACT

Stationary Monopole Search String (SMSS)
having come into operation on April 1986
at Lake Baikal is described. 50 days of SMSS
live time monopole search preliminary results
are given.

The aim of our physical investigations which are
carried out on Lake Baikal from 1980 [1] is to construct
large area ($\sim 4 \times 10^5 m^2 \cdot$ ster) Cherencov detector. After the
preliminary studies (in particular, the investigations of
characteristics of Baikal water luminescence) we can be
sure that realisation of this project is quite possible.

Stationary Short String (SSS) which was operating
during 50 days in spring of 1984 [2] , had been constructed. It has become one of the important stages of these
works. As the analysis of the SSS results had shown, it's
possible to carry out large-scale experiment on search
for superheavy magnetic monopoles on Lake Baikal.

On April 1986 at Lake Baikal the Stationary Monopole search String (SMSS) began to operate in regime of
long-term exploitation (Fig.1). This array contains 36
optical sensors on base of photomultiplier tube FEU-49B
with diametr of photocatode 15 cm. The main tasks of
SMSS there are following:
    -search for superheavy magnetic monopole by means
of the effect of barion decay catalysis [3,4],
    -search for relativistic magnetic monopole [5],
    -some methodic studies for more precise determi-

G. Domogatsky

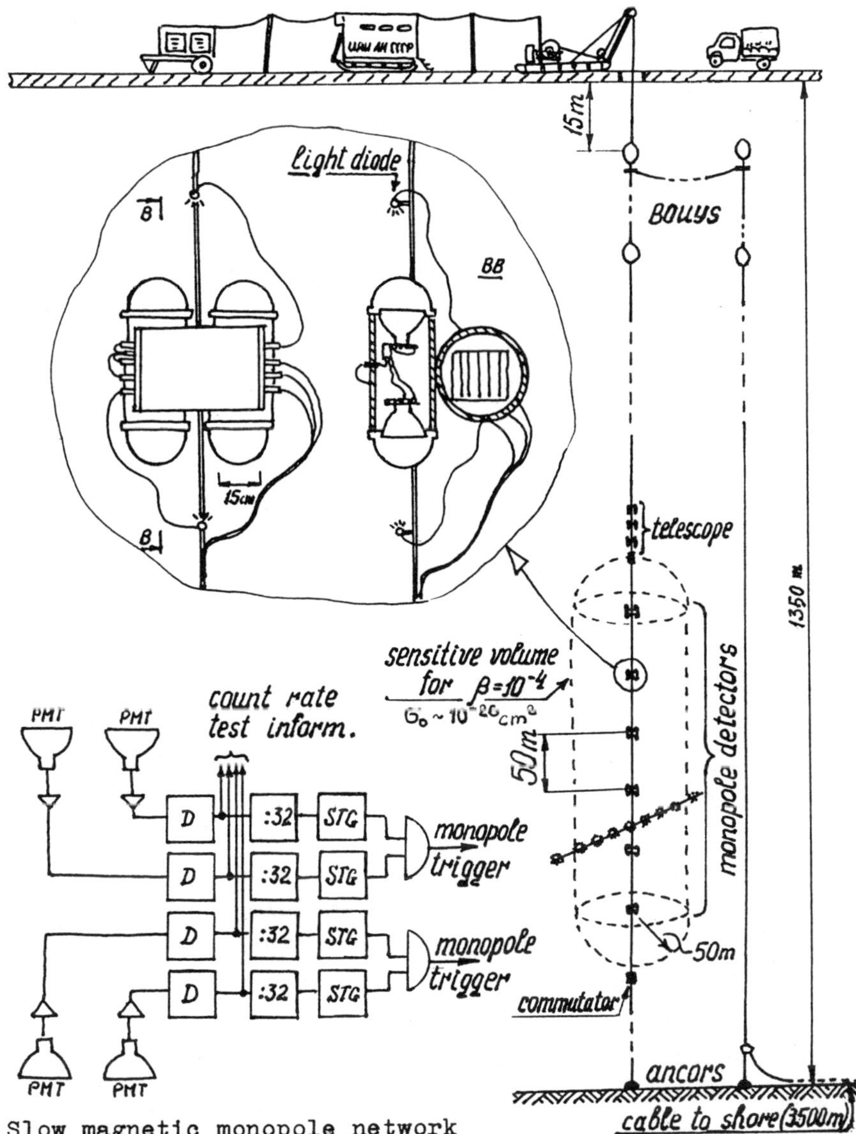

Slow magnetic monopole network
D – lowlewel discriminator
STG – sliding time gate (3 pulses for 0.5ms)
:32 – frequency divider

Fig.1.

nation of background conditions and criterions of event
selection in future experiments.

The SMSS consists of seven independent detectors.
One of them contains 12 PMTs and represents telescope
which detects atmospheric muons, showers, muons from
neutrino moving through the array upwards. This detector
also realizes the search for relativistic monopole. The
search for superheavy monopole is the main task of six
other detectors. The deployment of the SMSS was performed by using of the ise cover of the lake.

The procedure of SMSS signal processing is the
following. Trigger system of each detector permits to
store data in electronic module. Any detector can be connected with communication line by underwater commutator
which is operated by the shore computer. Then the stored
information in connected detector, count rate information
of each PMT and different kind of test information may
be sent to the shore station in reply to shore computer
request. After finishing of reply the electronic system
of this detector is initialized.

Telescope consists of six PMT-modules and an
electronic module (Fig.2). Each PMT-module contains two
PMTs. The electronic module consists of power supply telescope, discriminators, trigger system, systems of time
and amplitude analysis, memory and communication electronics. Telescope has three trigger conditions (each trigger can be settled at command from the shore):
1. Appearance of signal on any two pairs of PMTs
during $0.5\,\mu s$.
2. Appearance of signal on any two pairs of PMTs
from three pairs viewing downwards during $0.5\,\mu s$.

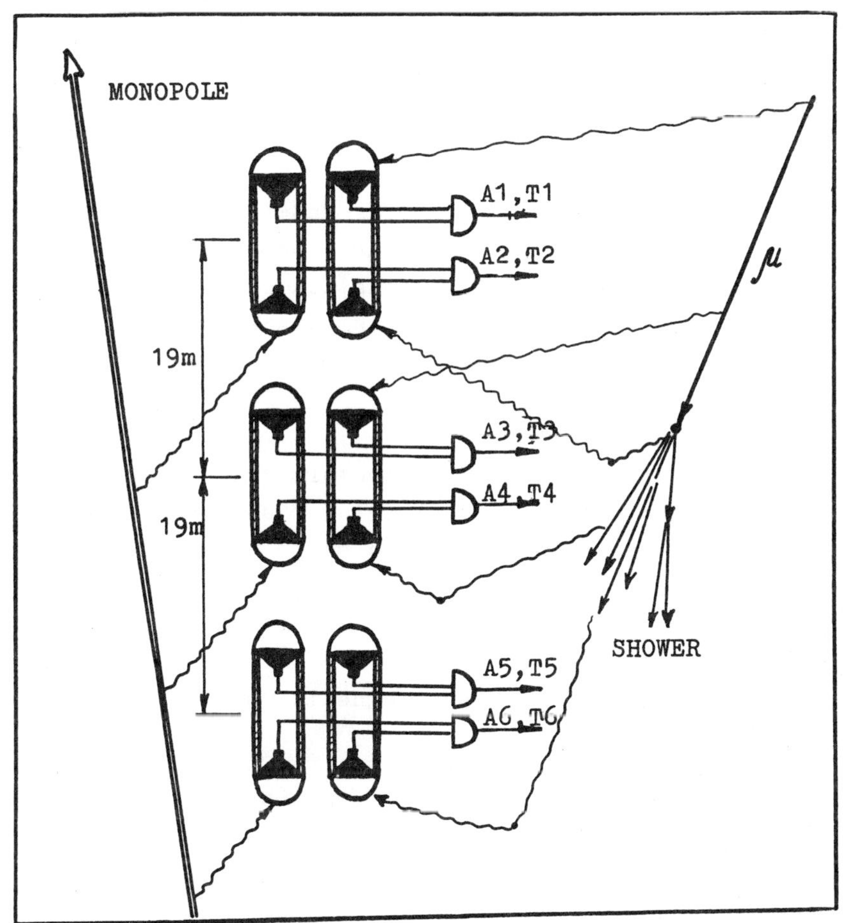

Fig.2. Telescope.

  3. Appearance of signal on the three pairs of PMTs viewing downwards during 0.5 μs.

  The main scientific tasks of the telescope are following:

  1.Registration of relativistic particles moving upwards, i.e., muons from neutrino and relatiwistic magnetic monopoles (β =v/c>0.75). Such monopoles (if they exsist) have mass $M_m < 6 \times 10^{10}$ GeV and are accelerated by galactic magnetic fields up to $E > 10^{20}$ eV [6]. Energy losses of such monopoles are less than ~10 GeV⋅cm^2⋅g^{-1} [7]. The main background process for upgoing muon or monopole is the shower which is produced by downgoing muons under telescope.

2. Measurement of counting rate of cosmic ray muons. These studies mace it possible to investifate the variations of optical characteristics of water medium.

3. Measurement of water luminiscence variations. It's the aim of periodical measurements of the counting rate of each PMT.

Each of six monopole detectors contain two PMT-modules and an electronic module. Power supply for detectors, discriminators, electronic sceme for selection of events from superheavy monopole, memory, communication electronics - that's what electronic modules contain.

Search for superheavy magnetic monopole is based on the following reaction:

$$\text{Mon} + p = \text{Mon} + e^+ + \text{pions} \qquad (1)$$

When such monopoles pass the array, PMTs counting rates ought to increase. The time duration of this increase depends on monopole velocity. Our detector have maximum sensitivity for velocity region $\beta = 10^{-5} \div 10^{-3}$. The slow monopole trigger is the appearance of counting rate increase during 0.5 ms simultaneously in two PMT viewing the same side.

The electronic scheme which is shown on Fig.1, realizes this trigger. The thresholds of discriminators are 0.3 p.e. The trigger condition is appearance of more than 96 pulses during 0.5 ms in each of two PMTs viewing the

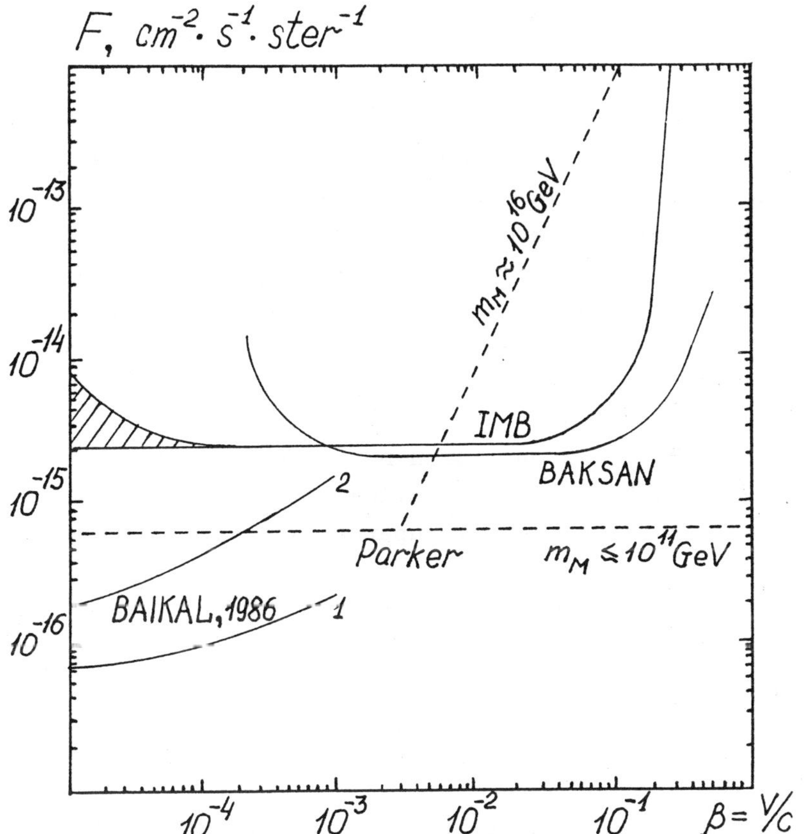

Fig.3. 9o%-C.L. upper limits on magnetic monopole flux as function of velocity β . Dash lines are astrophysical limits. Cross-section of
proton decay catalysis process is $\sigma = 0{,}17\cdot\sigma_0/\beta^2$
1 - $\sigma_0 = 10^{-26}$ cm^2, 2 - $\sigma_0 = 10^{-28}$ cm^2.

same side. Some additional information about velocity and direction of monopole moves is also registered.

For calibration the monopole detectors have been illuminated by special flash-tube from the distance of 50 meters. The light-diods which are used for test of electronic system can be operated at the command from the shore.

In contrast to underground experiments, acceptance of which is given mainly by their geometric dimensions the SMSS acceptance for slow monopole flux depends mainly on cross-section σ_c of proton decay catalysis process. The most interesting information the SMSS may give on monopole flux with velocity $\beta = 10^{-5} \div 10^{-3}$. In this case $\sigma_c = 0.17 \cdot \sigma_o \cdot \beta^{-2}$ may achieve the value of $\sigma_c \sim 10^{-21} cm^2$ under natural suggestion that σ_o is about $10^{-28} cm^2$. During 50 days of SMSS operation no monopole events has been observed. It gives the followings 90%-C.L. upper limits on isotropic flux of superheavy magnetic monopoles:

$$\text{for } \sigma_o \gtrsim 10^{-26} \qquad F_{Sm} \leqslant 6.5 \times 10^{-17} \; cm^2 \times s^{-1} \times sr^{-1}$$
$$\text{for } \sigma_o \gtrsim 10^{-28} \qquad F_{Sm} \leqslant 2.0 \times 10^{-16} \; cm^2 \times s^{-1} \times sr^{-1}.$$

References

1. G.V.Domogatsky, Rep. at Int. Conf. "Neutrino-82" Balatonfüred.
2. G.V.Domogatsky, Rep. at Int. Conf. "Neutrino-84" Dortmund.
3. V.A.Rubakov, JETPh Lett. 33, 644(1981).
4. V.A.Rubakov, M.S.Serebryakov, Nucl.Phys.B218,240(1983).
5. I.M.Frank, JINR Preprint P4-11777 (1978).
6. E.Goto, Prog.Theor.Phys. 30, 700(1963)
7. S.P.Ahlen, Phys.Rev. D17, 229 (1978).
8. M.S.Turner et al., Phys.Rev. D26, 1296 (1982).

9. J.Arafune, M.Fukugita, Phys, Rev. Lett. 50,1901(1983).

10 S.Errede et al., Phys. Rev. Lett. 51, 245(1983).

11. E.N.Alexeev et al., Proc. of 19th Int. C.R.Conf. v.5, 52(1985).

FUTURE PROJECTS IN ACCELERATORS

B.H.Wiik

II.Institut of Experimentalphysic, Univ. Hamburg, and

Deutsches Elektronen-Synchrotron DESY, Hamburg.

1. Introduction

It is by now widely believed that new physics will occur at a mass scale of 1 TeV or below. This expectation is based on the observation[1] that elastic scattering of longitudinally polarized vector bosons W_L or Z_L violates partial wave unitarity at 1.8 TeV unless new physics appears.

In the standard model[2] the infinities are removed by introducing elementary scalar fields[3] and generating vector boson masses by spontaneous symmetry breaking. In the simplest model there is one observable scalar particle - the Higgs boson.

All the relevant experimental data are described by the standard model using a consistant set of parameters; furthermore its predictions have been strikingly confirmed[4] by the discovery of the vector bosons $W^{\pm}$ and Z^0 at the CERN $p\bar{p}$ collider. Even so the standard model is not complete and it may simply represent the low energy limit of a more fundamental theory. The model does not contain gravity, a large number of free parameters are needed and there is a large number of elementary particles.

The very different mass scales occuring in the standard model - 100 GeV for the electroweak unification and 10^{15} GeV for the grand unification - poses a technical problem. It can be shown[5] that the Higgs mass would naturally be of the same order as the next mass scale - i.e. 0 $(10^{15}$ GeV). This result, in conflict with the requirement[6] that the Higg mass must be of order 1 TeV to preserve unitarity is avoided in the standard model by fine tuning constants to 26 decimal places.

A low Higgs mass is perhaps more naturally obtained in supersymmetric theories[7] in which every fundamental fermion and boson ac-

746

quires a partner which differs in spin by 1/2 a unit. Both partners contribute to the Higgs mass but with opposite sign resulting in a net contribution which is proportional to $|M-\tilde{M}|$, i.e. to the mass difference of the partners. Thus unitarity is saved if supersymmetric particles occur with mass below 1 TeV. Indeed many models predict the lightest supersymmetric particles to have a mass of order M_W.

It has been proposed[8], as an alternative to the standard Higgs mechanism, that the symmetry breaking arises dynamically from the gauge interactions themselves. In this model a new set of unbroken non-Abelian gauge interactions with a mass scale of the order of 1 TeV is introduced. This interaction gives rise to a complicated spectrum of technicolourless bound scalar technipion states with mass starting around 1 TeV.

The technicolour model must be extended in order to explain the mass of the fermions. Extended technicolour models[9] predict a rich spectrum of new particles with mass below a few hundred GeV.

It has been speculated that quarks and leptons are not elementary but composite[10] particles made of common building blocks bound by a new force. In this case quarks and leptons will have excited states and remanents of the new force may show up as a pointlike interaction well below the compositness scale.

Extensions of the standard model predict the existence of new vector bosons, for example as carriers of right handed currents.

Thus we expect new particles and new interactions to occur at a mass scale well below 1 TeV and perhaps be accessible to the next generation of particle accelerators. In the first part of my talk I'll describe the status of these projects and make a few comments on e^+e^- and $p\bar{p}$ physics at these facilities. In the second part I'll briefly review at the physics potential of the HERA ep collider.

2. Accelerator Facilities

Accelerators under construction or just completed are listed in Table I.

Table I - Facilities under construction or just completed

Type	E_{beam} (GeV)	E_{CM} (GeV)	Luminosity $(10^{30}cm^{-2}s^{-1})$	Facility	Physics
e^+e^-	2.8x2.8	5.6	20	BEPC, Beijing	1988
	30x30	60	80	TRISTAN, KEK, Tokyo	1986
	50x50	100	6	SLC, SLAC, Stanford	1987
	50x50	100	16	LEP I, CERN Geneva	1989
	100x100	200	-	LEP II, CERN Geneva	
$\bar{p}p$	315x315	630	4	ACOL/Sp$\bar{p}$S CERN, Geneva	1987
	1000x1000	2000	1	TEV I, FNAL, Chicago	1986
pm_p	3000xm_p	75	$5\times10^{12}p/s$	UNK I, Serpukhov Moscow	
pp	400x3000	2200	100	UNK II, Serpukhov Moscow	
ep	30x820	314	15	HERA, DESY Hamburg	1990
em_p	4xm_p	2.7	$1.3\times10^{15}e/s$	CEBAF Newport News	-

These machines represent a large step in energy compared to the energy available at present accelerators. The available cms energy in e^+e^- collisions will be increased from 46 GeV at PETRA to about 200 GeV with the advent of LEP II, in $p\bar{p}$ collisions the energy limit will grow from 660 GeV reached at the CERN Sp$\bar{p}$S via 2000 GeV soon to be exploited at FNAL to 2200 GeV which will be available with the completion of UNK II. The 30 GeV lepton-hadron cms energy available in secondary beams at FNAL will be increased by an order of magnitude to 314 GeV with the turn on of the ep collider HERA.

Besides the approved facilities several proposals for new accelerators have appeared in the past few years.

Table II - Proposed Facilities

Type	E_{beams} GeV	E_{CM} GeV	Luminosity $(10^{30}cm^{-2}s^{-1})$	Facility
pp	20000x20000	40000	1000	SSC, USA
	9000x 9000	18000	1500	LHC, CERN, Geneva
ep	65x 8100	1440	123	LHC, CERN, Geneva
AA (heavy ion)	100/amux100/amu	-	0.001	RICH, BNL Upton
	10/amux10/amu	-	0.0003	ORHIC, OakRidge
pm_p (Kaon factories)	45 x m_p	9.2	2.1×10^{14} p/s	LAMPF II Los Alamos
	30 x m_p	7.5	6.3×10^{14} p/s	TRIUMF II Vancouver
	30 x m_p	7.5	3.1×10^{14} p/s	SIN II Villigen

The large hadron colliders allow us to search[11] for new particles with masses in the range 1-10 TeV.

With the heavy ion colliders one might be able to create[12] a quark gluon plasma.

Experiments at the kaon factories with their extreme high intensities of secondary particles will lower the limits on rare decays by several orders of magnitude. For example, normalized to the 1984 AGS secondary beams the LAMPF II facility offers[13] an increase of a factor of 193 in the kaon and 302 in the $\bar{p}$ yield at 7 GeV/c whereas the ν_μ flux would be up by a factor of 112.

3. Electron - Positron Colliders

3.1 SLC

The SLC project[14] is interesting both as a physics machine and as a prototype linear collider.

The main machine parameters are listed below:

<u>Table III</u> - SLC Machine Parameters

Repetition rate	120/180 sec^{-1}
Beam energy	50 GeV
Energy spread ($\Delta E/E$)	$\pm$ 0.2%
Number of Klystrons	240
Radio frequency	2856 MHz
Peak voltage	315 kV
Peak output power	50 MW/Klystron
Accelerating gradient	17 MV/m
<u>Arcs:</u>	
Bending radius	279.4 m
β^*	0.75 / 0.50 cm
Bunch length σ_z	1.0 mm
Beam size at I.P. $\sigma_{x,y}$	1.5 μm
Particles per bunch	$5/7.2 \times 10^{10}$
Number of collision points	1
Luminosity	$0.6/6 \times 10^{30} \mathrm{cm}^{-2}\mathrm{s}^{-1}$

The layout of the SLC facility is shown in Fig. 1. The main components are the e^-/e^+ sources including the high gradient linac and the 2 km positron return line, the e^+/e^- damping rings, the upgraded linac with new 50 MW klystrons and a new control system, the two arcs made of small aperture combined function magnets and the final focus section designed to collide the micron sized beams.

Let us consider one collision cycle to better appreciate the complexity of the collider. The cycle starts with two bunches of electrons and two bunches of positrons stored in their respective damping rings. Each cycle one of the positron bunches and both of the electron bunches are extracted from the ring, compressed in longitudinal phase space and injected into the linac with the positron bunch leading. The two first bunches are accelerated to the end of the linac and transported around their respective arc and final focus transport to collide in the

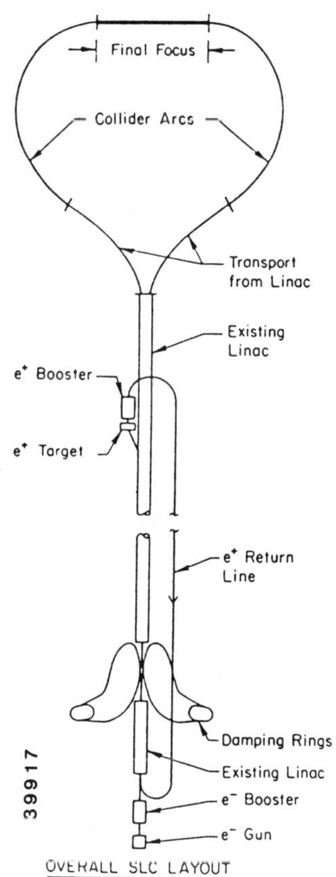

Fig. 1
Layout of the SLAC Linear
Collider Facility

interaction point. The third bunch is kicked out at the 2/3 point of the linac and targeted to produce positrons. The positrons are collected and accelerated to 200 MeV in a high gradient strongly focused linac. They are then sent down a 2 km long return transport to the beginning of the linac for subsequent acceleration to 1.21 GeV. Just prior to the arrival of the positrons the electron gun is fired producing two bunches to be accelerated and supplied to the electron ring.

The SLC project is now[15] nearing completion with first collisions expected early 1987. The e^+/e^- - sources, the two damping rings and the installation of the magnets in the arcs and the final focus transport system are either completed or will be completed this autumn. The most critical item is the replacement of the 230   34 MW klystrons downstream of the damping rings with new 50 MW klystrons. At the present production rate sufficient 50 MW klystrons will be available in the spring of 1981 to start 50 GeV operation.

The initial luminosity is expected to be around $6 \times 10^{29} \mathrm{cm}^{-2}\mathrm{s}^{-1}$ corresponding to the production of a few 10^6 Z^0's per day. The luminosity may later be increased to $6 \times 10^{30} \mathrm{cm}^{-2}\mathrm{s}^{-1}$ by an increase in re-

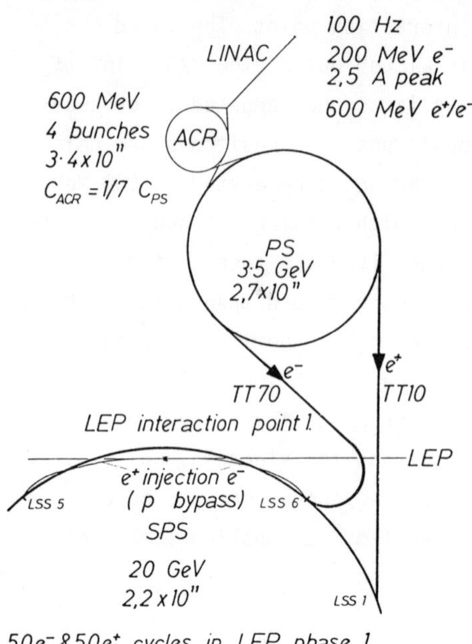

Fig. 2
Layout of the LEP Injector
Complex

petition rate, by a stronger final focus requiring superconducting quadrupoles (available 1988/89) and by increasing the number of particles per bunch.

Work on the polarized electron source is continuing and it is expected to become available in 1987.

3.2 LEP

LEP[16] with its circumference of 26.66 km is a factor of 4 larger than any circular accelerators built so far. The injection chain - shown in Fig. 2 - makes use of the PS and the SPS in addition to a newly built linear accelerator and a booster synchrotron. The LEP machine parameters are listed below.

Table IV - LEP I machine parameters

Machine energy	55 GeV
Energy spread ($\Delta E/E$) at 50 GeV	0.8×10^{-3}
Circumference	26.66 km
Bending radius of dipoles	3.096 km
Energy loss per turn at 55 GeV	260 MeV
Max. voltage per turn (55 GeV)	360 MV
Number of klystrons	96
Radio frequency	352 MHz
Peak output power	16 MW
Active cavity length	272 m
Accelerating gradient	1.47 MV/m
Number of collision points	4
Luminosity	$1.6 \times 10^{31} cm^{-2} s^{-1}$

The LEP project is now[17] well underway. More than 60% of the tunnel is completed. Series production of all major components is ongoing and machine installations have started. The first e^+e^- collisions are scheduled for the end of 1988 with the first physics runs in early 1989.

The full potential of LEP can only be explored by installing a large superconducting RF system. Two (of many) possible scenarios[18] are listed in Table V.

Table V - Superconducting RF system for LEP II

Cavities		Gradient MV/m	Power at 4.2 K (kW)	Beam Energy GeV
Number	Length (m)			
192	326	5	16	80
		7	32	87
384	652	5	32	95
		7	64	104

The superconducting RF programme has yet to be approved. It has been estimated that after approval four years are needed to produce the system.

3.3 Physics in e^+e^- Annihilation

Due to the well defined production mechanism and the absence of troublesome background e^+e^- annihilation is ideally suited to search for new phenomena.

The focuss of the next generation of large e^+e^- colliders will be on the production and decay of the Z^0 boson. These data will yield information[19] on:

- the mass and the width of the Z^0 with an accuracy of perhaps $\pm$ 50 MeV without a polarized beam and perhaps a factor 2-3 better with polarization. Note that each new generation of light neutrinos adds a 170 MeV to the Z^0 width.

- stringent tests of the standard model. It seems feasible to determine $\sin^2\theta_W$ with an error of perhaps less than 0.001.

- search for new fundamental particles including scalar quarks and scalar leptons for a mass up to 40-45 GeV. The missing link in the

standard model, the Higgs boson will presumably be detected in
the decay $Z^0 \rightarrow H \ell^+ \ell^-$ if its mass is below 50 GeV.

The search for bound tt-states will be an important task for the
e^+e^- colliders including TRISTAN. The level schemes and the decay modes
of the $t\bar{t}$-system provides stringent tests of QCD. Decays of $tt(1^{--})$
$\rightarrow$ 3 g give the possibility to study the fragmentation of gluon jets,
$t\bar{t}(1^{--}) \rightarrow 2gg\gamma$ should be a good hunting ground for gluonia and finally
the decay $t\bar{t}(1^{--}) \rightarrow H\gamma$ may be used to search for the Higgs boson.

P.ir production of charged vector bosons is one of the most interesting reactions which can be studied at LEP II. In this reaction one
may test the non-abelian nature of the electroweak interaction and
measure the coupling of 3 gauge bosons. The mass of the charged vector
boson can be extracted from these data with an uncertainty of perhaps
$\pm$ 100 MeV. Combined with the Z^0 mass this yields a high precision
measurement of the ρ parameter.

LEP II can be used to search for pair production of new particles
with masses up to 100 GeV. A new Z' will show up as a peak in the
cross section if its mass is below 2E = 200 GeV. A high mass Z' will
cause observable γ-Z' interference effects if its mass is below say
800 GeV.

Higgs bosons with masses below 100 GeV can be searched for in the
reaction $e^+e^- \rightarrow Z^0H$. A compositness scale of order 10 TeV will lead
to observable effects in $e^+e^- \rightarrow e^+e^-$ and $e^+e^- \rightarrow \mu\bar{\mu}$.

4. Hadron Colliders

4.1 ACOL / Sp$\bar{\text{p}}$S

In 1983 CERN decided to construct a new antiproton collection system ACOL[20] designed to improve the number of antiprotons accumulated
per hour by an order of magnitude to 6 x 10^{10} $\bar{\text{p}}$/hr. This would result
in a factor 10 increase in the peak luminosity to 4 x 10^{30}cm^{-2}s^{-1}.
This large increase results from an improved $\bar{\text{p}}$ conversion target using

high current lithium lenses to focuss the incident protons on to the production target and to collect antiprotons produced in a large solid angle.

The antiproton cooling is done in two stages: a fast initial cooling in ACOL with a rate matched to the repetition rate of the PS followed by storage and final cooling in the AA ring.

Test of the system will start[21] in spring of 1987 and routine operation for physics should begin in the fall of 1987.

4.2 The TEV I Collider

The basis for the $p\bar{p}$ collider is the 800-1000 GeV superconducting proton accelerator TEV II which has provided beams for the fixed target programme at FNAL since 1984. Limited by magnet performance TEV II has been operated at 800 GeV. Every minute the machine provides an average of 1.6×10^{13} protons distributed in a 20s long spill.

The layout[22] of the TEV I complex is shown in Fig. 3.

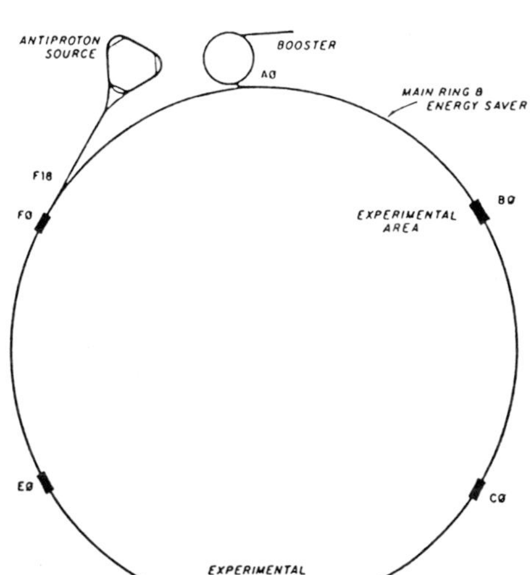

Fig. 3
Layout of the TEV I
project with the
antiproton source

In Table VI[23] the design parameters of the TEV I project are compared with the performance achieved late 1985 and the goals for the running period 86/87.

Table VI - TEV I collider history and goals

	Design	Oct.85	Goals winter 86-87
P/Bunch	6×10^{10}	2×10^{10}	4×10^{10}
$\bar{P}$/Bunch	6×10^{10}	few 10^{6}	10^{10}
Number of bunches	3x3	1x1	3x3
Transverse Emittance 95% normalized ($\pi\mu$radm)	24	15-50	24
Luminosity ($cm^{-2}s^{-1}$)	10^{30}	few 10^{24}	10^{29}
Accumulated $\bar{p}$ per hr	11×10^{10}	10^{9}	1.5×10^{10}
MR intensity	2×10^{12}	1×10^{12}	1.5×10^{12}
Target cycles/hr	1800	720	1200
Average proton flux on target (p/s)	10^{12}	0.2×10^{12}	0.5×10^{12}

As in the Sp$\bar{p}$S the luminosity is limited by the number of antiprotons. To obtain a high $\bar{p}$ accumulation rate the cooling is divided into two stages, a fast initial cooling system (debuncher ring) with a rate matched to the repetition rate of the Booster ring and a combined accumulation and cooling ring.

The layout of the p$\bar{p}$-complex is shown in Fig. 3. Every 2s a booster batch of 2×10^{12} protons is accelerated to 120 GeV in the main ring and focussed on to a production target. For 2×10^{12} protons incident on the target 7×10^{7} antiprotons with an energy of 8 GeV are collected using a high current lithium lens as a focussing element, transported and injected into the debuncher ring. Trading momentum spread and bunch length by a fast longitudinal bunch rotation the momentum spread $\Delta p/p$ is reduced from 3% to 0.2%. In the two seconds in which the antiprotons remain in the debuncher ring they are cooled from a transverse

emittance of 20 $\pi\mu$mrad to 7 $\pi\mu$mrad. The beam is then transferred to
the accumulator ring where the cooling is continued. A total of
4.3×10^{11} antiprotons with a transverse emittance of 2 $\pi\mu$mrad and
a longitudinal density of 10^5 antiprotons per eVolt will be accumulated
in four hours.

4.3 Physics with hadron colliders

Deep inelastic lepton-hadron interactions have shown that protons
are made of quarks bound by gluons and that its momentum is shared
in roughly equal parts between quarks and gluons. A proton-(anti)pro-
ton collision is thus equivalent to quark-quark, quark-gluon or gluon-
gluon collision and indeed data obtained at the CERN $p\bar{p}$ collider give
clear evidence for hard constituent collisions. The energy and the
momentum of the colliding constituents are not known, but typical constituents cms energies are of order 1/12 of the pp cms energy.

The mass limit[11] which can be obtained in $p\bar{p}$ collision depends
not only on production rate but also on background and final state
signature. This makes it difficult to predict the mass range which
can be exploited with the new colliders.

At TEV I the Higgs boson may be discovered if its mass is below
100 GeV. New vector bosons can be observed at TEV I if they occur with
a mass below 500-700 GeV and couples to quark and leptons with the
universal weak coupling strength.

Gluon fusion is a profilic source of quark production. New quarks
are produced at TEV I with detectable rates for a quark mass up to
100-200 GeV. Scalar quarks and gluons should be seen if they occur
at a mass below 100-200 GeV.

Composite scalars, predicted in technicoulor models, are produced
copiously in gluon-gluon collisions. The techni-eta should be seen
if its mass is below 200 GeV.

If quarks are composite objects made of common constituents the
Drell-Yan cross section will be modified due to contact interactions.
A compositness scale of a few TeV should lead to observable effects
in Drell-Yan production of massive lepton pairs.

The integrated luminosity collected at a dedicated proton-proton collider during one year of operation is presumably a factor of 100 to 1000 higher than the integrated luminosity collected at TEV I or the CERN p$\bar{\text{p}}$ collider. The discovery range - for a given cms energy - scales with luminosity to first approximation as $L^{0.2}$. Thus for most processes listed above the mass range can be extended by a factor of 2.5 to 4 by a dedicated 2 TeV pp collider.

5. Electron-proton colliders

5.1 HERA

The ep collider now under construction at DESY is built as an international collaboration modeled after the construction of large experimental facilities. Institutions in Canada, France, Netherlands, Israel, Italy, Peoples Republic of China, Poland, Switzerland, United Kingdom and the USA are contributing either technical components built in their home countries or skilled manpower working at DESY. The construction of HERA was authorized in April 1984 and the physics programme is scheduled to start in 1990.

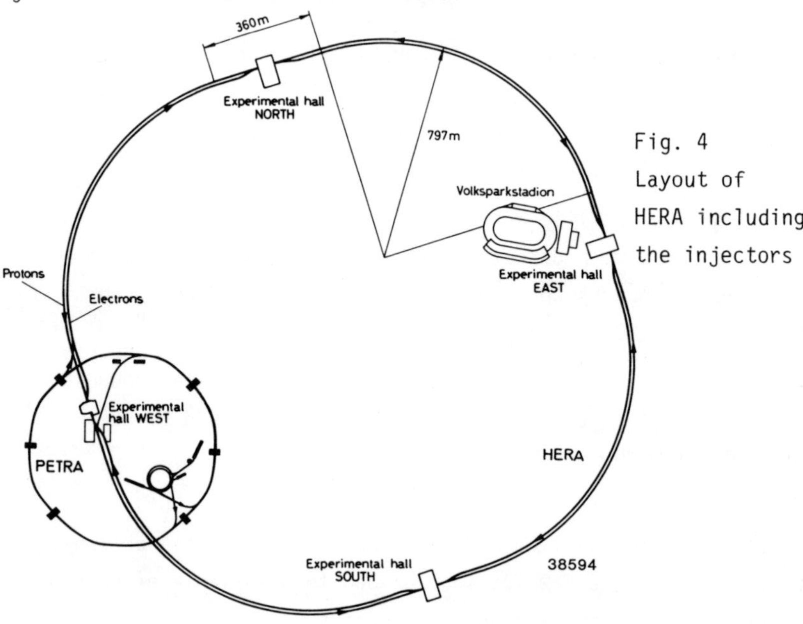

Fig. 4
Layout of
HERA including
the injectors

The general parameters of the machine are listed below in Table VII.
The layout is shown in Fig. 4.

Table VII - HERA machine parameters

	p-ring	e-ring	units
Energy range	820-300	30-10	GeV
Circumference	6336		m
Number of interaction points	4		
Length of each straight section	360		m
Free space for experiments	15		m
Polarization time (30 GeV)	27.8		min
Magnetic bending field	4.68	0.165	T
Injection energy	40	14	GeV
RF frequency	208.194/51.545	499.666	MHz
Max.circumferential volts	2.4/0.190	290	MV
Circulating current	163	58	m
Number of bunches	210		
Beta functions at the interaction point	10.0/1.0	2.0/0.70	m
Beam size at interaction points	0.293/0.0655	0.263/0.0693	mm
Luminosity	1.5×10^{31}		$cm^{-2}s^{-1}$
Installed refrigeration power at 4.3 K	20/60		kW/(g/s)
Installed refrigeration power at 40-60 K	60		kW

Construction of the tunnel and the four experimental halls is well advanced and will be completed by the end of 1987. Indeed DESY has now benificial occupancy of the first hall and installations of machine components in the tunnel will start in September of this year.

All the components of the electron ring have been ordered. The initial electron r.f. system will be transferred from PETRA to HERA. To boost the energy from 27.5 GeV provided by this system to 30 GeV or perhaps to 35 GeV will require superconducting cavities. A programme which will provide two industrially produced 8 cells 500 MHz cavities by spring of 1987 is underway.

Prototypes of all the different kinds of superconducting magnets for HERA have been tested. The field strengths are comfortably above the values required for 820 GeV operation and the field qualities are well within the specification. Series production of the dipoles and the correction magnets have begun in respectively Italy and Netherlands. The quadrupoles are out for tender.

Installation of the large helium plant is nearly completed and first operation is scheduled for the end of 1986.

Construction of the proton injection system is well underway.

Injection of positrons into the first HERA quadrant using the proton injection channel is scheduled for April 1987. Electrons will be injected into the second quadrant using the electron injection channel in July of 1987. First injection of protons will take place in summer of 1988.

6. Physics with HERA

6.1 Introduction

The kinematical region in Q^2 and ν available at HERA is shown in Fig. 5. The scale is set by the black dot in the left hand corner representing the region which can be explored using a 1 TeV muon or neutrino beam on a fixed target. The expected luminosity is sufficient to explore electroweak processes for Q^2 up to 30000 GeV^2.

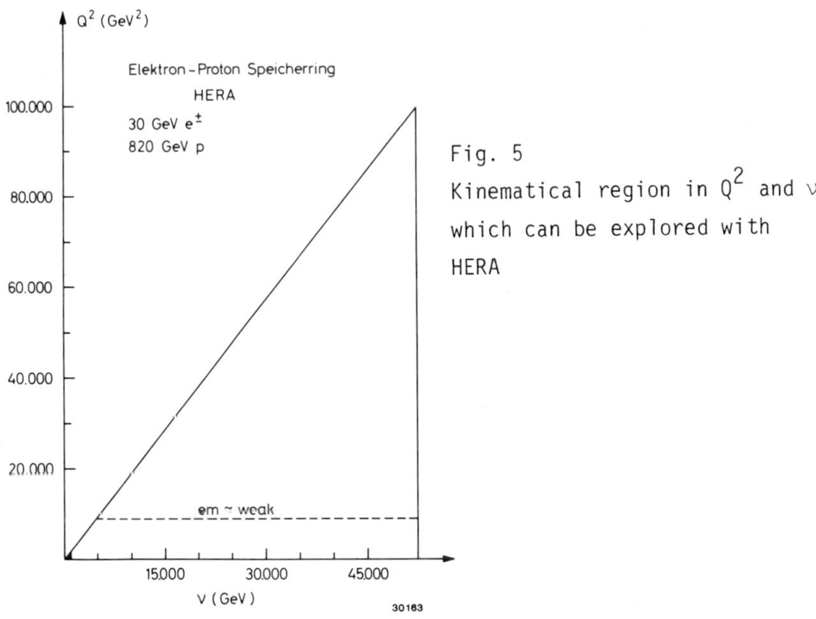

Fig. 5
Kinematical region in Q^2 and ν which can be explored with HERA

At very low values of Q^2 the electron beam is equivalent to a well collimated bremsstrahlungsbeam with an endpoint energy of 52 TeV. The untagged photon-proton luminosity is typically of the order of a few percent of the electron-proton luminosity yielding some 10^6 hadronic events per day.

The photon has a dual character, it may convert into a vector meson and interact like a hadron. In this case HERA is equivalent to a rho-proton collider exploring vector meson-proton interactions up to

314 GeV in cms. However, the photon through its pointlike part may induce hard processes like deep inelastic Compton scattering and the QCD analogues of Compton scattering and Bethe-Heitler processes.

For values of Q^2 above a few GeV^2 the incident electron will interact directly with one of the constituents in the proton - i.e. HERA is an electron-quark collider. The variables and the kinematic of deep inelastic interactions are defined in Fig. 6.

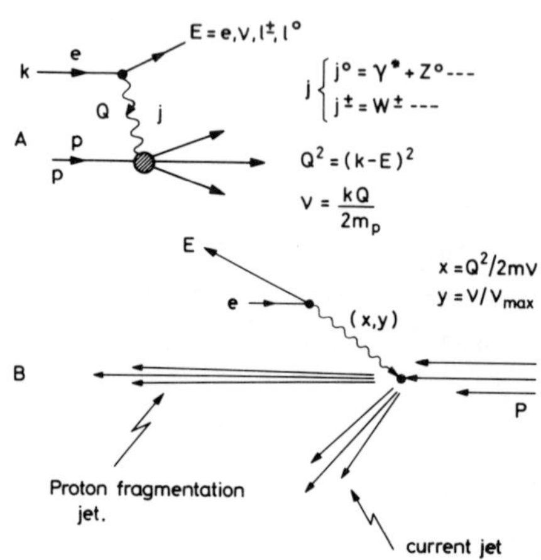

Fig. 6

Kinematic of deep inelastic electron-proton collisions

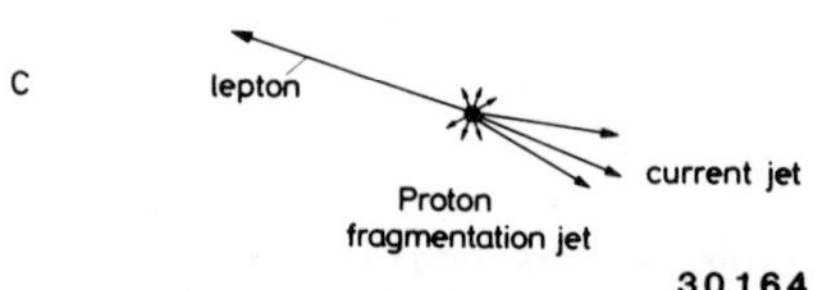

Hard scattering processes may be investigated at distances comparable to the Fermi scale

$$1/\Lambda_F = (\sqrt{2}\, G_F)^{-1/2} \sim (250\ \text{GeV})^{-1}.$$

This part of the physics programme[25,26] may be summarized as follows:

1) Determine the properties of spacelike electroweak currents at short distances. This will provide stringent test of our present understanding of the electroweak interaction.

2) Use the local, well defined electroweak current to explore the proton at distances down to 10^{-17} cm. Measurements of the structure function and an absolute determination of the strong coupling constant α_s will provide a crucial test of our present understanding of strong interactions.

3) New physics

   HERA is well suited to search for new particles with electron quantum numbers like scalar electrons, excited electrons, coloured electrons and leptoquarks. Search for new gauge bosons or pointlike residues of a new hypercolour confining force. HERA also offers a unique opportunity to search for right handed currents coupled to massive right handed neutrinos, to search for flavour violating transition like $e \to \mu$, free quarks etc.

The circulating e^+/e^- beams of HERA are expected to be polarized transversely to the plane of the accelerator. The transverse polarization can be turned into states of well defined helicity by a system of vertical and horizontal bending magnets located at the ends of the straight sections. Note that all helicity states will be available - i.e. e_L^-, e_R^-, e_L^+ and e_R^+.

The final state topology in deep inelastic electron-proton interactions is striking and easy to recognize. As indicated in Fig. 6B and Fig. 6C the scattered lepton appears at a large angle with respect to the beam axis and its transverse momentum is balanced by the struck quark which fragments into a jet of hadrons appearing at large angles on the opposite side of the beam axis. The remains of the proton give rise to a forward jet of hadrons focused along the proton beam axis with no net transverse momentum. Because of the imbalance between incident electron and proton momenta the particles will in general emerge in the forward hemisphere along the proton direction. The proton jet, the quark jet and the lepton defines a plane with small momenta transverse to the plane and large momenta in the plane.

Simply on the basis of topology it seems unlikely to confuse a deep inelastic electron-proton event with a background event such that the accessible Q^2-range appears to be limited by rate and not by background.

Note that particles from the lepton vertex and the quark vertex are kinematically well separated. In the standard model only a single neutrino or a single electron are allowed at the lepton vertex. The observation of any other type of particles or of jets emerging from the lepton vertex is a unique signature of new physics. The background resulting from photon-hadron interactions can be strongly reduced by a cut on transverse momentum.

The Q^2 range which can be explored depends therefore only on the rate which in turn is determined by the luminosity and the center-of-mass energy. To produce 100 charged current events with $Q^2 > 10000$ GeV^2 per year requires a luminosity of 1.5×10^{30} cm^{-2}s^{-1} - a factor of 10 below the HERA design luminosity.

6.2 Spacelike currents

6.2.1 Charged currents

The x, y distribution[27] of charged current events are plotted in Fig.7a in bins of dxdy = $(0.2)^2$ for a 30 GeV electron beam colliding with a 820 GeV proton beam and an integrated luminosity of 200 pb^{-1}. Given the distinct signature it seems possible to determine the properties of charged current events for values of Q^2 up to 30000 GeV^2.

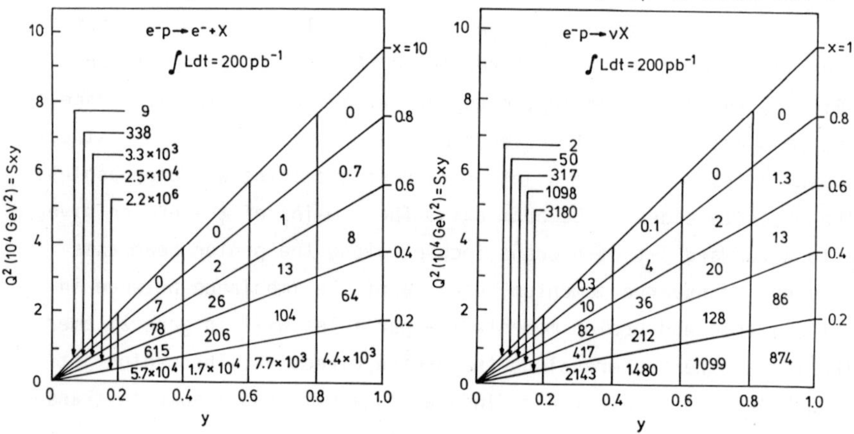

Fig. 7 - Number of deep inelastic events expected for an integrated luminosity of 200 pb^{-1}.

At $x > 0.4$ only the valence quarks contribute to the cross section - i.e. $e^- + u \rightarrow d + X$ and $e^+ + d \rightarrow u + X$ are the dominant processes. This makes it possible to study the fragmentation of quarks with well defined flavour and it may open the possibility of measuring the Kobayashi-Maskawa mixing angles.

6.2.2 Neutral currents

One photon exchange and Z^0 exchange contribute coherently to $e + p \rightarrow e' + X$ and both contributions are of similar strength at HERA energies. The number of neutral current events produced[27] in a bin $dxdy = (0.2)^2$ for an integrated luminosity of 200 pb^{-1} is plotted in Fig. 7b. Due to the characteristic topology of deep inelastic events HERA can extend the Q^2 range from the present few hundred GeV^2 out to perhaps 30000 GeV^2.

The presence of a weak current in the amplitude has clear signatures: 1) Parity violation, which leads to different cross sections for left handed and right handed electrons and positrons.

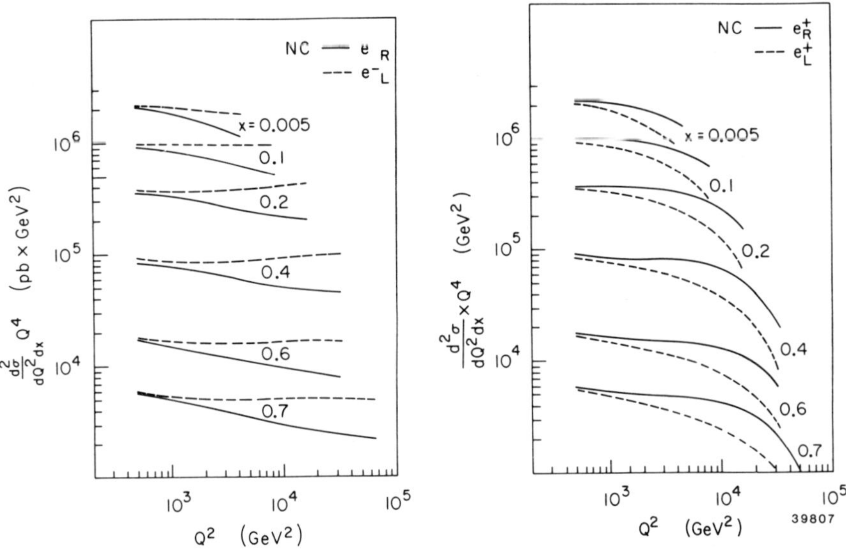

Fig. 8 - The cross section for longitudinaly polarized e^+ and e^-

2) Violation of charge conjugation, which leads to different cross sections for electrons and positrons.

3) The presence of a $1-(1-y)^2$ term in the cross section.

The neutral cross sections for left and right handed electrons and positrons are plotted in Fig. 8 as a function of Q^2 with x as a parameter. These asymmetries are rather large over the HERA range and depend sensitively on the values of the various weak coupling constants and the presence of additional neutral vector bosons.

6.3 New vector bosons

If vector bosons are composite objects then we may indeed expect to find a sequence of vector bosons with their coupling strengths adjusted such that the low Q^2 region is not affected:

$$\frac{G_F}{\sqrt{2}} = \frac{g^2}{8M_W^2} = \frac{g^2}{8m_1^2} + \frac{g^2}{8m_2^2} + \ldots$$

A new heavy vector boson will thus modify the cross section prediction at large values of Q^2. Altarelli et al. have evaluated[28] the change in cross section due to the exchange of two vector besons. The results are shown in Fig. 9 with $g_1 = g_2 = g$ and $m_1 = m_W$ and m_2 as a free parameter. It seems feasible to observe a second vector boson if its mass is less than 600 GeV to 800 GeV.

A similar result can be derived for additional Z^0's. Note that superstring theories predict the existence of a second heavy neutral vector boson.

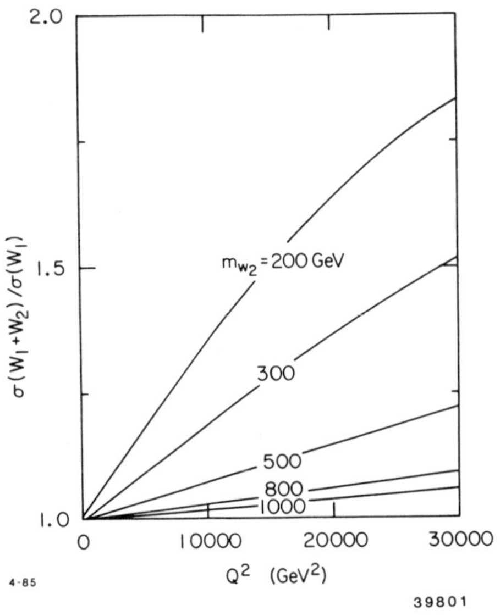

Fig. 9
Ratio $\sigma(W_1 + W_2)/\sigma(W_1)$ plotted versus Q^2 with the mass of W_2 as a parameter

6.4 Right handed currents

Right handed currents do not occur in the standard model. It has been conjectured[29] that the left - right handed symmetry is restored at some high energy by partnering the electron with a heavy right handed neutrino L_o. Measurements with right handed electrons and left handed positrons will reveal the existence of right handed currents if the mass of the propagator is less than 600 GeV and the longitudinal beam polarisation is at least 80% known to an accuracy of 1%. This mass limit - unlike the limit extracted from the β decay of the muon - is valid for massive right handed neutrinos.

If the L_o is a Majorana particle it will decay with equal probabilities into e^- and e^+. The observation of such spectacular events with positrons emerging from the lepton vertex may also serve as a signature for right handed currents. In this case one may be able to push the search for right handed currents up to 1 TeV in propagator mass.

6.5 Exploring the proton

The Q^2 evolution of the form factors can be unambigously computed in QCD and hence a determination of these form factors is a stringent test of the theory. Only a small correction factor is needed to extract the cross section from the raw data and the form factors can thus be measured with the required relative precision of a few percent over nearly the entire kinematical region. Furthermore the uncertainties caused by mass terms, higher twist operators and heavy flavour thresholds at present energies will be negligible small at HERA energies.

The strong QCD coupling constant $\alpha_s(Q^2)$, as evaluated[28] by Altarelli et al., is plotted in Fig. 10 versus Q^2 with the strong interation scale Λ as a parameter.

. Note that α_s varies by nearly a factor two over the HERA range and that the value of α_s at high Q^2 is rather sensitive to Λ. Thus at HERA one should be able to confirm or reject the predicted logarithmic variation of α_s.

At HERA α_s can be measured in various ways:
- from the evolution of the structure functions for Q^2 between 100 GeV^2 and 10^4 GeV^2
- from the properties of gluon bremsstrahlung events
- from the ratio of QCD to QED Compton events.

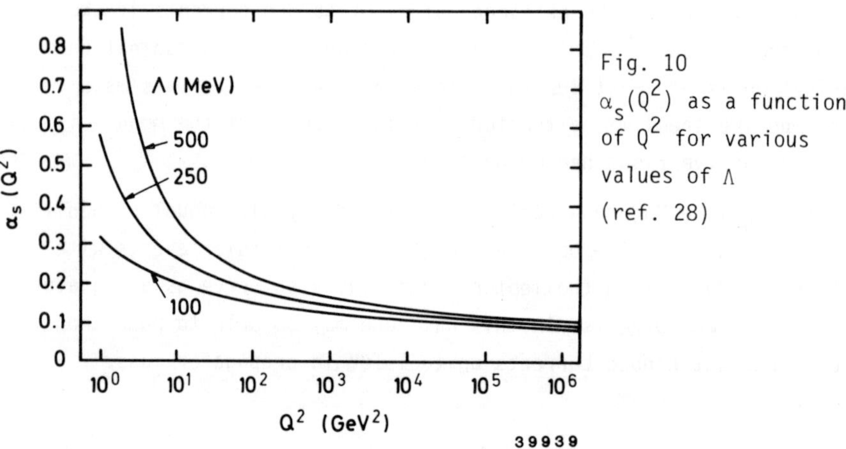

Fig. 10 $\alpha_s(Q^2)$ as a function of Q^2 for various values of Λ (ref. 28)

It is possible to determine the structure functions for valence
and sea quarks by combining data from all four reactions:

$$e^- + p \rightarrow e^- + X \qquad e^- + p \rightarrow \nu + X$$

$$e^+ + p \rightarrow e^+ + X \qquad e^+ + p \rightarrow \bar{\nu} + X$$

It may also be possible to measure the gluon structure functions
directly from a measurement of photon-gluon fusion[30] into light quark
pairs as shown in Fig. 11. Observation of scaling violations in the
gluon structure function would be direct evidence for the existence
of the three gluon vertex.

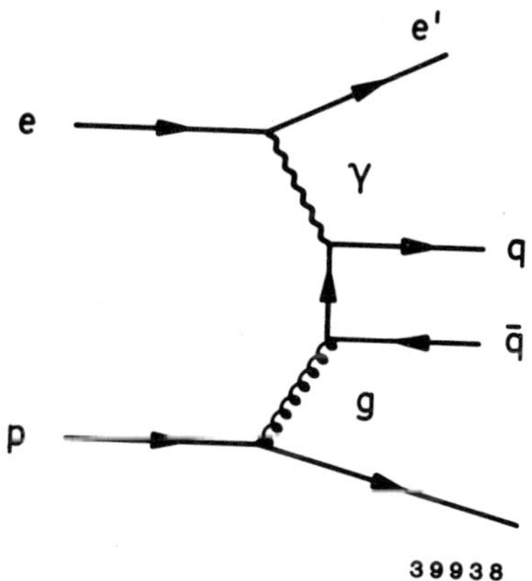

Fig. 11
Gluon fusion diagram

6.6 Production of new quarks and leptons

The photon-gluon fusion graph depicted in Fig. 11 is a strong source
of heavy quark pair production. The total cross section scales with
mass and the charge of the quark as:

$$\sigma(e\,q \rightarrow Q\bar{Q}X) \sim e_q^2\, M_Q^{-4}$$

For a t-quark of mass 50 GeV we expect a few hundred events per
year. The production of heavy quark pairs can be identified since

they yield a large number of final states particles distributed iso-
tropically in the plane perpendicular to the beam.

There may be new vector bosons connecting light and heavy constituents. A possible candidate would be the carrier of an right handed
current as discussed above. The number of $e + p \to L^0 + Q$ events has
been evaluated for m_W = 78 GeV and the normal weak coupling strength.
The rates are plotted in Fig. 12 as a function of the heavy quark mass
with the mass of the neutral lepton as a parameter.

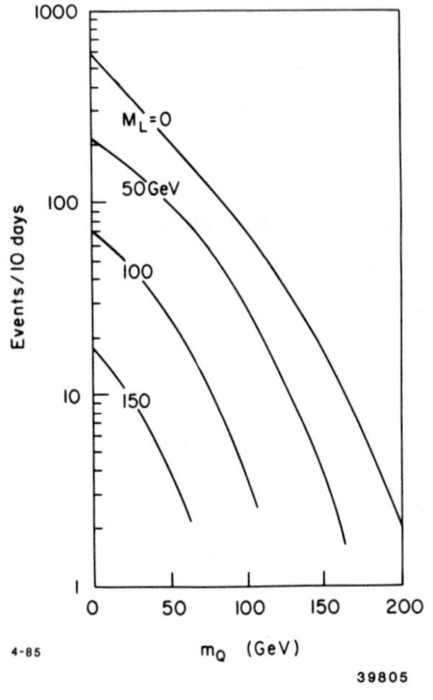

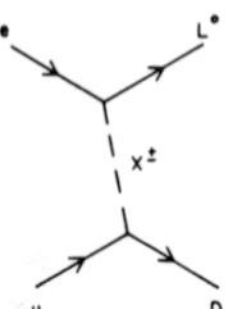

Fig. 12
The number of events per 10 days
for $e^- + p \to L^0 + Q$
assuming m_W = 78 GeV and the
normal weak coupling strength

Leptons and quarks with masses as heavy as 150-200 GeV can be
found in this way. The decay of the massive lepton is a source of
spectacular events; the single lepton emerging from the lepton vertex in the standard model will be replaced by $L_0 \to e^- Qq$ - i.e. a high
multiplicity jet containing leptons.

6.7 Composite leptons and quarks

Composite leptons and quarks would lead to an apparent violation
of scaling. The cross section would be modified by a form factor

$$F(Q^2) \; = \; \frac{1}{(1 \, + \, Q^2/M^2)}$$

giving rise to scaling violations which are very different from the
logarithmic violations expected QCD. A 10% measurement of the cross
section at $Q^2 = 4 \times 10^4$ GeV^2 will be sensitive to M $\leq$ 1 TeV. Note that
the form factors for composite quarks may increase or decrease depending
on the charges and the weak coupling constants of the new constituents.

Composite leptons will have excited states and ultimately such leptons will be directly produced. An excited lepton could decay into
e + γ, e + Z^0 and e + W leading to peaks in the invariant spectrum.
The topology of such final states with several particles emerging from
a lepton vertex makes it easy to find.

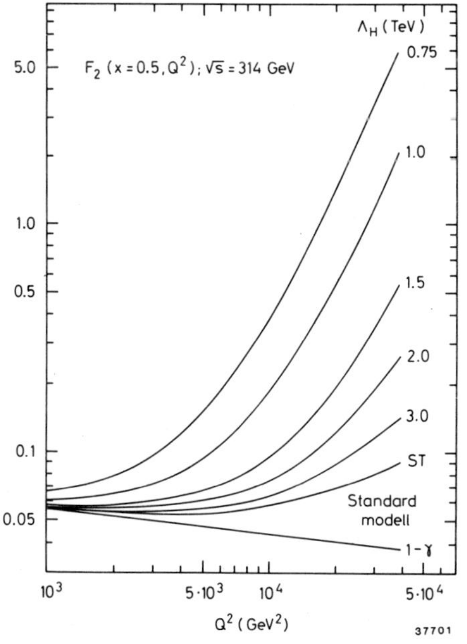

Fig. 13
$F_2(x = 0.5, Q^2)$ is plotted
versus Q^2 for e+p → e'+X.
The form factor is evaluated [32]
at $\sqrt{s}$ = 314 GeV for one
photon exchange, for the
standard model and for the
standard model including
contact interactions

If the quark and the lepton are made of common constituents[31] the normal spacelike interaction will be modified by a contact interaction resulting from constituent interchange. The form of the contact interaction may be written as:

$$\mathcal{L} \sim \frac{g^2}{\Lambda_c^2} \sum_{\alpha,\beta} \bar{e}_\alpha \gamma^\mu e_\alpha \; \bar{q}_\beta \gamma_\mu q_\beta$$

where α,β denotes states of definite helicity. Λ_c defines the composite scale and g is the coupling strength defined by $g^2/4\pi = 1$. The structure function[32] $F_2(x,Q^2)$ is plotted in Fig. 13 as a function of Q^2 for $x = 0.5$ and various values of Λ_c. The standard model predictions for F_2 with γ exchange only or with $\gamma + Z^0$ exchange are also shown in Fig. 13. The cross section including contact interactions is increased above the value predicted by the standard model. Using polarized beams HERA is sensitive[32] to values of $\Lambda_c \leq 5$ TeV.

6.8 Technicolor

The technicolour interaction[8,9] will result in leptoquarks, fundamental particles with combined lepton and baryon numbers and a mass predicted[33] to be around 160 GeV. The cross section for ep → (lq)x resulting from the Feynman graphs in Fig. 14a has been evaluated by Rudaz and Vermaseren[34] and is plotted in Fig. 14b versus the mass of the leptoquark. Roughly 0.1 event per day is expected for a leptoquark mass of 160 GeV and a suppression factor $\sin^2\theta_{et} = 0.05$. The topology of such an event is remarkable with a broad jet, resulting from the decay of the leptoquark (lq) → et, emerging from the lepton vertex.

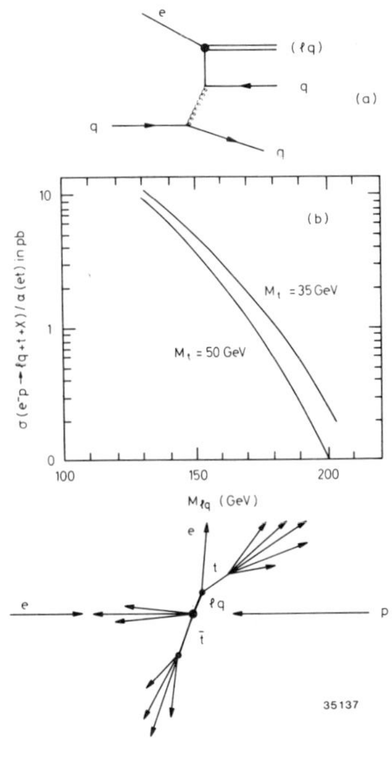

Fig. 14

a) The Feynman graph for producing a leptoquark in deep inelastic ep reactions

b) The cross section for ep → (1q) + X evaluated according to the Feynman graph in a

c) The final state topology

6.9 Leptoproduction of supersymmetrie particles

The presence of scalar quark and spin 1/2 gluons in the proton will strongly affect the form factors at asymptotic values of Q^2. In a conventional QCD 53% of the protons momentum is carried by the quarks as compared to 36% in a supersymmetric world. The problem is that the asymptotic values are approached very slowly for massive supersymmetric particles. Thus it may be difficult to convincingly demonstrate the existence of supersymmetric constituents from the observation of small scaling violations.

However, above threshold scalar quark and scalar leptons can be produced directly via the reactions:

$$e + q \rightarrow \tilde{e} + \tilde{q} \ ,$$

$$e + q \rightarrow \tilde{\nu} + \tilde{q} \ .$$

In standard neutral current events, however, the direction and energies of the scattered lepton and the current jet are strongly correlated. Since in supersymmetric neutral current events the electron and current jet are decay products, this correlation is destroyed and the signal can be enhanced by suitable cuts.

Estimates[26,35] show that one may produce 100 events a year of the type $e^- + q \to \tilde{e} + \tilde{q}$ with $\tilde{m}_q + \tilde{m}_e \approx 140$ GeV. This is easily detectable since the background from the standard neutral current events can be removed by cuts on coplanarity and on $y = \nu/\nu_{max}$ as evaluated from the electron and the current jet.

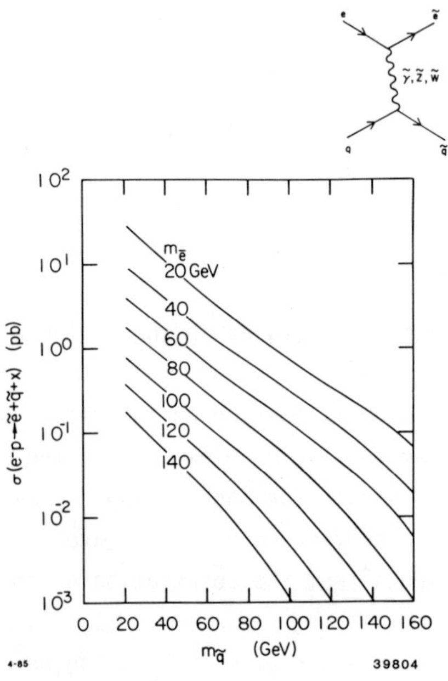

Fig. 15 Cross section for the reaction $e + p \to \tilde{e} + \tilde{q}$ as a function of scalar quark mass with the mass of the scalar electron as a parameter

Scalar leptons can also be searched for in the process[36] $e + p \to e^* + p \to \tilde{\gamma} + \tilde{e} + p \to e^- + p + 2\tilde{\gamma}$. The signature is very clear with only an electron and a uncorrelated proton in the final state.

References

1) J.M.Cornwall, D.Levin and G.Tiktopoulos, Phys.Rev.Lett. 30,
   1208, 1973, Phys. Rev. D10, 1145 (1970)
   C.H.Llewellyn-Smith, Phys.Lett. 46B, 233 (1973)

2) S.L.Glashow, Nucl.Phys. 22, 579, 1961
   S.Weinberg, Phys.Rev.Lett. 19, 1264 (1967)
   A.Salam, Proc. Nobel Symposium, ed. N.Svartholm (Almqvist and
   Wiksells, Stockholm, 1968) 367

3) P.W.Higgs, Phys.Rev.Lett. 12, 132, 1964 and 13, 508, 1964
   and Phys.Rev. 145, 1156, 1966

4) UA1 Collaboration, G.Arnison et al., Phys.Lett. 122B, 103 (1983),
   Phys.Lett. 126B, 398 (1983)
   UA2 Collaboration, M.Banner et al., Phys.Lett. 122B, 476 (1983);
   Phys.Lett. 129 B, 130 (1983)

5) S.Coleman and E.Weinberg, Phys.Rev. D7, 1888 (1973)

6) B.W.Lee, C.Quigg, and H.Thacker, Phys.Rev. D16, 1519 (1977)

7) Yu, A.Gol'fand and E.P.Likhtman, JETP Letters 13, 323 (1971)
   D.V.Volkov and V.P.Akulov, Phys.Lett. 46B, 109 (1973)
   J.Wess and B.Zumino, Nucl.Phys. B70, 39 (1974)
   for a recent review see:
   II.E.Haber and G.L.Kane, Phys.Rep. 117, 75 (1985)

8) S.Weinberg, Phys.Rev. D13, 979 (1976), D14 1277 (1979)
   L.Susskind, Phys.Rev. D20, 2615 (1979)
   For reviews see:
   E.Fahri and L.Susskind, Phys.Rev. 746, 277 (1981)
   J.Ellis, Pro. 1981 SLAC Summer Institute on Particle Physics,
   ed. A.Mosher, SLAC-245, 621 (1982)

9) S.Dinopoulos and L.Susskind, Nucl.Phys. B155, 237 (1979)
   E.Eichten and K.D.Lane, Phys.Lett. 90B, 125 (1980)

10) For a recent review including references see:
    W.Buchmüller in Proceedings of the XXIV Schladming School,
    CERN TH-4189, 185

11) J.Ellis, G.Gelmini and H.Kowalski, Proc. of the ECFA Workshop
    on the possibility of a large hadron collider in the LEP tunnel,
    DESY 84-071
    E.Eichten, I.Hinchliffe, K.D.Lane and C.Quigg,
    Rev. Mod. Phys. 56, 579 (1984)
    Fermilab Dedicated Collider, April 1983

12) For a review of the field see: Proc. of the Bielefeld Workshop,
    Quark Matter Formation and Heavy Collisions, World Scientific,
    Singapore 1982

13) A.A.Thiessen, IEEE Transactions on Culcear Science, Vol.NS-32 (1985)

14) SLC Design Handbook, Stanford Linear Accelerator Center,
    December 1984

15) B.Richter, private communication

16) CERN / ISR - LEP 79-33 and Addendum, CERN/SPC/472 LEP Phase 1

17) G.Plass, private communication

18) P.Bernard, H.Lengeler and E.Picasso, LEP Note 524

19) For an up to date review of the physics with large e^+e^- colliders
    see: Physics at LEP edited by J.Ellis and R.Peccei, CERN 86-02

20) Design Study of an Antiproton Collider (ALCOL)
    edited by E.J.Wilson, CERN 83-10, October 1983

21) E.Jones private communication

22) Design report Tevatron I project, September 1984
    J.Peoples, The Fermilab Antiproton Source, IEEE Trans. on Nucl.
    Science, Vol. NS-30, 1970 / 1983

23) H.Edwards private communication

24) HERA Proposal - DESY HERA 81-10 (1981)

25) Physics with ep colliders in view of HERA, DESY HERA 81-18 (1981)
    Proceedings of the Workshop Experimentation at HERA,
    NIKHEF, Amsterdam, June 9-11, 1983, DESY HERA 83-20 (1983)

26) R.J.Cashmore et al., Phys. Reports 122, 277 (1985)

27) The ZEUS Detector, Technical Proposal of the ZEUS Collaboration,
    March 1986

28) G.Altarelli, B.Mele and R.Rückl, CERN TH-3932

29) J.C. Pati and A.Salam, Phys.Rev. D10, 275 (1974)
    R.N.Mohapatra and R.E.Marshak, Phys.Rev.Lett. 44, 1316 (1980)
30) L.M.Jones and H.W.Wyld, Phys.Rev. D17, 759 (1978)
    H.Fritzsch and K.H.Streng, Phys.Lett. 72B, 385 (1978)
    J.Babcock, D.Sivers and S.Wolfram, Phys.Rev. D18, 162 (1970)
31) E.B.Eichten, K.D.Lane and M.E.Peskin, Phys.Rev.Lett. 50, 811 (1983)
32) R.Rückl., Phys.Lett. 129B, 363 (1983,
    Nucl. Phys. B 234, 91 (1984)
33) M.E.Peskin, Nucl.Phys. B 175, 197 (1980)
34) S.Rudaz and J.Vermaseren, CERN Preprint TH-2961 (1981)
35) S.K.Jones, C.H.Llewellyn-Smith, Nucl.Phys. B 217, 154(1983)
36) P.Salati and J.C.Wallet, Phys. Lett. B 122, 397 (1983)

Looking Beyond the Standard Model

R.D. Peccei

Deutsches Elektronen-Synchrotron DESY,
Hamburg, Fed. Rep. Germany

ABSTRACT

I discuss a variety of theoretical, and some experimental, reasons for
contemplating physics beyond the standard model. I illustrate this
general discussion with two concrete examples. One concerns the GSI
positron signals and their possible relation to variant axions. The
other is connected with superstring theories (which I motivate in a
heuristic manner) and the issue of neutrino masses in these theories.

I. Theoretical and Experimental Reasons to go Beyond the Standard Model

It is rather easy to make a long list of theoretical reasons for considering physics beyond the standard $SU(3) \times SU(2) \times U(1)$ model of the
strong and electroweak interactions. It is, however, very hard to find
any experimental reasons to do so. Phenomenologically the standard
model works marvelously well! Before discussing the few experimental
open questions that there are, let me thus concentrate on the theore-
retical open questions. These questions are essentially related to
structural aspects of the standard model and, instead of making a long
list of queries, it proves useful to categorize them into three broad
classes.

The first class of questions one can ask about the standard model is
why is it a gauge theory based on the group $SU(3) \times SU(2) \times U(1)$?
There are connected questions to this query. For instance, why at

$q^2 \sim 100$ GeV^2 are the strong and electroweak couplings so disparate in strength: $\alpha_3 \simeq 0.15$, $\alpha_2 \simeq 0.03$, $\alpha_1 \simeq 0.01$? Or why does the standard model (and we!) live in D=4 dimensions? Or where does gravity fit in the picture? Broadly speaking, we would like to have some idea really of why gravity, electromagnetism, the weak and the strong force are the forces we see in nature.

The second class of questions about the standard model is related to the kind of matter there is. We would really like to know why do the quarks and leptons transform as they do under the standard model. Of particular importance is trying to understand why do quarks and leptons appear in chiral asymmetric representations under the electroweak group? Related questions concern the U(1) assignments of the known fermions, which give third integral charge to the quarks but integral charge to the leptons. Another set of questions, of the same kind, concerns the reasons why a repetitive family structure appears and the issues of whether the number of repetitions stops at three and of whether families contain or not a right handed neutrino.

The third broad class of questions related to the structure of the standard model is concerned with the dynamics of mass generation. The spectrum of masses in the standard model is extensive and peculiar. The photon and gluon are massless and so are, for all intents and purposes, the neutrinos. The charged fermions we know, including a presumed top quark, range in mass over almost five orders of magnitude, from one half of an MeV to near 50 GeV. The W and Z masses are close to 100 GeV but we have as yet no evidence for where does the Higgs boson mass lie. There are associated questions here, too. For example, what physics fixes the Cabibbo Kobayashi Maskawa mixing angles? All elementary excitations we know get masses only after SU(2) x U(1) breakdown. That is

$$m \sim \Lambda_F \, , \tag{I.1}$$

where $\Lambda_F = (\sqrt{2}\, G_F)^{-1/2} \simeq 250$ GeV is the scale of this breakdown. What fixes Λ_F and why is Λ_F so disparate from the other two scales we know:

$$M_{Planck} \simeq 10^{19} \text{ GeV} \gg \Lambda_F \gg \Lambda_{QCD} \simeq 100 \text{ MeV} \tag{I.2}$$

where M_{Planck} is the scale associated with Newton's constant: $G_N \sim \dfrac{1}{M_{Planck}^2}$ and Λ_{QCD} is the dynamical scale of QCD. Eq. (I.2) is related to the famous naturalness question of the scalar sector of the standard model /1/.

In addition to these three broad classes of theoretical questions concerning the structure of the standard model there is a related, perhaps more technical, question: Why is the $\bar{\theta}$ angle so small? This question is associated with the, so called, strong CP problem. Because of the nature of the QCD vacuum /2/ and the existence of a chiral anomaly /3/ the Lagrangian of the standard model is augmented by a T, P and CP violating term

$$\mathcal{L}_{CP\,viol.} = \frac{\alpha_3}{8\pi}[\theta + Arg\,det\,M]\,\hat{F}_a^{\mu\nu}F_{a\mu\nu} = \frac{\alpha_3}{8\pi}\bar{\theta}\,\hat{F}_a^{\mu\nu}F_{a\mu\nu} \tag{I.3}$$

Here θ is the QCD vacuum angle and M is the quark mass matrix. The term in (I.3) can give rise to a nonvanishing electric dipole moment for the neutron and from the existing bound on this quantity /4/ one infers a bound on $\bar{\theta}$ /5/

$$\bar{\theta} \lesssim 10^{-8} - 10^{-9} \tag{I.4}$$

This is one of two examples I know where there is a cancellation between two, apparently, distinct sectors of the standard model. For $\bar{\theta} \simeq 0$ it is necessary that the QCD contribution, θ, should cancel against the electroweak contribution coming from Arg det M. The renormalizability of the electroweak theory is insured by a similar amazing cancellation. Due to the U(1) assignment of quarks and leptons one has that

$$\text{Tr } Q_{\text{quarks}} + \text{Tr } Q_{\text{leptons}} = 0 \qquad (I.5)$$

thereby assuring that no chiral anomalies /3/ spoil the renormalizability of the SU(2) x U(1) theory. It would be nice to know a reason why these interrelations obtain in the standard model.

In comparison to the theoretical reasons for wanting to go beyond the standard model - the issues of forces, matter and mass dynamics - there is a dearth of experimental reasons to do so. I could think only of one good reason, and two more speculative reasons, for contemplating physics beyond SU(3) x SU(2) x U(1) (plus gravity). The good reason is related to the matter-antimatter asymmetry of the universe. We know that the ratio of baryon minus antibaryons in the Universe is approximately /6/

$$\frac{n_B - n_{\bar{B}}}{n_\gamma} \simeq 10^{-10} \qquad (I.6)$$

Unless the Universe started that way - a peculiar initial condition - there must have been baryon number violating interactions at some level. However, in the standard model, baryon number is conserved (I will not worry here about possible non perturbative violation of baryon number in the standard model /2/ /7/). Hence to explain Eq. (I.6) it is necessary to have some physics beyond the standard model.

Less direct, but still quite compelling, evidence for physics beyond the standard model is related to neutrino masses and the solar neutrino problem, which has been the focus of much of this meeting. The discrepancy between the predicted rate /8/ for solar neutrinos in the Davis experiment /9/ and the measured rate:

$$R_{\text{Theory}} = 7.6 \pm 1.5 \text{ SNU}$$
$$R_{\text{Measur}} = 2.1 \pm 0.3 \text{ SNU} \qquad (I.7)$$

could well be due to our imperfect understanding of the sun. It is very
appealing, however, to suppose that this discrepancy is due to $\nu_e \rightarrow \nu_x$
conversion in the sun, as suggested recently by Mikheyev and Smirnov /10/
based on earlier work of Wolfenstein /11/. As Bethe has emphasized /12/,
the MSW effect can occur even for small mixing angles, provided
$\delta m^2_{\nu_e - \nu_x} \leq 5 \times 10^{-5}$ (eV)^2, where ν_x is a distinct neutrino from
ν_e. Barring accidental cancellations, this would require a tiny neu-
trino mass

$$m_{\nu_x} \lesssim 7 \times 10^{-3} \text{ eV} \tag{I.8}$$

Although a Dirac mass of this magnitude is not illegal in the standard
model, it is silly! Much more reasonable is to suppose that such a
small value for m_{ν_x} arises through the see-saw mechanism /13/:

$$m_{\nu_x} \sim \frac{m^2_{\ell_x}}{M} \tag{I.9}$$

where M is a large, lepton under violating, Majorana mass. If $x = \mu$
then Eq. (I.9) (which is a heuristic equation at any rate) implies
$M \sim 10^9$ GeV, while for $x = \tau$ one has $M \sim 10^{12}$ GeV. Lepton number violating scales of these orders of magnitude obviously necessitate
physics beyond the standard model.

The last potential evidence for physics beyond the standard model,
which I would like to mention here, has been discussed by Schweppe /14/
at this meeting. This is related to the anomalous positrons, and
correlated electrons, produced at GSI. This data can be understood, at
the kinematical level, by supposing that what results in the
heavy ion collision is a particle a, of mass near 1.7 - 1.8 MeV, which
is produced nearly at rest and then decays into e^+e^- pairs. It should
be mentioned that it is very difficult to understand /15/ how any such
particle could be produced almost at rest in these collisions. Nevertheless, the fact that M_a is so light suggests that a might be an
axion. Axions /16/ result from trying to solve the strong CP problem

by imposing an extra symmetry /17/ on the standard model, which guarantees that $\bar{\theta} = 0$. Thus if the GSI phenomena has anything to do with axions there is the exciting possibility that we have found some physics beyond the standard model which actually explains one of the open puzzles. This would be progress. Alas, as I shall now discuss this does not turn out to be the case!

II. Variant Axions: Birth and Untimely Death

The strong CP problem can be solved by augmenting the standard model by a global chiral symmetry: $U(1)_{PQ}$. One can show /17/ that such a symmetry locks the phase of the quark mass matrix to the QCD vacuum angle θ , thereby leading to $\bar{\theta} = 0$. To impose a $U(1)_{PQ}$ symmetry on the standard model it is necessary to have two Higgs doublets, so that chiral quark rotations can be compensated by Higgs phase rotations. The Yukawa Lagrangian in the Peccei Quinn model /17/:

$$\mathcal{L}^{PQ}_{Yukawa} = \Gamma^{u}_{ij}\, \bar{Q}_{Li}\, \Phi_1\, u_{Rj} + \Gamma^{d}_{ij}\, \bar{Q}_{Li}\, \Phi_2\, d_{Rj} + h.c.$$

$$(II.1)$$

where Q_{Li} are quark doublets of different families i, is clearly invariant under the chiral global $U(1)_{PQ}$ transformation:

$$u_{Rj} \to e^{i\alpha} u_{Rj} \qquad \Phi_1 \to e^{-i\alpha} \Phi_1$$

$$d_{Rj} \to e^{i\beta} d_{Rj} \qquad \Phi_2 \to e^{-i\beta} \Phi_2$$

$$(II.2)$$

(To rotate $\bar{\theta}$ away it is necessary that there be a net chiral transformation in the quark sector, so that $\alpha \neq -\beta$. /17/ It is for this reason that one needs two Higgs doublets, since with one Higgs doublet $\Phi_2 = \tilde{\Phi}_1 = i\tau_2 \Phi_1^*$ and thus $\alpha = -\beta$). Because of this extra symmetry, however, when the Higgs fields acquire vacuum expectation value:

$$\langle \Phi_i \rangle = \frac{1}{\sqrt{2}}\, f_i$$

$$(II.3)$$

with

$$f = \sqrt{f_1^2 + f_2^2} = (\sqrt{2}\, G_F)^{-1/2} = \Lambda_F \simeq 250 \text{ GeV} \qquad \text{(II.4a)}$$

being fixed by the scale of the SU(2) x U(1) breakdown and

$$x = f_2/f_1 \qquad \text{(II.4b)}$$

remaining a free parameter, also $U(1)_{PQ}$ breaks down. Associated with the breakdown of this global symmetry there arises a Goldstone boson – the axion /16/. In fact, since $U(1)_{PQ}$ is a chiral symmetry which suffers from an ABJ anomaly /3/ the axion is not really massless, but acquires a light mass /18/:

$$M_a \simeq m_\pi \frac{f_\pi}{f} \frac{(m_u m_d)^{1/2}}{(m_u + m_d)} N\left(x + \frac{1}{x}\right) \simeq 25\, N\left(x + \frac{1}{x}\right) \text{KeV}$$

$$\text{(II.5)}$$

where N is the number of generations of fermions. Other properties of, what has been called, the standard axion are also fixed by the value of x. For instance the axion couplings to quarks is given by

$$\mathcal{L}_{aqq} = \frac{m_q}{f}\, \bar{q}\, i \gamma_5 q\, a \left\{ \begin{matrix} x \\ x^{-1} \end{matrix} \right\} \qquad \text{(II.6)}$$

where the top line applies to charge 2/3 quarks and the bottom line to charge – 1/3 quarks.

The GSI excitation cannot be the standard axion since the standard axion has already been ruled out experimentally /19/! Actually, this analysis applies for an axion of mass less than $2m_e$ which has a very slow lifetime into two photons. It can, however, be repeated even in the case of an axion which decays rather rapidly into $e^+ e^-$ pairs. In particular, one of the processes $\psi \to \gamma a$ or $\Upsilon \to \gamma a$ badly violates the existing experimental bounds. One has /19/

$$B(\psi \to \gamma a) = (4.9 \pm 0.8) \times 10^{-5} \, x^2 < 1.4 \times 10^{-5}$$

$$\text{(II.7)}$$

$$B(\Upsilon \to \gamma a) = (2.7 \pm 0.7) \times 10^{-4} \, x^{-2} < 3 \times 10^{-4}$$

If $M_a \simeq 1.7 - 1.8$ MeV Eq. (II.5), for $N = 3$, tells us that x or x^{-1} is of order 20 and so one of the above decays would have given a spectacular signal, which has not been seen.

Although the quarkonia bounds are deadly for the standard axion, it is possible to consider a variant of the original Peccei Quinn model which neatly avoids these bounds. These variant axion models were originally discussed by Bardeen and Tye /18/ and were revived by Krauss and Wilczek /20/ and by Wu, Yanagida and myself /21/ as possible candidates for an explanation of the GSI phenomena. The idea behind these models is very simple. Instead of coupling Φ_1 to all u_{R_i} and Φ_2 to all d_{R_i} as in Eq. (II.1), in variant axion models one couples Φ_1 only to some u_{R_i}, with the rest of the u_{R_i} being coupled to $\tilde{\Phi}_2$. Effectively these models have an $U(1)_{PQ}$ symmetry only for some doublets but not for all N of them. The simplest model /20//21/, for instance has only u_R coupled to Φ_1, so that $N_{eff} = 1$, and only the u quark has a coupling proportional to x (cf. Eq. II.6). In view of Eq. (II.5), for $N_{eff} = 1$, x is very large $(x \simeq 70)$. Furthermore, one can arrange the models so that also electrons couple to Φ_1 ($\tilde{\Phi}_1$ to be precise) and have an enhanced coupling leading to a fast decay lifetime:

$$\tau(a \to e^+ e^-) = 3 \times 10^{-9} \, x^{-2} \, \sec \simeq 6 \times 10^{-13} \, \sec \qquad \text{(II.8)}$$

where the numerical value holds for a 1.7 MeV variant axion of the type described.

Variant axions appeared viable at the time of their proposal since they could escape the quarkonia bounds by construction and could, in view of their fast decay rate, avoid other bounds arising from beam dumps and nuclear deexcitation processes /19/. Because of the GSI signals they naturally elicited considerable interest. However, unfortunately,

these axions have also been ruled out experimentally now. Evidence
against them, arising from an electron beam dump experiment at KEK, has
been reported at this conference by Sasao /22/. I want to describe
similar negative results arising from hadronic experiments, which
really serve to rule out entirely any kind of variant axion model.

One can characterize variant axion models by their isovector and isoscalar mixing parameters λ_3 and λ_s, respectively /23/. (Axions do not
carry in general definite isospin and the parameters λ_3 and λ_s detail
their isospin couplings to hadrons). For the simplest variant model of
Ref. /20/ and /21/ one has

$$\lambda_3 \simeq 3/8 \; x \simeq 25 \; ; \quad \lambda_s = 0 \tag{II.9}$$

One can show, however, that for <u>any</u> <u>kind</u> of variant axion model the
difference between λ_3 and λ_s is fixed by the axion mass /23/:

$$\left(\lambda_3 - \lambda_s\right)^2 \simeq \left(\tfrac{3}{8} N x\right)^2 \simeq (25)^2 \tag{II.10}$$

where the numerical value applies for a 1.7 MeV variant axion. There
exist now data relevant for isovector axions which is in strong contradiction with the expectations of Eq. (II.9). Furthermore, there is also
a bound on the isoscalar mixing parameter λ_s, which taken in conjunction with the bound on λ_3, is at variance with the result (II.10). Thus
a combination of these bounds serves to rule out entirely all possible
variant axion models.

The isovector bounds come from an experiment at Caltech /24/ which
looked for axions in the deexcitation of the $2^+,1$ 9.17 MeV state of
^{14}N to its $1^+,0$ groundstate and from a search at SIN /25/ for axions
in π^+ decay: $\pi^+ \rightarrow a\, e^+\, \nu_e$. In both experiments axions of mass near
1.7 MeV decaying rapidly into e^+e^- pairs were looked for with negative
results. Bardeen, Yanagida and I /23/ have calculated the expectation of variant axion models for these decays and used the experimental
results to bound λ_3. The axion to photon rate for the ^{14}N transition,

using the results of Donnelly et al. /26/ to estimate the nuclear
matrix element ratio, is

$$\frac{\Gamma_a}{\Gamma_\gamma} = 2 \times 10^{-5}(\lambda_3)^2 \qquad\qquad \text{(II.10a)}$$

to be compared to the experimental limit /25/

$$\frac{\Gamma_a}{\Gamma_\gamma} < 4 \times 10^{-4} \qquad\qquad \text{(II.10b)}$$

For the π^+ decay an even stronger bound on λ_3 is obtained. Theoretically /23/ one predicts

$$B(\pi^+ \to a\, e^+ \nu_e) = 3 \times 10^{-9}(\lambda_3)^2 \qquad\qquad \text{(II.11a)}$$

while the SINDRUM experiment gives

$$B(\pi^+ \to a\, e^+ \nu_e) < (1\text{-}2) \times 10^{-10} \qquad\qquad \text{(II.11b)}$$

(The range above depends slightly on the exact lifetime of the axion).
Using Eq. (II.11) one has

$$\lambda_3 \lesssim 0.2 \qquad\qquad \text{(II.12)}$$

in strong disagreement with the prediction of the simplest variant axion
model given in Eq. (II.9).

The bound on the isoscalar mixing parameter λ_s comes from a reanalysis
by Calaprice et al. /27/ of an old internal pair conversion experiment
of Warburton et al. /28/. The presence of axions decaying into e^+e^-
pairs would have affected the measured angular distributions. For the
isoscalar transition in ^{10}B from the 3.59 MeV $2^+,0$ state to the $1^+,0$
state at 0.72 MeV the reanalysis of Calaprice et al. gives the bound
/27/

$$\frac{\Gamma_a}{\Gamma_\gamma} < 7.5 \times 10^{-5} \qquad\qquad \text{(II.13a)}$$

while our calculations yield

$$\frac{\Gamma_a}{\Gamma_\gamma} = 6.1 \times 10^{-4}(\lambda_s)^2 \qquad\qquad (II.13b)$$

Hence

$$\lambda_s \lesssim 0.29 \qquad\qquad (II.14)$$

Obviously, given Eq. (II.12) and Eq. (II.14), the constraint of Eq. (II.10) characteristic of variant axion models is badly violated and these models are ruled out. The possible remaining worry, connected with using an old experiment to bound λ_s, has actually now being removed by a direct bound /29/ on the ^{10}B transition of the 3.59 MeV 2^+,0 state to its 3^+,0 ground state. For this decay one predicts

$$\frac{\Gamma_a}{\Gamma_\gamma} = 7.9 \times 10^{-4}(\lambda_s)^2 \qquad\qquad (II.15a)$$

and the present bound at 90 % confidence level is /29/

$$\frac{\Gamma_a}{\Gamma_\gamma} < 7.2 \times 10^{-3} \qquad\qquad (II.15b)$$

giving

$$\lambda_s \lesssim 3.02 \qquad\qquad (II.16)$$

which, although not as strong as the bound of Eq. (II.14), suffices to rule out variant axions. My conclusion therefore is that the GSI signal remains mysterious but it has nothing to do with axions (or particle physics).

III. The Express Line to Superstrings

I want to return now to the central theoretical issues suggesting physics beyond the standard model - the questions of forces, matter and mass dynamics - and indicate, by means of a set of "non stringy" argu-ments, how the recently very popular superstrings theories /30/ give an extremely appealing answer to these questions. (In a subsequent section, though, I shall discuss also some of the troublesome issues in these theories.) The set of arguments I would like to put forward here involve three principal ideas. That of unification, that of supersymmetry and that of spontaneous compactification. Let me discuss these in turn.

The idea of <u>unification</u> of the existing forces into theories based on a simple gauge group /31/ is a natural extension of the process that led to the SU(2) x U(1) theory of electromagnetic and weak interactions. These, so called, GUT theories suppose that the standard model group is embedded in a larger group structure which, at some scale M_x, suffers a spontaneous breakdown, totally analogous to the breakdown of SU(2) x U(1) $\rightarrow$ U(1)$_{em}$ at Λ_F. The natural sequence for GUTs is the exceptional group sequence /32/, starting from the standard model SU(3) x SU(2) x U(1) $\cong E_3$ to E_8, the largest exceptional group. This sequence is detailed in Table III.1.

Table III.1: The Exceptional Group Sequence

Group	Number of Gauge Fields
$E_3 \cong$ SU(3) x SU(2) x U(1)	12
$E_4 \cong$ SU(5)	24
$E_5 \cong$ SO(10)	45
E_6	78
E_7	133
E_8	248

With GUTs the question of forces changes from why SU(3)xSU(2)xU(1) to, say, why SU(5)? However, some important conceptual advantages have been gained. For instance, now at energies above M_x there is just one coupling constant α_U. The disparate values of α_3, α_2 and α_1 at low q^2 values obtain from the different renormalization group behaviour of the coupling constants below M_x /33/. Indeed, requiring that these

couplings unify at M_x into a single coupling α_U - a non trivial requirement - yields $M_x \sim 10^{14} - 10^{15}$ GeV. Hence the physics of GUT
theories is characterized by energy scales much closer to the Planck
scale.

If the forces are unified in a GUT group, there is also some simplification in the matter assignments, since it is necessary that the fermions
of the theory themselves be in some GUT representation. These representations are much simpler than those under which the fermions transform
in the standard model. Furthermore, since $G \supset SU(3) \times SU(2) \times U(1)$ one
understands naturally why the particular $U(1)$ assignments of the quarks
and leptons (their charges, if you will) obtain. For instance, for
SU(5) the fermions of each family sit in a $\bar{5}$ plus a 10 representation.
For SO(10) all fermions of a given family are put into a 16, which requires the addition of a right handed neutrino $\nu_R \sim \nu_L^c$ to the theory
for each family. In E_6 quarks and leptons sit in the 27, which requires that there be another 11 extra states, etc.

Although unification of the forces brings some simplification in the
matter assignments, it brings no direct understanding of why we have in
nature these particular matter representations. From the point of view
of GUTs any fermion representation is ok for matter fields. To tie down
the matter representations we need in some way to unify matter with
forces. For that one needs <u>supersymmetry</u>, a symmetry which relates
bosons to fermions /34/. To follow the line of logic I am adopting here,
one wants the matter fermions to be partners of the gauge fields and
so to sit in a vector multiplet $V = (1, 1/2)$ of a supersymmetric GUT
theory. This is <u>not</u> the assumption one makes normally in these theories.
Usually matter sits in chiral multiplets $\Phi = (1/2, o)$ and gauge fields
have their own separate gauginos. The pairing I am suggesting here in
fact appears crazy as it runs into two fundamental problems:

i) Because the adjoint representation is real, pairing fermions to
gauge fields in a vector multiplet V necessarily puts the fermions in

real representations. However, we know that fermions are in chiral
asymmetric representations!

ii) To get the quantum numbers of the known fermions right the GUT group
must be at least E_6 and we need E_8 for 3 families. However, in all these
cases there are many other extra fermions, whose traces are yet to be
seen in nature!

Both of these points are illustrated by decomposing the 78 of E_6 into
its SO(10) subgroup and the 248 of E_8 into its E_6 x SU(3) maximal subgroup. One has

$$78 = 45 + 1 + \overline{16} + 16 \qquad\qquad\qquad \text{(III.1)}$$

$$248 = (78,1) + (1,8) + (\overline{27},\overline{3}) + (27,3) \qquad\qquad \text{(III.2)}$$

The matter fields of one family can be put in the 16 of (III.1) and
those of three families into the (27,3) of (III.2), including some
additional states. However, one sees from the above that in addition
to a 16 there is a $\overline{16}$ in (III.1) and analogously a $(\overline{27},\overline{3})$ in (III.2).
So the fermions in these theories have chirally symmetric partners -
mirror fermions - whose couplings with respect to the weak interactions
are V+A. In addition both (III.1) and (III.2) contain many more real
states besides the wanted matter fermions. The idea that force and
matter are unified via supersymmetry makes no sense unless the above
problems are somehow solved. Indeed, at this stage, also gravity is
missing.

The third concept, that of <u>spontaneous compactification</u>, makes the above
ideas potentially viable. In space time dimensions D = 4n + 2 it is
possible to have fermions which are chiral and yet obey reality conditions /30/. Furthermore if the extra dimensions beyond four compactify,
one may be able to obtain chiral fermions in four dimensions. The
number of the produced chiral fermions in D=4 is related to the geometrical characteristics of the compactified space K. Essentially this number
is the number of zero modes of the Dirac operator acting on the extra

dimensions /35/. If the properties of K are linked to the gauge fields present in the theory, and the theory is supersymmetric, then the chiral fermions that emerge in D=4 will have specific quantum numbers. Finally, since one is dealing with space time properties gravity is naturally included in these kind of considerations.

The above scenario is precisely what is thought to happen in superstring theories /30/. These theories are supersymmetric string theories in D=10 dimensions with a fixed gauge group which is either SO(32) or $E_8 \times E_8$ (Only this latter group is useful phenomenologically). It is not yet proven that superstring theories compactify to D=4, but if one assumes that this happens leaving an N=1 supersymmetry in four dimensions the compact manifold K is specified. This manifold, as Candelas, Horowitz, Strominger and Witten showed in a seminal paper /36/, is a so called Calabi-Yau space, which is Ricci flat, has a Kähler structure and has SU(3) holonomy. In particular, in these manifolds the gauge fields in an SU(3) subgroup of $E_8 \times E_8$ are identified with the spin connection, so that they are non vanishing. This means that in four dimensions the remaining gauge group is smaller than $E_8 \times E_8$. Assuming that this identification is made with respect to the SU(3) of the maximal $E_6 \times$ SU(3) subgroup of one of the E_8, the gauge structure present in 4 dimension is $E_6 \times E_8$. However, if K is not simply connected the E_6 group can be further broken by non vanishing flux loops /37/ to some subgroup of E_6. Since E_6 is one of the GUT groups one sees that superstrings offer the possibility of understanding why only certain forces appear. Note that in this respect the other E_8 provides a shadow world coupled to our world only gravitationally.

Superstrings offer also the possibility of understanding the matter content we see. Of the fermions present in the 248 adjoint representation of the non shadow E_8, only those that have a non trivial SU(3) content can emerge in four dimensions. Using Eq. (III.2) one sees that these are either in a 27 or a $\overline{27}$ of E_6. Now the 27 is precisely the right representation to contain the quarks and leptons! Furthermore, the number of 27 minus $\overline{27}$ which emerges is fixed by the geometry of K via an index theorem. One finds /36/ that

$$n_{27} - n_{\overline{27}} = 1/2 \, |\chi(\kappa)| \qquad\qquad \text{(III.3)}$$

where χ is the Euler number. Thus the number of families is determined by the topology of the Calabi-Yau manifold K, a remarkable result.

We see that even though in D=10 matter and forces were on equal footing, both sitting in the adjoint of $E_8 \times E_8$, in D=4 the emerging picture is rather disparate. Forces, acting on ordinary matter, are mediated by gauge fields which transform according to the 78 of E_6. Each of these gauge fields has an associated gaugino, so one has a vector multiplet V(1, 1/2) (There may be less gauge fields and gauginos if there is flux breaking). Matter on the other hand transforms as certain numbers of 27 and $\overline{27}$ of E_6, with the difference between 27 and $\overline{27}$ being fixed by the Euler number, Eq. (III.3). Each of these fermions will be accompanied by a complex scalar, since in the compactification the supersymmetry is preserved. Hence matter transforms as chiral multiplets ϕ (1/2,o). Clearly not much symmetry is apparent in D=4 between forces and matter. However, in superstring theories both had the same origin.

There is another aspect of superstring theories which is also very nice and which I want to mention since it addresses the matter of mass dynamics discussed earlier. In principle all Yukawa couplings are fixed in the compactification. They originate basically from vertices involving fermions and gauge fields in D=10. Now gauge fields along the compact dimensions become scalar fields in D=4. Hence, very roughly speaking, the Yukawa couplings in D=4 are nothing but gauge couplings in D=10 after integrating over the compact space. Provided one can generate the Fermi scale Λ_F, superstring theories appear to be able to predict in principle all fermion masses, since these are just proportional to the Yukawa couplings times Λ_F. However, superstring theories really set up the physics at a scale M_{comp}, which should be of the order of M_{Planck}. To understand whether these considerations have anything to

do with reality one must try to make connection with physics at
100 GeV and see if there are any problems, or indeed if any of the
predictions survive the transit.

IV. The Troublesome Neutrinos in Superstring Theories

I hope that my discussion in the last section has given you an idea of
why many theorists find superstrings so appealing. Here is a theory which
may provide the first example of a consistent quantum theory of gravity
and which, at the same time, has within it all the elements to illuminate
the structural mysteries of the standard model. This said, however, there
remain profound difficulties to overcome in superstring theory before
one can make a claim that it is at all connected to particle physics, as
we know it. (There are also very hard problems to solve connected with
the string aspects of superstrings, but these lie beyond both the scope
of this talk and my own competence). Let me list a few of these difficulties:

i) With Calabi-Yau compactification the theory is left with an unbroken
supersymmetry. Obviously this supersymmetry must somehow be broken but
at scales close to Λ_F, if one wants to appeal to supersymmetry as a
cure for the hierarchy problem /1/. How this breaking occurs and how Λ_F
itself is generated is, however, not too clear. One appeals to the hidden
sector of the theory to trigger the supersymmetry breaking /38/ and then
to radiative corrections to generate Λ_F /39/, but no convincing demonstration, ab initio, of these mechanisms starting from the superstring
theory has been given /40/.

ii) With flux loop breaking /37/ there are few workable models which
survive at low energy. If there are no intermediate scales, one can
show that these models must have an extra U(1) gauge group /41/. However, with intermediate scale breaking, it is possible that no such U(1)
appears /41/. Hence it is difficult to pin down precise predictions
arising from the superstring. Much of what can be seen at low energy

seems to depend more on the path taken than on the original 10 dimensional theory.

iii) Superstring models have generic problems with both proton decay and with neutrinos, which require careful monitoring.

Let me discuss in some detail the last point above and state more clearly both the nature of the problem and some of the proposed solutions. I indicated earlier that superstring theories give definite predictions for Yukawa interactions (and also Higgs self couplings) in D=4. Although one cannot yet compute these terms in detail, the overall structure of these interactions is known. The superpotential (from which one can derive both the Yukawa and Higgs self interactions) contains trilinear products of 27's or $\overline{27}$'s, which have an E_6 invariant form, although their coefficients need not respect E_6 /41/:

$$W \sim (27)^3 + (\overline{27})^3 \tag{IV.1}$$

Inside each 27 there is a full family of quarks and leptons, including a right-handed neutrino. In addition, there is an extra charge -1/3 quark and its antiquark, a lepton doublet plus its antidoublet and one more neutrino-like excitation. Specifically, decomposing the 27 in terms of its SU(5) content one has

$$27 = \{\bar 5 + 10 + 1\} + \{5 + \bar 5 + 1\} \tag{IV.2}$$

where the $\{5 + \bar 5 + 1\}$ are the extra states. The usual matter, written in terms of left-handed fields, reads

$$\bar 5 + 10 + 1 = \left(d^C, L\right) + \left(u^C, e^C, Q\right) + \nu^C \tag{IV.3a}$$

while the new matter reads

$$5 + \bar 5 + 1 = \left(g, H^C\right) + \left(g^C, H\right) + S \tag{IV.3b}$$

Here, L, Q, H^C and H are all SU(2) doublets and g ist the extra charge -1/3 quark.

R. D. Peccei

The presence of the extra 5 and $\bar{5}$ states in the 27 is the source of both
the baryon number and neutrino problem. One can check that $(27)^3$ contains
the following five terms involving g or g^c and the usual quarks and
leptons /42/

$$QQg \qquad u^c d^c g^c \qquad QLg^c \qquad u^c e^c g \qquad d^c \nu^c g \qquad\qquad \text{(IV.4)}$$

If all these terms have nonvanishing coefficients then one has very
rapid baryon number violation mediated by the scalar partners of g.
Typically one presumes that these objects have mass of $O(\Lambda_F)$ and so
the trouble is real! Two solutions have been proposed for this problem.
The first assumes that the compactified manifold K is so clever that the
actual coefficient for the first two, or the last three, terms in (IV.4)
vanishes, thereby restoring Baryon number as a global (accidental)
symmetry of the model. To my knowledge, however, no manifolds with this
property have been exhibited. A more appealing solution is that there
is an intermediate scale of breaking, below $M_{comp} \sim M_{Planck}$. In this
case it might be possible to give sufficient high mass to the g quarks
to avoid direct contradiction with experiment /41/ /43/.

The neutrino mass problem is of a similar nature. There are five potential "neutrino" helicity states per family in each 27, namely:

$$27: \quad \nu, \nu^c, N, N^c, S \qquad\qquad \text{(IV.5)}$$

where N and N^c are the neutral leptons in the doublets H and H^c. Because
of this it is quite natural to expect a Dirac mass for the neutrinos.
Indeed the superpotential (IV.1) gives the following two terms involving the fields in (IV.5)

$$\nu \nu^c N^c \qquad NN^c S \qquad\qquad \text{(IV.6)}$$

and one sees that a Dirac mass for the neutrino field ν ensues if N has
a vacuum expectation value. A possible solution to this problem is again
to appeal for the vanishing, for topological reasons, of the first term
in (IV.6). But again no examples exist where this really happens. Note
that it is not possible here to appeal directly to the see saw

mechanism /13/ to get small neutrino masses. To provide for a large
Majorana mass for the right handed neutrino one needs a Higgs multiplet
transforming as the 351 of E_6, but this multiplet is <u>not</u> part of theory.
However, as suggested by Deredinger, Ibañez and Nilles /44/ and by
Nandi and Sarkar /45/ such a term may effectively be generated from non
renormalizable terms in the superpotential. For instance, a term like

$$W_{NR} \sim \frac{(27)(27)\,\overline{27}\,\overline{27}}{M_{comp}} \qquad\qquad (IV.7)$$

contains an effective Majorana mass for ν^c of order of the expectation
value of $\frac{\langle\overline{27}\rangle^2}{M_{comp}}$. If there is an intermediate mass scale, so that some
component of the $\langle\overline{27}\rangle$ is large, this effective Majorana mass
may be sufficient to initiate the see-saw. Note that for this mechanism
to be feasible there must be some $\overline{27}$ in the theory. Indeed only in these
circumstances it is possible to contemplate the existence of intermediate mass scales because with only 27's the potential can never have
a minimum, if there are large vacuum expectation values /41/.

A second possibility to obtain reasonable neutrino masses is to marry
the neutrino field ν^c with some other E_6 singlet neutral fermions X, an
excitation which actually might be present after compactification /46/.
(An analogous mechanism using gauginos was discussed by Mohapatra /47/).
Then additional neutrino masses can arise from terms in the superpotential
of the structure

$$W \sim 27\,\overline{27}\,X \qquad\qquad (IV.8)$$

Intermediate scale breaking triggered by the $\overline{27}$ vacuum expectation
value can give large Dirac mass terms for the $\nu^c X$ combination. In conjunction with the usual $\nu\nu^c$ Dirac mass term one is left with a heavy
Dirac neutrino and a massless Majorana neutrino. This latter field can
then obtain a small mass from an induced supersymmetry breaking Majorana
mass for the X /48/. Clearly this is just one of possibly many scenarios
for generating neutrino masses. Unfortunately the relationship of these
"solutions" of the neutrino problem to the original superstring theory

becomes quite remote.

V. Concluding Remarks

Given the state of theoretical uncertainties of what might really be
the physics beyond the standard model, and the lack of any real experimental limits at present, all that one can really do now is to wait
for new experimental developments. Important clues on whether the superstring ideas I discussed are on the correct track should emerge in the
next five years, both from experiments at the accelerators now being
completed (Tevatron, SLC, LEP and HERA) and from non accelerator experiments. Of particular importance here are solar neutrino experiments, which may well give us positive evidence for the existence of an
intermediate mass scale. Accelerator experiments, on the other hand
could discover supersymmetric partners of ordinary matter which are of
fundamental importance for superstring theories (at least at the
psychological level - if for nothing else!). However, it is my belief
that theorists should not all migrate to the Planck mass. There are
still unclear aspects of the standard model and it might well be that
in trying to understand these - like the issue of strong CP - one might
find the clue for what physics really lies beyond the standard model.
Although the smart money is betting on supersymmetry and superstrings,
I am not quite ready to bet that the meeting in this series in 1992 will
change its name to Sneutrino 92.

References

/1/   For discussions see for example E. Witten, Nucl. Phys. **B188** (1981)
      513 or J. Ellis in Proceedings of the 1985 Lepton Photon Symposium,
      Kyoto, Japan

/2/   G. 't Hooft, Phys. Rev. **D14** (1976) 3432

/3/   S. Adler, Phys. Rev. **177** (1969) 2426; J.S. Bell and R. Jackiw,
      Nuovo Cimento **60A** (1969) 49; W.A. Bardeen, Phys. Rev. **184** (1969)
      1848

/4/   I.S. Altarev et al., Phys. Lett. **102B** (1981) 13

/5/ V. Baluni, Phys. Rev. D19 (1979) 2227;
    R. Crewther, P. di Vecchia, G. Veneziano and E. Witten, Phys.
    Lett. 89B (1979) 123

/6/ See for example, M. Turner in Proceedings of the 1983 NATO ASI
    "Quarks, Leptons and Beyond", Munich, Fed. Rep. Germany

/7/ V.A. Kuzmin, V. Rubakov and M. Shaposhnikov, Phys. Lett. 155B
    (1985) 36

/8/ J.N. Bachall et al., Ap. J. 292 (1985) 279

/9/ R. Davis Jr. in Proc. of the Neutrino Mass Miniconference, Telemark,
    Wisconsin (1980)

/10/ S.P. Mikheyev and A. Smirnov, Nuovo Cimento 9C (1986) 17

/11/ L. Wolfenstein, Phys. Rev. D16 (1978) 2369

/12/ H. Bethe, Phys. Rev. Lett. 56 (1986) 1305

/13/ T. Yanagida in Proceedings of the Workshop on Unified Theory and
    Baryon Number of the Universe, KEK, Japan 1979; M. Gell Mann,
    P. Ramond and R. Slansky in Supergravity (North Holland 1979)

/14/ J. Schweppe, these proceedings

/15/ See for example J. Reinhardt et al., Phys. Rev. 33C (1986) 194

/16/ S. Weinberg, Phys. Rev. Lett. 40 (1978) 223;
    F. Wilczek, Phys. Rev. Lett. 40 (1978) 279

/17/ R.D. Peccei and H.R. Quinn, Phys. Rev. Lett. 38 (1972) 1440;
    Phys. Rev. D16 (1977) 1791

/18/ W.A. Bardeen and S.-H.H. Tye, Phys. Lett. 74B (1978) 229

/19/ For a review, see for example A. Zehnder, Proceedings of the 1982
    Gif-sur-Yvette Summer School

/20/ L.H. Krauss and F. Wilczek, Phys. Lett. 173B (1986) 189

/21/ R.D. Peccei, T.T. Wu and T. Yanagida, Phys. Lett. 172B (1986) 435

/22/ N. Sasao, these proceedings

/23/ W.A. Bardeen, R.D. Peccei and T. Yanagida, DESY 86-054, Nucl.
    Phys. B to be published

/24/ M.J. Savage et al., Phys. Rev. Lett. 57 (1986) 178

/25/ R. Eichler et al., SIN Preprint PR 86-07

/26/ T.W. Donnelly et al., Phys. Rev. <u>D18</u> (1978) 1607

/27/ F.P. Calaprice et al., Internal Princeton Report 1986

/28/ E.K. Warburton et al., Phys. Rev. <u>B133</u> (1964) 42

/29/ F.W.N. de Boer et al., Groningen preprint, Phys. Lett. B to be published

/30/ For reviews see J.H. Schwarz, Phys. Rep. <u>89C</u> (1982) 233; M.B. Green, Surveys in High Energy Physics <u>3</u> (1982) 127

/31/ For a review see G.G. Ross <u>Grand Unified theories</u> (Benjamin, San Francisco 1985)

/32/ R. Slansky, Phys. Rep. <u>79C</u> (1981) 1

/33/ H. Georgi, H.R. Quinn and S. Weinberg, Phys. Rev. Lett. <u>33</u> (1974) 451

/34/ For a review see J. Bagger and J. Wess, <u>Supersymmetry and Supergravity</u> (Princeton Univ. Press, 1984)

/35/ For a discussion, see for example, P. Candelas, G.T. Horowitz, A. Strominger and E. Witten, in Proceedings of the Argonne Symposium on Anomalies, Geometry and Topology (World Scientific 1985)

/36/ P. Candelas, G. Horowitz, A. Strominger and E. Witten, Nucl. Phys. <u>B258</u> (1985) 46

/37/ Y. Hosotani, Phys. Lett. <u>126B</u> (1983) 309; <u>129B</u> (1983) 193; E. Witten, Phys. Lett. <u>149B</u> (1984) 351

/38/ M. Dine, R. Rohm, N. Seiberg and E. Witten, Phys. Lett. <u>156B</u> (1985) 55; J.P. Derendinger, L.E. Ibañez and H.P. Nilles, Phys. Lett. <u>155B</u> (1985) 65

/39/ L.E. Ibañez, Phys. Lett. <u>118B</u> (1982) 73; K. Inoue, A. Kakuto, H. Komatsu and S. Takeshita, Prog. Theor. Phys. <u>68</u> (1982) 927

/40/ M. Dine and M. Seiberg, Phys. Lett. <u>162B</u> (1985) 299

/41/ E. Witten, Nucl. Phys. <u>B258</u> (1985) 75; M. Dine, V. Kaplunovsky, M. Mangano, C. Nappi and N. Seiberg, Nucl. Phys. <u>B259</u> (1985) 549

/42/ J.D. Breit, B. Ovrut and G. Segrè, Phys. Lett. <u>158B</u> (1985) 33

/43/ F. del Aguila, G. Blair, M. Daniel and G.G. Ross, CERN TH 4336 (1985), Nucl. Phys. to be published

/44/ J.P. Deredinger, L.E. Ibañez and H.P. Nilles, CERN TH 4228
     (1985), Nucl. Phys. to be published

/45/ S. Nandi and U. Sarkar, Phys. Rev. Lett. $\underline{56}$ (1986) 566

/46/ E. Witten, Princeton preprint (1985)

/47/ R.N. Mohapatra, Phys. Rev. Lett. $\underline{56}$ (1986) 561

/48/ R.N. Mohapatra and J. Valle, Maryland preprint 86-127

Closing Words

George Marx
Secretary of the International Neutrino Commission

At the end of the NEUTRINO'86 conference one may ask: what have we learned? That the neutrinos behave, respecting the standard theory (the quantum flavor dynamics)! We may start our undergraduate lectures with the existence of Z and W (it is a fact like the Millikan experiment). We may express the electroweak mixing angle in terms of measured masses,

$$\sin^2\theta = 1 - m_W^2/m_Z^2 \ .$$

We can speak about the elementary processes

$$\nu_\mu e^- \rightarrow e^- \nu_\mu \qquad \text{and} \qquad \mu^- \rightarrow e^- \nu_\mu \bar{\nu}_e$$

like we have used to speak about Compton Scattering and the Klein Nishina formula. Experimentalists are getting able to measure electroweak radiative corrections (like Lamb shift was measured 40 years ago). QFD is getting complete like QED is, leaving the details to be clarified by quark people. (It has been noted at the Conference that the hadronic QCD radiative corrections get better known experimentally than theoretically!) But the number of input parameters of QFD is disturbingly large...

What to do next? Which way should search turn? This has been the central question at all Neutrino Conferences. We have mapped the photons, gluons, weak intermediate bosons, the lepton and quark peaks in the Sea of Fundamental Forces. We see now a Superpacific Ocean spreading far beyond our empirical horizons. The hypothetical Higgs archipelago is still hidden in fog, the promised Susy Continent is nowhere, the GUT and PLANCK peaks lay far beyond the range of our 20th century boats. Which way to sail in the vastness of 4 or 10 or more dimensions? Should we leave the crossing of the Superpacific to our children? Or are there unexpected coral reefs (may be, rich Southern Sea islands) quite near-by (at 28eV, at 1.8MeV or at 10^{-32} sec), waiting for the daring sailors? Neutrino research were always a case of serendipity, at the Neutrino Conferences we witnessed the discoveries of neutral currents, of tau lepton etc.

Our conferences started in Europe, at the Lake Balaton in 1972. Then we crossed the Atlantic repeatedly, moving to Pennsylvania, Caucasus, Aachen, back to Balaton, Bergen, Illinois, Sicily, Hawaii, Balaton again, Nordkirchen. Now we have crossed the whole Pacific to come to Sendai! Here we have enjoyed the unforgettable Japanese hospitality, accurate organization and expert reports. It is time to express our thanks to all the sponsors, to each member of the Local Organizing Committee, to the never tired staff of the Bubble Chambers Group at Tohoku University, and above all to the spiritual leader and main motor of Neutrino'86: Domo arigato, Kitagaki San!

In Sendai the International Neutrino Committee has decided to accept the repeated cordial invitation of Professor Jack Schneps: Neutrino'88 will be held in Massachusetts, on the Campus of Tufts University, June 5 - 11, 1988. The six sponsoring universities of the Boston region (Boston U, Brandeis, Harvard, MIT, Northeastern U, Tufts) are deeply involved in many of the forefront areas: neutrino interactions at the highest accelerator energies; collider physics to mass produce W's and Z's; a new wave of underground actions; developing GUTS and superstrings on

Earth or in the Sky.

Then the Neutrino Conference will return to the old Europe. We are happy about the several invitations for Neutrino'90. This is an indication that the interdisciplinary interest of particle physicists, nuclear physicists, geochemists and astronomers in neutrino science is alive and growing. So good luck for unexpected discoveries!

List of Contributed Papers

1. BEBC WA66 Collaboration - H. Grassler, W. Droge, U. Idschok,
   H. Kreutzmann, B. Nellen and B. Wunsch, A.M. Cooper-Sarkar,
   D.C. Cundy, H. Foeth, A. Grant, G.G. Harigel, H. Klein,
   D.R.O. Morrison, M. Nikolic, L. Pape, M.A. Parker, P. Schmid,
   H. Wachsmuth, M. Dris, E. Simopoulou, A. Vayaki, K.W.J. Barnham,
   D.B. Miller, M.M. Mobayyen, M. Talebzadeh, M. Aderholz, L. Deck,
   N. Schmitz, W. Wittek, P. Bostock, J. Krstic, G. Myatt,
   D. Radojicic, J. Guy, W. Venus, T. Bolognese, M.L. Faccini-
   Turluer, D. Vignaud, P.O. Hulth, K. Hultqvist, Ch. Walck :

   Prompt Neutrino Production in 400 GeV Proton Copper Interactions,
   Univ. of Stockholm Preprint   USIP Report 86-02

2. BEBC WA59 Collaboration - N. Armenise, M. Calicchio, O. Erriquez,
   G. Iaselli, G.T. Jones, R.P. Middleton, S.W. O'Neale,
   D. Bertrand, P. Marage, M.A. Parker, H. Wachsmuth, E. Simopoulou,
   A. Vayaki, E. Zevgolatakos, C. Vallee, E.F. Clayton, D. Miller,
   M.M. Mobayyen, T. Coghen, M. Aderholz, W. Wittek, K. Varvell,
   J. Wells, A.M. Cooper-Sarkar, J. Guy, P. Kasper, W. Venus,
   G. Gerbier, M. Lagraa, M. Neveu, T. Azemoon, F.W. Bullock,
   P.J. Fitch, R.A. Sansum, M. Berggren, P.O. Hulth :

   Evidence for Higher Twist Effects in Fast π^- Production by
   Antineutrinos in Neon, CERN Preprint   CERN/EP 85-195

3. N.N. Nikolaev :

   EMC Effect and Quarks and Gluons in Nuclei : Flavour Dependence
   as a Signature of Mechanisms of the Effect,
   Univ. of Tokyo Preprint   INS-Rep. -539

4. T.G. Rizzo and J.L. Hewett :

   The Stech and Fritzsh Hypotheses in the Leptonic Sector : An
   addendum to "Neutrino Mass Limits from the Fritzsch Mass Matrix",
   Iowa State Univ. Preprint   IS-J 2031

5.   BEBC WA21/WA59 Collaboration - M. Calicchio, O. Erriquez,
      M.T. Fogli-Muciaccia, G.T. Jones, R.W.L. Jones, B.W. Kennedy,
      R.P. Middleton, S.W. O'Neale, K. Varvell, E. Hoffmann,
      D. Bertrand, P. Marage, J. Sacton, H. Klein, M.A. Parker,
      D.R.O. Morrison, P. Schmid, H. Wachsmuth, E. Simopoulou,
      A. Vayaki, C. Vallee, K.W.J. Barnham, E.F. Clayton, F. Hamisi,
      D.B. Miller, M.M. Mobayyen, M. Aderholz, N. Schmitz, W. Wittek,
      G. Corrigan, G. Myatt, D. Radojicic, P.N. Shotton, S.J. Towers,
      J. Wells, T. Coghen, A.M. Cooper-Sarkar, J.G. Guy, P. Kasper,
      W. Venus, J.P. Baton, M. Lagraa, M. Berggren, P.O. Hulth,
      F.W. Bullock, P.J. Fitch, R.A. Sansum :

      Measurement of Total Cross Sections for Neutrino and Antineutrino
      Charged Current Interactions in Hydrogen and Neon, CERN Preprint
      CERN/EP 86-31

6.   BEBC WA59 Collaboration - N. Armenise, M. Calicchio, O. Erriquez,
      S. Natali, G.T. Jones, R.P. Middleton, S.W. O'Neale, K. Varvell,
      D. Bertrand, P. Marage, J. Sacton, H. Klein, D.R.O. Morrison,
      M.A. Parker, H. Wachsmuth, E. Simopoulou, A. Vayaki, V. Brisson,
      P. Petiau, C. Vallee, E.F. Clayton, F. Hamisi, D.B. Miller,
      M.M. Mobayyen, W. Burkor, T. Coghen, M. Aderholz, W. Wittek,
      P. Allport, J. Wells, A.M. Cooper-Sarkar, J. Guy, P. Kasper,
      W. Venus, J.P. Baton, G. Gerbier, M. Neveu, M. Berggren,
      P.O. Hulth, F.W. Bullock, P.J. Fitch, R.A. Sansum :

      Coherent Single Pion Production by Antineutrino Charged Current
      Interaction and Test of PCAC.

7.   M.B. Voloshin, M.I. Vysotsky :

      Neutrino Magnetic Moment and Time Variation of Solar Neutrino
      Flux, ITEP Moscow Preprint   ITEP-1

8.   L.B. Okun :

      On the Electric Dipole Moment of Neutrino, ITEP Moscow Preprint
      ITEP-14

9.   L.B. Okun, M.B. Voloshin, M.I. Vysotsky :

      Electromagnetic Properties of Neutrino and Possible Semiannual
      Variation Cycle of the Solar Neutrino Flux, ITEP Moscow Preprint
      ITEP-20

10. A.E. Asratyan, V.I. Efremenko, A.V. Fedotov, P.A. Goritchev,
    G.K. Kliger, V.Z. Kolganov, S.P. Krutchinin, M.A. Kubantsev,
    I.V. Makhlueva, V.I. Shekelyan, V.G. Shevchenko, V.V. Ammosov,
    V.S. Burtovoy, A.G. Denisov, G.S. Gapienko, V.A. Gapienko,
    V.I. Klyukhin, V.I. Koreshev, P.V. Pitukhin, V.I. Sirotenko,
    E.A. Slobodyuk, Z.U. Usubov, V.G. Zaetz :

    The Experimental Comparison of EMC Effect in the νn and νp
    Interactions in Neon Nucleus, ITEP Moscow Preprint   ITEP-115

11. L.M. Sehgal and M. Wanninger :

    Atomic Effects in Coherent Neutrino Scattering, PITH Aachen
    Preprint   PITHA 86/1

12. L.M. Sehgal :

    Differences in the Coherent Interactions of ν_e, ν_μ and ν_τ
    Phys. Letter 162B, 370 (1985)

13. L.M. Sehgal :

    Neutral Currents, Theory and Application, Progress in Particle
    and Nuclear Physics Volume 14

14. R.N. Mohapatra, S. Nussinov, J.W.F. Valle :

    Could Cyg X-3 Muons Indicate a Light Supersymmetric Particle ?,
    Phys. Letter 165B, 417 (1985)

15. R.N. Mohapatra, J.W.F. Valle :

    Neutrino Mass and Baryon Number Non-Conservation in Superstring
    Models, Preprint   Phys. Pub. No. 86-127

16. D. Chang, R.N. Mohapatra, S. Nussinov :

    Could Goldstone Bosons Generate an Observable 1/R Potential ?,
    Phys. Rev. Lett. 55, 2835 (1985)

17. Ernest Ma :

    Neutrino Mass Matrix and Nucleon Stability in a Superstring
    Context, Univ. of Hawaii report.

18. BIS-2 Collaboration - A.N. Aleev, V.A. Arefiev, V.P. Balandin,
    V.K. Berdyshev, V.K. Birulev, V.D. Cholakov, A.S. Chvyrov,
    I.I. Evsikov, T.S. Grigalashvili, B.N. Gus'kov, I.M. Ivanchenko,
    I.N. Kakurin, M.N. Kapishin, N.N. Karpenko, D.A. Kirillov,
    I.G. Kosarev, V.R. Krastev, N.A. Kuz'min, B.A. Kulakov,
    M.F. Likhachev, A.L. Lyubimov, A.N. Maksimov, P.V. Moisenz,
    A.N. Morozov, S. Nemecek, Nguyen Mong Zao, V.V. Pal'chik,
    A.V. Pose, L.V. Sil'vestrov, V.E. Simonov, L.A. Slepets,
    G.G. Sultanov, G.G. Takhtamyshev, P.T. Todorov, R.K. Trayanov,
    N.V. Vlasov, K. Hiller, H. Nowak, S. Nowak, H.E. Ryseck,
    A.S. Belousov, E.G. Devitsin, A.M. Fomenko, V.A. Kozlov,
    E.L. Malinovsky, V.V. Pavlovskaya, S.V. Rusakov, Yu.V. Soloviev,
    P.N. Shareiko, L.N. Shtarkov, A.R. Terkulov, Ya.A. Vazdik,
    M.N. Voichishin, M.V. Zavertyaev, E.D. Molodtsov, E.A. Chudakov,
    J. Hladky, M. Novak, A. Prokes, M.V. Tosheva, V.J. Zayachky,
    D.T. Burilkov, V.I. Genchev, I.M. Geshkov, P.K. Markov,
    N.S. Amaglobeli, V.P. Dzhordzhadze, V.D. Kekelidze, N.L. Lomidze,
    G.I. Nikobadze, R.G. Shanidze :

    The Λ_c^+ Production by 40-70 GeV Neutrons on Carbon, Z. Phys.
    C23, 333 (1984)

19. BIS-2 Collaboration - A.N. Aleev, V.A. Arefiev, V.D. Balandin,
    V.K. Berdyshev, V.K. Birulev, A.S. Chvyrov, I.I. Evsikov,
    T.S. Grigalashvili, B.N. Gus'kov, K. Hiller, I.M. Ivanchenko,
    I.N. Kakurin, M.N. Kapishin, N.N. Karpenko, D.A. Kirillov,
    I.G. Kosarev, V.R. Krastev, N.A. Kus'min, M.F. Likhachev,
    A.L. Lyubimov, A.N. Maksimov, P.V. Moisenz, A.N. Morosov,
    V.V. Pal'chik, A.V. Pose, A. Prokes, L.V. Sil'vestrov,
    V.E. Simonov, L.A. Slepets, M. Smizanska, G.G. Sultanov,
    G.G. Takhtamyshev, P.T. Todorov, N.V. Vlasov, H. Nowak, S. Nowak,
    H.E. Ryseck, A.S. Belousov, E.G. Devitsin, A.M. Fomenko,
    V.A. Kozlov, E.I. Malinovsky, V.V. Pavlovskaya, S.V. Rusakov,
    Yu.V. Soloviev, L.N. Shtarkov, A.R. Terkulov, Ya.A. Vazdik,
    M.N. Voichishin, M.V. Zavertyaev, E.A. Chudakov, J. Hladky,
    M. Novak, M. Vecko, M.V. Tosheva, V.J. Zayachky, D.T. Burilkov,
    P.K. Markov, P.K. Trayanov, V.D. Cholakov, N.S. Amaglobeli,
    V.P. Dzhordzhadze, V.D. Kekelidze, N.L. Lomidze, G.I. Nikobadze,
    R.G. Shanidze :

    Observation of $\bar{D}$-Mesons in nC Interactions at 40-70 GeV/c, JINR
    Dubna Preprint   E1-85-662

20. T. Hayashi, M. Tanimoto, S. Wakaizumi :

    Fourth Generation of Quarks and its Effect on CP Violation,
    Kure Tech. Coll./Hiroshima Univ. Preprint   KTCP-8501/HUPD-8505

21. M. Klein, H. Rupertsberger :

QCD-Corrections for Semiphenomenological Treatment of Heavy Quark
Decay, Univ. Wien Preprint   UWThPh-1986-2

22. A. Apostolakis, P. Ioannou, S. Katsanevas, J. Koutentakis,
C. Kourkoumelis, P. Pramantiotis, L.K. Resvanis, M. Vassiliou,
M. Baldo-Ceolin, F. Bobisut, E. Calimani, S. Ciampolillo,
H. Huzita, M. Loreti, G. Miari, G. Puglierin, C. Angelini,
A. Baldini, L. Bertanza, R. Fantechi, V. Flaminio, R. Pazzi,
B. Saitta, U. Camerini, W.J. Fry, R. Loveless, M. Procario,
D.D. Reeder :

Search for $\nu_\mu \longrightarrow \nu_e$ Oscillations using BEBC

23. A.I. Afonin, A.A. Borovoy, S.N. Ketov, A.N. Kheruvimov,
V.I. Kopeikin, L.A. Mikaelyan, M.D. Skorokhvatov, I.V.Kurchatov :

Neutrino Experiments at Rovno Power Reactor, Inst. of Atomic
Energy Preprint

24. J. Bouchez, M. Cribier, J. Rich, M. Spiro, D. Vignaud,
W. Hampel :

Matter Effects for Solar Neutrino Oscillations

25. M. Cribier, J. Gorry, B. Pichard, J. Rich, M. Spiro, D. Vignaud,
A. Besson, A. Bevilacqua, F. Caperan, G. Dupont, W. Hampel,
T. Kirsten :

Production of a High Intensity 746 KeV Neutrino Source for the
Calibration of Solar Neutrino Detectors

26. W.A. Mann, T. Kafka, M. Derrick, B. Musgrave, R. Ammar, D. Day,
J. Gress :

K-Meson Production by ν_μ-Deuterium Reactions near Threshold :
Implications for Nucleon-Decay Searches, ANL/Tuft Univ. Preprint
ANL HEP PR 86-04/TUFTS HEP 8601

27. N. Baumann, H. Gurr, Z. Greenwood, W. Kropp, M. Mandelkern,
L. Price, F. Reines, H. Sobel :

Two Position Results from the Irvine Mobile Neutrino Oscillatin
Detector, Univ. of Calif. Irvine Preprint   UCI Neutrino No.86-5

28. John M. LoSecco :

Comment of "Possible Explanation of the Solar Neutrino Puzzle",
Univ. of Notre Dame Preprint   UND-PDK-86-5a

29. J.M. LoSecco, R.M. Bionta, G. Blewitt, C.B. Bratton, D. Casper,
     P. Chrysicopoulou, R. Claus, B. Cortez, S. Errede, G. Foster,
     W. Gajewski, K.S. Ganezer, M. Goldhaber, T.J. Haines, T.W. Jones,
     D. Kielczewska, W.R. Kropp, J.G. Learned, E. Lehmann, H.S. Park,
     F. Reines, J. Schultz, S. Seidel, E. Shumard, D. Sinclair,
     H.W. Sobel, J.L. Stone, L. Sulak, R. Svoboda, J.C. Van der Velde,
     C. Wuest :

     Limits on the Neutrino Lifetime, Univ. of Notre Dame Preprint,
     UND PDK-86-6

30. J.M. LoSecco, R.M. Bionta, G. Blewitt, C.B. Bratton, D. Casper,
     P. Chrysicopoulou, R. Claus, B. Cortez, S. Errede, G. Foster,
     W. Gajewski, K.S. Ganezer, M. Goldhaber, T.J. Haines, T.W. Jones,
     D. Kielczewska, W.R. Kropp, J.G. Learned, E. Lehmann, H.S. Park,
     F. Reines, J. Schultz, S. Seidel, E. Shumard, D. Sinclair,
     H.W. Sobel, J.L. Stone, L. Sulak, R. Svoboda, J.C. Van der Velde,
     C. Wuest :

     Neutrino Physics with a Massive Underground Detector,
     ν'86 Contributed paper (Univ. of Notre Dame)

31. CHARM Collaboration - F. Bergsman, J.V. Allaby, U. Amaldi,
     G. Barbiellini, M. Baubillier, A. Capone, W. Flegel,
     F. Grancagnolo, L. Lanceri, M. Metcalf, C. Nieuwenhuis,
     J. Panman, R. Pain, R. Plunkett, C. Santoni, K. Winter,
     I. Abt, J. Aspiazu, A. Bungener, F.W. Busser, H. Daumann,
     P.D. Gall, T. Hebbeker, F. Niebergall, P. Schutt, P. Stahelin,
     P. Gorbunov, E. Grigoriev, V. Khovansky, A. Rozanov,
     A. Baroncelli, L. Barone, B. Borgia, C. Bosio, M. Diemoz,
     C. Dionisi, U. Dore, F. Ferroni, E. Longo, P. Loverre,
     L. Luminari, P. Monacelli, S. Morganti, F. De Notaristefanni,
     L. Tortora, V. Valente :

     A Precise Determination of the Electroweak Mixing Angle from
     Semileptonic Neutrino Scattering, CERN Preprint   WA18-INT/86-7

32. K. Sakurai :

     Short-Term Variation in the Solar Neutrino Flux and its Related
     Internal Processes of the Sun, ν'86 contributed paper
     (Kanagawa Univ.)

33. J. Arafune, N. Koga, K. Morokuma, T. Watanabe :

     Estimates of Molecular Effects on the Neutrino Mass Determination
     by Triton β Decay, Kobe Univ. Preprint   KOBE-86-02

34. H. Daniel, K.-H, Hidemann, O. Schwentker :

     Search for a Rest Mass of the Electron Antineutrino in the Beta-
     Decay of Tritium, ν'86 contributed paper
     (Technische Univ. Munchen)

35. J.B. Cole, S. Kunori, G.A. Snow, C.Y. Chang, D. Son,
P.H. Steinberg, D. Zieminska, R.A. Burnstein, J. Hanlon,
H.A. Rubin, T. Kitagaki, S. Tanaka, H. Yuta, K. Abe, K. Hasegawa,
A. Yamaguchi, K. Tamai, Y. Otani, H. Hayano, H. Sagawa,
Y. Yanokura, T. Kafka, W.A. Mann, A. Napier, J. Schneps :

The Non-Singlet Valence Quark Distribution from Neutrino-
Deuterium Deep Inelastic Scatterring

36. H.C. Ballagn, H.H. Bingham, T.J. Lawry, J. Lys, G.R. Lynch,
M.L. Stevenson, F.R. Huson, E. Schmidt, W. Smart, E. Treadwell,
R.J. Cence, F.A. Harris, M.D. Jones, A. Koide, M.W. Peters,
V.Z. Peterson, H.J. Lubatti, K. Moriyasu, E. Wolin, U. Camerini,
W. Fry, D. Gee, M. Gee, R.J. Loveless, D.D. Reeder :

Observation of Coherent ρ^+ Production in Neutrino Neon
Interactions, Univ. of Calif. Berkely Preprint   UCB PPG/860530

37. P. Kostka, W. Lange, M. Pohl, CH. Spiering, M. Walter, P. Wegner,
D. Kiss, E. Kiss, L.S. Barabash, Yu.A. Batusov, S.A. Bunyatov,
V.V. Chalyshev, O.Yu. Denisov, I.A. Golutvin, A.I. Grigoriev,
A.I. Ivanenko, A.G. Karev, A.V. Karpukhin, M.Yu. Kazarinov,
V.S. Khabarov, A.M. Kharin, V.S. Kurbatov, O.M. Kuznetsov,
V.V. Lyukov, E.I. Maltsev, L.P. Meszaros, B.A. Morozov,
A.A. Popov, S.N. Prakhov, A.M. Rozhdestvensky, V.P. Sarantsev,
V.V. Sidorkin, V.I. Snyatkov, A.Yu. Sukhanov, I.A. Tereshchenko,
V.I. Tretiak, V.N. Vinogradov, N.I. Zamyatin, A.A. Borisov,
N.K. Bulgakov, R.M. Fakhrutdinov, V.N. Goryachev, A.S. Kozhin,
V.I. Kravtsov, A.T. Mukhin, Yu.I. Solomatin, Yu.A. Sviridov,
V.L. Tumakov, A.S. Vovenko, V.A. Yarba :

IHEP-JINR Neutrino Detector, ν'86 contributed paper (IHEP-JINR)

38. M. Kawasaki, K. Sato :

Cosmological Effect of Radiative Decay of Neutrinos,
ν'86 contributed paper (Univ. of Tokyo)

39. N. Terasawa, M. Kawasaki, K. Sato :

Neutrino Decay and Primordial Nucleosynthesis, ν'86 contributed
paper (Univ. of Tokyo)

40. E636 Europe - US - India Collaboration - (Berkeley-Birmingham-
Brussels-CERN-Chandigarh-FNAL-Hawaii-IIT-Imp.College London-
Jammu-Munich-Oxford-Rutherford-Rutgers-Saclay-Tufts)

Observation of Coherent Charged Current Neutrino Interactions in
the 15' Bubble Chamber at the FNAL Tevatron, ν'86 contributed
paper

41.   BEBC WA59 Collaboration - (INFN Bari-Univ. of Birmingham-IUIHE
      Brussels-CERN-NRCD Athens-Ecole Polytechnique-Imp.Coll. London-
      INP Cracow-MPI Munchen-Oxford-RAL-CEN Saclay-Univ.of Stockholm-
      Univ.Coll. London)

      Study of Coherent Production of ρ^- Mesons by Charged Current
      Antineutrino Interactions in BEBC, ν '86 contributed paper

42.   SKAT - Collaboration - V.V. Ammosov, A.A. Ivanilov, P.V. Ivanov,
      V.I. Konyushko, V.M. Korablev, V.A. Korotkov, V.V. Makeev,
      A.G. Myagkov, A.Yu. Polyarush, A.A. Sokolov, H.-J. Grabosch,
      H.-H. Kaufmann, U. Krecker, R. Nahnhauer, S. Nowak,
      S. Schlenstedt :

      Preliminary Results for the Production of μe-pairs in neutrino
      Interactions in the Energy Range 3 - 30 GeV, ν'86 contributed
      paper

43.   H.-J. Grabosch, H.-H. Kaufmann, U. Krecker, R. Nahnhauer,
      S. Nowak, S. Schlenstedt, V.V. Ammosov, D.S. Baranov,
      A.A. Ivanilov, P.V. Ivanov, V.S. Konyushko, V.M. Korablev,
      V.A. Korotkov, V.V. Makeev, A.G. Myagkov, A.Yu. Polyarush,
      A.A. Sokolov :

      Observation of Internal Muon Bremsstrahlung in ν_μ- Freon
      Interactions, ν'86 contributed paper (IHEP Berlin, IHEP-
      Serpukhov)

44.   O.K. Manuel, G. Hwaung :

      Solar abundances of the Elements, Meteoritics. Vol. 18, No.3,
      209 (1983)

45.   J.M.D. MacElroy, O.K. Manuel :

      Can Intrasolar Diffusion Contribute to Isotope Anomalies in the
      Solar wind ?, Journal of Geophysical Research, Vol. 91, No.B4,
      D473 (1986)

46.   L.S. Peak, A.M. Bakich, P.R. Gerhardy, J. Malos :

      The Current Status of Sydney University Solar Neutrino Program,
      ν'86 contributed paper (Univ. of Sydney)

47.   CHARM Collaboration

      Parametrization of the Gluon Structure Function at Q^2=10 GeV^2,
      ν'86 contributed paper

48. R.N. Mohapatra, J.W.F. Valle :

Solar Neutrino Oscillations from Superstrings, Preprint   Physics
Publication No. 86-164 (Univ. Autonoma)

49. Bugey Collaboration - J. Bouchez :

Neutrino Oscillations : Bugey Experiment, Talk at Mariond Conf.
(1986) (CEN Saclay)

50. A.E. Asratyan, A.V. Fedotov, P.A. Goritchev, S.P. Krutchinin,
M.A. Kubantsev, I.V. Makhlueva, V.I. Shekelyan, V.G. Shevchenko,
V.V. Ammosov, V.S. Burtovoy, A.G. Denisov, G.S. Gapienko,
V.A. Gapienko, V.I. Klyukhin, V.I. Koreshev, P.V. Pitukhin,
V.I. Sirotenko, E.A. Slobodyuk, Z.U. Usubov, V.G. Zaetz :

Observation of the charmed strange pseudo vector meson cascade
radiative decay.
Institute of Theoretical and Experimental Physics, ITEP Moscow
Preprint   ITEP 86-77

51. A.E. Asratyan, A.V. Fedotov, P.A. Goritchev, S.P. Krutchinin,
M.A. Kubantsev, I.V. Makhlueva, V.I. Shekelyan, V.G. Shevchenko,
V.V. Ammosov, V.S. Burtovoy, A.G. Denisov, G.S. Gapienko,
V.A. Gapienko, V.I. Klyukhin, V.I. Koreshev, P.V. Pitukhin,
V.I. Sirotenko, E.A. Slobodyuk, Z.U. Usubov, V.G. Zaetz :

Observation in Nuclear Emulsion of a Charmed Σ_c^0 Baryon Decay to
$\Lambda_c^0 \pi^-$ with a Subsequent Λ_c^+ Decay to $\Sigma^+ \pi^- \pi^+$, ITEP Moscow
Preprint   ITEP-51

52. M.R. Krishnaswamy, M.G.K. Menon, N.K. Mondal, V.S. Narasimham,
B.V. Sreekantan, Y. Hayashi, N. Ito, S. Kawakami, S. Miyake :

K.G.F. Proton Decay Experiment, ν'86 contributed paper (Tata,
Osaka City, Kanagawa)

53. K. Lang, A. Bodek, F. Borcherding, N. Giokaris, I.E. Stockdale,
P. Auchincloss, R. Blair, C. Haber, S. Mishra, E. Oltman,
M. Ruiz, F.J. Sciulli, M. Shaevitz, W.H. Smith, R. Zhu, Y.K. Chu,
O.B. MacFarlane, R.L. Messner, D.B. Novikoff, M.V. Purohit,
D. Garfinkle, F.S. Merritt, M. Oreglia, P. Reutens, R. Coleman,
H.E. Fisk, Y. Fukushima, Q. Kerns, B. Jin, D. Levinthal,
T. Kondo, W. Marsh, P.A. Rapidis, S. Segler, R. Stefanski,
D. Theriot, H.B. White, D. Yovanovitch, O. Fackler, K. Jenkins :

Neutrino Production of Dimuons, ν'86 contributed paper

54. N.De Leener-Rosier, J. Deutsch, M. Lebrun, O. Naviliat-Cuncic,
     R. Prieels, A. Amsler, L. Van Elmbt, M. Schaad, P. Truol,
     Cl. Joseph, J.P. Perround, M.T. Tran :

     New Limits on Generation Mixing for Massive Neutrinos from
     $\pi \to e\nu$ decay, ν'86 contributed paper

Name	Institute	Country
Abe, Kazuo	KEK	JAPAN
Abe, Koya	Tohoku	JAPAN
Akagi, Takashi	Tohoku	JAPAN
Akiba, Tomoya	Tohoku	JAPAN
Alexeev, Evegnii	INR Moscow	USSR
Allkofer, O.C.	Kiel	F.R. GERMANY
Aprile, Elena	Columbia	USA
Arafune, Jiro	ICRR Tokyo	JAPAN
Bakich, A.	Sydney	AUSTRALIA
Baldo-Ceolin, Milla	Padova	ITALY
Barabash, Leonid S.	JINR Dubna	USSR
Barish, Barry C.	CALTECH	USA
Bellotti, Enrico	INFN Milano	ITALY
Bergsma, Felix	CERN	SWITZERLAND
Bingham, Harry H.	Berkeley	USA
Blair, Robert E.	Columbia	USA
Blecher, M.	Virginia PI	USA
Bodek, Arie	Rochester/KEK	USA
Boehm, Felix H.	CALTECH	USA
Bonn, J.	Mainz	F.R. GERMANY
Bouchez, Jacques M.	CEN Saclay	FRANCE
Bowles, T.	Los Alamos	USA
Brock, Raymond L.	Michigan State	USA
Bugg, W.M.	Tennessee	USA
Bunyatov, Stepan A.	JINR Dubna	USSR
Burnstein, Ray A.	IIT Chicago	USA
Busser, F.W.	Hamburg	F.R. GERMANY
Caldwell, David O.	Santa Barbara	USA
Carlini, Roger D.	Los Alamos	USA
Cavaignac, Jean F.	ISN Grenoble	FRANCE
Chen, Herbert H.	Irvine	USA
Chikawa, Michiyuki	Kinki	JAPAN

Ching, Cheng-rui	ITP Beijing	P.R. CHINA
Clark, Allan G.	CERN	SWITZERLAND
Cohn, H.O.	Oak Ridge	USA
Conforto, Gianni	INFN Firenze	ITALY
Costa, Giovanni	Padova	ITALY
Daniel, Herbert G.	Munchen	F.R. GERMANY
Dekerret, H.	LPC Paris	FRANCE
Domagatsky, G.V.	INR Moscow	USSR
Dore, Ubaldo	INFN Roma	ITALY
Ebata, Takeshi	Tohoku	JAPAN
Ejiri, Hiroyasu	Osaka	JAPAN
Fabbri, F.L.	INFN Frascati	ITALY
Fackler, Orrin	LLNL Livermore	USA
Ferroni, Fernando	Roma	ITALY
Fetscher, W.	SIN	SWITZERLAND
Fiorini, Ettore	Milano	ITALY
Fitch, Peter J.	College london	U.K.
Flaminio, Vincenzo	INFN Roma	ITALY
Frodesen, Anne G.	Bergen	NORWAY
Fujioka, Manabu	Tohoku	JAPAN
Fukui, Shuji	Nagoya	JAPAN
Furuno, Kouichiro	Tohoku	JAPAN
Gaidos, James	Purdue	USA
Galeotti, Piero	Torino	ITALY
Gotoh, Kasuo	Virginia PI/KEK	USA
Graham, Robert L.	Guelph	CANADA
Gunji, Masanori	Tohoku	JAPAN
Guyot, Claude	CEN Saclay	FRANCE
Haga, Katsuhiko	Tohoku	JAPAN
Haidt, Dieter W.	DESY	F.R. GERMANY
Hanada, Hiromitsu	Tohoku	JAPAN
Hara Yasuo	Tsukuba	JAPAN
Harton, John	MIT	USA
Hasegawa, Katsuo	Tohoku	JAPAN
Hayashi, Kooichi	Kinki	JAPAN

Hepp, Verker	Heidelberg	F.R. GERMANY
Hidaka, Keisho	Tokyo-Gakugei	JAPAN
higuchi, Masato	Tohoku-Gakuin	JAPAN
Hill, Henry	Arizona	USA
Hladky, Jan	Inst of Phys.Praha	CZECHOSLAVAKIA
Ho, Tso-hsiu	ITP Beijing	P.R. CHINA
Holzschuh, Eugen	Zurich	SWITZERLAND
Hoshi, Yoshimoto	Tohoku-Gakuin	JAPAN
Hubert, Philippe	CEN Bordeaux	FRANCE
Huzita, Humiaki	INFN Padova	ITALY
Ishihara, Sumio	Tohoku	JAPAN
Iso, Chikashi	Tokyo Inst of Tech	JAPAN
Ito, Nobuo	Osaka City	JAPAN
Jones, Goronwy T.	Birmingham	U.K.
Jovanovic, Drasko	FNAL	USA
Kabe, Seiji	KEK	JAPAN
Kafka, T.	Tufts	USA
Kaplan, I.G.	Karpov Inst.	USSR
Katayama, Jimitsu	Tohoku	JAPAN
Kato, Sadayuki	INS Tokyo	JAPAN
Kawasaki, Masahiro	Tokyo	JAPAN
Kayser, Boris	NSF Washington	USA
Kirsten, Till	MPI Heidelberg	F.R. GERMANY
Kitagaki, Toshio	Tohoku	JAPAN
Kleinknecht, K.	Mainz	F.R. GERMANY
Kobayakawa, Keizou	Kobe	JAPAN
Konuma, Michiji	Keio	JAPAN
Koshiba, Masatoshi	Tokyo	JAPAN
Kotani, Tsuneyuki	Osaka	JAPAN
Krogh, Juergen Von	Heidelberg	F.R. GERMANY
Kropp, William R.	Irvine	USA
Kundig, W.	Zurich	SWITZERLAND
Kunori, Shuichi	Maryland/FNAL	USA
Kurino, Hiroyuki	Tohoku	JAPAN
Learned, John G.	Hawaii	USA

Lee, Wonyong	Columbia	USA
Levy, J.M.	P&M Curie paris	FRANCE
Li, Ti-pei	IHEP Beijing	F.R. CHINA
Liang, Dong-qi	IAE Beijing	P.R. CHINA
LoSecco, John M.	Notre Dame	USA
Lu, Tan	Nanjing	P.R. CHINA
Lubimov, Valentin	ITEP Moscow	USSR
Mak, Hay-Bonn	Queens	CANADA
Manuel, Oliver K.	Missouri-rolla	USA
Marage, P.	Bruxelles	BELGIUM
March, Robert	Wisconsin	USA
Maruyama, Takashi	Wisconsin/SLAC	USA
Marx, George	Eoetvoes	HUNGARY
Maschuw, R.	Karlsruhe	F.R. GERMANY
Masuda, Hiroyuki	Tohoku	JAPAN
Merritt, Frank S.	Chicago	USA
Meyer, Hinrich	Wuppertal	F.R. GERMANY
Mikheyev, S.P.	INR Moscow	USSR
Miyake, Saburo	Kanagawa	JAPAN
Moe, Michael K.	Irvine	USA
Morales, Angel	Zaragoza	SPAIN
Mori, Shigeki	Tsukuba	JAPAN
Muciaccia, Maria T.	Bari	ITALY
Muraki, Yasushi	ICRR Tokyo	JAPAN
Mursula, Kalevi	Nordita	DENMARK
Murtagh, Michael J.	BNL	USA
Nagashima, Yorikiyo	Osaka	JAPAN
Nahnhauer, Rolf	IHEP Berlin	GERMANY DDR
Nakai, Satoshi	Tohoku	JAPAN
Nakajima, Takashi	Tohoku	JAPAN
Nakamura, Masayasu	Tohoku	JAPAN
Nath, P.	Northeastern	USA
Naumann, Robert A.	Perinceton	USA
Nolte, E.	Munchen	F.R. GERMANY
Numano, kazuaki	Tohoku	JAPAN

Oshima, Takayoshi	INS Tokyo	JAPAN
Otter, G.	RWTH Aachen	F.R. GERMANY
Ozaki, Satoshi	KEK	JAPAN
Panman, Jacob	CERN	SWITZERLAND
Peccei, R.	DESY	F.R. GERMANY
Perez, Patrice	CEN Saclay	FRANCE
Pessard, Henw C.	Lab. d'Annecy	FRANCE
Peterson, V.Z.	Hawaii	USA
Picciotto, Charles	Victoria/TRIUMF	CANADA
Pietshmann, H.	Wien	AUSTRIA
Plano, Richard J.	Rutgers	USA
Pomansky, Alexandr	INR Moscow	USSR
Pose, Dietrich	JINR Dubna	USSR
Price, LeRoy R.	Irvine	USA
Prieels, Rene	Louvain	BELGIUM
Reines, Frederick	Irvine	USA
Rjazhskaya, Olga G.	INR Moscow	USSR
Robertson, Barry C.	Queens	CANADA
Robertson, R.G.H.	Los Alamos	USA
Romanowski, T.A.	Ohio State	USA
Romero, Alessandra	INFN Torino	ITALY
Roos, Charles E.	Vandervilt	USA
Rosen, Simon Peter	Los Alamos	USA
Rossi, Antonio M.	INFN Bologna	ITALY
Rupertsberger, H.	Wien	AUSTRIA
Saavedra, Oscar	Trino	ITALY
Sakurai, Kunitomo	Kanagawa	JAPAN
Sanda, A.I.	Rockefeller	USA
Santoni, Claudio	INFN Sanito' Roma	ITALY
Sasaki, Makio	Tohoku	JAPAN
Sasao, Noboru	Kyoto	JAPAN
Sato, Humitaka	Kyoto	JAPAN
Sato, Katsuhiko	Tokyo	JAPAN
Sato, Minoru	Tohoku-Gakuin	JAPAN
Schmitz, N.	MPI Munchen	F.R. GERMANY

Schneps, Jack	Tufts	USA
Schweppe, John	Yale	USA
Sehgal, Lalit M.	RWTH Aachen	F.R. GERMANY
Seto, Richard	Columbia	USA
Shang, Ren-cheng	Tsinghua	P.R. CHINA
Shinkawa, Takao	KEK	JAPAN
Smith, Wesley H.	Columbia	USA
Sobel, Henry W.	Irvin	USA
Steigman, Gary	Deraware	USA
Stenger, Cictor J.	Hawaii	USA
Sugawara, Hirotaka	KEK	JAPAN
Sun, Han-cheng	IAE Beijing	P.R. CHINA
Suzuki, Atsuto	Tokyo	JAPAN
Suzuki, Hajime	Tohoku	JAPAN
Suzuki, Hideyuki	Tokyo	JAPAN
Suzuki, Yoichiro	Osaka	JAPAN
Takagi, Fujio	Tohoku	JAPAN
Takahashi, Kasuke	KEK	JAPAN
Takahashi, Koichi	Tohoku-Gakuin	JAPAN
Takayama, Tomoaki	Tohoku	JAPAN
Takeda, Gyo	Tohoku	JAPAN
Taketani, Hiroshi	Tokyo Inst of Tech	JAPAN
Takeuchi, Mine	Tohoku	JAPAN
Tamae, Kyoko	Tohoku	JAPAN
Tamai, Kunio	Tohoku	JAPAN
Tanaka, Syo	Tohoku	JAPAN
Tanimoto, Narimitsu	Ehime	JAPAN
Tenner, Armin G.	NIKHEF-H	NETHERLANDS
Terada, Susumu	KEK	JAPAN
Terasawa, Nobuo	Tokyo	JAPAN
Thomas, James	CALTECH	USA
Traweek, Sharon	MIT/KEK	USA
Tsai, Sheng-Yi	Nihon	JAPAN
Ukai, Kumataro	INS Tokyo	JAPAN
Valle, J.W.F.	Autonoma	SPAIN

Valov, Tzvetan	JINR Dubna	USSR
Vitale, Sandro	INFN Genova	ITALY
Vogel, P.	CALTECH	USA
Vrana, Jiri	LPC Paris	FRANCE
Wachsmuth, Horst W.	CERN	SWITZERLAND
Wakaizumi, Seiichi	Hiroshima	JAPAN
Watanabe, Tadashi	Kobe	JAPAN
Wiik, B.	DESY	F.R. GERMANY
Winter, K.	CERN	SWITZERLAND
Wittek, W.	MPI Munchen	F.R. GERMANY
Wolfenstein, L.	Carnegie-Mellon	USA
Yahil, A.	Stony Brook	USA
Yamada, Sakuei	INS Tokyo	JAPAN
Yamaguchi, Akira	Tohoku	JAPAN
Yamaguchi, Yoshio	Tokai	JAPAN
Yamakoshi, kazuo	ICRR Tokyo	JAPAN
Yamamura, Takeshi	Tohoku	JAPAN
Yamazaki, Toshimitsu	Tokyo	JAPAN
Yasumi, Sinjiro	Teikyo/KEK	JAPAN
Yoshiki, Hajimc	KEK	JAPAN
Yoshimura, Motohiko	KEK	JAPAN
Yuta, Haruo	Tohoku	JAPAN
Zatsepin, G.T.	INR Moscow	USSR

SENDAI
AOBANŌ
J. Maki
86